陈华癸论文著作集

——诞辰100周年纪念

张忠明　陈雯莉　周　启　主　编

科学出版社
北　京

内 容 简 介

中国科学院院士陈华癸教授是中国著名土壤微生物学家和农业教育家，在国内外土壤微生物学界享有崇高的声誉。陈华癸教授毕生精于科研、潜心教学，主张教学、科研和技术推广三者并举。陈华癸教授不仅在共生固氮微生物和水稻田土壤微生物两个科研领域上获得丰硕成果；同时大力推广在水稻田里种植草籽绿肥(紫云英)，为国家增加粮食产量作出了重大贡献，成为几代人学习的典范。

本论著集分为上、下两篇。上篇为论文集，收录了陈华癸教授在共生固氮微生物、水稻田土壤微生物和其他方面发表的101篇论文；下篇为著作集，收录了陈华癸教授编写或翻译的教材和专著15部。本论著集汇集了陈华癸教授一生的主要贡献，以纪念教授百岁诞辰，同时希望将教授为科研教学严谨献身的精神和为国家无私奉献的风节继承并发扬光大。

图书在版编目(CIP)数据

陈华癸论文著作集——诞辰100周年纪念/张忠明，陈雯莉，周啟主编.
—北京：科学出版社，2014.5

ISBN 978-7-03-040442-8

Ⅰ.①陈… Ⅱ. ①张… ②陈… ③周… Ⅲ. ①土壤微生物学-文集 Ⅳ. ①S154.3-53

中国版本图书馆CIP数据核字（2014）第076767号

责任编辑：李 悦 / 责任校对：钟 洋
责任印制：钱玉芬 / 封面设计：铭轩堂设计公司

科 学 出 版 社 出版
北京东黄城根北街 16 号
邮政编码：100717
http://www.sciencep.com
北京凌奇印刷有限责任公司 印刷
科学出版社发行 各地新华书店经销
*
2014年5月第 一 版 开本：889 × 1194 1/16
2014年5月第一次印刷 印张：37 插页：10
字数：1 100 000
POD定价： 248.00元
（如有印装质量问题，我社负责调换）

《陈华癸论文著作集——诞辰100周年纪念》编辑委员会

中国科学院院士陈华癸教授论文著作集

诞辰 100 周年纪念

(1914 年 1 月 11 日至 2014 年 1 月 11 日)

陈华癸教授
著名的土壤微生物学家、农业教育家
(1914 年 1 月 11 日至 2002 年 11 月 19 日)

简　介

陈华癸，1914 年 1 月 11 日出生于北京，祖籍江苏昆山。1935 年毕业于北京大学生物系，后留校任助教一年，师从张景钺教授，学习植物学和植物形态发生学。1936 年赴英国伦敦大学学习，一年后在著名的洛桑试验站开始攻读博士学位，研究豆科-根瘤菌共生固氮作用，同时利用该站的有利条件，学习土壤学和土壤植物营养学，奠定了与土壤微生物学有关的理论基础。1939 年取得哲学博士学位，时年 25 岁。1940 年回国，在西南地区工作期间，曾实地调查了云南、四川、陕西、广西、湖南的豆科绿肥生产和应用情况，随后着重研究紫云英根瘤菌。1948 年，陈华癸教授又开拓了一项新的研究课题，即水稻田土壤微生物的研究。

科学研究的成果

陈教授一贯主张高等农业院校既要重视教学、科研和技术推广三者并举，又要在科研上实行解决国民经济发展中的近期和长期问题并重的方针。正是在这种思想的指导下，从 20 世纪 30 年代开始，他坚持从生产实践出发的科研道路，锲而不舍地从事适合我国国情的土壤微生物的研究（包括共生固氮和水稻田土壤微生物两个领域），作出了重要贡献，取得了丰硕的科研成果，共发表论文 115 篇。陈教授是我国土壤微生物学领域的三位奠基人之一，在国内外的土壤微生物学界享有崇高的声誉。

一、共生固氮方面研究的成果：①早在 20 世纪 30 年代初就首次发现豆科植物根毛在根瘤菌感染之前的伸长和弯曲是与根瘤菌分泌生长素类物质的作用有关；②在研读博士期间，先后在英国《自然》、《英国皇家学会会刊》和《农业科学杂志》上发表 4 篇论文，其中对《有效根瘤菌株和无效根瘤菌株在寄主上结瘤的生长发育比较研究》一文，受到从事共生固氮研究的专家们的高度重视；③首次揭示紫云英是一个独立的互接种族，并定名为中国紫云英（*Astragalus sinicus* L.），结果发表在美国《土壤科学》杂志上，这一成果为紫云英根瘤菌人工接种大面积生产应用奠定了基础；④1957 年在中国-苏联稻作学术会议上，他提出水旱两作的水稻土生物循环中有 5 个特点的论述，在当时是有开拓性意义的；⑤关于豆科根瘤菌分子遗传学的研究共发表约 50 篇论文，其中一篇获得农业部科学技术进步奖一等奖；⑥80 年代中期在马桑 Frankia 菌的同一个根瘤中首次分离到 2 个完全不同的菌株——Frankia 菌和属于小单孢菌科的放线菌，到 2005 年，这一结果被国外学者陆续从不同 Frankia 菌的根瘤和豆科羽扇豆根瘤中得到完全证实。

二、水稻田土壤微生物方面研究的成果：①首先发现兼厌气性的硝化作用和硝化细菌；②分离获得亚硝酸细菌及其伴生菌的纯培养；③证实伴生菌不仅在好气和嫌气条件下均能进行反硝化作用，而且在完全没有有机碳化合物的亚硝酸培养基中也能单独生长和繁殖；④证实亚硝酸细菌和伴生菌共同生活在一起是造成水稻田土壤中无机态氮大量流失的主要原因。

教学事业的贡献

抗日战争胜利后，陈教授转入教育界，致力于农业教育。先后于北京大学农学院和武汉大学农学院任教授，并分别筹建和创建土壤系及农化系，任系主任。1952年院系调整后，他进入华中农学院（1985年更名为华中农业大学）继续任教授，创建土壤与农业化学系并任系主任。从此，他以渊博知识、勤奋好学、诲人不倦、教书育人和严谨治学的精神把自己的青春都奉献给了土壤与农业化学系，为华中农学院的发展作出了积极贡献。当时的土壤与农业化学系，师资、学科门类都十分有限，业务水平也亟待提高，他把培养青年教师作为迫切任务，采取以教学和科研任务相结合的方式，首先在教学上，鼓励青年教师制订教学大纲、建设实验室、参与编写全国统编教材和实验指导、组织外语学习（包括翻译）等，并尽早让他们担任主讲教师，成为各门课程的教学骨干。在科研上，从选题、制订试验方案、项目的具体实施到撰写论文，总是悉心指导，反复修改，以培养其学术思维、研究能力和严谨学风。这些措施非常奏效。至20世纪90年代初，已为华中农业大学培育了一支学科齐全、学力坚实、结构合理的教学、科研梯队，全系拥有40余名正副教授，3个博士点和3个博士后流动站；6名博士生导师；4个专业；4个研究室和1个开放实验室；1个重点学科。在他的领导下，土化系成为国际学术交流活跃的学系。同时教授还十分关心全国兄弟院校的师资培养，积极接受他（她）们来校进修，帮助提高他（她）们的业务水平。20世纪60年代初和70年代初，先后培养了一位越南留学生和一位美国留学生。在1965年前后，陈教授按省政府的要求，主持在华中农学院筹建紫云英根瘤菌肥料厂，并进行大量生产，产品除了满足湖北省各县、市的需要外，还销到广东、广西的部分地区。1989年，教授和英国约翰·英纳斯研究所合作，共同在华中农业大学举办了一期国际链霉菌遗传操作实践训练班，结果很成功，影响较大。陈教授一生中还编写了包括教材、专著和译著共计15部图书。

此外，他还非常重视和关心我国农业教学的改革，他单独或和同事合作，先后发表8篇文章，提出各种改革方案和宝贵意见，直到2001年他还与同事一起发表了他一生中的最后一篇论文。教授这种为祖国农业教学事业的发展和改革而奋斗终生的宝贵精神，值得人们永远学习。

陈华癸教授影集

陈华癸教授与夫人周如松教授的金婚合影

庆祝教授八十岁寿辰庆典

陈华癸教授在他 80 岁寿辰庆祝典礼上致答谢词（右座是夫人周如松教援）

代表向陈教授和夫人献花

教授和参加其 80 岁寿辰庆典的人员一起合影

参加教授 80 岁寿辰庆典的 49 级学生和老师一起合影

教授的教学和科研工作

讲授本科生微生物学课程

课后辅导

指导微生物学实验课

指导研究生和年轻教师的科研

担任农业微生物重点实验室学术委员会主任时与学术委员会成员合影

主持研究生学位论文答辩

主持国际链霉菌遗传操作实践训练班开学典礼

与美国留学生讨论问题

在灯光栽培室观察紫云英根瘤菌优良菌株选育实验的结果

教授的学术活动

全国豆科根瘤菌共生固氮学术讨论会和我国共生固氮研究 50 年纪念会代表合影（成都，1987）（前排就坐右起第八至第十人依次为我国土壤微生物学奠基人 陈华癸、张宪武、樊庆笙教授）

全国高等农业院校生物课程教学改革研讨会（1996）

访问英国约翰·英纳斯研究所。在遗传室观察链霉菌分子遗传实验结果（1982）

访问 43 年前曾在这里做博士论文的洛桑试验站，与该站国内访问学者合影（1982）

访问英国温室作物研究所。在温室观察应用真菌防治番茄蚜虫的良好效果（1982）

访问 Glasgow 大学，与该校教授座谈（1982）

教授的其他活动

农业部部长陈耀邦来学校看望他的两位老师杨新美教授和陈华癸教授

任华中农学院院长期间，教授向农业部部长何康汇报工作

教授和挚友樊庆笙教授

教授与武汉大学高尚荫教授

参观伦敦植物园，在大温室前留影（1982）

教授生病住院期间，他的夫人和学生一起来医院看望他

序

早在20世纪50年代初，我带领原北京农业大学（现更名为中国农业大学）的学生去长沙紫云英根瘤菌菌肥厂实习，从而了解到陈华癸教授多年来一直从事紫云英根瘤菌的研究，他带领他的团队还开展了紫云英根瘤菌菌肥的生产与推广应用，他们的产品不仅在国内推广使用，而且还销往东南亚国家。从此，我便开始认识并不断深入了解陈华癸教授。

陈华癸院士于1914年出生于北京，1935年毕业于北京大学生物系并留校工作，1936年赴英国伦敦大学攻读博士学位，在著名的洛桑试验站研究豆科-根瘤菌共生固氮的作用机制，1939年以优异的成绩获得博士学位并回国工作。在长达近70年的科学研究和高等教育的学术生涯中，他针对国家经济发展和生产实际的需求，瞄准世界科学技术的前沿，在根瘤菌的共生固氮、水稻田土壤微生物等领域开展了卓有成效的科学研究和推广应用工作，在国际上享有崇高的学术声誉，是我国土壤微生物学的奠基人之一，以他名字命名的“中慢生华癸根瘤菌”就是国内外学术同仁对他学术贡献的一种公认。同时，陈华癸院士在长期的高等农业教育实践中，慢慢形成了具有独特见解的教育思想，他不仅培养了数以千计的全日制本科生和研究生（包括留学生），而且针对生产实际和其他兄弟院校师资培养的急需，还培养了大量的进修生，为国家建设培养了众多的栋梁之才。

陈华癸院士具有高瞻远瞩的国际视野，对我国的高等农业教育和科学研究具有许多独到的见解，并利用各种机会建言献策，特别是他在担任华中农学院（现华中农业大学）院长、全国人大代表和中国科学院院士期间，向中央和国家有关部门提出了许多建设性的意见和独到的见解，对相关领域的计划制定和工作推进作出了重要贡献。

陈华癸院士不仅具有很高的学术造诣，而且具有严谨务实的学风和刚正不阿的人格，他敢于伸张正义、坚持真理，反对主观武断、弄虚作假，为良好的学风建设作出了重要贡献，成为我们的学习榜样。在我同陈华癸院士半个世纪的联系与合作过程中，他崇高的人格魅力、忘我的敬业精神和精湛的学术造诣深深地感染着我，一直鼓舞着我不断前行。在陈华癸院士诞辰一百周年之际，我们重新拜读他的论著，重温他的谆谆教诲，回顾他的人生轨迹，继承他的遗志，将他未尽的事业推向新的高度！我想这是我们作为学生和晚辈纪念和学习他的最好方式。

李季伦

中国科学院院士

中国农业大学教授

2013年8月15日于北京

前　言

中国科学院院士陈华癸教授是一位著名的土壤微生物学家和农业教育家。他从抗日战争胜利后进入教育界以来，一直主张高等农业院校既要重视教学、科研和技术推广三者并举，又要在科研上实行解决国民经济发展中产生的近期和长期问题并重的方针，正是在这种思想指导下，从 20 世纪 30 年代开始，他坚持从生产问题出发的科研道路，锲而不舍地从事适合我国国情的土壤微生物的研究（包括共生固氮微生物和水稻田土壤微生物两个领域），从而在科研上获得丰硕成果。此外，他还大力推广在我国中南地区的水稻田里种植草籽绿肥(紫云英)，扩大双季稻面积，为国家增加粮食产量作出了重大贡献。

陈教授不仅是著名的土壤微生物学家，也是我国三位土壤微生物学奠基人之一。在国内外土壤微生物学界中享有崇高的声誉。

从他在北京大学生物系学习期间首次发表论文开始，一直到 2001 年为止，共计发表论文 115 篇。在这些论文中，有将近一半是他单独和领衔撰写的，其他论文则是他指导研究生和青年教师以及与同事合作撰写的。值得指出的是，他在指导研究生进行科学研究时，从选题、制订试验方案、具体实验的贯彻到撰写论文，总是悉心指导，反复修改，以培养其学术思维能力、研究能力和严谨的学风，从而提高学生的学术水平。此外，教授在一生中还编写了 15 本著作。

在教学方面，他为本科生讲授微生物学课程，并在实验课中亲自进行具体指导，促使学生们掌握坚实的基础知识和具体的操作技能，为学生的科研能力进一步提高奠定了基础。此外，他还讲授过土壤学、肥料学和微生物遗传学等课程，他对研究生和青年教师的培养则采取教学和科研相结合的方式，积极鼓励他们制订教学大纲、建设实验室、参与编写全国统编教材和实验指导书、学习外语(包括翻译)等，让他（她）们勇挑教学、科研重担，早日成材。

40 多年来，陈教授为国家培养出具有土壤、农化和微生物方面知识的本科生、进修生、留学生和研究生的数量已近 3000 人，桃李满天下。这些高级人才活跃在全国各地的教学、科研和生产部门等，为祖国的建设事业发挥了重要作用。

教授为人爽直，顾全大局，秉公处事，认真不苟。他学风严谨，辛勤劳动，数十年如一日，耕耘在华中农业大学的校园里，他这种高尚的品德、渊博的知识、独有的学术和教学思想，以及对我国教学和科研事业的忠诚和重大的贡献等，一直受到年青一代的敬佩和爱戴，也为后人树立了光辉榜样，成为几代人学习的典范。

教授是我们的导师，也是将我们引入到科学研究园地的恩师，他对我们的成长乃至于整个人生励志有深远的影响。为了报答恩师对我们的期望，我们将恩师的一生功绩汇集成一本《陈华癸论文著作集》；一方面是纪念恩师的百岁诞辰，另一方面也希望本书为读者和后人所吸收和继承并发扬光大。

本书分为上、下两篇以及附件。上篇为论文集，下篇为著作集。论文集分为三部分：共生固氮微生物、水稻田土壤微生物和其他。每部分有一个提要，说明该部分论文的内容和重要结果等。每篇论文均注有该文原载杂志、卷号、期号、页数和发表时间，没有发表过的文章则注明未发表或未正式发表。著作集则分为主编教材、专著和译著。所有著作都配有封面照片、编著者姓名和著作内容(目录)等。

本论著集不是按发表或出版的时间顺序排列，而是根据内容分类排列的。

附件包括附件Ⅰ，即已在《微生物与农业》上刊登不再在本论著集中再重复登载的论文目录；附件Ⅱ，教授历届招收的研究生名单；附件Ⅲ，教授的简历以及编后记。

在本书编辑过程中，包括论文、著作和教授资料的收集，以及论文的打印、校对和照片的加工等，不少老师和同学都花费了很多精力和时间。除在此表示感谢外，还将他(她)们的姓名均列入本书的参编人员中。同时我们也要感谢对本书的出版予以关心和支持的领导和同仁。

由于时间仓促，编写和校对如有遗漏不妥之处，恳请作者和读者批评指正。

周 启

2013 年 6 月 10 日

目　录

上篇　论　文　集

第一部分　共生固氮微生物

第二项 非豆科-弗兰克菌(Frankia)

第三项　固氮螺菌

第二部分　水稻田土壤微生物

第一项　水稻土营养元素的生物循环

第二项　硝化作用和硝化细菌

第三部分　其　他

第一项　陈华癸教授独立撰写和由他指导研究生、青年教师，以及与同事合作撰写的论文

第二项 教育改革研究

下篇 著 作 集

（一）教 材

（二）专 著

（三）译 著

附 件

第一部分　共生固氮微生物

提　要　这是陈华癸教授研究最早、时间最长、发表文章最多的课题，研究内容有三项，共发表论文 65 篇[注]。

第一项　豆科-根瘤菌

早在 20 世纪 30 年代，陈华癸教授首次发现豆科植物根毛在被根瘤菌感染之前的伸长和弯曲与根瘤菌分泌的生长素类物质的作用有关。1936 年他在英国伦敦大学攻读博士学位时，在著名的洛桑试验站由 Thomton H.G.博士指导开展了豆科-根瘤菌共生固氮研究。他首次对豆科植物无效根瘤和有效根瘤的形态发育进行了比较研究，阐明了共生固氮有效性机理的一个重要方面。论文发表在英国皇家学会会刊（2 篇）和 *Joural of Agricultural Science* 上。该发现一直受到学术界的高度重视，其结论一直被 Stanier R.Y.等人著的《微生物世界》和 Becjevsen F.J.的论文及有关教科书所引用。Quispel A.在 1974 年还高度评价了他的研究结果的意义。他对共生固氮的另一贡献，是他在 1941 年与同事一起首先发现紫云英根瘤菌是一个独立的“互接种族”，并将它命名为中国紫云英（*Astragalus sinicus* L.）。论文发表在美国 *Soil Science* 杂志上。这项成果不仅丰富了根瘤菌和寄主植物共生结瘤固氮的特异性知识，而且为紫云英人工接种的大面积应用奠定了基础。20 世纪 50～60 年代，他发表的论文主要着重于根瘤菌剂生产和草籽绿肥推广应用方面，为南方扩大双季稻的栽培面积和粮食增产作出了贡献。在 70 年代后期，他带领同事开拓了共生结瘤固氮的分子遗传学研究的新领域。1984 年，论文发表在《中国科学》（B 辑）上，1986 年获得农业部科学技术进步奖一等奖。

第二项　非豆科-弗兰克氏菌（*Frankia*）

陈教授在 80 年代初和中国农业科学院油料作物研究所周平贞研究员合作开创尼泊尔马桑（*Coriaria nepalensis*）*Frankia* 菌的研究。除分离到大量纯培养外，还对根瘤形成发育过程中的内部结构进行了观察。文章先后发表在《武汉植物学研究》、《微生物学报》和《中国油料》上。直至 1992 年，一直在研究马桑 *Frankia* 菌的硕士来华中农业大学深造继续进行研究。他在周启教授和周平贞研究员共同指导下，研究马桑 *Frankia* 菌纯培养的生物学特性，获得的主要成果有：①揭示了马桑 *Frankia* 菌纯培养的生物学特性的多样性，为 *Frankia* 菌属的分类奠定基础。结果在《中国农业科学》上发表后，被编辑部选为优秀论文，并和其他优秀论文一起编译成一本英文论文集出版。②另一篇文章发表在《科学通报》后，被 SCI 收录并刊登出该文的摘要。

值得一提的是在 1987 年前后，研究小组首次在马桑根瘤中分离到一些与典型 *Frankia* 菌完全不相同的放线菌，并证实它们均为小单孢菌科的成员，由此得出结论，认为马桑根瘤中包含有 *Frankia* 菌和属于小单孢菌科的放线菌。直至 2005 年，国外研究者从不同 *Frankia* 菌的根瘤中和豆科羽豆根瘤中也都分离到小单孢菌科的放线菌，进一步证实了此前的研究结果和结论。

第三项　固氮螺菌

这是陈教授在中国科学院武汉微生物研究所（现为中国科学院武汉病毒研究所）兼任副所长时，指导该所研究人员进行研究的项目。

[注]：包括已在《微生物与农业》上刊登的论文，见附件 I。

第一项　豆科-根瘤菌

豆科植物之根瘤*

陈华癸

（北京大学）

（一）引言

豆科植物根上有许多瘤状突起，吾人名之曰“根瘤”。根瘤乃植物根薄膜组织因受某种细菌之侵害刺激而起之不正常生长。此种细菌名 *Bacillus radicicola*，Biejerinck，侵入根内而与之行互惠生活，盖彼吸收高等植物体内之碳水化合物而供给寄主以大量之氮化物也。此种共生现象不仅豆科植物为然，其它科目中亦多有之。兹文则仅以豆科为范围。

（二）根瘤

细菌自根毛之尖端侵入，达于皮层，则被侵害之皮层细胞及其紧邻乃受一定刺激而变为生长组织，加速分裂以构成根瘤，根瘤大部分由内外两层之薄膜组织构成，内部之薄膜组织即所谓含菌组织（Bacterioid tissue），具薄膜及大量之蛋白体，蛋白体为细菌状之小棍体，T 形，Y 形，或分支状，即类菌体（Bacterioids）也。外层之薄膜组织含淀粉粒极富。为养分之储藏所。在薄膜组织之间有管系通入司养分及氮素之运输。根瘤之外则被以木栓层。因含菌组织、管系分布，生长方向之不同，Spratt 氏[1]将豆科植物之根瘤分为四型：

（Ⅰ）Genisteae 型　根瘤初发生时，生长组织为一团，后来大部分之组织成熟，不再分裂，只余许多分散的小块的生长组织，结果根瘤乃成鸡冠状；在瘤之底部有一束管系伸入；含菌组织为薄层的不含菌组织分隔成几块。Lupinus，Genista，Spartium 等植物之根瘤属之。

（Ⅱ）　Trifoleae 型　似前者，生长组织幼时为一团，但很快即变为几个顶生长组织，因顶生长之结果，根瘤遂成树枝状；含菌组织不为不含菌组织所隔；管系之分布与前者相似。Phaseolus，Trifolium，Lotus，Coronilla 等植物之根瘤属之。

（Ⅲ）Viceae 型　根瘤分枝或否，含菌组织完整，自瘤之底部有两束管系通入。Vicia，Pisum，Cotutea 等植物之根瘤属之。

（Ⅳ）Mimosoideae 型　前三者皆为一年生植物之根瘤，此则为多年生植物之根瘤状态，根瘤亦可生活多年，其生长组织之寿命与根瘤同长，在 Acacia 等植物，其含菌组织之外为一层新鲜活泼之生长组织所包，在 Sophoia，Robinia 等植物则仅有一团顶生长组织。

（三）根瘤菌

根瘤之发现也，在十五世纪之中叶，而根瘤菌之发现则远在三百年后，盖细菌之第一次发现不过百余年事尔。十九世纪之末，植物病理学家 Frani 氏见根瘤中有反光性极强之细丝，以为是高等菌类之菌丝，而 Brunchorst 氏则以为不过是类蛋白体（Proteids）。至 1888 年，Beijerinck 氏[2]始认为属裂殖菌类（Schizophyte）而名之曰 *Bacillus radicicola*，是为根瘤菌得名之始。

根瘤菌为球形或短棍状之小裂殖菌属 Bacteriaceae，其游动时期（motile stage）生一根或多根之鞭

---

*原载于《中国植物学杂志》，2(1):497～500, 1935.

毛，在根瘤内则多成 Y 或 T 型之类菌体，性嗜酸，生活于 pH=4-7 之间。无数之细菌胶连成一侵蚀丝（infection thread）以侵入根内，是即 Frank 氏所见之反光性极强之丝也。凡豆科植物之根瘤菌皆属此种。Hanson 氏[3]观察其动时期之不同，分之为两型：甲，豇豆、毛豆、落花生等植物之根瘤菌在动时期只有一根鞭毛；乙，金花菜、甜金花菜、豌豆、苜蓿等植物之根瘤在动时期有数根鞭毛。更因生理性质之不同，*B. radicicola* 又可分为不同的 Biotypes，关于此点以 Wright 及 Steven 二氏之研究为最多，精深繁杂，兹从略。

根瘤菌只能自根毛侵入寄主体内，侵入后即四散蔓延于皮层中，此时皮层细胞之细胞浆立即加浓，核亦立即加大而成生长组织之状态，开始分裂，长成根瘤。侵蚀丝每入一皮层细胞内即散布无数细胞于细胞中而成含菌组织，此时在侵蚀丝外寄生细胞分泌一层纤维质包围之，并用以防止细菌之散布也。侵蚀丝既被纤维质膜所包围，容积不能加，而细菌增殖不已，结果乃将膜一部涨起成胞（Cyst），涨至不能再涨，则膜破而细菌仍可散布细胞浆内。散布细胞浆中之细菌多膨胀而成 Y 或 T 形之类菌体。生长组织至一定时期，则大部分之细胞再分裂，仅余一小部分之顶生长细胞，各种细胞开始分化构成管系，于是细胞乃得充分之养分，繁殖鼎盛，生活活泼，供给寄主以大量之氮素。

秋落之后，植物不再制造碳水化合物，细菌之力源供给日减，生气衰微，氮素之固定日少，终至毫无裨益而成寄生生活，待夫植物体内之储藏养料不继，细菌即以寄生细胞浆为食物，细胞浆逐渐收缩，细胞中之类菌体亦破为小粒，随收缩之细胞浆共赴灭亡，根瘤腐败，而散布于细胞之间之细菌入土中休眠，以待来年。

（四）共生现象

根瘤为共生现象；高等植物供给细菌以碳水化合物而维持其生活，情形单简，不多赘，兹略伸述细菌之氮气固定对寄主之裨益于后：

农作之主要肥料有三，曰氮，曰钾，曰磷，三者不可一缺，固吾人之所习之也。Wagner 氏[4]以燕麦及豌豆做实验，各以之植于三种不同之土壤中：甲为不施肥料者；乙为有钾及磷而缺氮素者；丙为三者皆丰富者。结果燕麦之生长恰如意料，只丙类之生长旺盛，前二者俱不能长大；而豌豆生长则出乎意料，不施氮肥料之植物亦能有极佳之生长与丙类者无异。一八八八年 Hellriegel 及 Wilfarth 两氏[5]又做系统实验，亦见豆科植物能生存于缺氮之土壤中，唯消毒之缺氮土壤则不能生存；两氏又见根瘤只生于未消毒土壤之植物根上，而消毒者无之，于是乃以为根瘤之造成乃豆科植物与下等植物之共生现象，根瘤且直接影响豆科植物之氮素吸收。同年 Beijerinck 及 Prazmovskii 两氏做成 *B. radicicola* 之单独培养，且进而探求其生活史，于是三百年来，根瘤之谜遂破。一八九四年 Winogradsky 氏更证明细菌能固定大气中游离之氮素，根瘤之直接影响与植物对大气中氮之吸收亦得证明，而吾人对根瘤之知识乃渐趋完善。

犹有一言愿为读者告知：细菌与豆科植物之互惠生活为条件的而非绝对的，如寄主不能履行其碳水化合物之供给，则细菌不能制造氮化物，且更噬食寄主细胞之细胞浆以取营养。第三节之末，所述根瘤之破败即其明证。一九二五年 Brenchley 及 Thornton 两氏[6]植植物于缺硼之土壤中，则根瘤菌之氮素制造极少且以寄主细胞为食料，详细观察，乃见植于缺硼土壤中之植物，其管系生长不正常，无维管伸于根瘤中；细菌缺养料之供给不得不行此下策也。蓋“朽腹从公”为天下绝无之事！

参 考 书

[1]Spratt, E. Ann of Bot; Vol. Xxxiii; pp. 189-99. 1919

[2]Beijenuck，M.W. Bot. Ztg；Vol. 46. 1888

[3]Hansen，R. Science n.s. Vol. 50；pp. 568-9

[4]Wagner，P. Ergebnisse von Dunungungsversuchen. 2te Auff. Dornstadt，1891

[5]Hellriegel，H. and Wilfarth, H. Beilege Zeilege. Rubenzucker Indust. d. deutsch. Reich. Berlin，1888

[6]Brenchley，W.E. and Thornton, H.G. Roy. Soc. Ser. B；London. Vol.98.pp.373-399

Production of Crowth-Substance by Clover Nodule Bacteria*

H. K. CHEN.

(Bacteriology Department, Rothamsted Experimental Station, Harpenden. Sept. 7.)

THIMANN[1], using the standard *Avena* technique, showed that a growth-substance is produced in considerable amount in root nodules. He furthermore claimed that the growth-substance produced is not derived from the meristematic tip of the nodule, but comes directly from the bacterial tissue.He found that the symptoms induced by 3-indole-acetic acid upon roots closely resembled those produced, in Molliard's work, by the action of sterile filtrate of nodule bacteria upon pea roots, and he consequently believed that the bacteria cultivated in laboratory media produce growth-substance in considerable amount.

Using Went's pea test technique[2], I have confirmed Thimann's view that nodule bacteria do produce a good deal of growth-substance in a culture provided with a small amount of tyrptophane in the medium. The filtrates of four weeks old cultures of strains of clover nodule bacteria grown in a yeast-water medium containing 0.02 per cent tryptophane were tested against pea shoots prepared according to Went's method. The results of a typical experiment are shown in the accompanying table.

	pH	Dilution				
		1/4	1/8	1/16	1/32	1/64
Uninoculated control	8.2	0	0	—	—	—
Strain 2057 in medium without tryptophane	8.4	0	0	0	0	0
Strain 2057	7.9	+	+	+	±	0
Strain 2027	8.0	+	+	+	±	0
Strain 202	7.8	+	+	±	0	0
Urine	—	+	+	±	—	—

+ Positive reaction; 0 No reaction

± Reaction doubltful; —Not tested

It appears that strains that are effective in fixing nitrogen in the plant produce in this tryptophane medium very little if any more growth-substance than do the non-beneficial strains that are not effective in fixing nitrogen. The old laboratory strains which have lost their virulence,that is,are unable to produce nodules when supplied to the plant,were found in most experiments to produce less growth-substance,as illustrated by the strain "202" in the table.

*原载于 *Nature*. 142: 753～754, 1938.

1. Thimann,K.V., *Proc.Nat.Acad.Sci.*,22,511-514（1936）.
2. Went,F.W.,and Thimann,K.V., "Phytohormones" ,54-55（New York,1937）.

The Structure of 'Ineffective' Nodules and Its Influence on Nitrogen Fixation*

H. K. CHEN AND H. G. THORNTON

Bacteriology Department, Rothamsted Experimental Station, Harpenden

(Communicated by Sir John Russell, F.R.S.-Received 4 April 1947)

[Plates 13, 14]

1. The anatomy and cytology of nodules produced on clover, peas and soy beans by 'effective' and 'ineffective' strains of *Rhizobia* were investigated, with especial reference to the changes in volume of the active infected tissue during the life of the nodule.

2. In clover the mean volume of this active bacterial tissue is about three times as great in 'effective' as in 'ineffective' nodules. This is due to an early arrest of growth in nodules produced by ineffective strains.

3. In all nodules the active bacterial tissue eventually disintegrates, but in effective clover nodules it remains without disintegration for about six times as long as in ineffective nodules.

4. In an experiment to test the nitrogen fixation by clover inoculated with an effective and an ineffective strain, the difference between the strains in the amounts of nitrogen fixed could be accounted for by the differences in volume and in duration of the active bacterial tissue.

5. In peas, nodules produced by an effective strain were nearly twice the length of those produced by an ineffective strain, and their bacterial tissue remained without disintegration for about twice as long.

6. In soy beans the mean volume of bacterial tissue was 4.75 times as great in effective as in ineffective nodules and the percentage of that volume composed of infected cells was twice as great.

7. In ineffective soy bean nodules disintegration of the bacterial tissue began when the plant was 4 weeks old and was practically complete by the twelfth week, at which time no disintegration could be found in effective nodules.

8. The difference in amount of nitrogen fixed by soy bean plants bearing each type of nodule could be accounted for wholly by the factors mentioned above.

9. Thus in both clover and soy bean nodules the volume and duration of the active infected tissue are the main, if not the only, factors determining differences in nitrogen fixation amongst the strains tested.

A. INTRODUCTION

The fact that strains of nodule bacteria differed in their ability to benefit the host legume was realized as early as the nineties of last century. It was, however, the careful work of Stevens (1925) working with lucerne, and of Wright (1925a, b) with soy beans, that definitely showed the wide differences in nitrogen fixed in the same species of host legume when infected with different strains of bacteria. Their work has been amply confirmed and has been extended to other host plants by numerous workers such as Helz, Baldwin and Fred (1927) working with peas, Eckhard, Baldwin and Fred (1931) with lupins, and Baldwin and Fred (1929) with clover.

In some cases, especially amongst clover nodule bacteria, strains have been found that fix very small

*原载于 *Proc. Roy. Soc. B.* 129:208～229, 1940.

amounts of nitrogen and confer scarcely any benefit on their host. Such strains are here referred to as 'ineffective' although this term is relative and it is doubtful whether strains exist that fix no nitrogen at all.

Various authors have attempted to find some correlation between ineffectiveness and such other characters as serological behaviour and cultural features shown *in vitro*. Stevens（1925）and Wright（1925a）had some success in relating ineffectiveness to agglutination reactions, while Baldwin and Fred（1927）found that strains of lucerne nodule bacteria showed differences in respect to mannitol fermentation that agreed with the determinations of effectiveness in nitrogen fixation found by Stevens（1925）. But, on the whole, little correlation has been found between effectiveness in the plant and behaviour of the organisms in laboratory culture.*

It has, nevertheless, been natural to suppose that bacteria of ineffective strains fix less nitrogen in the host plant owing to some defect in the mechanism of nitrogen fixation possessed by them. This is not self-evident, however, since crude nitrogen determinations of host plants inoculated with different strains do not provide a measure of the nitrogen-fixing efficiency of the bacteria unless an estimate can be made of the quantities fixed by unit masses of bacteria in unit time. To obtain this estimate, corrections must be made for such factors as the number of nodules, the volume of active tissue containing bacteria within these nodules, and duration of its activity. Adequate data for making even approximate corrections of the last two factors have not hitherto been available. The need for such data has, however, long been apparent from the frequently recorded fact that both the numbers and mean size of nodules show large differences as between strains, those that are ineffective tending to produce more numerous but much smaller nodules.

Surprisingly few observations have been recorded in which the anatomy and cytology of nodules produced by effective and ineffective strains have been compared. Elizabeth McCoy（1929）studied the anatomy of nodules on *Phaseolus*. Earlier workers such as Frank（1890）and Schneider（1892）considered that nodules on *Phaseolus* as a class were not beneficial to the host plant. Benefits from inoculation have been recorded with *Phaseolus* by Wilson and Leland（1929）but it does seem that ineffective strains are particularly prevalent with this host plant. In any case, the nodules studied by McCoy were apparently on an ineffective type. She found that the formation of the nodule was due rather to the multiplication of infected cells than to that of uninfected cells which were later invaded by the bacteria, as is more usual in legume nodules. The former process, however, has been found by Milovidov（1926, 1928）to be typical of lupin nodules, an observation confirmed by us, and is hence more likely to be conditioned by the type of host plant than by the particular strain of the invading organism. McCoy also found her *Phaseolus* nodules to be unusual in the following characters.（1）There were relatively few infected cells in the central tissue.（2）The nodules contained an abundance of starch associated with a great development of mitochondria.（3）The bacteria were rod-shaped and did not change into swollen or branched 'bacteroids'.

Marie löhnis（1930）studied the anatomy of pea and clover nodules, in each case produced by an effective and an ineffective strain. She found that the two types of pea nodules differed in the amount and distribution of starch, the ineffective nodules containing more starch, more widely distributed through the nodule. Her effective nodules also contained a peculiar type of bacterial cell not found in the ineffective nodules. In clover nodules, however, she records no differences in anatomy, in amount and distribution of starch or in the shape of the bacteria as between effective and ineffective nodules. This similarity suggested that the unusual characters found in ineffective nodules on *Pisum* and *Phaseolus* are not necessarily associated with inefficiency.

---

* Dr Hugh Nicol tested the strains of nodule bacteria that form the subject of this paper as regards their growth and the change of reaction produced in media containing a wide range of carbohydrates and higher alcohols. No differences were found that could be correlated with effectiveness towards the host plant.

It therefore seemed desirable to study the anatomy and cytology of nodules produced on several host plants by effective and ineffective strains, first, in order to determine what characters were common to ineffective but not found in effective strains, and, secondly, to obtain quantitative data from which it might be possible to form some estimate of the relative amount of nitrogen fixed by a unit mass of bacteria in unit time in nodules of each class. For without such an estimate it was impossible to determine how much of the ineffectiveness of a given strain was due to relative inability of the individual bacteria to carry out the process of nitrogen fixation, and how much to poor growth of the bacteria within the host tissues.

B. MATERIAL AND METHODS

The nodules on clover, peas and soy beans were chosen for this investigation, because these are widely different types of host plant from which strains of nodule bacteria have been isolated that differ greatly in effectiveness. The following bacterial strains were used:

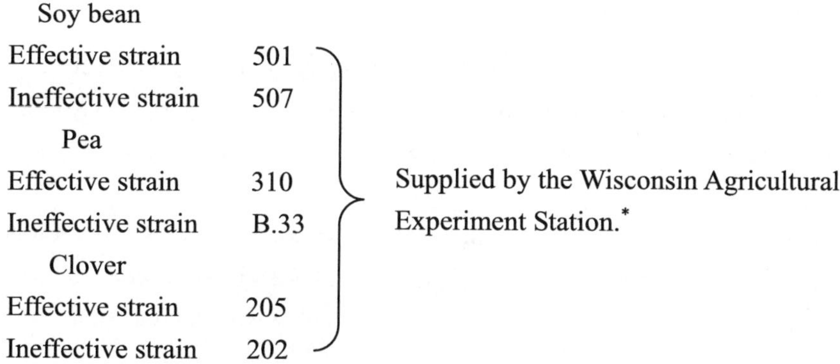

Soy bean		
Effective strain	501	Supplied by the Wisconsin Agricultural Experiment Station.*
Ineffective strain	507	
Pea		
Effective strain	310	
Ineffective strain	B.33	
Clover		
Effective strain	205	
Ineffective strain	202	

Effective strain A, supplied by the Agricultural Experiment Station, Stockholm.

Ineffective strain Coryn, obtained from hill pasture on Coryn Mountain, Aberystwyth.

Strains 205 and 202 are the same that formed the subject of Marie Löhnis' observations of clover nodules.

The original strains 205 and 202, obtained from Wisconsin, lost their power of producing nodules during the course of the work, although strain 205 has since recovered this power without loss of effectiveness. In the meantime, however, a second active culture of each strain was obtained from Wisconsin in 1937, and these are here distinguished by the numbers 2057 and 2027.

The anatomical descriptions given below are based on the study of nodules on clover grown in agar and on peas and soy beans grown in sand.

The clover plants used for this study were grown in agar media in test tubes $1\frac{1}{4}$ in. in diameter. Three methods were employed. The first was the usual one of growing the plants in agar blocks made by pouring about 30 ml. of the melted agar medium into each tube and allowing it to solidify with the tube kept upright. Seeds were sown near the glass so that a good proportion of the root system should remain visible. The second method was to pour 10 ml. of melted medium into each tube and to solidify it by rotating the tube under a stream of cold water, as is done in the 'roll-tube' method of plating bacteria. Seeds were sown at the upper edge of the film of medium. Such cultures have the advantage that the whole root system remains visible through the glass, so that the appearance and growth of the individual nodules can be followed. It was used in obtaining growth curves of nodules and to obtain nodules of known age for anatomical study. The method suffered from a tendency of the agar film to dry up. This drying could be delayed by adding a flew ml. of sterile water to each tube, but it was later replaced by a third method in which 12 ml. of agar medium were added to each tube and allowed to solidify and form a slope, the seed being sown at the upper end.

For the first, or agar block method, the following medium was employed:

K_2HPO_4	0.5g	NaCl	0.1g	$FeCl_3$	0.01g
KH_2PO_4	0.5g	$Ca_3(PO_4)_2$	2.0 g	Agar	10.0g
$MgSO_4 \cdot 7H_2O$	0.2g	$FePO_4$	0.5g	Tap water	1 L

For the roll and slope cultures, the medium was modified as follows in order to stiffen the agar gel and to reduce the opaque phosphate precipitate:

K_2HPO_4	1.0 g	$FeCl_3$	0.01 g
$MgSO_4 \cdot 7H_2O$	0.2 g	Agar	20.0 g
$CaH_4(PO_4)_2 \cdot 2H_2O$	0.5 g	Tap water	1 L
NaCl	0.1 g		

The tubes of agar were autoclaved for 20 min. at 15 lb. pressure. For inoculation, the tubes were cooled to 42℃ and a loopful of the appropriate culture was added to the melted agar and mixed by gentle shaking. This method gives quicker and heavier infection of the plant than that of adding the bacteria after the agar has solidified.

The seeds were externally sterilized by shaking for 3 min. in absolute alcohol, for another 3 min. in 0.2% $HgCl_2$ and washing in four changes of sterile water. Two to four seeds were planted in each tube with a flamed platinum loop. During growth, the tubes were supported in blocks of wood drilled with suitable holes of such a depth as to keep the root system shaded.

For nitrogen determinations, clover was grown in quart milk bottles each containing 800g. of sand and 100 ml. of the following food solutions:

K_2SO_4	0.9 g	$MgSO_4 \cdot 7H_2O$	0.5 g	Boric acid	0.02 g
K_2HPO_4	0.5 g	NaCl	0.5 g	$MnSO_4$	0.02 g
$CaH_4(PO_4)_2 \cdot 2H_2O$	0.5 g	$FeCl_3$	0.02 g	Tap water	1 L

The bottles of sand wore autoclaved for 2 hr. and the food solution separately autoclaved for 15 min. at 15 lb. pressure. The inoculum was suspended in the food solution and the suspension added to the sand with a sterile pipette. Seven to ten externally sterilized seeds were sown in each bottle.

Peas and soy beans were grown in glazed earthenware pots, the former in small pots containing 3 kg., and the latter in large pots holding 12 kg. of sand. The pots of sand for peas were rendered free from a contamination of nodule bacteria by blowing steam upwards through the sand for 14 min. via the hole at the base of the pot. In the case of soy beans the pots and sand were not sterilized, as these were found to be free from soy bean nodule bacteria.

The same food solution was used for peas and soy beans as for the bottle cultures of clover 200 ml. of this solution was added to each of the small pots and 1000 ml. to each of the large pots. More food solution was added as required during the course of the experiments. The solution used in growing the peas was sterilized and watering was done with boiled water. The inocula were added to the first dose of food solution before adding this to the pots. Seeds of both peas and soy beans were externally sterilized and sown at a uniform depth. In all experiments with clover, peas and soy beans, uninoculated controls wore grown and these remained free from nodules.

For anatomical study, nodules wore fixed at intervals during the growth of the host plants and in the case of clover, nodules of known age were taken. They were usually fixed in Allen's modification of Bouin's solution （Allen's P.F.A.3） and occasionally in Flemming's medium solution （as given by McCoy 1929）. They were brought through alcohols into chloroform, imbedded in paraffin wax and cut into sections 6-10 μ thick. Most of the sections were stained with Heidenhain's iron haematoxylin with or without a counter-stain of erythrosine or orange G.

For studying the distribution of starch, slides were treated with a mordant of 2% aqueous solution of tannin for 12 hr., stained for 2 min. with 1% aqueous gentian violet and differentiated in 95% alcohol. Excellent results were obtained when this method was combined with the haematoxylin stain. The slides were then treated as above described after the haematoxylin had been differentiated with iron alum, and were finally counter-stained with orange G.

C. THE DEVELOPMENT AND STRUCTURE OF CLOVER NODULES PRODUCED BY DIFFERENT STRAINS OF BACTERIA

In agar cultures of clover, the nodules induced by the strains here studied, first appeared on plants about 8 days old, which had produced their first true leaves. Nodules visible to the naked eye appeared on the same day in which infected root hairs were first observed, indicating that the bacteria pass down the root hair and induce proliferation of the root cells within 24 hr. the early development of the nodule follows the course described by Thornton （1930a） for lucerne nodules. The first stage consists of a mass of proliferating cells （figure 1, plate 13） mostly in the cortex but penetrating into the pericycle. The central cells soon swell and most of them become infected with bacteria brought into them by the infection threads（figure 2, plate 13）. At this stage also, vascular strands begin to be differentiated along the sides of the nodule. By the time the nodule is a week old the cytoplasm of the infected cells in the central region becomes filled with bacteria. This central tissue is referred to below as the 'bacterial tissue'（figure 3, plate 13）.

At the distal end of the nodule a cap of cells remains meristematic. In all clover nodules studied, the bacteria in recently infected cells dose behind this meristem cap are rod shaped, but in older parts of the bacterial tissue they change into branched or swollen forms, often referred to as 'bacteroids'. The shape of these bacteroids differs considerably in different strain, but these differences are not correlated with the effectiveness of the strain. The bacteroids in nodules produced by strains 2057, A and 2027 are usually pear-shaped or branched, an observation already made for the two American strains by Marie Löhnis （1930）. In the Coryn strain the bacteria undergo a type of change never observed by the authors in any other nodules （figure 4）. Swellings appear either at the end or in the centre of the young rod-shaped cells and increase until they absorb the whole of the rod, which is thus converted into a sphere. At first the staining material is distributed throughout the spherical cell but later the staining material is distributed throughout the spherical cell but later this became collected to form a deeply staining granule. Final disintegration releases these granules from the cells.

The amount and distribution of starch varies very greatly in individual clover nodules but seems to bear no relation to the effectiveness of the strain.

The early course of nodule formation is common to all the strains studied, whether effective or ineffective. It results in the formation of an approximately spherical nodule about 0.5 mm. in diameter. From this stage, however, the course of events differs in effective and ineffective nodules.

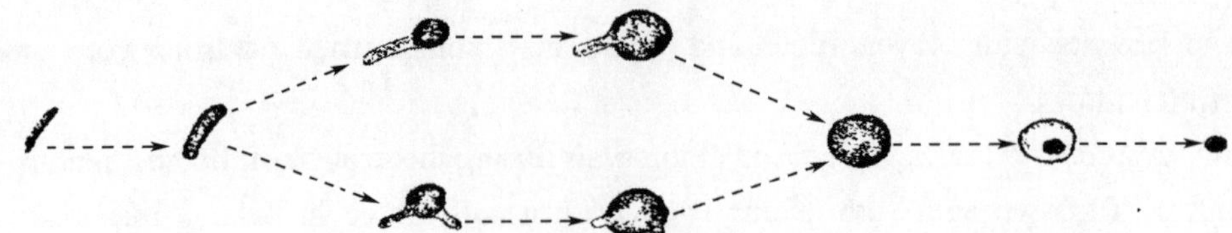

FIGURE 4 Formation of spherical forms by Coryn strain bacteria in the nodule.

In nodules produced by the effective strains 205，2057 and A，the apical meristem continues its activity（figure5，plate13）and causes the nodules to grow in length so that，when 7 weeks old，they have a mean length of about 2 mm（figure 6）.* The mean length of such nodules of all ages on plants ten weeks old is 1.45mm. In nodules produced by the ineffective strains Coryn and 2027，the meristem cap ceases to function after about 7 days so that the nodules remain small and round，having a mean diameter of 0.8 mm. In these nodules there is therefore a very small volume of bacterial tissue （figure 3，plate 13）. The volume of this tissue in clover nodules bears a fairly constant relationship to their overall length，as was also found to be the case with lucerne（Thornton and Nicol 1936）. This relationship，found for nodules of strain 2057 and Coryn，is shown graphically in figure 7，which is based on measurements of nodules from plants of varying age grown in agar and on the assumption that the bacterial tissue is cylindrical in shape. The mean length of strain 2057 nodules on plants 10 weeks old corresponds to a bacterial tissue volume of 0.22 cu.mm.，whereas that of Coryn nodules corresponds to a bacterial tissue volume of only 0.05 cu. mm.

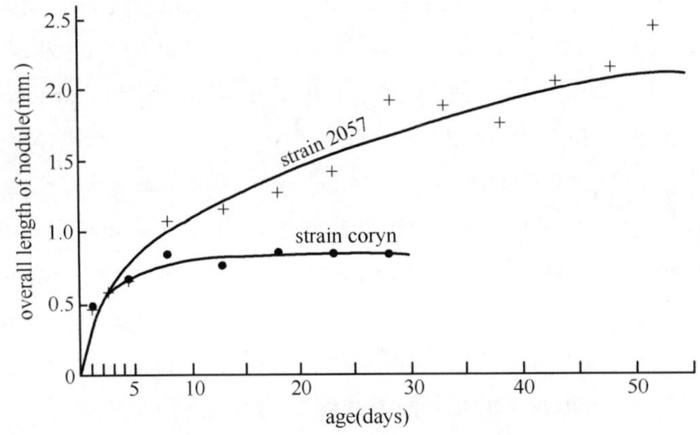

FIGURE 6. Growth in length of effective and ineffective clover nodules.

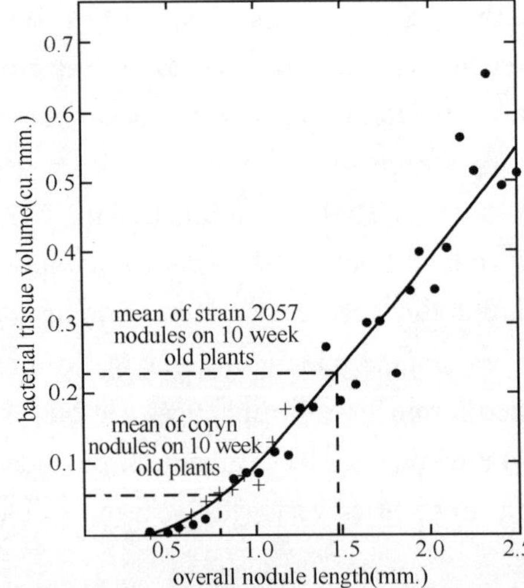

FIGURE 7. Relation of overall length to volume of bacterial tissue in clover nodules. Each point gives the mean volume of bacterial tissue in from 3 to 10 nodules of the same overall length. Dots refer to strain 2057，crosses to the Coryn strain nodules.

* The data plotted in figure 7 are derived from the agar cultures described below in § D（P.218）. The volume of organized bacterial tissue plotted in figure 10 was also obtained from these cultures.

In clover nodules of all the strains studied, the bacterial tissue eventually disintegrates owing to parasitic attack on the tissues by the bacteria, as has been described in the nodules on clover and lucerne by Thornton (1930b). But one of the most striking differences between effective and ineffective nodules is the time at which this disintegration takes place. In nodules produced by the effective strains, the bacterial tissue begins to disintegrate at the base, when the nodule is about a month old (figure 8, plate 13).

DESCRIPTION OF PLATES

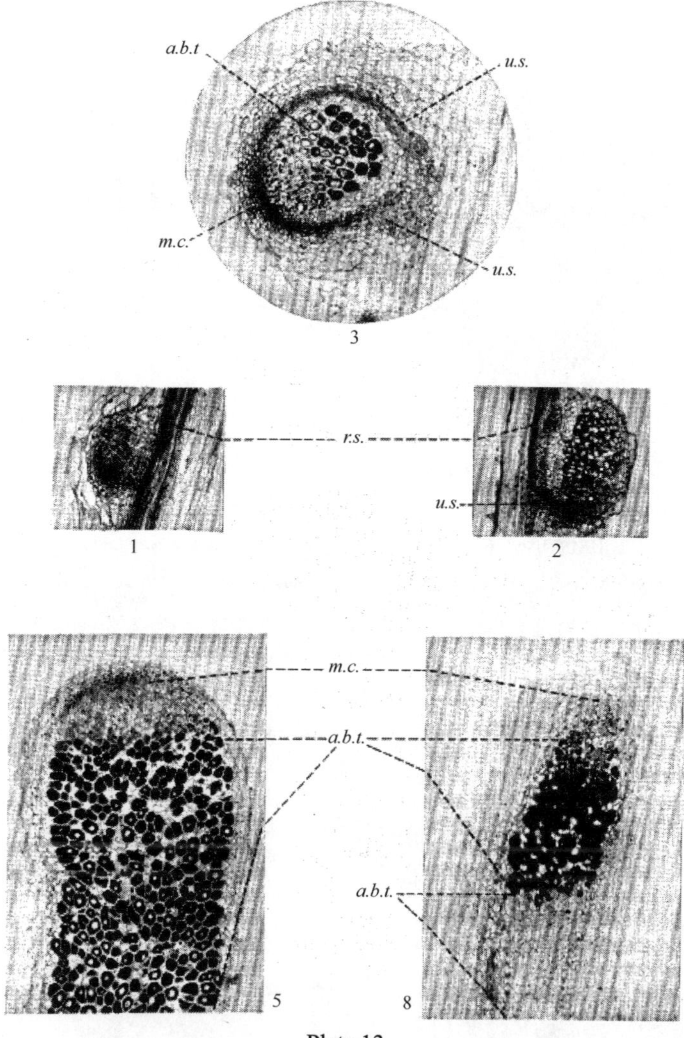

Plate 13

FIGURES 1 to 3. Development of an ineffective clover nodule produced by the Coryn strain.

FIGURE 1. Nodule 1 day old, composed of meristem tissue.

FIGURE 2. Nodule 3 days old, showing swelling of the central tissue cells, and formation of a vascular strand.

FIGURE 3. Nodule 6 days old, and about fully grown, showing the small amount of bacterial tissue.

FIGURE 5. Effective clover nodule one month old produced by strain 205, showing large development of bacterial tissue.

FIGURE 8. Six weeks old nodule produced by strain 205. Disintegration of the bacterial tissue commencing at the base.

m.c. meristem cap; *o.b.t.* organized bacterial tissue; *d.b.t.* disintegrated bacterial tissue; *v.s.* vascular strands; *r.s.* stele of root.

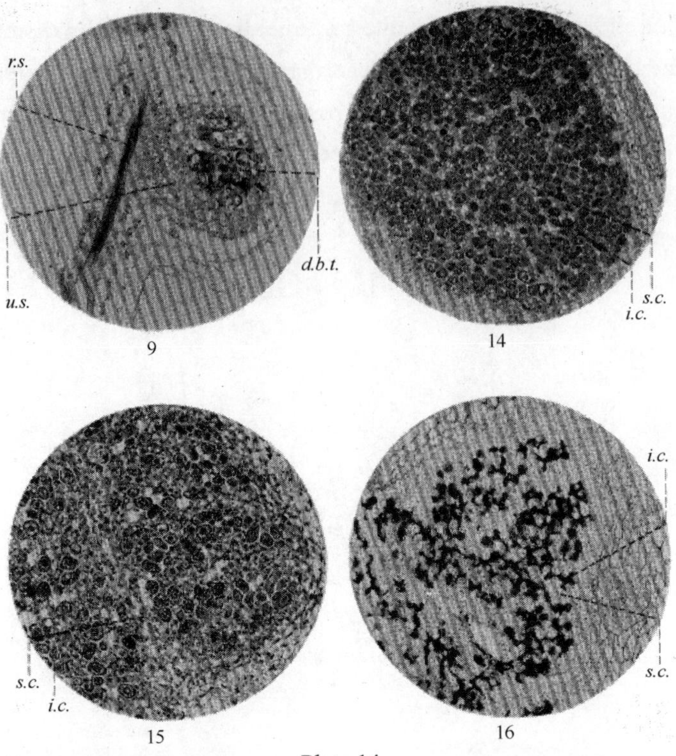

Plate 14

FIGURE 9. Clover nodule, 15 days old, produced by the Coryn strain, showing disintegration of the bacterial tissue.

d.b.t. disintegrated bacterial tissue; *v.s.* vascular strands; *r.s.* stele of root.

FIGURES 14 to 16. Sections of the bacterial tissue of soy-bean nodules.

FIGURE 14. Strain 501 (effective), showing large proportion of infected cells, *i.c.*

FIGURE 15. Strain 507 (ineffective), showing small numbers of infected cells and large proportion of uninfected cells, *s.c.* containing starch.

FIGURE 16. Bacterial tissue of strain 507 nodule undergoing disintegration through collapse of the infected cells, *i.c.*, the sterile cells, *s.c.*, remaining turgid.

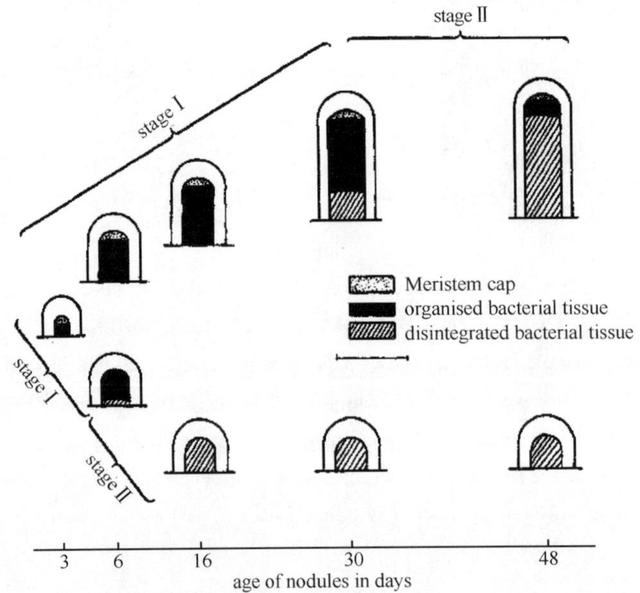

FIGURE 10. Growth and decay of the bacterial tissue in clover nodules.

After about eight weeks the centre of the nodule usually shows complete disintegration and is filled with collapsed and necrotic cells. In effective nodules of strains 2027 and Coryn, disintegration begins when the nodule is only about 7 days old, that is, almost as soon as the bacterial tissue is developed, and is complete by about the 15th day（figure 9，plate 14）.

Measurements of the volume of total bacterial tissue such as are plotted in figure 7, therefore give an incomplete picture of the differences between effective and ineffective nodules, for not only does the volume attained by the bacterial tissue differ in the two types of nodule, but also the duration of its active life prior to disintegration. It is the differences in volume of organized, or undisintegrated bacterial tissue throughout the life of the nodules that are of importance. These differences are illustrated diagrammatically in figure 10. The history of the bacterial tissue in a nodule can be divided into two stages. During stage I, which lasts from its first appearance until disintegration commences, the volume of the organized bacterial tissue is increasing. During stage II, the progress of disintegration from the base upwards causes a decrease in the volume of the organized bacterial tissue, which is in the end completely destroyed. In effective nodules the bacterial tissue continues to grow, so that on plants 10 weeks old its mean volume per nodule is 0.22 cu. mm. The nodule does not begin to show disintegration till it is nearly a month old and this process is not complete until it is about 8 weeks old, the organized bacterial tissue, formed by the end of the first week, having an active life of about 7 weeks. In ineffective nodules the bacterial tissues ceases to grow after about a week when its mean volume is about 0.05 cu.mm and it is completely disintegrated by about the fifteenth day having thus an active life of little more than a week.

The cells of the bacterial tissue contain almost the entire bacterial population of the nodule, and this tissue, prior to its disintegration, may therefore be assumed to be the centre of nitrogen fixation. The activity of the nodule in fixing nitrogen must therefore be greatly affected by the differences above described, and should bear a relation to an integration of the volume of organized or undisintegrated bacterial tissue over the period covered by stages I and II.

D. RELATION OF VOLUME AND DURATION OF THE BACTERIAL TISSUE TO NITROGEN FIXATION IN CLOVER NODULES

A series of clover cultures in agar was made in order to obtain more exact data as to the changes in volume of organized and disintegrated bacterial tissue during nodule growth. The plants were grown in wide test tubes by the agar slope method above described. Half of the tubes were inoculated with the effective strain 2057 and half with the effective inoculated with the effective strain 2057 and half with the ineffective Coryn strain. The tubes were kept in a warm glasshouse and moisture maintained by the occasional addition of sterile culture solution.

Nodules were individually marked as they appeared, so that their age at the time of sampling might be known. Two tubes inoculated with each strain were taken at weekly intervals and the overall lengths of their nodules were recorded and measurements were made of the length and width of the organized and disintegrated bacterial tissue in these nodules, from free-hand sections, stained with 0.1% thionin made up in a 5% phenol solution. From these measurements the volumes of bacterial tissue were calculated, assuming a cylindrical shape. The results are shown graphically in figure 11, based on the examination of 152 nodules of strain 2057 and 71 Coryn nodules. If individual bacteria of the two strains are equally active in fixing nitrogen in the nodule, the amount of nitrogen fixed by a nodule should be proportionate to the product, vt, of the mean volume of organized bacterial tissue and the time during which it acts. This product was obtained from the data by calculating the term $\{S(x)/n\}t$, where x represents the volume of organized bacterial tissue found on n

occasions over a growth period of t days. The calculation expresses the result in units of 1 cu.mm. bacterial tissue acting for 1 day （'eu. mm.-days'）.

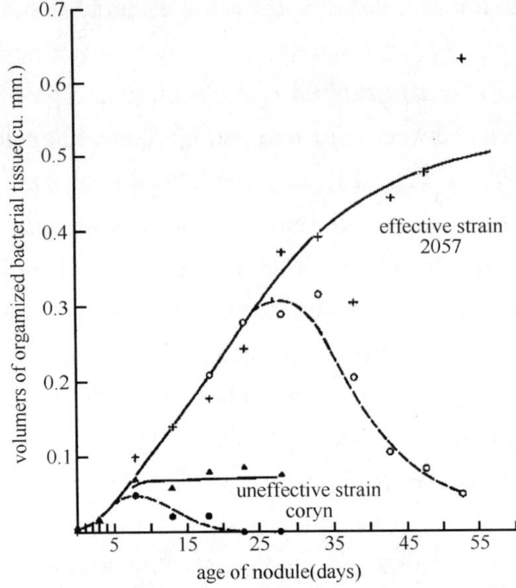

FIGURE 11. Changes in volume of total and of organized bacterial tissue in effective and ineffective clover nodules— Total bacterial tissue; ---organized bacterial tissue.

The value of *vt* was found to be 8.25 cu. mm.-days per nodule for the effective strain 2057 and 0.42 for the ineffective Coryn strain. The difference in the factor *vt* should therefore account for a difference of nearly twenty-fold in the nitrogen fixed by individual nodules of the two strains. To determine the true nitrogen fixing efficiency of the bacteria of each strain it was thus necessary to estimate the amount of nitrogen fixed per unit volume of the organized bacterial tissue per day of each strain.

An experiment was made to determine the quantities of nitrogen fixed by red clover, grown in quart milk bottles of sand as described in the section on methods. The bottles were divided into three sets, one uninoculated, one inoculated with strain 205 and one with the Coryn strain. After a growth period of about 3 months, the number of nodules in each bottle was counted and the nitrogen determined by the Kjeldahl method. The results are set out in table 1. The amount of nitrogen fixed per bottle was obtained by subtracting the amounts round in the uninoculated set.

Table 1 Nirogen-fixing efficiency of clover bacteria Exp. I, 1936

	Uninoculated	Strain Coryn	Strain 205
No. of replicate bottles	12	12	16
Nitrogen per bottle（mg.）	0.91±0.136	2.42±0.53	6.80
Nitrogen fixed per bottle（mg.）	—	1.51±0.547	5.89
Mean volume×duration of organized bacterial tissue per nodule in cu.mm. days（*vt*）		0.42	8.25
No. of nodules Per bottle（*d*）		771	185
e=mg. nitrogen fixed per ml.active tissue Per day= 1000*f*/*vtd*		4.59	3.90

Applying *e*=3.90 to data for Coryn.

Expected nitrogen fixed per bottle should be 1.27 mg.

The amount fixed per ml, per day, *e*, was calculated from the formula *e*=1000*f*/*vtd* where *f* is the nitrogen fixed and d the mean number of nodules per bottle.* The value of for strain 205 is 3.9 mg of nitrogen fixed per c.c.-day and that for the Coryn strain is 4.59. If one assumes that the latter strain has the same efficiency as strain 205，a calculation can be made of the nitrogen that should be fixed per bottle by the Coryn strain if the value of *e* for this strain were also 3.9. This calculation gives the expected amount of nitrogen fixed per bottle as 1.27 mg. which does not differ significantly from the observed value of 1.51 whose standard error is $\pm$ 0.547，with 11 degrees of freedom. Thus，although the plants with strain 205 nodules fixed nearly four times as much nitrogen as those bearing Coryn nodules，the whole of this difference can be accounted for by the smaller volume and shorter duration of the organized bacterial tissue in nodules of the ineffective strain.

E.THE STRUCTURE OF PEA NODULES PRODUCED BY EFFECTIVE AND INEFFECTIVE STRAINS

Observations were made on the anatomy of nodules produced by the effective strain 310 and by the ineffective strain B33 on peas grown in small pots of sand. Roots were washed at different stages in the growth of the host plant，the nodules measured and sample nodules of various sizes were fixed and sectioned.

There was a marked difference in the size of nodules by the two strains. On plants 10 weeks old，strain 310 nodules of all ages had a mean length of 1.8 mm. and this included a number of long shaped and branched nodules. On plants of the same age，B33 nodules had a mean length of 1.08 mm. and were spherical in shape.

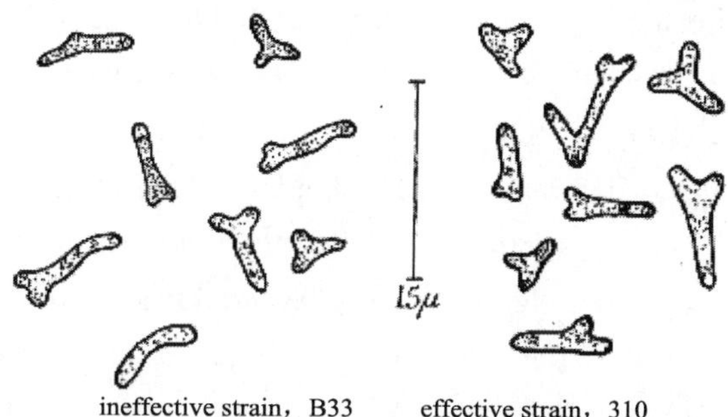

FIGURE 12. Bacteroid strains from pea nodules.

The course of nodule development in the pea generally resembles that in lucerne and clover. Sections showed that，on plants 10 weeks old，even large nodules of the 310 strain retained an active meristem cap and that very few showed any disintegration of the central tissue. Most of the nodules of B33 strain，even on plants six weeks old，showed no active meristem cap and had their bacterial tissue completely disintegrated.

The appearance of the bacteria was particularly examined in view of the statements by Nobbe and Hiltner（1893）and Marie Löhnis（1930），that ineffective nodules contained bacteria in the rod stage which did not change into typical 'bacteroids'. In our material typical swollen and branched 'bacteroids' were found in the organized bacterial tissue of nodules produced by both the effective and ineffective strains（figure 12）. These disappeared as usual during the process of disintegration of the bacterial tissue and so were absent in most nodules of the B33 strain from plants over 6 weeks old，since by that time disintegration was usually complete.

Nobbe and Hiltner mention that their ineffective nodules showed no meristem cap and it therefore seems likely that they，at any rate，based their description on nodules in a state of disintegration，such as quickly follows the arrest of meristem activity. Marie Löhnis（1930）also describes certain unusual bacterial cells，referred to as 'brown bacteroids'，in effective nodules. No such forms could be identified in our material，in

* In this experiment，the data used for calculating volumes of bacterial tissue were derived from agar cultures but applied to the analytical results obtained from sand cultures. This procedure seemed to be justified because the mean lengths of 205 and Coryn nodules were in approximately the same ratio in the agar and sand cultures.

which no constant differences in the shape of the bacterial cells could be found to distinguish the strains.

Young nodules produced by either strain contained considerable amounts of starch. Disintegrated bacterial tissue seldom contains any starch so that the earlier onset of disintegration caused an earlier disappearance of the starch in ineffective than in effective nodules.

Thus the principal differences found in effective and ineffective pea nodules are the smaller size of the latter and the shorter duration of the bacterial tissue in them. These differences are similar to those found in clover nodules.

F.THE COURSE OF GROWTH AND STRUCTURE OF SOY BEAN NODULES PRODUCED BY EFFECTIVE AND INEFFECTIVE STRAINS

The material used for studying the anatomy of Soy bean nodules was obtained from sand cultures of the plants in large pots made in 1937，1938 and 1939. The plants were inoculated with the effective strain 501 and with the ineffective strain 507，and nodules of various sizes were taken and fixed at different stages in the growth of the plant.

The development of nodules on the soy bean differs in some important respects from that in clover and pea nodules. Its general course has been described by Bieberdorf （1938） with whose account the present authors agree.

The nodules remain approximately spherical but become slightly flattened with age. Nodules of strain 501 continue to grow until they attain a considerable size. In the 1939 experiment，nodules of this strain，on plants four months old，attained a mean diameter，taken at right angles to the root，of 2.95 mm. Nodules of strain 507 stop growing quite early and remain small. The mean diameter of such nodules on four months old plants from the same experiment was only 1.9 mm. There is，in consequence，a large difference in volume of the bacterial tissue contained in the two types of nodule. Measurements were made of the bacterial tissue from freehand sections of 100 nodules of each strain taken from plants of varying ages grown in 1939. From these data the volumes were calculated，assuming the shape of the bacterial tissue to be an ellipsoid. The volume of bacterial tissue is plotted against the overall diameter of the nodule in figure 7. The relation of the two characters differs in the two strains，an effective nodule of given length having more bacterial tissue than an ineffective one. The volume of bacterial tissue，corresponding to the mean diameters given above are 11.4 cu. mm for strain 501 and 2.3 cu. mm for strain 507.

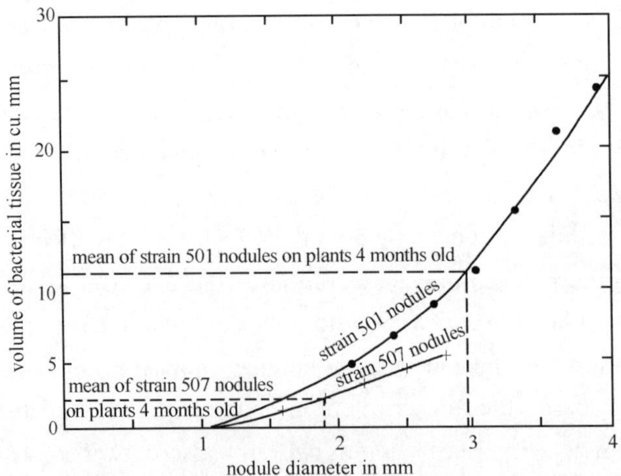

FIGURE 13. Relation of overall diameter to volume of bacterial tissue in soy-bean nodules. Each point gives the mean volume of bacterial tissue in from three to thirty-two nodules of the same overall length.

The bacterial tissue of soy-bean nodules differs in appearance from that of clover and pea nodules. In the latter the great majority of cells are filled with bacteria both in effective and ineffective nodules, so that the volume of this tissue in the two types gives a comparable estimate of the relative numbers of bacteria present. But in soy-bean nodules the bacterial tissue contains a relatively large proportion of uninfected cells and this proportion differs strikingly in effective and ineffective nodules studied (figures14, 15, plate 14). Estimates of the percentage volumes of infected cells in the bacterial tissue were obtained by measuring the areas occupied by infected and by uninfected cells in random microscope fields in microtome sections from a number of nodules of each type. The results are shown in table 2. Measurements made from soy-bean nodules grown in 1938 showed that 82.99% of the volume was occupied by infected cells in strain 501 nodules and 43.44% in strain 507 nodules. In the material grown in 1939, the percentages were 94.19 for strain 501 and 49.16 for strain 507. The percentages for the two strains in each year were in the same ratio, namely 1.91 : 1, but they were different in the two seasons, this seasonal difference being significant in the case of strain 501. In the infected cells of both types of nodules, the bacteria lie close-packed in the cytoplasm, so that a comparison of the percentage volumes of infected cells should provide an estimate of the relative numbers of bacteria per unit mass of the bacterial tissue in the two types of nodule.

TABLE 2

	501 nodules		507 nodules	
	1938	1939	1938	1939
No. of nodules examined	5	11	5	11
Percentage volume of infected cells (mean)	82.99	94.19	43.44	49.16
Standard error	±3.15	±1.05	±5.19	±2.05

In soy-bean nodules the infected cells were found to be free from starch, but the uninfected cells contained many starch grains. Since these latter cells were much the more numerous in inefficient nodules, the amount of starch in these nodules was proportionately greater. The 507 nodules somewhat resembled the ineffective *Phaseolus* nodules, described and illustrated by Elizabeth McCoy (1929), in having many uninfected cells filled with starch in the central tissue.

The bacteria in infected cells in soy-bean nodules remain in the rod stage and show scarcely any change toward the ‘bacteroid’ condition in any part of the nodule. There was no noticeable difference in their appearance in the organized bacterial tissue in nodules produced by the two strains.

The process of disintegration of the bacterial tissue in soy beans differs in important respects from that in clover and pea nodules. It does not always begin at the base but may start at any point in the central tissue. In the early stages, the bacteria break up into granules and the infected host cells lose their turgidity. The uninfected cells, however, remain turgid and their pressure on the infected cells causes these to collapse. When this process is far advanced the appearance of a section suggests, and was, indeed, at first mistaken for intercellular infection (figure 16, plate 14). It probably accounts for the statement of Kâs (1930) that soy-bean nodules show both intra- and intercellular infection, and may also explain the ‘intercellular’ type of infection described in *Serradella* by Milovidov (1926, 1928). Bacterial tissue showing this collapse of the infected cells is here referred to for convenience as ‘disintegrated’, although there is no general collapse of the tissue as occurs in clover and pea nodules. This is clearly due to the large proportion of uninfected cells which remain turgid, and support the tissue.

There is a marked difference between nodules of strains 501 and 507 in the age at which disintegration takes place. In the 1939 nodules of strain 507 the process began when the host plant was only 4 weeks old and was nearly complete by the twelfth week. Strain 501 nodules on plants twelve weeks old did not yet show any disintegration, which only began to appear when the plants were at the end of their growth period, at an age of about 17 weeks.

Effective and ineffective soy-bean nodules thus show differences in their growth and decay of the same general type as those found in clover and pea nodules. Effective nodules grow to a mean diameter 50% greater than that of ineffective nodules. They contain more than five times the volume of bacterial tissue, which bears twice the percentage by volume of infected cells, and this tissue lasts more than four times as long as in ineffective nodules before becoming disintegrated.

It was, however, necessary to follow the course of growth and decay of the bacterial tissue throughout the growing period of the plant before the quantitative effect of these factors on nitrogen fixation could be estimated.

The size of the plant made it impracticable to use the agar culture method, used with clover for obtaining individual nodules of known age. The measurements here recorded are therefore the means from batches of nodules taken from successive reapings of plants of recorded ages taken during the pot-culture experiment of 1939. These data did in fact give a good measure of the growth and decay of the nodules because, as is usual in soy beans, nearly all the nodules appeared when the plants were quite young and so did not vary greatly in age within each batch.

The soy beans were grown in thirty large pots, twelve of which were inoculated with strain 501, twelve with strain 507, while six were left uninoculated. Three or four plants inoculated with each strain were removed at weekly intervals, and, from these, twenty nodules of each type were examined by means of hand sections stained with carbol thionin. In each section, measurements were made of the diameters of the nodules and of the area occupied by bacterial tissue, and an estimate was made of the percentage of that area showing disintegration. From these measurements, the mean volumes of organized and of disintegrated bacterial tissue in each batch of twenty nodules were calculated. The volumes of total (that is, organized plus disintegrated), bacterial tissue have been plotted against nodule diameter in figure 13, already discussed. Changes in volume of the organized bacterial tissue, with time are shown in figure 7, and discussed below. At the end of the growth period, the remaining plants, then 4 months old, were reaped, the nodules counted and nitrogen determinations made by the Kjeldahl method.

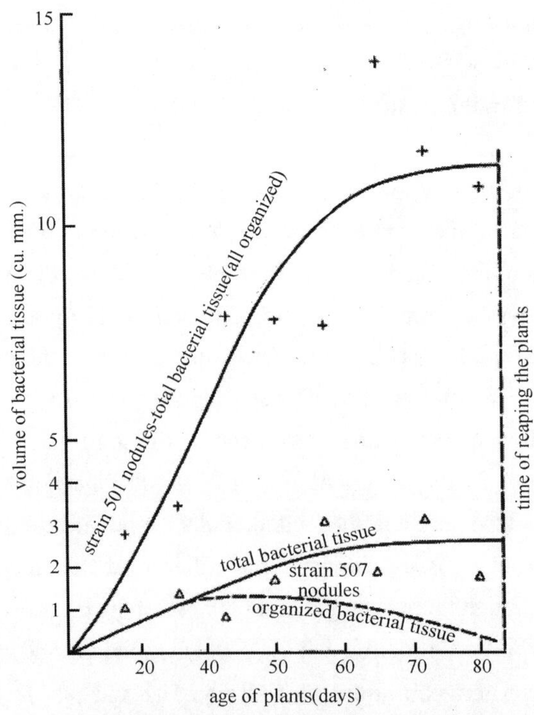

FIGURE 17. Changes in volume of organized and total bacterial tissue in effective and ineffective soy bean nodules. Each point represents the mean of 20 nodules.

In figure 17, the calculated volumes of total and of organized bacterial tissues, taken from the weekly measurements, are plotted against the age of the plants. From these data the integrated cu.mm.-days, *vt*, are calculated according to the formula

$$vt = \frac{S(b - by)}{n100} t,$$

where b is the mean volume of total bacterial tissue found on each occasion and y the estimated percentage of that tissue showing disintegration.

The values of *vt* thus calculated are 83.95 for strain 507 and 647.4 for strain 501. These are in units of 1 cu.mm of *organized* bacterial tissue acting for one day, but they do not represent the relative volumes of active tissue until further corrected for the percentage, *c*. of infected cells within the bacterial tissue of each strain. When this correction is made, using the mean percentages obtained from 1939 data shown in table, the value of *vtc*/100 for strain 507 is 41.27 and for strain 501 it is 609.79. The average 501 nodule should thus fix nearly fifteen times as much nitrogen as the average nodule of strain 507, by reason of its larger content of infected cells.

The data with regard to nitrogen fixation are set out in table 3. Plants with 501 nodules fixed about six times as much nitrogen as did those with 507 nodules. The quantity e of nitrogen fixed per ml. of active infected cells per day was calculated from the formula $e = 1000f/vt\frac{c}{100}d$ where f and d are means per plant for nitrogen fixed and nodule numbers respectively.

Table 3 Nitrogen-fixing efficiency of soy-bean bacteria (1939 experiment)

	Uninoculated	Strain 507	Strain 501
No. of plants	40	33	18
Nitrogen per plant (mg.)	16.69	19.34	31.70
Nitrogen fixed per plant (mg.)		2.65	15.01
Mean volume×duration of organized bacterial tissue per nodule in cu.mm.-days (*vt*)		93.95	647.4
Infected cells as percentage of organized bacterial tissue (C)		49.16	94.19
No. of nodules per plant (*d*)		45.2	21.1
e=mg. nitrogen fixed per c.c. active infected cells per day= $1000f/vt\frac{c}{100}d$		1.42	1.17

Applying e=1.17 to data of strain 507.

Expected nitrogen fixed per plant should be 2.18.

The value of e, for the two strains 507 and 501, were 1.42 and 1.17 respectively. If the value for strain 507 were the same as for the other strain, namely 1.17, the expected nitrogen fixed per plant would be 2.18 mg, actually a fraction less than the observed figure. Thus the whole of the large difference in nitrogen fixed by soy-bean plants bearing the two types of nodules can be accounted for by differences in the total volume of infected cells and in the duration of their activity.

DISCUSSION

The course of the growth and decay of nodules produced by effective and ineffective strains of bacteria on clover, peas and soy beans thus shows that there are two quantitative characters closely correlated with effectiveness, namely the combined volume of the infected cells and the length of time that elapses before they collapse and disintegrate. Other features of nodule anatomy, unless causally connected with the above two characters, were found to show no association with effectiveness. Attempts were made with clover and soy beans to give quantitative values to these factors of volume and duration of the organized tissue containing bacteria，and to apply these values as corrections in estimating，from the results of analysis，the quantities of nitrogen fixed by a unit volume of infected cells in unit time. These attempts were so far successful that in experiments with both host plants，the whole of the large differences between the amounts of nitrogen fixed by 'effective' and 'ineffective' strains respectively could be accounted for by the differences in volume and duration of the infected nodule cells. The data showed no evidence that the 'ineffective' strains were really less efficient in fixing nitrogen per unit of bacterial mass in unit time，than the 'effective' strains.

The problem of ineffective nodules thus resolves itself into the need for explaining why the volume of infected cells in them remains so small and why these cells disintegrate so soon. The small volume of bacterial tissue is related to the early cessation of apical meristem growth in the nodule，but since this meristem activity is doubtless stimulated by the bacteria，both the short growth period of the nodule and the early decay of the infected cells are indications of poor bacterial growth within the tissues.

Strains of nodule bacteria differ in the rapidity of their growth on artificial media，but there is no correlation between this character and their effectiveness in the host plant. There is thus no reason to believe that ineffective strains are naturally less vigorous in growth than others，when given a suitable medium. Hence the poor growth of these strains within the nodule indicates that the tissues of the host plant provide an environment that is less suited to the ineffective than to the effective strains.

This may either mean that some unfavourable factor is normally present in the host tissues，or that such a factor appears in them as a consequence of infection by the ineffective strain. The latter alternative is rather suggested by the fact that ineffective nodules commence their growth quite normally and only later show arrested development.

REFERENCES

[1]Baldwin，I. L. and Fred，E. B. 1927 *Soil. Sci.* 24，217-230.

[2]Baldwin，I.L. and Fred，E.B. 1929 *J. Bact.* 17，17-18.

[3]Bieberdorf，F. W. 1938 *J. Amer. Soc. Agron.* 30，375-389.

[4]Eckhard，M. M.，Baldwin，I. L. and Fred，E. B. 1931 *J. Bact.* 21，273-285.

[5]Frank，E. 1890 *Landw. Jb.* 19，523-640.

[6]Helz，G. E.，Baldwin，I. L. and Fred，E. B. 1927 *J. Agric. Res.* 35，1039-1055.

[7]Kâs，V. 1930 *Vestn. Csl. Akad. Zemed.* 6，1073-1078.

[8]Löhnis，Marie P. 1930 *Zbl. Bakt.*，2 Abt. 80，342-368.

[9]McCoy，Elizabeth F. 1929 *Zbl. Bakt.*，2 Abt. 79，394-412.

[10]Milovidov，P. F. 1926 *Zbl. Bakt.*，2 Abt. 68. 333-345.

[11]Milovidov，P.F. 1928 *Rev. gen. Bot.* 40，193-205.

[12]Nobbe，F. and Hilther，L. 1893 *Landw. Versuchw.* 42，459-478.

[13]Schneider，A. 1892 *Bull. Torrey Bot. Cl.* 19，203-218.

[14]Stevens，J. W. 1925 *Soil. Sci.* 20，45-66.

[15]Thornton，H. G. 1930*a Ann. Bot.*，*Lond.*，44，385-392.

[16]Thornton，H.G.1930*b Proc. Roy. Soc.* B. 106. 110-122.

[17]Thornton，H. G. and Nicol，H. 1036 *J. Agric. Sci.* 26，173-188.

[18]Wilson，J. K. and Leland，E. W. 1929 *J. Amer. Soc. Agron.* 21，574-586.

[19]Wright，W. H. 1925*a Soil Sci.* 20，95-129.

[20]Wright，W.H. 1925*b Soil Sci.* 20，131-141.

The Growth of Nodule Bacteria in the Expressed Juices from Legume Roots Bearing Effective and Ineffective Nodules*

H.K.CHEN， HUGH NICOL AND H.G.THORNTON

(Bacteriology Department，Rothamsted Experimental Station)

Strains of pea and soy-bean nodule bacteria，differing in their effectiveness in benefiting the host legume，were grown in media containing the unheated root juices from uninoculated host plants and from host plants bearing effective and 'ineffective' nodules and their growth was measured.

The growth of the different bacterial strains on root juice from uninoculated plants was not correlated with their effectiveness.

The juice from roots with effective nodules produced significantly better growth of the bacteria than juice from roots with ineffective nodules in twenty-seven comparisons out of forty-four，the differences in the remaining comparisons being insignificant.

The juice from roots with effective nodules produced significantly better growth than the juice from uninoculated roots in ten comparisons out of twenty-five，and significantly poorer growth in three comparisons.

The juice from roots with ineffective nodules produced significantly poorer growth than the juice from uninoculated plants in eleven comparisons out of twenty-five，and better growth in only one comparison.

The production，as a result of infection of soluble substances affecting growth of the bacteria，affords an explanation of those differences in nodule growth that determine the effectiveness or ineffectiveness of the different strains of bacteria as regards nitrogen fixation within the host.

INTRODUCTION

Different strains of bacteria-producing nodules on a given host legume differ greatly in their ability to benefit that host by fixing nitrogen. Some strains are so 'ineffective' that they produce no appreciable improvement in growth of the legume and fix nitrogen in amounts difficult to detect.

The ineffectiveness of such strains was at first attributed to some defect in the process of nitrogen fixation.

The recent study of clover，pea and soy-bean nodules made by Thornton and Chen （1940），however，has shown that nodules produced by an ineffective strain contain a much smaller volume of infected cells filled with nodule bacteria than do effective nodules，and also that in them，the central tissue containing these infected cells disintegrates after a relatively very short active life. Estimates were made of the mean volume and duration of this central 'bacterial tissue' in nodules of effective and ineffective types. Plants of both clover and soy beans showed large differences in nitrogen content according to which type of nodule they carried，but the quantities of nitrogen fixed by a unit mass of infected nodule cells in unit time was found to be the same for effective and ineffective nodules. The whole of the large differences in total nitrogen fixed could be accounted for by differences in the total volume of infected nodule cells and the duration of these cells prior to disintegration. The further elucidation of the cause of ineffectiveness thus requires that the relatively small total volume and the short life of the infected cells in ineffective nodules be explained.

---

*原载于 *Proc. Roy. Soc. B.* 129:475～491，1940.

Nodules produced by effective and ineffective strains develop during their early stages in the same manner, and begin to grow at the same rate. But after a very short while (with clover in about 7 days), the ineffective type of nodule stops growing, while nodules produced by an effective strain continue their growth. This arrested growth is the principal cause of the small volume of infected cells contained in an ineffective nodule, although in soy beans the smaller proportion of infected to sterile cells in the central tissue is a contributory factor. Arrested growth is due to the stopping of cell division in the apical cap of nodule cells. Presumably the initiation and maintenance of the apical meristem is due to growth substances produced by the bacteria. The production of growth substance by the nodule bacteria has been demonstrated by Link (1937) and Thimann (1936, 1939), but Chen (1938) found that effective and ineffective strains of clover bacteria when growing equally vigorously in liquid culture produced similar amounts of growth substance, as assayed by Went's split pea stem test. The stopping of growth in an ineffective nodule would thus seem to be a consequence of the early arrest or decrease in growth of the contained bacteria. Weaker growth of the invading bacteria in ineffective nodules is also suggested by the observation that in soy-bean nodules the central tissue contains a smaller proportion of infected cells in ineffective than in effective nodules.

There is considerable variation in the growth rates of various strains of nodule bacteria in laboratory media, but no correlation could be found between this character and effectiveness towards the host plant. So that the striking difference between the growth of effective and ineffective strains of bacteria within the nodule is not due to any inherent difference in their powers of multiplication, but must indicate that the latter type of strain finds the host tissues an unsuitable environment for its growth.

Disintegration of the central bacterial tissue with its contained bacteria seems eventually to take place in all nodules by whatever strain they are produced, although in effective nodules it does not normally occur until the nodule is getting old. It can, however, be induced in quite young nodules, even of effective strains, by imposing abnormal conditions on the host plants, as by growing these in a boron-deficient medium (Brenchley and Thornton 1925), in the dark (Thornton 1929), or with an excessive nitrate supply (Thornton and Rudorf 1936). In nodules produced by ineffective strains on normally-grown host plants, the disintegration of the central tissue and bacteria also takes place in quite young nodules, and is closely similar to that which can be produced by artificially changing the physiology of the host plant. Like the stopping of nodule growth, the disintegration indicates a maladaptation between host plant and nodule organism.

This maladaptation of inefficient strains to their host plant may be due to any of the following three causes: (a) Some factor may be normally present, in the host root system, that is specifically less favourable to the ineffective than to the effective strains. (b) The presence of an effective strain within the tissues may induce a change favourable to the growth of the bacteria, which is not induced by the presence of the ineffective strain. Here the action of the induced factor need not specifically favour either type of strain. (c) The presence of invading bacteria of an ineffective strain may induce some change in the host tissue that is detrimental to growth of the bacteria. Here again the induced factor need not act specifically against ineffective strains, but these strains must be the specific cause of its appearance.

The substances in solution within the nodule tissue must supply food material to the bacteria so that it seemed natural to search in the root and nodule juice for the factors affecting their growth. On our first hypothesis, (a), the root juice of the host plant, whether uninoculated or bearing either type of nodule should be specifically less favourable to the growth or ineffective than to effective strains of the bacteria. On the second hypothesis, (b), juice from roots bearing effective nodules should be more favourable to the growth of the bacteria than that from roots bearing ineffective nodules or from uninoculated roots. On the third hypothesis, (c), juice from roots bearing nodules containing an ineffective strain should be less favourable to the growth than juice from roots bearing effective nodules or from uninoculated roots.

To test these hypotheses, plants were grown in sterilized sand without inoculation, and inoculated with effective and ineffective strains; the root juices were extracted from each set, sterilized by filtration and their effects on the growth of nodule bacteria of either type were tested *in vitro*. Clover was found to be unsuitable for this work owing to its small root development in the young state. Peas and soy beans were used because these provide ample root material, and since strains showing wide differences in effectiveness have been isolated from their nodules (Helz, Baldwin and Fred 1927; Wright 1925). The cultures of pea and soy-bean nodule bacteria used in this work were supplied by the Wisconsin Agricultural Experiment Station, to whom the authors are gratefully indebted.

TECHNIQUE

The plants were grown at Rothamsted in sand in glazed earthenware pots, the peas in small pots holding 3 kg, and the soy beans in large pots holding 12 kg of sand. In the experiments with peas, the sand was rendered free from live nodule bacteria by blowing superheated steam through the pots of sand via the aperture at the base. The sand used for experiments with soy beans was not sterilized, since it was found not to contain bacteria capable of producing nodules on soy beans. In all the experiments described below, whether with peas or soy beans, the uninoculated sets remained free from nodules.

In the experiments carried out in 1936 and 1937 the following food solution was added to the sand, to about 15% of its dry weight, and fresh supplies were given during the growth of the plant:

K_2SO_4	0.87 g	$CaSO_4$	0.2 g
K_2HPO_4	0.3 g	$FeCl_3$	0.04 g
KH_2PO_4	0.3 g	$MnSO_4$	0.04 g
NaCl	0.5 g	Boric acid	0.04 g
$MgSO_4 \cdot 7H_2O$	0.5 g	Tap water	1 litre

For the 1939 experiments this food solution was modified as shown below, so as to give a better calcium supply.

K_2SO_4	0.9 g	$FeCl_3$	0.04 g
K_2HPO_4	0.5 g	$MnSO_4$	0.04 g
$CaH_2(PO_4)_2 \cdot 4H_2O$	0.5 g	Boric acid	0.04 g
$MgSO_4 \cdot 7H_2O$	0.5 g	Tap water	1 litre
NaCl	0.5 g		

The food solutions were sterilized before addition to the pots, and in the experiments with peas all watering was with boiled water.

In all experiments, inoculation was accomplished by mixing a suspension of the appropriate organism with the food solution before this was poured into the sand, the uninoculated pots receiving an equal volume of sterile water in place of the bacterial suspension. The seeds were externally sterilized by immersion in absolute alcohol followed by 0.2% aqueous $HgCl_2$ for 3 min., and then washed in four changes of sterile water. They were sown immediately after application of the food solution to the pots. When the plants reached the flowering stage, the roots were shaken free from sand, washed, and the surplus water drained off them. They were then minced, ground in mortar and the juice squeezed through muslin. This juice was then passed through two thicknesses of Whatman no.4 filter-paper, and further cleared by filtration under reduced pressure through a pad 2 in. thick of compressed filter-paper fragments set up in a wide glass cylinder.* To prepare this pad, wet pieces of filter-paper about 1 cm. square were compressed with a glass ramrod. In the 1937 experiment with soy beans this filter was replaced by a Seitz filter.

After clearing, the juice was rendered sterile by filtration successively through an L1 Chamberland filter-candle, and through a sterilized L5 candle set up in a sterile vacuum flask.

The effect of the various root juices upon the growth of the nodule bacteria *in vitro* was tested by two different methods. In the 1936 experiments with soy beans and peas the juices were added to a melted agar medium which was poured into petri dishes and allowed to set. A number of point inoculations of the test organism were made on each plate, and the mean area of the resulting colonies was taken as the measure of growth.

In later experiments the test organisms were grown in liquid media containing the root juices, and their growth was measured by haemocytometer counts. As the details of technique varied somewhat in the different experiments they will be described with each experiment.

EXPERIMENT WITH SOY BEANS, 1936

Five sets of soy beans were grown in pots of sand. One set was uninoculated and the remaining sets inoculated with nodule bacteria of the effective strains 501 and 505, and of the ineffective strains 502 and 507. Six parallel pots, each with eight plants, were set up and growth carried out from 1 September till 26 October.

The roots were washed and the juice extracted and sterilized by filtration. The medium used for testing the effects of the juices on bacterial growth had the following composition:

K_2HPO_4	1.0 g	$FeCl_3$	0.02 g
KH_2PO_4	1.0 g	Sucrose	10.0 g
$MgSO_4 \cdot 7H_2O$	0.4 g	Yeast water*	100 ml
NaCl	0.2 g	Tap water	900 ml
$CaCO_3$	1.0 g	Agar	20.0 g

* Yeast water was made by boiling 10% yeast in water and filtering.

This basal medium was divided into five portions which were autoclaved, cooled to 42° C and, to each portion, one of the juices was added with a sterile pipette to make 18% of the final volume. From each of the four media containing juice from the inoculated plants, sixteen petri dish plates were poured, each of 10 m1. of medium. These were allowed to set and incubated at 25° C for 24 hr. to allow the agar surface to dry. Each batch of sixteen plates were then divided into four sets of four replicate plates, one set being inoculated with each of the four strains of soy-bean nodule organisms 501, 505, 502 and 507.

In the case of the uninoculated roots, however, the small supply of juice enabled only three replicate platings to be inoculated with each test organism.

By the above plan each of the four strains was grown on media containing each of the live juices, namely, from uninoculated plants and from plants that had borne nodules produced by each of the same four strains (see Table 1).

The inoculation of the agar plates was carried out as follows. Ten needles were set about 1.5 cm. apart in a large cork, with their points projecting and carefully adjusted so that the points were level. The culture to be

* The apparatus used is illustrated in *Proc. Roy. Soc. B* (1936), 119, figure 1, p. 478.

used as inoculum was suspended in sterile saline solution, and about 10 ml. were poured into a sterile petri dish. The needles, sterilized by flaming, were lowered into the suspension, and their points then gently brought down on to the surface of the agar medium in the plate. In this way each plate was given ten point inoculations. After incubation for 14 days at 25℃, the colony areas were measured. The results are shown in table 1, which gives the mean colony areas in square millimetres. The mean colony area on each plate was taken as the unit in calculating the standard error. The number of degrees of freedom for each set is given in the table. This is not always three, because less than four replicate plates with juice from uninoculated plants were tested with each strain, and because some plates in other sets were lost from various causes. In particular, the set in which strain 502 was tested on juice 501 was spoilt owing to water condensing on the agar surface. In order to obtain a valid standard error for the whole experiment, a hypothetical value was allotted to this set, calculated by Yates's method for missing plots （Yates 1933）. The standard error for the differences between any two individual set means is ±1.38. The bottom row gives the mean colony areas for the four test organisms combined in the case of each juice tested. The standard error for the differences between any two of these juice means is ±0.69.

Table 1 Experiment with soy beans（1936）

Colony areas of four strains of soy-bean nodule bacteria on agar media containing root juice from variously treated plants

Mean colony area sq.mm.（*m*） Degrees of freedom		Medium containing root juices from				
		Uninoculated plants（control juice）	Plants inoculated with strain			
Strain	（*n*）		501	505	502	507
501	*m*	4.17	6.75	11.35	0	0
	n	1	3	3	3	3
505	*m*	8.58	6.06	14.04	6.27	8.17
	n	2	3	3	2	2
502	*m*	8.49	（11.52）	19.56	6.22	10.96
	n	1	0	4	3	2
507	*m*	6.28	7.07	7.99	5.18	4.37
	n	0	3	2	3	3
All strains	*m*	7.02	7.85	13.23	4.41	5.87

S.E. of difference between any two individual means =±1.38.

S.E. of differences between juice means (bottom row)=±0.69.

The relative effect on the bacterial growth of root juice from plants bearing effective nodules, as against root juice from plants with ineffective nodules, can be tested in fourteen comparisons, excluding, of course, the hypothetical value. Juice from plants bearing the effective strain gave significantly the better growth in ten of these comparisons.

The root juice from plants bearing 501 nodules did not differ significantly from the juice of uninoculated roots （described below as 'control juice'） in its effects on any of the strains tested. Root juice from plants bearing the other effective strain 505, however, gave significantly better growth than did control juice with three of the test organisms and no significant difference with the fourth.

Juice from roots, bearing ineffective nodules, of strains 502 and 507 completely prevented growth of strain 501, but with the other three test, organisms these two juices produced growth not significantly different from control.

In this experiment the results are suggestive, but the difference in action of the two juices from plants with effective nodules and the different response of the four test organisms to the juices from plants with ineffective nodules，makes the comparison with control juice inconclusive.

EXPERIMENT WITH PEAS，1936

In this experiment peas，inoculated with the effective strains 310 and 317，with the ineffective strains 313 and B33 and uninoculated，were grown in small pots of sand as described above. The seeds were sown on 1 September and the plants reaped on 21 October. The root juices were extracted，and the growth of the four strains of pea-nodule bacteria on agar media containing each of the juices was tested by the same method as in the previous experiment. Each organism was，in addition，grown on the basal agar medium in which the root juice was replaced by distilled water. The results are shown in table 2 in which the mean colony areas are derived from three to five replicate plates，each bearing ten colonies. The standard error for the difference between any two individual means is ±18.38，while that for the difference between the juice means given in the bottom row is ±9.19.

In this experiment，juice from plants with efficient nodules gave significantly better growth than juice from plants with inefficient nodules in eight out of the sixteen possible comparisons and insignificant results in the remainder.

In comparison with the uninoculated control juice，that from roots bearing one of the effective strains（310），has given variable results，and its mean effect on all four test organisms does not differ significantly from that of the control juice. Roots bearing the other effective strain （317） have produced a juice giving significantly better growth of two test organisms，as compared with control juice，but insignificant results with the other two. Both the root juices from plants bearing ineffective nodules （strains 313，B33） produced poorer growth than control juice with every test organism，although these differences were significant in only three out of the eight comparisons with individual test organisms. But when the growths of the four test organisms are combined，the growth on these two juices differs from that on control juice with a high degree of significance. In only one comparison out of the eight did these two juices produce significantly better growth than the medium without any root juice at all.

Table 2 Experiment with peas（1936）

Colony areas of four strains of pea-nodule bacteria on agar media with and without root juices

Strain	Mean colony area sq.mm.（*m*） Degrees of freedom（*n*）	Medium containing root juices from: Uninoculated plants（control juice）	Plants inoculated with strain: 310	317	313	B33	Medium without root Juices
310	*m*	153.4	198.6	190.2	106.3	151.3	102.5
	n	3	3	3	3	4	4
317	*m*	159.4	114.2	145.3	104.3	112.2	94.0
	n	2	3	3	3	2	3
313	*m*	105.6	103.3	156.7	92.5	94.3	90.0
	n	2	3	3	3	3	4
B33	*m*	130.6	124.1	148.1	94.7	113.6	112.6
	n	3	3	3	3	3	3
All strains	*m*	137.3	135.0	162.3	94.4	117.8	99.7

S.E. of difference between individual means =±18.38.
S.E. of differences between juice means (bottom row)=±9.19.

This experiment thus confirms the conclusions of the first experiment in that plants bearing one of the effective strains（317）produced a juice stimulating bacterial growth compared with control juice，while the juice from plants bearing either of the ineffective strains was definitely less good than control juice as a medium for the growth of the test bacteria.

EXPERIMENT WITH PEAS，1937

The method of estimating bacterial growth by colony area was found to be not altogether satisfactory owing to uneven growth of the colonies；it was，therefore，decided to estimate growth by means of haemocytometer counts of the test organisms growing in liquid culture.

Peas were grown in small pots of sand，sets of eight replicate pots being inoculated with strains 310，317 and B33，and one set left uninoculated. The peas were grown from 8 March till 6 June，and the root juices extracted and sterilized by filtration. The test organisms were grown in the following liquid medium：

K_2HPO_4	0.5g	$CaCO_3$	0.5 g
KH_2PO_4	0.5g	Sucrose	10.0 g
$MgSO_4 \cdot 7H_2O$	0.2g	Yeast water （10% yeast）	100 ml
NaCl	0.1g	Tap water	900 ml
$CaSO_4$	1.0g		

The medium was put in 20 ml. portions in small flasks，sterilized in the autoclave and 5 ml. of juice added aseptically，to each flask. Duplicate flasks of medium containing each type of juice were inoculated with standard loopfuls of a suspension of B33 strain of pea-nodule bacteria. After 18 days' growth at 25℃，the numbers of bacterial cells were counted on a haemocytometer ruled in $\frac{1}{400}$ sq. mm. The results are set out in table 3，where each value given is the mean count of duplicate flasks. The standard error for the difference between any two means is +25.9.

Table 3　Experihent with-peas（1937）

Growth of pea organism，strain B33. Bacterial cells per ml. in liquid media containing root juices，means of duplicate cultures

Medium containing root juices from peas			
Uninoculated （control juice）	Inoculated strain 3 10	Inoculated strain 3 1 7	Inoculated strain B 33
1828	2083	1915	1608
S.E. of difference between any two means= ±25.9，n=4			

Both the efficient strain root juices gave significantly better growth than did the control juice. The juice from roots bearing the inefficient B33 nodules produced less growth than any of the other juices，the significance of the differences having in each case a probability of less than 1%.

EXPERIMENT WITH SOY BEANS，1937

Soy beans were grown in large pots with one set uninoculated and other sets inoculated with strains 501，502，505 and 507. The seed was sown on 10 April and the plants reaped on 10 July. Root juices were extracted and sterilized as before. A liquid medium was made up similar to that used in the last experiment but with the sucrose replaced by dextrose，which was found better for the growth of soy-bean bacteria. This medium was put up in 15 ml. portions in wide test-tubes, to each of which, after sterilization, 5 ml. of juice to be tested was added aseptically. Batches of six replicate tubes were given control juice, and juices from plants bearing 505 and 507 nodules respectively. Two of each of these tubes were inoculated with strain 501, two with strain 502 and two with strain 507. The yield of juice from plants bearing 501 and 502 nodules was insufficient, so that only two strains of test organisms, 501 and 502, were grown on these juices, each in duplicate tubes.

After 12 days' incubation at 25℃, the growth of the bacteria was estimated from haemocytometer counts. The results of this experiment are shown in table 4.

Table 4 Experiment with soy bean (1937)

Growth of soy-bean organisms. Bacterial cells per ml. in liquid media with root juices, mean of duplicate cultures

Test Organism	Medium containing root juice from soy beans				
	Uninoculated (control juice)	Inoculated with strain			
		501	505	502	507
507	1251	—	790	—	668
S.E. of difference between any two means, 65.1, n=3					
501	1835	1938	1533	403	960
502	1680	2122	1263	1328	660
Means of both strains	1757	2030	1397	866	810
S.E. of the difference between individual means= ±210.7, n=10					

On account of the incompleteness of the test with 507 bacteria, separate standard errors have been calculated for this series and for the remaining two series inoculated with 501 and 502 bacteria. With the former the standard error for a difference between the means is ±65.1, while for the latter the error for differences between individual means is ±210.7.

When growth on juices from plants bearing efficient and inefficient strains is compared for the three test organisms, the inefficient strain juices give significantly the poorer growth in seven, and insignificant differences in two of the nine comparisons.

Five comparisons can be made between growth of the test organisms on juices from plants bearing effective nodules of strains 501 and 505 with their growth on juice from uninoculated plants. In one comparison the former type of juice gave significantly better and, in a second, significantly poorer growth. The other differences were not significant. The two root juices from plants bearing inefficient nodules of strains 502 and 507 gave significantly poorer growth than did the control juice in four comparisons out of five, and gave an insignificant result in the remaining comparison.

This experiment thus confirmed the depressing action of juices from roots with inefficient strains, but did not bear out the previous experiments as regards the improved growth on juices from roots with the efficient nodules as compared with control juice.

EXPERIMENT WITH PEAS, 1939

Peas were grown in small pots of sand with one set uninoculated and others inoculated with strains 317 and 313. The plants were grown from 4 April till 26 June, and the root juices extracted and filtered as before. Small flasks were set up, each containing 50 m. of the following medium in which the yeast extract was replaced by asparagin as a nitrogen source, so that the root juice should supply the sole source of accessory growth substances.

K_2HPO_4	0.5 g	CaSO4	0.5 g
KH_2PO_4	0.5 g	$FeCl_3$	0.02 g
$MgSO_4 \cdot 7H_2O$	0.2 g	Asparagin	0.5 g
NaCl	0.1 g	Dextrose	5.0 g
$CaCl_2$	0.1 g	Tap water	1000 ml

The flasks of medium were sterilized, and to each 10 ml of root juice were added aseptically. Each flask was inoculated with 1 ml of a suspension of the test organism. Eight flasks of control juice and six of each of the other types of juice were set up, and half the flasks with each juice were inoculated with 317 and half with B33 organisms. Two of the flasks inoculated with strain B33 and one inoculated with strain 317 were lost by contamination, so that the number of replicates counted varies between the different sets as shown in table 5. The results of the haemocytometer counts after 10 days' incubation at 25℃ are shown in table 5.

The growth of strain 317 was poor on all media, and did not make significantly different growth on any of them. Organism B33 grew better. Its growth was much poorer on juice from roots bearing inefficient 313 nodules than on control juice, but growth was also significantly poorer on the 317 juice. Indeed, the two juices from inoculated roots did not differ significantly from each other in their effects in this experiment.

Table 5 Experiment with peas (1939)

Growth of pea-nodule bacteria, strains 317 and B33.Bacterial cells per ml. in media with pea-root juices

Test organisms	Bacterial numbers (m) Degrees of freedom (n)	Media containing root juices from peas		
		Uninoculated (control juice)	Inoculated with strain	
			317	313
317	m	169	145	156
	n	3	2	2
	S.E. of difference between two means= ±12.73, n=7			
B33	m	2079	1516	1295
	n	2	1	2
	S.E. of difference between two means= ±129.8, n=5			

EXPERIMENT WITH SOY BEANS, 1939

Soy beans were grown in large pots of sand with one set uninoculated and other sets inoculated with strains 501 and 507 respectively. The seed was sown on 4 April and after 10 weeks' growth the roots were washed and the root juices extracted and sterilized by filtration. The same liquid test medium was employed as in the last experiment with peas; duplicate flasks were set up with each type of juice 10ml. of juice being added aseptically to 50 ml of sterilized medium. All the flasks were inoculated with 1 ml of a suspension of strain 507 and incubated at 25℃ for 3 weeks growth of the bacteria being slow. Haemocytometer counts of the resulting growth gave the mean figures shown in table 6.

Table 6 Experiment with soy beans (1939)

Growth of soy-bean nodule bacteria, strain 507. Bacterial cells per m1. in media containing soy-root juices. Means of four replicate cultures

Medium containing root juice from plants		
Uninoculated	Inoculated with strains	
(control juice)	501	507
471	1980	1127
S.E. of difference between means= ±129.8, n=9		

The juice from roots bearing the inefficient 507 nodules gave significantly poorer growth of the test organism than did the juice from roots with the efficient 501 nodules, but the control juice behaved in an unusual manner and gave much poorer growth than either the other juices. It is possible that this unusual result was due to the omission of yeast from the basal medium. This experiment thus provides the only case in which juice from plants bearing nodules of an inefficient strain gave better growth than juice from nodule-free roots.

DISCUSSION

The composition of root juice is clearly liable to be affected both by growing conditions of the plant and by details in the method of extraction and filtration that are very difficult to control. It is to be expected, therefore, that somewhat variable results will be obtained in experiments of the type described above. The experiments must be considered together in order to assess the validity of conclusions derived from them. They comprise six experiments, three with peas and three with soy beans. The fact that the technique of testing the effect of the juices was varied from one experiment to another will strengthen the conclusions common to all the experiments. In most of the experiments several test organisms were used, and in all of them several types of root juices were compared.

It was pointed out in the introduction (p.477) that the maladaptation to their host plant shown by ineffective strains might be due to some factor normally present in the root system, that acted specifically against such strains. The growth of the different bacteria in root juice from uninoculated plants, shown in tables 1, 2, 4 and 5, is not correlated with the effectiveness or otherwise of the strain used as test organism. This shows that the root juice does not normally contain any factor that acts specifically against ineffective strains.

On the other hand, there is evidence that the type of nodule on the root system affects the ability of the juice from roots and nodules to support growth of the test organisms. Analyses of variance were made for each experiment, and the standard errors of differences between means obtained. The latter are given for each experiment. From them has been calculated the significance of differences in growth of the nodule bacteria on juices from plants bearing effective and ineffective nodules, and on juice from nodule-free roots. These comparisons have been summarized in table 7, in which a 'plus' sign means that there has been significantly better, and a 'minus' sign significantly poorer growth, while '0' means that the difference did not attain significance. The third column shows the comparative growth of test organisms on juices from roots bearing effective nodules as against growth on juices from roots with ineffective nodules. Out of the forty-four possible comparisons, there were twenty-seven in which the former juice gave significantly the better growth. In all remaining seventeen cases the differences were insignificant.

This effect of root juice upon bacterial growth affords a simple explanation of the observed appearances in effective and ineffective nodules. The rapid growth and long duration of activity amongst the bacteria in effective nodules and the slower growth and early onset of disintegration of the bacteria in ineffective nodules is just the result that might be expected from the different environment provided by the root juice in each case.

The difference in effect of juice from roots bearing effective and ineffective nodules might itself be due to a stimulating action upon bacterial growth of the former type of juice, or to a depressing action of the latter type. These two effects can, to some extent, be separated by comparison with the effects of control juice from nodule-free roots.

Twenty-five comparisons have been made between the effects of juices from roots bearing effective nodules and those of control juice (column 4 of table 7). The former gave significantly better growth of bacteria than control juice in ten cases and significantly poorer growth in three cases.

Twenty-five comparisons were made between growth on juices from plants with ineffective nodules and on control juice (column 5, table 7). In eleven of these the juice from roots with ineffective nodules gave growth significantly poorer than control juice, and in only one case was this result reversed.

The evidence taken as a whole thus indicates that there is an increased stimulating effect from juices of roots bearing effective nodules, possibly connected with the products of nitrogen fixation, while the presence of ineffective strains in the roots induces the appearance of some factor in the root juice which is depressing in its action on bacterial growth.

With regard to the nature of this depressing factor, it should be noted that the test organisms made growth on all the juices that were tested except on two in the first experiment. In no instance could a transmissible lyric agent be found in the juice. This provides strong evidence against the hypothesis that any of the strains tested owe their ineffectiveness to the presence of bacteriophage. Had bacteriophage been present in a concentration sufficient to produce the symptoms observed in the ineffective nodules, the technique of growing the same strain in the presence of filtrates of the crushed nodules and roots should have revealed its presence. The same technique of juice extraction has, in fact, been successfully used to demonstrate the presence of bacteriophage in nodules of peas grown in garden soil.

The present results provide no evidence as to whether the factors affecting the growth of nodules and of their contained bacteria are localized in those nodules or distributed throughout the root system. But it seems probable that they are localized in or near the nodules; for, when nodules of both effective and ineffective types develop on the same root system, the characteristic size and appearance of each type of nodule is not affected by the presence of the other type on the same root.

Table 7　Summary of significant differnces

		3	4	5
		Comparative growth of bacteria on root juices from plants inoculated with		
Experiment	Test organism	Effective strain-Ineffective strain	Effective strain-uninoculated	Ineffective strain-uninoculated
Soy 1936 (table 1)	501	++++	+0	—
	505	++0 0	+0	0 0
	502	++	+	0 0
	507	++0 0	0 0	0 0
Pea 1936 (table 2)	310	++++	+ +	-0
	317	+0 0 0	—0	—
	313	++0 0	+0	0 0
	B33	0+0 0	0 0	0 0
Pea 1937 (table 3)	B33	++	+ +	—
Soy 1937 (table 4)	501	++++	0 0	—
	502	+++0	+ 0	-0
	507	0	—	—
Pea 1939 (table 5)	317	0	0	0
	B33	0	—	—
Soy 1939 (table 6)	507	0	+	+
Totals:				
Significantly greater		27	10	1
Significantly less		0	3	11
Insignificant		17	12	13
Total comparisons		44	25	25

A plus sign means that the first-named root juice produced significantly better, and a minus sign significantly poorer growth than the second-named juice. '0' indicates an insignificant difference.

The evidence as a whole suggests that a soluble substance is formed with in the nodules produced by an ineffective strain, that is harmful to the growth of the bacteria. The formation of such a substance would account for the early arrest in nodule growth and for the rapid onset of disintegration of the contained bacteria.

REFERENCES

[1]Brenchley, W.E. and Thornton, H.G.1925. *Proc. Roy. Soc.* B, 98, 373.
[2]Chen, H.K.1938. *Nature, Lond.*, 142, 753.
[3]Chen, H.K. and Thornton, H.G.1940. *Proc. Roy. Soc.* B, 129, 208.
[4]Helz, G.E., Baldwin, I.L. and Fred, E. B. 1927. *J. Agric. Res.* 35, 1039.
[5]Link, G.K.K. 1937. *Nature, Lond.*, 140, 507.
[6]Thimann, K.V. 1936. *Proc. Nat. Acad. Sci.*, *Wash.*, 22, 511.
[7]Thimann, K.V. 1939. *Trans. Third Commission Int. Soc. Soil Sci.* A, 24.
[8]Thornton, H.G. 1929. *Proc. Roy. Soc.* B, 106, 111.
[9]Thornton, H.G. and Rudorf, J. 1936. *Proc. Roy. Soc.* B, 120. 240.
[10]Wright, W.H. 1925. *Soil Sci.* 20, 95, 131.
[11]Yates, F. 1933. *Empire J. Exp. Agric.* 1, 129.

The Limited Numbers of Nodules Produced on Legumes by Different Strains of *Rhizobium*[*]

H. K. CHEN

(Soil Microbiology Department，Rothamsted Experimental Station，Harpenden)

In the field a legume crop usually obtains its nodules from a mixed population of nodule bacteria including a variety of strains doubtless varying in their effectiveness towards the host plant. It is therefore of practical importance to deter mine what are the factors that determine which strains will infect the plant and in what proportions. Nicol & Thornton （1941） found that an important factor controlling this was the competition that took place between bacteria of different strains outside the roots of the host plant. Where one was markedly dominant in competition，this became the determining factor controlling infection. But otherwise the relative infectivity of the strains determined the proportion of the total nodules contributed by each of them.

The numbers of nodules produced on a legume by a given strain of *Rhizobium* will clearly be conditioned by a number of factors，some arising in the root surroundings，others from the general physiology of the host plants，and others more specifically related to the strain of invading bacterium. It is with the latter more specific relationships that the present paper deals. Obvious differences exist between strains of *Rhizobium* not only as regards the mean size but also the mean number of nodules characteristically produced by them under conditions of host-plant cultivation rendered as carefully standardized as possible. These differences might be due to different rates of successful infection,[1] or to the establishment of a limiting equilibrium due to ability of a given strain to produce only a limited number of nodules on a given mass of the root system. If such a limiting equilibrium exists，the number of nodules per gram of root should attain a constant level characteristic of each strain. This number should remain constant on a growing root system though the absolute number of nodules increases with the growth of the roots. On a root system that makes its growth over a short period，the further production of nodules should stop when the roots cease growing or when the specific limiting number of nodules per gram of root has been reached，whichever occurs last.

On this latter type of root system，if the limiting number of nodules per gram of root is quickly reached the nodules on the young plant will comprise a large fraction of the number finally possible and hence will greatly reduce the formation of further nodules by the same or by a different strain. A number of authors have recorded such an inhibiting effect of nodules. In particular，Dunham & Baldwin （1931） found that early nodules produced by one strain could entirely inhibit the formation of nodules by a different strain. Nicol &Thornton（1941），however，found that with peas and soy beans the inhibiting effect of the early nodules acted with equal intensity against the same as against a different strain. In this work which was designed to study competition between strains，the second strain was applied to sand already populated with the first，so that the results were complicated by competition between strains outside the plant. To measure the effect of the early nodules upon subsequent infection it is necessary to remove the bacteria，derived from the first inoculum，that remain in the root surroundings，and to transplant the roots into a medium populated only by the second inoculum. Experiments of this kind were made with the object of determining whether the nodule

[*]原载于 *The Journal of Agricultural Science*. 31:479～487，1941.

1 A large part of the root hair infections must fail to result in nodules（see McCoy，1932）. The term ‘successful infection’ is here used to designate infections that result in the formation of nodules.

numbers per unit mass of roots attained a liming equilibrium and how the establishment of this equilibrium affected subsequent infection by the same and by a different strain. These experiments were made with clover, whose root system continues its growth over a long period, and with soy beans, whose roots make most of their growth during early stages of culture.

EXPERIMENTS WITH RED CLOVER, 1939

In this experiment two strains of clover *Rhizobium* were used-the efficient strain 205 obtained from Wisconsin[2] and the inefficient Coryn strain, whose nodules were described by Chen & Thornton (1940). The rates of nodule appearance due to these strains on clover seedlings were found by Nicol & Thornton (1941) to be markedly different, while their experiment in which clover was grown in sterilized sand showed that the absolute number of nodules produced in three months' growth by the two strains also differed characteristically.

In the present experiment, Montgomery red clover was sown in wide test-tubes on slopes of agar medium of the following composition:

K_2HPO_4	0.5 g	NaCl	0.1 g	$FeCl_3$	0.01 g
KH_2PO_4	0.5 g	$Ca_3(PO_4)_2$	2.0 g	Agar	10 g
$MgSO_4 \cdot 7H_2O$	0.2 g	$FePO_4$	0.5 g	Water	1 L

The tubes of media were sterilized in the autoclave. Twenty replicates were left sterile, twenty supplied with strain 205 and twenty with the Coryn strain. The bacteria were mixed with the melted agar cooled to 42℃ before making the slopes. Two seeds, externally sterilized by immersion for 3 min in absolute alcohol and for 3 min. in 0.2% $HgCl_2$ and washed with sterile water, were sown at the top of each slope. The seeds were sown on 13 February, and on 27 March the seedlings were removed from the tubes and their nodules counted. They were then replanted in small pots each containing 3 kg. of nitrogen-deficient sand, sterilized by blowing superheated steam through each pot for half an hour. 175 ml of the following sterilized food solution was added to each pot:

K_2SO_4	0.9 g	$FeCl_3$	0.02 g
K_2HPO_4	0.5 g	Boric acid	0.02 g
$CaH_2(PO_4)_2 \cdot 4H_2O$	0.5 g	$MnSO_4$	0.02 g
$MgSO_4 \cdot 7H_2O$	0.5 g	Tap water	900 ml
NaCl	0.5 g	Lucerne root extract	10 ml

Twenty replicate pots were supplied with a heavy inoculum of strain 205 and twenty with one of the Coryn strain, the bacteria being mixed with the food solution before addition. One seedling bearing strain 205 nodules, one bearing Coryn nodules and two plants without nodules were planted in each pot, each plant being separately labelled. On 10 July the roots were washed, the nodules on each plant were counted and the dry weights of individual root taken. The results are shown in Table 1.[3] The experiment was so designed as to test whether any of the following factors had any effect upon the final nodule numbers per gram of root: (1) time at which the bacteria were first applied; (2) size of the root system as modified by the efficient strain applied at by seeding time; (3) a possible inhibiting action of the early nodules against the same or a different strain.

2 The author's thanks are due to the staff of the Wisconsin Agricultural Experiment Station for supplying cultures of this strain and of the four strains of soy bean nodule bacteria used in the second experiment, described below.

3 The nodules per gram of root were separately calculated for each plant and the means of the figures so obtained are those shown in the table. They differ from those derivable from the mean nodule numbers (column 51 and the mean root weights (column 6). The same was followed for the corresponding figures in Table 3.

Table 1 Effect of strain of *Rhizobium* on nodule numbers in red clover

Set	Strain applied		Mean nodule numbers Per plant		Final root dry wt mg Means per plant	Final nodules per g.root	n
	At Sowing time	After Trans planting	When Trans planted	At end			
1	—	Coryn	—	357.2 ± 42.8	118	3402.4±475.8	17
2	Coryn	Coryn	38.7±4.4	385.5± 74.0	163	2770.9±558.3	10
3	205	Coryn	7.3 ±0.8	904.1±181.4	381	2273.7±360.9	14
4	—	205	—	182.1 ± 30.8	355	544.8 ± 56.8	15
5	Coryn	205	50.0 ±1.4	169.3 ± 38.4	239	694.5 ±101.3	10
6	205	205	9.1 ±1.5	331.1 ± 50.7	573	652.6 ±121.3	14

At the time of planting out, seedlings that had grown for 6 weeks on agar already showed differences in nodule numbers characteristic of the strain supplied at seeding time（column 4). When removed from the agar the seedling root system showed no differences in size according to the culture supplied-the efficient nodules not having had time to produce increased growth.

After transplanting into the pots some plants died. The numbers surviving can be deduced from the degrees of freedom, *n*, shown in the last column. These made considerable growth before harvest with a large increase in nodule numbers. The final root weights shown in column 6 were much increased where 205 nodules, effective in nitrogen fixation, had developed on the seedling while growing in agar. This appears in comparing set 1 with 3 and set 4 with 6.

The absolute number of nodules at the end of the experiment(column 5)were not significantly affected by the presence of Coryn nodules on the seedlings but the presence of 205 nodules at the time of transplanting greatly increased subsequent nodule formation by either strain. This effect was in fact due to the enlargement of the root system resulting from nitrogen fixation by the early-formed efficient strain 205 nodules.

The nodules per gram of root（column 7）show no significant differences between sets receiving different treatments in their early growth but later grown in pots supplied with the same strain. Thus the time at which the bacteria were first supplied to the roots was without final effect on the nodules per gram of root. Nor were there very large differences in size of the root system produced by the early formed 205 nodules. There were, on the other hand, large differences in the mean number nodules per gram of root according to the strain in the sand which was in contact with the root system during the period when it made most of its growth. This mean number reached a definite limit characteristic of the strain present in the sand. This limit was apparently attained quite early in the plant's growth. The mean figure for plants grown in pots containing Coryn bacteria was 2816, and that for plants in pots containing strain 205, only 631 nodules per gram of root. These figures are in the ratio of 4.6：1. This ratio can be compared with that between the actual nodule numbers developed by the two strains in agar（column 4), because during this early period the size of the root systems was similar in all sets. The Coryn strain developed a mean of 44.4 nodules per seedling, and strain 205 a mean of 8.2, at the time of transplanting. These figures are in the ratio of 5.4:1. So that the two strains produced nodules whose numbers per unit of the root system were in approximately the same ratio both on seedling roots grown in agar and subsequently on plants grown in pots of sand. Thus the limit of infection for a root system of given size is characteristic of each strain and is quickly reached. But the absolute nodule numbers of nodules increased *pari passu* within the growth of the root system, which in clover continues over a long period. This explains why the presence of nodules on the seedling did not stop further nodule formation, which took place on a growing root system, and why the final number of nodules per gram of root was that characteristic of the second applied strain, which was in contact with the root system while this was making most of its growth.

The following experiment，similar in general design to the first，was made with soy beans，whose root system makes most of its growth when the plant is quite young. It was designed to determine what specific limits of nodule numbers per gram of root were possessed by four strains of soy bean *Rhizobium* and to test the influence of the early nodules upon later infection by the same and by different strains.

EXPEEIMENT WITH SOY BEANS，1939

Soy beans were grown in glazed earthenware pots each containing 12 kg of sand and 1L. of food solution similar in composition to that used in the first experiment. Five seeds externally sterilized were sown in each pot on 21 June. Eight replicate pots were left uninoculated and eight each were supplied with each of the following strains of *Rhizobium*:

Wisconsin 501 } Effective
” 505
” 502 } Ineffective
” 507

The plants were grown for 9 weeks and their roots were thoroughly washed and the nodules counted. They were then replanted in the pots in such a way that each pot whose sand contained an effective strain（501 or 505） received one plant bearing nodules produced by the same strain，one plant bearing nodules produced by each of the ineffective strains and one uninoculated plant. Similarly each pot whose sand contained an ineffective strain（502 or 507） received one plant bearing nodules produced by the same strain，one plant bearing nodules produced by each of the effective strains and one plant without nodules. Each plant was separately labelled. The scheme of transplanting is shown in Table 2. After a further 14 weeks' growth the nodules were recounted and dry weights of the roots were taken. The results are shown in Table 3. The plants which bore nodules produced by strains 501，502 or 505 before transplanting did not show any significant increase in nodule numbers after transplanting （sets 1-9，columns 4 and 5）. Thus the limit of nodule numbers attainable on the root systems in these sets had been reached within the first 9 weeks' growth. The number of nodules per gram of root was specific to the strain of Rhizobia（column 7）. The mean number of nodules per gram of root in sets 4，5 and 6 which bore nodules produced by strain 502，was 284.7，a figure significantly higher than the mean numbers，191.7 and 163.4 of the sets whose nodules were produced by strains 501 and 505 respectively（sets 1-3 and 7-9）.

Table 2 Soybean experiment，scheme of transplanting

Plants with nodules，when transplanted，of strain	Transplanted into pots whose sand contained strains 501 Set	502 Set	505 Set	507 Set
501	1	2	—	3
502	4	5	6	—
505	—	7	8	9
507	10	—	11	12
No nodules	13	14	15	16

The plants without nodules at the time of transplanting made considerably greater root growth during the second growth period，probably because they were smaller at the time of transplanting and suffered less check. These plants in sets 13，14 and 157 planted in sand containing bacteria of strains 501，502 and 505 respectively，developed nodules whose numbers per gram of root did not differ significantly from those on plants that had received the corresponding strain at the time of sowing （compare sets 1 and 13，5 and 14，8 and 15）. Thus the specific limit of nodules per unit mass of root system was attained regardless of the total mass of the root system，which varied widely，or the time at which the infection took place. This latter point shows that the number of nodules is determined by the size of the root system and not vice versa，since most of the root growth in sets 13，14 and 15 took place before the plants had developed any nodules. Strain 507 has a much higher level of nodule numbers than the other three strains. The mean final nodule numbers per gram of root for sets 10，11 and 12 was 444.4，a figure significantly higher than that for any other set or group of sets. This high figure was not reached during the period of 14 weeks' growth in set 16 which first received the bacteria at the time of transplanting. Nor was the full number of nodules reached during the first 9 weeks of seedling growth in sets 10，11 and 12，which developed more nodules after transplanting. In sets 10 and 11 these additional nodules were in fact produced by the strains 501 and 505 respectively as was shown by examining the nodules. These later-formed nodules were comparatively few and the number of nodules per gram of root finally reached was that characteristic of strain 507. The figures for sets 10 and 11，451 and 446，do not differ significantly from that of 436.2 for set 12 which received strain 507 both at sowing time and after transplanting.

Table 3 Effect of strain of *Rhizobium* on nodule numbers in soy beans

Set	Strain applied		Mean nodule numbers per plant		Final root dry wt. mg. Mean per plant	Final nodules per g. root	n
	At Sowing time	After Trans planting	When Trans planted	At end			
1	501	501	17.7	16.1 ± 2.7	94	206.3 ±30.2	9
2	501	502	17.2	18.5 ± 1.5	162	149.8 ±38.6	4
3	501	507	22.4	22.6 ± 3.4	144	219.1 ±68.4	6
	501	Mean				191.7 ±25.9	21
4	502	501	29.1	29.4 ± 4.8	123	269.1 ±44.7	8
5	502	502	35.9	37.8 ± 4.4	158	284.3 ±38.1	7
6	502	505	36.3	36.5 ± 6.1	131	300.8±46.9	9
	502	Mean				284.7 ±24.6	26
7	505	502	18.9	21.0± 1.7	170	134.4 ±20.9	6
8	505	505	22.5	19.6 ± 2.0	114	184.1 ±29.9	7
9	505	507	18.4	19.3± 2.0	185	171.7 ±50.6	6
	505	Mean				163.4 ±19.6	21
10	507	501	28.0	34.6 ± 4.7	77	451.0 ±56.5	6
11	507	505	29.6	44.6 ± 8.9	101	446.0 ±49.4	7
12	507	507	32.1	56.6±10.1	141	436.2 ±53.5	8
	507	Mean				444.4 ±29.4	23
13	—	501	—	30.9 ± 7.4	179	211.4±26.0	7
14	—	502	—	70.3±13.4	303	233.0 ±39.7	8
15	—	505	—	43.0 ± 8.8	244	186.4 ±27.4	7
16	—	507	—	57.0±13.4	253	225.3 ±43.6	7

DISCUSSION

In the experiments described above the number of nodules n divided by the dry weight of the roots m was found to reach a limiting figure that was constant and specific for each strain of *Rhizobium*，$n = mk$. If a plant's roots are exposed in succession to pure cultures of two strains of *Rhizobium* having the limiting constants k_1 and k_2 and if each strain is allowed time to reach its limit，the number of nodules n_1，produced by the first strain will be m_1k_1 where m_1 is the mass of the roots developed while in contact with this strain. On the simplest supposition, the number n_2 produced by the second strain will be m_2k_2 where m_2 is the additional mass of roots developed in contact with it. The total nodules developed by the two strains will therefore be $n_1+n_2=m_1k_1+m_2k_2$. In the experiment with clover, nearly all the root growth took place after transplanting so that m_2 was very large relatively to m_2. Consequently the number of nodules was determined by k_2 and was that characteristic of the second-applied strain. In the soy bean experiment, all or nearly all the root growth took place in the presence of the first applied strain, whose specific constant, k_1, determined the limit of nodule numbers reached.

It would be interesting to investigate the condition where the host is removed into the presence of the second strain before the first strain had reached its limit of infection for a given mass of roots, and to discover to what extent the first strain can then impose its specific limit on further nodule formation in these same roots by the second strain. Thus if the number of nodules produced by the first strain, $n_1=m_1k_1-x$, would the number x, produced by the second strain, be determined by, the constant k_1 or by a constant k_2 specific to the second strain? The answer to this question might throw light on the mechanism of nodule limitation. The evidence from sets 10 and 11 in the soy bean experiment suggests the continued operation of the constant k_1 specific to the first strain, but this evidence is insufficient to form any basis for discussion.

SUMMARY AND ABSTRACT

Pot experiments were made with red clover and with soy beans to determine how far the number of nodules developed was a specific character of the strain of *Rhizobium* supplied.

The number of nodules per gram of root was found to reach a limit specific to each strain. This limiting equilibrium was attained regardless of the size of the root system or the age of plant at which the culture was first supplied, provided enough time were allowed for the limit to be reached.

When two different strains were applied to the root surroundings in succession, the final number of nodules was determined by the limit specific to the strain in contact with the roots while these were making most of their growth. In clover this was the second and in soy beans the first applied strain.

REFERENCES

[1]CHEN, H. K. & THORNTON, H. G. (1940) . *Proc. Roy. Soc. B*, 129, 208-29.

[2]DUNHAM, D. H. & BALDWIN, I. L. (1931) . *Soil Sci.* 32, 235-9.

[3]MCCOY, ELIZABETH (1932) . *Proc. Roy. Soc. B*, 110, 514-33.

[4]NICOL, H. & THORNTON, H. G. (1941) . *Proc. Roy. Soc. B*, 130, 32-59.

NOTE CONCERNING AUTHORSHIP

The work described in this paper was carried out by Dr Chen shortly before his departure for Central China. Some difficulty in communication due to wars has made it necessary for the undernamed to write the paper from Dr Chen's notes and data without his having the opportunity to see it before publication. The writer thinks he has drawn conclusions from the data in agreement with Dr Chen's opinions, but he accepts full responsibility for these conclusions and for the actual writing, although credit for the work is due solely to Dr Chen.

H. G. THORNTON.

(*Received* 23 *May* 1941)

Note on the Root-Nodule Bacteria of *Astragalus Sinicus* L.*

H. K. CHEN AND M. K. SHU

(National Agricultural Research Bureau, China)

On the basis of a fairly extensive cross-inoculation study of the wild leguminous plants in Wisconsin, Bushnell and Sarles (2), reported that none of the four species of *Astragalus* studied [*A. alpinus* L., *A. canadensis* L., *A. caryocarpus* Ker., and *A. neglectus* (T.&G.) Shelden] was assigned to any tested cross-inoculation group, and pointed out that "these results, although they do not warrant the establishment of a new cross-inoculation group for *Astragalus* spp., do lend weight to the probability that *Astragalus* spp. make up their own select cross-inoculation group." This is the only published record relating the cross-inoculation characteristics of the genus *Astragalus*, notwithstanding the fact that *A. sinicus* is one of the major green manuring and forage crops in China and Japan. It is not mentioned in Itano's book on soil microbiology (3) or in Junk-Oppenheimer-Weisbach's *Tabulae Biologicae* (5) .

The present note is a summary of inoculation tests carried out during the last 3 years on the strains of root-nodule bacteria isolated from *A. sinicus* and other plants and their nodule-forming properties.

Morphologically and culturally, our strains of *A. sinicus* bacteria conform well with those described by Aso and Ohkawara (1) and by Itano and Matsuura (4) and with Bushnell and Sarles' bacteria of other *A. stragalus* species (2) . They are all gram-negative, uniformly stained when young, are in the form of coccoid rods smaller than pea bacteria, and are motile. On yeast-mannitol agar slants, they grow more slowly than pea bacteria and more rapidly than most cowpea and soybean bacteria. A moderately abundant growth is often obtained after 1 week's incubation at 25-30℃. On yeast-mannitol congo-red agar plates, the colonies are round, raised, whitish, moist, gummy, and unstained by congo-red. On calcium-glycerol- phosphate agar, they grow moderately abundantly and are moist and opaque. On Clark and Lub's medium, they are turbid with gummy deposit. They turn litmus milk alkaline and give a clear serum zone. On potato strip medium, they grow moderately abundantly and are milky white, watery, and glistening.

In the inoculation tests, plants of *A. sinicus* were grown in sterilized sand containing modified Crone's solution and were inoculated with various strains of root-nodule bacteria. The sterileness of the culture medium was checked in each case with uninoculated controls. A total of 30 strains of bacteria of known cross-inoculation groups representing the seven major groups, and also six strains of bacteria isolated from root nodules of *A. sinicus*, were used in the tests. The results are summarized in table 1. Repeated experiments showed consistently that *A. sinicus* formed nodules only with bacteria isolated from the same plant species and with a strain of cowpea bacteria isolated from nodules of *Desmodium heterophllum*.

Table 1 Summary of root-nodule formation of different strains of bacteria on *A. sinicus*

Strain	Source of isolation*	Gross-inoculation group	Nodule formation
107	*Vicia hirsuta*	Pea	—
108	*V.sativa*	Pea	—
110	*Vicia* sp.	Pea	—
114	Sweden commercial	Pea	—
201	*Trifolium pratense*	Clover	—

*原载于 *Soil Science*. 58 (4) :291～294, 1944.

续表

Strain	Source of isolation*	Gross-inoculation group	Nodule formation
205	*Trifolium* sp.	Clover	—
301	*Melilotus indica*	Alfalfa	—
306	*M.indica*	Alf 以 fa	—
302	*Medicago hispida*	Alfalfa	—
303	*M.sativa*	Alfalfa	—
304	*M.sativa*，Wisconsin 107	A1falfa	—
305	*M.sativa*.New Jersey A2	A1falfa	—
401	*Soja max*	Soybean	—
S1	*S.max*	Soybean	—
S2	*S.max*	Soybean	—
501	*Cajanus cajan*	Cowpea	—
507	*Lespedeza striata*	Cowpea	—
508	*Pueraria hirsuta*	Cowpea	—
510	*Indigofera suffrutieosa*	Cowpea	—
511 B	*Phaseolus aureus*	Cowpea	—
516	*P.aconitifolius*	Cowpea	—
512	*Vigna sinensis*	Cowpea	—
515	*Crotalaria striata*	Cowpea	—
517	*Alysicarpus vaginalis*	Cowpea	—
520	*A.vaginalis*	Cowpea	+
519	*Atylosia scarabeoides*	Cowpea	—
521	?	Cowpea	—
522	*Desmodium heterophyllum*	Cowpea	+
523	*Derris elliptica*	Cowpea	—
701	*Phaseolus vulgaris*	Bean	—
601	*Astragalus sinicus*		+
602	*A.sinicus*		+
603	*A.sinicus*		+
604	*A.sinicus*		+
605	*A.sinicus*		+
606	*A.sinicus*		+

*The writers acknowledge the assistance of C. J. Wang，who kindly identified the host plants.

SUMMARY AND CONCLUSIONS

The present investigation forms the counterpart of Bushnell and Sarles' work，in which bacteria isolated from *Astragalus* spp. were tested on plants of known cross- inoculation groups. The results agree，in that root-nodule bacteria of *Astragalus* do not produce nodules on other genera of leguminous plants，nor do bacteria isolated from other sources，except *Desmodium heterophyllum*，form nodules on *Astragalus* plants. Since it is well known that bacteria of the cowpea group often infect plants of different groups，the nodulation of *A. sinicus* by *D. heterophyllum* bacteria should not be regarded as invalidating the conclusion that *Astragalus* and its root-nodule bacteria must be considered as a select cross-inoculation group.

REFERENCES

[1]Aso，K.，AND OHKAWARA，S. 1928. Studies of the nodule bacteria of Genge. *Proc. and Papers 1st Internatl. Cong. Soil Soi. Comn.* 3:183-184.

[2]BUSHNELL，O. A.，AND SARLES，W.B. 1937. Studies on the root-nodule bacteria of wild leguminous plants in Wisconsin. *Soil Sci.* 44:409-423.

[3]ITANO，A.，1934. Soil Microbiology，ed. 2.（Japanese.）

[4]ITANO，A.，AND MATSUURA，A. 1934. Nodule bacteria of *Astragalus sinicus*（Genge）: II. III. *Ber. O'hara Inst. Landw. Forsch.* 6:259，341.

[5]WILSON，P. W.，AND SXRLES，W.B. 1939. Root nodule bacteria. *Tabulae Biologicae*. 17:338-367.

草籽绿肥、根瘤细菌和人工接种*

陈华癸

（武汉大学农学院农业化学系）

一、施 肥 增 产

要大力开展丰产运动，要每亩田地的生产量增高，人人都晓得，必须要培育土壤的肥沃性要供给大量的肥料。肥料从何而来？来源畅不畅？够不够？主要的肥料来源有：

（1）农家储粪积肥

（2）城市粪尿

（3）肥料和化学肥料

（4）绿肥

解放以后，经过剿匪反霸、土地改革、换工互助，农民生产的积极性大大提高了，依靠自己储粪积肥，增施肥料，两年来已经得到了优良的成绩。但这是有限度的，要求将农业生产再提高一步，只从农家自己储粪积肥是不够的。城市粪尿的供给量有限，供给范围也有限，对靠近城市的农家有些帮助，量也不多。商品肥料，最主要的是籽饼，价格很高，量也不够。化学肥料要等化学肥料工业发达以后才能大量供给，目前的生产量很小，价格也高，只能重点试用，累积经验，对于当前的丰产运动是不能起主要作用的。

因此，在现有的储粪积肥所能达到的水平之上，展开丰产运动，主要的要依靠搞绿肥。

什么是绿肥？概而言之，把各种植物，压翻在土壤中，促使它腐烂，产生肥田效果，就是搞绿肥。山青、湖草、各种绿肥作物，都可以用（请参阅陈华癸著《绿肥》，武汉通俗图书出版社，1950 年）。

二、 草 籽 绿 肥

水稻田冬季栽种草籽绿肥，在本省和整个的长江流域都具有悠久的历史，在农业生产制度中，占有重要的地位。在某些地带，农民根据了祖上累积的经验，利用草籽绿肥，获得了水稻丰产效果。但这些地带的生产经验并没有普遍的展开，没有普遍展开的原因是多方面的，主要是原因是：

（1）由于农业生产社会的闭塞性，鸡犬之声相闻而老死不相往来，好的生产知识，不能传开。

（2）由于各地气候和土壤的性质不同，甲地的优良生产知识不能机械的介绍到乙地去利用，必须要予以必要的改变，使它和乙地的情况配合，才能收效。

（3）由于封建的生产关系，限制了生产手段的充分开展。就以水稻田种草籽绿肥而论：种草籽的好处是增收水稻。在解放以前，许多地方，秋收主要是交给地主的，春收才轮到农民自己。种了草籽，自己无利，白白地为了地主。在这种情形之下，农民没有种草籽的道理。

现在封建的生产关系已经打倒了，第三个原因已不存在。在共产党和人民政府的领导之下，通过农会，农民已经组织起来了，换工互助，生产合作，已经开始从农业社会形式逐渐向有组织的工业社会形式发展，农业社会的闭塞性也打破了，限制了良种美法的传播和提高主要是由于上述第二个原因。

甲乙两地的土质不同，气候不同，生产制度也有些不同，甲地的生产经验介绍到乙地不能机械的搬去，那是会碰壁的，行不通的，必须要做一番消化和变更的工作，将一时一地的感性知识（甲地的生产经验），提高为理性知识（生产经验的科学理论），再与另一时地的实际情况相结合，才能成为适用于乙

---

*原载于《新科学》，3:33～38，1952.

地的生产知识。这类工作是我们农业科学工作者、各地农事试验场的负责和技术干部的工作。

三条障碍已打开了两条，剩下这一条是我们的责任。今天演讲的主题是整个问题的一部分。

三、种草籽绿肥的好处

种草籽绿肥为什么可以肥田？种草籽绿肥的好处很多，主要的是：

种了草籽的田地，土壤泡松，耕犁省工、省力。这是一般绿肥都有的好处，冬天种一季绿肥作物，连根带叶可以生产五、六千斤有机物质（植物和动物的体质，化学上称为有机物质），这些有机物质压青（翻入旱地土壤中称为压青）或泡青（翻入水田中称为泡青）后在土壤中腐烂，增加了土壤腐殖质（土壤中的有机质），就是使土壤泡松的效果。种一坵绿肥完全用在本坵，或者分施二三坵都可以。

种了草籽或其他豆类绿肥还有一项主要的好处，它能供给土壤中十分缺乏而植物很需的氮质营养料。农作物生长需要从土壤中吸收许多种营养料；需要最多，而土壤中很缺乏的就是氮质营养料。种水稻、玉米、麦类、蔴类等农作物，只靠土壤中的氮质营养料是不够的，必须施用氮质肥料。籽饼、粪尿、厩肥、硫酸铔肥田粉等都含有很丰富的氮质营养料。前面讲到过，要展开丰产运动这些肥料还不够用。要购买的商品肥料太贵，并且也不够分配。种了草籽绿肥作物或它种豆类作物能够大量的增加土壤中的氮质营养料。

四、氮质固定作用

空气中有80%氮气，可惜一般的农业作物都没有吸收氮气，作为氮质营养料的功能。空气中氮气虽多，如果土壤中没有足够的氮质营养料（这不是氮气，而是别种含氮化合物，以硝酸化合物和铵质化合物为主），一般的农业作物还是长不好。豆类植物则不然，它有吸收利用氮气的功能，蓝花草籽和红花草籽都是豆类植物，它们也有这功能。冬季种了草籽，草籽吸收了氮气，变为它体质中的氮质成分（以蛋白质为主）。到春天翻入土壤中，充分腐烂，由于土壤中各种微生物的腐解作用，将草籽体质中的氮质成分分解为硝酸和铵质化合物，存在土壤中。再种水稻，土壤中就有丰富的氮质营养料供给水稻吸收，达到增产效果。

这样，通过栽培草籽绿肥作物（它种豆类植物也可以），将空气中无用的氮气变为土壤中有用的氮质营养料，称为“氮质固定作用”，种一季草籽，一亩田地可以产五、六千斤植物体质，以每亩5000斤计算，可生产20斤有价值的氮质，等于400斤籽饼，或100斤硫酸铔肥田粉。买籽饼做肥料，最便宜1斤饼1斤壳；请计算一番，种一季草籽，可以省下多少肥料本钱。

五、根瘤和根瘤细菌

专靠豆类植物本身，没有氮质固定作用。随便挖掘一株草籽或它种豆类植物，会发现根系上有许多疙瘩，这种疙瘩为根瘤（图1）。

将根瘤切开，做成显微镜片，在显微镜下观察，会发现许多细菌，这种细菌称为根瘤细菌（图2）。

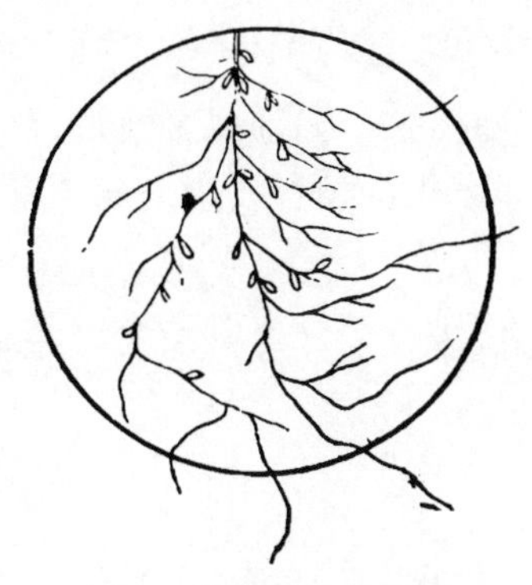

图1 苕籽的根和根瘤

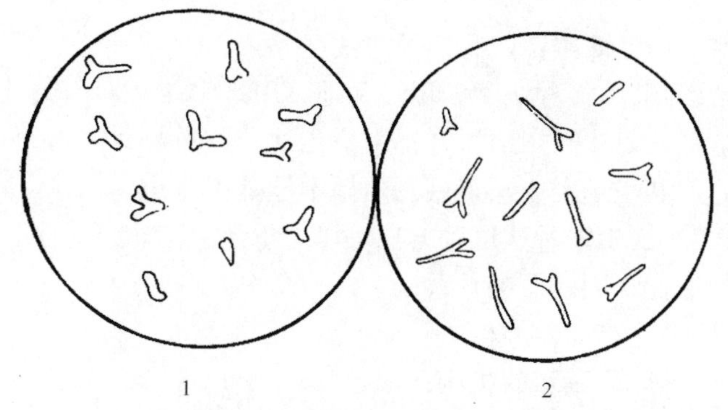

图2 从根瘤中取出的根瘤细菌在显微镜下观察的形象

人类利用豆类绿肥作物肥田，自古已然，现在已不知是何时开始了。然而，将这自古已然的经验知识，提高为科学的理性知识，只是近几十年的事。1888 年两位德国科学家，黑尔格利智尔和威尔法尔斯，研究证明，豆类植物受根瘤细菌沾染以后，才能生根瘤，生了根瘤的豆类植物才能吸收利用空气中的氮气，变为体质中的氮质成分。因此，氮质固定作用是豆类植物和根瘤细菌两种生物（一种植物，一种微生物）共同生活的作用，称为“共生作用”，种草籽绿肥，要它产生高度的肥田效果，不但要保证草籽本身长得好，而且要保证它受根瘤细菌的沾染，产生根瘤才好（图 3）。

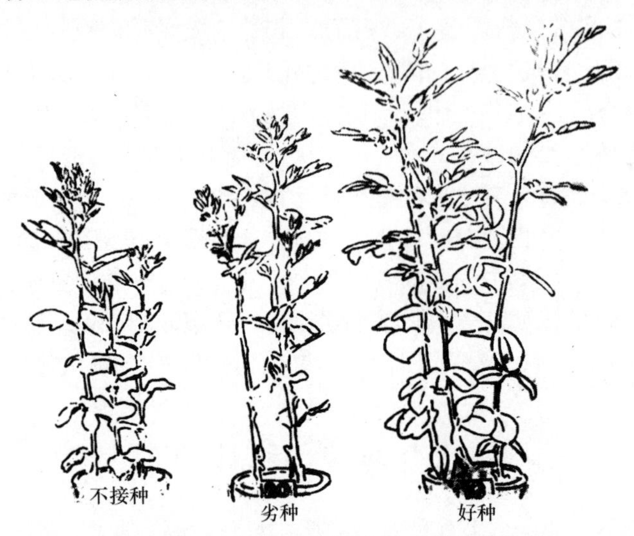

图 3 蚕豆不接种、接了劣种或好种的比较

许多地方，土壤里有根瘤细菌，只要种草籽，草籽根上就生长根瘤，就能固定氮质。但是也有许多地方，土壤里没有根瘤细菌，要种草籽，不但要播种草籽种子，而且还要播细菌种籽。如果专播草籽种子，土壤中没有根瘤细菌，草籽根上就不会长根瘤，没有共生现象，不能固定氮质。这样，草籽绿肥就和非豆类植物一样，不能增加土壤中的氮质营养料，没有这主要的肥田效果。

什么土壤中有根瘤细菌，什么土壤中没有根瘤细菌？根瘤细菌是微生物，个体太小，我们无法掘开土壤来查看。一般而论，如果某地种过多年草籽，每年种草籽，草籽根上根瘤生得都很好，足以证明土壤中有很多的根瘤细菌。没有种过草籽的地带，土壤中不一定有适宜的根瘤细菌，要在这地带推广草籽绿肥，就应当设法保证根瘤细菌的沾染，播草籽种，同时也要播细菌种，今年省农林厅在各农场试种草籽绿肥，也随带播细菌种，就是这个意思。在中和性和石灰性土壤上，推广草籽绿肥，每年同时播草籽种和细菌种，几年以后土壤中有了足够的根瘤细菌，就无需再下细菌种了。在酸性的土壤上，根瘤细菌怕酸，不能在土壤中自然繁殖，种草籽绿肥，就有每年播细菌种的必要。因为根瘤细菌怕酸，在酸性土上种草籽，最好同时施用石灰。

六、 人 工 播 种

田中播根瘤细菌，有两种方法：

（1）移土接种法譬如甲乙两地，甲地多年种草籽，土壤中根瘤细菌很多，乙地新种草籽，要播根瘤细菌种，可以自甲地挑几担土到乙地，与草籽种子混合，一起播种，这样种的方法称为移土接种法，显然的，利用这种方法，甲乙两地不能隔得太远。

（2）人工接种法这种方法可以大量的供应很广大的区域，一省，一大行政区，甚至全国。实行人工接种法，需要三个条件：

（1）生产菌苗的机关；

（2）分配菌苗的机构；

（3）在正确的技术指导下，举行人工接种。

一省，一大行政区，或者全国要成立生产菌苗的制作机关。这机关负责培养优良的根瘤细菌菌种，并大量的制造菌苗。制造菌苗的过程，简而言之是这样的：第一步用细菌学的纯种培养法，从壮盛的豆类植物根瘤中培育根瘤细菌的纯系，用隔离栽培法，比较所培养的根瘤菌纯系培养体，然后，采用一种或数种纯系根瘤细菌，大量培养，制成菌苗，供给各地需要。

菌苗从生产机关发出，经过分配运输，送到要用的地方。由于菌苗是生活的，不可受热或破坏污染，或受其他毒害。

达到要用的地方，按照最有效的施用方法，将农民组织起来，在草籽播种以前一两天举行集体的人工接种，简而易行的人工接种方法可以采用下列手续：

（1）菌苗收到后必须放置阴凉之处，在利用以前不可开封。开封太早，可能受杂菌污染有所伤害。

（2）按照最有利的比例（在菌苗包装外应有说明）准备菌苗、种子和干土壤。干土壤最好是中和性的砂质壤土。如果是酸性土壤要加点石灰。如果土壤黏重要加些河砂。

（3）充分混合土壤、种子和菌苗，加适量的冷水（最好是冷开水）。水不可加多，以保持潮润而不流失为度。

（4）混合后立即播种或隔一、二日播种均可，但不可耽搁太久，也不可受日晒雨淋。

七、根瘤细菌的种类

根瘤细菌不止一类，譬如说，从豌豆根瘤中培育而得的纯系根瘤细菌，接种给大豆，就不能使大豆生根瘤。科学工作者试验比较各种豆类各种根瘤细菌，将能够互相接种的称为同一“互接种族”，不能互相接种的称为非同一互接种族，通常栽培的豆类可以归纳成为下列几个互接种族：

（1）苜蓿族包括紫苜蓿、黄花苜蓿、草木樨等属于苜蓿和木樨两属植物都归这一族，这族之间，各种植物的根瘤细菌可以互相通用。

（2）三叶草族包括三叶草属植物。

（3）豌豆族包括豌豆、苕籽（蓝花草籽）、蚕豆等植物。

（4）四季豆施包括四季豆、扁豆等植物。

（5）羽扁豆族。

（6）大豆族包括大豆一种植物，一切的大豆品种，黄豆、黑豆、泥豆、禾根豆等都是。

（7）豇豆族包括的植物很杂，农艺上重要的有豇豆、花生、绿豆、赤豆、小豆、猪屎豆、胡枝子等。

（8）紫云英族：紫云英（红花草籽）单归一族。

检查上述，显然的，举行蓝花草籽和红花草籽人工接种不能用同一种菌苗。

八、好种和劣种

根瘤细菌是一种微小的生物，它和别的生物一样（例如，水稻、棉花、牛、马等）有好种也有劣种。好种所生成的根瘤能固定多量的氮质，劣种所生成的根瘤不固定氮质，或只固定极少的氮质（参看图3）。讲者在1938年曾举行了一个试验，比较两种三叶草根瘤细菌的好坏。将三叶草种在大玻璃瓶里，有20瓶接种了好种根瘤细菌，12瓶接种劣种根瘤细菌，每瓶中植物所固定的氮质相差四倍多：平均每瓶固定的氮质好种0.654克，劣种0.151克，因此，培育各种根瘤细菌，菌种必须要选择优良的品系，举行根瘤细菌培育和制备工作应当和作物选种一样，经常的选择培育好种，淘汰劣种。

九、田间考验

举行了人工接种，并不就算保证草籽绿肥长得好。人工接种可能失败，就算人工接种成功，可能所接的是劣种。同时，在一个新区试种草籽，气候、土宜和栽培方法不一定就适宜，试种草籽绿肥，必须随时考验。

试种草籽绿肥，看样就算成功？最后的考验，自然要看来年水稻是否生长得好，但是，如果观察得仔细，今年秋末，已经可以看出好坏了，种得好的草籽，在秋末应当植株旺盛，叶也暗绿。掘出一株植物，根上应当有许多健旺的根瘤，根瘤挺壮，呈粉红色，切开根瘤，里面应当是红色。

如果植株矮小，叶色浅绿或黄绿，基呈红色。掘出根来，上面没有根瘤，或者只有几个或一、二十个根瘤，这是人工接种没有做好的缘故。

如果植株矮小，叶色浅绿或黄绿，其呈红色。掘出根来，根瘤很多，但是个体小，颜色不正，青白色或是黄色，切开根瘤，里面没有红色，这是接种了劣种。

人工接种没有做好，或是接种了劣种，应当立即报告有关机关，进行检查，以便明年再做时可以改善。

如果植株矮小，但是掘出根来，根瘤是健旺正常的，这可能是土壤不宜种草籽，或者栽培的方法不对。强酸性土壤（pH 低于 5.5）上草籽可能种不好，要种草籽应当施用石灰。中南区有些土壤，除氮质营养料以外，磷质和钾质营养料也缺乏，种草籽可能因为缺乏磷质或钾质而长得不好，应当多施用骨粉和草木灰。排水很困难的土壤，潮湿太甚，土壤通气不畅，草籽也长不好。碰到上列情形，应当针对问题，寻求改善土宜，这和栽培一般农作物一样，而不与根瘤菌人工接种相干。

以上情况，不仅对草籽绿肥的新推广区而言，在种了多年草籽绿肥的地方，也可能有同样的情况发生，负责一地方的农业干部应当多多注意。

Application of Astragalus Rhizobium Inoculants*

LI FU-DI，CAO YAN-ZHEN，ZHOU QI，ZHOU JUN-CHU，ChEN HUA-KUI

（Huazhong College of Agriculture，Wuhan，Hubei）

1. BACKGROUND

Autumn sown astragalus（*Astragalus sinicus*）is an important green manure crop preceding “two crop of rice per year” in Southern China. The green yield （shoots） of astragalus amounts generally to 4000-6000 jin per mou， and is ready for cutting or plowing under a week or so before planting of early season rice. A mou of astragalus in good stand satisfies the need of 1-2 mou of rice as basic manure.

In the fifties and decades before that， most rice fields in Hubei were planted to one crop of rice （of middling growth season） per year， supplemented by a winter crop of wheat， barley or rapeseed for seeds， or vetch for green manure. In the sixties， there was a drastic expansion of “two crop of rice per year” crop system， which needed more manure crop for early season rice， because it needed longer growth period than permissible. Consequently there was a corresponding increase of astragalus （see Table 1）.

Table 1 Expansion of acreages under early season rice and astragalus in Hubei

Year	Acreage（$\times 10^4$ mou）	
	Early season rice	Astragalus
1949	114	—
1957	361	439
1962	369	461
1965	666	886
1966	848	ca.965
1970	1019	ca.1009
1973	1502	1334
1974	1648	1470
1975	1638	1594

（1 mou=1/15 hectare）

In the field on which astragalus rhizobium in soil（see Table 2）. No rhizobium isolated from other genera of legumes has yet been found to be able to nodulate *Astragalus sinicus*. Consequently， the effective artificial inoculation of astragalus rhizobia becomes the key for successful growth of astragalus in soils newly grown to it， provided soil moisture and soil phosphorous are suitable （see Figure 1）.

Table 2 Number of Rh. astragali in soil （rhizobia/g. dry soil）

Date	Soil，astragalus never grown	Field condition No. of rhizobia	Soil，astragalus grown	Field condition No. of rhizobia
Feb.2	Fallow	0	Astragalus in winter	9.6×10^4
Apr.9	Field flooded	3.6	Astragalus in full blossom	2.9×10^4
Jun.11	Under rice，field flooded	0	Under rice，field flooded	3.0×10^4

（1）Numbers of rhizobia were tested by “plant-dilution infection technique” in 1964.

*原载于《中澳生物固氮双边学术研讨会论文集》，1986，未正式发表。

Fig.1 Astragalus in field; uninoculated front, inoculated at back

In the sixties and later on, we had isolated a good many pure cultures of astragalus rhizobia in Hubei and adjacent provinces, and tested them for relative efficiency in pot experiments as well as field experiments. We had shown that inoculation of efficient astragalus rhizobia preparations significantly promoted the growth of astragalus and resulted in increased green yield, which in turn provided condition for bumper crop of early season rice.

The experiments carried out in Hubei on the effectiveness of inoculation of efficient strains of astragalus rhizobia and some related problems are briefly presented in the paper.

2. THE EFFECT OF INOCULATION ON THE GROWTH OF ASTRAGALUS IN EARLY STAGE

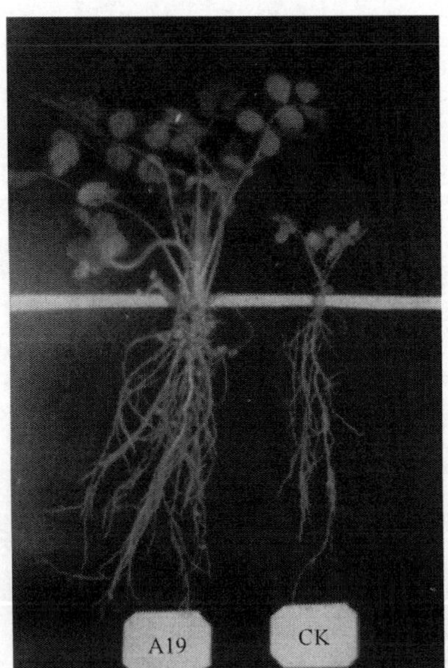

Fig.2 Difference shown in field experiment; astragalus effectively nodulated (left) and uninoculated (right)

With the application of efficient astragalus rhizobia, the growth of astragalus seedlings is significantly promoted. Table 3 gives the result of a plot experiment. The experiment showed, in comparison with uninoculated control plots, the inoculated plants grew higher and greener. Examined 80 days later, most plants in inoculated control plots remained branchless, while plants in inoculated plots possessed 1.1 to 1.4 branches on average. Number of nodules per plant in inoculated plots averaged 4.7-9.3 times of nodules per plant in control plots. The nodules were big, reddish and located mostly on upper parts of root system(see Figure 2). Stimulation of seedling growth was due chiefly to the effect of nitrogen fixation. The pot experiments gave similar results, as shown in figure 3. Increase of nitrogen content of inoculation is given in table 4.

Astragalus is an autumn sown legume, healthy growth before winter freeze is requisite for good growth in the next spring. Hence, inoculated with efficient rhizobia not only promotes nitrogen fixation but also increases cold resistance.

Table 3 Influence of efficient strains on growth of astragalus seedling

Strain	Seedling weight (cm)	No. of branches (branches/plant)	No. of nodules (nodules/plant)
A 16	5.5	1.4	19.5
A 19	4.8	1.1	9.8
A 21	5.3	1.4	13.7
A 26	5.5	1.4	17.0
CK	3.3	0.2	2.1

(1) Each value is the mean of 4 plots. 20 seedlings were examined per plot.

(2) Observed on 80 days after sowing.

Fig.3 Difference shown in pot experiment; CK being uninoculated

Table 4 Comparison of nitrogen content of uninoculated plant and plant inoculated with efficient strains

Treatment	N%
Uninoculated plant	1.50
Inoculated plant: A 16	3.80
A 19	3.79
A 21	3.34
A 26	3.85

(1) Aerial parts of oven dried plant were assayed.

(2) A 16, A 19, A 21, A 26 being strains of rhizobia inoculated.

3. EFFECT OF ARTIFICIAL INOCULATION ON GREEN YIELD OF ASTRAGALUS

Repeated experiments and extensive field survey showed that artificial inoculation of efficient astragalus rhizobia was an important factor for good yield of astragalus green manure crop, especially when astragalus was sown on the field where it had never been grown. Results of extensive field survey are given in Table 5. Table 5 shows, in old region (where astragalus had grown for many years) as well as in new region (where astragalus had never been grown before), most of the artificially inoculated fields yielded more than 4, 000 jin per mou, while uninoculated fields never attained 4, 000 jin per mou, with more than half then below 2, 000 jin.

On many field newly sown to astragalus, artificial inoculation resulted in most significant increase in green yield (see Table 6).

Table 5 Effect of inoculation on fresh weight of astragalus grown in field （jin/mou）

Inoculated fields				Uninoculated fields				
No. of field	<2000	2000-4000	>4000	No. of field	<1000	1000-2000	2000-3000	>3000
		0				0	0	
21	3	8	10	9	3	3	3	
New region	14%	38%	48%		33%	33%	33%	
13		2	11	9	1	3		5
Old region		15%	85%		11%	33%		56%

（1）Estimation of fresh weight cut at full blossom stage in 52 fields in 11 counties of Hubei province.

（2）1 jin = 1/2 kilogram; 1 mou = 1/15 hectare.

Table 6 Effect of inoculation on fresh weight of astragalus in field experiment*

Treatment	Fresh weight of shoot（jin/mou）	Percent increment
Exp.I Uninoculated	1400	
Inoculate	3100	121.43
Exp.II Uninoculated	2800	
Inoculate	5000	78.57

* The experiments were conducted in Jingmen County of Hubei Province. Growth was examined at full blossom stage of astragalus. Each value was the mean of 4 plots.

In old region，artificial inoculation with efficient preparations was also profitable. Table 7 gives results of field experiment comparing the effectiveness of efficient strains of rhizobia（A106 and A107）with that of an inefficient strain （A9）.

Table 7 Effect of different strains on fresh weight of astragalus shoots*

Strain	Fresh weight of shoot（jin/mou）	Percent increment or decrement with control with control
A106	6090	+31.8
A107	5310	+15.0
A7	4350	-6.0
CK	4620	

* The experiment was conducted in the State Dongxihu Farm. Each value was the mean of 4 plots.

The results clearly showed that selection and application of efficient rhizobium strain was an important technical link of growing astragalus.

4. ON QUALITY OF INOCULANTS

In Hubei，astragalus rhizobium inoculants were generally manufactured in summer at scattered local stations. Production condition was relatively poor and cold storage was not available. In the effort of the extension of artificial inoculation，it was found that storage temperature was an important factor affecting the quality of the preparations.

The effect of storage time and temperature of the peat preparation was studied. Viable count was carried out by plating method. Nodulation was tested in plants grown in test tube. The preparations were diluted to the range of 10^{-1} - 10^{-8} or 10^{-1} - 10^{-10}. Tubes were inoculated with inoculants of different degree of dilution. Each dilution was replicated 4 times. The test tubes were kept in growth chamber and illuminated artificially （16 hours illumination and 8 hours darkness per day）. Table 8 shows the nodulation effectiveness of preparations stored at 28℃ for various periods.

Table 8 Effects of nodulation of inoculants of different storage time

Storage time (day)	No. of rhizobia (10^8/g)		Level of dilution of original inoculant sample							
			10^{-1}	10^{2}	10^{-3}	10^{-4}	10^{-5}	10^{-6}	10^{-7}	10^{-8}
		(1)			4	4	4	4	4	1
1	52.4	(2)			1.8×10^6	1.8×10^5	1.8×10^4	1.8×10^3	1.8×10^2	18
		(3)			4.0	3.4	4.1	6.3	4.8	0.7
		(1)	4	4	4	4	4	4	2	
30	8.6	(2)	2.9×10^7	2.9×10^6	2.9×10^5	2.9×10^4	2.9×10^3	2.9×10^2	29	
		(3)	2.7	5.7	3.5	4.3	3.4	3.6	0.4	
		(1)	4	4	4	1	0	0	0	
90	1.7	(2)	5.7×10^6	5.7×10^5	5.7×10^4	5.7×10^3	5.7×10^2	5.7×10	5.7	
		(3)	11.1	13.6	5.3	1.3	0	0	0	

(1) Dilution nodulated and number of nodulated tubes.

(2) Number of rhizobia per seed. The values were extrapolated from data determined by plate viable count of inoculants sample.

(3) Nodules per plant.

On storage, not only the viable count of rhizobia decreased, but also results in decrease of nodulation ability. As shown in Table 8, inoculation of fresh preparation (storage time 1 day) diluted to 175 viable bacteria per seed resulted in 100% nodulation (i.e. nodulation occurred in all 4 replicates). Inoculation of preparation stored for 90 days at 28℃ required a concentration of 57000 bacteria per seed to ensure 100% nodulation. Storage for 30 days showed no signification harmful effect on nodulation. The experiment also showed that the amount of inoculated bacteria exceeding the requirement for 100% nodulation for 100% nodulation produced no more nodules pre plant.

Storage temperature was an even more important factor affecting quality of inoculants. Table 9 shows the results in change of nodulation-ability of preparations stored for 3 months at different temperatures.

Table 9 Effects of nodulation by inoculants stored under different temperatures*

Storage temperature	No. of rhizobia (10^8/g)		Level of dilution of original inoculant sample								
			10^{-1}	10^{-2}	10^{-3}	10^{-4}	10^{-5}	10^{-6}	10^{-7}	10^{-8}	10^{-9}
		(1)				4	4	4	4	3	1
Below 12℃	10.8	(2)				36000	3600	360	36	3.6	0.36
		(3)				3.7	4.6	3.6	4.7	1.9	0.08
		(1)			4	4	4	4	4	3	0
25℃	4.5	(2)			150000	15000	1500	150	15	1.5	
		(3)			2.9	3.2	3.3	2.4	3.2	2.0	
		(1)	0		0	0	0	0	0		
35℃	0.01	(2)	33000		3300	330	33	3.3			
		(3)	0		0	0	0	0			

*Legend being the same as in table 8.

From Table 9 can be seen that the preparations stored for 3 months at temperature below 25℃ preserved high nodulation-ability. Only 15 or 36 bacteria per seed were required for 100% nodulation. On the other hand, preparations stored for 3 months at 35℃ resulted in complete loss of nodulation-ability. No nodule was found with inoculation of as much as 35, 000 bacteria per seed.

5. ON NODULATION-ABILITY OF ASTRAGALUS RHIZOBIA

Loss of nodulation-ability is a fairly common phenomenon when pure cultures of rhizobia are maintained and repeatedly transferred in laboratorial conditions. In our experience, the declination and loss of nodulation-ability of astragalus rhizobia occurred very frequently. The repeated isolation of new efficient strains to take the place of declined strains proved to be a necessity for the production of artificial inoculants. Since the late fifties, we have already changed rhizobial strains three times in succession. The loss of nodulation-ability has been regarded as the result of combined effects of natural mutation followed by selection of some environmental factors, high glycine concentration for instance. Although some different selection factors have been reported, nothing has been settled yet.

As already presented above, the raise of storage temperature proved to be a marked factor leading to declination or even to lose of nodulation-ability. It seemed that the nodulation-ability mutants had been more tolerant to high temperature than the nodulating wild types with the result that the selection pressure favored the dominance of the non-nodulating mutants.

Isolated colonies had been picked from dilution plates of strain A21. 100 colonies were picked in random. Every picked colony was tested for its nodulation-ability. Only 52 colonies formed nodules on host seedling, 48 colonies failed to form nodules in one experiment. If a mutant occurred in a rhizobial population, and if the selection pressure favored the said mutant, in the course of maintenance and transfer, the mutant would have to dominant and finally complete replace the wile type.

The relatively high frequency of loss of nodulation-ability of rhizobium has been attributed probably to its genetic determinants being located on plasmid instead of on the chromosome. Considering the practical significance of the property of nodulation-ability in maintaining and raise of nitrogen fixing efficiency, this problem is certainly valuable both in theory and practice.

REFERENCES

[1]陈华癸.（Chen，H.K.，1962），微生物学实验.

[2]Beringer，J.E.（1976），The demonstration of conjugation in *Rhizobium leguminosarum*，In Symbiotic nitrogen fixation in plant，pp.91-97.

[3]Date，R.A.（1976），Principles of *Rhizobium* Strain selection，ibid. pp.137-150

[4]Kapusta，G.，Rouwenhorst，D.L.（1973），Influence of inoculum size on *Rhizobium japonicum* serogroup distribution frequency in soybean nodules，Agron. J. 65:916-919

[5]Kunelius，H.T.，Gupta，U.C.（1975），Effects of seed inoculation methods with peat-based *Rhizobium meliloti* on alfalfa，Can. J. Plant Sci.，55:555-563

[6]Schwinghamer，E.A.（1977），The genetic aspects of nodulation and dinitrogen fixation by legume: The microsymbiont. In A Treatise on Dinitrogen fixation，section III: Biology，pp.577-622

[7]Tuziumra，K.，Watanabe，I.（1961），Ecological studies of Rhizobium in soil by a noduletation -dilution frequency method，Soil Sci. Plant Nutr.，7:61-65

[8]Vicent，J.M.（1971），A manual for the Practical Study of Root-Nodule Bacteria.

结晶紫在根瘤菌肥料生产中的应用*

曹燕珍 周俊初 陈华癸

（华中农学院）

在我院根瘤菌肥料生产中，经常发生搪瓷培养罐污染，对生产造成严重的影响。因此，减少培养罐污染是提高菌肥产量的一个亟待解决的问题。

结晶紫有抑制杂菌的作用。我们在 1963～1964 年的研究工作中，曾进行过高浓度结晶紫对紫云英根瘤菌影响的试验，结果表明：紫云英根瘤菌具有很强的抗结晶紫性能，能耐 1∶1000 高浓度结晶紫。在加入 1∶100 000 和 1～50 000 结晶紫培养基中进行培养的紫云英根瘤菌，经过冻粉管及砂培试验，均证明其结瘤能力不受结晶紫的影响。根据这一试验结果，在以后的选种工作中，我们分离根瘤菌时，就采用 1∶100 000 浓度结晶紫抑制杂菌，特别是抑制芽孢细菌。

1972 年为了降低根瘤菌肥料生产中培养罐污染率，我们计划在培养罐培养基中加 1∶100 000 浓度结晶紫，以防止杂菌污染。

但是，要将结晶紫应用到生产中去，必须充分肯定下面几个问题：

（1）能抑制杂菌特别是芽孢细菌；

（2）对根瘤菌的繁殖没有抑制作用；

（3）不影响根瘤菌的结瘤性能；

（4）不影响根瘤菌的固氮能力；

现将试验初步结果和目前生产情况报告如下。

一、结晶紫对紫云英根瘤菌繁殖的影响

1．试验方法

用 A16，A19 作为试验菌株，每一菌株分别用加十万分之一浓度结晶紫和不加结晶紫的豆芽汁培养基培养，每一处理培养 5 罐，每罐装 17 000 毫升培养基。

每罐接种种子菌液 200 毫升，种子菌培养液不加结晶紫。接种后，进行通气培养，于 1 小时、24 小时、48 小时、72 小时、96 小时分别用稀释平板法进行测数，比较根瘤菌在加结晶紫和不加结晶紫培养基中菌数的繁殖情况。

2．试验结果

五次测数结果列于表 1：

表 1　结晶紫对紫云英根瘤菌繁殖的影响

菌株	处理	菌　数（亿/毫升）				
		1 小时	24 小时	48 小时	72 小时	96 小时
A_{16}	加结晶紫	0.21	47	136	159	66
	不加结晶紫	0.22	71	138	76	88
A_{19}	加结晶紫	0.08	31	125	71	92
	不加结晶紫	0.03	26	176	87	70

从上表看出，在加结晶紫和不加结晶紫的培养基中，培养 24 小时后，菌数均达到几亿个；48 小时后，菌数达到一百多亿，两种处理没有明显的差异，说明 1∶100 000 浓度结晶紫对紫云英根瘤菌的繁

*原载于《华农科技》，第 1 期:1～3，1972。

殖并没有影响。

二、结晶紫对紫云英根瘤菌结瘤能力和固氮作用的影响

1. 试验方法

用 A16 和 A21 作为试验菌株，接种在加结晶紫（1∶100 000）和不加结晶紫的培基中，振荡培养三天，取菌液进行稀释平板测数的同时，取 10^{-5}，10^{-6}，10^{-7} 三个稀释度的菌液，每一级稀释液接种 7 管砂培紫云英植物，每管加 1 毫升，并用同一稀释度接种 3 个砂罐，每罐三穴，每穴接种 7 毫升稀释液。

2. 试验结果

砂管和砂罐结瘤试验和植株全氮量测定结果列于表 2 和表 3。

表 2 结晶紫对紫云英根瘤菌结瘤能力和固氮作用的影响（砂管试验）

菌株	稀释度	接种菌数（万/毫升）	处 理	株高	叶片数	叶色	每株瘤数	含氮量（%）
A16	10^{-5}	1.57	加结晶紫	8.0	5	绿	2.1	3.978
		1.78	不加结晶紫	7.3	4	绿	1.9	3.740
A21	10^{-5}	3.36	加结晶紫	8.6	5	绿	3.6	3.376
		0.30	不加结晶紫	8.4	4	绿	2.9	3.642
	10^{-6}	0.33	加结晶紫	8.7	5	绿	3.1	3.372
		0.03	不加结晶紫	7.9	4	绿	2.5	3.624
	10^{-7}	0.03	加结晶紫	9.2	5	绿	3.8	3.637
		0.003	不加结晶紫	8.3	4	绿	2.5	3.476
对照				8.0	4	绿	0.0	2.148

从表 2 可以看出，加结晶紫后，每株结瘤数反而略有增加。菌株 A21 加结晶紫后，植株含氮量略低于不加结晶紫的处理，但菌株 A16 则相反。因此总的来看，加结晶紫对固氮能力基本没有影响。

表 3 结晶紫对紫云英根瘤菌结瘤能力和固氮作用的影响（砂罐试验）

菌株	稀释度	接种菌数（万/毫升）	处 理	株高	叶片数	叶 色	含氮量（%）
A16	10^{-6}	0.15	加结晶紫	9.5	5	绿	3.909
	10^{-6}	0.17	不加结晶紫	9.0	5	绿	3.817
A21	10^{-6}	0.03	加结晶紫	9.0	4	绿	3.663
	10^{-6}	0.33	不加结晶紫	8.5	4	绿	3.794

表 3 中所列砂罐固氮试验结果与砂管试验结果相似，而植株含氮量则更加相近。

三、结晶紫对防止培养液污染的效果

经过试验，肯定了结晶紫（1∶100 000）对紫云英根瘤菌的繁殖、结瘤能力、固氮能力都没有影响，今年菌肥生产在培养罐的培养液中均加入了 1∶100 000 浓度结晶紫。从近一个月的生产情况与 1971 年相比较，结晶紫对防止培养罐污染有明显的效果（见表 4）：

表 4 结晶紫对防止培养液污染的效果

时 间	生产罐数	污染罐数	合 格 率
71 年（6～8 月）	1495	966	35.3%
72 年（24 天）	425	148	65.2%

从表 4 可以看出，加结晶紫后，有效地防止了杂菌污染，因而大大地提高了培养罐的合格率。

1971 年造成培养罐严重污染的细菌主要是革兰氏正反应的细菌，今年培养罐由于加了结晶紫，革兰氏正反应的细菌已经基本上被抑制了。

目前培养罐污染的细菌主要是革兰氏负反应的小杆菌（因结晶紫对这类细菌没有抑制作用），如果进一步严格操作，加强管理，这种负反应的小杆菌所造成的污染也是可以克服的。

湖北省稻田绿肥*

陈华癸

（华中农学院）

一、水稻田的绿肥耕作制

“水田靠青”是我省水稻产区用地、养地和增施肥料的一项主要经验，与此相应，水稻田绿肥耕作制是我省水稻产区最主要的轮作耕作制度。这个制度的特点是：

（1）以水稻为主作物；

（2）水旱作物轮换种植，每年3～7个月水田，5～9个月旱地，和相应的排灌和土壤耕作技术；

（3）通过种植肥料（主要是越冬绿肥）来培肥土壤，并做到肥料基本自给，使得土壤生产力长期地保待在较高的水平上；

（4）尽可能混种越冬作物或晚秋旱作物，以提高复种指数，提高年亩产量；

（5）随着人地比率和土地利用集约程度的变化，年亩产量300～500斤（一人3～4亩地，两种一熟或更多一些）到600～800斤（一人1～2 亩地，两种二熟或更多一些）。

这些年来，由于生产关系和生产技术的猛烈变革，原来的水稻田绿肥耕作制受到了较大的扰动，茬口打乱了，必要的土壤耕作放松了，绿肥或其他种植肥料的面积大幅度下降。例如：应城县1949～1956年稻田绿肥都在20～24万亩之间，占全部水稻田面积的40%～60%，而1957～1960年间只有5.6～7.4万亩，只占水稻田面积的12%～15% （根据县统计资料，实际面积可能要高一些）。又例如：孝感专区江南几县的水稻田一向主要是中稻和草籽、荞麦水旱轮作。前几年强调多种麦，少种或不种草籽。结果由于肥料不足，年亩产量反而下降了。社员们说：“想吃几十斤麦子，不让种绿肥，丢了更多的谷”（通山县农业局调查资料）。又例如：在黄陂、孝感水稻田秋种泥豆的面积，解放以来大幅度的下降，孝感县1956年泥豆面积只有1951～1953年的1/3强，1958年只有1/4～1/5，1960～1961年只剩5%弱（根据县统计资料）据说，在解放前，泥豆的种植面积，比1951～1953年还要大些。

绿肥或其他种植肥料的面积大幅度下降，后果是严重的。以田养田，肥料自给的水平大大的降低了，对于商品肥料的依赖性大为增高，超过了供应能力，并埋下了地力逐渐衰退的隐患，这个隐患更由于这几年的连续灾害而暴露了出来。

这是我省恢复和发展农业生产必须解决的问题。

二、多种并存，湖北省稻田绿肥的主要种类

建国以来，我省在绿肥方面做了一些科学研究工作，取得了一些成绩，较完整地总结和提炼了草籽绿肥的增产技术；在一些试种地区确立了草籽绿肥的种植技术和对水稻的增产效益；反复地证明了磷肥拌种对于草籽绿肥的大幅度增产作用；反复地证明了根瘤菌拌种对于草籽绿肥增产作用和保证作用。

然而，由于科学实践工作的局限性和研究者们在见解上的片面性，在稻田绿肥的生产实践和稻田绿肥的科学研究工作之间，总还是存在着一定程度的隔阂，最突出的表现是：科学研究者们几乎是全力以赴的研究草籽绿肥，而在生产实践中，稻田绿肥确实是多种并存的。

显然，沟通这项隔阂是十分必要的，在这个问题上，省农业厅的工作者们做出了有益的贡献。他们

---

*原载于《湖北农业科学》，2:11～15，1962。

反复指出，湖北省稻田绿肥是多种多样的，各有其特殊的优缺点和适应条件，而片面地强调草籽绿肥，不重视其他绿肥，甚至于采取否定的态度，对于恢复和发展稻田绿肥以及开展有关的科学研究工作都是不利的。

我省稻田绿肥和相应的水稻田绿肥耕作制，有下列几个主要种类：

大麦绿肥：是我省中部面积最大，适应性最广的稻田绿肥，相应的耕作轮作制有：

中稻～炕土～大麦收获或沤青……………………………………………………………………[1]

中稻～秧荪～大麦收获或沤青……………………………………………………………………[2]

早稻～双季晚稻～大麦收获或沤青………………………………………………………………[3]

早稻～泥豆～大小麦收获或沤青…………………………………………………………………[4]

在以上各种耕作轮作制中，一般大麦收获都是次要的（面积小），沤青是主要的（面积大）。

大麦绿肥有许多优点。首先是收、沤两可，或收或沤不必在冬前播种时决定，而是在春暖时决定，这样就提供了根据春季气候（主要是雨水）条件灵活决定的有利因素，这对于我省年雨量1000毫米左右，不仅有秋旱，也常有春旱的地区，这种两来全的作物安排对于保产稳收十分重要。雨水好，就大麦沤青，争取获得秋季丰收稻谷；雨水不好，就大麦收获，保证一季夏收。其次是大麦绿肥前后茬口宽松，农活好安排，秋播从十一月初一直到十二月中，翻耕浴田在四月底就可进行，有从容的时间插早、中稻。再次是对土质、地形要求不严格，凡是勉强能水旱两作的稻田，都能种大麦绿肥。第四是种子用量不多，种源方便。

大麦绿肥的主要缺点是肥效较低，生长得茂盛，一亩田也只能肥一亩田，生长得不很好，肥效就很低了。同时，大麦绿肥没有固定空气中的氮气，提高土壤含氮量的作用，只有提供有机质，转化土壤中养料成分的作用。

大麦可收可沤，究竟在什么条件下收，什么条件下沤？决定收沤的条件大致如下（华中农学院调查资料）：①看雨水，雨水好就沤，不好就收；②看塘堰水，塘堰水满就沤，不满就收；③看生长好坏，长的好就收，不好就沤；④看田底子，田底子好就收，不好就沤；⑤看田远近，远田送肥困难就沤，近田送肥容易就收。换言之，大麦收、沤主要决定于水肥和劳力条件。

草籽绿肥：我省的主要草籽绿肥是红花草籽（紫云英）和蓝花草籽（苕籽），两者之间有共同性，在有些情况下，可以互相代替；也有各异性，分别适应于不同的耕作轮作制度。我省草籽绿肥在面积上和在增产总收益上都占绿肥的第一位。草籽绿肥肥效高，一般生产量，一亩可肥两亩，草籽是豆类，生根瘤，能进行共生固氮作用，丰富土壤中的氮素养料含量。

蓝花草籽主要是中稻的后茬，相应的水稻田绿肥耕作制是：

中稻～蓝花草籽。……………………………………………………………………………[5]

中稻～蓝花荞麦混播。………………………………………………………………………[6]

中稻～蓝花荞麦混播～中稻～炕土或秧荪～大小麦或其他越冬作物收获。………………[7]

蓝花草籽播种早，产量高，播种晚了，不仅产量下降，而且很不稳定。为了提早秋播，同时争取一季晚秋，提高复种收获指数，提高年亩产量，在我省多与秋荞混播，晚秋收50～100斤荞麦，并不影响蓝花草籽的生长。

值得指出的是，我省过去在试种蓝花草籽的地区，大都是一方面强调早播的必要性，一方面又只是单播蓝花草籽，不和荞麦混播。这样，在有种植晚秋作物习惯的地区，种一季蓝花草籽就不仅顶掉了一季越冬作物，还顶掉了一季晚秋作物。这是不必要的。技术上的简单化对于推广草籽绿肥实际上起了阻碍的作用。

蓝花草籽对土质要求也不严格，大致能种大麦的田地，也能种蓝花草籽。

蓝花草籽春季翻耕沤青，沤青的时间也和大麦沤青时间相同。

可以这样说，有必要在春间才决定是收是沤的田地，大麦绿肥有它显然的优点，而有条件在秋播就决定做绿肥的田地（主要是水源保证问题）则以种蓝花草籽绿肥为宜，蓝花草籽达至一般水平（翻耕的地面青草量在3000斤左右）就可以一亩肥两亩，从而还保证了一亩田地可以夏收。

蓝花草籽能不能在秧荪、一季晚或双季晚收获前套种在生板田内？在技术上有争论，就有些种植经验来说是成功的。这样就创造了将它作为中稻秧荪、一季晚稻以及双季晚稻的后茬绿肥：

中稻～秧荪～蓝花草籽……[8]

一季晚稻～蓝花草籽……[9]

早稻～双季晚稻～蓝花草籽……[10]

蓝花草籽在中稻秧荪以后生板套种，争取了一季晚秋作物，对于复种指数要求高的地区是重要的。在一季晚稻（尤其是收获期较早的一季晚稻品种，如412等）田内生板套种，前后茬口都好，由于蓝花草籽生长期比较长，翻得晚些，青草产量更高些，肥效更大些。双季晚稻套种蓝花草籽一般不如红花草籽适宜（见后）。

如果种植得好，种一亩蓝花草籽肥两亩，上述[8]、[9]两种制度都可以发展为：

中稻～秧荪～蓝花草籽～中稻～炕土～大小麦或油菜……[11]

一季晚稻～蓝花草籽～一季晚稻～小麦……[12]

后者是在水情好的田地上的一种高产，肥料基本自给，茬口宽松，节省劳力的稻麦两熟制。

红花草籽比蓝花草籽播种期晚些，翻耕期可早些。除一般地可以代替蓝花草籽外，红花草籽更适宜于做双季稻的后茬和早稻的前茬，适应的耕作轮作制度有：

中稻～炕土（或不炕）～红花草籽……[13]

中稻～炕土（或不炕）～红花草籽～中稻炕土或秧荪大小麦或油菜……[14]

中稻～双季晚稻～红花草籽……[15]

红花草籽也不很择土地，一般能种大麦的田地也能种红花草籽。种双季稻大麦的田，一般水情好，可以保证有沤青的水，与其大麦沤青，不如红花草籽沤青。

较普通地有这样的看法，在新区初种红花草籽，头两年生长不好，要三、五年后才生长得好。这究竟是什么原因？湖南省推广红花草籽的经验指出，如果进行根瘤菌拌种，第一年就可以生长好。究竟是不是这样，值得认真研究解决。

红花草籽可以和本地油菜混种。或者在保收一季油菜（亩产50～80斤菜籽）的基础上，另得1000斤红花草籽绿肥，一亩肥一亩，每年两种两收：

中稻～炕土（或不炕）～油菜加红花草籽……[16]

或者在主要是红花草籽的田中加一些油菜，亩收20～40斤菜籽，红花草籽产量并不比单独播种低，一般水平，一亩肥两亩：

中稻～炕土（或不炕）～红花草籽加油菜～中稻～坑土或秧荪～大小麦……[17]

后者也适合于双季稻田。

不论蓝花草籽或红花草籽，种子供应问题和就地留种问题都是亟待研究解决的。我省一向是草籽种子的产区，并运销外省，但似乎在产种技术和种子供销问题上，该做的许多工作都还没有做。

蚕豆绿肥：蚕豆做中稻绿肥一般在五月中下旬收青豆沤青，生长得好，每亩所收青豆可当干豆100斤，也就是说，蚕豆是既有收获，又能肥田的作物。一亩可肥田一亩到两亩，相应的耕作轮作制度是：

中稻～蚕豆～中稻～炕土或秧荪～大小麦……[18]

蚕豆择地较严，在有些田地只生茎叶，不结荚，在有些田地上茎叶也长不好。在什么土上长得好，在什么土上长不好，我们还不甚了解，蚕豆田对水情要求比大麦、草籽都要严格些，能种大麦、草籽的田地水情不一定适宜种蚕豆。

由于病害严重，蚕豆不能重茬。蚕豆绿肥用种量大，收青豆不能做种。

由于上述各种原因，蚕豆绿肥虽好，在稻田中蚕豆绿肥所占的面积是有很大的局限性的。

蚕豆和油菜混播，收油菜，收青豆，沤青作蚕豆绿肥，是中稻的好后茬，相应的耕作轮作制度是：

中稻～蚕豆油菜混播～中稻～秧籽或坑土～大小麦……[19]

泥豆：严格来说，泥豆不是绿肥，但为一种种植肥料。早水、晚旱种泥豆。泥豆后茬种麦，等于施了肥，收获的泥豆主要也是用做肥料，相应的耕作轮作制度是：

早稻或早熟中稻～泥豆～大小麦 [20]

早稻～泥豆～小麦（收获）～中稻～秧荪或炕土～大麦（收获或沤青） [21]

在黄陂、孝感一带泥豆原本是早熟中稻（如等泡齐）后的晚秋旱作近些年来胜利籼代替等泡齐等早熟品种收获期晚 5～7 天，就下适宜种泥豆了。这样泥豆就转变为晚熟高产的早稻（如南特号） 的好后茬。

在黄、孝一带，年雨量 1000 毫米左右，是秋旱较频繁的地带，早水晚旱接越冬作物是十分重要的作物安排形式。也是以田养田，肥料基本自给的主要手段。

最近的田间试验结果指出，泥豆在春季播种，可以做中稻或一季晚稻的绿肥，在孝感县，清明前播种，立夏后翻耕，每亩可得青草产量 2000 斤左右。

湖草：滨湖田地，打湖草做肥料，可以做到以湖养田，湖草不是一种种植肥料，但确实是一种绿肥，在离湖近、劳力多、水情好的田地，以湖草肥田，可以确实的做到一年三种三熟。

早稻～双季晚稻～大小麦收获……[22]

年产量大面积地稳定在 900 斤以上。

对有些田地，田青也起和湖草类似的作用。

三、调整和发展稻田绿肥和其他种植肥料的几项技术问题

提高单产，无论哪一种绿肥或种植肥料，单产总是产生肥效的最主要的因素。如何提高单产，是生产实践和科学研究的当前问题。总起来说，决定绿肥或其他种植肥料单产的主要因素是：

（1）土宜、地形相应的水情对于各种绿肥和种植肥料的土宜，科学研究工作很不充分。各地的生产实践经验是比较丰富的，但还没系统地总结出来，并说明其究竟。

（2）适时播种。

（3）整地播种质量，包括整地质量、播种方法和火候、播种量等，以及相应的出苗质量。

（4）越冬绿肥，春季开沟排渍的工作质量和实效。

（5）以小肥养大肥。

（6）适时翻耕沤青。

这里着重讨论以小肥养大肥这个单产因素。不论是哪种绿肥或种植肥料，合理施肥总是提高单产的一项有效措施。对于越冬绿肥，施用小量种肥（或盖籽肥）、腊肥和春肥，都有显著的增产效果 （中南绿肥座谈会，1962；湖北省绿肥磷肥座谈会，1962）。

研究工作指出，对各种豆类绿肥来说：①根瘤菌肥料拌种对于青草产量有增产或保证作用；②施磷肥有显著的增产作用；③根瘤菌肥料和磷肥相结合，增产效果更为显著。

根瘤菌肥料用量不多，成本低，在农业生产的成本上不成什么问题。

磷肥用量较多，成本较高，如何使用较少量的磷肥，用在刀口上，争取较大的绿肥增产效应和相应的稻谷增产效应，是合理使用磷肥，以小肥养大肥之需遵循的技术路线。

过磷酸钙用做底肥、种肥、春肥以及后期喷施，对于蓝花草籽或红花草籽都有增产效果（湖北省绿肥磷肥座谈会，1962）。从实验成效、磷肥的本质以及作用机制来看，都指出晚施不如早施。过磷酸钙用做蓝花草籽或红花草籽的种肥，与根瘤菌肥料结合使用，看来可能是最适当的施肥方法。研究工作表明，施用过磷酸钙对于蓝花草籽或红花草籽的根系发育以及根瘤的数量、大小和质量都有显著的刺激作用，这样就：①用做种肥，有利于秋季扎根，在冬前形成健旺的根系，而冬前形成健旺的根系对于越冬抗寒以及开春开旺都打下了良好基础；②用做种肥，对于根瘤的形成有良好作用，因此不仅发挥了磷肥的直接肥效，而且通过对根瘤形成的刺激作用，同时提高了共生固氮作用，制造更多的氮肥。田间试验结果证明，红花草籽用少量过磷酸钙拌种（每亩 2～3 斤）和根瘤菌肥料拌种结合，每亩增产青草产量 1000 斤以上。

在恢复中有发展，恢复水稻田绿肥耕作制的实施面积。在八权下放，以及认真地贯彻了各项农业政

策以后，各种绿肥和种植肥料的面积也确实在积极恢复着。

一个特定的地区，有一定的作物安排，其中包括一定的绿肥或其他种植肥料种类以及相应的轮作换茬制度。这是在长时期生产实践中形成的，不能轻易否定。

但是，这决不意味着，在这个问题上，就只有生产实践的恢复而没有科学研究和技术改革的任务了。恰好相反，在恢复中发展、调整和变革，从而提出了一系列的科学研究和技术改革任务。

在原本没有种植绿肥习惯的地区有必要试种和推广绿肥，从而提高当地以田养田，肥料基本自给的水平。这几年来，我省蓝花草籽和红花草籽的种植区域正在从南向北发展着。在这从南向北的发展过程中，除有一般的技术传授和推广任务外，还必须研究和制定适合于新地区的种植技术及引种驯化适宜的品种。总的来说，生长季节缩短了，冬寒加强了，原本适合于我省南部，生长期较长，冬季较温暖的绿肥品种不一定完全适合于新的生活环境。在南部，原本比较宽松的前后茬口，在中北部也变得不很宽松了。在这个问题上，科学研究和技术改革跟不上，将会限制绿肥在新区的健康发展。

在原本有种植绿肥习惯的地区，由于生产条件的变化，在绿肥品种上也要求有所改革，如何尽可能地用豆类绿肥代替非豆类绿肥，从而尽可能地发挥共生固氮作用，以及如何尽可能提高肥、粮（油）兼收的效果和种植面积，都是当前的重要课题。

例如，在原本是双季稻、大麦沤青的地区和田地，改种红花草籽来部分地代替原本可收可沤的大麦，以提高以田养田，肥料基本自给的水平，并争取提高收获指数。如果，红花草籽种得好，一亩肥两亩，那么，再隔一年收一年大麦，复种收获指数为 2.5。即早稻～双季晚稻～红花草籽～早稻～双季晚稻～大（小）麦收获。

在确实有春旱威胁的地区和田地上，收、沤两可的作物安排是不可忽视的。在这种情况下：

（1）尽可能扩大蚕豆或蚕豆油菜混播来代替可能代替的大麦沤青或收获。研究种蚕豆的土宜，防治蚕豆病害的方法，克服不能重茬的缺点，争取蚕豆保产稳收等都是十分重要的科研课题。

（2）用大麦、瘪豌子间作来代替单播大麦也是一项值得认真研究的技术。瘪豌子是豆类，有根瘤，能固定氮素，同时也是可收可沤的作物。雨水好，和大麦一起沤青；雨水下好，两者都可以收获，做粮食用。这个种植制度目前还只在很狭窄的地区内应用着（钟祥县）。

前面提到，泥豆是早水、晚旱加越冬作物，一年三种的水稻轮作制度中的一项重要种植肥料，目前正在积极恢复中。然而，由于水稻品种的变化，种植泥豆也就不是单纯的恢复问题，而是创造一套适合于新的水稻品种的轮作、耕作制度，以及相应的全套生产技术问题。

各种绿肥的间作、套作、混播制度，如荞麦、蓝花草籽混播，油菜、红花草籽混播，蚕豆、油菜混播，大麦、瘪豌子间作等等，其效果都不只是两种植物的机械相加，而是在时间上、土地面积上提高了利用效率，而且，研究工作也指明，两种作物之间还可能存在着互利的作用。这类混合种植的原理和技术是农业科学的重要研究课题，而且在生产实践中存在着很大的推广潜力，不少间作、套作、混播制度还只是在局部地区进行着，没有理由认为，它们只适合于现在的特定地区，而不能在更大的范围内加以利用。

在某种意义上说，研究两种或更多的大田作物的间作、套作和混播，研究同时同地两种和更多作物的相互有利关系的原理和技术是我国（和我省）农业科学的独特问题。在这方面我国（和我省）的生产经验很丰富，从而向理论研究提供了丰富的经验知识。尤其是目前人们很注意的农作物群体结构的研究，从单一作物向两种或两种作物以上的群体结构上发展，是值得研究者们关切的。

做好发展根瘤菌肥料的科学技术工作*

陈华癸

（华中农学院）

根瘤菌肥料，或称根瘤菌剂，是一项现代农业技术。根瘤菌肥料的作用在于将适宜的活根瘤菌接种在豆科植物上（主要是采用拌种的方法，即将活根瘤菌与种子拌和，一起播种），保证豆科作物结有效根瘤，发生共生固氮作用，从而改善豆科植物的氮素营养，提高当季和后茬作物的产量，并培肥土壤。这是一项现代农业技术，这项技术的特点在于生产、贮运和施用活细菌，还要求活细菌在土壤和植物中旺盛生活、繁殖和起作用。

植物生长需要氮素营养。但就大多数植物种类来说，它们只能吸收利用土壤和水中的氮素化合物，而不能吸收利用空气中的氮气。只有能进行固氮作用的生物（称为固氮生物），才能吸收利用空气中的氮气。而在固氮生物之中，根瘤菌和豆科植物的共生固氮作用既是十分特殊的，又是十分重要的。

豆科植物和根瘤菌的共生固氮作用的特殊性在于它们在独立生活时（即豆科植物不结瘤时，或根瘤菌在土壤中或人工培养中生活时）都没有固氮能力，只有在共生情况中（即豆科植物结根瘤，根瘤中含有生活的根瘤菌，根瘤实际上是根瘤菌钻进豆科植物根中结的瘤子）才能进行固氮作用（即吸收利用空气中的氮气）。

豆科植物和根瘤菌共生固氮作用的重要性在于：（一）它是使得空气中氮气有效化（对植物营养而言）的一项重要环节，并且就现有的知识来说，是各种生物固氮作用中的最强的一种。一般估计，每亩豆科植物可固氮五斤到十斤左右，种四百万亩到八百万亩豆科植物的固氮总量就等于年产十万吨氮肥的化学肥料工厂（以硫铵计）的固氮量（制造化学氮肥也是将空气中氮气有效化）。（二）在农业利用上效果最为突出。就农业生产实践经验而言，不论国内国外，它就有两千多年的历史了。自从现代自然科学（尤其是农业化学和微生物学）兴起之后，它又发展成为一项重要的现代农业科学技术。

种植豆科植物除本身需肥较少外，还对后茬有益，能肥田。豆科绿肥是后茬作物的主要肥料，肥劲大。牧草中要有适宜的豆科植物比例，才能保证草场肥沃，放牧价值高。这是国内外农牧业的普遍知识。在很多地方，还从长期的生产经验中，朴素地知道了，只有结瘤的豆科植物才生长得好，才有上述各种好处。有些地方还掌握了客土接种根瘤菌、保证结瘤的方法。

自从十九世纪末期，农业化学和微生物学阐明了豆科植物和根瘤菌的共生固氮作用的实质之后，在二十世纪初期，人们就设法人工培养根瘤菌，制造根瘤菌肥料，用来和豆科植物种子拌种，保证豆科植物结瘤好，共生固氮作用旺盛。在有些国家中，这项技术在三十年代就在生产实践中推广了。第二次世界大战以后，它更进入了世界范围推广的阶段。

关于这项技术，我国在解放前几乎是空白的，而解放以后却开展得较早，十多年来，不论是在根瘤菌肥料的生产技术上、使用效果上和理论研究上，都取得了一定的成绩。现在已经可以肯定，对花生、大豆和几种豆科绿肥进行根瘤菌肥料人工接种，确有保证结瘤、提高产量的作用。我们已经创办了根瘤菌肥料制造工厂。

然而，要把这项现代科学技术在生产实践中巩固下来，并且不断地提高质量、扩大成果，则还必须在生产技术上和科学原理上开展大量的科学研究工作。

发展根瘤菌肥料需要解决的科学技术问题很多，当前紧要的有以下四个问题。

---

*原载于《人民日报》，1964. 5. 19。

菌种问题

这问题有两个方面。一个方面是研究什么豆科植物需要什么根瘤菌问题，也就是植物—细菌群的关系问题；另一个方面是选出生产用的优良菌株问题。

在植物—细菌群关系方面，现代科学已经积累了不少具体知识，掌握了一些科学规律，能为合理制造和施用根瘤菌肥料提供有益的科学依据。扼要介绍如下：（一）豆科植物和根瘤菌的共生关系（结根瘤和进行共生固氮作用）是有专性的，不是所有的豆科植物种类都能和所有的根瘤菌菌株共生。（二）可以将豆科植物种类和根瘤菌菌株（就已研究过的来说）分为若干植物—细菌群，属于同一个植物—细菌群的植物和细菌，一般都能够发生共生关系，不属于同一个植物—细菌群的就不能够，可是这种关系并不是绝对的，也有例外。（三）在同一植物—细菌群中，产生结瘤作用和共生固氮作用的关系又可以分为两类：一类是能结瘤又能发生共生固氮作用，一类是只能结瘤但不能发生共生固氮作用，或只有微弱的共生固氮作用。当然，农业生产中需要的是前一类。在前一类之中，不论是结瘤作用或共生固氮作用，都还有程度上的差别，农业上追求的是结瘤作用好、固氮能力高的共生关系。（四）植物和细菌间的关系，既决定于它们的遗传性，也受环境条件的影响。不论是植物的遗传性或细菌的遗传性，都有决定性的作用。因此，必须选育优良的根瘤菌株，还必须针对着特定的豆科植物种类和品种选育适合于它们的优良根瘤菌株。而且在一定程度上，还需要针对着特定的环境条件开展菌株的选育工作。

上述几点科学结论是十分宝贵的，但是要把这几点科学结论化作为生产技术，则还必须掌握足够的具体知识。就世界范围来说，已经获得了不少科学资料，这些资料是提出上述科学结论的基础，同时也是有些国家的生产实践的数据根据。但是，关于我国自己栽培的豆科植物和品种的植物—细菌群关系的科学资料，迄今为止，还是比较缺乏的。我国的豆科植物种类十分丰富，栽培历史久，地理分布广，品种多，因此，系统地进行这方面的研究工作，是发展根瘤菌肥料的重要科学基础。

选育优良根瘤菌株的重要性是无须解说的。值得着重指出的是，由于上述植物—细菌群关系（亦即共生固氮作用的专性）的复杂性，以及我国豆科植物种类、品种的丰富性和地理分布的广泛性，我们的任务不是选育一个或少数几个优良菌株，而是选育出适合于不同豆科植物群的许多优良菌株。

十几年来我国已选出了少数比较优良的菌株，在试验中或在生产实践中证实了它们的有效性，但是太不够了。例如，地区分布广、品种丰富的花生和大豆等豆科植物，现已掌握的少数菌株只是在部分地区和少数品种的使用上经受过考验。我们还说不出，在哪些地区，对哪些品种，已经有了比较优良的菌株，哪些地区哪些品种还没有。又例如，对于在南方水稻田中广泛种植的紫云英和苕子绿肥，迄今我们也还只有极少数经过部分地区考验过的菌株；在这极少数中，有的还表现得不够好。至于其他豆科植物，大多数还缺乏经过实践检验过的有效菌株。

显然，按地区、作物种类和品种等不同条件进行选育优良菌株的工作是亟待展开的，而且工作量相当大。如何将全国的专业力量组织起来，形成一个选育网，是一项十分必要的工作。并且，在广泛选育菌株的基础上，还需要形成几个专门中心，承担决选和保存优良菌株的任务。

根瘤菌肥料的生产技术问题

生产根瘤菌肥料的难处在于不仅要求在工厂中生产大量活细菌，而且成品也是活细菌，并且要求在较长时期内保持旺盛的活力。这在生物制品工业中也是比较特殊的。尤其是作为一种农用制品，产品价格需要尽可能地降低，包装储运条件也需要尽可能地降低（同医药制品比较而言）。这些要求又进一步地增加了根瘤菌肥料在生产和贮运上的困难。

比较有利之点是，根瘤菌生长繁殖所需求的生活条件不甚苛刻。因此，根瘤菌肥料的生产，上马比较容易，而巩固却比较困难，产品质量难以保证和提高。我国这几年的生产经验也正是如此，产品质量很不稳定。

这个问题也包括两个方面：一个是解决生产和供应（包括保存、运输和分配）中存在的技术困难问题，另一个是判定和颁布产品规格问题。这两方面提出的科学研究任务基本上是相同的。产品规格必须与生产技术水平相符合，而且只有在产品质量较高而且比较稳定的基础上，才有可能制定和颁布产品规格。

从菌肥制造的角度来说，质量标准主要是两条，一条是制品中含有大量的生活根瘤菌（例如每克几亿或几十亿），另一条是制品中不含或只含极少杂菌（例如，不超过百分之十）。问题不在于新成品的质量如何（新成品的质量比较容易保证），而在于经过几个月（不能少于三个月）的贮存、运输和分配时期之后还能保持较高的制品质量。这对于含活细菌的制品来说是特别困难的，尤其是就我国目前的仓库和运输条件来说，还不能要求几个月的贮存、运输和分配（主要是在农村）期都在人造低温条件下度过。

同时，究竟是把菌数稳定在每克几千万、几亿或几十亿，是十分重要的。每克菌数多，就可以一瓶当十瓶或当百瓶用，在贮存和运输上也都方便得多。需不需要在完全无菌操作条件下生产根瘤菌肥料？目前国内外的意见都还不一致。为了能经过高温时期不生杂菌或杂菌不多，根瘤菌肥料制造技术的发展趋势是采取无菌技术。经验证明，杂菌多了，根瘤菌就会大量死亡。同时并不是说，能防止杂菌的大量繁殖，就能保证根瘤菌不大量死亡；如果制品中污染了杂菌，那么能防止杂菌大量繁殖的条件多半也就是不适于根瘤菌生活的条件，根瘤菌也就随着大量死亡了。

施用技术问题

适宜的施用法和施用量是保证根瘤菌肥料发挥最大作用的一项关键性的环节。结瘤作用的优劣和根瘤菌肥料的施用量之间的关系是比较复杂的。当然，足够的活根瘤菌数是保证充分结瘤的必要条件。但是，究竟什么是足够的活根瘤菌数？施用菌肥之后，结瘤之前，根瘤菌有几天或几十天生活在土壤中，在这个时期中，根瘤菌的增加或减少，决定于土壤条件。因此，施用时的数量只是起一个好的开端的作用。

可以这样说，在豆科植物和根瘤菌的相互关系中，目前我们了解得最不够的一个方面就是，在豆科植物的根际土壤中，根瘤菌究竟是如何发展着的。在这个问题上，盆栽试验的经验不能等比例地应用到大田中去，而大田生产或试验经验又很不容易完善地总结整理出来。用灭菌土壤进行的试验的参考价值是不大的，因为根瘤菌在灭菌土壤中的变化情况和在没有灭菌的土壤（自然土壤实际上是这样的）中的变化情况显然不是一回事。但是迄今为止，一旦将根瘤菌加入没有灭菌的自然土壤中后，还没有什么可靠和简便的方法来直接测定它们的变化情况。因此，这是一项比较困难的研究课题。

特别是这个问题牵涉到土壤条件的影响。在酸性土壤中（例如大多数红壤），根瘤菌的死亡率很大。要保证有足够的根瘤菌起结瘤作用，除研究适宜施用量问题外，还须研究能克服酸性土壤对根瘤菌的有害影响的施用方法。例如，有些国家采用种子石灰颗粒化和拌种相结合的方法，这在我国还没有试验过。

正确的施用量和施用法也是节省根瘤菌肥料的关键因素之一，这也意味着更充分地发挥一定的根瘤菌肥料生产设备的作用的问题。用过大的施用量来保证优良的结瘤作用，意味着等比例地降低菌肥生产设备的实际作用和等比例地提高了施用根瘤菌肥料的费用。

充分的田间比较试验资料

在数量上足够的、在地区分布上广泛的、在科学质量上严格的田间比较试验资料，既是对这项现代生产技术作出正确评价的必要依据，也是推广这项技术的有效手段。目前，我们还拿不出多少符合现代田间试验技术要求的根瘤菌肥效的田间比较试验来。这是急待大规模地系统化地开展的工作。在充分的田间比较试验的基础之上，还必须开展更大规模中间生产性质的样板田试验工作。

从以上四个问题的内容可以看出，要做好这项工作，既需要多学科（主要是微生物学、植物生理学、土壤学和植物栽培学）的协作，还需要多种工作方式的配合；有些是要在专业研究单位中进行的，有些则是要求在广大的生产实践环境中进行的。由于根瘤菌肥料的中心环节是根瘤菌，土壤微生物学研究单位就有责任承担起主要的项目，并争取其他专业科技单位和广大生产基层的协作。

豆科植物与根瘤菌共生关系的形成与发展*

在河南省绿肥学术讨论会上的报告摘要

陈华癸

（华中农学院）

我从豆科绿肥根瘤菌方面，讲讲近年来的进展。

一、种植豆类作物不是一个单一的植物，而是种一个共生体

豆科植物在根部能形成根瘤，但只有根毛分泌物或根毛脱落物对某一种根瘤细菌的繁殖有利，某一种根瘤菌才会旺盛繁殖并侵入根毛。就是说豆类根际需要形成一个相应的生态环境，某种根瘤菌才能与某一特定豆类植物形成一个共生体。科学试验证明：利用根瘤菌接种，形成根瘤的结瘤部位不是在播种的种子部位，而是在根系的分布层。在形成根瘤的部位要有细菌生长和繁殖。在种子发芽十天左右，根瘤菌侵入根部，它不是靠细菌的机械运动，也不是靠细菌的简单扩散，主要是靠细菌的繁殖，以及根际环境能否满足根瘤菌繁殖的要求。根瘤菌在根际的繁殖叫根际效应。根际效应是很强的。如放一粒豌豆在 200 毫升的矿质营养液中，经过 2～3 周生长过程，矿质营养液中的细菌数量从原加入浓度的每毫升 10^2 发展到 10^8 其所需要的养料，都来源于豌豆根与根毛的分泌物和根毛脱落物。随着种子发芽，在根附近提供好的条件，使根瘤菌旺盛生长。人工接种根瘤菌是否能旺盛生长和导致形成根瘤，是接种的第一个环节。只有在根际效应满足细菌繁殖，才可能保证结瘤。我们的研究和生产任务是要保证植物结瘤由优良菌种所造成的。优良菌种接种后，能否排斥土生菌种？现在要研究的问题是在特定条件下寄主植物对根瘤菌的根际效应何如？何如保证接种后所结瘤是优良菌种所形成的？

二、共生关系的形成是有选择性的，其决定因素可能是豆类植物根系表面的血凝素成分

根据近十几年研究表明：不同豆类植物表面所具有的血凝素成分不同，决定双方接触的选择性。血凝素的结合对象是细菌多醣。不同互接种族的血凝素与细菌表而多醣的关系的差异，决定了它们能否造成第一步接触。如能够接触就有可能形成根瘤，反之，就不能形成根瘤。细菌接触后对植物根毛产生作用，刺激根毛细胞向内凹陷，细菌繁殖、侵入、刺激使之进一步凹陷，最后形成一“侵入线”。侵入线尖端的分泌物刺激使靠近中轴地方的内皮层细胞，由不能分裂繁殖的细胞转化为可分裂繁殖的生发组织细胞，类似侧根，但不是侧根。类似侧根的生发组织细胞进一步分化伸长，在两个维管束中间及分生组织后面根瘤菌进入植物细胞，细菌逐步长大成为畸形，叫类菌体。含类菌体的植物细胞是红色的，这种细胞才具有共生固氮作用。

三、具有固氮酶活的类菌体是一个活体

试验证明：把类菌体的细胞组织，从根瘤菌中分离出来，在一定的条件下可以固氮，从这点讲它是活的，但它不能繁殖，因而又是死的。最近两、三年研究发现，由于类菌体细胞壁构造不完全，抗渗透

---

*原载于《河南农林科技》，8:1～2，1981。

压作用小，如把它放在等渗压或亚等渗压条件下，也可以繁殖，证明类菌体是活的。固氮酶催化的固氮作用是还原作用，其化学反应是吸热反应，它必须与放热反应相偶联，才能进行。植物细胞通过光合作用提供足够的三磷酸腺苷（ATP），把放热反应与吸热反应偶联起来。其反应式：

$$\left.\begin{array}{l}\text{ATP} \rightarrow \text{ADP} + \text{Pi} + \text{热} \\ \text{三磷酸腺苷} \quad \text{二磷酸腺苷} \quad \text{无机磷} \\ N_2 + 6H + 6e \rightarrow 2NH_3 - \text{热}\end{array}\right\} \text{二者偶联}$$

生物固氮作用的强弱与植物光合作用强弱密切联系。固氮酶的活性要求一个重要条件，即在高氧压下固氮酶失活，在低氧压下固氮酶有活性，在0.5%～2%含氧量为合适。植物供给充足的ATP是在好气作用下的产物。因此，固氮酶活性与供给足够的ATP存在着需氧与厌氧的矛盾，豆科共生固氮是通过豆血红蛋白（即在根瘤中所见到的红色物质）来克服这个矛盾。它是吸氧和放氧的缓冲剂。

豆科植物从发芽生长—感染—共生固氮，经历一个复杂的过程，这个过程是由既决定于植物和根瘤菌的基因型，也决定于环境条件是否允许有关基因的表达，试验证明：根瘤菌的结瘤基因不是在染色体上，而是在质粒上，因而它很不稳定，易使结瘤性能发生变异，但对杂交和改造菌种是有利的。

（陈婉华根据录音整理并加标题）

Differentiation and Viability of Nodule Bacteria in Host Cells*

CAO YANZHEN ZHOU JUNCHU AND CHEN HUAKUI

(Laboratory of Biological Nitrogen Fixation, Huazhong Agricultural College, Wuhan)

ABSTRACT In the mature bacteroidal cells of nodules formed by fast-growing rhizobia and host legumes (*Astragalus sinicus*, *Medicago denticulata*, *Trifolium repens*, *T. subterranian* and *Vicia* sp.), there are two distinct morphological classes of bacterial cells. One class consists of large T-shaped, Y-shaped, and club-shaped differentiated bacteroids considered to be active nitrogen fixing forms which are not viable (not multiplicable); the other class consists of small, non-metamorphic rod-shaped bacteria similar to rhizobia grown in culture medium which are viable.

In soybean nodules, on the other hand, there is no apparent morphological differentiation. The cell shape and size are similar to the rhizobia grown in culture medium, and most of the cells are viable.

I. INTRODUCTlON

It is generally accepted that a bacterial cell is viable if it multiplies in a suitable environment. The inability of bacteroids in root nodules to form colonies on common culture media was first reported by Almon[1] in 1933. Bergersonr[2] in 1968 reported that only 0.02% of soybean bacteroids was viable if plated on standard yeast extract mannitol agar (RMY). Sutton et al.[3] reported that the viability of bacteroids of *Rh. lupine* varied with nodule age and was dependent on osmotic protection of the culture medium. Tsien et al.[4] found almost 100% of viability of bacteroids were prepared from nodules of soybean and frenchbean of different ages. Gresshoff et al.[5] showed that under osmotic protection, 60%~90% of the bacteroids in bacteroidal cells of clover nodules were multiplicable. Gresshoff and Rolfe[6] reported that soybean bacteroids were multiplicable under definite experimental conditions. Their experiments showed that soybean bacteroids could multiply only when they were plated on culture medium containing high concentration of mannitol (B^+Man medium). However, they pointed out that the effect of high mannitol concentration was probably not due to osmotic protection, because the passing of the bacteroids through water or protoplast dilution buffer (PDB) did not affect the multiplicability of the rhizobia. All investigators mentioned above regarded all the bacteria present in bacteria-containing cells as bacteroids, regardless of their morphology. At the Fourth International Conference on Biological Nitrogen Fixation, this concept was clearly stated by Pankhurst who suggested "that the term (bacteroid) be applied to all rhizobium cells found within the central tissue cells of legume root nodules, without regard to morphology, or physiology" [7].

The work reported in this paper was undertaken in order to verify the form differentiation and viability of bacterial cells present in the bacteroidal cells of mature were studied (Table 1).

Table 1 Legume-rhizobia symbiotic systems

Legume	Rhizobial strain
Fast-growing types	
Astragalus sinicus L.	Ra 106
Medicago denticulate	M3

*原载于 *Science in China*, *Ser. B*. 6: 593~600, 1984.

续表

Legume	Rhizobial strain
Fast-growing types	
Trifolium repens	T8
Vicia sp.	Infected naturally
Slow-growing Type	
Glycine Max L.	SM31

II. MATERIAIS AND METHODS

1）Legume-rhizobia symbiotic systems. Five legume-rhizobia symbiotic systems were studied（Table 1）.

2）Culture media used （Table 2）

Table 2 Culture media used in experiments

Ingredient	RGY b)	RMY b)	RSY b)	B^+Man (b)
$Na_2HPO_4 \cdot 12H_2O$				0.36g
$K_2HPO_4 \cdot 3H_2O$	0.5g	0.5g	0.5g	
$MgSO_4 \cdot 7H_2O$	0.2g	0.2g	0.2g	0.08g
$FeCl_3 \cdot 6H_2O$				0.003g
$CaCl_2 \cdot 2H_2O$	0.1g			0.04g
$CaCO_3$		3.0g	3.0g	
NaCl	0.1g	0.1g	0.1g	
Mannitol		10g		37.0g
Sucrose			10g	
Glycerin	10ml			
Sodium glutamate				0.5g
Yeast		5.0g	5.0g	0.5g
Trace elements a)	4ml	4ml	4ml	4ml
Vitamin H	0.001g			
Vitamin B_6	0.005g			
Ca pantothenate	0.005g			
Agar-ager c)				

a） Trace elements：Mixed solution of 0.5% H_3BO_3 and 0.5% Na_2MoO_4 in equal volume.

b） RGY，RMY and RSY are routine media used in this laboratory.

c） As needed.

3）Protoplast dilution buffer（PDB），according to Gresshoff et al.[5]

4）Nodulation test，according to Chen，H. K.[8]

5）Isolation of protoplast from bacteria-containing nodule cells. The surface of detached nodules was sterilized by routine procedure and transferred to sterile PDB（0.5M）. The apical one third of the nodule was cut off to eliminate the cells newly infected by the rhizobia from infection threads. The nodules were sliced into segments of 0.2mm thickness and washed again in sterile PDB. Each segment was incubated individually in 0.5ml of 4% filter-sterilized cellulase dissolved in PDB（0.5M）in a 5ml vial for 1 hr at 36℃. The digested segment was transferred into a drop of PDB （0.5M） on a glass slide and examined under microscope. The single protoplasts thus obtained were sucked into a sterile capillary tube with the help of a manipulater and transferred onto a new glass slide. The protoplast was examined and transferred into fresh PDB drops several times to remove bacteria possibly attached to the surface of the protoplast and examined under phase-contrast microscope to be certain that there were no free bacteria present outside the protoplast.

6) Release of bacteria from the bacteroidal protoplast. Single bacteroidal protoplasts were transferred from 0.5M PDB to 0.25M PDB or water. Within 2~3 min, the protoplast swelled and then burst. Bacterial cells were thus released gradually.

7) Count of bacteria per protoplast. A hemocyte counter was used under phase-contrast microscope.

8) Count of viable bacteria per protoplast. A single protoplast was transferred into a sterilized 5ml vial with 2 glass beads containing 1ml sterilized distilled water (or 0.25M PDB). In order to affect the burst of protoplast and the complete releasing of bacteria, the solution was shaken for 2~3 min. The viable bacteria were counted by serial dilution and plating.

9) Microchamber culture. A loop of suspension containing bacteria released from a bacteroidal protoplast was placed on a cover slip. A drop of melted 55℃ agar medium (B^+Man or RSY) was added and mixed immediately. The cover slip was placed invertedly onto a hollow slide and sealed with vaseline. The microchamber was placed in a petri dish with a piece of moist filter paper and incubated at 28℃. The multiplication of bacteria was examined in the microchamber at a fixed location and regular intervals for 8~10 days under phase-contrast microscope.

10) Separation of form-class of bacteria in bacteroidal protoplast by filtration. A large number of bacteroidal protoplasts obtained by the above methods were plasmolysed and filtered through a sterile fritted glass filter (G5 porosity 1.5~2.5 micron). The viability of the bacteria present in the filtrate as well as larger bacteroids left on the filter was determined by plating on RSY agar.

III. RESULTS

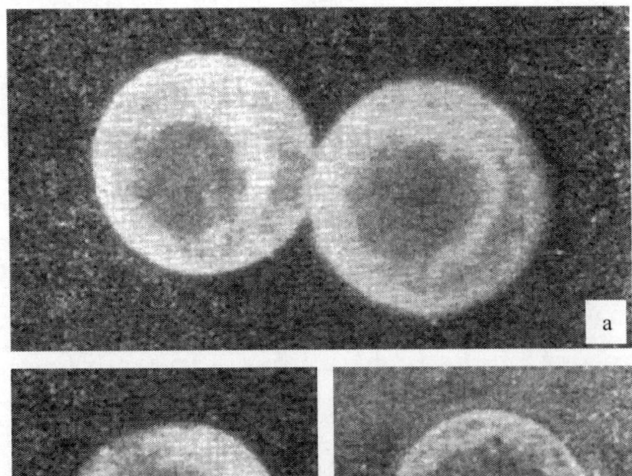

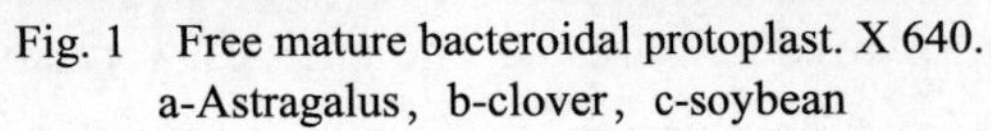
Fig. 1 Free mature bacteroidal protoplast. X 640.
a-Astragalus, b-clover, c-soybean

1) Mature bacteroidal protoplast. Free mature bacteroidal protoplasts in 0.2M PDB are shown in Figs. la (*astragalus*), lb (*clover*) and lc (*soybean*).

2) Releasing of bacterial cells from protoplast. Bacteroidal protoplasts of swelling, bursting and releasing bacterial cells are shown in Figs.2a, 2b, 2c (*astragalus*) and 3 (*soybean*). Fig 2d shows the two types of cell morphology. In 0.25M PDB solution, the protoplast sometimes had to be helped by mechanical pressing to burst.

3) The form differentiation and viability of the bacterial cells in bacteroidal protoplast of fast-growing type nodules. The bacterial cells released from mature bacteroidal protoplasts of *astragalus* and other fast-growing types could be clearly separated into two morphological classes. One consisted of large T-shaped, Y-shaped and club-shaped barteroids. The other class consisted of small rod-shaped bacteria similar to those grown on conventional culture media (Fig. 2d)

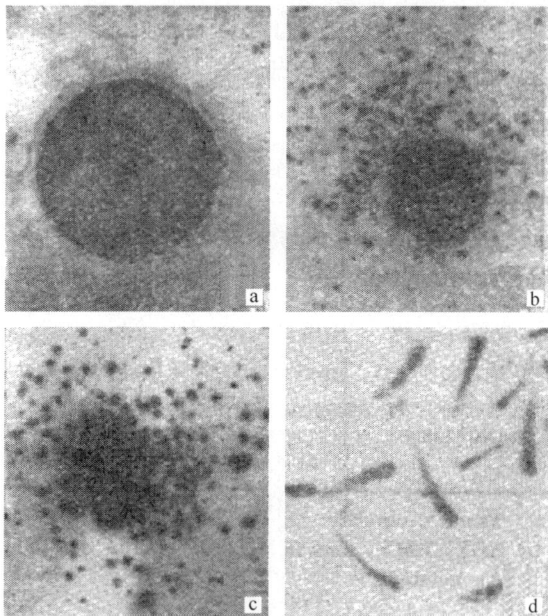

Fig. 2 Bacteroidal cell released from protoplast of astragalus nodule.

a-Protoplast swelling, ×640; b-protoplast bursting and bacteria releasing, ×640; c-protoplast collapsed, ×640; d-two morphologieal types of bacterial cell, ×3200.

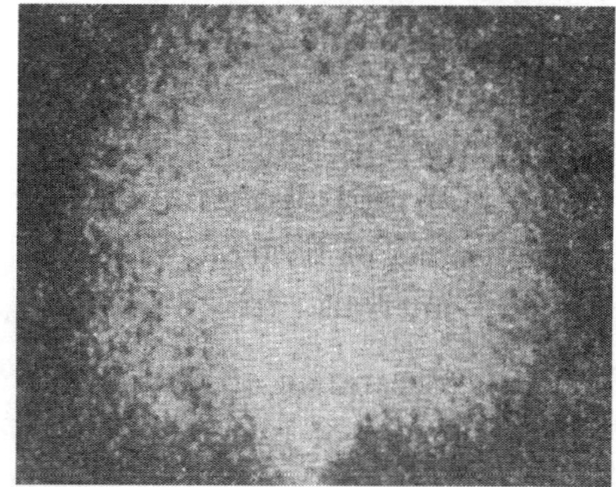

Fig. 3 Bacterial cell relasing from protoplast of soybean nodule.

The average number of bacterial cells per astragalus bacteroidal protoplast was 23.4×10^3 (average of 10^3 protoplasts).

The microcultures were examined 3 times a day at the fixed locations of the culture. It was observed that the small rod-shaped bacteria multiplied gradually into micro-colonies, while the large bacteroids remained undivided during the total period of observation of 9 days (Fig. 4).

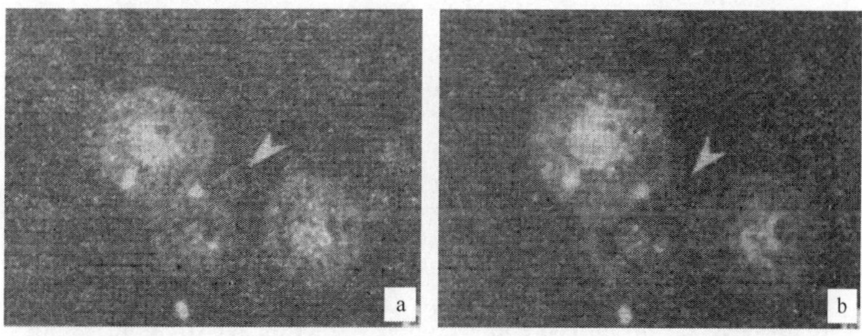

Fig. 4 Bacteria released from protoplast incubated in mieroehamber.

a-small rod bacteria developed into microcolonies and non-viable large bacteroids (arrowhead), b-microcolonies developing and covering part of bacteroid (arrowhead).

The small rod-shaped bacteria were separated from the large bacteroids by filtering through a fritted glass filter (G5). When small rod-shaped bacteria in the filtrate were inoculated onto RSY medium, they multiplied and formed normal colonies (Fig.5). When the bacteroids left on the surface of the filter were inoculated onto the same medium, no growth was observed.

Three hundred clones were obtained from the small rods released from the mature bacteroidal protoplasts of astragalus nodules. Nodulation tests were performed with each clone, ninety-two percent of the clones were able to form nodules on the host plant *Astragalus sinicus*.

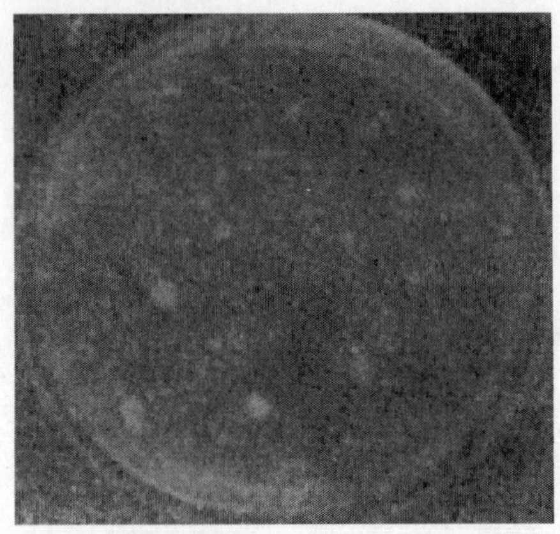

Fig. 5 Normal colonies formed by filtered small rods (half of natural size).

Experiments with other legume-rhizobia systems of fast-growing type, e.g. Lucerne clover and vetch, gave similar results.

4) Total and viable bacterial cells present in a protoplast. Table 3 shows the total number and viable number of bacteria cells present in a mature bacteroidal protoplast of 5 legume-rhizobia systems.

The total number of bacterial cells in each protoplast was more than 2×10^4, with only 0.09%~0.27% viable.

5) The relationship between the abundance of viable bacteria and acetylene reduction activity of bacteroidal protoplasts of varying nodule ages. Acetylene reduction activities of astragalus nodules of different nodule ages were determined. Bacteroidal protoplasts were isolated from the nodules used for acetylene reduction determinations. The results showed that both the acetylene reduction activity and the abundance of viable bacterial cells varied with nodule age, and were negatively correlated (Table 4).

Table 3 Number of total and viable cells in bacteroidal protoplast

Plant	Method of cunting	Number of potoplast examined	Number of vable bcteria per protoplast	Percentage of viable bacteria per protoplast
Astragalus	Plant count[a)]	17	63.9	0.27%
	Total cell[b)] count	103	23375.5	
Alfalfa	Plant count	28	26.5	0.11%
	Total cell count	15	22755.0	
White clover	Plant count	15	17.7	0.09%
	Total cell count	11	19509.0	
Subterranean clover	Plant count	29	19.3	0.11%
	Total cell count	15	17813.0	
vetch	Plant count	28	29.9	0.12%
	Total cell count	15	23973.0	

a) Grown in B^{+}Man medium;
b) Counted by hemoeyte counter.

Table 4 Number of viable bacteria per protoplast and acetylene reduction activity of astragalus nodule of different ages

	Nodule age (Days)			
	14	21	45	180
Viable cells/protoplast	1600	154	145	810
nMC_2H_4/mg of fresh weight/hr	1.05	7.61	8.55	0.38

6) The form and viability of bacterial cells in the mature bacteroidal cells of nodules of the soybean-rhizobia (SF31) system. There was no morphological differentiation of rhizobia in bacteroidal cells in the red mature, *N*-fixing nodules of soybean and other slow-growing symbiotic systems. All bacteria in bacteroidal cells were rod-shaped, similar to the bacteria grown on culture media. By the same method of preparing protoplasts, the viability of released bacteria was 81% (Table 5). In microchamber culture, virtually all the bacteria multiplied and formed micro-colonies (Fig. 6).

Table 5 Number of bacteria released from bacteroidal protoplast of soybean nodule

Method of counting	Number of protoplasts examined	Number of bacteria per protoplast	Percentage of viable bacteria per protoplast
Plate count	52	50903	81%
Total cell count	216	62781	

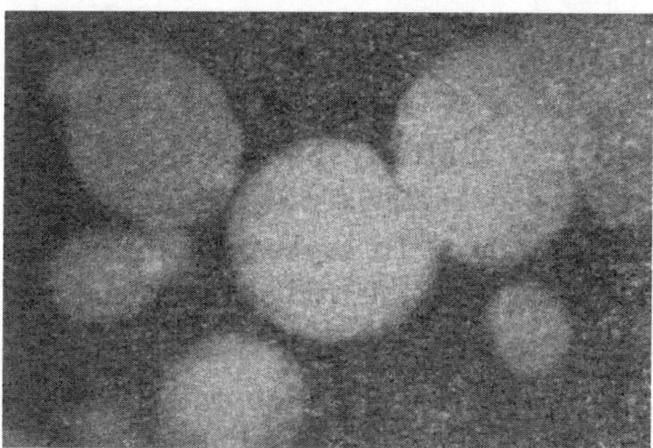

Fig. 6 Microcolonies formed by bacterial cells released from protoplast of soybean nodule. ×245

7) The influence of osmotic protection on the viability of bacteriods. Swelling in water or in 0.25*M* PDB did not affect the number of viable cells in each protoplast, as tested with clover and lucerne (Table 6).

Table 6 Effect of swelling by water or 0.25M PDB on the number of viable cells

Plant	Solution used	Number of protoplast examined	Viable cells/protoplast	S.D.
Clover	Water	10	14.5	±6.2
	0.25M PDB	10	14.25	±5.3
Alfalfa	Water	6	20.37	±7.6
	0.25M PDB	6	27.62	±11.6

The difference in osmotic pressure of culture media had no significant effect on the number of viable bacterial cells in each protoplast (Table 7).

Table 7 Influence of media of different osmotic pressures on the viability of bacteria

Plant	Media	Number of protoplast examined	Viable cells/protoplast	S.D.
Alfalfa	RSY	12	24.0	±10.4
	B^+Man	12	22.1	±9.5
Soybean	RSY	12	25.6×1000	±25.5×1000
	B^+Man	12	34.1×1000	±22.4×1000

IV. DISCUSSION

The questions raised in the 1930's were raised again in the late 1970's and have not yet been satisfactorily resolved.

1)Do the mature, pink-coloured, nitrogen-fixing nodules contain only one morphological class of bacterial cell, or are the bacterial cells differenciated into different morphological classes? The definition of bacteroid given by Paukhurst[7] is applicable to the soybean-rhizobia symbiotic system, but does not conform to the legume-rhizobia symbiotic systems of the fast-growing type, as reported in this paper. In the nodules of astragalus, clover, lucerne and vetch, there are distinct differentiations in form and function. There are two distinct morphological classes present in the mature bacteroidal cells: (i) large T-shaped, F-shaped, and club-shaped bacteroids which have lost the ability to multiply but are biochemically functional, i.e. *N*-fixing[9]; （ii） normal small rod-shaped bacteria possessing the ability to multiply, hence playing the role of clonal continuation. In soybean nodules, there is no such differentiation of formation and function, and the rod-shaped bacterial cells present in bacteroidal plant cells are viable as well as nitrogen-fixing.

2）Is the osmotic protection essential for the viability of the bacteroid? Previous reports gave contradictory data and opinions. The experiments reported in this paper show that neither water nor 0.25M PDB affects the viability of the rod-shaped bacterial cells in bacteroidal protoplasts. In the soybean bacteroidal protoplast 81% of the rod-shaped bacteria were viable. With regard to the legume-rhizobia systems of the fast-growing types, the viability of the rod-shaped bacterial cells was not significantly harmed by water or 0.25M PDB.

In the senescent nodules of the fast-growing types, the bacteroidal tissues begin to rot and many rod-shaped bacteria are found. It has been considered that these rods are transformed from the large bacteroids within the host cells. The results of our experiments indicate that the high number of rod-shaped bacteria present in the senescent bacteroidal cells results from the multiplication of rod-shaped bacteria which are always in the bacteroidal cells, though in low numbers during the active N-fixing stage. The senescent bacteroidal cell probably provides the necessary saprophytic condition suitable for the multiplication of the rod-shaped bacteria. The high proportion of rod-shaped bacteria in senescent bacterioidal cells is probably not the result of the conversion of bacteroids to rod-shaped bacteria.

REFERENCES

[1]Almon, L., *Planta*, 142（1978）, 329-333.

[2]Bergersen, F., *Trans. 9th Int. Congr. Soil Sci.*, Adelaide, 2（1968）, 49-63.

[3]Sutton, W. et al., Plant Physiol., 59 （1977）, 741-744.

[4]Tsien, H. C. et al., Appl. and Environ. Microbiol., 34（1977）, 804-856.

[5]J Gresshoff, P. M. et al., Plant Science Letters, 10（1977）, 299-304.

[6]Gresshoff, P. M. & Rolfe, B. G., *Planta*, 142（1978）, 329-333.

[7]Pankhurst, C. E., in Current Perspectives in Nitrogen Fixation（Eds. Gibson, A. H., & New-ton, W. E.）, Australian Academy of Science, Canberra, 1981, 304.

[8]陈华癸. 微生物学实验. 农业出版社. 1962, 119-122.

[9]Mulder, E. G., in Nitrogen fixation by Free-living Micro-organisms（Ed. Steward, W.D.P.）, Cambridge University Press, 1975, 3-28.

根瘤菌在寄主细胞内的分化及存活性*

曹燕珍　周俊初　陈华癸

（华中农学院土壤化学系）

摘　要　本文证明在快生型紫云英、苜蓿、三叶草和苕子根瘤菌的成熟根瘤内的类菌体组织细胞中，根瘤细菌有两种形态，一种为已分化发育的类菌体，一般呈棒状、T 状或 Y 状，即通常认为活跃固氮的形态，它们不能生长繁殖；另一种为未经分化的杆状菌，与在培养基上生长的形态一样，它们能够生长繁殖。

在慢生型的大豆根瘤含菌组织中，类菌体形态分化不明显，菌体的形状大小同在培养基上生长的类似，它们绝大多数都是能够生长繁殖的。

通常将能否进行繁殖作为细菌细胞的生存界限。先前（Almon，1933 年）认为豆类植物根瘤中的类菌体是不能繁殖的[1]，即失去了存活性。Bergensen （1968 年） 报告说大豆根瘤的含菌细胞中的类菌体只有 0.02%能在甘露醇酵母洋菜（RMY）上生长繁殖[2]。Sutton 等（1977 年）发现羽扇豆根瘤的类菌体的繁殖能力随瘤龄而变化，并认为其繁殖能力依赖于培养基渗透压的保护作用[3]。Tsien 等（1977 年）发现不同年龄的大豆和菜豆根瘤中的类菌体几乎 100%具有活力[4]。Gresshoff 等（1977 年）的研究结果表明，三叶草根瘤的类菌体在渗透压保护下有 60%~90%能繁殖[5]。Gresshoff 和 Rolfe （1978 年）报道说大豆根瘤类菌体在一定的试验条件下能繁殖[6]。他们的实验表明大豆根瘤内的类菌体只有在含高浓度甘露醇的培养基（B^{+}Man）上能够繁殖。但高浓度甘露醇的作用是否起了渗透保护作用，不能肯定。因为在处理过程中菌体经过等渗液（PDB）或水都不能影响菌体繁殖能力。这些研究者们都把根瘤中的含菌细胞内所有的细菌个体全部看做是类菌体，而不论它们表现为什么形态。在第四次国际生物固氮会议上，Pankhust （1980 年）建议，“只把存在于成熟的，有固氮活性的含菌植物细胞中的细菌称为类菌体，以避免混乱”[7]。明确地表达了这种观念。

本文介绍根瘤菌在寄主细胞内的分化及存活性研究工作，对含菌细胞中细菌的形态分化及繁殖能力提供了直接的验证。

一、材料和方法

1. 豆类一根瘤菌共生体　本实验采用了 5 种豆科植物及相应的根瘤菌（表 1，其中快生型根瘤菌为紫云英、苜蓿、三叶草和苕子，慢生型根瘤菌为大豆）。

表 1　试验豆类-根瘤菌共生体

豆类	接种菌株	豆类	接种菌株
紫云英	Ra 106	苕子	自然感染
苜蓿	M3	大豆	SM31
三叶草	T8		

*原载于《中国科学 B 辑》，14（3）：237~243，1984.

2. 培养基的种类和成分（表 2）

表 2 培养根瘤菌的培养基成分

	RGY*	RMY*	RSY*	B^+Man **[6]
$Na_2HPO_4 \cdot 12H_2O$				0.36g
$K_2HPO_4 \cdot 3H_2O$	0.5g	0.5g	0.5g	
$MgSO_4 \cdot 7H_2O$	0.2g	0.2g	0.2g	0.08g
$FeCl_3 \cdot 6H_2O$				0.003g
$CaCl_3 \cdot 2H_2O$	0.1g			0.04g
$CaCO_3$		3.0g	3.0g	
NaCl	0.1g	0.1g	0.1g	
甘露醇		10g		37.0g
蔗糖			10g	
甘油	10ml			
谷氨酸钠				0.5g
酵母		5.0g	5.0g	0.5g
微量元素+	4ml	4ml	4ml	4ml
洋菜**				
维生素H	0.001g			
维生素B_6	0.005g			
泛酸钙	0.005g			

+微量元素：0.5% H_3BO_3；0.5% Na_2MoO_4 等量混合液。
*RGY、RMY、RSY 均为华农生物固氮研究室培养根瘤菌常用培养基。
**洋菜用量按需要添加。

3. 等渗溶液（PDB） Gresshoff 等[5]用的 PDB （0.25M 甘露醇，0.25M 山梨醇，10mM K_2HPO_4，2mM $CaCl_2$，pH 5.8）等渗溶液，

4. 结瘤试验 按本研究室的常规洋菜管结瘤和砂培结瘤法[8]进行.

5. 根瘤含菌细胞原生质体的分离 根瘤按常规方法表面灭菌后，转入灭菌 PDB 液（0.5M）中，切去根瘤前端，以排除侵入线新侵染的细胞，然后将剩余根瘤切成 0.2 毫米左右的薄片，再用 PDB 液洗一次。每个根瘤薄片分别放入容量为 5 毫升的小瓶中；加入用 PDB 液（0.5 M）配制的 4%纤维素酶（Cellulase“ONO-ZUKA”R-10 Yakult Biochemicals，Japan） 0.5 毫升，于 36℃下酶解一小时，用吸管将根瘤薄片吸入放有一滴 PDB 液（0.5 M）的载玻片上，在低倍显微镜下观察，用显微镜操作器和无菌毛细管将单个原生质体吸出移到另一载玻片上，并用 PDB 液洗涤几次，洗去原生质体外可能存在的游离细菌，然后将单个原生质体放在相差显微镜下检查，确证原生质体外无游离细菌。

6. 从单个含菌原生质体中释放出细菌 将表面无游离细菌的原生质体，从 0.5 M PDB 液移到 0.25M PDB 中或水中，在 2-3 分钟内即可观察到原生质体先吸水涨圆，随后崩裂，原生质体内含的细菌逐渐释放出来。

7. 单个原生质体中总菌数的测定 在相差显微镜下用血球计数器测数。

8. 单个原生质体中活菌数的测定 在含 1 毫升无菌蒸馏水（或 1 毫升 0.25M PDB 液）的 5 毫升无菌小玻璃瓶（瓶中放 2 粒小玻珠）中，移入一个原生质体，振荡 2~3 分钟，使原生质体破裂，释放出所含细菌，对此菌悬液进行平板培养测数。

9. 微室培养 从原生质体破裂后释放出的细菌悬液中取一环放在盖玻片上，随即加一小滴融化并冷至 55℃的洋菜培养基（B^+Man 或 RSY），用接种环迅速混匀，将盖玻片反扣在凹玻片上，用凡士林密封，将此凹玻片放入培养皿（皿底放湿滤纸保湿）中，28℃保温培养，定位和定时观察微室培养中菌

体的繁殖情况，连续观察 8~10 天。

10. 原生质体中大小菌体的过滤分离 用纤维素酶处理快生型根瘤获得大量含类菌体的原生质体，待胀破原生质体释放出细菌，用灭菌过滤器过滤（重熔玻璃过滤器 G5，孔径 1.5~2.5 微米），将过滤液和未过滤物悬浮液分别接种在 RSY 平板上培养，检查菌落生长情况。

二、试验结果

共观察了根瘤中分离出来的 3000 多个单原生质体。

1. 成熟的含类菌体原生质体 在 0.5M PDB 液（等渗液）中，含类菌体原生质体的外形如图 1a（紫云英）、1b（三叶草），lc（大豆）。

2. 含菌原生质体中释放出细菌 将原生质体悬浮在蒸馏水或 0.25M PDB 液（低渗压）中，在相差显微镜下，可以清晰地观察到原生质体的膨胀和崩裂情况，崩裂后细菌逐步地从原生质体中释放出来。在 0.25M PDB 液中有时需要挤压后，原生质体才破裂。图 2a，2b，2c 显示紫云英根瘤内含菌原生质体胀圆、崩裂，释放出细菌；图 2d 是释放出来的细菌个体形态，图 3 显示大豆根瘤含菌原生质体释放出细菌。

3. 快生型根瘤含类菌体原生质体中菌体形态的分化和存活性 从紫云英和其他快生型根瘤的成熟的含类菌体原生质体中释放出来的细菌在形态上明显地分为两类，一类是大的、棒状、Y 状或 T 状的类菌体，一类是与常用培养基上生长的小杆菌在大小、形态上相同（图 2d）。

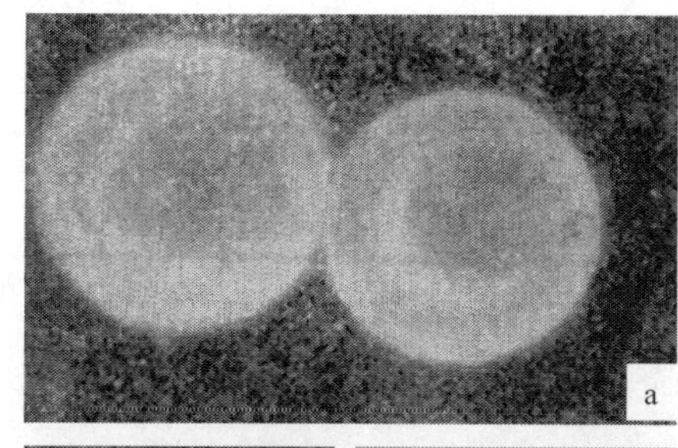

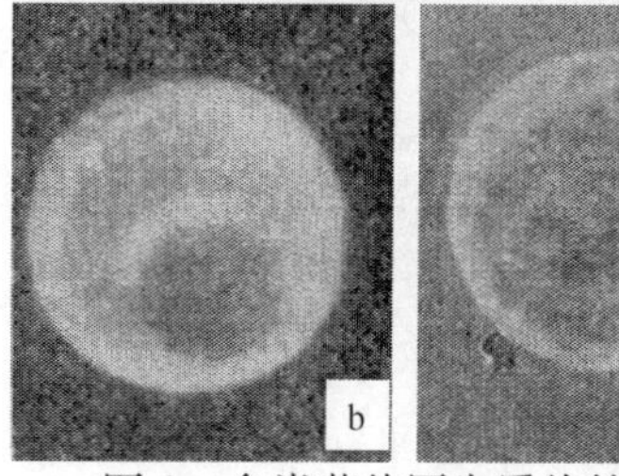

图 1 含类菌体原生质体的外形

a.紫云英约×560，b.三叶草约×560，c.大豆约×560

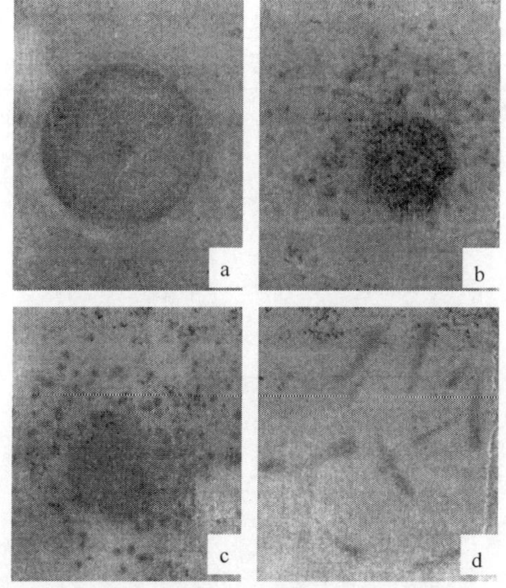

图 2 紫云英根瘤内含菌原生质体胀圆崩裂释放出细菌

a.原生质体胀圆约×640，b. 原生质体开始崩裂释放出细菌约×640，c. 原生质体完全崩裂约×640，d. 原生质体中释放出来的两类细菌形态约×3200

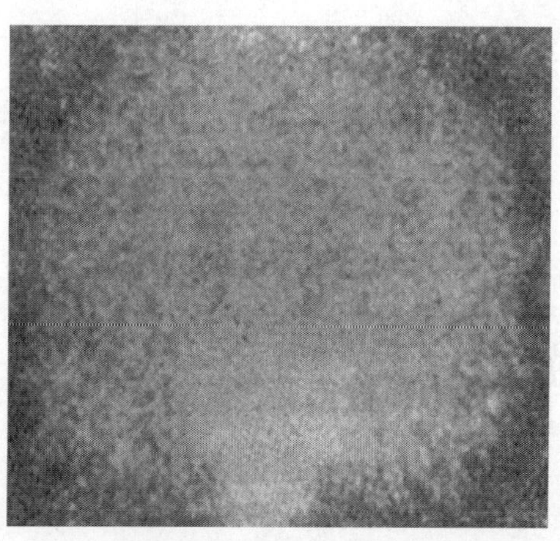

图 3 大豆根瘤含菌原生质体释放细菌

测定了单个原生质体中释放出的细菌个体的数目。103 个紫云英根瘤含类菌体原生质体中平均含细菌细胞数为 2.34 万个。紫云英的含类菌体原生质体释放出的细菌接种到 RSY 或 B^{+}Man 培养基上，进行微室培养，在培养过程中定点观察，每日观察三次，小的杆菌逐渐繁殖形成微菌落，大的类菌体经九天都没有分裂繁殖（图 4）。

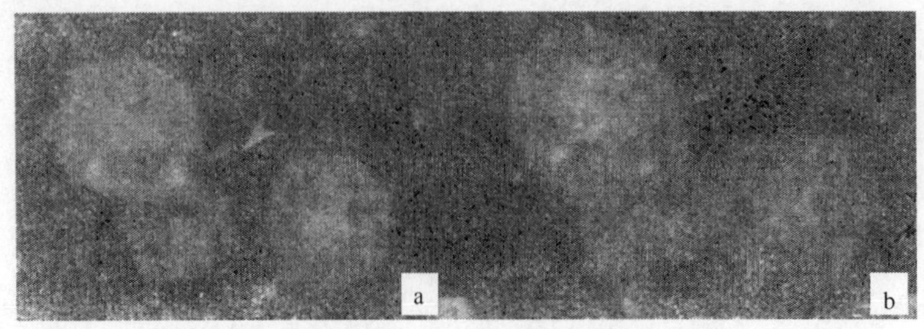

图 4 紫云英根瘤原生质体释放出的根瘤细菌在微室培养中的情况

a.小杆菌已繁殖成微菌落，大类菌体（箭头所指）不能繁殖；b. 微菌落进一步发展，并覆盖了部分类菌体（箭头所指）

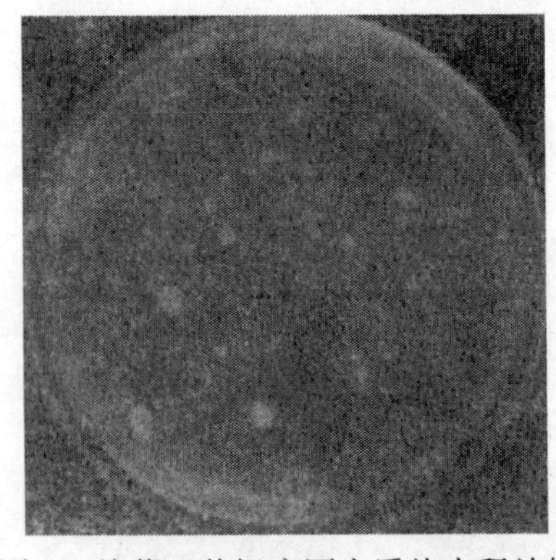

图 5 从紫云英根瘤原生质体中释放的细菌中过正常菌落（为正常大小的 1/2）

用重熔过滤器 G5 过滤分离紫云英根瘤含菌原生质体中释放出来的细菌悬液。小杆菌可以滤过，大的类菌体不能滤过，将滤过的小杆菌接种于 RSY 培养基上，能够繁殖形成正常菌落（图 5）。过滤器上面留下的大的类菌体经接种在同样培养基上，不能繁殖形成菌落。

从紫云英根瘤成熟的含类菌体原生质体中释放出的小杆菌取得了 300 个培养系，对每个培养系进行了结瘤试验，结果有 277 个培养系在紫云英根上结瘤。

用其他快生型豆类-根瘤菌共生体来进行平行实验，包括苜蓿、三叶草和苕子，所得试验结果相同。

4. 快生型共生体的含菌细胞中的菌体总数和活菌体数 5 个豆类-根瘤菌共生体中每个成熟的类菌体细胞所含总菌数和活菌数的测定结果：见表 3。每个原生质体中的细菌总数均在两万个上下，其中活菌数只占 0.09%~0.27%。

表 3 测定从紫云英、苜蓿、三叶草和苕子根瘤含菌原生质中释放出的菌数

植物	测数方法	测定的原生质个数	平均每个原生质体中菌数（个）	每个原生质体中活菌数百分比
紫云英	平板测数*	17	63.9	0.27%
	直接测数**	103	23375.5	
苜蓿	平板测数	28	26.5	0.11%
	直接测数	15	22755.0	
白三叶草	平板测数	15	17.7	0.09%
	直接测数	11	19509.0	
地三叶草	平板测数	29	19.3	0.11%
	直接测数	15	17813.0	
苕籽	平板测数	28	29.9	0.12%
	直接测数	15	23973.0	

*用 B^{+}Man 培养基平板测活菌数

**用血球计数器测总菌数.

5. 紫云英不同瘤龄的根瘤原生质体中的活菌数与根瘤乙炔还原活性的反相关 随着紫云英植物不同瘤龄的变化，我们发现不同瘤龄的单个原生质体中的活菌数和乙炔还原活性有明显反相关（见表 4）。

表 4 紫云英不同瘤龄的单原生质体中的活菌数和根瘤乙炔还原活性

	瘤龄（天）			
	14	21	45	80
活菌数（个/单个原生质体）	1600	154	145	810
乙炔还原活性（毫微克分子乙烯/毫克鲜重/小时）	1.05	7.61	8.55	0.38

6. 大豆-根瘤菌（菌株 SM31）含类菌体细胞中的菌体形态和存活性 大豆根瘤菌和其他慢生型共生体系中的红色的、成熟的、有固氮功能的含类菌体细胞中的菌体并没有形态分化、全都是杆状细菌，大小也和培养基上生长的菌体相似。

用同样的方法制备原生质体，释放出的细菌，成活率很高。每个原生质体中所含活菌数的百分比为 81%（见表 5）。进行微室培养几乎所有细菌都能繁殖形成微菌落（图 6）。

表 5 大豆根瘤含菌原生质体中释放出的菌数

测数方法	测定的原生质体个数	平均每个原生质体含菌数	每个原生质体中活菌数百分比
平板测数	52	50903	81%
直接测数	216	62781	

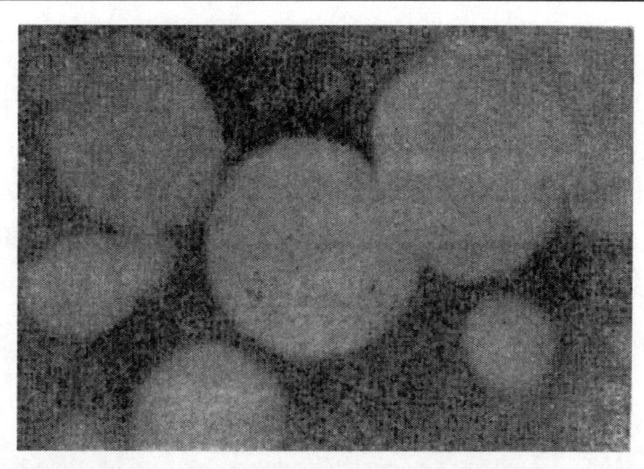

图 6 从大豆根瘤原生质体中释放的细菌进行微室培养形成的微菌落约×245

7. 渗透压保护对类菌体活性的影响 表 6 说明用水或用 0.25M PDB 胀破原生质体时，对于每个原生质体内的活细胞数没有不同的影响．表 7 说明培养基渗透压不同，并不明显地影响每个原生质体中活细胞的百分数。

表 6 水胀破或 0.25M PDB 胀破对细菌存活性的影响

植物	胀破液	测定的原生质体数	活菌数*/原生质体	平均数标准差
三叶草	水	10	14.5	±6.2
	0.25M PDB	10	14.25	±5.3
苜蓿	水	6	20.37	±7.6
	0.25M PDB	6	27.62	±11.6

*活菌数用 RSY 平板培养测定

表 7 不同渗透压培养基对细菌存活性的影响

植物	培养基	测定的原生质体数	活菌数*/原生质体	平均数标准差
苜蓿	RSY	12	24	±10.4
	B^+Man	12	22.1	±9.5
大豆	RSY	12	25.6×1000	±25.5×1000
	B^+Man	12	34.1×1000	±24.0×1000

三、讨　论

这个在三十年代提出并进行过初步研究的问题（Almon，1933 年），在七十年代后期又重新提出，并且报道了一些互相矛盾的实验结果和见解。

第一个问题是，在豆类-根瘤菌共生的根瘤中的红色的、成熟的、有固氮功能的含菌细胞（通常认为是含类菌体细胞）中的细菌有没有分化成为在形态上和功能上有明显区别的细菌体？

Pankhurst（1980 年）对于类菌体所下的定义，对大豆—根瘤菌共生体系而言是适合的，但是根据我们的研究结果，对于快生型共生体系是不适合的。在这类含菌细胞中的细菌有明显的形态和功能的分化；①棒状、Y 状、T 状的大类菌体有固氮功能（类菌体有固氮功能是早已验证的，见 Mulder，1975 年[9]）、但失去了繁殖能力；②正常的小杆菌保持有繁殖能力。也就是说，形态和功能都有明显差别。可以认为，从根瘤中分离培养根瘤菌，实质上是这种小杆菌的后代，对于大豆-根瘤菌共生根瘤中含菌细胞中的细菌，则并没有这种形态和功能分化。

第二个问题是：从根瘤含菌细胞中取出细菌细胞，渗透压保护是不是必需的条件？过去的结论也是分歧的。本文报道的试验表明，它并不是必需的条件，在大豆—根瘤菌共生体中，不论是在水中或 0.25M PDB 中都有 80%左右细菌能繁殖，并同时是固氮的功能者；在快生型细菌的共生体中，小杆菌的繁殖能力也不受水或 0.25M PDB 的影响。

在衰老的快生型根瘤中，通常发现含菌组织形态腐烂，其中充满了大量的小杆菌。一直认为这些小杆菌是由大的类菌体转化的，是根瘤菌在寄主细胞中的生活史的一个阶段。这篇报道的试验结果表明，在衰老的含菌组织中的大量小杆菌是本来存在于健康的含菌细胞中的小杆菌的大量繁殖（衰老的植物细胞实际上提供了根瘤菌腐生生活的条件）结果，而不是类菌体向杆菌的形态转化。

参 考 文 献

[1]Almon. L.，Cited from Gresshoff. P. M. & Rolfe，B. G.，*Planta*，142（1978），329-333.

[2]Bergersen，F.，*Trans. 9th Int. Congr. Soil Sci.*，Adelaide.（1968），2，49-63.

[3]Sutton，W.，Jepsen，N. & Shaw. B.，*Plant Physiol*，59（1977），741-744.

[4]Tslen. H. C.，Oain，P. S. & Schmidt，E. L.，*Appl. and Environ. Microbiol.*，34（1977），854-856.

[5]Gresshoff. P. M.，Skotnicki，M. L.，Eadie. J. F. & Rolfe. B. G.，*Plant Science Letters*，10（1977），299-304.

[6]Gresshoff，P. M. & Rolfe，B. G.，*Planta*. 142（1978），329-333.

[7]Pankhurst，C. E..in *Current Perspectives in Nitrogen Fixation*（Ed. Gibson，A. H. & Newton，W.E.），Australian Academy of Science. Canberra，（1981），304.

[8]陈华癸，微生物学实验，农业出版社，1962，119-122.

[9]Mulder，E. G.，in *Nitrogen fixation by free-living micro-organisms*（Ed. Steward，W. D. P.）. Cambridge University Press，（1975），3-28.

紫云英根瘤菌的不结瘤（Nod^-）突变*

李仲贤+　张春生++　李阜棣　陈华癸

（生物固氮研究室）

摘　要　以紫云英根瘤菌 A106L（A106 菌系的一株）为出发菌株，在正常培养条件下得到的 170 个分离系中，有 19 个分离系对紫云英无结瘤能力；在吖啶橙处理条件下得到的 150 个分离系中有 40 个无结瘤能力，在热处理条件下得到的 165 个分离系中有 30 个无结瘤能力。常规形态，生理鉴定和抗原–抗血清凝集试验表明这些不能结瘤的分离系都是 A106L 的后代，是 A106L 的 Nod^- 突变体。由于分离系总数不多，以及吖啶橙与热处理的方法多样，得不出两种处理能显著提高 Nod^- 突变率的统计分析数据。从所得 Nod^- 突变分离系中，随机抽出 20 个分离系，考查它们不能在紫云英根上结瘤的症结所在，发现障碍全部在结瘤过程早期，即不能使根毛“卷曲”（Cur^-），不能侵入根毛形线成入线（Thr^-）。试验还表明，Nod^- 突变与对链霉素和卡那霉素（10 微克/毫升）的自然抗性的丧失同时发生：$Nod^+ \cdot Str^R \cdot Kan^R \rightarrow Nod^- \cdot Str^S \cdot Kan^S$。

在紫云英根瘤菌的研究工作和接种剂的生产实践中，丧失结瘤能力的突变出现的频率很高（李阜棣等，1979；Li，1981）。曹燕珍、李阜棣（1979）从紫云英根瘤菌 A21 菌株的一个培养体分离出 100 个单菌落，分别进行结瘤试验，其中 48 个菌落没有结瘤能力。

本文报道了以菌株 A106L 为出发菌株所进行的一些试验，包括 $Nod^+ \rightarrow Nod^-$ 的自然突变，经吖啶橙处理和热处理得到的 $Nod^+ \rightarrow Nod^-$ 突变，Nod^- 突变株的表现型特点，以及与 $Nod^+ \rightarrow Nod^-$ 突变同时丧失的自然抗药性。

材料和方法

1. 出发菌株：华中农学院保藏的紫云英根瘤菌株 A106。将 A106 斜面稀释培养在 YMA 平面上，分别从 50 个单独菌落得到 50 个分离系，通过四次回接试验，选出一株现瘤早，瘤数多的菌株 A106L，作为试验用的出发菌株。

2. 紫云英种子：地方品种。

3. 酵母汁、甘露醇、洋菜培养基（YMA）：$K_2HPO_4 \cdot 3H_2O$ 0.5 克，$MgSO_4 \cdot 7H_2O$ 0.2 克，NaCl 0.1 克，酵母汁 100 毫升，甘露醇 10 克，微量元素（B、Mo）液 4ml，$CaCO_3$ 0.3 克（液体培养基不加），洋菜 20 克（液体培养基不加），水 900 毫升，pH 固体培养基不调整；液体培养基调整到 7.2，灭菌 121℃，30 分钟。

4. 无氮植物培养基（SA）：K_2SO_4 0.9 克，$K_2HPO_4 \cdot 3H_2O$ 0.5 克，$MgSO_4 \cdot 7H_2O$ 0.5 克，$CaHPO_4 \cdot 2H_2O$ 0.5 克，NaCl　0.5 克，$FeCl_3$　0.2 克，H_3BO_4　0.2 克，$Na_2MoO_4 \cdot 2H_2O$ 0.2 克，洋菜 10 克（采用 Fahraeus 方法时不加），pH 7.0~7.2，灭菌 121℃，30 分钟。

5. 结瘤试验：紫云英种子表面灭菌，在无菌条件下萌发，播入灭菌后的 SA 斜面试管内，每管两颗，接种供试细菌，在光照室内培养，观察结瘤现象。回接试验重复三次，每次每分离系接三管。重复三次均不结瘤者确认为不结瘤菌株（Nod^-）。

6. 吖啶橙（2，8-双二甲基氨基吖啶，简称 AO）贮备液：每毫升灭菌蒸馏水含 500 微克 AO，盛于棕色瓶中，在室温下贮存一周后按需要稀释使用。

---

*原载于《华中农学院学报》，2（1）：10~18，1983.江苏省淮阳地区农业科学研究所朱铭富同志等为本项研究制备 A106L 的抗血清，特此致谢。

+现在湖南省微生物研究所。

++现在中国农业科学院原子能研究所。

7. 链霉素（Str）、卡那霉素（Kan）选择培养基：每毫升 YMA 培养基内含 10 微克链霉素或卡那霉素。

8. 乳酸石碳酸棉蓝染色液（LPCB）：乳酸（比重 1.21）10 毫升，石碳酸 10 克，棉蓝 0.02 克，甘油 20 毫升，水 10 毫升。

9. 紫云英根瘤菌 A106L 的抗血清：委托江苏省淮阳地区农业科学研究所微生物组制备。

10. 根瘤菌侵入根毛的观察：

a. SA 斜面试管培养紫云英幼苗，接种或不接种，将斜面置于显微镜载物台上直接观察。

b. Fahraeus（1957）法。

c. 将载玻片埋在 SA 培养基平面内，播入萌发的紫云英种子，接种根瘤菌或不接种，将平面竖立，在光照室内培养，用显微镜直接观察平面上生长的根毛，或取出载玻片在显微镜下观察根毛。后一种情况下，用乳酸石炭感棉蓝染色后，可清晰地观察侵用线在根毛内的发展。

试 验 结 果

1. A106 菌系的 Nod⁻自发突变　以 A106L 为出发菌株，先后共进行七次试验，实际上每次都是吖啶橙处理或热处理试验的对照。七次试验共得 170 个分离系，分别进行结瘤试验。170 个分离系中，151 个能结瘤（Nod^+），19 个不能结瘤（Nod^-），占总数的 11.2%。试验数据和 Nod^- 分离系编号见表 1。

表 1　紫云英根瘤菌 A106L 的 Nod^- 自发突变

试验日期	培养温度、时间、方法	接种菌量细胞/毫升	终止菌量细胞/毫升	测定分离系数 a)	Nod^- 突变分离系数
1981.10.7-9	28℃，2 天，振荡	1.2×10^8	32×10^8	20	3
1981.10.7-9	28℃，2 天，静止	1.2×10^8	1.3×10^8	20	0
1981.10.7-14	28℃，7 天，静止	1.2×10^8	23.2×10^8	20	0
1981.11.12-14	28℃，2 天，静止	0.58×10^8	5.1×10^8	25	1
1981.11.12-20	28℃，8 天，静止	0.56×10^8	17×10^8	25	2
1981.11.20-27	28℃，7 天，振荡	54×10^8	24.1×10^8	30	7
1981.11.20-27	28℃，7 天，静止	1.03×10^8	44.7×10^8	30	6

a）终止培养后，稀释平面培养，随机桃取 20-30 个单菌落移入试管斜面培养，得分离系。挑取的总分离系数 170 个，Nod^- 离系 19 个，占总数的 11.2%。

菌系编号：0—12，0—18，2—28—24，8—28—7，8—28—4，3—0—1，3—0—9，3—0—16，3—0—2，3—2—28，3—0—25，3—0—20，7—28—4，7—28—1，7—28—9，7—28—12，7—28—30，7—28—31，在另—次 28℃振荡培养 7 天的试验中（ 未列入上表） 得 1 个 Nod^- 分离系，编号是 7—0—20。

2. AO 处理获得的 Nod^- 分离系　共进行两次试验，两次试验的 AO 处理尝试都分别在 0、5、10、20 微克/毫升。第一次试验采用大接种量（1.2×10^8 细胞/毫升），28℃振荡培养 2 天；第二次试验采用较小的接种量（54×10^4 细胞/毫升），28℃振荡培养 7 天。终止培养后，稀释平面培养，从单菌落得分离系，测定结瘤能力。从 150 个 AO 处理分离系中得 40 个 Nod^- 分离系。试验结果和 Nod^- 分离系编号见表 2。从每毫升处理液含 AO 20 微克、28℃振荡培养 7 天的处理中得到的 Nod^- 分离系最多，试验的 30 个分离系中有 22 个 Nod^- 分离系。但由于供试的分离系总数不多，处理多样，得不出 AO 处理能否显著提高 Nod^- 突变率的统计分析数据。

表 2　吖啶橙处理紫云根瘤菌 A106L 得到的 Nod^- 突变分离系

	AO浓度微克/毫升	接种菌量细胞/毫升	终止菌量细胞/毫升	测定分离系数	Nod^- 分离系数
试验一	0	1.2×10^8	31.8×10^8	20	3[b)]
1981.10.7-9	5	1.2×10^8	17.3×10^8	20	0
28℃，2天	10	1.2×10^8	2.9×10^8	20	0
振荡	20	1.2×10^8	0.33×10^8	20	0

续表

	AO浓度微克/毫升	接种菌量细胞/毫升	终止菌量细胞/毫升	测定分离系数	Nod^-分离系数
试验二	0	54×10^4	24.1×10^8	30	7[b)]
1981.11.20-27	5	54×10^4	12.1×10^8	30	9[a)]
28℃，7天	10	54×10^4	1.4×10^8	30	9[a)]
振荡培养	20	54×10^4	$0.15\times.0^8$	30	22[a)]

a）AO 处理获得的 Nod^-突变分离系 40 个，编号如下：

3—5—25，3—5—13，3—5—31，3—5—10，3—5—1，3—5—33，3—5—22，3—5—32，3—5—10—4，3—20—8，3—20—7，3—20—24，3—20—23，3—20—19，3—20—20，3—20—11，3—20—35，3—20—36，3—20—14，3—20—30，3—20—29，3—20—9，3—20—31，3—20—18，3—20—17，3—20—25，3—20—26，3—20—5，3—20—21，3—20—22，3—20—33

在另一次于 28℃，AO 为 5 微克/毫升，振荡培养天的实验中，得 3 个 Nod^-分离系（未列入上表中），编号是：7—5—24，7—5—17，7—5—13。

b）对照处理得到的 Nod^-突变分离系已列入自发突变范围内，见表 1。

3. 热处理获得的 Nod^-分离系　共进行了五次热处理试验：第 1 次，37℃保温培养 2 天；第 2 次，35℃或 37℃保温培养 7 天；第 3 次，37℃保温培养 2 天；第 4 次，35℃或 37℃保温培养 8 天；第 5 次，35℃保温培养 7 天。以上试验均为大接种量，静止培养，每天定时摇动 3 次。终止培养后，稀释平面分离，共得 165 个分离系，其中有 30 个 Nod^-突变分离系。在一次 35℃保温培养 8 天的热处理试验中，25 个供试分离系内有 14 个 Nod^-突变分离系。试验数据和 Nod^-突变分离系编号见表 3。同样由于分离系总数少，处理方法多样，得不出热处理显著提高 Nod^-突变率的统计分析数据。

表 3　热处理紫云英根瘤菌 A106L 得到的 Nod^-突变分离系

试验日期	处理温度和时间[b)]	接种苗量细胞/毫升	终止菌量细胞/毫升	测定分离系数	Nod^-突变分离系[c)]
1981.10.7-9	37℃，2天	1.2×10^8	1.3×10^8	20	2
1981.10.7-14	35℃，7天	1.2×10^8	5.8×10^8	20	2
1981.10.7-14	37℃，7天	1.2×10^8	0.38×10^8	20	1
1981.11.12-14	37℃，2天	0.56×10^8	0.16×10^8	25	7
1981.11.12-20	35℃，8天	0.56×10^8	0.24×10^8	25	2
1981.11.12-20	37℃，8天	0.56×10^8	0.0068×10^8	25	14
1981.11.20-27	35℃，7天	1.03×10^8	0.47×10^8	30	2

a）28℃保温培养（对照）的 Nod^-突变体已列入自发突变之内，见表 1，

b）各处理均为静止培养，每天定时摇动 3 次。

c）热处理得到 30 个 Nod^-突变分离系，编号如下：2—37—19，2—37 一 20，7—35—13，7—35—15，7—37—6，2—2—37—14，2—2—37—18，2—2—37—24，2—2—37—11，2—2—37—13，2—2—37 一 7，2—2—37—2，8—35 一 8，8—35—10，8—37—20，8—37—26，8—37—4，8—37—7，8—37—19，8—37—14，8—37—17，8—37—22，8—37—8，8—37—27，8—37—2，8—37—1，8—37—9，8—37—16，3—7—35—25，3—7—35—12。

4. Nod^-分离系的血清学鉴定　对得到的 Nod^-突变分离系，除了进行形态、生理常规检查，表明与出发菌株 A106L 一致以外，还进行了血清学鉴定。以 A106L 为抗原的抗血清是委托江苏省淮阳地区农科所微生物室制备的。凝集反应（欧阳谅，1980）效价是以抗原与抗体发生“+ +”的凝集作用的免疫血清稀释倍数之倒数来表示的（即凝集作用中等强度，管内液体半浑浊，管底可见明显的凝集块）。鉴定结果汇总于表 4。从表 4 可以看出，所有考查过的 Nod^-分离系的凝集反应都和“A106L”及“1—0—16”（同属 A106 菌系）两个 Nod^-菌株的相同。因此，可以认定所得 Nod^-分离系确为 A106L 的突变体。

从 93 个 Nod^-突变体中随机抽取 20 个突变体，它们是 0—12，0—5，0—18，2—37—1 9，2 —3 7 —2 0，7 —3 5—1 3，7—3 5—1 5，7 —37 —6，7 —5 —24，7—5—17，7—5—13，2 —2 —28 —24，2—2—37—14，2—2—37—1 8，2—2 —37—24，2—2—37—13，2—2—37—7，2—2—37—2，8—28—7，2—2—37—11，作为供试材料。

表4 紫云英根瘤菌 A106 L 的 Nod⁻突变分离系对 A106 L 抗血清的凝集效价 [a]

凝集原b	凝集效价
A106 L（出发菌株，Nod^+）	6400
1—0—16（A 106L的后代，Nod^+）	6400
41 个Nod^-分离系	6400
47 个Nod^-分离系	3200
另有5个Nod^-分离系	未测定

a） 凝集效价以抗原与抗体发生“+ +”的凝集作用（即凝集作用中等强度，管内液体半浑浊，管底可见明显的凝集块）的免疫血清稀释倍效的倒数来表示。

b） 共有 93 个 Nod^-分离系，其中有 5 个因故未鉴定；有 41 个的凝集效价是 6400；有 47 个的凝集效价是 3200，因此，血清学反应证明，鉴定过的 Nod^-突变体都是出发菌株 A106 L 的后代。

5. 阻碍 Nod^-突变体侵染过程发展的关键环节 采用 SA 斜面试管法，Fahraeus 法和 SA 平面载玻片法培养幼苗，接种根瘤菌，在显微镜下观察 Nod^-和 Nod^-菌株的早期侵染过程。用乳酸石碳酸棉蓝染色检查有无侵入线形成。观察结果如下：

1）观察了 60 株未接种的紫云英幼苗根尖上的根毛形态（图 1A），未发现类似 Nod^+菌株开始侵染时所引起的根毛“卷曲”形象。(图 1D)。

2）观察了 20 个 Nod^-突变体接种的 200 个紫云英植株幼苗根尖上的根毛形态，这些紫云英根毛也是直的，与未接种的紫云英根毛的表现形态完全相同（图 1A，B)。

3）观察了 28 个 Nod^+分离系接种的 37 个紫云英植株幼苗的根毛形态，在随机观察的 12018 条根毛中，有 1953 条呈现“卷曲”形象（图 1C)，占根毛总数的 16.5%。

4） 观察了用 A106L（出发菌株，Nod^+）接种的 17 株紫云英幼苗的根毛形态，在计数的 1881 条根毛中，有 270 条“卷曲”根毛，占根毛总数的 14.4 %，在 270 条卷曲根毛中，有 25 条侵入线，有侵入线的根毛占卷曲根毛的 9.2 % （图 1D)。

上述观察结果表明，随机抽取的 20 个 Nod^-突变体都在同一侵染环节上遇到阻碍，即在侵染的最早期不能使根毛“卷曲”，不能侵入根毛形成侵入线。

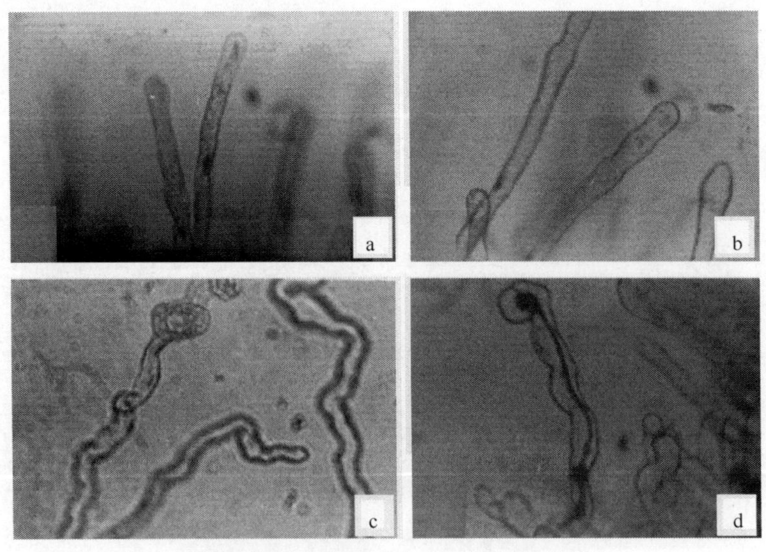

图 1 接种或未接种情况下的根毛形态

a. 未接种根瘤菌；b. 接种 Nod^-根瘤菌；c. 接种 Nod^+ 根瘤菌，根毛伸长“卷曲”
d. 接种 Nod^+ 根瘤菌. 根毛内形成侵入线

6. 与 Nod^-突变同时丧失的自然抗药性 测定了 A106L 菌系的 Nod^+菌株和 93 个 Nod^-突变体对 8 种抗生素（链霉素、卡那霉素、艮他霉素、四环素、氯霉素、利福平、萘啶酮酸、红霉素）的自然抗药性，测定浓度为每毫升培养基内含抗生素 10 微克。结果见表 5。表 5 表明，在测定过的 26 个 Nod$^+$菌株（均属 A106 菌系，其中包括 A106L）中，有 25 个抗链霉素（Str^R），27 个 Nod$^+$菌株全部抗卡那霉素（Kan^R）。与此相反，93 个 Nod$^-$突变体全部丧失了对链霉素和卡那霉素的抗性。

表 5　A106 L 菌系的 Nod$^+$菌株和 Nod$^-$突变株的自然抗药性

抗生素 / 供试菌株	链霉素浓度 10微克/毫升	卡那霉素浓度10 微克/毫升
测定过的Nod$^+$菌株数[a)]	26	27
抗药性的Nod$^+$菌株数	2	27
测定过的Nod$^-$菌株数[b)]	93	83
抗药性的Nod$^-$菌株数	0	0

a）均为 A 106L 菌系的后代，包括 A106 L。
b）A106 L 的 Nod$^-$突变株。

讨　　论

豆类植物-根瘤菌共生固氨作用的形成，既决定于寄主植物的遗传因子，也决定于根瘤菌的遗传因子。寄主植物和根瘤菌的遗传因子的遗传和变异是独立发展的。为了研究工作的方便，我们可以针对特定的根瘤菌的基因型，研究寄主植物的遗传和变异对共生固氮作用的影响（Nutman，1981），也可以针对特定寄主植物的基因型，研究根瘤菌的遗传和变异对共生固氨作用的影响（Beringeretal，1950）

能否结瘤和能否固氮是豆类植物-根瘤菌共生固氮关系最容易辨认的两个表现型，前者常用 Nod$^+$或 Nod$^-$表示，后者常用 Fix$^+$或 Fix$^-$表示。Nod$^+$和 Fix$^+$的充分表现要求一系列必需的生命活动的进行都不受阻碍。这既要求必要的外部因素，又要求必需的遗传因子的存在和表达；从本质上讲，后者是决定性的。针对特定寄主植物的基因型，根瘤菌能否侵入，能否形成形态上完整的根瘤、能否固氮，要求根瘤菌具备这一系列过程所需要的阳性（＋）遗传因子（*nod* 基因群和 *nif* 基因群），其中任何一个遗传因子从阳性变为阴性（－），就会成为共生固氮作用的不可逾越的障碍。整个过程包括：①寄主植物根际对根瘤菌生活和繁殖的影响（rhizospheric effect）；②互相辨认（Rec$^+$或 Rec$^-$）；③根毛，“卷曲”（Cur$^+$或 Cur$^-$）；侵入线形成与延伸（penetration of infection thread，Thr$^+$或 Thr$^-$），⑤根瘤组织的增殖和分化（prolification and differentiation of nodule structure，Pro$^+$或 Pro$^-$），⑥根瘤菌侵入细胞内寄生（intracellular infection，Intr$^+$或 Intr$^-$）；⑦类菌体形成（formation of bacteroids，Roid$^+$或 Roid$^-$）；⑧豆血红蛋白的形成（formation of Leghaemoglobin，Lb$^+$或 Lb$^-$）；以及全部 *nif* 基因所表达的固氮酶的形成和功能的发挥（陈华癸，1979；Rolfe et al，1981；Roberts and brill，1981）。

紫云英根瘤菌自发丧失结瘤能力的频率很高。从本文报告的试验结果来看，一个出发菌株（A106L）的 170 个单菌落分离系中出现 19 个自发的 Nod$^-$突变株，占总数的 11.2%。这和 Higashi（1967）早先报道的三叶草和菜豆根瘤菌的不结瘤自然突变率 12%接近。AO 处理和热处理的效应，由于试验分离系不多，处理条件和方法各异，未得到统计分析方面的显著效果。

最近的研究表明，根瘤菌细胞内含有多个大小不一的质粒（plasmids），质粒携带一些基因（Beringer et al.，1980；Ruvkun and Ausubel，1981；Nuti，1981，Denarie et al.，1981）。Beringer 等（1980）总结现有知识，提出根瘤菌质粒具有的遗传功能有：①中等细菌素的产生；②小细菌素的抑止；③结瘤能力；④寄主专一性；⑤根瘤功能；⑥固氮酶；⑦细胞壁多糖；⑧色素产生；⑨转移功能；⑩不相容性。从本文报道的试验结果看，结瘤能力和对链霉素与卡那霉素的自然抗性紧密相连着。那么根瘤菌质粒的遗传功能还应当加上一项，即对链霉素与卡那霉素的自然抗性。当然这项功能又是和转移功能联系着的，R 质粒通常同时携带着抗药性因子和致育因子。

如果确实是这样；那么，质粒上携带的结瘤能力遗传因子是包括从辨认（rec）到形成完整根瘤的全部遗传因子呢，还是只有其中的各个别因子呢？由于质粒的丢失或部分缺失，有关因子的丧失必将阻碍根瘤的最后形成。随机抽取的 20 个 Nod^-突变体都表现为 Cur^-、Thr^-，没有一个是 Cur^+或 Thr^+的，指出了两种可能性：①质粒上携带全部结瘤基因群（nod 基因群），质粒的丢失，首先表现出 Cur^-、Thr^-（Rec 未测定），其他后继因子无从表达，②质粒上只携带结瘤基因群的先行因子，"卷曲"因子、侵入线伸入因子，它们的丢失，导致结瘤过程不可能发生。目前我们对于 nod 基因群的知识实在是很贫乏的。

参考文献

[1]曹燕珍，李阜棣. 1979. 紫云英根瘤菌的诱变，微生物学通报，6（4）：7-9

[2]陈华癸，王子芳，曹燕珍，李阜棣. 1979. 在三个水平上的生物固氮研究。中国微生物学会学术年会，莫干山

[3]李阜棣，李燕珍，周启，周俊初. 1979. 紫云英根瘤菌接种剂的应用，中国生物固氮座谈会，北京

[4]欧阳谅. 1980. 微生物学实验法，148-155，南昌：江西人民出版社

[5]Beriner J.E.，Brewin N.J. and Johnston A.W.B.，（1980） The Genetic Analysis of Rhizobium in Relation to Symbiotic Nitrogen Fixation. Heredity 45：161-186

[6]Denarie J.，Rosenberg C.，Boistard P.，Truchet G. and Casse-Delbart F.，（1981） Plasmid Control of Symbiotic Properties in *Rhizobium meliloti*，Current Perspectives in Nitrogen Fixation 137-141 Australian Academy of Science，Canberra

[7]Fahraeus，G.（1957） The Infection of Clover Root Hairs by Nodule Bacteria Studied by a Simple Glass Slide Technique，J. Gen. Microbiol. 16：374-381

[8]Higashi，S.（1967） Transfer of Clover Infectivity of *Rhizobium trifolii* to *R. phaseoli* as Mediated by an Episomic Factor，J. Gen. Microbiol. 13：391-403

[9]Li Fudi （1981） The Production of Lagume Inoculants in China，In Biological Nitrogen Fixation Technology for Tropical Agriculture，Ed. P.H.Graham and S.C.Harrris，Publication CIAT，Cali，Columbia，U.S.A.

[10]Nuti M.P.，Ledeboer A.M.，Lepidi A.A and Schilperoort R.A. （1977） Large Plasmids in different *Rhizobium spp.*，J. Gan，Microbiol. 100：241-248

[11]Nutman P.S. （1981） Hereditary Host Factors Affecting Nodulation and Nitrogen Fixation，Current Perspectives in Nitrogen Fixation，194-204，Australian Academy of Science，Canberra

[12]Roberts G. P. and Brill W.J. （1981） Genetics and Regulation of Nitrogen Fixation，Ann. Rev. Microbiol.，35：207-235

[13]Rolfe B.G.，Djerdjevic M.，Scott K.F.，Hunghes J.E.，Badenoch Jones J.，Gresshoff P.M.，Cen Y.，Dudman W.F.，Zurkowshi W. and Shine J.，（1981.） Analysis of the Nodule forming ability of Fast-growing Rhizobium Strains. Current Perspectives in Nitrogen Fixation，142-14，Australian Academy of Science，Canbrra

[14]Ruvkun G.B. and Ausubel F.M（1981）Physical Mapping of Symbiotic Nirogen Fixation Genes in *Rhizobium meliloti*，Current Perspectives in Nitrogen Fixation，137-141，Australian Academy of Science，Canberra

易位子 Tn5 引入紫云英根瘤菌的研究*

李阜棣　曹燕珍　黎若娅　陈华癸

（华中农学院生物固氮研究室）

前　　言

易位子（transposon） 是可转移的遗传单位，它可以插入细菌染色体或质粒的不同位点；借助载体（一般是可转移性质粒）通过细菌杂交可被转入受体细胞。如果载体是自杀性质粒，则易位子进入受体细胞后可随机插入染色体或质粒上。被插入位点的基因受到损伤，不能表达功能，即细菌发生突变。与物理和化学等诱变因素不同的是，它只引起一个基因突变；而且它本身带有抗药性标记，便于选择。所以易位子是很有效的分子遗传学研究手段。

易位子 Tn5 具有抗卡那霉素标记。它进行根瘤菌分子遗传学的研究，七十年代后期以来取得了很大进展。发现了一些快生型根瘤菌的共生基因存在于质粒上，目前正在进行结瘤基因和固氮基因的结构分析。本文是我室根瘤菌遗传学研究工作的一部分，目的是用 Tn5 来获得大量紫云英根瘤菌突变株，作为遗传学研究的材料。

材料和方法

菌株：本研究所用菌株列于表1中。

表 1　细菌菌株

菌株	性状	来源
埃氏大肠杆菌给体		
JB1830	Pro met，带 pJB4JI	Beringer 等（1978）
RL29	thr　leu　thi，带 pR64：：Tn5	李和 Beringer（1981）
RL30	thr　leu　thi，带 pR136：：Tn5	李和 Beringer（1981）
RL32	thr　leu　thi，带 pR199：：Tn5	李和 Beringer（1981）
紫云英根瘤菌受体		
Ra6	$Nod^+ Nal^r sm^r$	本研究室
Ra18	$Nod^+ Nal^r$	本研究室

培养基：用修改的 Vidaver 氏（1967）培养基（NBY）培养大肠杆菌，成分如下（克/立升）：牛肉膏 3.0；蛋白胨 5.0；酵母膏 2.0；K_2HPO_4 0.5；葡萄糖 5.0；$MgSO_4 \cdot 7H_2O$ 0.25。葡萄糖 （10%，W/V）和 $MgSO_4 \cdot 7H_2O$（1M）分别灭菌后加入。

用 TY 完全培养基 （Beringer，1974） 培养根瘤菌。Y 基础培养基 （ Beringer，1974） 添加下列成分作为选择性培养基 （Fy）：甘露醇 10.0 克；KNO_3 1.0 克；生物素 0.75 毫克；硫胺素 0.75 毫克；DL 泛酸 0.75 毫克。卡那霉素（Km）浓度为 50 微克/毫升，萘啶酮酸 （Nad）浓度为 20 微克/毫升。上述各成分先配制成储备液，分别灭菌，倒平板前按量加到基础培养基中。

*原载于《华中农学院学报》，3（2）：52~54，1984.

杂交方法：给体菌株接种在 NBY 中于 37℃静止培养过夜。第二天上午用温 NBY（37℃）稀释 5 倍，置 37℃再培养 2 小时。受体菌株接种在 TY 斜面上于 28℃培养 2 天，用时每管加入 5 毫升生理盐水（或无菌水）。杂交时每对组合的给体和受体各取 0.5 毫升菌悬液混合，收集在 0.45 微米孔径滤膜（Millipore 公司出产）上。然后将含菌滤膜平放在 TY 平板上于 28℃培养 20 小时。把培养好的滤膜转移到 10 毫升 VS 液体培养基（Vincent，1970）中振荡并稀释。将不同稀释度的菌悬液分别涂抹在 FY 和 TY 平板上。3 天后（或待菌落出现时）进行检查，统计菌落数后，挑起所需菌落移入斜面上培养。

结果和讨论

细菌杂交的抗卡那霉素标记的转移频率列于表 2 中。

表 2 转移频率

埃及大肠杆菌给体	紫云英根瘤菌的抗卡那霉素菌落频率	
	Ra6	Ra18
JB1830	1.3×10^{-6}	3×10^{-8}
RL29	4.3×10^{-7}	—
RL30	5.3×10^{-7}	1×10^{-7}
RL32	5.6×10^{-7}	4.4×10^{-7}

受体菌株的自发突变频率低于 10^{-9}。因此表 2 说明易位子 Tn5 确实通过杂交从埃氏大肠杆菌转移到了紫云英根瘤菌中。获得了两类转移杂交体，即 $Nod^{+}\ km^{r}$ 和 $Nod^{-}\ km^{r}$。它们正在用于进行两类实验。用紫云-英根瘤菌的结瘤缺陷突变株（$Nod^{-}km^{r}$）来分析形成共生关系的步骤。Vincent（1980）描述了 11 个步骤，Rolfe 等（1981）在研究白三叶草固氮根瘤的形成时又增加了两个步骤。为了对根瘤菌进行改造，必须确切地了解根瘤菌的那些性状对于结瘤和固氮是最主要的。易位子诱变对于遗传分析和菌株改建都是很方便有用的技术。它也能应用于研究菌肥生产和使用中遇到的许多问题，例如结瘤能力丧失的原因，这是在实验室条件下保存根瘤菌纯培养时常遇到的现象，尤其是在反复传代的情况下（李等，1979）。我们已获得了稳定表型的 $Nod^{+}km^{r}$ 突变株。由于这一特性，可以很容易地在培养基上测定结瘤能力的丧失，而不必要进行大量费时的植物测定。这两类实验目前都在进行中。

参 考 文 献

[1]Beringer，J.E.，1974，J. Gen. Microbiol. 84：188-198.

[2]Beringer，J.E. et al.，1978，Nature，276：633-634.

[3]Li，F. et al.，1979，Chinese-Australian Symposium on Biological Nitrogen Fixation，Beijing.

[4]Li，F. and Beringer，J.E.，1981，Rothamsted Report for 1981，part I. 215-216.

[5]Rolfe，B.G. et al.，1981，In Current Perspectives in Nitrogen Fixation，142-145.

[6]Vidaver，A.K.，1967，Appl. Microbiol.15：1523-1524.

[7]Vincent，J.M.，1970，A Manual for the practical Study of Root-Nodule Bacteria.

[8]Vincent，J.M.，1980，In Nitrogen Fixation，Vol. II，103-129.

紫云英根瘤菌有效根瘤形成的初步遗传分析*

郑 玲 陈华癸 李阜棣

（华中农学院土化系生物固氮研究室）

前言

根瘤菌和豆科植物的共生关系是非常复杂的，涉及共生二者的一系列双边关系的发展，就根瘤菌方面说，从根毛感染到固氮作用的发挥可能包括许多基因。要有的放矢地对根瘤菌进行遗传改造，首先就必须确切地知道根瘤菌的结瘤基因（*nod*）和固氮基因（*fix*）的结构及其功能。而研究*nod*和*fix*基因的前提条件是筛选大量的共生缺陷突变株。过去是通过物理和化学方法如紫外线、亚硝基胍来获得突变株。然而，两个主要的问题妨碍了共生的缺陷基因的遗传图绘制和操作，第一，缺陷的表型可能是多点突变的结果；第二，唯一可识别的表型是不能形成具有充分功能的根瘤，因为由于诱变成活的每一菌落必须在植物上检测，以鉴定其共生表型，这是一件费时费力的工作。长期以来，证实共生缺陷的表型就成了遗传研究的主要障碍。

携带抗药性基因的易位子（transposon）对于克服这两个问题是理想的诱变剂。易位子是一种能在细胞染色体或质粒的不同基因间转移的DNA序列，它有三大主要特点：在每个细胞中只发生一次插入，如果插入是在一个基因内则只是被插入的基因失活，避免了多重突变，所以它是研究基因功能的有效手段；易位子本身带抗性标记，因而有利于在平皿上选择突变体，大大减少了费时的植物试验的次数；基因突变的位置就是易位子插入的位置，便于检定受阻基因的DNA序列。易位子Tn5是根瘤菌遗传研究上应用较多的一种。70年代后期以来，正是由于易位子技术的应用，使根瘤菌遗传学研究取得了很大进展，最大突破之一是证实了许多快生型根瘤菌的共生基因定位在质粒上。

本项研究工作目的是应用易位子Tn5诱变紫云英根瘤菌（*Rhizobium astragali*），筛选共生缺陷菌株。通过研究它们与寄主植物的形态学关系，进行结瘤作用的遗传分析。

材料和方法

一、菌　株

供体和受体菌株及其有关性状如下。

菌抹	性状	来　源
供体菌株：*E.coli* RL30	thr-leu-thi-Km^r （pR1363：：Tn5）	李阜棣（1981）
受体菌株：*R astragali* Ra6-3	$Nod^+Fix^+Nal^r$	本研究室（1982）

二、培 养 基

*E. coli*培养基：BPY[8] *R. astragali*培养基：TY[8] 、FY[8]、VS[1]。植物培养基：无氮植物培养基[1]，Fahracus载玻片培养基。

*原载于《华中农学院学报》，4（2）：36~40，1985.

三、抗 生 素

卡那霉素（Km）和萘啶酮酸（Nal），在培养基中的终浓度为50μg/ml（即50r）。

四、易位子Tn5诱变技术[3]

供体菌株RL30在BPY中37℃静止培养过夜。次晨用预热的BPY稀释5倍，置37℃再培养4小时。受体菌株Ra6-3在TY斜面上于28℃培养2天（对数生长期），加入5毫升无菌水制成菌悬液。混合供体和受体菌悬液（细胞浓度5：1），混合后收集在0.45微米孔径滤膜上。将滤膜平放在TY平皿上于28℃培养20小时。将滤膜转移到5毫升VS液体培养基中，充分振荡后进行稀释，取0.1毫升稀释液涂抹在添加50r Km和50r Nal的FY平皿上，长出的菌落可能为Tn5插入引起的杂交子。

五、植物结瘤试验

采用琼脂培养和砂培两种方法，筛选Tn5插入引起的共生缺陷菌株。

六、Fahraeus载玻片法研究Nod^-菌株与寄主根毛的形态学

准备内置一块载玻片的培养皿，将已融化的Fahraeus培养基倒入此培养皿内，使覆盖在载玻片上，琼脂厚度不超过1毫米，然后用无菌镊子轻轻地把幼苗胚根插入琼脂下面，根尖朝下，置灯光室中培养2天，然后接种菌悬液，继续放回灯光室培养。培养一段时间后从培养皿内取出载玻片，置光学显微镜下观察根毛的形态变化，此载玻片仍返回灯光室培养，以后连续观察。

结果

一、分离易位子描入的Nod^-和Nod^+Fix^-共生缺陷菌株

经过15次杂交试验，共获得了1341株Km^r的杂交子。易位子Tn5插入频率为1.4×10^{-8}将这些杂交子通过紫云英植物筛选共生缺陷菌株。经反复植物筛选均不能结瘤的菌株初定为Nod^-；通过肉眼观察根瘤外形、颜色及现瘤时间的早晚，初步为Nod^+Fix^-菌株。Nod^+Fix^-菌株。Nod^+Fix^-菌株形成的根瘤体积小，根瘤颜色白，现瘤时间比Nod^+Fix^+的晚15-20天。这样初筛出3株Nod^-和6株Nod^+Fix^-共生缺陷菌株。编号为Ra6-3-936、Ra6-3-156、Ra6-3-152、Ra6-3-1032、Ra6-3-842、Ra6-3-430、Ra6-3-391、Ra6-3-335。易位子Tn5诱变*R. astragali*菌株的共生缺陷株频率为0.67%。

初筛的3株Nod^-共生缺陷菌株又经3次琼脂试管和3次砂培试管复证，每次每菌株重复10管，它们都不能使植物结瘤。肉眼判断的6株Nod^+Fix^-共生缺陷菌株进行了两次气相色谱测定，均无乙炔还原反应。

通过以下四个方面的考察，确证获得的9株共生缺陷菌株是亲本菌株Ra6-3的后代。

1. 血清学鉴定，以亲本受体菌株Ra6-3的免疫血清为抗体，以9株共生缺陷菌株为抗原，采取试管凝集和荧光抗体标记两种测定方法，同时设亲本受体菌株Ra6-3抗原对照。实验表明它们与亲本受体菌株对照一样，其试管凝集效价达2560倍。

2. 培养特征观察表明，它们与亲本受体菌株一样，在碳水化合物酵母汁培养基上生长的孤立菌落为圆形，边缘整齐，无色，具光泽，产黏液。

3. 形态和染色特征观察表明，它们与亲本受体菌株一样，小短杆菌，革兰氏染色负反应，细胞内含有折光性的聚β-羟基丁酸颗粒，使细胞染色不均匀。

4. 抗性测定表明，它们在含50r Km和50r Nal的根瘤菌培养基上生长好，而亲本受体菌株不能生长。

综合上述培养特征，形态特征，抗性和血清学鉴定结果，这9株共生缺陷菌株虽然丧失了在紫云英根部结瘤或者固氮能力，但仍保留着紫云英根瘤菌Ra6-3的形态学、生理学和免疫学特征，它们不是杂菌，而是受体菌株Ra6-3的共生缺陷后代。

二、*Nod*⁻共生缺陷菌株与寄主根毛的形态学研究

正常情况下，未接种的根毛是直的（图版-1），虽然可以偶尔看见2%的根毛弯曲；但这可能是根毛在载玻片上生长发育过程中遇到阻碍或者根毛间相互接触引起的物理变形。

接种亲本菌株Ra6-3的根毛，8小时内出现根毛弯曲反应（图版-2），当培养时间延长到48小时，根毛弯曲数目增多。统计了接种时未出现根毛区域内的根毛弯曲比例，所调查的5840根根毛中，变形根毛达2680根，根毛变形比例为45.89%。

三株*Nod*⁻共生缺陷菌株经8次载玻片试验，从接菌后1~4天内观察根毛反应，它们均不能引起根毛变形，根毛直（图版-3），与未接菌的根毛反应几无差异。

三、Nod^+Fix^+株与Nod^+Fix^-共生缺陷菌株形成根瘤的内部结构比较

紫云英植物的根瘤属无限根瘤，外形是长枣形，分生组织在顶端皮层内。成熟的根瘤具有下列主要结构：根瘤皮层；含菌组织；分生组织；维管束；胞间空隙。其中含菌组织（即含类菌体细胞组织）是固氮作用的部位，类菌体在其中还原分子态氮。类菌体是根瘤菌生活在根瘤中的形态，紫云英根瘤内的类菌体呈犁状，棒杆状，幼小根瘤内极少数分枝杆状。

选出其中一株Nod^+Fix^-共生缺陷菌株Ra6-3-1032的25天瘤龄的根瘤和亲本受体菌株Ra6-3的25天瘤龄的根瘤，经石蜡包埋切片后进行扫描电镜观察。从电镜照片看出，Ra6-3的根瘤体积大（图版-4），含菌细胞大，充满类菌体，类菌体呈犁状或棒杆状（图版-5）；相比之下，Ra6-3-1032的根瘤体积小（图版-6），其中部组织中空，有的甚至破裂，没有看见类菌体形态，只看见极少数小杆菌(图版-7)。Ra6-3-1032形成根瘤的压碎汁观察，也没有看见类菌体形态。

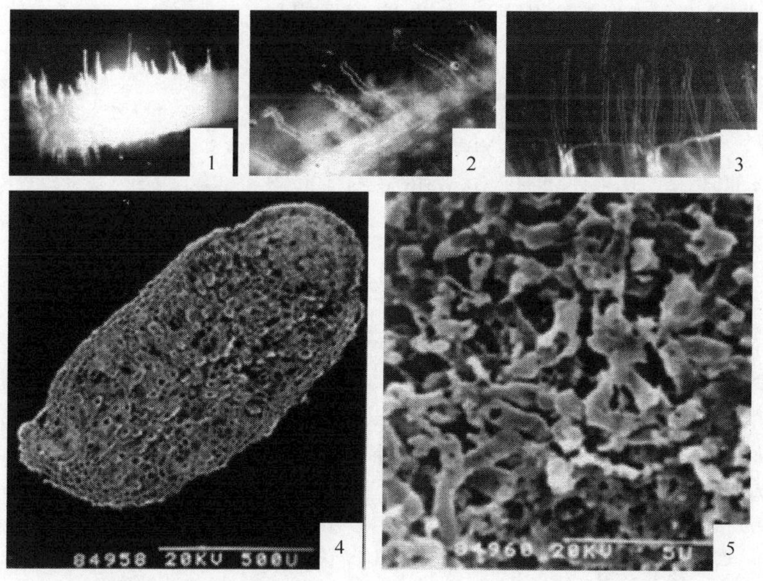

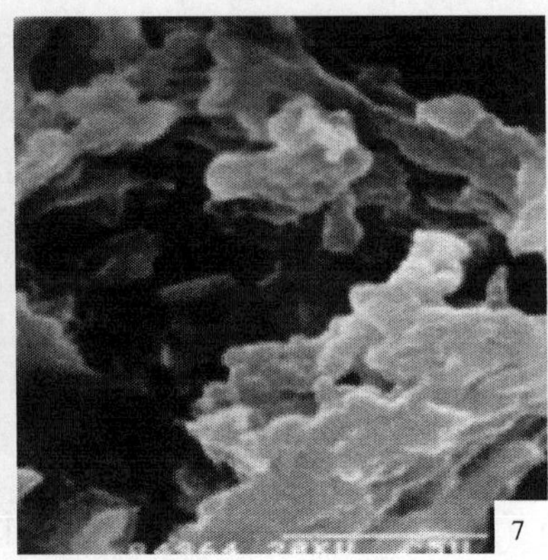

图版说明

1. 未接种根瘤菌的根毛是直的；2. 接种出发菌株 Ra6-3 的根毛发生弯曲；3. 接种 Nod$^-$共生缺陷菌株根毛是直的；4. Ra6-3 菌株形成的根瘤纵切面（×80）；5. 含菌细胞的类菌体（×7000）；6. Nod$^+$ Fix$^-$菌株 Ra6-3-1032 形成无效根瘤的纵切面；7. 无效根瘤含菌细胞的类菌体（×7000）。

讨论

1980年，Vincent[9]根据形态变化，将结瘤作用的全过程分成11个步骤：根瘤菌在根面繁殖；吸附根毛；根毛分枝；根毛卷曲；根毛形成侵染线；多倍体分生组织发育，根瘤发育和分化；根瘤菌从侵染线释放到细胞内；类菌体充分发育；固氮酶固氮；生理和生化功能互补根瘤功能持续。Rolfe （1980年）在此基础上又添加了两步：侵染线分枝和根瘤开始。这些不同的步骤是由根瘤菌基因控制，还是植物基因控制？每一基因控制一个步骤，还是一个基因控制几个步骤？全世界范围内正在研究这些问题，从目前研究的进展看，只报道了一些初步结果，都不能作肯定性结论。

本实验从Tn5插入的1341株杂交体中，筛选出3株*Nod*$^-$共生缺陷菌株，它们都不能引起根毛卷曲，说明它们被阻碍在同一共生环节上。这一结果至少说明引起根毛弯曲的基因由根瘤菌基因控制。对于引起根毛弯曲和形成根瘤两种表型是否由相同基因控制的问题，这3株*Nod*$^-$菌株不引起根毛弯曲表型和不形成根瘤表型连锁发生，可能此两种表型由相同基因控制。但不可排除没有分离出能引起根毛弯曲而不能结瘤的*Nod*$^-$共生缺陷菌株。

从亲本受体菌株Ra6-3和*Nod*$^+$*Fix*$^-$的共生缺陷菌株Ra6-3-1032的根瘤电镜照片看出，在相同寄主、相同瘤龄、相同培养条件的情况下，两者根瘤内部结构差异明显，Ra6-3-1032的根瘤内看不见类菌体形态，只看见极少数小杆菌，可能根瘤菌的类菌体形成基因受到损伤。

世界范围内，对共生关系的分子遗传学实质的深入研究只是在不久前才开始的，但它是一个非常活跃，进展很快的研究领域。自从1983年Dreyfus报道了快生型田菁根瘤菌能离体固氮[8]，就充分肯定了根瘤菌具备有效性的全部基因。现在还不知道*nod*和*fix*基因的全部结构，但已证明它们具备与*K. pneumonia*的固氮酶结构基因KDH同源的DNA片段。快生型根瘤菌的*nod*和*fix*基因一般存在质粒上。

主要参考文献

[1]芬森特编，1970，根瘤菌实用研究手册，上海人民出版社，上海植物生理研究所固氮室译.

[2]郑国昌主编，1978，生物显微技术，17-18 页，人民教育出版社.

[3]Beringer，J.E.，Beynon，J. E. et al.，1978，Transfer of the drug-resistance transposon Tn5 to Rhizobium，Nature，276：633-34.

[4]Bohlool，B.B.，Schmidt. E. L. and May. S.N.，Serological techniques in microbial ecology，in Training Manual，University of Hawaii.

[5]Dart.，1977，Infection and development of leguminous nodules. In A Treatise on Dinitrogen Fixation section 3 ed. Hardy，R. W.F. and Silver. W.S.，pp367-472

[6]Dreyus，B.L.，Elmerich，C. and Dommerzues，Y.，1983，Free-living Rhizobium strain able to grow on N_2 as the sole nitrogen source，App，Enviromn. Microbiol. 45：711-713.

[7]Fahraeus，G.，1957，The infection of clover root hairs by nodule bacteria studied by a simple glass slide techniques. J. gen. Microbiol. 16：374-386.

[8]Li，F. D. et al.，1984，Introduction transposon Tn5 into *Rhizobum astragali.* in Advances in Nitrogen Fixation Research，Ed. Veeger and Newton.

[9]Rolfe，B.G.，Djordjevic，M. et al.，1980，Analysis of the nodule forming ability of fast-growing Rhizobium strains，in Current Perspective in Nitrogen Fixation，pp 142-145

紫云英根瘤菌质粒的研究Ⅰ、五个菌株大质粒的分离纯化和菌株7653质粒的内切酶酶切图谱*

张忠明　周俊初　许耀才　陈华癸　李阜棣

（华中农业大学生物固氮研究室）

范云六

（中国农业科学院分子生物学研究室）

用修改的 Hirsch 法从紫云英根瘤菌的五个菌株中分离出 1~2 个大质粒，比较研究了不同蛋白酶和菌液浓度对质粒分离效果的影响。用 CsCl-EB 密度梯度离心从菌株 7653 中提取出质粒 DNA，得到其用不同的限制性内切酶作用的酶切图谱。

近年来，共生固氮遗传学研究表明，一切快生型根瘤菌中存在着大质粒[1-5]，结瘤基因和固氮基因（至少是部分基因）位于大质粒上。三叶草根瘤菌（*R. trifolii*）、菜豆根瘤菌（*R. phaseoli*）、豌豆根瘤菌（*R. leguminosarum*）的大质粒约为 100~200Md，苜蓿根瘤菌（*R. meliloti*）的大质粒可能达 800~1000 Md[6]。本文对几株紫云英根瘤菌的质粒进行了分离研究。

材料和方法

一、菌　　株

Ra27、Ra6、Ra65 均为本室分离和保存的菌株；Sm^r52（以下简称 S_{52}）由中国科学院病毒研究所惠赠；菌株 7653 由江苏扬州地区农科所供给。以上菌株均能在紫云英上结瘤，固氮。

二、培　养　基

TY 培养基：胰蛋白胨 5 克；酵母粉 5 克；水 1000 毫升；pH7.0。

TA 培养基：按 Hirsch[7]法稍作修改：以胰蛋白胨代替胃蛋白胨。胰蛋白胨 4 克，$MgSO_4 \cdot 7H_2O$ 0.6 克，水 1000 毫升，pH7.0。

三、琼脂糖凝胶电泳

按 Hirsch[7]方法，稍作修改；菌种接种于 TY 琼脂斜面，有 28℃培养两天，挑取一环菌苔接种于 80 毫升 TA 培养液中，在 28℃振荡培养两天。4000r/min 离心 20 分钟，收集细胞，用 10 毫升 TE 缓冲液（Tris 0.05M，EDTA 0.02M，pH8.0）洗涤一次，再在同样条件下离心收集细胞。沉淀物再悬浮于 TE 缓冲液中。取 4 毫升菌悬液，加入 10% SDS（0.5 毫升），再加入 0.5 毫升（10 毫克/毫升）链霉蛋白酶 E（Pronase E），链霉蛋白酶 E 在加入前，有 37℃预处理 1 小时。混匀后在 37℃水浴中保温 1 小时左右，菌悬液转变为透明液。加入 3M NaOH 0.125 毫升（pH 调至 12.4），在室温放置 20 分钟，然后加入 2M Tris-HCl （pH7.0）0.375 毫升，混匀后再加入 5M NaCl 1.375 毫升，在冰箱（4℃）中置放 4~6 小时，取出在 4℃、17000r/min（约 23000g）离心 20 分钟。去沉淀，取上清液加入 1.6 毫升 50%的聚乙二醇-6000

*原载于《华中农业大学学报》，5（4）：326~331，1986.

（PEG）。混匀后置冰箱（4℃）过夜。次日取出，在 4℃，15000r/min 离心 15 分钟，弃去上清液，沉淀物加入 80 微升 TE 缓冲液。置冰箱中保存备用。

取 60 微升样品，加入 10 微升含有 20%的（Ficoll-400）0.05M EDTA，0.125%克溴酚蓝混合液。混匀后加到 0.7%水平琼脂糖凝胶样品槽中，在室温下，70 毫安电泳 4~6 小时。电泳缓冲液为 89mM Tris，89mM 硼酸，2.5M EDTA，pH8.3（TBB），电泳结束后，将胶取下，放入含有 0.5 微克/毫升的溴化乙锭（EB）溶液中，染色约 20 分钟，转入蒸馏水中脱色 5~10 分钟，有紫外灯下观察和照相。

四、CsCl-EB 浮力密度梯度离心

培养 2000 毫升菌液，按上述制备质粒 DNA 粗提物的方法（各试剂用量按比例放大），所得质粒 DNA 粗提物溶于 1.45 毫升 TE 缓冲液中，加入 3.3 毫升 TE 缓冲液饱和的 CsCl 和 0.5 毫升的溴化乙锭 5 毫克/毫升。转入离心管中，用超速离心机（Backman L80）20℃，60000r/min 离心 16 小时。离心结束后，将离心管小心取出，并在紫外灯下观察、照相。随后收集质粒 DNA，用饱和正丁醇脱去溴化己锭，转入透析袋中，置于 2000 毫升 TE 缓冲液（10mM Tris，1mM EDTA，pH8.0）透析两天，每八小时换一次透析液。透析后，取出样品，置冰箱保存备用。

五、限制性内切酶切裂试验

内切酶：*Hind* II、*Bgl* I、*Xba* I （中国医学科学院基础医学研究所生化室制品）；*Bgl* I、*Xba* I（中国科学院生物物理所生物化学试剂厂制品）。

酶切按常规方法进行，不同内切酶缓冲溶液按厂方要求配制。质粒 DNA 量视浓度而定，作用条件均为 37℃，1 小时，然后进行电泳，观察结果。

试验结果

一、不同蛋白酶的溶菌效果试验

以菌株 R27 为试验菌株，比较了五种酶（其浓度均为 200μg/ml）的溶菌效果和后续的质粒分离的效果。表 1、图 1 表明，链霉蛋白酶 E（pronase E）的溶菌效果及以后质粒 DNA 带和染色体 DNA 分离效果均最好，溶菌酶（lysozyme）几乎无溶菌效果，其他几种蛋白酶溶菌效果也不理想，后续的质粒分离效果比较差。因此，以后的工作均采用链霉蛋白酶 E。

表 1　五种蛋白酶的溶菌效果

酶名称	溶菌效果	电泳检测出的质粒带数
链霉蛋白酶 E（pronase E）	透亮	2
溶菌酶（lysozyme）	混浊	0
蛋白酶 K（proteinase k）	较透亮	1
蛋白酶（proteinase）	微透亮	1（微弱）
胰蛋白酶（typsinase）	较混浊	0

二、溶菌时的适宜菌液浓度

溶菌时的菌液浓度对质粒的分离效果影响很大，浓度太低（推算的 OD_{660} 低于 0.5），只能看见染色体 DNA 带，看不见质粒 DNA 带。菌液浓度太高（推算的 OD 大于 4），不能完全溶菌，得不到透亮的

细胞裂解物，而且染色体 DNA 及其他细胞碎片一起被除掉。此时，在琼脂糖凝胶电泳板上也检测不到质粒 DNA 带。为了获得正确的溶菌时的菌液浓度，我们将在 TA 中的培养物首先测定光密度，然后离心收集细胞，用 TE 缓冲液洗涤后，按光密度值计算加入适量的 TE 缓冲液，使推算的 OD 值为 1。在此浓度下容易获得透亮的裂解物，而且琼脂糖凝胶电泳检测质粒的效果也很好。

三、紫云英根瘤菌的大质粒带

分离了菌株 Ra6、Ra65、Ra27、S_{52}和 7653 的电泳质粒带[图 2]，Ra6、Ra65 和 7653 只有一条质粒带，Ra27 和 S_{52}有两条明显的质粒带，分子质量均在 200Md 左右，它们的相差电泳位置稍有差异。

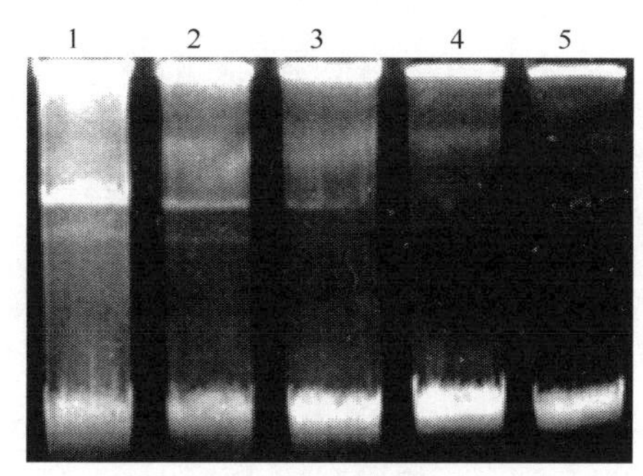

图 1　五种蛋白酶的溶菌效果

1.链霉蛋白酶 E；2.溶菌酶；3.蛋白酶 K；4.蛋白酶；5.胰蛋白酶

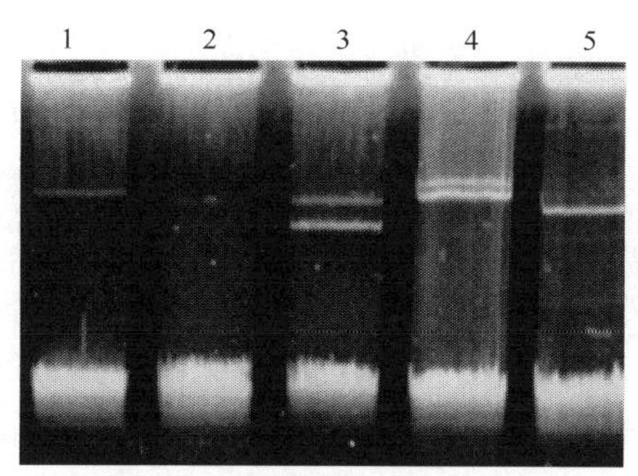

图 2　不同菌株的大质粒电泳图

1. Ra6；2. Ra65；3. Ra27；4. S_{52}；5. 7653

四、CsCl-EB 梯度密度离心提纯菌株 7653 的大质粒

获得的大质粒带见图 3。分别收集经梯度密度离心的质粒带的染色体碎片，通过萃取和透析，再进行琼脂糖凝胶电泳，得图 4。图 4 中 1，2 为染色体碎片 DNA 带，3 为空白对照，4，5 为质粒 DNA 带。4、5 中仍见有少量“染色体 DNA”碎片带，可能是在操作中部分质粒 DNA 断裂的人为结果。

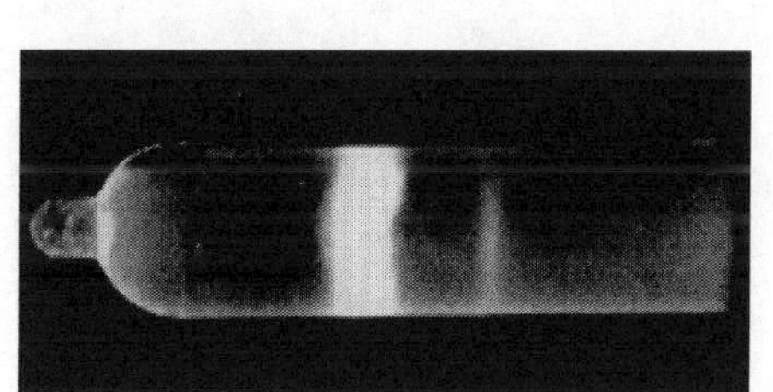

图 3　紫云英根瘤菌 7653 大质粒的 CsCl-EB 梯度密度离心结果

宽带为染色体 DNA；窄带为质粒 DNA

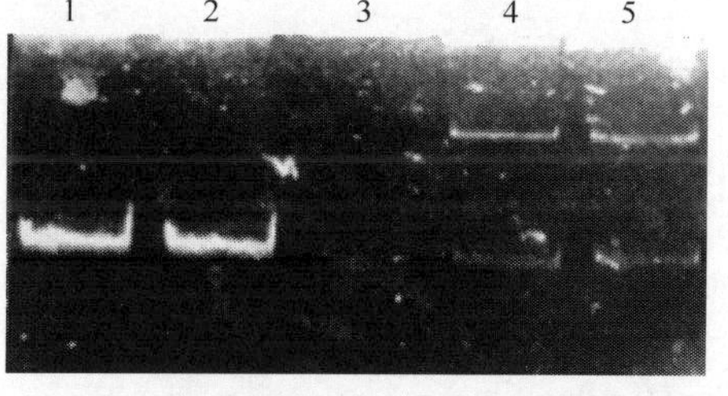

图 4　CsCl-EB 梯度密度离心管中收集不同带的琼脂糖电泳

1，2.染色体 DNA 带，4-5.质粒 DNA

五、质粒 DNA 的酶切图谱

将 CsCl-EB 梯度密度离心提纯的菌株 7653 的质料 DNA，用 *Hin*d II、*Bgl* I、*Bgl* II、*Xbo*I、*Xba* I 内切酶切裂，得各种内切酶切裂带的琼脂糖电泳图谱，见图 5。*Xbo* I 的酶切谱分为 20 条带；*Bgt* II 的酶切谱分为 13 段；*Xba* I 的酶切谱分为 5 段，其中 2、3、4 段 DNA 带几乎是紧挨着的；*Bgl* I 的酶切谱分为 10 段；*Hin*d II 的酶切谱分为 6 段，第二段量最大并与第一段挨着。

图 5 质粒 DNA 的酶切图谱

1-*Hind*III；2-*Bgl* I；3-*Xba* I；4-*Bgl* II；5-*Xho* I

讨论

紫云英根瘤菌在培养过程中产生大量荚膜和黏液，它们严重影响溶菌处理的效果。用 Hirsch 的 PA 培养基可以降低黏液的产生，但紫云英根瘤菌生长不良，菌浓度太低。改用胰蛋白胨代替 Hirsch 原配方的胃蛋白胨，细菌生长良好，黏液也不多，在 28℃摇床振荡培养（约 180rpm）48 小时，OD 值可达 0.1 左右，溶菌效果较好。

大质粒在操作过程中容易由于机械因素断裂，必须避免过度振动。溶菌过程中在三角瓶中进行，只需轻微摇动，便可将加入的试剂与细胞裂解充分混匀，并使细胞碎片，蛋白质等聚合物在 NaCl 中完全沉淀而析出，使 DNA 充分游离出来；DNA 在混匀的聚乙二醇中也能充分沉淀，并不至于破坏质粒 DNA 整体，同样在 CsCl-EB 密度梯度离心分离纯化质粒时也必须十分轻缓的操作，以防止质粒 DNA 整体的断裂。

本试验中提取的质粒带大小与三叶草、豌豆、菜豆根瘤菌中分离出的质粒的分子量大小接近（100~200Md），但没有分离出苜蓿根瘤菌那样的大分子质量的质粒（800~1000Md）[6]。根据最近根瘤菌分类的研究[8]，三叶草、豌豆、菜豆根瘤菌同属一个种（*Rhizokium leguminosarum*）；苜蓿根瘤菌属另一个种（*R. meliloti*）。根据陈文新的聚类分析研究，紫云英根瘤菌与 *R. leguminosarum* 和 *R. meliloti* 在种的级别上又属于另一个聚类群。至于试验中分离出的质粒是否携带结瘤（*nod*）或固氮（*fix*）基因，尚待进一步研究。

参 考 文 献

[1]Nuti MP，Ledeboer AM，Lepidi AA and Schilpcroort RA，（1977） Large plasmids in different Rhizobium Species. Journal of General Microbiology 100：241-248.

[2]Hirsch PR，（1979） Plasmid-determined bacteriocin production by *Rhizobium leguminosarum*. J. Gen. Microbiol. 113：219-228.

[3]Johnston AWB，（1980） Rhizobium Genetics，Current perspectives in Nitrogen Fixation. Proceedings of Fourth international Symposium on Nitrogen Fixation held in Canberra Australia，1 to 5 December.

[4]Beynon JL，Beringer JE and Johnston AWB，（1980）Plasmids and Host-range in *Rhizobium leguminosarum* and *Rhizobium phaseoli*. J. Gen. Microbiol. 120：421-429.

[5]Casse F，Boucher C，Jullot JS，Michel M and Denarif J （1979） Identification and characterization of large plasmids in *Rhizobium meliloti* using agarose gel electrophoresis. J. Gen. Microbiol. 113：220-242.

[6]Toro N and Olivares J（1985） Book of Abstracts for other Inter. Sym. On Nitrogen Fixation. 9-46.

[7]Hirsch PR，Van Montagu M，Johnston AWB，Brewin NJ and Schell J，（1980） Physical identification of Bacteriocingenic，nodulation and other Plasmids in Strains of *Rhizobium leguminosarum*. J. Gen. Microbiol. 120：403-412.

紫云英根瘤菌质粒的研究Ⅱ、紫云英根瘤菌（*Rhizobium astragali*）经高温和吖啶橙处理后结瘤和固氮突变株的产生和质粒的消除*

周俊初　张忠明　黄诚金　李阜棣　陈华癸

（华中农业大学生物固氮研究室）

用高温和吖啶橙处理紫云英根瘤菌菌株 7653^r 和 S_{52} 分别获得了表型性状为不结瘤（Nod^-）和结瘤而不固氮（Nod^+、Fix^-）突变株，与亲本菌株具有的两条大质粒带相比，上述突变株均丢失了其中的一个较大的质粒。这一结果初步表明，菌株 7653^r 的 *nod* 基因和 S_{52} 的 *nif* 或 *fix* 基因可能定位在已消除的那一个大质粒（pRa7-2 和 pRa5-2）上。

关键词　根瘤菌；紫云英根瘤菌；质粒

前言

质粒是染色体外控制生物次要性状的遗传因子。Nuti 等[1]对根瘤菌产物进行沉降分析和密度梯度离心，直接证明三叶草、豌豆、大豆和菜豆等根瘤菌中存在分子质量为 70~400Md 的大质粒。在自然条件下，质粒常可由于自发或诱发突变而缺失部分基因或完全丢失。Beynon 等[2]报导菜豆根瘤菌 1233 由于其中一个质粒自发地中间缺失而失去了在菜豆上结瘤的能力。Denarie 等[3]发现三个 Nod^-的苜蓿根瘤菌突变株均在一个大于 300Md 的大质粒上发生了缺失。Scott[4]发现三叶草根瘤菌的自发 Nod^-突变株也丢失了一个分子质量为 120Md 的大质粒。吖啶橙或高温处理已经多次被证明可以在很高的频率（10^{-1}~10^{-2}）上消除质粒。Higashi[5]用吖啶橙处理三叶草和菜豆根瘤菌，使 30%~70%的分离株失去了侵染能力。Parijskaya[6]和 Zurkowski 等[7]也报导用吖啶橙取得了较好的效果；但 Dunican 和 Cannon[8]和 Zurkowski[9]却报导说明吖啶橙处理的 Nod^-突变率很低。对于高温处理的文献报导比较一致，Zurkowski[9]用 35℃处理三叶草根瘤菌的 10 个供试菌株，分别获得 1%~75%的 Nod^-突变率。Hooykaas[10]用 37℃处理三叶草根瘤菌，也得到了 Nod^-突变体。

李仲贤[11]用吖啶橙处理紫云英根瘤菌，当接种量为 10^4/ml 时，得到 24.1%~45.5%的 Nod^-突变率。用 35℃，7 天处理的 Nod^-突变率为（8.2±1.7）%；用 37℃，2~8 天处理可使 Nod^-突变率提高到（24.8±23）%。他选取 20 个 Nod^-突变株进行的根毛感染试验表明都不能导致根毛卷曲或形成侵入线（Hac^-），但没有进一步研究高温的吖啶橙处理对质粒的影响及质粒消除与 *nod* 或 *nif* 基因的关系。宁林夫等[12]用高温处理紫云英根瘤菌，获得了消除质粒的 Nod^-突变株，并且接合转移方法，实现了 *nod* 基因的诱动转移，使 Nod^-突变株恢复为 Nod^+。

材料和方法

一、供 试 菌 株

Rhizobium astragali 7653：Nod^+，Fix^+，来自江苏扬州地区农科所。凝胶电脉检测有一个大质粒带，编号为 pRa7-1。

7653^r：本室从菌株 7653 分离得到的自发抗链霉素（Str）突变株，Nod^+，Fix^+，凝胶电脉检测表明除有一个与 7653 相同的质粒带以外，还有一个稍大的质粒带，编号为 pRa7-2。

Sm^rS52：Nod^+，Fix^+，来自中国科学院武汉病毒所，凝胶电脉检测有两条质粒带，其中分子量小

*原载于《华中农业大学学报》，6（2）：156~164，1987.

的编为 pRa5-1，大的编号为 pRa5-2（以下简写为 S_{52}）。

二、培 养 基

RSY：修改的 YMA（13），以蔗糖代替甘露醇。

TY：Tryptone （Difco）：5g，Yeast Extract （Difco）：3g，蒸馏水：1000ml，pH6.8~7.0。

RSY+Str：RSY 加链霉素使其终浓度为 500μg/ml。

三、高温处理法

选择 RSY 和 RSY+Str 两种培养基，采用平皿法和振荡法两种处理方式：

1. 平皿法 将供试菌在 RSY 中于 28℃经 48h 振荡培养后系列稀释至 10^{-8}，分别用 10^{-6}~10^{-8} 三个稀释度的菌悬液各 0.1ml 涂 RSY 平板，三次重复，于 28℃培养供活菌计数用；另取 10^{-1}~10^{-4} 四个稀释度 0.1ml 涂 RSY 和 RSY+Str 平板，各作 10 次重复，于 37℃中保温培养 7 天后，转入 28℃，待单菌落长出后取出计数并挑取代表性单菌落，移植斜面，编号保存。

2. 振荡法 将在 RSY 中生长 2 天的供试菌培养液 10ml 转入 80ml 新鲜 RSY 培养液中，在摇床（~120r/min）上于 37℃振荡培养 7 天后取出，系列稀释至 10^{-8}，取 10^{-3}~10^{-8} 稀释液各 0.1ml 涂 RSY 和 RSY+Str 平板，6 次重复，于 28℃培养 4~5 天后取出，计算活菌数并挑取单菌落，移植斜面，编号保存。

四、吖啶橙处理法

先用灭菌蒸馏水配制 500μg/ml 吖啶橙贮备液，在棕色瓶中保存，临用前，取需要量经 0.22μm 滤膜过滤后使用。供试菌转接 TY 培养液后于 28℃振荡培养 24h，同时取样作显微镜直接测数和稀释平板测数，并根据直接计数结果将培养液含菌量稀至 10^5 个每毫升。将此菌悬液 1ml 分别加入已预先配好的吖啶橙含量为 20μg/ml 及 40μg/ml 的 18ml TY 培养液中，经 28℃振荡培养 7 天后，系列稀释至 10^{-8}，取 10^0~10^{-8} 各稀释液 0.1ml 涂抹 TY 平板，三次重复，于 28℃保温培养出单菌落后，计数并挑取单菌落，编号保存。

五、质粒 DNA 的检测

采用修改的 Hirsch 碱变性-PEG 沉淀法。供试菌经链丝蛋白酶（Pronase E，Merck 公司）和 SDS 在 37℃水浴中溶菌，调 pH 至 12.4 使染色体 DNA 变性，再用 Tris-HCl 回调 pH 至中性，加入 NaCl 以沉降蛋白质和细胞碎片，于 4℃，17000r/min 条件下离心 20min，上清液再加聚乙二醇（PEG MW=6000）沉淀质粒 DNA。

琼脂糖凝胶电泳为 160V，70mA，4~5h，凝胶板经溴化乙锭染色，在紫外检测仪下观测并拍照。具体操作见文献[18]。

六、植物结瘤试验

取紫云英（*Astragatus sinicus* L.）种子经 0.1% $HgCl_2$ 表面灭菌，洗涤催芽后，用 Fahraeus 植物无氮营养液[13]，分别以半固体琼脂斜面或砂土管栽培，在光照室进行人工光照培养，温度 15~25℃，光照：8~10h/d，每供试菌 5 次重复，一月后检查结瘤和生长情况。

七、根瘤固氮酶活性测定

取等重量的新鲜根瘤于 18×180mm 大试管中，注入 10%新生乙炔，于 28℃保温 3h 后，抽取 1ml 气体，在 100 型气相色谱仪（上海分析仪器厂）上测定乙烯产生量。

八、供试菌株的抗血清和荧光抗体制备及检测

供试菌抗原抗体的制备，抗体 IgG 的提纯及荧光标记和荧光抗体检测的具体方法参见文献[13]。

试验结果

一、稀释分离对菌株 7653 结瘤和固氮能力的影响

为了进一步纯化供试菌株，对 7653 进行了稀释分离。分别涂抹 RSY 和 RSY+Str 两种平板，经 28℃培养，从 RSY 上挑取 10 个单菌落，分别编号为 7653-1，2，3，4，5，6，7，8，9，10；从 RSY+Str 平板上仅挑得一个单菌落，编号为 7653^r。

洋菜和砂培结瘤试验表明：除 7653-7，8，9，10 和 7653^r 仍保持其结瘤和固氮性能外，765 3-1，2，3，4，5，6，已失去在紫云英上结瘤的能力（图 1），其 Nod^-的百分率为 54.5%。

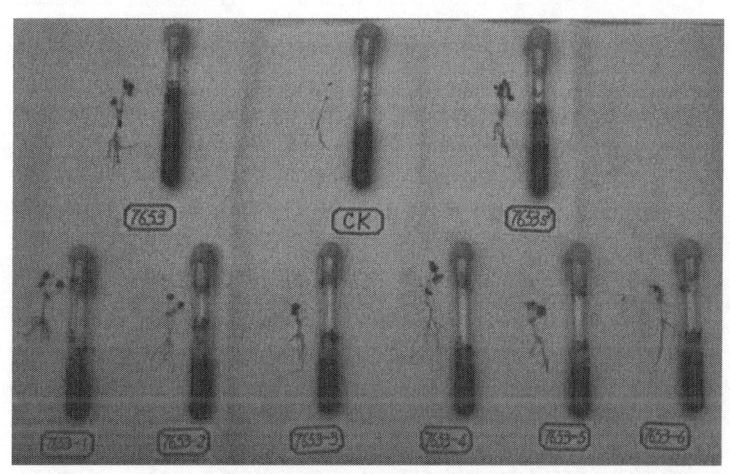

图 1　稀释分离对 7653 结瘤和固氮能力的影响（7653^r_s 即本文中的 7653^r）

二、7653 及其稀释分离菌株质粒带的检测

用修改的 Hirsch 法对以上分离菌株的质粒检测结果见图 2。除 7653^r 有 2 条质粒带（见图 5）以外，其余供试菌株均只有一条 pRa7-1 质粒带，考虑到 Nod^-的 7653-1，2，3，4，5，6 等 6 个菌株仍然具有与 7653（Nod^+，Fix^+）相同的质粒带，可以推论 pRa7-1 不是带有结瘤或固氮基因的共生质粒，因而决定选择 7653^r 作为质粒消除试验的供试菌株之一。

三、高温处理对紫云英根瘤菌生长、结瘤和固氮能力的影响

高温对 7653^r 和 S_{52} 两个出发菌株，用两种不同的培养基，采用两种处理方式取得的全部试验结果汇总于表 1 之中。

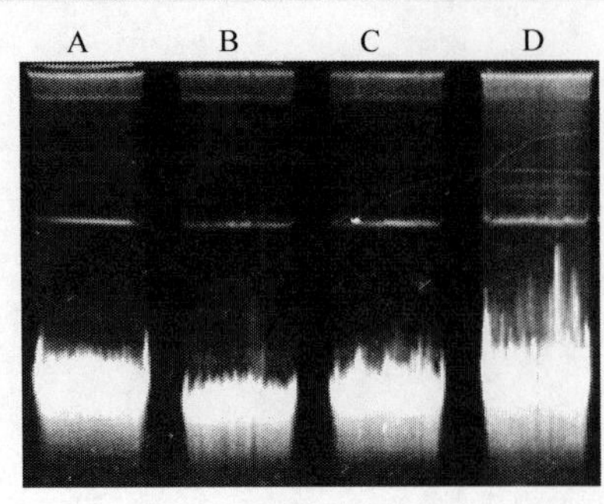

图 2 7653 及其稀释分离菌株的质粒图谱

（7653r 的另一大质粒隐若可见，碱变性抽提法）A-7653-1；B-7653s；C-7653；D-7653-r

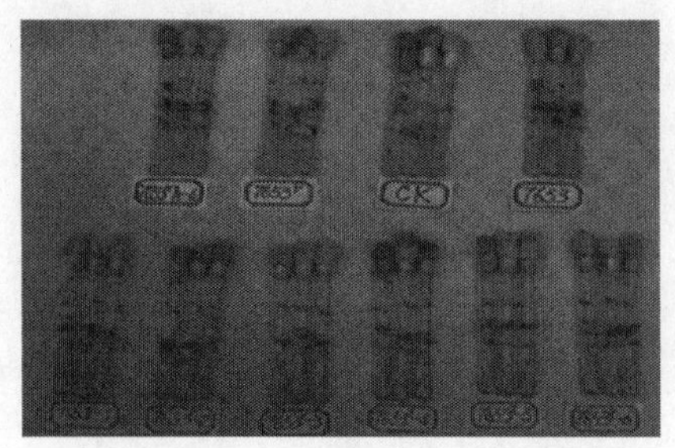

图 3 高温对 7653r 结瘤和固氮能力的影响

图中的 7653r-1，2……6 即本文中的 HRa7-1……6，为 Nod⁻；7653，7653r 和 7653rh-6 为 Nod+ Fix+

由表 1 可见高温对供试的两菌株的作用有明显的差异：从 7653r 的四种不同处理中选择了 40 个分离株，结瘤试验证明，其中有 8 株为不结瘤（Nod⁻）突变株，其系统编号为 HRa7-1，2，3，4，5，6，7，8。用同样的方法处理菌株 S_{52}，尽管高温处理的存后率已降至 10^{-3} 以下，仍然未能得到 Nod⁻突变株；唯一有趣的是从高温平皿法在 RSY+Str 平板上得到的全部分离株，虽然仍能结瘤，却失去了正常的固氮能力（Nod+，Fix⁻），其系统编号为 HRa5-1，2，3，4，5，6，7，8，9，10。

表 1 高温对紫云英根瘤菌生长、结瘤和固氮能力的影响

供试菌株	处理方式	培养基	菌数（个/毫升）		存活率	结瘤与固氮能力		
			起始	终止		供试菌林数	Nod⁻数	Fix⁻数*
7653r	平皿法	RSY	59.2×10^8	$<10\times10^5$	$<10^{-3}$	10	0	0
		RSY+Str	59.2×10^8	$<10\times10^5$	$<10^{-3}$	10	1	0
7653r	振荡法	RSY	2.6×10^8	1.5×10^8	5.77×10^{-1}	10	4	0
		RSY+Str	2.6×10^8	1.1×10^8	4.23×10^{-1}	10	3	0
S_{52}	平皿法	RSY	199×10^8	$<3\times10^5$	$<10^{-5}$	3	0	0
		RSY+Str	199×10^8	$<10\times10^5$	$<10^{-4}$	10	0	10
S_{52}	振荡法	RSY	3.1×10^8	9.6×10^5	3.1×10^{-4}	10	0	0
		RSY+Str	3.1×10^8	2.8×10^4	9×10^{-5}	10	0	0

*为除 Nod⁻以外的不固氮菌株数

高温对菌株 7653r 结瘤和固氮能力的影响见图 3，对菌株 S_{52} 结瘤和固氮能力的影响见图 4。对 S_{52} 和 HRa5-1~10 等 10 个分离株的根瘤乙炔还原活性的比较测定结果见表 2。

表 2 高温平皿法处理对紫云英根瘤菌 S_{52} 结瘤和固氮能力的影响

供试菌株和处理	CK	S_{52}	HRa5-									
			1	2	3	4	5	6	7	8	9	10
根瘤数/株	0	3.8	0.6	3.0	3.6	2.2	2.6	3.6	3.0	5.4	3.2	1.8
乙烯峰高*（mm）	6.0	83.0	9.0	13.6	9.0	9.0	5.0	6.0	6.0	8.0	8.0	6.0
质粒带 pRa5-2		+	-	-	-	-	-	-	-	**未测	**未测	**未测

*选取中等大小的根瘤 10 个，于 18×180mm 大试管中，注入 10%乙炔，在 28℃保温 3 小时后，抽取 1ml 气体用 100 型气相色谱仪测定，固氮酶活性以所产的乙烯的峰高值来表示。

**因在保存中污染丢失，未能作质粒带检测。

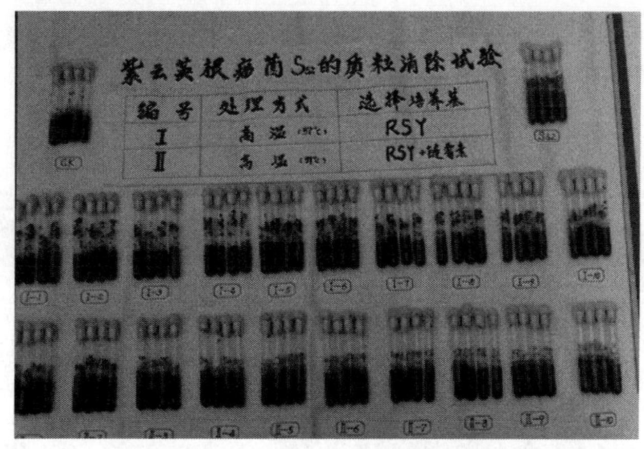

图4　高温和链霉素处理对 S_{52} 结瘤和固氮能力的影响

图中的 II-1，……II-10 即本文中的 HRa5-1，……HRa5-10，为 $Nod^{+} Fix^{-}$；I-11……I-20 即本文中的 HRa5-11……HRa5-20，为 $Nod^{+} Fix^{+}$

四、吖啶橙处理对紫云英生长、结瘤和固氮能力的影响

供试菌株用两种不同浓度吖啶橙处理的全部结果列于表 3 之中。表 3 的数据说明：40μg/ml 的吖啶橙处理浓度过高，7 天后使两种供试菌全部死亡，未得到分离株。在 20μg/ml 的吖啶橙处理的突变分离株中，来自 S_{52} 的全部结瘤固氮，仅从 7653^{r} 的 10 个分离株中获得了 4 个 Nod^{-} 突变株，分别编号为 ARa7-1，2，3，4。

表 3　吖啶橙处理对紫云英根瘤菌生长和结瘤能力的影响

供试菌株	吖啶橙浓度（μg/ml）	菌数（个/毫升）		存活率	结瘤能力	
		起始	终止		供试菌株数	Nod^{-} 株数
7653^{r}	20	2.2×10^{5}	4.5×10^{2}	2×10^{-3}	10	4
	40	2.2×10^{5}	0	0		
S_{52}	20	2.3×10^{5}	1.6×10^{2}	7×10^{-4}	10	10
	40	2.3×10^{5}	0	0		

五、高温处理对根瘤菌大质粒的消除效果

用修改的 Hirsh 碱变性法和快速检测法（王常霖等，未发表）对获得的部分 7653^{r} Nod^{-} 突变株和 S_{52} Nod^{+} Fix^{-} 突变株的质粒进行了检测，结果发现：（1）7653^{r} 的 Nod^{-} 突变株均丢失了其中的一条大质粒带 pRa7-2（图 5 中 HRa7-1，2，3）。（2）S_{52} 的 Nod^{+} Fix^{-} 突变株也丢失了其中的一条大质粒带（图 5 中 HRa5-1，2 和图 6 是的 HRa5-3，4），而 Nod^{-}，Fix^{+} 突变株则仍然保持着与出发菌株相同的两条粒带（图 5 中的 HRa5-11 和图 6 中的 HRa5-12）。

六、供试菌与分离突变株的荧光抗体检测

用制备的 7653^{r} 和 S_{52} 的荧光抗体对供试菌及其代表性分离突变株的检测结果均表现为强的正反应，表明这些分离株不是杂菌，它们确系这两个出发菌株的分离后代。

讨论

上述结果表明，紫云英根瘤菌的不同菌株对高温和吖啶橙处理的反应有着明显的差异：7653^{r} 对处理敏感，很容易分离到 Nod^{-} 突变株；而 S_{52} 用两种处理虽也获得了很高的致死率，却未能得到 Nod^{-} 突

变，只在 RSY+Str 平板上经高温处理获得了 Nod^+ Fix^-突变株。菌株对处理的这种特异性可以解释以往文献中关于吖啶橙处理效果结论不一致的原因。

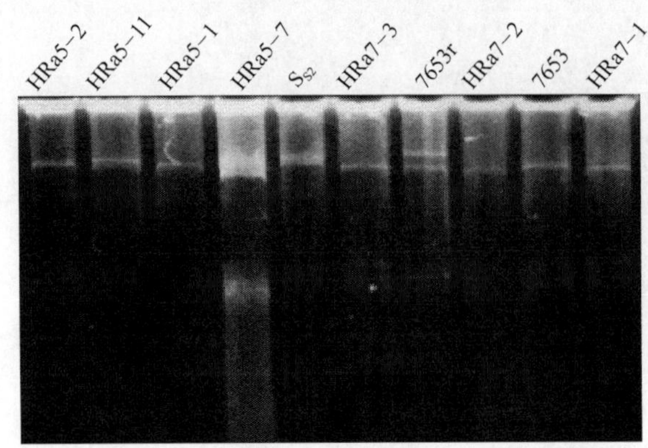

图 5 7653^r，S_{52} 及其热处理突变株的质粒图谱（快速检测法）

其中 HRa5-11，S_{52}，7653^r，7653 为 Nod^+ Fix^+；HRa5-1，HRa5-2，HRa5-7 为 Nod^+ Fix^-；HRa7-1，HRa7-2，HRa7-3 为 Nod^-

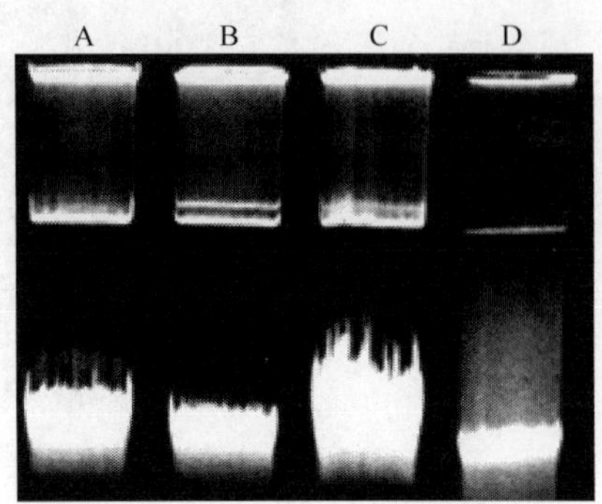

图 6 S_{52} 及其热处理突变株的质粒图谱（碱变性抽提法）

A-HRa5-3；B-S_{52}；C-HRa5-12；D-HRa5-4；A，D 为 Nod^+ Fix^-；B，C 为 Nod^+ Fix^+

7653^r的 Nod^-突变株丢失了其中的一条大质粒带，这表明该菌株的 *nod* 基因可能定位在该质粒上，这与宁林夫等最近报道的用热处理后，证明紫云英根瘤菌 SR72、531 和快生型大豆根瘤菌 USDA205 的 *nod* 基因也定位在一个大质粒上的结论相似[12]。

用高温平皿法从菌株 S_{52} 获得的 Nod^+ Fix^-突变株也丢失了其中的一个大质粒，说明该菌株的 *nod* 基因与 *nif* 基因（或与固氮作用有关的其他基因）没有定位在同一个大质粒上。Masteron 等[17]新近也报导快生型大豆根瘤菌 USDA205 的 *nod* 和 *nif* 基因分别定位在分子质量为 195Md 和 112Md 的两个不同的大质粒上。

十分有趣的是，菌株 7653^r的 Nod^-稀释分离菌株仍然具有与出发菌株相同的质粒带（pRa7-1），这一结果也从另一个方面证明它不是根瘤菌的共生质粒。胡学俊[14]最近采用与本文类似的方法处理紫云英根瘤菌 634 菌株，从 60 个分离株中获得了 23 株 Nod^-突发株，对其中的 20 个菌株进行的质粒检测表明：除一个 Nod^+菌株的质粒带丢失和一个 Nod^-菌株的质粒发生缺失以外，其中 18 个 Nod^-菌株的质粒带丢失和一个 Nod^-菌株仍然具有同出发菌株相同的质粒带，这一结果也表明 *nod* 基因不在该质粒上。Hynes 等[16]最近报道已从苜蓿根瘤菌（*R. meliloti*）中分离到分子质量为 800~1000Md 的巨大质粒，说明我们还需要进一步改进现有的大质粒提取和检测方法，并配合采用 Tn5 易位子转座和基因探针技术，才能彻底查清根瘤菌共生固氮的 *nod*、*nif* 和 *fix* 等基因的确切位置，并为固氮基因的转移做好准备。

参 考 文 献

[1]Nuti MP et al.，（1977） Large plasmids in different *Rhizobium* species，J. Gen. Microbiol，100：241-248.

[2]Beynon JL et al.，（1980） Plasmids and host-range in *Rhizobium leguminosarum* and *R. phnseoli*，J. Gen Microbiol，120：421-429.

[3]Denarie J et al.，（1981） Plasmid control of symbiotic properties in *Rhizobium meliloti*，Current Perspectives in Nitrogen Fixation （ed. By Gibson AH and Newton WE） 137-141.

[4]Scott DB，（1981） Characterization of plasmids essential for nodulation in *Rhizobium trifolii*，Current Perspectives in Nitrogen Fixation （ed. By Gibson AH and Newton WE） 405.

[5]Higashi S，（1967） Transfer of cover infectivity of *Rhizobium teifolii* to *R. phaseoli* as mediated by an episomic factor. J. Gen. Appl. Microbiol. 13：391-403.

[6]Parijskaya AH，（1973） The effect of acridine orange and mitomycin C on the Symbiotic properties of *R. meliloti*. Microbiologiya，42：119-121

[7]Zurkowskji W et al.，（1973） Effect of aeriflavine and sodium dodecyl sulphate On infectiveness of *R. trifolii*，Acta Microbiologica Polonica，5（2）：55-60.

[8]Dunican LK and Cannon FC,（1971） The genetic control of symbiotic properties In *Rhizobium*: evidence for plasmid control, Plant and Soil, special volume，73-79.

[9]Zurkowskii W et al.,（1978） Effective method for the isolation of non-nodulating Mutants of *Rhizobium trifolii*，Genetical Research，32：311-314.

[10]Hooykaas PJ et al.,（1981） Sym-plasmid of *Rhizobium trifolii* expressed in Different rhizobial species and *Agrobacterium tumefaciens*, Nature，291：351-353.

[11]李仲贤，（1985） 紫云英根瘤菌结瘤能力的研究，微生物学通报，12（4）：151-153。

[12]宁林夫等，（1986） 几类快生型根瘤菌质粒的研究，微生物学报，26（3）：271-276。

[13]周俊初，（1984）三叶草根瘤类菌体的繁殖能力，华中农学院学报，3（4）：33-45。

[14]胡学俊，（1986）紫云英根瘤菌结瘤基因的消除和 RP4 质粒在根瘤菌和其他革兰氏阴性细菌之间的转移，华中农业大学土化系农业微生物硕士学位论文。

[15]Hynes MF et al.,（1985）Characterization of the two megaplasmids of *Rhizobium meliloti*, Book of Abstracts for both Inter Sym.on Nitrogen Fixation，9-22.

[16]Toro N et al.,（1985） Genetic analysis of one no-pSym plasmid of *Rhizobium meliloti*，Book of Abstracts for both Inter. Sym. on Nitrogen Fixation，9-45.

[17]Masterson RV et al.,（1985） Conservation of symbiotic nitrogen fixation gene sequences in *Rhizobium japonicum* and *Bradyrhizobium japonicum*，J. of Bacteriol，163（1）：21-26.

[18]张忠明等，（1986） 紫云英根瘤菌质粒的研究 1.五个菌标大质粒的分离，纯化和菌株 7653 质粒的内切酶图谱，华中农业大学学报，5（4）：326-331.

紫云英根瘤菌寄主专性结瘤的基因和固氮基因的鉴定*

张忠明　周俊初　陈华癸

（华中农业大学生物技术中心）

摘　要　用植物筛选法从紫云英根瘤菌 7653R 基因文库中分离到能互补 7653R+1（原菌株 7653R 消除共生质粒的突变株）共生缺陷功能的重组质粒 pRaZ15 质粒克隆有紫云英根瘤菌共同结瘤基因簇。采用三亲交配—植物试验法，将 pRaZ15 质粒转移到豌豆根瘤菌菌株 T83K3 和菜豆根瘤菌菌株 3622−15 中，用以回接紫云英植株，并在紫云英上结瘤，证实了 pRaZ15 质粒上克隆有紫云英根瘤菌寄主专性结瘤基因。将 pRaZ15 质粒再次转移到紫云英根瘤菌不结瘤突变株 7653R+1 中，回接紫云英植株，用乙炔还原法测定出所结根瘤的固氮酶活性。用 α-^{32}P-dCTP 标记克隆有肺炎克氏杆菌发源的 *nif* HDK 基因的 pSA30，以此作为探针，与 pRaZ15 质粒的 *Eco*RI 片段杂交，证实了紫云英根瘤菌 *nif* HDK 基因也存在于 pRaZ15 质粒上。

关键词　紫云英根瘤菌；寄主专性结瘤基因；固氮酶基因；pRaZ15 质粒；三亲交配

引　　言

紫云英（*Astragalus sinicus* L.）是中国南方水稻田中广泛种植的一种绿肥植物。从 40 年代以来，我国广泛开展了对紫云英根瘤菌的研究[1]。众多的研究结果表明，紫云英根瘤菌属于一独立的互接种族[2]，是种级水平上的一个独立聚类群（陈文新，私人通讯）。然而对于紫云英根瘤菌分子遗传学的研究则起步不久。几年来，继证实了紫云英根瘤菌广泛存在大质粒[3]和共生基因存在于大质粒上[4]后，又构建了紫云英根瘤菌基因文库[5]，随后我们首次从基因文库中分离到克隆有完整功能的紫云因根瘤菌结瘤基因簇的重组质粒 pRaZ15[6]。

本文进一步证实了 pRaZ15 质粒上不仅克隆有共同结瘤基因（*nod* DABC），而且也含有紫云英根瘤菌寄主专性结瘤基因（*hsn*）和固氮酶结构基因（*nif* HDK）。

1　材料和方法

1.1　细菌菌株与质粒

见表 1

表 1　细菌菌株与质粒

菌株与质粒	有关性状	来源及文献
紫云英根瘤菌:		
7653R	野生型 $Nod^{+}Fix^{+}$	本室收集
7653R+1	7653R 消除共生质粒突变株 $Nod^{-}Tc^{s}$	[7]
7653R+1（pRaZ15）	$Nod^{+}Tc^{r}$	本工作
豌豆根瘤菌:		
$T_{83}K_3$	在豌豆上能结瘤 Km^{r}	[9]
$T_{83}K_3$（pRaZ15）	能在紫云英上结瘤 Km^{r} Tc^{r}	本工作

*原载于《华中农业大学学报》，8（4）：285~290，1989.

续表

菌株与质粒	有关性状	来源及文献
菜豆根瘤菌：		
3622-15	在菜豆上能正常结瘤固氮 Km^r	[9]
3622-15（pRaZ15）	能在紫云英上结瘤 Km^r Tc^r	本工作
质粒：		
pRaZ15	pLAFR1 上克隆有 25kb 的紫云英根瘤菌 DNA 片段（包括有 *nod*、*hsn*、*nif* 等）Tc^r	[6]
pRK2013	协助转移质粒 Km^r	[10]
pSA30	pAcyc184 上克隆肺炎氏杆菌的 *nif* HDK 基因	[11]

1.2 三亲本接合转移方法

将根瘤菌受体接种于 5ml TY 培养液中，28℃振荡培养 24~36 小时。含 pRaZ15 质粒的大肠杆菌和含 pRaZ15 质粒的大肠杆菌分别接种于 5ml LB 中（分别含相应的四环素和卡那霉素），37℃振荡培养 12 小时，按 1/5 的接种量转移到新鲜的 LB 中，37℃继续振荡培养 2~3 小时。将三个培养好的菌株等比例混合后，用无菌注射器注射到醋酸纤维素滤膜上（Φ25mm，孔径为 0.45μm），然后将膜置于已制备好的 TY 平皿上，28℃培养 16~24 小时。用无菌水洗下菌苔，所洗下的菌悬液涂布到含四环素和卡那霉素（需要时加）的 SM 培养基上，0.2 毫升/皿，置 28℃培养 3~5 天。供体和受体菌在培养基上均不能生长，只有杂交子才能生长，从平皿上分别挑取 10 个单菌落，混合在一起接种紫云英植株。

1.3 紫云英结瘤试验

按常规方法进行紫云英种子表面灭菌、催芽，并将种芽转移到含无氮植物营养液的砂土管中，置光照室培养 3~4 天后接种根瘤菌菌悬液，继续培养 30 天左右观察结果。其培养基配方和培养条件见文献[6]。

1.4 有关项目的测定方法和杂交方法

根瘤菌质粒检测按 Hirsch 法[12]进行。

DNA 操作及分子原位杂交。在大肠杆菌质粒 DNA 的制备按 Birnboin 法[13]进行。限制性内切酶消化反应按 New England Biolabs 所示条件进行。各种酶类均购自华美生物工程公司。

DNA-DNA 膜杂交按 Southern[14]的方法，采用缺口翻译法标记克隆有肺炎克氏杆菌 *nif* HDK 基因的 pSA30 作为探针。α-^{32}P-dCTP 为北京福瑞同位素公司的产品。

根瘤的固氮酶活测定。剪取鲜瘤置于 15ml 小瓶中，加盖橡皮塞，注入 2ml 乙炔，置 28℃下 2 小时，用气相色谱测定乙炔还原能力。

2 结果

2.1 三亲交配及 pRaZ15 质粒的转移

克隆有紫云英根瘤菌结瘤基因的 pRaZ15 质粒，在协助转移质粒 pRK2013 的作用下，分别与豌豆根瘤菌 T83K3、菜豆根瘤菌 3622-15 和不结瘤紫云英根瘤菌突变株 7653R+1 进行三亲交配，在含有四环素的 SM 培养基上选择杂交子并计算其转移频率，结果见表 2。

表 2 pRaZ15 向不同根瘤菌中转移的频率

供体	受体	转移频率
pRaZ15（HB101）	$T_{83}K_3$	3.14×10^{-4}
	3622−15	9.89×10^{-3}
	7653R+1	4.87×10^{-3}

2.2 紫云英结瘤实验——寄主专性结瘤基因（*hsn*）的鉴定

本试验从 3 个杂交组合中分别挑取 10 个杂交子混合接种紫云英植株，置光照室（光照 12 小时，黑暗 12 小时）培养 1 个月后，它们都能结瘤，结果见表 3 和图版。

表 3 不同杂交子在紫云英上的结瘤试验

供试菌株	培养基	结瘤状况（紫云英）
$T_{83}K_3$（pRaZ15）	SM+Km+Tc	+
$T_{83}K_3$	SM+Km	-
3622-15（pRaZ15）	SM+Km+Tc	+
3622-15	SM+Km	-
7653R+1（pRaZ15）	SM+Tc	+
7653R+1	SM	-
7653R	SM	+
ck（空白）		-

紫云英植株均生长在砂培管中。

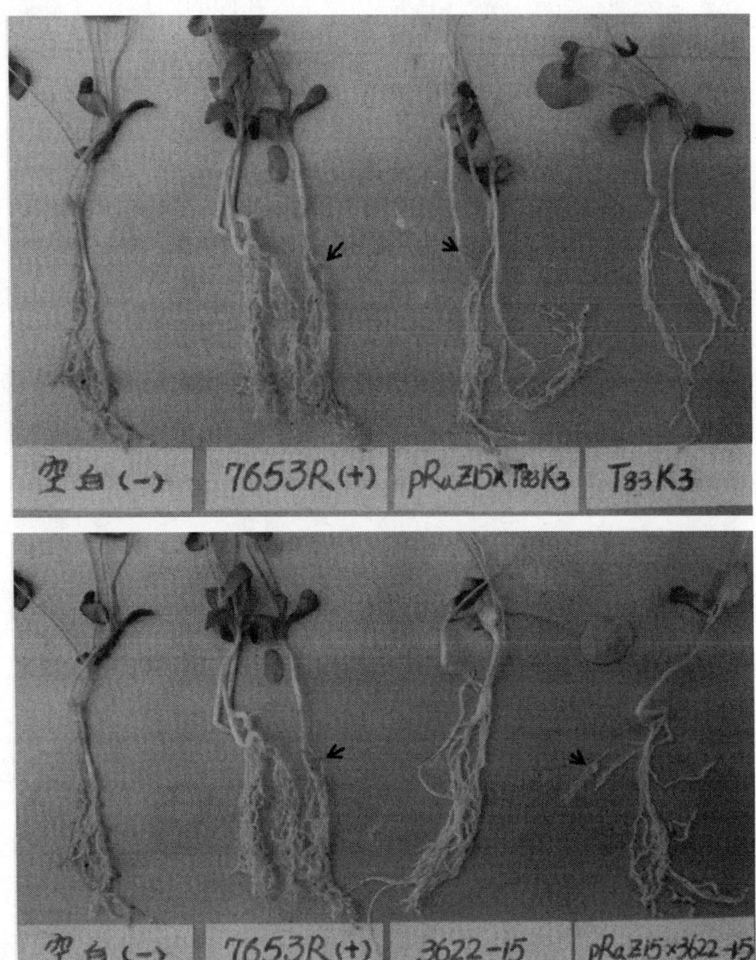

图版说明：不同杂交子在紫云英上结瘤：上图中 $T_{83}K_3$ 为豌豆根瘤菌接种紫云英植株；下图中 3622-15 为菜豆根瘤菌接种紫云英植株。pRaZ15×$T_{83}K_3$ 为转移接合子接种紫云英植株，pRaZ15×3622-15 为转移接合子接种紫云英。箭头所示处为紫云英根瘤。

由表 3 和图版可知，豌豆根瘤菌、菜豆根瘤菌与紫云英不属于同一互接种族，因此这两种根瘤菌在紫云英上不能结瘤，而将 pRaZ15 质粒转移到这两种根瘤菌后，它们在紫云英上表现出结瘤能力。说明了重组质粒 pRaZ15 上克隆有识别紫云英植株的寄主专性结瘤基因（*hsn*）。

2.3 pRaZ15 质粒上固氮酶基因的鉴定

制备和纯化 pRaZ15 质粒 DNA，分别用限制性内切酶 *Eco*RI 和 *Sal*I 消化，走琼脂糖凝胶电泳，结果见图 1（左）。并将其转移到硝酸纤维膜上，用 α-^{32}P -dCTP 标记克隆有肺炎克氏杆菌 *nif* HDK 基因的 pSA30，以此作为探针 DNA，与上述的膜作 DNA-DNA 分子原位杂交，结果见图 1(右)。除载体 pLAFR1 带上有显影外，在另外的三条片段上也有显影。说明 *nif* HDK 基因存在于该质粒上。同时将 pRaZ15 质粒转移到 7653R+1 菌株中，通过前述的植物试验，取鲜瘤通过乙炔还原法测定固氮酶活，结果表明（表 4），供试菌株均具有乙炔还原活性。

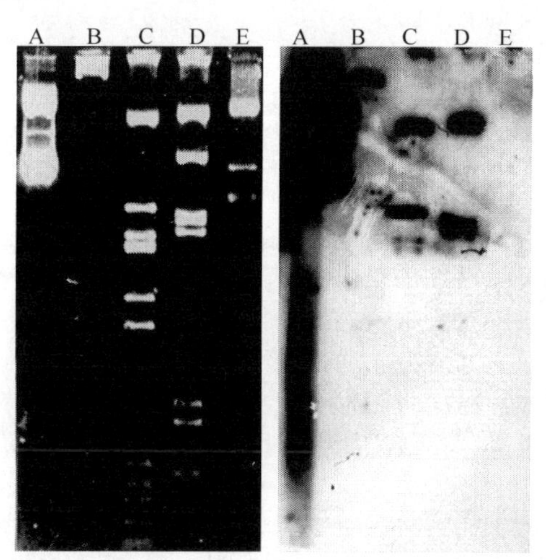

图 1 Southern 法分子滤膜杂交的放射自显影图

用 ^{32}P 标记克隆有肺炎克氏杆菌 *nif*HDK 基因的 pSA30 DNA，与经 Southern 法转移固定在硝酸纤维素滤膜上的 DNA 进行杂交，得到的放射自显影图（右）和用 0.7%琼脂糖电泳图谱（左）。其中：A，pSA30 质粒 DNA；B，pRaZ15 质粒 DNA；C，pRaZ15-Sal1；D，pRaZ15-EcoRⅠ；E，αDNA-*Eco*RⅠ。

表 4 紫云英-7653R+1（pRaZ15）共生根瘤固氮酶活测定

供试菌*	结瘤管数	鲜瘤数	鲜瘤数（g）	固氮酶活**（nmol 乙烯/h/克瘤）
R11	2/3	20	0.015	24.7
R12	2/3	13	0.008	12.0
R13	3/3	22	0.02	28.0
R15	3/3	20	0.01	25.6
R16	3/3	22	0.02	22.0
R17	3/3	2	0.003	26.7
R18	3/3	18	0.02	24.0
R110	3/3	10	0.01	54.4
7653R（ck）	3/3	13	0.01	18.4

* R11-R110 供试菌均为 pRaZ15 转移到消除共生质粒的 7653R+1 突变株中的杂交子，7653R 为野生型菌株。

** 固氮酶活都很低是因生长条件有限，植株生长在砂土管中，根瘤也较小。

3 讨论

pRaZ15 质粒进入豌豆和菜豆根瘤菌能够使它们在紫云英植株上结瘤。这一结果说明了 pRaZ15 上含有识别紫云英植物的寄主专性结瘤基因（*hsn*）；同时说明了克隆的紫云英 *hsn* 基因能够在豌豆和菜豆根瘤菌中表达，从而增加了它们的寄主范围。关于这一点，Long[15]曾提出不是所有的根瘤菌获得另一种根瘤菌的共生质粒或克隆的共生基因后都能在新的寄主上结瘤。如苜蓿根瘤菌在接受了三叶草根瘤菌的共生质粒或共生基因后，仍只能在苜蓿上结瘤，而不能在三叶草上结瘤。pRaZ15 能在其他根瘤菌中表达，为我们利用紫云英根瘤菌共生基因来改造其他根瘤菌的寄主范围、结瘤和固氮效应提供了可能。

pRaZ15 质粒能够恢复消除了共生质粒而不能在紫云英上结瘤的突变株 7653R+1 的结瘤和固氮的能力。本文通过分子杂交试验也证明了 *nif* HDK 基因存在于这个克隆质粒上，这一结果表明紫云英根瘤菌的结瘤基因和固氮酶结构基因是紧密连锁的。同时根瘤的固氮酶活测定表明，该质粒上也可能含有与固氮有关的*fix*基因。在其他快生型根瘤菌中，有许多报道[16~19]虽都指出克隆的根瘤菌结瘤基因簇能恢复消除共生质粒突变株的结瘤功能，并含有*nif*基因，但没有说明有固氮酶活性。而紫云英根瘤菌共生基因（包括*hsn*、*nod*、*nif*、*fix*）聚在一条25kb片段上，它具有决定紫云英根瘤菌寄主专性、结瘤和固氮的功能，这样为我们在整体功能上对紫云英根瘤菌共生基因分析和进一步的遗传操作提供了极大的方便。

参 考 文 献

[1]Chen HK et al.，Note on the Root-Nodule Bacteria of *Astragalus sinicus* L.，Soil Sci.，1944，51：291-293.

[2]陈华癸，草籽绿肥根瘤细菌和人工接种，新科学，1952，3：33-38.

[3]张忠明等，紫云英根瘤菌质粒研究：Ⅱ 五个菌株大质粒的分离、纯化和7653质粒的内切酶酶切图谱，华中农业大学学报，1986，5：326-331.

[4]宁林夫等，紫云英根瘤菌共生质粒的鉴定及结瘤功能的诱动转移，遗传学报，1986，13：1-10.

[5]张忠明等，紫云英根瘤菌基因文库的构建及结瘤基因的分离，生物技术在农业上的应用论文集。上海科学技术出版社，1987，30-32.

[6]张忠明等，紫云英根瘤菌基因文库的构建及完整结瘤基因的重组质粒pRaZ15的分离。（寄生物工程学报，待发表）

[7]周俊初等，紫云英根瘤菌质粒的研究：Ⅱ.紫云英根瘤菌经高温和吖啶橙处理后结瘤和固氮突变株的产生和质粒的消除，华中农业大学学报，1987，156-164.

[8]Johnston AWB et al.，Two transmissible plasmids in *Rhizobium leguminosarum* strain 300，1982，128：85-93.

[9]王常霖等，转座子Tn5-Mob对菜豆根瘤菌共生基因的诱变、转移和初步定位，遗传学报，1988，15：25-33.

[10]Diuta G et al.，Broad host range DNA cloning system for gram negative bacteria：Construction of a gene bank of *Rhizobium meliloti*，proc. Natl. Acad. Sci. USA，1980，77：7347-7351.

[11]Cnnon FC et al.，Overlapping Sequences of *Klebsiella pneumoniae nif* DNA cloned and Characterized，Mol. Gen. genet. 1979，174：59-66.

[12]Hirsch PR，Physical identification of Bacteriocinogenic，nodulation and other plasmids in strain of *Rhizobium Ieguminosarum*，J. Gen. Microbiol. 1980，120：403-412.

[13]Birnboim HC et al.，A rapid alkaline extraction for screening recombinant plasmid DNA，J. Nucleic Acids Res.，1979，7：1513-1523.

[14]Southern EM et al.，Detection of specific sequences among DNA fragments separated by gel electrophresis，J. Mol. Boil. 1975，98：503-517.

[15]Long SR，Genetic Advances in The Study of *Rhizobium* Nodulation，Genetic Enginering，1987，8：135-150.

[16]Downie JA et al.，Cloning of the symbiotic region of *Rhizobium leguminosarum*：the nodulation genes are between the nitrogenase genes and *nif* A-like gene，1983.

[17]Schofield P et al.，Hostspecific nodulation is coded on a 14kb DNA fragment in *Rhizobium trifolii*，Plant Mol. Boil. 1984，3：3-11.

[18]Kondorosi E et al.，Physical and genetic analysis of a symbiotic region of *Rhizobium meliloti*：identification of nodulation genes，Mol. Gen. Genrt. 1984，193：445-452.

[19]Fisher RW et al.，Conserved nodulation genes in *Rhizobium meliloti* and *Rhizobium trifolii*，Appl. And Environ Microbiol，1985，49：1432-1435.

Tn5–Mob 转座系统对紫云英根瘤菌 SR72 的诱变和诱动作用*

韩素贞　周俊初　陈华癸

（华中农业大学土化系生物固氮研究室）

摘　要　通过接合作用将携带有转座子 Tn5–Mob 的“自杀”性载体质粒 pSUP5011 引入紫云英根瘤菌 SR72，得到卡那霉素抗性（Km^r）菌落的频率为 6.99×10^{-6}，测得受体菌的 Km^r 自发突变频率 $<10^{-8}$。从对 1071 个 Km^r 突变体进行的植物砂培结瘤中筛选出结瘤不固氮（Nod^+，Fix^-）突变株 17 个，不结瘤（Nod^-）突变株 4 个。另外，还从近 3000 个 Km^r 突变体中选出腺苷营养缺陷型突变株 3 个。通过 Tn5 探针进行的菌落原位杂交试验证明：这 21 个共生固氮突变株中均含有 Tn5 序列，进一步通过接合作用将协助质粒 RP4–4（Tc^r）引入 Nod^+，Fix^+ 突变株，获得了含有 Tn5-Mob 和 RP4-4 的新突变株 SR72ZR（Km^r，Tc^r），但试图通过它们的协同作用将 SR72 中的大质粒诱动转移到根癌农杆菌 A136 的试验未获成功。

关键词　紫云英根瘤菌；转座子；质粒诱动转移

转座子（transposon）是带有抗性基因的一类插入因子，能在宿主染色体或质粒 DNA 的许多位点上插入或易位，转座子不但能使受体菌被插入部位的基因失活，还同时使受体菌产生新的抗药性，从而给插入受体菌的筛选提供了标记。Tn5 是带有卡那霉素抗性的转座子，广泛应用于根瘤菌共生固氮的遗传学研究中。Beringer 等[1]首先将带有 Tn5 的质粒 pJB4JI 引入豌豆、三叶草和菜豆根瘤菌并获得了共生固氮突变株。Johnston 等[2]也在同期用 Tn5 转座技术证明豌豆根瘤菌的共生固氮基因定位在 pRL1JI 质粒上。Banfalvi 等[3]用带 Tn5 的菌株进行热处理，发现 Km^r 的消失总与 Nod^-或 Fix^-表型同步，并进一步用固氮酶基因（*nif*）探针技术证明苜蓿根瘤菌的 *nif* 基因定位在大质粒 pRme41b 上。Kondorosi 等[4]用 Tn5 进行饱和诱变作出了苜蓿根瘤菌的共生固氮基因图，发现了共同结瘤基因（Common nod）A、B、C、D 和寄主专性结瘤区域（hsn）。Lamb 等[5]用菜豆和豌豆根瘤菌的 *nod* 基因探针和 Tn5 诱变证明慢生型大豆根瘤菌的结瘤和固氮基因连锁。Evans 等[6]用 Tn5 诱变技术找到了 *nod I* 基因。

除豌豆根瘤菌的共生质粒 pJB5JI [2]和产细菌素质粒 pRL1JI、pRL3JI 和 pRL4JI[7]及三叶草根瘤菌的共生质粒 pRtr5a[8]等少数质粒具有较高的自我转移能力外，大多数根瘤菌的共生大质粒难以通过接合作用转移。Scott 等[9]利用 Inc P1 群质粒 pR68.45 将三叶草根瘤菌的共生质粒 pRtr514a 诱动到 *nod*⁻突变株中。Kondorosi 等[4]将质粒 RP4 的 *mob* 片段插苜蓿根瘤菌的共生质粒 pRme41b 上，然后再克隆到 pBR325 上并成功地将其导入其它根瘤菌和根癌农杆菌中。Simon 等[10，11]巧妙地构建了一种 Tn5-Mob 诱动系统，他们先后构建了一系列同时带有 Tn5 和 RP4 的诱动基因片段（Mob）的“自杀性”载体质粒，该类质粒一旦进入根瘤菌等革兰氏染色负反应（G^-）细菌后即停止复制而解体，从中释放出的 Tn5-Mob 片段可以“随机”地插入受体菌 DNA 中而引起基因突变。如再借助于导入另一个带有转移基因（tra）的协助质粒 RP4-4 的联合作用，可以将插有 Tn5-Mob 片段的大质粒以较高的频率（达 10^{-5} 数量级）诱动转移到其他受体细菌之中。宁林夫等[12]也利用质粒 RP1：：Tn501 诱动转移了紫云英根瘤菌 SR72 的结瘤基因，他们虽然未能在受体菌中找到诱动转入的大质粒，但却发现一个新的、分子质量为 40Md 的质粒。

*原载于《中国微生态学杂志》，2（1）：56~63，1990.

材料与方法

一、菌株和质粒

本试验所用的菌株和质粒的特性及其来源见表 1。全部菌株均用冻干或 30%甘油于-20℃低温下保存。大肠杆菌（*E.coli*）用 LB，紫云英根瘤菌（*R. astragali*）和根癌农杆菌（*Agrobacterium tumefaciens*）用 TY 培养基培养。

表 1 供试菌株和质粒

菌株	质粒	基因型或表现型	来源
Rhizobium astragali			
SR72	两条质粒带	Thi^-，Nod^+，Fix^+	浙江农科院
SR72Z218，409，420，443，434，447，491，509，866，808，802，821，827，846，882，940，946	" "	Thi^-，Nod^+，Fix^-，Km^r	本研究
SR72Z496，662，669，942	" "	Thi^-，Nod^-，Fix^-，Km^r	"
SR72MQ-2，3，5	" "	Ade^-，Nod^+，Fix^+，Km^r	"
E.coli			
S17-1	pSUP5011	pBR325：：Tn5-Mob，Cm^r，Km^r，Pro^-	Simon[11]
RR1	pKan2	pBR325：：Tn5，Km^r	Scott[20]
SM10	RP4-4	Tc^r，Km^r	Hedges[9]
Agrobacterium tumefaciens			本研究
A136（Str）		△Tc^r，Str^r	

二、培养基和试剂

YMA[13]、TY[11] 、LB[14]和 SM[15]培养基，Fahraeus 无氮植物营养液[13]，质粒检测用 PA 培养基及有关试剂[17]，以及菌落原位杂交所用的试剂[17]均参见所列的文献。

三、抗生素及其浓度

硫酸卡那霉素（Km）：50μg/ml；盐酸四环素（Tc）：10μg/ml；氯霉素（Cm）：50μg/ml；硫酸链霉素（Str）：100μg/ml。

四、植物砂培结瘤试验

紫云英砂培结瘤试验按文献[13]的方法用大试管进行，植物播种后 1~2 天后接种根瘤菌，在人工光照条件（温度 20~25℃，光照 12 小时/天）下培养一月后观察结果，项目包括结瘤和生长状况及根瘤的固氮酶乙炔还原活性测定等。

五、Tn5-Mob 插入试验-供体菌 S17-1 与受体菌 SR-72 的接合作用

将在相应培养基上活化至对数期的 S17-1 和 SR-72 分别制成菌悬液，各取 1ml 混合均匀后吸取 0.1ml 加至已事先灭菌并贴放在 TY 平板上的微孔滤膜（0.22μm）上，于 28℃培养过夜，用无菌水将滤膜上的菌体洗下，将洗脱液作 10 倍系列稀释至 10^{-8}，取洗脱原液和低倍稀释液各 0.1ml 涂于 SM+Km（50μg/ml）平板以筛选 Tn5-Mob 接合转移的突变株；用高倍稀释液（10^{-6}~10^{-8}）各 0.1ml 涂 SM 平板以计算受体菌数量。为彻底淘汰供体菌 S17-1，获得的接合突变株需进一步在 SM+Km 平板上划线纯化三次。

按上法同样操作但不加供体菌 S17-1，即可测得受体菌产生自发 Km 抗性的突变频率。

六、质 粒 检 测

用修改的碱变性法进行[18]。供试菌先在 TY 斜面上活化后转接 PA 培养基振荡培养 1 天（对 *E.coli*）或 2 天（对根瘤菌）。离心收集菌体后用 TE 缓冲液洗涤并离心再沉淀，依培养液光密度 OD 值大小计算应加入的 TE 缓冲液用量，取 4ml 菌悬液于 125ml 三角瓶中，加入 0.5ml 10% SDS、0.5ml Pronase E（10mg/ml）于 37℃水浴上溶菌至清亮，加入 0.125ml 3mol/L NaOH 使 pH 至 12.4，于室温下保持 20′ 使染色体 DNA 变性，滴加 0.375ml 2mol/L Tris-HC1（pH7.0）使 pH 降至约 8.5，然后加 1.375ml 5mol/L NaCl 于 4℃下 4 小时以上以沉淀蛋白质和染色体 DNA。将上述沉淀液于 17000r/min 下离心，上清加 1.6ml 50% PEG 使终浓度为 10%，于冰箱过夜后于 15000r/min 下离心，沉淀用少量 TE 溶解即为质粒 DNA。提取的质粒经电泳后，于溴化乙锭液中染色，即可在紫外透射仪上观察。

七、菌落原位杂交

用上法从 RR1 中提取带 Tn5 的质粒 pKan2，经 CsCl 密度梯度离心后进一步纯化，用从英国 New England Nucllear 公司订购的 α-^{32}P dCTP 进行缺口转移，制成的 Tn5-探针用于对经接合转移获得的 Tn5-Mob 插入突变代表菌株作菌落原位杂交，具体操作方法见文献[17]。

八、质粒的诱动转移

第一步是依方法五所述的方法以 SM10 为供体菌，选取经菌落原位杂交证实已含有 Tn5-Mob 片段的 SR72 Nod$^+$Fix$^+$突变株为受体菌，在 SM+Km+Tc 的抗性平板上筛选出转移有 RP4-4 质粒的 SR72 双抗性突变株。第二步是用自发突变的方法从根癌农杆菌 A136 中筛选并获得抗链霉素（Str^r）的突变株为供体菌，A136（Str）。第三步以双抗性的 SR72 突变株为供体菌，A136（Str）为受体菌，用上述接合转移的方法，以 YMA+Km+Str、YMA+Tc+Str 和 YMA+Km+Tc+Str 等三种选择平板筛选 Tn5-Mob、RP4-4 单独和共同转移的接合子，并对纯化的接合子作质粒检测。

九、Tn5-Mob 诱变 SR72 营养缺陷型突变株的筛选、鉴定和回变分析

1. 营养缺陷型突变株的筛选和鉴定　选取在方法五中 S17-1 与 SR72 接合转移的 Km^{r}突变株，分别点种在完全培养基 YMA 和基本培养基 SM 上，经培养后在 YMA 上生长而在 SM 上不生长的是初筛的营养缺陷型菌株。将初筛菌株在 YMA 液体培养基中培养；取菌悬液点种在含有不同营养因子的 10 组不同的 SM 平板上（表 2）[11]，经 28~30℃培养三天后观察结果并判定其所属的营养缺陷型类型。

2. 回变分析 将用上法筛得的营养缺陷突变株用YMA液体培养后稀至10^{-3}后取0.1ml涂SM平板，以计数回变的野生型菌落数；继续稀到10^{-8}，取10^{-6}、10^{-7}、10^{-8}三个稀释度各0.1ml涂YMA平板，以计数缺陷型菌株总菌数和回变频率。

表2 鉴定营养缺陷型用培养基分组及其营养成分（以SM为基础）

分组号	1	2	3	4	5
6	腺苷 Adenosine	鸟苷 Guanosine	半胱氨酸 Cysteine	甲硫氨酸 Methionine	
7	组氨酸 Histidine	亮氨酸 Leucine	异亮氨酸 Isoleucine	赖氨酸 Lysine	缬氨酸 Valine
8	苯丙氨酸 Phenylalanine	酪氨酸 Tyrosine	色氨酸 Tryptophan	苏氨酸 Threonine	脯氨酸 Proline
9	谷氨酰胺 Glutamine	天门冬氨酸 Aspartic acid	尿嘧啶 Uridine	天门冬氨酸 Aspartic acid	精氨酸 Arginine
10	胸腺嘧啶 Thymine	丝氨酸 Serine	谷氨酸 Glutamic acid	二氨基庚二酸 Diaminopimelic acid	甘氨酸 Glycine

结果

一、Tn5-Mob 转座频率的测定

以S17-1为供体菌，SR72为受体菌进行的两次接合转移试验结果见表3。由表3可见：Tn5-Mob的转座频率一般比自发突变频率高出上百倍，表明Tn5-Mob片段可能已进入到受体菌SR72之中。为了进一步在分子水平上确证这一点，我们随机从中挑取并进一步在SM+Km平板上用划线分离法纯化出1071个Km^r突变株单菌落并编号为SR72 Z1-1071。

表3 Tn5-Mob 转座频率的测定

试验次数	SR72 自发突变			Tn5-Mob 转座			
	总菌数（$\times10^8$/ml）	突变Km^r菌落数（个/ml）	频率	受体菌总数（$\times10^8$/ml）	Km^r接合子菌落数（个/ml）	转座频率	平均频率
1	4.65	0	$<4.65\times10^{-8}$	1.43	1210	8.46×10^{-6}	6.99×10^{-6}
2	4.65		$<4.65\times10^{-8}$	1.98	1008	5.5×10^{-6}	

二、转座突变株的结瘤性能和固氮酶活性测定

对上述1071个转座突变株全部进行了砂管结瘤试验，经初筛获得结瘤不固氮（$Nod^+ Fix^-$）突变株166个，不结瘤（Nod^-）突变株91个，其余814个仍保持了结瘤和固氮能力。从上述共生固氮突变株中再选出Nod^+，Fix^-突变株35个，Nod^-突变株25个作了进一步的结瘤试验，表4是其中部分菌株的测定结果。

表4 *R.astragali* SR72 与代表性转座突变株的固氮酶活性

供试菌株	表型性状	根瘤重量（g）	固氮酶活性（$nMolc_2h_4$/g/h）
CK			0
SR_{72}	Nod^+Fix^+	0.0345	2.7×10^4
$SR_{72}Z_{492}$	Nod^+Fix^+	0.0306	1.35×10^4
$SR_{72}Z_{509}$	Nod^+Fix^-	0.0584	0
$SR_{72}Z_{21}8$	Nod^+Fix^-	0.0370	0

三、菌落原位杂交

按第二批结瘤试验的结果，从 1071 个划线分离株中选出 184 个菌株点种在硝酸纤维微孔滤膜上，经贴于 TY 培养基上长出后，依次经 10%SDS-变性液-中和液处理后于 80℃下烘干。Tn5 探针用自 RR1 提取的 pKan2 质粒经 α-^{32}P dCTP 缺口转译而成，探针 DNA 与硝酸纤维滤膜的杂交在 68℃下 18 小时，菌落原位杂交的放射性自显影结果见图 1。由图可见：Tn5-Mob 的供体菌 S17-1 为强阳性，受体菌 SR72 为阴性；在全部 184 个供试转座突变株中有 81 个呈阳性反应，占 44%。其中表型为不结瘤（Nod$^-$）的有 4 个，即 SR72 Z496、662、669 和 942 菌株；表型为结瘤不固氮（Nod$^+$Fix$^-$）的有 17 个，它们是 SR72 Z218、409、420、434、443、447、491、509、866、808、802、821、827、846、882、940 和 946 菌株，其余 60 个均为结瘤固氮的菌株。

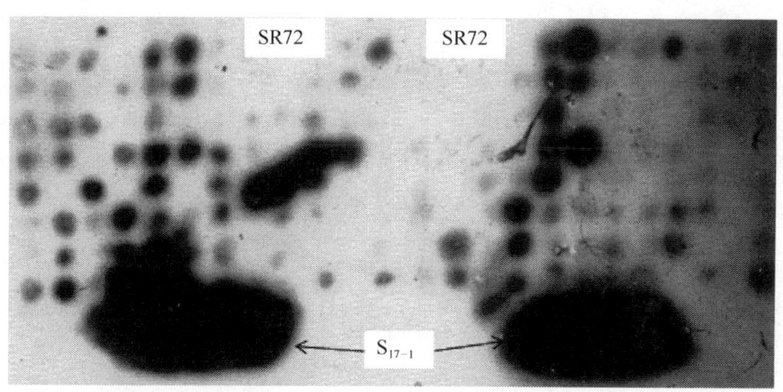

图 1　*R.astragali* SR72 的 Tn5-Mob 转座突变株的放射自显影图

四、营养缺陷型 Tn5-Mob 转座突变株的检出和鉴定

用 YMA 完全培养基和 SM 基础培养基比较点种法从近 3000 个供试 Tn5-Mob 转座突变株中初筛获得 5 个营养缺陷菌株（图 2），并编号为 SR72MQ-1、2、3、4 和 5。

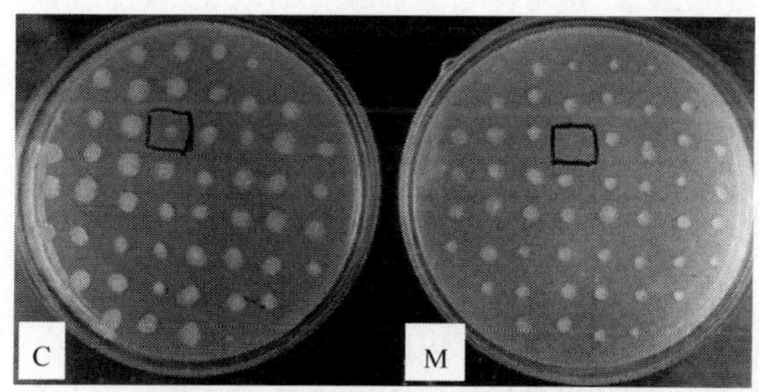

图 2　营养缺陷突变株初筛试验

C：YMA 培养基 M：SM 培养基，方框中标明为营养缺陷突变株

用修改的 Davis[11]分组方案，因 SR72 系维生素缺陷型。需在 SM 基础培养基中加入硫胺素、泛酸钙和尼克酰胺等维生素，所以本实验中取消了原方案中的第 11 组，即维生素组。五个初筛菌株的测定结果见表 5。

由表 5 可见：MQ-2、3 和 5 三个突变株均只在第 1 和 6 养料组中生长。查表 2 可知，它们的共同成分是腺苷，因此可以判定这三株都是腺苷缺陷型。MQ-1 和 4 在供试的 10 组培养基中均能生长，说明它们不是真正的营养缺陷型，进一步的复证也表明它们也能在 SM 基础培养基上生长。

表 5 营养缺陷型突变株的鉴定

供试菌株	培养基分组									
	1	2	3	4	5	6	7	8	9	10
SR_{72}	+	+	+	+	+	+	+	+	+	+
SR_{72}MQ-1	+	+	+	+	+	+	+	+	+	+
SR_{72}MQ-2	+	−	−	−	−	+	−	−	−	−
SR_{72}MQ-3	+	−	−	−	−	+	−	−	−	−
SR_{72}MQ-4	+	+	+	+	+	+	+	+	+	+
SR_{72}MQ-5	+	−	−	−	−	+	−	−	−	−

+：生长；−：不生长

五、营养缺陷型突变株的回复突变和结瘤试验

为了测定 Tn5-Mob 片段引起的营养缺陷型突变株的稳定性，我们用 YMA 和 SM 平板对 MQ-2、3 和 5 三个突变株的回复突变进行了测定，结果见表 6。由表 6 可见：三个腺苷缺陷型菌株在分裂过程中均十分稳定。回复突变为野生型的频率均低于 10^{-8}。

表 6 营养缺陷型菌株的回复突变

供试菌株	在 YMA 平板上的活菌数（$\times10^8$/ml）	在 SM 平板上的菌落数（个/ml）	回变频率
SR_{72}MQ-2	3.60	0	$<3.6\times10^{-8}$
SR_{72}MQ-3	3.52	0	$<3.5\times10^{-8}$
SR_{72}MQ-5	3.20	0	$<3.2\times10^{-8}$

六、转座突变株的质粒检测

用碱变性法检测了部分代表性转座突变株的质粒图谱(图 3)。结果表明:突变株均不含供体菌 S17-1 所有的小质粒 pSUP5011，除 SR72Z218 具有与受体菌 SR72 相同的二条大质粒带以外，其余 5 个突变株还具有一条新的质粒带，它们可能是转座子与染色体 DNA 作用使部分 DNA 脱落而产生的。

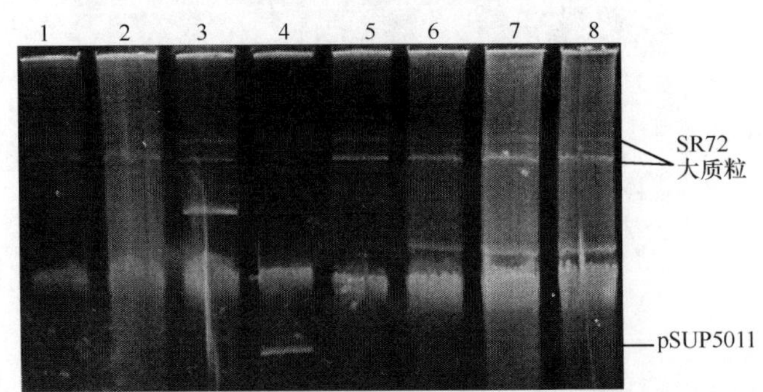

图 3 Tn5-Mob 转座突变株的质粒图谱

1：SR72；2：SR72Z218（Nod^+Fix^-）；3：SR72Z313R（Nod^+Fix^+）；4：S17-1；5：SR72Z496（Nod^-）；6：SR72Z509（Nod^+Fix^-）；7：SR72Z725（Nod^+Fix^+）；8：SR72Z942（Nod^-）

七、诱动转移试验

选择经菌落原位杂交证实 Tn5-Mob 片段插入的 Nod^+Fix^+转座突变株 SR72Z214、234、253、302、313、442、500、516、725 和 815 等 10 个菌株为共生固氮基因的供体，试图通过 Tn5-Mob 系统实现共

生固氮基因向根癌农杆菌的转移。该试验分两步进行：首先将 RP4-4 协助质粒从 *E.coli* SM10 引入上述供体菌，通过接合转移在 SM+Km+TC 选择平板上获得了大量双抗突变株（表 7），测得 RP4-4 的平均转移频率高达 4.8×10^{-3}。

表 7 Tn5-Mob 转座突变株的 Tc^r 自发突变和与 SM10 的接合转移频率

供试菌株	自发突变			接合转移（与 SM10）			
	总菌数（$\times10^8$/ml）	突变 Tc^r 菌落数（个/ml）	频率	受体菌总数（$\times10^8$/ml）	Tc^r 接合子菌落数（个/ml）	接合转移频率	平均频率
SR72Z253	5.58	0	$<5.58\times10^{-8}$	7.93	10×10^{-4}	1.3×10^{-2}	
SR72Z302	10.30	0	$<1.63\times10^{-9}$	15.80	2.57×10^{-4}	1.7×10^{-4}	4.8×10^{-3}
SR72Z213	16.30	0	$<1.63\times10^{-9}$	15.80	2.45×10^{-4}	1.6×10^{-4}	

用快速检测法对其中部分双抗突变株进行了质粒检测，结果表明 RP4-4 确实已从供体菌 SM10 进入到转座突变株之中（图 4）。

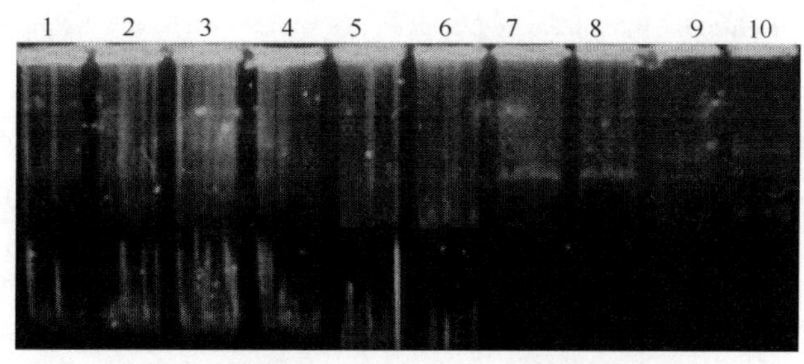

图 4 诱动转移部分接合突变株的质粒图谱

1：A136；2：A136；3：A136×SR72Z313R；4：A136×SR72Z302R；5：SM10；6：SM10；7：SR72；8：SR72；9：SR72Z313R；10：SR72Z302R

第二步是用在第一步已获得的带有 RP4-4 和 Tn5-Mob 的突变株为供体与根癌农杆菌 A136（Str^r）进行接合转移，以 YMA+Str+Tc；YMA+Str+Km；YMA+Str+Km+Tc 等三种选择平板筛选接合转移子。本试验共选用了 SR72Z214R 等 10 个双抗转座突变株作供体，A136 为受体反复多次试验，仅在 YMA+Str+Tc 平板上筛选到接合子，表明 Tn5-Mob 片段未能在 RP4-4 诱动下向农杆菌转移。

讨论

Tn5-Mob 与 Tn5 转座子一样，是一种有效并常用的诱变因素，Tn5-Mob 的优点是有可能在协助质粒的帮助下实现插入部位基因或质粒的转移。Forrai 等[19]报道用 Tn5 诱变苜蓿根瘤菌产生的共生固氮突变株的频率为 0.3%；Sherman 等[20]用 Tn5-Mob 诱变大豆根瘤菌产生的共生固氮突变株的频率为 3%。本研究从 1071 个 Tn5-Mob 突变株中最终筛选出不结瘤突变株 4 个，结瘤不固氮突变株 17 个，占总数的 2%。

菌落原位杂交试验表明上述 21 个共生固氮突变株中含有 Tn5 序列，但仍不足以表明 Tn5-Mob 的具体插入位置。马庆生等[21]在用 Tn5 诱变获得的 4 株 nod^- 表型突变株中发现有 1 株不是由于 Tn5 插入引起的。Long 等[22]也发现 Meade 等[14]先前报道的某些 nod^- 突变株实际上不是 Tn5 插入而是苜蓿根瘤菌本身的插入因子易位造成的。Forrai 等[19]也表明苜蓿根瘤菌的共生固氮突变株中约有 40%与 Tn5 插入无关。因此对本研究获得的 21 个共生固氮突变株还需作进一步的试验研究。

宁林夫等[12]曾报道用 RP1∷Tn501 质粒将紫云英根瘤菌 SR72 的结瘤基因诱动转移到根癌农杆菌中。胡学俊等[23]在试图用 RP4 质粒诱动 SR72 大质粒转移的研究中只发现 RP4 从供体菌向受体菌转移。本研究采用 Simon[10]的 Tn5-Mob 系统在 RP4-4 的协助下也未能实现 SR72 大质粒的转移而只发现 RP4-4 能以较高的频率转移，这一结果初步表明：本实验获得的 Nod^+Fix^+ 转座突变株的 Tn5-Mob 片段可能没有插在大质粒上而是插在染色体上，因而难于实现诱动转移。

参 考 文 献

[1]Beringer JE，et al.，Transfer of the drugresistance transposon Tn5 to Rhizobium. Nature，1978，270：633.

[2]Johnston AWB，et al.，High frequency transfer of nodulating ability between strains and species of Rhizobium. Nature，1978，276：634.

[3]Banfalvi Z，et al.，Location of nodulation and nitrogen fixation genes on a high molecular weight plasmid of *R. meliloti*. Mol. Gen. Genet. 1984，184：318.

[4]Kondorosi EV，et al.，Physical and genetic analysis of a symbiotic region of *Rhizobium meliloti*；Identification of nodulation genes. Mol. Gen. Genet. 1981，193：445.

[5]Lamb JW，et al.，In *Bradyrhizobium japonicum* the common notulation genes，nod ABC，are linked to nif A and fix A. Mol. Gen. Genet. 1986，202：512.

[6]Evans J，et al.，The nod I gen product of *Rhizobium leguminosarum* is closely related to ATP-binding bacterial transport proteins，Nucleotide sequence analysis of the nod I and nod J genes. Gene 1986，43：93.

[7]Hirsch RP，et al.，Plasmid-determined bacteroicin production by *Rhizobium leguminosarum*. J. G. Microbiol. 1979，113：219.

[8]Hooykaas PJJ，et al.，Sym-plasmid of *Rhizobium trifolii* expressed in different rhizobial species and *Agrobacterium tumefaciens*. Nature，1981，291：351.

[9]Scott PB，et al.，Identification and mobilization by cointegrate formation of a nodulation plasmid in *Rhizobium trifolii*. J. Bacteriol. 1982，151：36.

[10]Simon R，et al.，A broad host range mobilization system for in vitro genetic engineering：transposon mutagenesis in Gram-negative bacteria，Biotechnology，1986，1：784.

[11]Simon R，et al.，High frequency mobilization of Gram-negative bacteria replicans by the in vitro constructed Tn5-mob transposon. Mol. Gen. Genet. 1984，196：413.

[12]宁林夫等，紫云英根瘤菌共生质粒的鉴定及结瘤功能的诱动转移. 遗传学报. 1986，13（1）：1。

[13]Vinvent JM. A manual for the practical study of root-nodule bacteria. Inter. Biol. Programme Handbook No：15. Blackwell Oxford，Edinberg：1970.

[14]Meade HM，et al.，Physical and genetic characterization of *Rhizobium meliloti* induced by transposon Tn5 mutagengsis. J. Bacteriol. 1982，149：114.

[15]周俊初等，快生型根瘤菌营养生理的研究. 华中农学院学报. 1981，3：44.

[16]Eckhardt TT，et al.，A rapid method for the identification of plasmid deoxyribonucleic acid in bacteria. Plasmid，1978，1：584.

[17]Maniatis T，et al.，Molecular cloning，Cold Spring Harbour Laboratory，1983.

[18]Hepbum AG. The art of coarse genetic engineering. Gentics Department，The John Innes Institute，Colneg Lane，Rorwich.

[19]Forrai T，et al.，Localization of symbiotic mutantions in *Rhizobium meliloti*. J. Bacteriol. 1983，153：635.

[20]Sherman SM，et al.，Transposon Tn5-Induced mutagenesis of *Rhizobium meliloti*. J. Bacteriol. 1984，159：335.

[21]Ma SQ et al. Molecular genetics of mutants of *Rhizobium leguminosarum* which fail to fix nitrogen. Mol. Gen. Genet. 1982，187：166.

[22]Long SP，et al.，Cloning of *Rhizobium meliloti* nodulation genes by direct complementation of nod mutations. Nature，1982，298：485.

[23]胡学俊，紫云英根瘤菌抗药性菌株的获得与初步应用. 华中农业大学土化系硕士生毕业论文，1986.

紫云英根瘤菌基因文库的构建及含完整结瘤基因的重组质粒 pRaZ15 的分离*

张忠明　陈华癸　李阜棣

（华中农业大学生物固氮研究室）

范云六

（中国农业科学院分子生物学研究室）

摘　要　以紫云英根瘤菌菌株 7653R 为材料，制备总 DNA，经 *Eco*RI 限制酶部分酶解，通过 10%~50%蔗糖梯度离心，分离到 20~30kb 的 DNA 片段。利用能在革兰氏阴性菌中转移和复制的广谱寄主载体 pLAFR1 质粒，构建了紫云英根瘤菌基因文库。通过与苜蓿根瘤菌 1021 菌株中 8.7kb 的共同结瘤基因（作探针 DNA）杂交，从基因文库中分离到紫云英根瘤菌共同结瘤基因片段。以紫云英根瘤菌不结瘤突变株 7653R+1（7653R 消除共生质粒）为受体、构建的 7653R 基因文库（*E. coli* C600）为供体，通过协助转移质粒 pRK2013（LE392）进行三亲交配，在含四环素的根瘤菌合成培养基（SM）上选择接合转移子。将得到所有接合转移子混合在一起接种植物，通过植物结瘤试验，分离到含紫云英根瘤菌结瘤基因的重组质粒 pRaZ15。将该质粒用 *Eco*R Ⅰ完全酶切，得到 25kb 左右的外源 DNA 片段，该片段携带完整的结瘤基因簇。

关键词　紫云英根瘤菌；结瘤基因；基因文库

紫云英（*Astragalus sinicus* L.）是我国南方水稻田里的主要豆科绿肥，对增加粮食产量、保持土壤肥力起着重要作用。陈华癸[1, 2]、范云六[3]等对紫云英根瘤菌进行过许多研究。

近几年来的遗传学研究表明，紫云英根瘤菌和其他一些快生型根瘤菌一样，也广泛存在着内生大质粒[4, 5]，其共生基因也存在于特定的大质粒上[8, 7]。迄今为止，尚未见到关于紫云英根瘤菌完整结瘤基因簇的分离和克隆的报道。本文利用能在革兰氏阴性菌中转移和复制的广谱寄主载体-pLAFR1[8]构建了紫云英根瘤菌 7653R 菌株的总 DNA 基因文库，并通过与苜蓿根瘤菌 8.7kb-*Eco*R Ⅰ片段的共同结瘤基因杂交，从基因文库中分离出含有紫云英根瘤菌共同结瘤基因片段。同时利用不结瘤（*nod*⁻）突变株 7653R+1，进行三亲交配，用接合转移子接种紫云英植株，从 7653R 基因文库中分离出具有完整功能的紫云英根瘤菌结瘤基因簇。

材料与方法

（一）材料

1. 细菌和质粒　见表 1。

2. 培养基

（1）TY 培养基（g）：胰蛋白胨 5，酵母粉 5，$CaCl_2 \cdot 6H_2O$ 1.3，蒸馏水 1000ml，pH7.0。

（2）TA 培养基（g）：胰蛋白胨 4，$MgSO_4 \cdot 7H_2O$ 0.6，蒸馏水 1000ml pH7.2。

（3）SM 培养基：蔗糖 10g，KNO_3 0.5g，K_2HPO_4 0.5g，$CaCl_2$ 0.1g，NaCl 0.1g，H_3BO_3 20mg，Na_2MoO_4 20mg，硫胺素 0.1mg，烟酰胺 0.1 mg，泛酸钙 0.1mg，生物素 1mg，蒸馏水 1000ml，pH7.0。

*原载于《生物工程学报》，7（3）：213~219，1991.

表 1 细菌和质粒

Table 1 Bacterial strains and plasmids

菌株与质粒 Strains & plasmids	有关性状 Relevant characteristics	文献及来源 Reference or source
R.astragali		
7653R	Nod^+ Fix^+ （wild type）	（7）
7653R+1	Nod^- mutant of 7653R devoid of Sym-plasmid	（7）
R15	Nod+ （7653R+1 contained recombinant plasmid pRaZ15）	This work
E.coli		
LE392	F^- hedk514 （rk^- mk^-） sup44，SupF58，LacYl or Δ（LaclZY）6 galK2，galT22，matB1，trpR55 λ^-	中国农科院分子生物学研究室
C600	F^- thx-1，thi-1，LeuB6，LacY1，tonA21，supE44，λ^-	Lad.Mol.Biology,
BHB2688	N205，recA-（λimm434，cIts´，b2，red3，Eam4，Sam7） λ	Chinese Acad.
BHB2690	N205，recA-（λimm434，cIts´，b2，red3，Dan15，Sam7） λ	Agri.sci.
R.leguminosarum		
T83K3	Nod^+ Fix^+ Km^r	Johnston（1978）
Plasmids		
pLAFR1（HB101）	Cosmid derivative of pRK290，Tc^r	（8）
pRmSL26（HB101）	*R.meliloti* nod gene cloned in pLAFR1	（14）
pBR325（HB101）	Amp^r，Tc^r，Cm^r	This Lab.
pRmZ1（HB101）	*R.meliloti* 8.7kb common *nod* gene cloned in pBR325	This work
pRa370（C600）	*R.astragali* common *nod* gene cloned in pLAFR1	This work
pRaZ15（HB101）	*R.astragali nod* gene cluster cloned in pLAFR1	This work
pRK2013（LE392）	Tra^+ ColEl	Ditta.et al.（1980）

（二）方法

1. DNA 的分离及操作 根瘤菌总 DNA 的分离方法参考 Ma，Q.S.等[9]的方法。提取的总 DNA 经 *Eco*R I 部分酶解后，经 10%~50%蔗糖梯度，25000rpm（Beckman SW28）20℃离心 24h，然后收集 20~30kb 的 DNA 片段，装入透析袋内，在 TE 缓冲液（10mmol/L Tris-HCl，1mmol/L EDTA，pH7.8）中透析 24h，每隔 6h 换一次缓冲液，取出样品，加入 1/10 体积的 3mol/L 醋酸钾（pH8.0），加入 2 倍体积的−20℃冷却的无水乙醇，置−20℃过夜。离心（15000r/min，4℃，20min）沉淀 DNA，加入 70%冷乙醇洗涤一次，以同样的条件离心再沉淀 DNA，最后 DNA 悬浮于 200μl 的 TE 缓冲液中。

载体 DNA（pLAFR1）的制备以及重组质粒 DNA 的检测按 Maniatis 等[10]的碱性裂解法进行。

DNA 的连接采用 T_4DNA 连接酶（日本 Zeon 公司）在 12℃连接 16h。

2. λ噬菌体包装物的制备及效价测定 包装物的制备参照 Puhler A.[11]的方法。效价测定指示菌的制备参照 Boehringer Mannheim 公司产品目录上介绍的方法[12]，指示菌为 LE392。

3. 重组 DNA 的体外包装及转染 重组 DNA 的体外包装及转染按 Puhler A.[11]的方法进行，重组子在含四环素（15μg/ml）的 LB 选择性培养基上筛选，受体菌为 C600。

4. 缺口转译及菌落原位杂交 缺口转译药盒为 BRL 公司产品，α-^{32}P-dCTP 为 Dupont 公司产品，被标记的质粒 pRmZ1 DNA 为碱解法制备物，反应条件按 BRL 公司缺口转移药盒的说明进行。

菌落原位杂交：硝酸纤维素膜（Serva 公司）的制备处理按 Maniatis 等[10]的方法进行。膜杂交按下述方法进行：膜在杂交前，用 50mmol/L Tris-HCl，pH 8.0，0.1% SDS，1mmol/L EDTA，1mol/L NaCl 在 42℃处理

30min，然后置于滤纸上吸干，按 0.2ml/cm² 膜的量加入预杂交液（1∶1 甲酰胺，5×SSPE，0.3% SDS，0.1mg/ml 热变性的鲑精 DNA），42℃预杂交 2.5h，倒去预杂交液，再按 1/4 原体积的量加入杂交液，热封口，置 42℃水浴中杂交 24h。然后取出膜置于 42℃预热的 0.5% SDS、2×SSC 中漂洗 30min，再转入 42℃预热的 0.1% SDS、2×SSC 中漂洗 30min。膜放在滤纸上晾干，并置于 X-光暗匣中，压上 X-光片，在−20℃曝光 24h。

5. 三亲交配及植物结瘤试验 取已转染的大肠杆菌 0.5ml（40%甘油保存）接种于 5ml 含四环素（15μg/ml）的 LB 培养液中，37℃振荡培养 12h。同时接种含质粒 pRK2013 的菌株于 5ml 含卡那霉素（50μg/ml）的 LB 培养液中，37℃培养 12h。紫云英根瘤菌不结瘤突变株 7653R+1 接种于 5ml TY 培养液中，28℃振荡培养 24~36h。将三个培养好的菌株等比例混合后，用无菌注射器注射到醋酸纤维素滤膜上（Φ25mm，孔径为 0.2μm），然后将膜置于已制备好的 TY 固体培养基上，28℃培养 16~24h，用无菌水洗下洗下菌苔，所洗下的菌悬液全部涂布到含四环素（10μg/ml）的 SM 培养基上。0.2ml/皿，用 7653R 的基因文库和 7653R+1 同时涂皿作对照，28℃培养 3~5 天后，长出的菌落用无菌水全部洗下，混合起来制成菌悬液，作为接种物进行植物结瘤试验。

紫云英种子经表面灭菌和催芽后，播入无氮植物营养液配制的琼脂栽培管中，置 25℃光照室生长 2 天后，将上述制得的菌悬液接种于琼脂栽培管中，0.5~1ml/每管，植物继续在光照室生长 30~40 天后观察结果。

结果与讨论

（一）紫云英根瘤菌基因文库的构建

紫云英根瘤菌 7653R 菌株的总 DNA 用 *Eco*R I 部分酶解后，通过蔗糖梯度离心，得到 20~30kb 的 DNA 片段，将这种 DNA 片段连接到经 *Eco*R I 酶切的 pLAFR1 质粒上，参照 Friedman 的方法[8]，连接时外源 DNA 与载体 DNA 的比例约为 8∶1，即根瘤菌 DNA 为 400μg/ml，载体 DNA 为 50μg/ml。高浓度的外源 DNA 可以降低载体 DNA 自身连接的比例，酶连结果见图版 I-A。这种连接物经体外包装并转染 *E. coli* C600，在含四环素（15μg/ml）的 LB 培养基上筛选四环素抗性克隆，结果见图 1。四环素抗性克隆数的得率为 7.4×10^4/μg DNA 连接物。

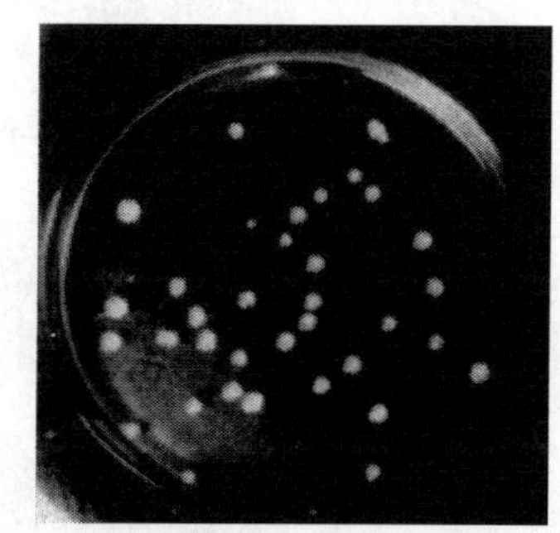

图 1 四环素抗性克隆的筛选

培养基为含四环素（15μg/ml）的 LB 培养基

Fig.1 Selection of tetracyline resistance clones

LB medium containing tetracycline (15μg/ml)

为了检验四环素抗性克隆中载体-外源DNA连接和载体自身连接的比例，我们随机挑取了 50 株四环素抗性克隆分离质粒作为初检，结果表明 50 中有 15 株含有载体与外源 DNA 连接的重组质粒，其余均为载体自连的 pLAFR1 质粒。结果见图版 I-B。图版 I-B 是 50 株四环素抗性克隆质粒检测中的部分结果（5 块电泳凝胶中的 1 块），从图版 I-B 可以看出：3，4，6，7，8 号样品的质粒带型与 1 号样品（载体 pLAFR1 质粒）完全一样，而 5，9，10，11，12 号样品带型与 1 号样品不一样，所以我们初步认为后者不一样的是重组质粒，即在 pLAFR1 上克隆有外源 DNA 片段。值得说明的是，本文采用的是碱性裂解法抽提质粒，而 pLAFR1 质粒及克隆有外源 DNA 片段的重组质粒都含有“cos”位点，“cos”位点在碱性条件下很容易连在一起，使质粒形成多聚体。从图版 I-B 可以看出除 2 号样品（无质粒对照菌株 C600）外，其余样品在染色体带以上均隐约可见多聚体带，根据 pLAFR1 的单体质粒带以及多聚体带的位置，我们初步确定 3，4，6，7，8 样品的质粒为载体自连的结果，其他均为重组质粒。通过这种判断方法，在上述 50 株中有 15 株是重组质粒。为了进一步确证这 15 株是否克隆有外源 DNA 片段。我们从 15 株中随机取 6 株制备质粒 DNA，用 *Eco*RI 完全酶切。并将这 6 株不同的质粒分别命名为 pRa1、pRa2、pRa3、pRa4、pRa5、pRa6。酶切分析结果见图版 I-C。结果表明这 6 个质粒均含有外源 DNA 片段，而且不同的质粒克隆的外源 DNA 片段都不相同，说明克隆有多种多样的外源 DNA 片段。其外源 DNA 片段的大小在 18.8—29.8kb 范围之间。因此，重组克隆的质粒在基因文库中的比例约为 30%。根据 Maniatis 等[10]介绍的公式：$N=\dfrac{\ln(1-p)}{\ln(1-f/g)}$ 根瘤菌基因组 DNA 为 1×10^{7}bp（g），插入片段以 2.5×10^{4}bp（f）计算，构建基因文库能以 99%的概率（P）覆盖根瘤菌整个 DNA，理论上需要的重组子数（N）可以通过上式计算为 1.8×10^{3}。本文所得基因文库的重组子数达到理论值，满足了建库的要求。

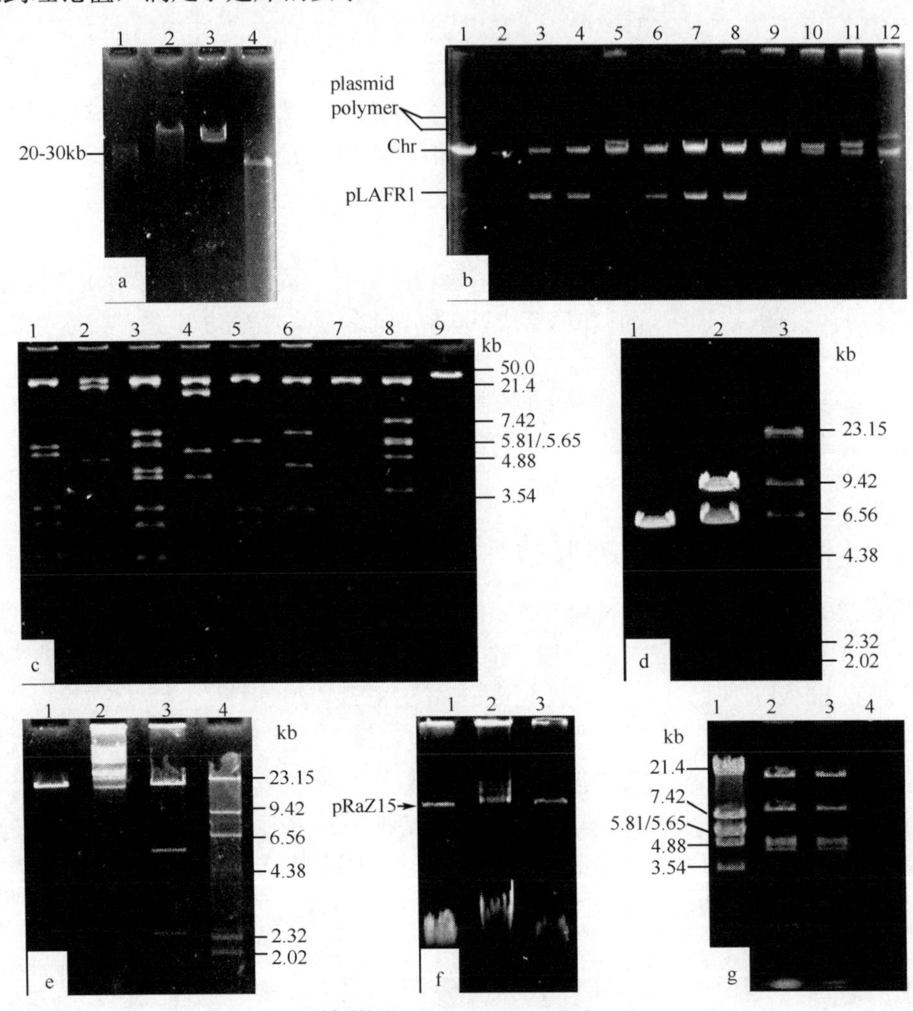

图版说明：Explanation of Plate I

A. 酶连结果 Ldentification of ligation of ligation：1. Foreign DNA fragments （20-30kb），2. Ligates of *Eco*RI-digested pLAFR1 and foreign DNA，3. λ DNA （-50kb），4. *Eco*RI-digested pLAFR1 DNA （21.6kb）

B. 四环素抗性克隆质粒检测的电泳图谱：1. Vector pLAFR1 plasmid；2. Recipient strain C600；3-12. Plasmid of tetracycline resistance clones

C. 六株不同重组子 DNA 的酶切鉴定 Identification of *Eco*RI digested patterns of 6 different recombinants：1. pRa1/*Eco*RI DNA fragments；2. pRa2/*Eco*RI DNA fragments；3. pRa3/*Eco*RI DNA fragments；4. pRa4/*Eco*RI DNA fragments；5. pRa5/*Eco*RI DNA fragments；6. pRa6/*Eco*RI DNA fragments；7. pLAFR1/*Eco*RI DNA fragments；8. λ DNA/*Eco*RI DNA molecular weight marker；9. λ DNA

D. pRmZ1 质粒的酶切鉴定 Identification of pRmZ1 plasmid digested with *Eco*RI：1. Vector pBR325/*Eco*RI DNA fragment；2. pRmZ1/*Eco*RI DNA fragments；3. λ DNA/*Hin*dIII DNA molecular Weight marker

E. pRa370 质粒的酶切鉴定 Identification of recombinant plasmid pPa370 DNA digested with *Eco*RI：1. *Eco*RI-digested vector pLAFR1 DNA；2. pRa370 plasmid DNA；3. *Eco*RI-digested pRa370 DNA；4. λ DNA/*Hin*d I DNA molecular weight marker

F. 转移接合子中重组质粒的检测 Detection of recombinant plasmid pPa370 DNA digested with *Eco*RI：1. transconjugant strain R15；2. *Rhizobium leguminosarum* strain T83K3 （containing 3 megaplasmids MW. 150，200，250kb）；3. recipient strain 7653R+1 （containing a megaplasmid 130kb）；

G. 重组质粒 pRaZ15 的 *Eco*RI 酶切鉴定 Identification of recombinant plasmid pRaZ15 digested with *Eco*RI：1. λ DNA/*Eco*RI DNA fragments（M W. marker）；2，3. pRaZ15/*Eco*RI DNA fragments；4. pLAFRI/*Eco*RI DNA fragment（21.6kb）.

（二）紫云英根瘤菌共同结瘤基因的分离

将根瘤菌基因文库中分离出的2000个菌落复印到硝酸纤维膜上（膜处理及杂交见材料和方法），从pRmSL26质粒上切下含有苜蓿根瘤菌8.7kb的共同结瘤基因片段，克隆到pBR325的*Eco*RI位点上，构成pRmZ1质粒，pRmZ1的酶切分析见图版I-D。将该质粒作为探针DNA，从上述菌落中分离到含紫云英根瘤菌共同结瘤基因片段的重组质粒，定名为pRa370，菌落原位杂交结果见图2。pRa370质粒经*Eco*R I完全酶切得到2条外源DNA片段，其分子质量大小分别为6kb和3kb（结果见图版I-E）。

（三）用互补Nod⁻突变体功能的试验分离紫云英根瘤菌结瘤基因簇

以紫云英根瘤菌不结瘤突变株（Nod⁻）7653R+1（消除了共生质粒）为受体，紫云英根瘤菌7653R基因文库（C600）为供体，以pRK2013（LE392）为协助转移质粒，进行三亲交配后得到的所有接合转移子，用无菌水洗下并混合在一起接种紫云英植株。在无氮植物营养液紫云英琼脂栽培管中，100管有9管结瘤，7653R正对照4/5管结瘤，7653R+1负对照5管都不结瘤（结果见图3）。从接合转移子所结的根瘤中分离并纯化根瘤菌，检测该根瘤菌是否具有四环素抗性。实验结果表明，分离的菌株都具有四环素抗性。按Hirsch[13]的方法分离质粒并作电泳检测，表明从瘤中分离出的根瘤菌增加了一个50kb左右的质粒，该质粒定名为pRaZ15（图版I-F），含该质粒的根瘤菌定名为R15。在图版I-F中的1号样品，其pRaZ15质粒也呈现多条带现象，我们认为也是由于碱解法引起的质粒多聚体。这种多聚体不影响转化和酶切，酶切后多聚体完全消失。我们将pRaZ15质粒的抽提物转化*E. coli* HB101，在含四环素的LB培养基（15μg/ml）上筛选转化子，然后从转化子中分离，制备质粒DNA，用*Eco*RI进行完全酶切。结果表明重组质粒pRaZ15除含有载体pLAFR1质粒DNA带（21.6kb）外，还含有7条外源DNA酶切带，累计外源DNA片段的分子量约为25kb（结果见图版I-G）。通过进一步的植物试验证明，该片段具有恢复紫云英根瘤菌不结瘤突变株7653R+1在紫云英植株上的结瘤功能。7653R+1是一株消除了共生质粒的突变株，因此试验表明该片段含有完整功能的结瘤基因簇（或7653R+1原丢失的共生质粒上的结瘤基因簇）。

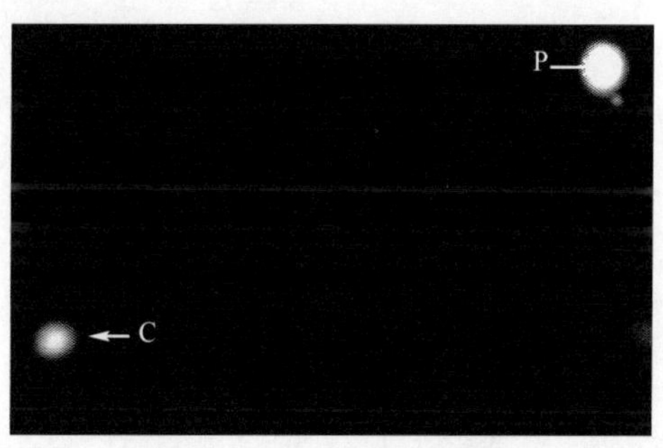

图2　菌落原位杂交结果

Fig.2　The result of colony hybridization

P.（pRa370）；C.（pRmSL26）　used as positive LB control，pRmZ1 used as a probe

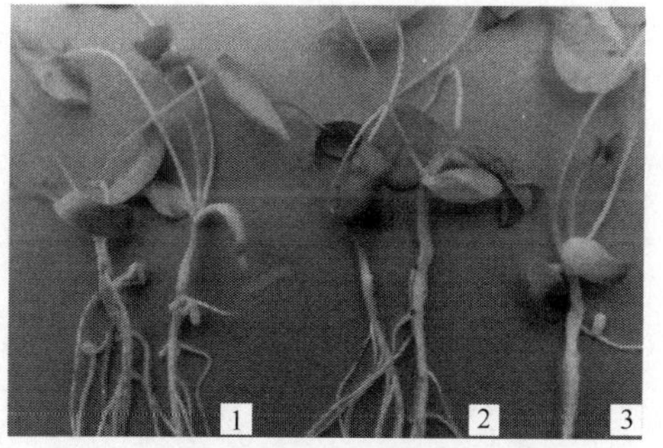

图3　紫云英植物结瘤试验的结果

Fig.3　The result of nodulation test on *A. sinicus*.
1. Inoculated with 7653R strain　(positive control)；2.Inoculated with 7653R+1 strain (negative control)；3. Inoculated with mixture of triparental mating transconjugants　(7653R gene library，pRK2013 and 7653R+1)．

参 考 文 献

[1]Chen H.K. et al.，*Soil. Sci.*，51：291-293，1944.

[2]陈华癸：新科学，3：33-38，1952.

[3]范云六：土壤通讯，3：44-45，1965.
[4]张忠明等：华中农业大学学报，5（4）：326-331，1986.
[5]王常霖等：华中农业大学学报，7（1）：15-21，1988.
[6]宁林夫等：遗传学报，13（1）：1-10，1986.
[7]周俊初等：华中农业大学学报，6（2）：156-164，1987.
[8]Friedman，A.M. et al.：*Gene*，18：289-296，1982.
[9]Ma，Q.S.et al.：*Mol.Gen.Gent.*187：166-171，1982.
[10]Maniatis，T. et al.：Molecular Cloning：A Laboratory Manual，Cold Spring Harbor Laboratory，New York，1982.
[11]Puhler，A.et al.：Advance Molecular Genetics，Spering-Verlag Berlin Heidelberg，pp.190-201，1984.
[12]Boehringer M.：Biochemicals for Molecular Biology，Eds：Boehringer Mannheim GmbH- Biochemica，Printed in Western Germany，pp.12，1980.
[13]Hirsch，P. R.：*J. Gen. Microbiol.*，120：403-412，1980.
[14]Long，S.R. et al.：*Nature*，298：485-488，1982.

Characteristics, Distribution, Ecology, and Utilization of *Astragalus* Sinicus–Rhizobia Symbiosis*

H.K. CHEN, F.D. LI, AND Y.Z. CAO

1. INTRODUCTION

Astragalus sinicus L. is a symbiotic nitrogen-fixing leguminous green manure crop traditionally grown in rice fields in the lower Yangtze Valley, southeastern China and Japan. In the lower Yangtze Valley, after the rice crop is harvested in September/October, *A. sinicus* seeds are sown to the drained field. The crop is plowed under at full blossoming stage in April/May; after that the field is flooded and prepared for transplanting rice seedlings. A fair yield of the green manure crop amounts to thirty to forty metric tons per hectare of fresh material, including tops and roots, containing approximately four kilograms nitrogen per metric ton, if well nodulated by efficient rhizobia. *A. sinicus* is grown yearly or in alternation with winter wheat, barley or rapeseed.

2. *Rhizobium huakei* （Chen et al. 1990）: THE *A. sinicus* Rhizobia

Detailed taxonomic studies of the *A. sinicus* rhizobia were carried out only very recently （Chen 1990）. A tentative name, *Rhizobium astragali*, was given to the *A. sinicus* rhizobia, which first appeared in a textbook of microbiology in 1959 （Chen 1959） following the then current emphasis on cross-inoculation at the species level. Chan et al. （1988） placed the organism in the genus *Bradyrhizobium.* A new species, *Rhizobium huakui*, was proposed by Chen et al. （1990） for the rhizobia isolated from *A. sinicus* L., based on numerical classification studies comparing two hundred phenotypic features of nine strains of rhizobia isolated from *A. sinicus* with those of forty eight strains of *Rhizobium* spp., *Bradyrhizobium* spp., *Sinorhizobiumfredii* and *Agrobacterium* spp., as well as polyacrylamide gel electrophoresis patterns of cell protein, DNA G + C content and DNA-DNA hybridization. All nine isolates from *A. sinicus* are closely related, clustering as one phenon within the genus *Rhizobium.* The morphology and colony characteristics of *R. huakui* are similar to those of other *Rhizobium* spp., except that it is motile by a single polar or subpolar flagellum （Fig. 1）, which is unique in the genus. *R. huakui* grows well in mannitol yeast extract medium but does not or very poorly grow in beef extract nutrient medium, similar to most other *Rhizobium* spp., Potassium nitrate or urea serves satisfactorily as sole nitrogen source. One or more vitamins （thiamin, nicotinic amide, etc.） are necessary for different strains （Zhou and Cao 1981）.

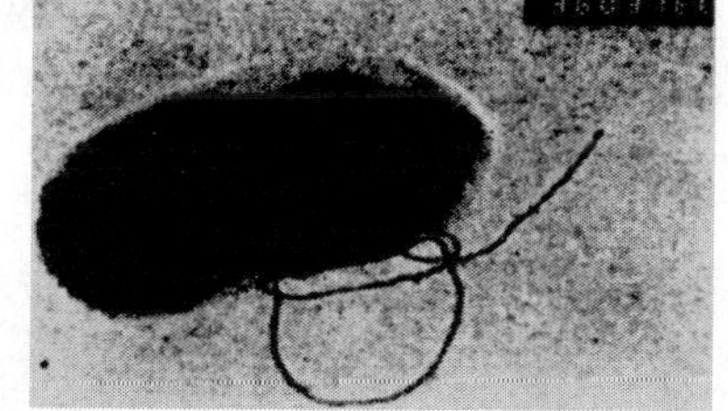

Fig.1

A synthetic medium was formulated for large-scale production of liquid cultures of *R. huakui* and compared favourably against yeast extract-based or soybean sprout extract-based liquid medium.

*原载于 *The Nitrogen Fixation and Its Research in China.*
Editor: Guo-fan Hong C. Springer-Verlag Berlin Heidelberg, pp439~455, 1992.

3. HOST-RHIZOBIA SPECIFICITY OF *A. sinicus* SYMBIOSIS

Astragalus is the largest genus in Leguminosae，containing 1500 to 2000 species. Allen and Allen （1981，pp. 78-80） listed 91 nodulated species. Results of plant-infection tests in early years suggested that symbiotic associations within the genus *Astragalus* were exclusively *inter se* （Chen and Shu 1944；Ishizawa 1954）；though there were reports to the contrary （Wilson 1939；Wilson and Chen 1947）. Allen and Allen mentioned in their monograph （1981，p. 77） that the host plant-rhizobia relationships of 18 species of *Astragalus* were markedly versatile. Not all *Astragalus* rhizobia nodulated *Astragalus* species tested. Five strains nodulated 2 species of *Phaseolus*，but not P. *lunatus*，3 strains nodulated *Medicago sativa* and *Mililotus* sp.，and 3 strains nodulated *Vigna* sp. All 18 strains were non-infective on *Lupinus*，*Glycine*，*Trifolium*，*Pisum*，*Lathyrus* and *Vicia.* Chen and Shu （1944） tested the nodule-forming ability of rhizobia isolates from nodules of twenty four species of eighteen genera of Leguminosae on *A. sinicus* host. All five isolates of *A. sinicus* rhizobia and one isolate from nodules of *Desmodium heterotyllum* formed nodules on *A. sinicus* host. All the remaining isolates did not form nodules on *A. sinicus* host，namely，isolates from *Vicia hersuta*，*V. sativa*，*V.* sp.，*Trifolium pratenese*，*T.* sp.，*Molilotus indica*，*Medicago hispida*，*M. sativa*，*Soja max*，*Cajanus cajan*，*Lespedeza striata*，*Indigofera sujfruticosa Phaseolus aureus*，*P. vulgaris*，*P. acoritifolius*，*Vigna sinensis*，*Crolatoria strata*，*Alysicarpus vaginalis*，*Atylosia scarabeoides*，*Derris elliptica*，and an unidentified legume. Ning etal. （1988） found that rhizobia isolated from *A. sinicus* did not nodulate *A. adsurgens* and vice versa. Three *A. sinicus* rhizobia isolates tested by Chen etal. （1990） nodulated *A. adsurgens* and *A. alginosus* but not *A. membranacens* and *A. mongholicus*；they did nodulate *Vicia villosa*，*Phaseolus vulgaris* and *Sesbania* sp.，but not *Hedysarum mongolicum*，*Pisum sepium*，*Trifolium repens*，*Melilotus albus*，*Lotos corniculalus*，*Glycine max*，*Lupinus* sp. and *Vigna sinensis.* It is known that there is no clear-cut demarcation dividing host-rhizobia cross-inoculation groups. All evidence considered，the *A. sinicus* rhizobia symbiosis may be taken as a monospecific cross-inoculation system，although interspecific and intergeneric cross-inoculation relations do happen. Observations of extension experts have confirmed this justification. On fields where nodules are naturally formed on *Vicia* spp.，*Faba faba*，*Glycine max* or other legumes，*A. sinicus* sown for the first time fail to nodulate if not properly inoculated.

4. THE SYMBIOTIC PLASMIDS

Same as in other fast-growing rhizobia，the nodulation and nitrogenase gene **determinants** are located on megaplasmids. One to four megaplasmids are present in the *A*，*sinicus* rhizobia examined. Table 1 presents the number of megaplasmids，phenotypic expression of nodulation and nitrogen fixation of wild and mutant strains of *A. sinicus* rhizobia. Sone strains contain only one megaplasmid and are positive of nodulation and nitrogen fixation. Strain 7653R contains two megaplasmids. The heat-cured mutant，strain 7653R1，has lost one of the wild-type megaplasmids and becomes uninfective to *A. sinicus.* The acridine orange-cured mutant strain 7653R5 has also lost one megaplasmid and behaves likewise. The wild-strain S52 contains also two megaplasmids and is nodulation positive and nitrogen fixation positive. Its heat-cured mutant，strain S52S2，has lost the larger megaplasmid（pRaS52b）and becomes nodula¬tion positive but nitrogen fixation negative. Radioprobing shows that the nod genes are located on the smaller megaplasmid （pRaS52a），responsive positively to 32P-labelled pRmSL42 probe containing nodDABC，and the nif genes are located on the larger megaplasmid（pRaS52b），responsive positively to 32P-labelled pIJ 1242 probe containing m/KDH（Wei and Li 1989）.

Table 1 Number of megaplasmids and the properties of nodulation and nitrogen fixation of different strains of *A. sinicus* rhizobia

Strain	Number of megaplasmid	Nodulation	Nitrogen fixation	Reference
7653[a]	1	+	+	Zhang et al. 1986
7653R[a]	2	+		Zhou et al. 1987
7653R1[b]	1	-		Zhou et al. 1987
7653R5[c]	1	-		Zhou et al. 1987
76531[a]	4	+	+	Wang et al. 1988
S52[a]	2	+	+	Wang et al. 1986
S52S2[b]	1	+	-	Zhou et al. 1987
HR104[a]	3	+	+	Wang et al. 1988
HR1042[b]	2	+	+	Wang et al. 1988
HR107[a]	3	+	+	Wang et al. 1988
HR112[a]	2	+	+	Wang et al. 1988
Ra31[a]	1	+	+	Wang et al. 1988
Ra1[a]	1	+	+	Wang et al. 1988
Ra74[a]	1	+	+	Wang et al. 1988
Ra81[a]	1	+	+	Wang et al. 1988
CZ74[a]	2	+	+	Wang et al. 1988
SR72[a]	2	+	+	Lin et al. 1986
Ra27[a]	2	+	+	Zhang et al. 1986
HR101[a]	2	+	+	Wang et al. 1988

Notes:

[a] wild type （isolated from nodules of field host）

[b] heat cured

[c] acridine orange cured

[d] all references except this one are given members of this laboratory

Zhang et al. （1989） prepared a gene bank of *A. sinicus* rhizobia strain 7653R. By triparental mating （*E. coli* HB101 gene bank as donor，*E. coli* HB101 （pK2013） as helper and *A. sinicus* rhizobia Nod$^-$ strain 7653R-1 as receptor） and selection by plant nodulation test，are constructed plasmid pRaZ15 was obtained. Strain 7653R1 （pRaZ15） regains abilities of the nodulation and nitrogen fixation on *A. sinicus.* pRaZ15 was transferred into *Phaseolus vulgaris* rhizobia strain 3622-15 and *Pisum sativum* rhizobia strain T83K3，and the resulting 3622-15 （pRaZ15） and T83K3（pRaZl5） nodulated *A. sinicus* （Fig. 2）. Probed by ^{32}P-labelled pRmZl containing R.m. nodDABC or ^{32}P-labelled pSA30 containing in *K. p. nif*KDH，pRaZ15 responds positively，indicating that the pRaZ15 contains both *nod* and *nif* genes，which is in agreement with the nodulation test.

5. INFECTION AND NODULE DEVELOPMENT

The infection and nodule development of *A.* smicus-rhizobia symbiosis is similar to the *R. leguminosarum* type. Infection of the root hair takes place after the elongation and curling of the root hair stimulated by the presence of rhizobia. Nodule development is visible on the seedling by the naked eye in six or seven days after inoculation of the rhizobia to germinated seeds. Mature nitrogen-fixing nodules are oblong to cylindrical in shape. The dividing meristem lies at the terminal region of the nodule. Reddish infected bacteroidal tissue develops progressively behind the meristem until the nodule stops developing. Two strands of vescular bundles lie between the infected bacteroidal tissue and the cortex surrounding the nodule.

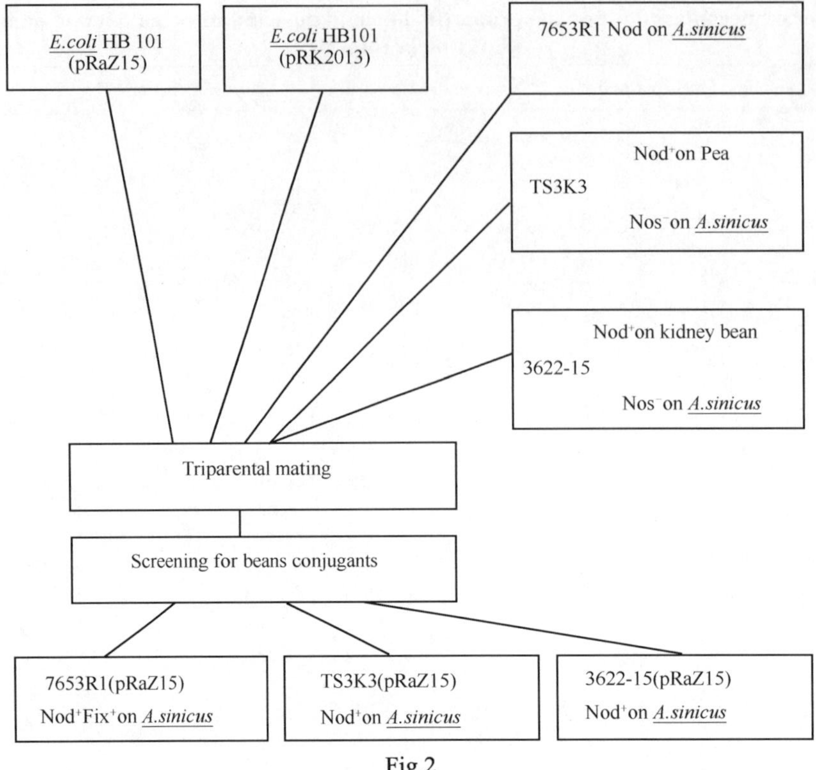

Fig.2

The bacteroids are mostly club-shaped, approximately forty to fifty times larger than the unswelled rod-shaped bacteria present in the infection threads or in in-virtro culture. The bacteroidal host cell contains vast amounts of swelled bacteroids and a small number of rod-shaped bacteria. Microculture of bacteroids and bacteria released from bacteroidal host protoplast （Fig.3） reveals that only the rod-shaped bacteria multiply （Fig.4）, while the club-shaped bacteroids remain undivided （Cao et al. 1984; zhou et al. 1985）. This is common to *A. sinicus* nodules and the nodules of alfalfa, clover and vetch, but not with soybean nodules. The 'bacteroids' in the soybean bacteroidal tissue are rod-shaped, unswelled, and almost all multiply themselves when transferred to the culture medium. Table 2 presents the comparisons of viable counts and direct counts of organisms released from bacteroidal host protoplasts of *A. sinicus*, *Medicago sativa*, *Trifolium repens*, *T. subterraneum*, *Vicia crocca* and *Glycine max* （Cao et al. 1984）.

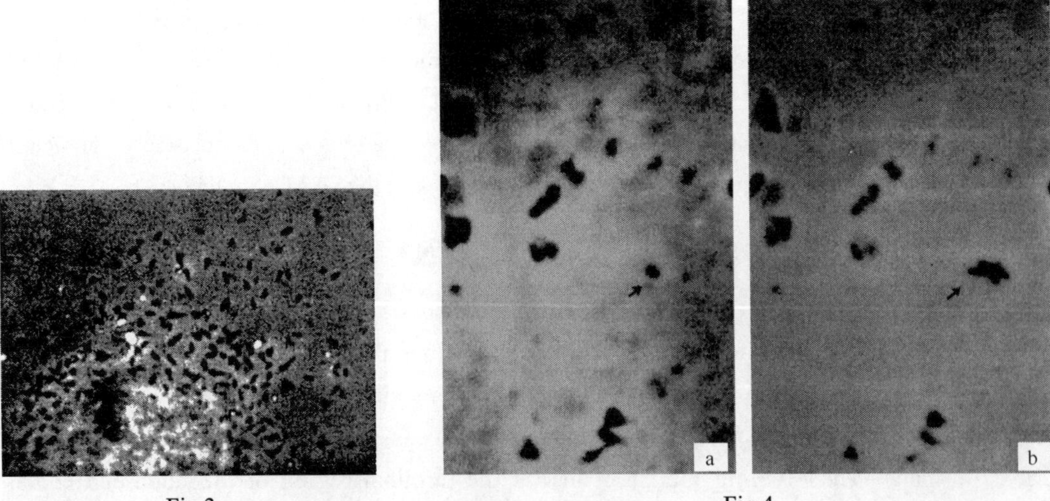

Fig.3　　Fig.4

Table 2 Viable and total counts of the rhizobia from bacteroidal host protoplasts from nodules of different legumes (Cao et al. 1984)

Legume	Protoplasts counted		Average number of rhizobia per protoplast	Ratio V：T
A. sinicus	Viable	17	63.9	
	Total	103	23375.5	0.27：100
Medicago	Viable	28	26.5	
sativa	Total	15	22755.0	0.11：100
Trifolium	Viable	15	17.7	
repens	Total	11	19509.0	0.09：100
Trifolium	Viable	29	19.3	
subterraneum	Total	15	17813.0	0.11：100
Vicia crocca	Viable	28	29.9	
	Total	15	23973.0	0.12：100
Glycine max	Viable	52	50903	
	Total	216	62781	81.08：100

6. DISTRIBUTION AND EXPANSION OF *A. sinicus*

In the Yangtze Valley and the southern provinces, the autumn-sown legumes, cereals and rapeseeds are grown over winter in properly drained rice fields. Among the legumes, *A. sinicus* and *Vicia crocca* are grown for green-manuring, and *Faba faba* is grown chiefly for seeds. Under the condition of energy stress and short supply of chemical nitrogen fertilizers, nitrogen-fixing green manure crops not only contribute a good deal of nitrogen nutrient for the succeeding rice crop, but also organic matter to the soil. However, the legumes can only grow in properly irrigated and drained rice fields. In the rain-fed rice fields, the fields have to be left fallow in order to receive and conserve water for the next rice crop. In the lowland, without properly developed irrigation and drainage system, the fields are flooded all year round, no upland winter crop could grow. Owing to the progress of construction of reservoirs and irrigation systems in the past forty years, the acreage of the winter legumes and other winter crops were very much expanded in the rice-growing region.

Traditionally there was an interboundary cut roughly at 112 degrees east longitude right across the Hubei and Hunan provinces. *A. sinicus* was grown chiefly in the east up to the sea coast. *Vicia crocca* was grown chiefly to the west and the southwestern provinces. Since the late nineteen fifties, double-cropping rice systems, i.e., an early rice crop followed by a late crop rice in the same year, is extended in the Yangtze Valley. *A. sinicus* fits better than *V. crocca* as green manure crop for the double-cropping rice system. Therefore, *A. sinicus* has taken the place of *V. crocca* in the traditional *V. crocca* region whereever the double-cropping rice system is practised.

For both reasons mentioned above, the acreage of *A. sinicus* expanded significantly in the rice-growing region in the sixties and seventies. The expansion of the acreage of *A. sinicus* in Hubei province between 1957 and 1975 increased approximately three-fold, according to a provincial sensus:

	hectares (10^4)
1957	29
1962	29
1965	59
1970	67
1973	89
1975	106

7. OCCURRENCE AND INTRODUCTION OF *A. sinicus* Rhizobia IN RICE FIELDS OILS

In fields where *A. sinicus* has never been sown, the soil is generally devoid of *A. sinicus* rhizobia. Table 3 presents the result obtained by comparing the occurrence of *A. sinicus* in the soil of two rice fields on the same farm at Shizishan, Wuhan, developed on Yangtze alluvial. Estimation of the quantity of *A. sinicus* rhizobia in soil was carried out by a modified Brokwell's nodulation method （Brokwell 1968）. Surface-sterilised seeds of *A. sinicus* were seeded to sterilised plant culture tubes. Serial dilutions of soil samples were introduced into the tubes as inoculant. The presence of *A. sinicus* rhizobia in the soil dilutions was indicated by the nodules formed on the seedlings grown for five weeks in the -growth chamber. The number of *A. sinicus* rhizobia were estimated in five replicates by the MPN method. In a field where *A. sinicus* had never been grown, no *A. sinicus* rhizobia were detected. In a field where *A. sinicus* was established for years, soil samples taken in different seasons contained thousands to hundred thousands *A. sinicus* rhizobia per gram of dry soil.

Table 3 Occurrence of *A. sinicus* rhizobia in two rice field soils at the same farm developed on Yangtze alluvial. Shizishan. Wuhan. 1964 （Cao and li, unpublished）

Date of sampling	Field Ⅰ[a]		Field Ⅱ[a]	
	Field condition	Rhizobia[b]	Field condition	Rhizobia[b]
Feb. 2	*A. sinicus*	94 000	Fallow	0
Mar. 31	*A. sinicus* blossoming	29 000	Water-logged	3.6
Jan. 11	Rice planted	30 000	Rice planted	0
Sept. 14	Soil plowed	1 800	Soil plowed	0
Dec. 16	*A. sinicus*	180 000	Rapessed	0

Notes:

[a] field I *A. sinicus* established for years; field Ⅱ *A. sinicus* never grown before

[b] Number of *A. sinicus* rhizobia per gram dry soil, estimated by the plant nodulation method, see text for brief description

For a field where *A. sinicus* was sown for the first time, proper inc culation of *A. sinicus* rhizobia not only ensures successful nodulation and symbiotic nitrogen fixation, but also ensures the integration of the rhizobia as a relatively stable member of the microbial population of the soil. Table 4 presents data of an experiment carried out at Xing'andu, Wuhan. A rice field where *A. sinicus* has never been sown before was divided into two parts. On one part *A. sinicus* was sown without inoculation, and on the other part *A. sinicus* was sown and properly inoculated. At the first sampling, before *A. sinicus* seeds were sown, soil in both parts contained no *A. sinicus* rhizobia, as shown by the plant nodulation method described above. At the second sampling, when the *A. sinicus* had grown for a month, the quantity of soil *A. sinicus* rhizobia in the inoculated parts amounted to 33 thousand per gram of dry soil. In the uninoculated part, the soil contained only 10 *A. sinicus* rhizobia per gram of dry soil, most probably from carried-over seeds. During the entire growth period *of A. sinicus* the quantity of *A. sinicus* rhizobia measured up to hundred times in both parts.

Table 4 Occurrence and abundance of *A. sinicus* rhizobia in rice field where *A. sinicus* was sown for the first time with or without artificial in inoculation at Xingandu, Wuhan, 1964 （Cao and Li, unpublished）

Date of sampling	Field condition	Rhizobia per gram of dry soil[a]	
		Uninoculated	Inoculated
Sept. 6	Ratoon rice Field flooded	0	0

续表

Date of sampling	Field condition	Rhizobia per gram of dry soil[a]	
		Uninoculated	Inoculated
Nov. 18	*A. sinicus*	10	35 000
Jan. 16	*A. sinicus*	4 480	145 000
Mar. 15	*A. sinicus* in full blossom	1 380	163 000
May. 17	Seeds ripe	318 000	10 100 000

[a] Estimated by the plant nodulation method

Field surveys carried out by this laboratory and other institutions offered confirming observations. Where *A. sinicus* were sown for the first time, no nodule was formed or only a few seedlings were nodulated. Once a population of the *A. sinicus* rhizobia is established in the rice field soil, it would persist for years even if no succeeding crop of *A. sinicus* is grown. Pot experiments were carried out at Xiaolingwei, Nanjing (Huang 1983). Sterilised pots were filled with soil taken from a rice field where *A. sinicus* was grown yearly before 1975 but no *A. sinicus* was grown afterwards. Surface-sterilised seeds of *A. sinicus* were sown to the pots with or without artificial inoculation. After 92 days of culture, an average of 14.2 nodules per plant were formed in the uninoculated pots and 52.4 nodules per plant were formed in the inoculated pots. The result proved that *A. sinicus* rhizobia did exit for seven years as an integrated member of microbial population of the soil (Table 5).

Table 5 Pot experiment on the nodulation of *A.sinicus* on soil taken from a field where no *A.sinicus* was grown for seven years but was grown yearly before, Xiaolingwei, Nanjing (Huang 1983)

	Uninoculated	Inoculated
Fresh weight (top and roots) per pot (g)	27.15	55.70
Number of nodules per plant	12.2	52.4

Note: Average value of 6 replicated pots, 15 plants per pot

8. ISOLATION OF EFFICIENT STRAINS

Isolation and selection of *A. sinicus* rhizobia were carried out at this laboratory and other institutions, mostly in the late fifties, sixties and seventies. No centralised culture collection institute was established until the establishment of the China Committee for Culture Collection of Microorganism (CCCCM) and China Centre for Type Culture Collection (CCTCC) in 1979. Efficient strains of *A. sinicus* rhizobia were obtained and lost. Table 6 presents a representative field experiment related to the performance of strains selected by this laboratory for the purpose of selecting strain used in inoculant preparation.

Table 6 Field performance of *A.sinicus*-rhizobia symbiosis of four *A.sinicus* rhizobia examined 80 days after sowing. Honghu, Hubei. 1973 (Cao and Li, unpublished)

	Uninoculated control	Inoculated with strain			
		A16	A19	A21	A26
Dry weight per plant (g)	3.3	5.5	4.8	5.5	5.5
Branching	0.2	1.4	1.1	1.4	1.4
Nodules per plant	2.1	19.5	9.6	13.7	17.0
%N content of oven-dry matter of aerial growth	1.5	3.8	3.75	3.34	3.85

Note: 4 replicated plots were laid out for each treatment; 20 plants were taken randomly from each plot. Figures shown are average values.

9. PREPARATION AND QUALITY OF INOCULANT

Most inoculant preparations are peat based. Neutral peat is sterilised in an autoclave and packed in bottles or polychloroethylene bags. Liquid cultures of rhizobia are produced in aerated ferment. Aliquotes of liquid culture are added to sterilised peat bottles or bags aseptically. The preparations are kept for five days and examined for quality. All unproperly packed, puddled or visibly molded bottles or bags are discarded. Samples are taken from each batch for bacteriological examination. Samples arc serially diluted and plated on agar media. A batch is discarded if the samples are contaminated or of very low count.*A. sinicus* is sown in September/October. The inoculants have to be prepared, stored and distributed in summer season. Quality of the preparation is determined by the process of production as well as, or even more by the condition and duration of storage and distribution. At high temperature, the viable counts of rhizobia drop drastically. Figure 5 presents data of changing trends of the viable counts of inoculant preparations during four months of storage at 12℃, 28℃and 35℃. The viable count of inoculant preparation kept at a much higher level at 12℃than those at higher temperature. At 35℃, the viable count droped to a level of under one million rhizobia per gram of preparation.

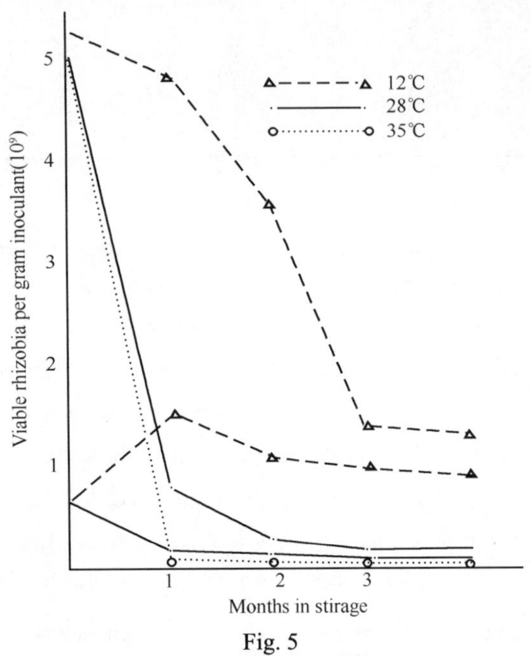

Fig. 5

10. APPLICATION OF INOCULANT

In agricultural practice, 40g peat-based inoculant preparation is mixed with every kg of *A. sinicus* seeds, consisting of approximately 280000 seeds. Three to four kg seeds are sown to one hectare of drained rice field, either on prepared land after single cropping rice, or 1 to 2 weeks before harvest of the second rice crop of the double-cropping system, between rows of rice stand.

Abundant inoculation is emphasised. Sand experiments were carried out to assess the number of viable rhizobia in inoculant preparation required per seed for satisfactory nodulation. In an experiment carried out with inoculant stored at 20℃for different durations, the minimum number of viable rhizobia per seed required for satisfactory nodulation was 175 and 200 viable rhizobia per seed for inoculant preparations stored for 1 and 30 days, respectively (Table 7). In another experiment, inoculant preparations were stored at different temperature (12℃ and 25℃) for the duration of 90 days, the minimum number of viable rhizobia per seed for satisfactory nodulation was 36 and 15, respectively (Table 7). No significant improvement in nodulation was observed, even hundred thousand times more viable rhizobia than the above minimum was added.

Table 7 Sand pot experiments estimating nodulation effectiveness of *A.sinicus* seed-rhizobia mixture (Zhou, Hu, Cao and Li 1965, unpublished)

Inoculant		Nodulation effectiveness						
28℃, 1 day	A/1.75	10^6	10^5	10^4	10^3	10^2	10^1	10^0
Viable rhizobia per gram	B	4/4	4/4	4/4	4/4	4/4	4/4	—

续表

Inoculant	Nodulation effectiveness							
52.4×10^3	C	4.0	3.4	4.1	6.3	4.8	0.7	—
28℃，30 days	A/2.90	10^6	10^5	10^4	10^3	10^2	10^1	10^0
Viable rhizobia per gram	B	4/4	4/4	4/4	4/4	4/4	2/4	—
10.8×10^8	C	5.7	3.5	4.5	3.4	3.6	0.4	—
12℃，90 days	A/0.36		10^5	10^4	10^3	10^2	10^1	10^0
Viable rhizobia pergram	B		4/4	4/4	4/4	4/4	3/4	1/4
10.8×10^8	C		3.7	4.6	3.6	4.7	1.9	0.08
25℃，90 days	A		10^5	10^4	10^3	10^2	10^1	10^0
Viable rhizobia pergram	B		4/4	4/4	4/4	4/4	3/4	0/4
4.5×10^8	C		3.2	3.3	2.4	3.2	2.0	0

Note：A-Number of viable rhizobia per seed inoculated，B-Pots inoculated in 4 replicates，C-Nodules per plant，3 seed hills per pot，3 seeds per hill

11. BENEFIT OF ARTIFICIAL INOCULATION TO *A. sinicus*-rhizobia SYMBIOSIS

Many field experiments were carried out by this laboratory and other institutions on the benefit of artificial inoculation of *A. sinicus* rhizobia to the *A. sinicus*-rhizobia symbiosis. Table 8 cites two field plot experiments carried out by this laboratory at Jingmen County，Hubei Province，in 1965，on rice fields where *A. sinicus* was grown for the first time. Statistics of pot and field experiments，carried out by various institutions showed that，similar to the cited experiments artificial inoculation of *A. sinicus* rhizobia was almost invariably necessary for *A. sinicus* grown for the first time in rice field soils. On the other hand，on fields where *A. sinicus* was carried out in farmers' fields，the vegetative growth of *A. sinicus* can only be estimated by pooling samples taken at random on a plot. The standard deviations of mean values of replicated plots so obtained are high. Table 9 presents data of an experiment，in which 28 percent increase of mean value of four replicated plots inoculated by an efficient strain of *A. sinicus* rhizobia over the mean value of uninoculated control plot is statistically not significant at 5 percent probability level. Table 10 presents a field survey of fifty two fields in twelve counties in Hubei Province for the effect of artificial inoculation of *A. sinicus* rhizobia in fields where *A. sinicus* were grown for the first time or for some years previously. Although factors beside artificial inoculation affect the yield and nitrogen content of *A. sinicus*，it is justified to conclude that in terms of practical farming，artificial inoculation is necessary to rice fields new to *A. sinicus* as well as to fields where *A. sinicus* has been grown previously.

Table 8 Effect of rhizobia inoculation on fresh weight （tops only） of *A.sinicus* grown for first time on field at Jingmen county，Hubei Province. 1965 （Hu and Li，unpublished）

Experiment	Treatment	MT per hectare[a]	% increment
I	Uninoculated	10.50	
	Inoculated	23.25	179**
II	Uninoculated	21.00	
	Inoculated	37.50	79*

[a] Mean of 4 replicated plots arranged randomly

Table 9 Field experiment on the effect of inoculation of *A.sinicus* rhizobia strains on field where was grown successfully in previous years at Qianjiang, Hubei, 1975-1976 (Li et al. unpublished)

Treatment	Freash wt. MT per hectare[a]	S.D.
Unino. contral	23.92	±1.995
Ino.st.A16	21.43	±2.487
Ino.st.A1105	25.99	±1.429
Ino.st.A1106	30.62	±2.487
Ino.st.A1107	25.20	±3.679
Ino.st.A1108	25.90	±3.506

[a] Mean value of 4 replicated plots for each treatment

Table 10 Survey of the effect of artificial inoculation on *A.sinicus* of 52 fields in 12 counties in Hubei Province, 1966 (Li, Hu and Zhou, unpublished)

Uninoculated		Inoculated	
Fresh wt. MT per hectare	Fields	Fresh wt. MT per hectare	Fields
A.sinicus grown for the first time			
Below 7.5	3		
7.5-15	3	Below 15	3
7.5-22.5	3	15-30	8
		Over 30	10
A.sinicus grown previously			
Below 7.5	1		
7.5-15	3	Below 15	0
7.5-22.5	5	15-30	2
		Over 30	11

12. AGRONOMIC CONSIDERATIONS

Proper agronomic management is demanded for high yield of the crop and high nitrogen-fixing activity of *A. sinicus* rhizobia symbiosis. *A. sinicus* cultivar, growth season, water-air regime, phosphorus supply and nodulation by efficient rhizobia are the predominant factors.

There are two major types of *A. sinicus* cultivars. The large-leaflet-type cultivars grow more vigorously but blossom two or three weeks later than the small-leaflet-type cultivars. Gross yield and succulence considered, full blossom stage is the most suitable time for plowing under the green matter for manuring. Choice of cultivar type is determined by the time demanded for plowing under the green matter, which, in turn is determined by the time of transplanting rice seedlings.

The *A. sinicus* crop has two active growth periods before and after the severe winter. In the winter, when the daytime temperature falls below 14℃, there is little or no aerial vegetative growth. The length of the dormant period varies with locality and winter severity of the year.

Water or air stress retards the development of *A. sinicus*. *A.* sinicus-rhizobia symbiosis suffers more than the host plant. In agricultural practice, in the diked rice fields, impeded drainage often causes soil air stress, especially during the wet spring season. In fields containing enough native rhizobia or properly inoculated, soil air stress causes the formation of small nodules containing little bacteroidal tissue which is colourless or only very lightly pink coloured, due to suppressed development of leghemoglobin. Moreover, the life span of the nodules is short and decays early. Shallow ditches dug at the periphery of and across the rice field is practised to improve the soil air regime. On the other hand，water stress is often met during the autumn drought. Light irrigation is practised to relieve water stress. Table 11 presents data of a pot experiment on the effect of the water-air regime on the development of *A. sinicus* and nodulation. Seeds were inoculated and sown to pots containing heavy loam. At the time of sowing，the water content was held at the water-holding capacity of the soil. In one treatment，the water content was left to dry naturally down to 40 percent of its water-holding capacity. In other treatments，the water contents were kept at 60，80 or 100 percent of their water-holding capacity. At the lowest water content level，the plant growth suffered slightly，but no nodule was formed at all. On the other extreme，in pots in which the water content was held at the soil's water-holding capacity，plant growth did not suffer but the number of nodules formed was very much reduced.

Table 11 Pot experiment on the effect of the water-air regime on *A.sinicus* nodulation （Cao，unpublished，1965）

water-air regime	Average value per plant		
（% water holding capacity）	Leaves	Plants nodulated （%）	Nodules
100% down to 40%	3.1	0	0.0
100% down，kept at 60%	3.3	69	2.7
100% down，kept at 80%	3.7	97	9.6
kept at 100%	3.7	80	3.4

Note：Soil （heavy loam） pot experiment （4 replicates），seeded and inoculated，grown for 24 days.

Except soils developed on purple shale soils，which are rich in total and available phosphorus，most soils in the Yangtze Valley are developed on weathering materials of granite，schist，tertiary，quarternary or recent deposites，which are poor in total and available phosphorus. *A. sinicus*，similar to other legumes，is very responsive to phosphorus fertiliser. Table 12 presents data of a field experiment carried out by a farming team on their own communal field，at Macheng，Hubei. The combined effect of inoculation and phosphorus fertiliser was much stronger than that when they were treated separately.

Table 12 Effect of superphosphate and inoculation on the yield of *A. sinicus*

Treatment	Green matter MT per hectare
Uninoculated，P unfertilised	4.58
Inoculated，P unfertilised	6.08
Uninoculated，P fertilised	8.61
Inoculated，P fertilised	12.22

Note：Field experiment carried out by the commune farming team of Macheng，Hubei，1965. *A.sinicus* grown for the first time 300kg of superphosphate applied per hectare. Yield estimated in late March at the stage before blossoming，under the superviseon of members of the Laboratory of Soil Microbiology，Huazhong Agriculture College.

REFERENCES

[1]Allen ON，Allen E （1981） The leguminosae. A source book of characteristics，uses and nodulation.University of Wisconsin Press，Madison，USA

[2]Brockwell J （1968） Accuracy of a plant-infection technique for counting populations of *Rhizobium trifolii.* Appl Microbiol 11：377

[3]Cao YZ，Zhou JC，Chen HK （1984） Differentiation and viability of nodule bacteria in host cells. Scientia Sinica （Series B） 27：583

[4]Chan CL，Lumpkin TA，Root CS （1988） Characterisation of *Bradyrhizobium* sp. *{Astragalus sinicus* L.） using serological agglutination，intrinsic antibiotic resistance，plasmid visualization and field performance Plant Soil 109：85

[5]Chen HK.Shu MK （1944） Notes on the root nodule bacteria *of Atragalus sinicus* L. Soil Sci 58：291

[6]Chen HK （ed） （1959） Microbiology. Higher Education Press，Beijing （in Chinese）

[7]Chen WX，Li GS，Qi YL，Wang ET，Yuan HL，Li JL （1990） *Rhizoboum huakui* sp.nov. isolated from the root nodules of *Atragalus*

sinicus. Int J Syst Bacteriol （accepted for publication）

[8]Huang LK（1983）Inoculant efficacy of green manuring rhizobia in the traditionally legume grown area（in Chinese）. Jiangsu Agr Sci 1983（1）：47

[9]Ishizawa S （1954） studies on the root nodule bacteria of leguminous plants II. part 1. Cross-inoculation test （in Japanese，English summary）. J Soil Manure （Japan） 24：297

[10]Lin LF，Gong HY，Zhou LM，Cen YH （1988） Study on the plasmid profiles from fast-growing rhizobia （In Chinese，English summary）. Acta Microbiol Sinica 26：271

[11]Ning KZ，Li YF，Huang YL （1988） Nodulation behavior of *Astragalus adsurgens* and *Astragalus sinicus* （in Chinese） Microbiol Magazine 8：56，Shengyang，China

[12]Wang CL，Chen J B （1988） The plasmids pattern and endoantibiotic resistence of *Rhizobium astragali* strains （in Chinese，English summary）. J of Huazhong Agr Univ 7：15

[13]Wei H and Li FD（1989）Physical evidence for plasmid-borne symbiotic genes in *Rhizobium astragali* （in Chinese，English summary）. J of Huazhong Agr Univ 8：121

[14]Wilson JK （1939） Leguminous plants and their associated organisms. Cornell Univ Agric Exp Sta Mem 221

[15]Wilson JK，Chin JH （1947） Symbiotic studies with isolates from nodules of species of *Astragalus.* Soil Sci 63：119

[16]Zhang ZM，Zhou JC，Xu YC，Chen HK，Li FD，Fan YL （1986） Studies on plasmids of *Rhizobium astragali.* Isolation，purification and enzyme-digestion of plasmids DNA from *Rhizobium astragali* （in Chinese，English summary）. J of Huazhong Agr Univ 5：326

[17]Zhang ZM，Chen HK，Li FD，Fan YL （1989） Construction of gene library and isolation of *nod* genes of *Rhizobium astragali.* In：Li Shuxuan （ed） Application of biotechnology in agriculture pp 30-32. Shanghai Scientific and Technical Publ

[18]Zhang ZM，Zhou JC，Chen HK （1989） Identification of *hsn* and m/HDK genes on pRaZ15 incorporated to *Rhizobium astragali* （in Chinese，English summary）. J of Huazhong Agr Univ 8：285

[19]Zhou JZ，Cao YC （1981） Nutrient requirement of the fast growing *Rhizobium* J of Huazhong Agr Univ 3：44 （in Chinese，English summary）

[20]Zhou JC，Tchen YT，Vincent JM （1985） Reproduction capacity of bacteroids in nodules of *Trifolium repens* L and *Glycine max* （L） Merr Planta 163：473

[21]Zhou JC，Zhang ZM，Huang CJ，Li FD，Chen HK （1987） Studies on *Rhizobium* plasmids II. Tests of plasmid elimination of *R. astragali*，J of Huazhong Agr Univ 6：156 （in Chinese，English summary）

紫云英根瘤菌与豌豆根瘤菌共生质粒间的相互作用*

张学贤　周俊初　张忠明　李阜棣　陈华癸

（华中农业大学生物技术中心）

摘　要　将豌豆根瘤菌共生质粒 pJB5JI 导入紫云英根瘤菌含共生质粒的 7653R 菌株和消除了共生质粒的 7653R+1 菌株中。7653R+1 接合子获得了在豌豆上结瘤的能力，7653R 接合子却不能在豌豆上结瘤。从而揭示出紫云英根瘤菌的共生质粒能抑制导入的豌豆根瘤菌共生质粒 pJB5JI 功能的表达。同时 pJB5JI 的导入使 7653R 接合子在其正常宿主紫云英上的结瘤固氮能力较亲本 7653R 显著提高。试验同时考察了 pJB5JI 导入紫云英根瘤菌后的稳定性。

关键词　紫云英根瘤菌；共生质粒；豌豆根瘤菌；接合转移

在自然界，一种根瘤菌只能与一种或几种相应的豆科植物形成共生固氮体系，即豆科植物构成互接种族关系。近年来，随着根瘤菌分子遗传学的发展和基因工程技术的应用，通过转移根瘤菌共生质粒或共生基因能够打破这种互接种族的范围[1]。如 Beynon 等[2]将豌豆根瘤菌 T83K3 菌株中的共生质粒 pJB5JI 转移到菜豆根瘤菌中，使后者获得了在豌豆上结瘤的能力，王常霖等[3]将 pJB5JI 转移到三叶草根瘤菌，获得的转移接合子既能在三叶草上结瘤，也能在豌豆根上结瘤。

紫云英根瘤菌是我国的一种重要的共生固氮资源，在互接种族中独属一族[4]。最近，陈文新等[5]用数值分类法，将其定名为 *Rhizobium huakuii*。自八十年代以来，国内外开展了对紫云英根瘤菌共生固氮作用的遗传学研究，已有的研究结果表明，紫云英根瘤菌含有 1-4 个大质粒[6]，而且共生固氮基因定位在质粒上[8, 9]。周俊初等[10]用高温和吖啶橙处理野生型紫云英根瘤菌菌株 7653R，获得消除了共生质粒而不能在紫云英上结瘤的 7653R+1 菌株（Nod^-）。张忠明等[11]从 7653R 菌株的文库中调出了 pRaZ15 重组质粒，其中的 7653R DNA 片段含有 *nod*、*nif* 和 *fix* 基因，转入 7653R+1 菌株，能恢复其结瘤固氮功能。

本研究将豌豆根瘤菌 T83K3 菌株的共生质粒 pJB5JI 分别导入紫云英根瘤菌 7653R 和 7653R+1 菌株，比较研究了 pJB5JI 在紫云英根瘤菌不同遗传背景条件下的行为和表达，以探讨紫云英根瘤菌与豌豆根瘤菌共生质粒间的相互作用。

1　材料与方法

1.1　细菌菌株与培养条件

细菌菌株与质粒见表 1。根瘤菌培养基：液体培养采用 TY[10]，固体培养采用 SM[11]，质粒检测采用 PA[6]；大肠杆菌培养基采用 LB[12]。培养温度：根瘤菌 28℃，大肠杆菌 37℃。抗生素使用浓度：四环素（Tc）15 r；链霉素（Str）300 r；卡那霉素（Km）50 r。

表 1　供试细菌菌株与质粒

菌株与质粒	有关性状	来源
紫云英根瘤菌		
7653R	野生型 Nod^+Fix^+ Str^r	本室
7653R+1	7653R 消除共生质粒突变株 Nod^-	见文献[10]
7653R（pRK404）	Str^r Tc^r	本研究
7653R+1（pRK404）	Str^r Tc^r	本研究

*原载于《华中农业大学学报》，11（2）：175~181，1992.

续表

菌株与质粒	有关性状	来源
7653R+1（pJB5JI）	$Str^r Km^r$	本研究
7653RA1……A10	分别从 7653R （pJB5JI）在紫云英上所结根瘤中分离	本研究
7653R+1Pa，Pb	分别从 7653R+1（pJB5JI）在豌豆上所结根瘤中分离	本研究
豌豆根瘤菌		
JI6015	$Str^r rif^r Nod^-$	见文献[3]
T83K3	JI6015（pJB5JI） $Nod^+ Fix^+$	见文献[3]
质粒		
pJB5JI	共生质粒 pRL1J1 上插入 Tn5，Km^r	见文献[3]
pRK404	$mob^+ tra^-$、Tc^r	见文献[13]
pRK2013	$Tra^+ Km^r$	见文献[13]

1.2 植物结瘤试验

采用本室改进的双层钵进行。砂土装在塑料杯中，塑料杯再放在装有 Fahraeus 无氮植物营养液的玻璃瓶中，通过纱布将营养液吸入砂土中。种子表面灭菌及催芽皆按常规方法进行，豌豆每瓶种一株，紫云英每瓶种 4 株。人工光照栽培，每天光照 14 小时，光强 4500~6000 lx，温度 15-24℃。

1.3 根瘤固氮酶活性测定

剪取鲜瘤（带一段小于根瘤宽度的根段）置于 20ml 小瓶中，加盖橡皮塞，马上抽气 2ml，注入乙炔 2ml，28℃下反应 2 小时，用日立 163 型气相色谱仪测定乙炔还原值。

1.4 其他方法

质粒的接合转移采用滤膜杂交法[11]，质粒快速检测采用修改的 Eckhardt 方法[12]。

2 试验结果

2.1 紫云英根瘤菌受体菌株的标记

由于供体菌 T83K3 为 Km^r 和 Str^r，紫云英根瘤菌受体菌株 7653R 和 7653R+1 也为 Str^r，为了便于选择转移合子，本试验通过三亲本杂交方法将 pRK404 质粒分别导入 7653R 和 7653R+1 中，使其具有 Tc^r，以便在杂交中淘汰供体菌 T83K3。pRK404 质粒具有在大多数革兰氏阴性细菌中转移和复制的能力，但在根瘤菌中，在无选择性压力条件下很容易丢失[13]。因此含有 pRK404 质粒的紫云英根瘤菌 7653R（pRK404）和 7653R+1（pRK404）在杂交前一直在含四环素的培养基上培养，杂交后的转移接合子则在不含四环素下培养，经转接两次，就淘汰了 pRK404 质粒，丧失 Tc 抗性。

2.2 豌豆根瘤菌共生质粒 pJB5JI 的转移及转移接合子的鉴定

豌豆根瘤菌 T83K3 菌株中的共生质粒 pJB5JI 具有自身转移的能力[16]。本研究将 T83K3 分别与含共生质粒的紫云英根瘤菌菌株 7653R（pRK404）和消除了共生质粒不结瘤突变株 7653R+1（pRK404）杂交，在含四环素和卡那霉素的 SM 培养基上筛选转移接合子。结果表明，豌豆根瘤菌共生质粒 pJB5JI 向紫云英根瘤菌中转移的频率约为 10^{-5}。随机从两个杂交组合中各挑取 5 个转移接合子，经在选择性培养基上纯化 3 次后，再将这些转移接合子在不加任何抗生素的培养基上转接 2 次，然后各挑取 100 个菌落分别点种于含卡那霉素和含四环素的培养基上，结果表明全部具有卡那霉素抗性，而丧失了四环素抗性，说明转移接合子中含有 pJB5JI，并具有稳定的遗传特性，而 pRK404 质粒被淘汰，质粒电泳结果见图 1。

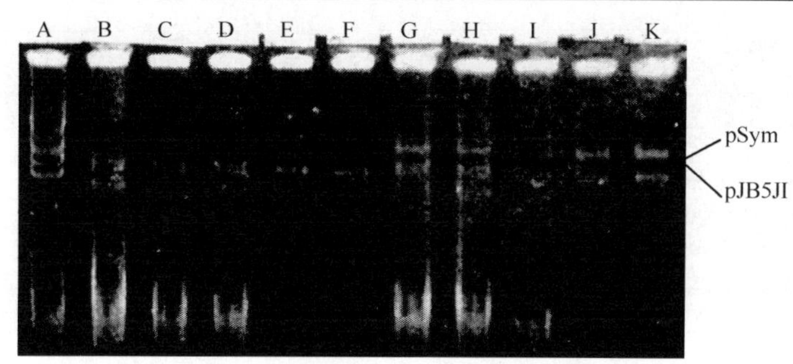

图 1　转移接合子的质粒电泳图谱

A.T83K3；B.C.D. 7653R+1 接合子；E.F. 7653R+1；G.H.I. 7653R 接合子；J.K. 7653R

2.3　pJB5JI 在紫云英根瘤菌中的共生特性

1）在紫云英植株上的共生特性

将转移接合子 7653R（pJB5JI）和 7653R+1（pJB5JI）分别接种紫云英植株，结果表明：7653R+1（pJB5JI）不能在紫云英上结瘤；而 7653R（pJB5JI）在紫云英上形成固氮的有效根瘤，而且根瘤的固氮酶活性极显著高于原菌株 7653R 所形成的根瘤（结果见表 2 及表 3）。随机从转移接合子 7653R（pJB5JI）所形成的根瘤中挑取 30 个根瘤分离根瘤菌作卡那霉素抗性和质粒检测。结果表明：它们都具有卡那霉素抗性，但大部分检测不到 pJB5JI 质粒，只有两株检测到 pJB5JI 质粒（结果见图 2）。这种现象表明转移接合子经过与紫云英共生之后，外源共生质料 pJB5JI 表现出不稳定的遗传特性。上述从根瘤中分离的转移接合子再接种紫云英植株，根瘤的固氮酶活性仍显著高于 7653R 所形成根瘤的固氮酶活性（结果见表 3、表 4）。

表 2　转移接合子的共生表型

接种物	紫云英		豌豆	
	重复	表型	重复	表型
7653R+1	5	Nod^-	5	Nod^-
7653R+1（pJB5JI）	5	Nod^-	10	Nod^+Fix^-
7653R	10	Nod^+Fix^+	5	Nod^-
7653R （pJB5JI）	10	Nod^+Fix^+	10	Nod^-
CK	10	Nod^-	10	Nod^-

表 3　7653R 转移接合子在紫云英上的共生效率

次数	接种物	地上部分干重（克/株）	固氮酶活性（10^{-2}μmol 乙烯/小时·株）								
			重复					均值	t 值	$t_{0.01}$	显著性
第一次	7653R	0.1074	26.4	12.65	11.00	23.10	4.40	15.80	3.70	2.88	极显著
			20.90	15.00	14.30	20.35	9.90				
	7653R（pJB5JI）	0.1203	17.90	20.35	51.25	32.45	31.90	30.65			
			14.30	34.10	35.20	32.45	36.60				
	CK	0.0337									
第二次	7653R	0.219	9.07	24.80	13.70	15.10	13.00	14.2	4.68	2.88	极显著
			12.60	12.60	11.90	13.70	15.50				
	7653RA2[1)]	0.28	27.70	33.10	26.30	18.40	22.70	25.1			
			29.90	32.40	17.30	16.60	26.60				
	CK	0.037									

1）7653RA2 为 7653R（pJB5JI）在紫云英上所结根瘤的分离物，检测到 pJB5JI

图 2 7653R（pJB5JI）在紫云英上所结根瘤分离物质粒检测图谱

A-I.为 7653R（pJB5JI）根瘤分离物；J.K. 为 7653R 根瘤分离物

2）在豌豆植株上的共生特性

分别将转移接合子 7653R（pJB5JI）和 7653R+1（pJB5JI）接种豌豆植株，结果表明：尽管 7653R（pJB5JI）含有豌豆根瘤菌和全套的紫云英根瘤菌结瘤和固氮基因，并不能在豌豆植株上结瘤；而消除了紫云英根瘤菌共生质粒的 7653R+1（pJB5JI）却能在豌豆上结瘤，但不能固氮（结果见表 2 及图 4）。从转移接合子 7653R+1（pJB5JI）在豌豆上形成的根瘤中分离出的根瘤菌都具有卡那霉素抗性和大部分含有 pJB5JI 质粒，结果见图 3。将这些根瘤分离物再回接豌豆植株时，未检测到 pJB5JI 的 7653R+1Pa 与检测到 pJB5JI 的 7653R+1Pb 都同样具有在豌豆上结无效根瘤的能力（结果见表 4）。

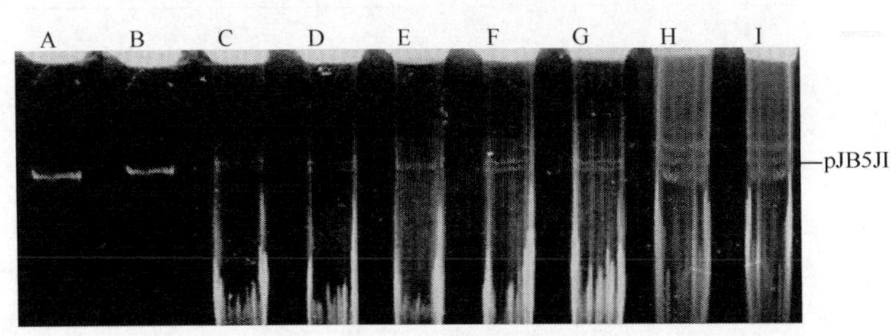

图 3 7653R+1（pJB5JI）在豌豆上所结根瘤分离物质粒检测电泳图

A.B.为 7653R+1 根瘤分离物；C-G. 为 7653R+1（pJB5JI）根瘤分离物；H.I. T83K3 根瘤分离物

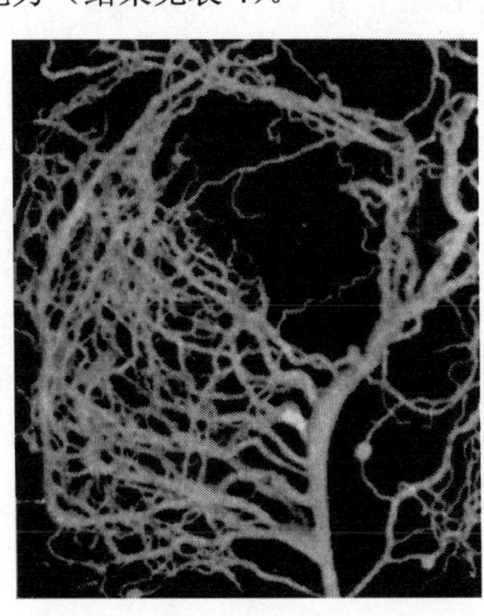

图 4 7653R+1（pJB5JI）在豌豆上结的根瘤

表 4 转移接合子所结根瘤的分离物共生表型

接种物	紫云英		豌豆	
	重复	表型	重复	表型
7653R+1	5	Nod^-	5	Nod^-
7653R+1Pa[1)]	5	Nod^-	10	Nod^+Fix^-
7653R+1Pb[1)]	5	Nod^-	10	Nod^+Fix^-
7653R	10	Nod^+Fix^+	5	Nod^-
7653RA2[2)]	10	Nod^+Fix^+	5	Nod^-
7653RA9[2)]	10	Nod^+Fix^+	5	Nod^-
CK	10	Nod^-	5	Nod^-

1）7653R+1（pJB5JI）在豌豆上所结根瘤的分离物，其中 7653R+1Pa 未检测到 pJB5JI，7653R+1Pb 检测到 pJB5JI

2）7653R（pJB5JI）在紫云英上所结根瘤的分离物，其中 7653RA2 检测到 pJB5JI，7653RA9 未检测到 pJB5JI

3 讨论

据已有的文献报道，豌豆根瘤菌共生质粒 pJB5JI 导入到三叶草根瘤菌[3]和菜豆根瘤菌[2]，能使其中绝大多数菌株具有在豌豆上结瘤的能力，而导入到亲缘关系较远的苜蓿根瘤菌[15]，快生型大豆根瘤菌[14, 16]、农杆菌[15]后，转移接合子则都不能在豌豆上结瘤。本研究将 pJB5JI 导入到紫云英根瘤菌含共生质粒的 7653R 菌株和消除了共生质粒的 7653R+1 菌株。结果表明：7653R+1（pJB5JI）能在豌豆上结无效根瘤，而 7653R（pJB5JI）却不能在豌豆上结瘤。从而首次揭示出 pJB5JI 在紫云英根瘤菌遗传背景中能够表达在豌豆上结瘤的功能，而且紫云英根瘤菌内在的共生质粒能够抑制 pJB5JI 这种功能的表达。由此可见紫云英根瘤菌共生固氮体系的遗传学背景与其它根瘤菌共生固氮体系的背景存在较大区别。

7653R+1 菌株由于消除了共生质粒，缺少紫云英专性结瘤基因，因而导入 pJB5JI 并不能恢复其在紫云英上结瘤的能力。而 7653R 菌株具有紫云英根瘤菌的全套共生基因，能在紫云英上正常结瘤，因而 7653R（pJB5JI）在其正常宿主紫云英上能够结有效根瘤，而且共生质粒间相互作用导致 7653R（pJB5JI）在紫云英上的结瘤固氮能力显著高于原菌株 7653R。至于这种共生质粒间相互作用的分子机理是什么，尚有待进一步深入研究。

本研究对 pJB5JI 在紫云英根瘤菌中的稳定性进行了考查，结果表明转移接合子在人工培养基上繁殖时，pJB5JI 能稳定存在，而经过与植物（紫云英、豌豆）共生之后则表现出不稳定的遗传特性，即部分根瘤分离物检测不到 pJB5JI。这与王常霖等[3]报道的 pJB5JI 在三叶草根瘤菌中的稳定性相类似，他们将这种宿主植物影响外源共生质粒稳定性的现象解释为功能不相容性（functional incompatibility）。7653R（pJB5JI）和 7653R+1（pJB5JI）分别与紫云英、豌豆共生之后的部分根瘤分离物虽然检测不到 pJB5JI 质粒，但它们都具有 Km^r 标记，据此推测 pJB5JI 可能部分或全部整合到紫云英根瘤菌的染色体上去了。其中未检测到 pJB5JI 的 7653R+1Pa 还具有在豌豆上结无效根瘤的能力，这就从功能上也证明了这一点。

参 考 文 献

[1]Long S R. Genetic Advances in the study of *Rhizobium* Nodulation. Genetic Engineering. 1987，8：135-150

[2]Beynon J L et al. Plasmids and Host-range in *Rhizobium leguminosarum* and *Rhizobium phaseoli*. J. Gen. Microbiol. 1980，120：421-429

[3]Wang C L et al. Host Plant Effects on Hybrids of *Rhizobium leguminosarum Biovars viceae* and *trifolii*. J. Gen. Microbiol. 1986，132：2063-2070

[4]陈华癸. 草籽绿肥根瘤细菌和人工接种. 新科学. 1952，3：33-38

[5]Chen W X et al. *Rhizobium huakui* sp. nov. Isolated from the Root Nodules of Astragalus sinicus. Int. J. Syst. Bacteriol. 1991，41：275-280

[6]宁林夫等. 几类快生型根瘤菌质粒的研究. 微生物学报. 1986，26：271-276

[7]王常霖. 陈金标. 紫云英根瘤菌质粒特征及内源抗药性. 华中农业大学学报. 1988，7：15-21

[8]宁林夫等. 紫云英根瘤菌共生质粒的鉴定及结瘤功能的诱动转移. 遗传学报. 1986，13（1）：1-10

[9]魏辉，李阜棣. 紫云英根瘤菌结瘤基因的定位研究. 微生物学报. 1990，30：330-335

[10]周俊初. 紫云英根瘤菌质粒的研究 II：紫云英根瘤菌经高温和吖啶橙处理后结瘤和固氮突变株的产生和质粒的消除. 华中农业大学学报. 1987，6：156-164

[11]张忠明等. 紫云英根瘤菌寄主专性结瘤基因（*han*）和固氮基因（*nif HDK*）的鉴定. 华中农业大学学报. 1989，8：285-290

[12]王常霖，赫茨. 转座子 Tn5-mob 对菜豆根瘤菌共生基因的诱变、转移和初步定位。遗传学报，1988，15（1）：25-33

[13]Gary Ditea et al. Plasmids Related to the Broad Host Range Vector. PRK290. Useful for Gene Cloning and for Monitoring Gene Expression. Plasmid. 1985，13：149-153

[14]Michale J et al. Differential Expression of the Pea Symbiotic Plasmid pJB5JI in Genetically Dissimilar Backgrounds Symbiosis. 1985，1：125-138

[15]Michale A Djerdjevic et al. Sym Plasmid Transfer to Varlous Symbiotic Mutants of *Rhizobium trifoll*，*R. leguminosarum* and *R. meliloti*. J.Bactoriol. 1983，156（3）：1038-1045

[16]Ruiz-Sainz J E et al. Transfer of a host range plasmid from *Rhizobium leguminosarum* to fast-growing bacteria that nodulate soybeans. J. Appl. Bacterio. 1984，57：309-315

紫云英根瘤菌重组质粒 pRaZ15 的亚克隆系统及限制酶图谱*

胡福荣　农　广　张忠明　陈华癸

（华中农业大学）

摘　要　在分离到能互补7653R-1重组质粒pRaZ15的基础上[1]，将pRaZ15用*Eco*R Ⅰ进行完全酶切，得到除载体pLARF1外的7个*Eco*R Ⅰ酶切片段，然后以pBR325为载体构建了7个*Eco*R Ⅰ酶切片段的亚克隆体系。同时利用不同的限制性内切酶进行单酶切、双酶切，构建了pRaZ15的限制酶的酶切图谱。

关键词　紫云英根瘤；亚克隆；酶切图谱

紫云英根瘤菌（*R. huakuii*）感染黄芪属（*Astragalus*）豆科植物紫云英（*A.sinica*）后能形成有效固氮根瘤。紫云英根瘤菌是一种快生型根瘤菌[1]，在互接种族中独属一簇[2]，是我国一种重要的共生固氮资源。近几年来，共生固氮遗传学研究表明，一切快生型根瘤菌中存在着大质粒，共生固氮基因（至少是部分基因）位于大质粒上[3, 4]。然而，迄今尚未见到有关紫云英根瘤菌共生固氮基因结构、功能及其遗传特性的报道。本项研究是在分离到能互补不结瘤的突变株7653R-1的重组质粒pRaZ15的基础上，开展对紫云英根瘤菌共生固氮基因结构、功能及遗传特性的研究，并在基因水平上比较紫云英根瘤菌与其他根瘤菌共生固氮基因间的异同，阐明紫云英根瘤菌共生固氮基因的结构特点，以期为揭示紫云英根瘤菌共生固氮体系的遗传学关系提供资料，为改造和提高共生固氮体系的共生关系和固氮效率提供依据和途径。

一、材料与方法

1. 质粒与菌株

见表1。

表1　供试质粒与菌株

菌株与质粒	有关性状	来源
大肠杆菌：		
HB101		本室收集、保存
pRaZ15（HB101）	pLARF1上克隆有一段紫云英根瘤菌DNA片段	文献[5]
pBR325（HB101）	Amp^r　Tc^r　cm^r	本室收集、保存

2. 质粒DNA分离、纯化及酶切　质粒DNA的分离及纯化按彭秀玲等[6]介绍的碱性裂解法进行。限制性内切酶均购自华美生物工程公司，酶切缓冲液购自Promega公司。单酶切、双酶切反应均在37℃反应2~3h，然后置68℃水浴10min变性，双酶切的公用缓冲液参照Promega公司提供的参考说明选取酶切效果为75%~100%之间的缓冲液。

3. DNA连接及转化　将pRaZ15和pBR325质粒DNA分别用*Eco*R Ⅰ完全酶切，然后将pRaZ15和pBR325酶切物按3∶1混合，加入T_4-DNA连接酶，在12℃中反应16h以上。将连接物转化*E. coli* HB101，在含四环素（20μg/ml）和氨苄青霉素（50μg/ml）的LB培养基平板上选择含有pRaZ15的*Eco*R Ⅰ完全酶切的外源片段的pBR325转化子。

4. 亚克隆片段的分离及鉴定　从含四环素和氨苄青霉素的LB平板上挑出若干转化子，分别于LB

*原载于《湖北农业科学》，8：4~6，1993.

肉汤中（含四环素）37℃培养过液，按 Maniatis[6]的碱性裂解法分离质粒 DNA，然后用 *Eco*R Ⅰ进行完全酶切。同时将 pRaZ15 也用 *Eco*R I 进行完全酶切。同时将 pRaZ15 也用 *Eco*R I 进行完全酶切。以 0.7%的琼脂糖凝胶，在室温下，40V 电泳 6~8h，电泳后，经 EB 染色，在紫外灯下观察，对照 pRaZ15 的七条不同酶切片段在凝胶上的位置，找出对应的带有 pRaZ15 的 *Eco*R I 完全酶切的亚克隆质粒。

5. pRaZ15 限制性内切酶图谱的分析　以常用的限制酶 *Eco*R I、*Hind* III、*Bam*H I、*Bgl* II、*Sac* I 分别对质粒 pRaZ15 进行了单酶切和不同组合的双酶切。然后将不同的酶切物与标准 Maker 同时电泳。琼脂糖凝胶上 DNA 片段的大小（kb）根据文献[7]的方法所衍生的计算机程序[8]而确定。再参照文献[9]所介绍的双酶消化法作出 pRaZ15 的图谱。

二、结果与分析

1. *Eco*R I 亚克隆的建立　用 *Eco*R I 对 pRaZ15 以及 pBR325 分别进行完全酶切，连接后经筛选得到的 7 个亚克隆体系如图 1。

2. pRaZ15 的酶切分析及作图　pRaZ15 总的分子质量大约等于 50kb 左右。为了确定 pRaZ15 上 *Hind* III、*Bam*H I、*Bgl* II、*Sac* I、*Eco*R I 各酶切位点的相互位置，我们对该质料分别进行了单酶切及各种组合的双酶切。不同酶切的片段数以及各片段的分子质量大小（kb）数列于表 2 中，各酶切片段在凝胶电泳上的位置如图 2。

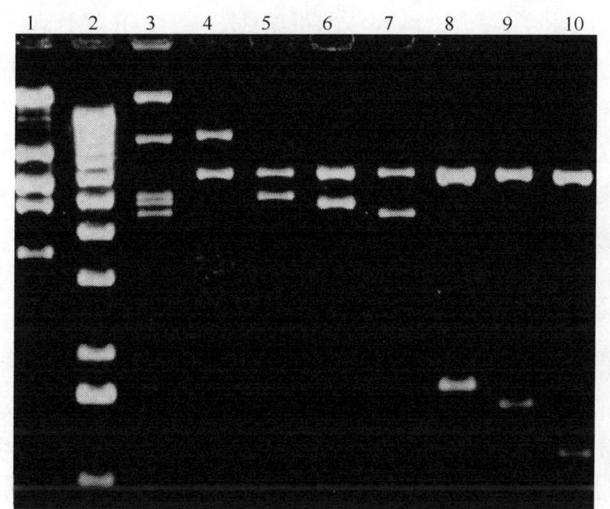

图 1　pRaZ15 7 个亚克隆的琼脂糖凝胶电泳图

1：λ +*Eco*R I Marker；2：1 kb DNA Ladder Marker；3：pRaZ15+*Eco*RI；4：亚克隆 7-325；5：亚克隆 6-325；6：亚克隆 5-325；7：亚克隆 4-325；8：亚克隆 3-325；9：亚克隆 2-325；10：亚克隆 1-325

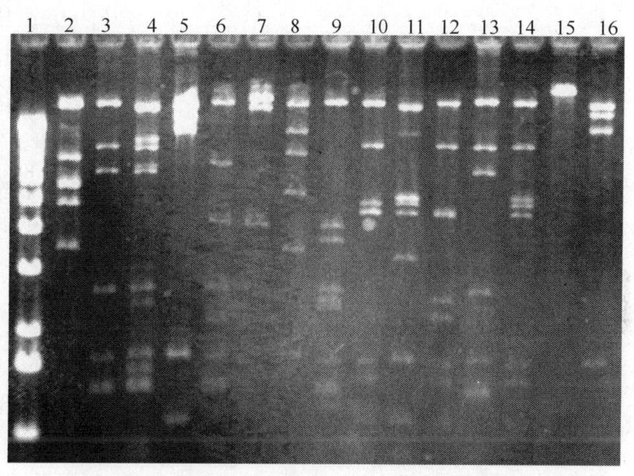

图 2　pRaZ15 各酶切片段的凝胶电泳图

1：Marker；2：Z15+*Hind*III；3：Z15+*Bam*HI；4：Z15+*Bgl*II；5：Z15+*Sac*I；6：Z15+*Eco*RI；7：Z15+*Eco*RI+*Hind*III；8：Z15+*Eco*RI+*Bam*H1；9：Z15+*Eco*RI+*Bgl*II；10：Z15+*Eco*RI+ *Sac*I；11：Z15+*Hind*III+ *Bam*HI；12：Z15+*Hind*III+ *Bgl*II；13：Z15+*Hind*III+*Sac*II；14：Z15+*Hind*III+ *Bgl*II；15：Z15+ *Bam*H1+*Sac*I；16：Z15+*Bgl*II+ *Sac*I

表 2　不同酶切的片段数及分子质量

限制酶	片段数	各片段分子质量（kb）							
Hind III	1	50							
*Bam*H I	2	32	18						
Bgl II	4	19.7	16.5	12	1.5				
Sac I	5	33.4	5.4	4.8	3.8	2.6			
*Eco*R I	8	21.6	9.3	5.3	4.9	4.6	1.7	1.6	1.1
Hind III/ *Bam*H I	3	28	18	4					
Hind III/ *Bgl* II	5	19.7	14	12	2.6	1.7			

续表

限制酶	片段数	各片段分子质量（kb）											
*Hin*d III/ *Sac* I	6	33.4	4.8	3.8	2.8	2.6	2.6						
*Hin*d III/ *Eco*R I	9	21.6	9.3	4.9	4.6	4.6	1.7	1.6	1.1	0.7			
*Bam*H I / *Bgl* II	6	19.7	10.1	8.0	6.5	3.8	1.8						
*Bam*H I / *Sac* I	7	25	8.4	5.4	4.8	2.6	2.6	1.2					
*Bam*H I / *Eco*R I	10	21.6	9.3	4.6	4.6	2.6	2.3	1.7	1.6	1.1	0.7		
Bgl II/ *Sac* I	9	19.7	11.7	4.8	4.8	3.8	2.6	1.8	（?）0.5	（?）0.2			
Bgl II/ *Eco*R I	12	18.7	5.7	5.3	4.9	4.6	3.5	1.8	1.7	1.2	1.2	1.1	（?）0.4
Sac I / *Eco*R I	13	21.6	5.0	4.9	4.2	3.1	2.6	2.2	1.7	1.1	1.0	（?）0.8	（?）0.6	（?）0.4

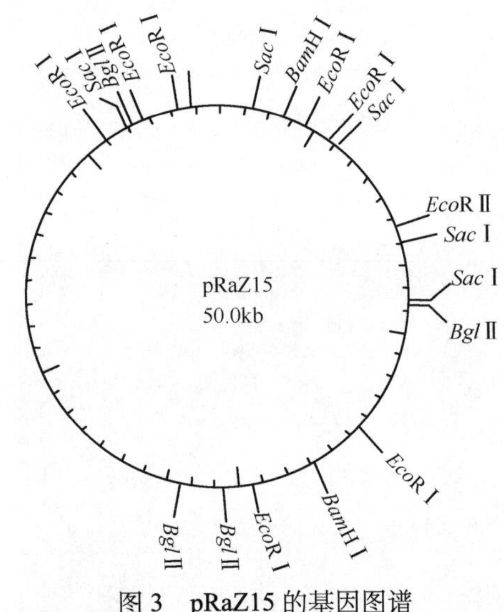

图 3 pRaZ15 的基因图谱

根据片段的大小，将每一对酶切的结果分别清理，除去在单酶和双酶消化时共同产生的片段，对照电泳图的带型，再按照分子量大小，找出新生小片段和一个消失片段之间的关系。如：在单酶切中，*Hin*d III 将 pRaZ15 切成一条 50kb 的片段，说明 *Hin*d III 在 pRaZ15 上只有一个切点。*Bam*H I 则将该质粒切成 32kb 和 18kb 二条片段，说明 *Bam*H I 在 pRaZ15 上有二个切点。而用 *Bam*H I + *Hin*d III双酶切后产生 28kb、18kb、4kb 三条片段。从酶切电泳图上看到双酶切后保留了原始 *Bam*H I 的 18kb 这条片段，而 32kb 大片段消失，增加了 28kb 和 4kb 二条片段，因此可以推定 *Hin*d III 的这个切点一定在 *Bam*H I 的 32kb 这条片段上。

如果按照不同酶切的情况，进行一一分析，找出相邻各片段的位置，作出基因图谱如图 3。其中细线部分是载体 pLAFR1 为 21.6kb，带有 Tc^r 抗性；而粗线是外源 DNA 片段约为 28.4kb。研究结果看出，pRaZ15 的外源片段中包括 1 个 *Hin*d III 片段，2 个 *Bam*H I 片段，4 个 *Bgl* II 片段，5 个 *Sac* I 片段和 7 个 *Eco*R I 片段。

参 考 文 献

[1]H.k.Chen et al. Soil.sci.1944，51：291~293
[2]陈华癸. 新科学，1952，3：33~38
[3]张忠明等. 华中农业大学学报，1986，5（4），326~33
[4]宁林夫等. 遗传学报，1986，13（1），1~10
[5]张忠明等. 生物工程学报，1991，7（3）：213~211
[6]彭秀玲等. 基因工程实验技术，长沙，湖南科学技术出版社，1987，p5~14
[7]Southern，E.M. Ann. Biochem. 1979，100：319~320
[8]Kieser，T. Nucleic Acids Research. 1984，12：679~688
[9]齐义鹏等. 基因工程原理和方法，成都，四川大学出版社，1989，p114 ~ 118

紫云英根瘤菌 7653R 基因文库的 5 个互补结瘤菌株的重组质粒分析*

农 广 胡福荣 张忠明 陈华癸

（华中农业大学农业微生物部级重点开放实验室）

摘 要 对紫云英根瘤菌 7653R 基因文库的 5 个重组质粒，即 pNR102、pNR103、pNR108、pNR203、pNR213 进行各种内切酶的酶切分析。结果证实它们均含有 1.7kb 的 *Eco*R I−*Bgl* II 双酶切片段和 1.9kb 的 *Eco*R I-*Sac* I 双酶切片段，而 pNR213 的外源片段最大，包含有其余质粒所具有的各种片段，以 pNR108 的外源片段作探针，进行 DIG 杂交，发现探针与其余 4 个重组质粒的外源片段有强的同源杂交；与 7653R 的内源质粒杂交还表明，pNR108 的外源片段来源于共生质粒 pRa7-2。

关键词 重组质粒；内切酶；DIG 杂交

紫云英根瘤菌 7653R 是能够在紫云英有效结瘤固氮的野生型菌株，具有两个内源质粒，即隐蔽质粒 pRa7-1 和共生质粒 pRa7-2。通过消除质粒获得 Nod⁻表型的菌株 7653R-1，证实它只有隐蔽质粒 pRa7-1 而消除了共生质粒 pRa7-2 [1]。以广谱寄主范围载体 pLAFR1，构建 7653R 总 DNA 的 *Eco*R I 酶切片段的基因文库，通过三亲本接合，重组质粒转入到受体根瘤菌 7653R-1，将四环素抗性筛选所获得的转移接合子进行植株结瘤试验，从根瘤中获得 1 个互补结瘤菌株，并证实含有结瘤基因的重组质粒 pRaZ15[2]，同时还从不同的根瘤中分离得到 5 个互补结瘤菌株 R102、R103、R108、R203 和 R213。本实验对这 5 个菌株的重组质粒进行了酶切分析，现报道如下。

1 材料与方法

1.1 供试材料

紫云英根瘤菌有效结瘤菌株 7653R，及其消除质粒的不结瘤突变株 7653R-1 由本室周俊初分离获得[1]。7653R 总 DNA 基因文库导入受体 7653R-1，从根瘤所分离到的互补结瘤菌株 R102，R103，R108，R203 和 R213 由本室张忠明提供。

1.2 根瘤菌质粒检测

按 Eckhardt 方法[3]进行。

1.3 反三亲本接合转移

将 5 个互补结瘤菌株 R102、R103、R108、R203 和 R213 接种于 TY 液体培养基，28℃振荡培养 24h。接种受体菌 HB101 和辅助转移菌株 ED8767（pRK2073）于 LB 液体培养基，分别含 $50\mu g \cdot ml^{-1}$ 链霉素和 $50\mu g \cdot ml^{-1}$ 壮观霉素，37℃振荡培养 12h。将 3 种菌按 1∶1∶1 混合，涂布于预先放置于 TY 固体培养基上的醋酸纤维素滤膜上（孔径 0.45μm），于 28℃培养 24h，再用无菌水洗脱，菌悬液涂布于含 $15\mu g \cdot ml^{-1}$ 四环素的 LB 固体培养基上，选取单菌落，用于质粒提取、纯化及酶切分析。

1.4 凝胶电泳 DNA 片段测定

利用 Kieser 的 DNAGEL 计算机程序[4]，以 BRL 1 kb Ladder Marker 为标准，在电磁感应图板上测定照片上的谱带迁移图距，计算机根据公式 kg（值）=A+B/（D-DC）直接输出所测 DNA 片段的 kb 值。

1.5 DNA 酶切分析

选择常用的内切酶，酶切反应于 37℃进行 2h。双酶切选用宝来曼公司的 Incubation Buffer，选择使两种酶都能达到 100%酶切效率的缓冲液。以酶切电泳照片测定各酶切片段的 kb 值。以 *Eco*R I 和其余各种酶的双酶组合，确定这些内切酶在重组质粒外源片段上有无位点以及所产生的酶切片段大小。再根

*原载于《华中农业大学学报》，13（6）：531~537，1994.

据在载体上有已知其 kb 值的内切酶 *Bgl* II、*Bst*E II 和 *Pvu* II，与在外源片段上有酶切位点的内切酶进行双酶切分析，确定各酶切位点及与载体的排列方向，从而确定重组质粒外源片段酶切图谱。

1.6 *DIG* 杂交

按宝来曼公司 DIG Kit 所附的说明书操作。

2 结果与讨论

2.1 重组质粒的分离与 *Eco*R I 酶切分析

将菌株 R102，R103，R108，R203 和 R213 进行根瘤菌质粒检测，电泳结果见图 1。

电泳结果证实，互补结瘤菌株 R102、R103、R108、R203 和 R213 均以 7653R-1 为受体，各携带有外源质粒多聚体。

进一步进行实验证实这些外源质粒为含有外源片段的重组质粒，将这 5 个菌株分别进行反三亲本杂交，使重组质粒转移到大肠杆菌 HB101，并在四环素抗性平板上筛选，各挑取单菌落，进行质粒提取和纯化，重组质粒经 *Eco*R I 完全酶切，显示分别来自这 5 个根瘤菌株的重组质粒都含有一个 *Eco*R I 酶切片段，电泳结果见图 2，同时将这 5 个重组质粒相应其所来源的根瘤菌命名为 pNR102、pNR103、pNR108、pNR203 和 pNR213。

根据图 2 照片，以 1 kb Ladder Marker 为标准，进行 DNAGEL 法酶切片段 kb 值测定。pNR102、pNR103、pNR108、pNR203 和 pNR213 的外源 *Eco*R I 酶切片段的 kb 值依序为：4.68，5.01，5.04，5.10 和 9.38。

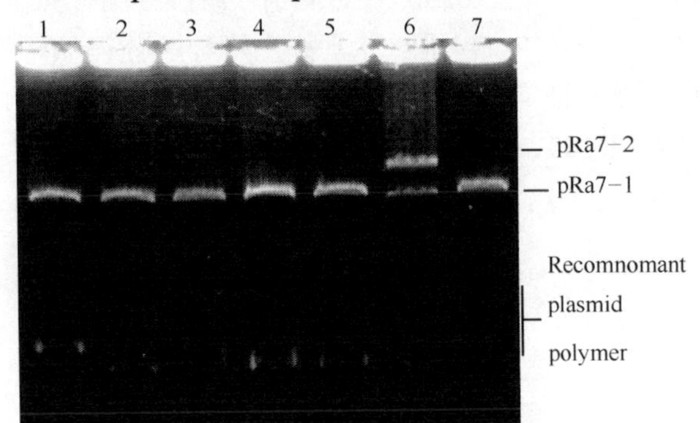

图 1 紫云英根瘤菌大质粒快速检测

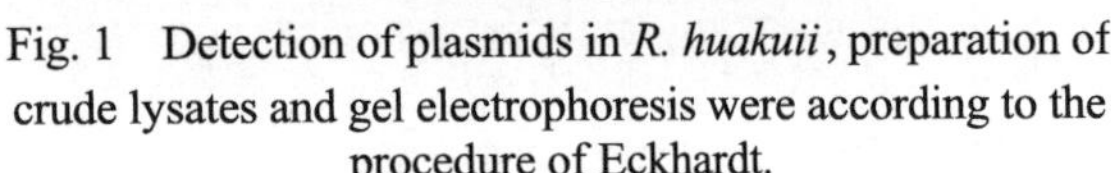

Fig. 1 Detection of plasmids in *R. huakuii*, preparation of crude lysates and gel electrophoresis were according to the procedure of Eckhardt.

1：R213；2：R203；3：R108；4：R 103；5：R102；6：7653R；7：7653R-1

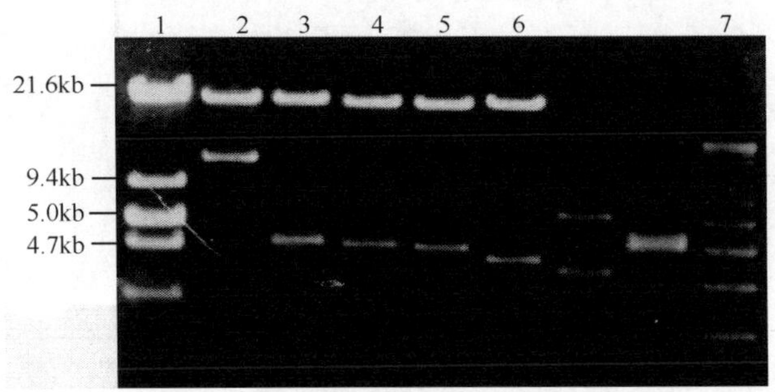

图 2 重组质粒的 *Eco*R I 酶切分析

Fig. 2 Analysis of recombinant plasmids digested with *Eco*R I

1：λ +*Eco*R I Marker；2：pNR213+*Eco*R I；3：pNR203+*Eco*R I；4：pNR108+*Eco*R I；5：pNR103 +*Eco*R I；6：pNR102+*Eco*R I；7：BRL 1 kb Ladder Marker

2.2 外源片段的同源杂交

以 pNR108 的 5kb 外源片段作探针，对其余 4 个重组质粒的外源片段和 7653R 大质粒进行 DIG 杂交，电泳和杂交结果见图 3 和图 4

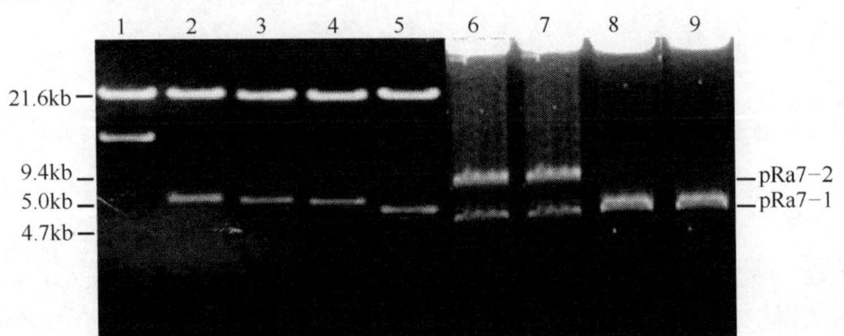

图 3 重组质粒的酶切分析及紫云英根瘤菌的质粒快速检测

Fig.3 Digestion of recombinant plasmids with *Eco*R I and detection of plasmids in *R. huakuii*

1：pNR213+*Eco*R I；2：pNR203+*Eco*R I；3：pNR108+*Eco*R I；4：pNR103+*Eco*R I；5：pNR102+*Eco*R I；6-7：7653R；8-9：7653R-1

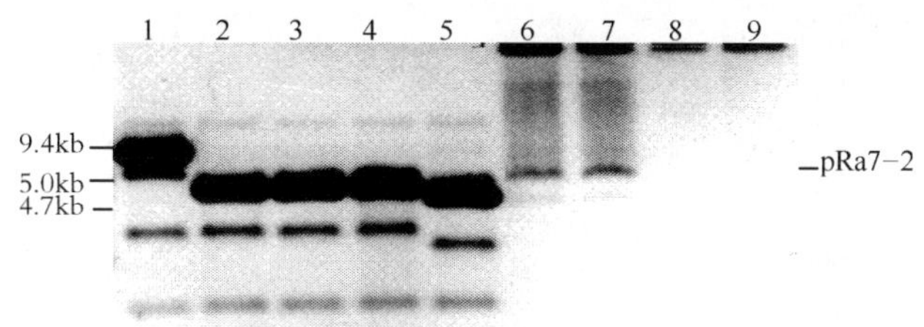

图 4　紫云英根瘤菌 7653R 内源质粒和重组质粒间的同源杂交分析

Fig. 4　Hybridization of DIG-Labeled probe，a 5.0 kb foreign fragment of pNR108，with other recombinant plasmids and the plasmids in *R. huakuii.*

1：pNR213+ *Eco*R Ⅰ；2：pNR203+*Eco*R Ⅰ；3：pNR108+*Eco*R Ⅰ；4：pNR103+*Eco*R Ⅰ；5：pNR102+*Eco*R Ⅰ；6-7：7653R；8-9：7653R-1

图 3 电泳分为两部分，6~9 孔为 7653R 及 7653R-1 的质粒快检，在电泳 5~6h 后，将 5 个重组质粒的 *Eco*R I 酶切溶液点样，继续电泳 2h，再将凝胶进行 Southern 转移，进行 DIG 杂交。大质粒快检与酶切片段大小无对应关系。

图 4 杂交结果显示：①pNR108 的外源片段与其余 4 个重组质粒的外源片段有极强的同源杂交，表明了这些重组质粒之间的同源性。②pNR108 与 7653R 的共生质粒 pRa7-2 有明显的同源杂交，表明重组质粒的外源片段来源于共生质粒。③pNR108 与其余 4 个重组质粒间的高度同源性以及 pNR108 与 7653R 共生质粒间的同源性提示我们，其余 4 个重组质粒的外源片段也可能是来源于共生质粒。④杂交照片上还出现了电泳照片中所没有的，比外源片段还小的杂交带。这是由于 DIG 杂交具有高灵敏度，其将 *Eco*R I 卫星酶切活性所产生的极低浓度的酶切片段，检测出来所造成的现象。

2.3　重组质粒酶切图谱

利用内切酶 *Eco*R I，*Bam*H I，*Bst*E II，*Xba* I，*Bgl* II，*Hind* II，*Kpn* I，*Pst* I，*Pvu* II，*Sac* I 和 *Xho* I 进行双酶切和单酶切分析。酶切结果表明重组质粒 *Eco*R I 酶切片段上没有 *Bam*H I 和 *Bst*E II 的酶切位点。而 *Eco*R I 和其余内切酶的双酶切所产生的片段及大小则可决定片段的可能排列方式。

再利用载体上已知酶切片段 *Bst*E II-*Eco*R I 为 1.84 kb，*Bgl* II-*Eco*R I 为 1.2kb，*Bgl* II-*Bgl* II 为 1.78 kb，*Pvu* II-*Eco*R I 为 1.4 kb，*Pvu* II-*Pvu* II 为 3.7 kb，并且 *Bgl* II、*Bst*E II 和 *Pvu* II 再与载体上无切点而外源片段上有酶切位点的内切酶组合，进行双酶切，并补充 *Bgl* II，*Hind* II 和 *Pvu* II 单酶切，由此确定片段间的排列及与载体的连接方式。由此确定各重组质粒酶切图谱如图 5-A、B、C。其中，重组质粒 pNR103 的酶切图谱及与载体的连接方式与 pNR108 相同；而 pNR203 酶切位点与 pNR108 相同，但载体的连接方向与 pNR108 相反。

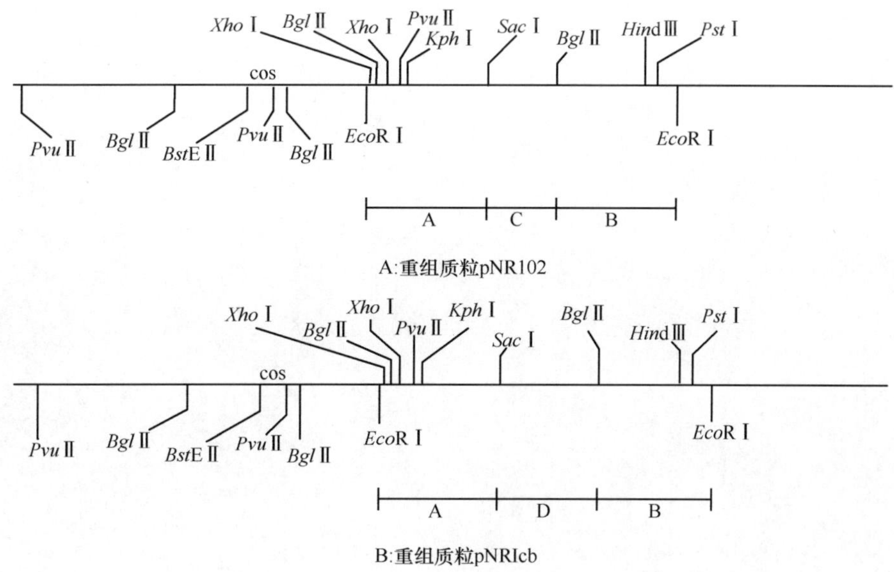

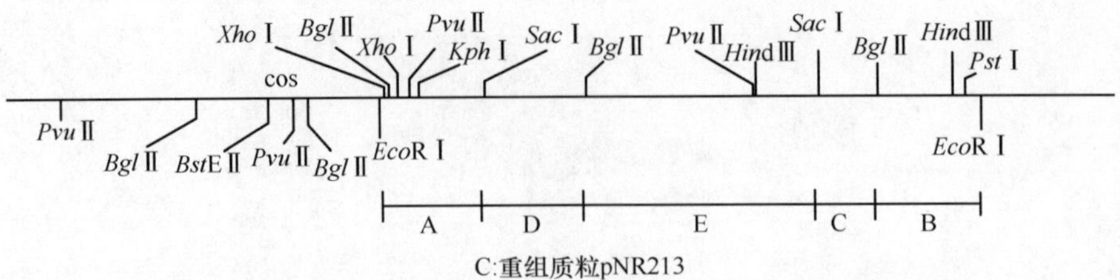

C:重组质粒pNR213

图 5 重组质粒酶切图谱分析及外源片段与载体的连接方向

Fig.5 Restriction enzyme cleavage maps of recombinant plasmids and the directions of ligation with vector

从重组质粒的酶切图谱可看到，它们的外源片段上都有 1.9 kb 的 A 片段（*Eco*R I-*Sac*I 酶切片段）和 1.7kb 的 B 片段（*Eco*R I -*Bgl* II 酶切片段）。这可说明 pNR108 外源片段与各个重组质粒外源片段极强同源杂交的原因。pNR213 除了具有 A 片段和 B 片段，还与 pNR102 共有相同的 1.2kb 的 C 片段（*Sac* I -*Bgl* II 酶切片段），与 pNR108 有相同的 1.4 kb 的 D 片段（*Sac* I -*Bgl* II 酶切片段）。并且 pNR213 还具有一个自身特有的 3.3 kb 的 E 片段（*Sac* I -*Bgl* II 酶切片段），这显示出 pNR213 包含其余 4 个重组质粒所具有的各种 DNA 片段。

考虑到这 5 个重组质粒都具有 A 片段和 B 片段。因此，可以推测这 2 个片段在功能上可能会具有重要作用，针对这 2 个片段开展功能研究是有意义的。

参 考 文 献

[1]周俊初，张忠明，陈华癸等. 紫云英根瘤菌质粒研究 II. 紫云英根瘤菌经高温和吖啶橙处理后结瘤和固氮突变株的产生和质粒消除. 华中农业大学学报. 1987.6（2）：156~164

[2]张忠明，范云六，陈华癸等.紫云英根瘤菌基因文库的构建及含完整结瘤基因的重组质粒 pRaZ15 的分离. 生物工程学报. 1991：7（3）213~219

[3]Eckhardt T. A rapid method for the identification of plasmid desoxyribonucleic acid in bacteria. Plasmid. 1978.（1）：584~588

[4]Kieser T. DNAGEL a computer program for determining DNA fragment sizes using a small computer equipped with a graphics tablet. Nucleic Acids Res. 1984.（12）.679~688

紫云英根瘤菌5个互补结瘤菌株所分离到的重组质粒外源片段的酶切图谱*

农 广 张忠明 胡福荣 陈华癸

（华中农业大学生命科学技术学院）

李立家

（武汉大学生命科学学院）

摘 要 对来源于紫云英根瘤菌7653R总DNA基因文库的5个互补结瘤菌株所分离到的重组质粒pNR102、pNR103、pNR108、pNR203和pNR213，*Eco*R Ⅰ酶切表明它们分别携带有一个4.68、5.01、5.04、5.10或9.38kb的外源DNA片段，选用11种内切酶进行单、双酶切分析，建立了重组质粒外源片段的酶切图谱，5个质粒都具有1.7kb的*Eco*R Ⅰ-*Bgl* Ⅱ片段和1.9kb的*Eco*R Ⅰ-*Sac* Ⅰ片段。

关键词 重组质粒；内切酶；酶切图谱

根瘤菌与植物的共生结瘤作用涉及其基因组内的二十几个基因。Long[1]和 Dowine[2]等分别从苜蓿根瘤菌和豌豆根瘤菌的总DNA基因文库中分离到能互补Nod⁻菌株结瘤的重组质粒，并证实结瘤基因分别位于8.7kb或约10kb的DNA片段上，张忠明等将紫云英根瘤菌7653R菌株的总DNA进行*Eco*R Ⅰ酶切，并克隆到广谱载体pLAFRl的*Eco*R Ⅰ位点，构建基因文库，并导入不结瘤突变株7653R-1，通过植株结瘤试验，从根瘤中分离到5个互补结瘤菌株R102、R103、R108、R203和R213。笔者从这些菌株中分离到5个重组质粒：pNR102、pNR103、pNR108、pNR203和pNR213[3]。本研究通过对这5个重组质粒进行单酶切和双酶切分析，建立了重组质粒外源片段的酶切图谱，并发现这5个重组质粒的外源片段上都有1.7kb的*Eco*R Ⅰ-*Bgl* Ⅱ片段和1.9kb的*Eco*R Ⅰ-*Sac* Ⅰ片段。酶切图谱的建立为进一步的功能分析和亚克隆研究奠定了良好的基础。

1 材料和方法

1.1 实验材料

重组质粒pNR102、pNR103、pNR108、pNR203和pNR213由笔者从菌株R102、R103、R108、R203和R213分离获得。

1.2 DNA酶切分析

选择常用的11种内切酶，进行单酶切和双酶切，酶切反应于37℃进行2h。双酶切反应选用Boehringer Mannheim公司的Incubation Buffer，使所选择的两种酶能达到100%的酶切效率，然后进行电泳。

根据电泳照片上各酶切片段的电泳迁移距离，利用Kieser的DNAGEL计算机程序[4]，以BRL 1kb Ladder Marker为标准，选择所要测定的样品酶切片段所处在的区间内的至少4个标准Marker片段建立标准曲线，从而获得曲线的截距A和斜率B，再测定样品酶切片段的迁移距离D，计算机依据公式：分段长度=A+B/（D-DC），直接输出所测酶切片段的kb值，公式中的DC为距离校正因子。

2 结果和分析

2.1 5个重组质粒的酶切分析

对从互补结瘤菌株R102、R103、R108、R203和R213所分离到的重组质粒pNR102、pNR103、

*原载于《武汉大学学报（自然科学版）》，41（4）：469～474，1995

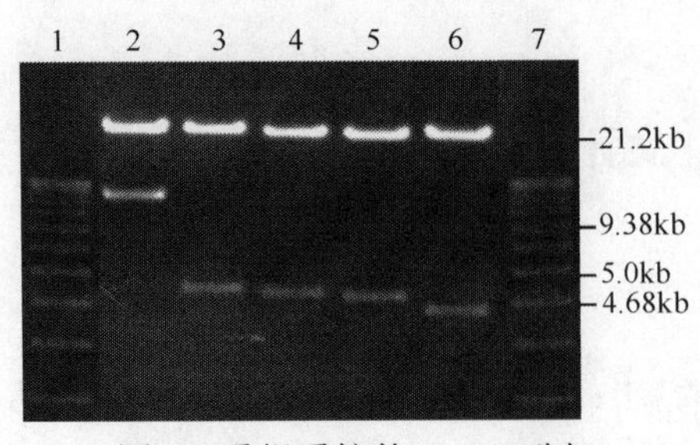

图 1 重组质粒的 *Eco*RI 酶切

1、7：BRL.1kb Ladder Marker；2：pNR213+E；3：pNR203+E；4：pNR108+E；5：pNR103+E；6：pNR102+E；

pNR108、pNR203 和 pNR213 进行 *Eco*R Ⅰ 酶切，酶切结果如图 1 所示。

酶切结果显示，5 个重组质粒各有一个外源片段，并且都有 21.6kb 的载体 pLAFRl 片段，表明了这些重组质粒来源于基因文库。经多次酶切证实，重组质粒各只含有一个外源片段，以 BRL 1kb Ladder Marker 为标准，根据 DNAGEL 方法[4]，从计算机测定了这 5 个重组质粒 pNR213、pNR203、pNR108、pNR103 和 pNR 102 外源片段 kb 值分别为 9.38、5.10、5.04、5.01 或 4.68。

2.2 外源片段的酶切位点分析

选用 11 种常用的内切酶 *Eco*R Ⅰ，*Bgl* Ⅰ，*Hind* Ⅲ，*Kpn* Ⅰ，*Pst* Ⅰ，*Pvu* Ⅰ，*Sac* Ⅰ，*Xho*Ⅰ，*Bst*E Ⅱ，*Bam*H Ⅰ 和 *Xba* Ⅰ。通过 *Eco*R Ⅰ与其余 10 种内切酶分别进行双酶切组合，所得电泳结果经 DNAGEL 方法测定各酶切组合中酶切片段的数目及大小，结果见表 1。

从表 1 各双酶切组合的结果发现：①内切酶 *Bst*E Ⅱ，*Bam*H Ⅰ和 *Xba* Ⅰ这 3 种酶在外源片段上没有切点。②其余内切酶各有 1～3 个酶切位点。由于切点较少，因此很容易能确定这些片段间的可能排列方式。需要说一点，表中各片段 kb 值总和可能与整个外源片段的理想值会有一定的误差，这是电泳及片段 kb 值测定过程中容许出现的，重要的是，每个酶切组合中的双酶切片段数目都必须由电泳照片上肯定，再根据各片段之和来确定它的可能性及合理性，由此确定酶切位点的存在。

表 1 重组质粒外源片段双酶切所产生的片段及大小（单位：kb）

内切酶组合	pNR102	pNR103/pNR 108/pNR203	pNR213
	0.14	0.14	0.14
EoR Ⅰ + *Bgl* Ⅱ	1.70	1.70	1.73
	2.92	3.10	3.15
			4.50
	0.50	0.50	0.50
EoR Ⅰ + *Hind* Ⅲ	4.12	4.58	3.18
			5.70
EoR Ⅰ + *Kpn* Ⅰ	0.62	0.62	0.62
	4.10	4.50	9.07
EoR Ⅰ + *Pst* Ⅰ	0.30	0.30	0.30
	4.35	4.79	9.20
	0.50	0.50	0.50
EoR Ⅰ + *Pvu* Ⅰ	4.00	4.50	3.70
			5.1
	1.93	1.95	1.93
EoR Ⅰ + *Sac* Ⅰ	2.66	3.08	2.66
			4.94
	0.075	0.075	0.075
EoR Ⅰ + *Xho* Ⅰ	0.29	0.29	0.29
	4.32	4.63	9.02
EoR Ⅰ +*Bst*E Ⅱ	4.65	5.06	9.38
EoR Ⅰ +*Bam*H Ⅰ	4.63	5.03	9.40
EoR Ⅰ +*Xba* Ⅰ	4.61	5.10	9.39

2.3　外源片段酶切位点的排列及载体的连接方向

选取在载体上有已知切点的内切酶 *Bst*E II，*Bgl* II 和 *Pvu* II，这 3 个酶的切点都集中于 *Eco*R I 位点的一侧，其中 *Bst*E II-*Eco*R I 片段为 1.84 kb，*Bgl* II-*Eco*R I 片段为 1.2kb，*Bgl* II-*Bgl* II 片段为 1.78kb，*Pvu* II-*Eco*R I 片段为 1.4 kb，*Pvu* II-*Pvu* II 片段为 3.7 kb，*Pvu* II-*Eco*R I 片段为 1.4 kb，*Pvu* II-*Pvu* II 片段为 3.7 kb。利用这些在载体上有切点的酶及在外源片段上有切点的酶进行单、双酶切分析，所得的电泳结果经 DNAGEL 方法处理，得到各片段及其大小值，结果见表 2。

表 2　重组质粒酶切段 kb 值

内切酶	pNR102	pNR103/pNR108	pNR203	pNR213
Bgl II	1.37	1.37	1.74*	1.40
	1.74*	1.76*	2.84	1.76*
	2.83	2.94	3.01	3.00
				4.30
Bgl II+*Pst* I	1.36	1.35	1.35	1.35
	1.42	1.43	1.53	1.42
	1.75*	1.78*	1.78*	1.76*
	2.61	3.02	3.02	3.00
				4.33
Bgl II +*Xho* I	1.40	1.40	1.73*	1.40
	1.75*	1.75*	2.85	1.75*
	2.52	2.91	3.00	3.00
				4.35
Bgl II+*Hind* III	1.24	1.29	1.30	1.30
	1.37	1.40	1.57	1.42
	1.70*	1.74*	1.77*	1.74*
	2.50	2.97	3.00	1.81
				2.40
				3.00
Hind III	—	—	—	3.14
*Bst*E III+*Kpn* I	2.42	2.42	6.23	2.49
*Bst*E II+*Sca* I	3.9	3.9	4.97	3.9
				4.42
Pvu I	1.84	1.84	3.7*	1.86
	3.70*	3.66*	5.8	3.73*
				5.04

*为载体的酶切片段

根据表 1 和表 2 所列片段以及已知的载体上酶切片段的 kb 值，利用新出现或消失片段的和或差关系，可逐个确定酶切位点的排列顺序和位置。如：pNR102 经 *Bgl* II 酶切后，表 2 中剔除*号所示的载体片段后，考虑到没有利用 *Eco*R I，因此载体上 1.2kb 的 *Bgl* I -*Eco*R I 部分必须与外源片段上的一部分构成一个新的 *Bgl* II-*Bgl* II 片段，联系表 1 考察比较，新出现片段 1.37kb=1.2+0.14，而表 1 中任何一个值与载体 1.2kb 相加都大于 1.37kb，因此可以断定靠近 *Eco*R I 多切点一侧 0.14kb 处有一 *Bgl* II 切点，而表 1 中的 3 个片段必有一个是 *Bgl* II-*Bgl* II 片段，考察表 1 和表 2 可以判断表 1 的 2.9 kb 片段为 *Bgl* II-*Bgl* II 位点，由此确定 *Bgl* II 切点在外源片段上的位置及与载体的连接方式，依理类推，最后将所确定的各个酶的酶切位点重合在一起，即得到一个完整的酶切图谱，见图2。

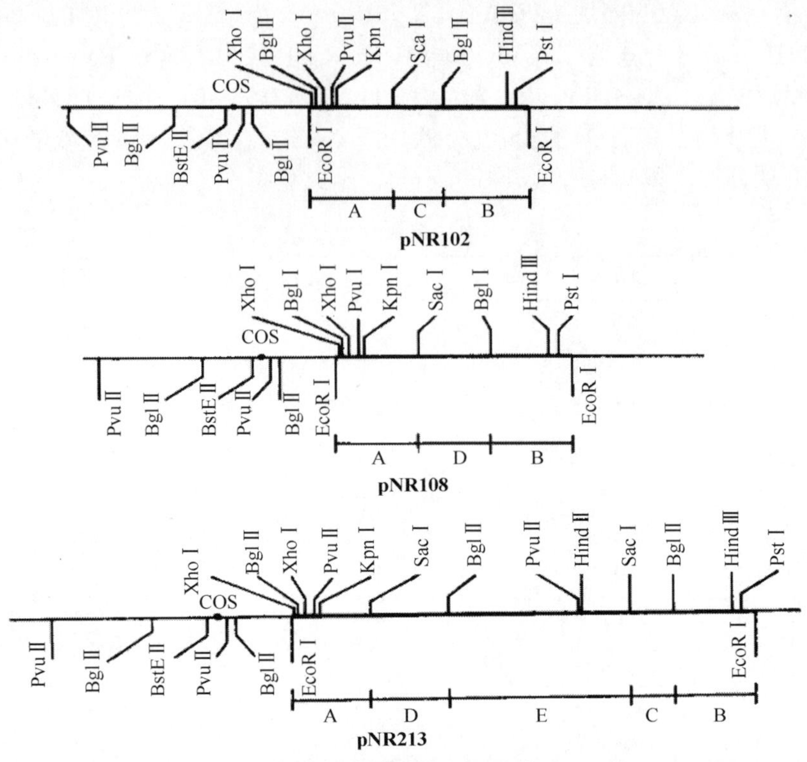

图2 重组质粒酶切图谱

对图2首先要说明的是，pNR013，pNR108和pNR203的外源片段上所得到的酶切位点是相同的，虽然它们分别来自不同的菌株，但含有相同的外源片段，并且pNR103和pNR108与载体的连接方式也是相同的，但pNR203与载体的连接方向与pNR108相反，而这并没有影响其互补结瘤功能，说明外源片段的连接方式对功能并没有决定作用。

进一步比较这5个重组质粒的酶切图谱可发现：它们都共有一个1.9 kb的*Eco*R I -*Sac* I片段（A片段）和一个1.7 kb的*Eco*R I –*Bgl* II片段（B片段）；pNR213与pNR102还共有一个相同的1.1 kb的*Bgl* II-*Sac* I片段（C片段），而pNR213与pNR108等共有一个相同的1.4kb *Bgl* II-*Sac* I片段（D片段），此外，pNR213还拥有一个3.3kb的*Bgl* II-*Sac* I片段（E片段）。酶切图谱显示了这些重组质粒DNA的共同性，也表现了pNR213在DNA片段上的包含性，酶切图谱的建立，为进一步的功能分析奠定了良好的基础。

酶切图谱所显示的重组质粒间片段的共同性也许正揭示了它们在DNA来源及功能方面的相似性，笔者曾以pNR108的5kb外源片段为探讨进行同源杂交分析[3]，结果显示：5个重组质粒外源片段有强的同源杂交，这与酶切图谱显示的片段共同性是相吻合的；杂交还发现，pNR108的外源片段与7653R菌株的共生质粒有同源杂交，这揭示这些质粒的外源片段的来源，又说明了重组质粒互补消除共生质粒突变菌株结瘤的功能基础。重组质粒共同具有的A片段和B片段也许正表明其功能关键所在，对其进一步研究将有助于揭示其功能区域及基因本质。

参 考 文 献

[1]Long S R，Buikema W J，Ausubel F M. Cloning of *Rhizobium meliloti* nodulation genes by direct complementation of Nod mutants. Nature，1982，298：485-488

[2]Downie J A，Johnston A W B，Rossen L，et al. Cloned nodulation genes of *Rhizobium leguminosarum* determine host-range specificity. MGG，1983，190：359-365

[3]农广，张忠明，陈华癸，紫云英根瘤菌7653R基因文库的5个互补结瘤菌株的重组质粒的分析，华中农业大学学报，1994，13（6）：531-537

[4]Kieser T. DNAGEL：a computer program for determining DNA fragmenr sizes using a small computer equipped with a graphics tablet. Nucleid Acids Res，1984，12：678-688

外源共生基因对紫云英根瘤菌共生效率的影响*

张学贤　周俊初　李阜棣　陈华癸

（华中农业大学生物固氮研究室）

摘　要　将豌豆根瘤菌共生质粒 pJB5JI、含紫云英根瘤菌共生基因的重组质粒 pRaZ15 及含苜蓿根瘤菌共生基因的重组质粒 pRmSL26 导入紫云英根瘤菌野生型菌株 7653R，然后对 3 种转移接合子的共生固氮效率进行了比较全面的测定。结果表明：转移接合子 7653R（pJB5JI）的共生固氮能力较出发菌株 7653R 大幅度提高，而且表现出较高的竞争结瘤能力，转移接合子 7653R（pRmSL26）在固氮酶活性、植物干重及结瘤数方面亦显著高于对照菌株 7653R；转移接合子 7653R（pRaZ15）虽然在固氮酶活性及植物干重方面没有发现显著性差异，但结瘤数显著减少。

关键词　紫云英根瘤菌；共生基因；共生效应；接合转移

紫云英（*Astragalus sinicus* L.）是我国南方广泛种植的一种稻田绿肥植物。能在紫云英上结瘤并进行共生固氮作用的紫云英根瘤菌是我国的一种重要的共生固氮资源，紫云英根瘤菌在互接种族中独属一族[1]。1991 年陈文新等[2]采用数值分类和 DNA 杂交等分子生物学鉴定方法，将其定名为 *Rhizobium huakuii* C.。多年来，我国学者在紫云英根瘤菌的应用方面已进行了许多卓有成效的研究[8]，但育种方法一直局限在自然选育及诱变育种。1982 年 De Jong 等[10]率先把根瘤菌共生质粒的转移应用于育种研究中。他们通过导入外源共生质粒使受体根瘤菌的结瘤固氮能力大幅度提高。1983 年 Brewin 等[7]研究证明导入外源共生质粒还可以提高根瘤菌的竞争结瘤能力，因此本研究将 3 种不同来源的外源共生质粒（基因）导入到紫云英根瘤菌野生型菌株 7653R 中，以考察外源共生基因对受体菌共生效率的影响，并探讨构建具有生产应用前景的优良菌株的可能性。

1　材料与方法

1.1　细菌菌株与培养条件

细菌菌株与质粒见表 1。根瘤菌培养基：液体培养采用 TY[2]，固体培养采用 SM[2]，质粒检测时采用 PA[2]，大肠杆菌培养基采用 LB[6]。培养温度：根瘤菌 28℃，大肠杆菌 37℃。抗生素使用浓度：四环素（Tc）15μg/ml，链霉素（Str）300μg/ml，卡那霉素（Km）50μg/ml。

1.2　植物结瘤试验

采用本室改进的双层钵进行。上层是塑料杯，内装粗砂，下层为装有 Fahraeus 无氮植物营养液[11]的玻璃瓶，塑料杯插在玻璃瓶中，通过纱布条将营养液吸人沙土中。每瓶种 4 株紫云英，种子表面消毒及催芽按常规方法进行。人工光照培养，每天光照 14 小时，光强为 4500～6000 lx，温度 15～24℃。

1.3　根瘤菌固氮酶活性测定

剪取待测植株的根系，去掉无瘤的部分后，置于 20ml 小瓶中，加盖橡皮塞，马上抽气 2ml，注入乙炔 2ml，28℃反应 1～2 小时，用日立 163 型气相色谱仪测定乙炔还原值。

1.4　植株地上部分含氮量的测定

采用半微量凯氏定氮法。

1.5　根瘤菌质粒的快速检测

采用修改的 Eckhardt 方法[6]。

*原载于《中国农业科学》，28（4）：14～19，1995.

表1 细菌菌株与质粒

Table 1 Bacterial strains and plasmids

菌株与质粒 Strain/plsamid	有关性状 1) Relevant characteristics	来源及文献 Source of reference
紫云英根瘤菌 （*Rhizobium huakuii*）		
7653 R	野生型 $Nod^{+}Fix^{-}Str^{r}$	[2]
7653 R（pJB5JI）	Str^{r} Km^{r}	本研究（This study）
7653 R（pRaZ15）	Str^{r} Tc^{r}	本研究（This study）
7653 R（pRmSL26）	Str^{r} Tc^{r}	本研究（This study）
7653 R（pLAFR 1）	Str^{r} Tc^{r}	本研究（This study）
7653 R（pRK404）	Str^{r} Tc^{r}	本研究（This study）
7653 RA 2	7653R（pJB5JI）在紫云英上所结根瘤的分离物 Isolate from a nodule form by 7653R（pJB5JI）on *Astragalus sinicus*	本研究（This study）
豌豆根瘤菌 （*R.leguminosarum biovar viceae*）		
$T_{83}K_3$ 质粒（piasmid）	Str^{r} Km^{r}	[3]
pJB5JI	豌豆根瘤菌共生质粒 Km^{r} Symbiotic plasmid of Rhizobium leguminosarum biovar viceae	[3]
pRaZ15	含紫云因根瘤菌共生基因重组质粗 Recombinant plasmid containing symbiotic gens of Rhizobium huakuii	[5]
pRmSL26	含苜蓿根瘤菌共生基因 Recombinant plasmid containing symbiotic gens of Rhizobium huakuii	[3]
pRK404	Mob^{+}，tra^{-}，Tc^{r}	[3]
pRK2013	Tra^{+}，Km^{r}	[3]

Km^{r}：卡那霉素抗性 Kanamycin rsistanc；Tc^{r}：四环素抗性 Tatracycline resistance；Str^{r}：链霉素抗性 Streptomycin resistance；Nod^{+}：结瘤 Nodulation：Fix^{+}：固氮 Nitrogen fixation；Mob^{+}：诱动功能 Mobilization：Tra^{+}：转移功能 Transfer.

2 结果

2.1 共生质粒（基因）的转移

pJB5JI 具有自我转移的能力，可以用供、受体菌的二亲本杂交进行质粒转移，但 pJB5JI 的供体菌 T83K3 及受体菌 7653R 都具有链霉素抗性，缺乏选择性标记。为了便于选择转移接合子，笔者对受体菌进行了质粒标记，即先通过三亲本杂交的方法在 pRK2013 的协助下，将 pRK404 质粒导入受体 7653R，使其具有四环素抗性（Tc^{r}）标记，以便在杂交中淘汰供体菌 T83K3。pRK404 质粒具有在大多数 G^{-}菌中转移和复制的能力，但在根瘤菌中，在无选择压力的条件下很容易丢失，因此含有 pRK404 质粒的紫云英根瘤菌 7653R（pRK404）在杂交前一直在含有四环素的培养基上培养，杂交后的转移接合子则在不含四环素的培养基中培养，经转接两次即可选择到四环素敏感（Tc^{s}）单菌落，从而淘汰了 pRK404。pJB5JI 质粒向 7653R（pRK404）中转移的频率（以受体计算）为 6.21×10^{-5}。

重组质粒 pRaZ15 及 pRmSL26 载体质粒都是 pLAFR1，不能自行转移，因此它们需要在协助质粒 pRK2013 的帮助下，通过三亲本杂交的方法进行接合转移，pRaZ15 及 pRmSL26 向 7653R 中转移的频率（以受体计算）分别为 4.1×10^{-3} 和 7.4×10^{-3}。

2.2 转移接合子的鉴定

随机从上述杂交组合中个挑取 5 个转移接合子，在各自的选择性平板上纯化 3 次后，利用快速检测法进行质粒检测。结果表明所有转移接合子都含有导入的外源质粒，见图 1 及图 2。

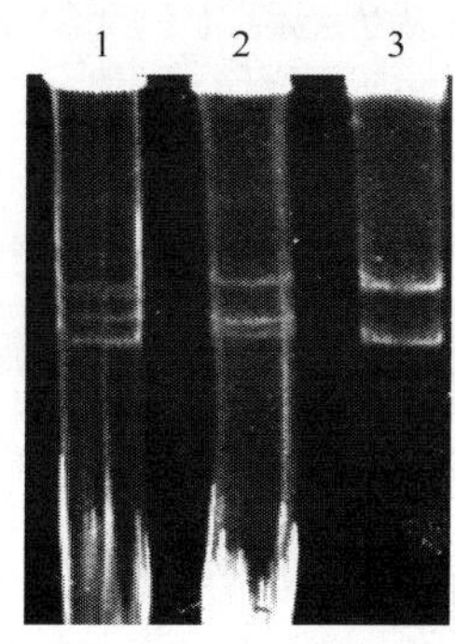

图 1 转移接合子 7653R（pJB5JI）的质粒检测

Fig.1 Eckhardt agarose gel electrophoresis of Plasmid DNA of transconjugant 7653R（pJB5JI）

1. *R. leguminosarum* biovar *viceae* T83K3; 2. *R. huakuii* 7653R（pJB5JI）; 3. *R.huakuii* 7653R

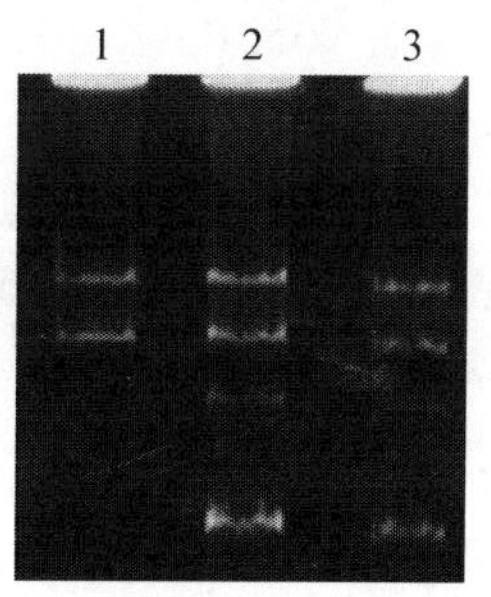

图 2 转移接合子 7653R（pRaZ15）及 7653R pRmSL26 的质粒检测

Fig.2 Eckhardt agarose gel electrophoresis electrophoresis of plasmid DNA of transconjugant 7653R（pRaZ15）and 7653R（pRmSL26）

1. 7653R; 2. 7653R（pRaZ15）; 3. 7653R pRmSL26

2.3 外源共生质粒的稳定性

以 pRmSL26 为代表来分析 pLAFR1 重组质粒的稳定性。将纯化后的 5 个转移接合子 7653R（pRmSL26）在 YMA 平板上转接 2 次，再分别稀释得到单菌落，各挑 100 个单菌落分别点种在 YMA+Tc 和 YMA 平板上，结果表明只有 87.3%保持 Tc^r，可见在人工培养条件下，pRmSL26 是不稳定的。

从 7653R（pRmSL26）所结根瘤中挑选 5 个，将根瘤压碎汁稀释，在 YMA 平板上得到单菌落，各挑 100 个点种在 YMA+Tc 和 YMA 平板上，结果表明，四环素抗性菌落仅占 60.7%，可见经过与紫云英共生之后，pRmSL26 也是不稳定的。转移接合子 7653R（pJB5JI）在人工培养条件下，pJB5JI 能稳定存在，而经过紫云英共生之后，pJB5JI 表现出不稳定的遗传特性。该结果已另文论述[4]。

2.4 转移接合子 7653R（pJB5JI×）共生效应的测定

2.4.1 共生固氮能力的测定

转移接合子 7653R（pJB5JI）在紫云英上形成有效根瘤，共生固氮能力的测定表明：7653R（pJB5JI）所结根瘤的固氮酶活性较原菌株 7653R 增加 94%，而且在 α= 0.01 水平达到极显著。接种 7653R(pJB5JI）的植株地上部分干重亦较 7653R 增加 12%（见表 2）。从转移接合子 7653R（pJB5JI）在紫云英上所结的根瘤中分离根瘤菌。选择含 pJB5JI 的根瘤分离物 7653RA2 作重复验证实验。结果表明：7653RA2 所结根瘤的固氮酶活性较 7653R 增加 76.8%，而且在 α=0.01 水平达到极显著。接种 7653RA2 的植株地上部分干重及含氮量也高于接种 7653R 植株（见表 2）。

表 2 转移接合子 7653R（pJB5JI）共生效应的测定

Table 2 Symbiotic Properties of transconjugant 7653R（pJB5JI）

接种物 Inoculants	乙炔还原活性 Acetylene reduced activity（$\mu mol.g^{-1}\cdot h^{-1}$）	地上部分干重 Shoot dry weight（g/plant）	含氮量 Percentage of N in shoot（%）
7653R	15.8	0.1074	
7653 R（pJB5JI）	30.65**	0.1203	
CK		0.0337	
7653R	14.2	0.219	3.54
7653RA2	25.1**	0.280	4.60
CK		0.037	1.03

**在 α=0.01 水平差异极显著 Significant difference at 0.01 level

2.4.2 竞争结瘤能力的测定

从 7653R（pJB5JI）及 7653R 所结根瘤中各随机挑选 50 个。将根瘤压碎汁在 YMA+Km 平板上划线。结果表明：所有 7653R（pJB5JI）形成的根瘤均能检出 Km^r 根瘤菌，而 7653R 形成的根瘤均不能检出 Km^r 根瘤菌。因此 Km^r 可以作为测定 7653R（pJB5JI）占瘤率的标记。将 7653R（pJB5JI）及 7653R 以细菌数 1∶1 的比例混合均匀，然后接种紫云英植物，从所结根瘤中随机挑选 50 个，将根瘤压碎汁涂 YMA+Km 平板，结果从其中 40 个根瘤中分离出 Km^r 根瘤菌。因此 7653R（pJB5JI）的占瘤率为 80%，而 7653R 的占瘤率为 20%。可见转移接合子 7653R（pJB5JI）的竞争结瘤能力也高于原菌株 7653R。

2.5 转移接合子 7653R（pRaZ15）及 7653R（pRmSL26）共生效应的测定

重组质粒 pRaZ15 及 pRmSL26 的载体质粒是 pLAFRl。在盆栽条件下比较转移接合子 7653R（pLAFRl）与 7653R 的共生固氮能力，结果表明：载体 pLAFRl 对 7653R 的共生固氮效率没有影响。转移接合子 7653R（pRaZ15）与 7653R 相比，在固氮酶活性及接种植株地上部分干重方面没有显著差异，但结瘤数量减少。转移接合子 7653R（pRmSL26）不但在固氮酶活性及接种植株地上部分干重方面显著高于 7653R，而且结瘤数量显著增多（见表 3）。重复验证仍得到类似结果（数据未列）。

表 3 转移接合子 7653 R（pRaZ15）和 7653R（pRmSL26）的共生效率

Table 3 Symbiotic properties of transconjugants 7653R（pRaZ15）and 7653R（pRmSL26）

接种物 Inoculants	瘤数 Nodule number（per plant）	乙炔还原活性 Acetylene reduced activity（$\mu mol \cdot g^{-1} \cdot h^{-1}$）	地上部分干重 Shoot dry weight（g/plant）
7653R	35.0	24.37	0.1025
7653R（pLAFR1）	31.3	27.36	0.1067
7653R（pRaZ15）	23.3[a]	31.53	0.1024
7653R（pRmSL26）	58.0[c]	40.53[b]	0.1259[a]
CK			0.0455

a：在 α=0.1 水平差异显著 Significant difference at 0.1 level b：在 α=0.05 水平差异显著 Signcant difference at 0.05level c：在 α=0.01 水平差异显著 Significant difference at 0.01 level

3 讨论

本研究将豌豆根瘤菌共生质粒 pJB5JI 导入紫云英根瘤菌野生型菌株 7653R 之后，植物盆栽结瘤实验表明转移接合子 7653R（pJB5JI）不但结瘤固氮能力较出发菌株 7653R 显著提高，而且表现出较高的竞争结瘤能力，从而具有潜在的应用价值。该结果表明通过导入异源共生质粒（基因）的方法是可以提高根瘤菌的共生固氮效率的。

自 20 世纪 70 年代以来，国内外的研究者进行了很多共生基因的转移工作，但其目的多数在于研究外源共生基因的表达，很少涉及导入的外源共生基因对受体菌共生固氮效率的影响，即两套共生基因间的相互作用。对于在育种工作中能否有效地利用共生基因间的相互作用来改变根瘤菌的共生特性还知之甚少。从本研究及笔者以前的工作[8]来看，根瘤菌共生基因间的相互作用是多方面的，向受体根瘤菌中导入外源共生基因，既可以产生正效应，也可以产生负效应，或者没有显著性影响。因此在针对特定根瘤菌的共生性状进行遗传改造时，导入足够多的不同来源的大片段共生基因，从中选出有价值的遗传组合是一条值得更进一步探索的技术途径。

参 考 文 献

[1]陈华癸.草籽绿肥根瘤细菌和人工接种.新科学.1952，3：33-38

[2]张学贤等.紫云英根瘤菌与豌豆根瘤菌共生质粒间的相互作用.华中农业大学学报.1992，11（2）：175-181

[3]张学贤等.导入外源共生基因对大豆根瘤菌共生效率的影响.农业生物技术进展与展望.中国科技大学出版社，1993，133-141

[4]张学贤等.外源共生质粒 pJB5JI 在紫云英根瘤菌（*Rhizobium huakuii*）中的稳定性，华中农业大学学报.1994，13（1）：1-8

[5]张忠明等.紫云英根瘤菌基因文库的构建及含完整结瘤基因的重组质粒 pRaZ15 的分离.生物工程学报.1991，7（3）：213-219

[6]王常霖，赫茨.转座子 Tn5-mob 对菜豆根瘤菌共生基因的诱变、转移和初步定位.遗传学报.1988，15（1）：25-33

[7]Brewin N J，et a1.，Natural variation in Rhizobium plasmids. In Molecular Genetics of the Bactaria-plant Interaction，Puhler A.（ed.）springer-velag，Berlin Heidelberg New York Tokyo. 1983，113-120

[8]Chen H K，et al.Characteristics，distribution，ecology and utilization of *Astragalus sinicus- rhizobia* symbiosis. In Nitrogen Fixation Research in China，G. .F. Hong（ed.）springer-Verlag. 1992，439-455

[9]Chen WX，et al.*Rhizobium huakuii* sp. nov. isolated from the root nodules of *Astragalus sirricus*. Int. J. Sys. Bacteriol. 1991，41（2）：275-280

[10]Dejong T M，et al.Rhizobium strain improvement by plasmid transfer. J. Gen，Microbiol. 1982，128：1829-1838

[11]Fahraeus G.，The infection of clover root hairs by nodule bacteria studied by a simple glass slide technique. J. Gen. Microbiol. 1957，16：374-381

Behavior of Plasmid pJB5JI in *Rhizobium huakuii* Under Free–Living and Symbiotic Conditions*

XUE-XIAN ZHANG，JUN-CHU ZHOU，ZHONG-MING ZHANG，HUA-KUI CHEN，FU-DI LI

（Department of Microbiology，Huazhong Agricultural University，Wuhan，430070，P.R. China）

ABSTRACT The *Rhizobium leguminosarum* biovar *viceae* host-range plasmid pJB5JI was transferred into *Rhizobium huakuii* strains，both wild-type 7653R and its sym plasmid-cured mutant 7653R-1. Transconjugant 7653R-1（pJB5JI）acquired the ability to form ineffective nodules on pea plants，whereas transconjugant 7653R（pJB5JI）could not do so，indicating that the indigenous symbiotic plasmid could restrict the functional expression of pJB5JI. On the other hand，transconjugant 7653R（pJB5JI）showed higher nitrogenase activity on *A.sinicus* and higher shoot dry weight than the recipient strain 7653R. The alien plasmid pJB5JI in both kinds of transconjugants remained stable during frequent transfer on culture media，but in part of the isolates from nodules formed by them the pJB5JI was not visualized on gel by the Eckhardt procedure. Southern hybridization with Tn5 and nod gene probes showed that these isolates still reserved，at least in part，DNA of pJB5JI，which was probably integrated onto the chromosome of cells.

The host-specific interaction between *Rhizobium* bacteria and leguminous plants leads to the formation of nitrogen-fixing root nodules. The host range，in addition to nodulation and nitrogen-fixing abilities，is determined by genes，which are harbored on plasmids in most of the *Rhizobium* strains. Some large *Rhizobium* plasmids carrying symbiotic genes，such as pJB5JI，can be transferred between various *Rhizobium* species，but their behaviors are different in the new host strains，owing to the dissimilarities of genetic background[1,8,10,15,18].

Rhizobium huakuii is the newly assigned name to some fast-growing strains isolated from nodules of *Astragalus sinicus* L.[4]. It forms nodules mainly on *A.sinicus* in a very host-specific manner；*A.sinicus* is an important legume crop for green manure，traditionally grown in rice fields in lower Yangtze River Valley and South China. The biology，physiology，and ecology of the *A.sinicus-R.huakuii* symbiosis，as well as its application in agriculture，have been extensively studied for several decades in China [3]. Studies on its genetics have been conducted only lately. The plasmid number of studied strains varied from 1 to 5（this laboratory，unpublished），and the symbiotic genes are located on one of the large plasmids. Wild-type strain 7653R harbors two large plasmids，pRh7653Ra and pRh7653Rb. Southern hybridizations with nodDABC and nifHDK probes indicated that the tested genes located on the larger plasmid pRh7653Rb. Subsequently，pRh7653Rb was cured by heat treatment，and the mutant strain obtained，7653R-1，was found to be Nod^{-}[20]. The purpose of this research was to determine whether the *R. leguminosarum* biovar *viceae* host-range plasmid pJB5JI could enable *R. huakuii* wild-type strain 7653R and its symbiotic plasmid-cured mutant strain 7653R-1 to form nodules on pea plants. We are interested in studying the possible interaction of pJB5JI with the endogenous symbiotic plasmid of strain 7653R. The stability of pJB5JI in *R. huakuii* strains was also investigated.

*原载于 *Current Microbiology*. 31:97～101，1995.

MATERIALS AND METHODS

Bacterial strains and plasmids. Bacterial strains and plasmids used in this study are listed in Table 1. Rhizobial strains were maintained and cultured on yeast extract mannitol（YEM） agar [17]. *Escherichia coli* strains were grown in LB medium [7].

Plasmid transfer and visualization. Plasmid pJB5JI transfer from *R. leguminosarum* strain T83K3 to *R. huakuii* strains 7653R and 7653R-1 was separately carried out by membrane crosses as described by Christensen and Schubert [5]. Transconjugants were selected on defined medium SM[19] supplemented with tetracycline（15μg/ml）, streptomycin（300μg/ml） and kanamycin（50μg/ml）. Plasmids were detected on agarose electrophoresis by the procedure of Eckhardt[9].

Table 1 Bacteria and plasmids used in this study

Strains/plasmids	Relevant characteristic ^{a}	Reference or source
Rhizobium huakuii		
7653R	Wild-type Str^r	[20]
7653R-1	sym plasmid-cured mutant of 7653R Nod^-	[20]
7653R（pJB5JI）	Str^r，Km^r	This study
7653R-1（pJBSJI）	Str^{r}· Km^r	This study
7653RA2，A9	Isolates from nodules formed on *A. siaicus* by 7653R（pJB5JI）	This study
7653R-1Pa，Pb	Isolates from nodules formed on pea by 7653R-1（pJB5JI）	This study
R. leguminosarum biovar viosas		
JI6015	Str^r，Rif^r，Nod^-	[11]
T83K3	JI6015（pJB5JI）	
	Nod^+Fix^+	[11]
Agrobacterium tumefaciena		
GMI9023	Str^r，RiP^r，plasmid free strain	[14]
GMI9023（pJB5JI）	Str^r，Rif^r，Km^r	This study
Plasmids		
pJB5JI	pea sym plasmid with a Tn5 insertion in the genes for bacteriocin production	[11]
pKan2	pBR322 containing 3.5-kb *Hin*dII fragment of Tn5	[6]
pRmSL26	nod segment of *R. maniloti*，including nodDIABC in pLAFR1	[12]

Symbiotic assays. Test plants were grown in an assembly pot similar to the Leonard jar designed by this laboratory. Sand was placed in a plastic cup with a small hole at the bottom through which a wick was anchored. The cup was loaded onto a glass bottle containing Fahraeus nitrogen-free solution. The whole device was sterilized at 121℃ for 120 min. *Pisum sativum* and *A. sinicus* seeds were pre-immersed in 95% ethanol for 5 min，sterilized with 0.1% $HgCl_2$ for 5 min，and rinsed with sterile water ten times. Surface sterilized seeds were germinated on a water agar plate for 2 days in darkness at room temperature. The seedlings were transferred onto the growth assembly and were inoculated when the cotyledons opened. Inoculated plants were grown in an illuminated chamber under 4500-6000 lux at 15-22℃ with a 14-h photoperiod. Uninoculated plants were set up as a control.

Acetylene-reduced activity of nodules was detected at 30 days（*P. sativum*） or 40 days（*A. sinicus*） after inoculation. Shoot dry weight was determined on the same day.

Strain re-isolation. Nodules were excised from the roots of *P. sativum* and *A. sinicus* and surface sterilized as described above. The excised nodules were crushed and streaked on SM medium supplemented with streptomycin.

Southern hybridizations. Dot hybridization was carried out as described by Sambrook et al. [16]. Southern blot hybridization was done as described by Christensen and Schubert [5]. Tn5 probe was the 3.5-kb *Hind*III fragment of plasmid pKan2 [6], and the nod genes probe was the 8.7-kb *Eco*RI fragment of plasmid pRmSL26 known to contain *R. meliloti* nodD_1ABC [12].

RESULTS

Transfer of plasmid pJB5JI and its stability in recipient strains. Crosses were done with *R. leguminosaturn* biovar *viceae* T83K3 as the donor of pJB5JI to *R. huakuii* wild-type strain 7653R and its symbiotic plasmid-cured mutant strain 7653R-1. The transfer frequency (per recipient) of Km^r marker was 10^{-5} to both 7653R and 7653R-1. The spontaneous mutation for kanamycin resistance of the recipient was less than 10^{-9}. Five single colonies were taken from each cross and purified by streaking three times successively on selective medium. For a test of the stability of plasmid pJB5JI during cultural transfers on medium, all ten transconjugants were subcultured three times on SM medium without kanamycin. One hundred colonies for each were picked up, cultured separately on SM medium with kanamycin, and SM medium as well. None was found to have become kanamycin-sensitive. Subsequently 20 isolates of each cross were chosen randomly for plasmid visualization. They all harbored pJB5JI. So its presence in both 7653R and 7653R-1 of *R. huakuii* was stable in free-living conditions.

Symbiotic properties of transconjugant 7653R-1 (pJB5JI). *R. huakuii* mutant strain 7653R-1 is Nod^- on *A. sinicus* because the symbiotic plasmid pRh7653Rb has been cured. Introduction of the pea host range plasmid pJB5JI into it could not recover its nodulation ability on *A. sinicus*, whereas transconjugant 7653R-1 (pJB5JI) acquired the ability to nodulate *Pisum* sativum, but no acetylene reduction activity was detected (Table 2). All isolates from 10 nodules formed on *P. sativum* by 7653R-1 (pJB5JI) were Km^r, and plasmid pJB5JI was demonstrated except for two of them (Fig. 1). Two isolates, one 7653R-1Pb carrying pJB5JI and the other 7653R-1Pa without pJB5JI, were selected to inoculate both *A. sinicus* and *P. sativum* plants separately. Results showed that both isolates could not nodulate *A. sinicus*, but could form ineffective nodules on *P. sativum* (Table 2).

Table 2 Symbiotic properties of transconjugants

Inoculant	Plant response		Inoculant	Plant response	
	Astragalus	*Pigeon activum*		*Astragalus*	*Pigeon activum*
7653R	Nod^+Fix^+ (10) [a]	Nod^- (5)	7653RA2	Nod+Fix+ (10)	Nod^- (10)
7653R (pJB5JI)	Nod^+Fix^+ (10)	Nod^- (10)	7653RA9	Nod+Fix+ (10)	Nod^- (10)
7653R-1	Nod^- (5)	Nod^- (5)	7653R-1	Nod^- (5)	Nod^- (5)
7653R-1 (pJB5JI)	Nod^- (5)	Nod^+Fix^- (10)	7653R-1Pa	Nod^- (5)	Nod^+Fix^- (10)
T83 K3	Nod^- (5)	Nod^+Fix^+ (5)	7653R-1Pb	Nod^- (5)	Nod^+Fix^- (10)
7653R	Nod^+Fix^+ (10)	Nod^- (5)	T83K3	Nod^- (5)	Nod^+Fix^+ (5)

[a]Numbers in parentheses refer to the number of plants tested.

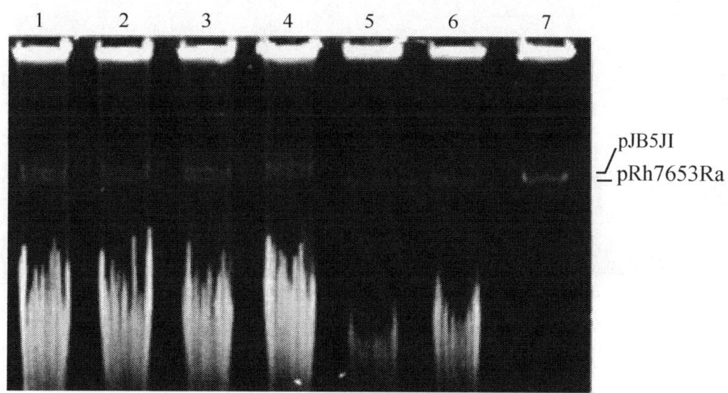

Fig.1 Eckhardt agarose gel electrophoresis of plasmid DNA of *Rhizobium huakuii* strains. Lanes 1-6, isolates from the nodules formed by transconjugant 7653R-1 (pJB5JI) on pea; lane 7, 7653R-1.

Symbiotic properties of transconjugant 7653R(pJB5JI). *R. huakuii* wild-type strain 7653R is Nod^+ on *A. sinicus*. Transconjugant 7653R (pJB5JI) not only could form effective nodules on *A. sinicus*, but nodules also showed much higher acetylene reduction activity compared with those nodules formed by strain 7653R (Table 3). On the other hand, it did not nodulate pea (Table 2). All isolates from 30 nodules formed on *A. sinicus* by transconjugant 7653R (pJB5JI) were kanamycin resistant, but no plasmid pJB5JI could be visualized on gel except two of them (Fig. 2). Two of the isolates, 7653RA2 carrying pJB5JI and 7653RA9 without pJB5JI, were chosen to inoculate both plant species. They formed nodules on *A. sinicus*, but not on *P. sativum* (Table 2). Furthermore, nodules on *A. sinicus* formed by 7653RA2 showed significantly higher acetylene reduced activity than those formed by its parent strain (Table 3).

Table 3 Nitrogen fixation abilities of transconjugant 7653R (pJB5JI) on *Astragalus sinicus*

Inoculant	Acetylene reduced activity (μ mol C_2H_4/g/hour)	Shoot dry weight (g/plant)	Percentage N In shoot (%)
7653R	15.8	0.1074	ND[b]
7653R (pJB5JI)	30.7[a]	0.1203	ND
CK		0.0337	ND
7653R	14.2	0.219	3.54
7653RA2	25.1[a]	0.280	4.60
CK		0.037	1.03

[a] The difference compared with the parent strain 7653R is significant at level of p=0.01

[b] ND, not detected

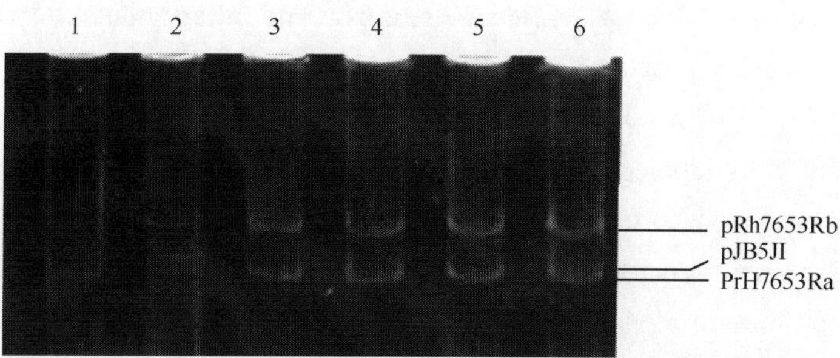

Fig.2 Plasmid profiles in strains. Lane 1, T83k3; 2-5,isolates from the nodules formed by transconjugant 7653R (pJB5JI) on *A. sinicus*; 6, 7653R.

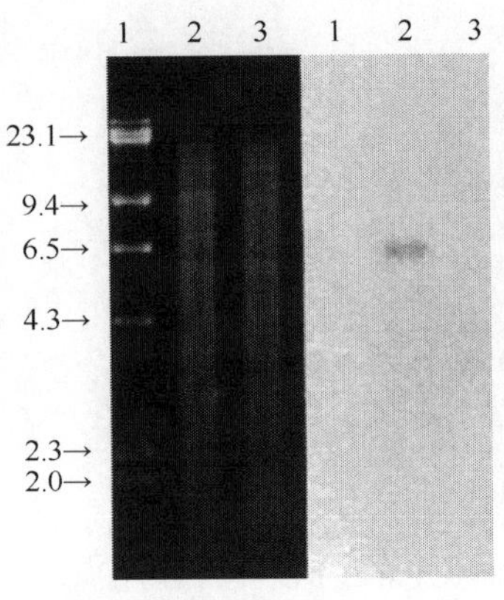

Fig. 3 *Eco*RI-digested total DNA（A）and the corresponding autoradiogram of ^{32}P-labeled nod gene probe（B）, 1, λ DNA/*Hin*dIII; 2, 7653R-1Pa; 3, 7653R-1.

DNA hybridizations using Tn5 and nod gene probes. A plasmid corresponding in size to pJB5JI was absent in isolate 7653R-1Pa, suggesting a loss of pJB5JI. But 7653R-1Pa was still kanamycin resistant and able to nodulate pea plants conferred by the presence of pJB5JI. For determination of the fate of pJB5JI, *Eco*RI-digested total DNA from 7653R-1Pa was hybridized with ^{32}P-labeled Tn5 and nod gene probes. Hybridization tests showed that there was a single homogeneous band of approximately 11 kb and 6.6 kb respectively (Figs. 3 and 4). Southern blot analysis of an Eckhardt agarose gel probed with ^{32}P-labeled Tn5 DNA showed no hybridization to the plasmid of 7653R-1Pa（Fig. 5）. The hybridization data suggested that at least part of pJB5JI was still present in 7653R-1Pa, and it was probably integrated onto the chromosome.

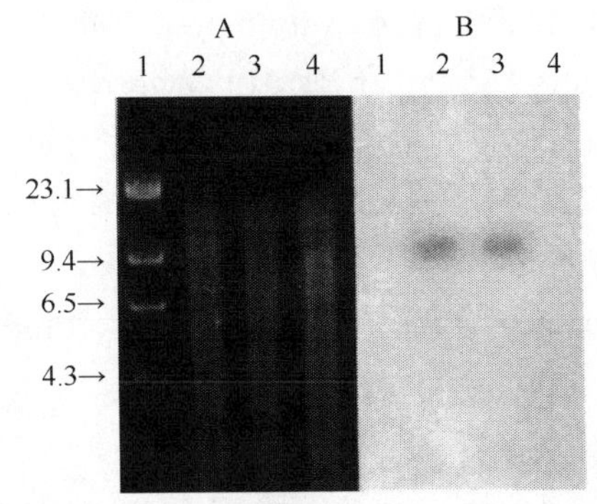

Fig. 4 *Eco*RI-digested total DNA（A） and the corresponding autoradiogram of ^{32}p-labeled Tn5 probe（B）.

1, λ DNA/*Hin*dIII; 2, T83k3; 3, 7653R-1Pa; 4, 7653R-1.

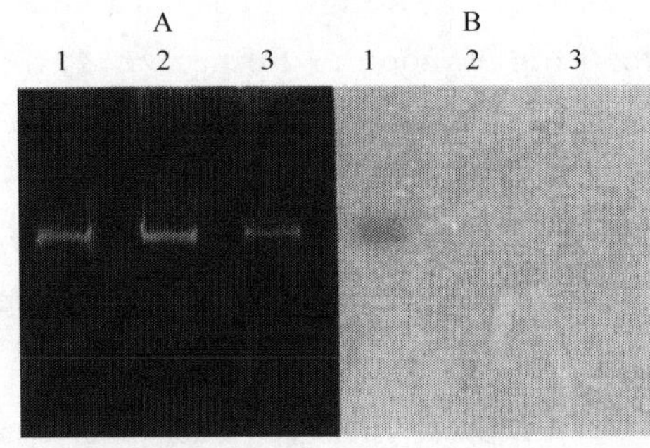

Fig. 5 Eckhardt agarose gel electrophoresis（A） and the corresponding autoradiogram of ^{32}P-labeled Tn5 probe（B）.

Lanes: 1, GMI9023（pJB5JI）; 2, 7653R-1Pa; 3, 7653R-1.

DISCUSSION

The host range plasmid pJB5JI of *R. leguminosarum* biovar *viceae* has been transferred to other species of rhizobia, and loss or deletion of plasmid pJB5JI after plant selection was frequently observed [2,15,18]. This paper is the first to report its behavior in *Rhizobium huakuii.* Results indicated that the host plant also influenced the stability of pJB5JI in *R. huakuii* strains. The alien plasmid pJB5JI in both kinds of transconjugants was stable during growth of the rhizobia in laboratory media. But in part of the isolates from nodules formed by them, pJB5JI could not be detected by the Eckhardt procedure. Subsequent Southern hybridization analysis with ^{32}P- labeled Tn5 and nod gene probes showed that at least part of pJB5JI was present in these isolates, and it was probably integrated onto the chromosome.

Rolfe et al. [13] once transferred pJB5JI into Rhizobium strain ANU237 and found that there was some kind of interaction between the introduced pJB5JI and the endogenous symbiotic plasmid of strain ANU237, because plasmid pJB5JI coded for pea nodulation and strain ANU237 was unable to nodulate clover, whereas transconjugant ANU237（pJB5JI） could produce nodules on white clover. In this study, pJB5JI was transferred into *R. huakuii*, both wild-type strain 7653R and its symbiotic plasmid-cured mutant strain 7653R-1. Transconjugant 7653R-1（pJB5JI） acquired the ability to form ineffective nodules on pea plant, but transconjugant 7653R（pJB5JI） could not do so. This indicated that the endogenous symbiotic plasmid pRh7653Rb restricts the functional expression of introduced pJB5JI. Little is known about its mechanism. One possible explanation that should be further confirmed is that the Nod factor produced by pJB5JI was modified by part of the nod genes of pRh7653Rb and could not be recognized by *Pisum sativum*.

ACKNOWLEDGMENTS

This research was supported in part by the National Natural Science Foundation of China（NSFC）, and also a IFS grant C/2305-1 to X.-X. Zhang.

LITERATURE CITED

[1]Beynon JL, Beringer JE, Johnston AWB（1980） Plasmid and host-range in *Rhizobium leguminosarum* and *Rhizobium phaseoli.* J Gen Microbiol 120：421–429

[2]Brewin NJ, Wood EA, Johnston AWB, Dibb NJ, Hombrecher W（1992） Recombinant nodulation plasmids in *Rhizobium leguminosarum.* J Gen Microbiol 128：1817–1827

[3]Chen HK, Li FD, Cao YZ（1992） Characteristics, distribution, ecology and utilization of *Astragalus sinicus*-rhizobia symbiosis. In：Hong GF（ed） Nitrogen fixation research in China. Berlin, Heidelberg, New York：Springer-Verlag, pp 439–455

[4]Chen WX, Li GS, Qi YL, Wang ET, Yuan HL, Li JL（1991） *Rhizobium huakuii* sp. nov. isolated from the root nodules of *Astragalus sinicus.* Int J Syst Bacteriol 41：275–280

[5]Christensen AH, Schubert KR（1983） Identification of a *Rhizobium trifolii* plasmid coding for nitrogen fixation and nodulation genes and its interaction with pJB5JI, a *Rhizobium leguminosarum* plasmid. J Bacteriol 156：592–599

[6]Chua KY, Pankhurst LE, Macdonald PE, Hopcroft DH, Jarvis BDW, Scott BB（1985） Isolation and characterization of transposon Tn5-induced symbiotic mutants of *Rhizobium loti.* J Bacteriol 162：335–343

[7]Ditta G, Stanfield S, Corbin D, Helinski DR（1980） Broad host range DNA cloning system for Gram-negative bacteria：construction of a gene bank of *Rhizobium meliloti.* Proc Natl Acad Sci USA 77：7347–7351

[8]Ditta G, Schmidhauser T, Yakobson E, Lu P, Liang XW, Finlay DR, Guiney D, Helinski DR（1985） Plasmids related to the broad host range vector, pRK290, useful for gene cloning and for monitoring gene expression. Plasmid 13：149–153

[9]Eckhardt T（1978） A rapid method for the identification of plasmid deoxyribonucleic acid in bacteria. Plasmid 1：584–588

[10]Espuny MR, Ollero FJ, Bellogin RA（1989） Selection and symbiotic properties of *Rhizobium leguminosarum* biovar *phaseoli* strains harboring pRtr5a. Curr Microbiol 19：179–181

[11]Johnston AWB, Beynon JL, Buchanan-Wollaston AV, Setchell SM, Hirsch PR, Beringer JE（1978） High frequency transfer of nodulating ability between strains and species of *Rhizobium.* Nature 276：635–636

[12]Long SR, Jacobs T, Beebe D, Egelhoff T（1983） Early nodulation genes of *Rhizobium meliloti.* In：Puhler A（ed.） Molecular genetics of the bacteria-plant interaction. Berlin, Heidelberg, New York, Tokyo：Springer-Verlag, pp 79–87

[13]Rolfe BG, Djordjevic MA, Morrison NA, Plazinski J, Bender GL, Ridge R, Zurkowski W, Tellam JT, Gresshoff PM, Shine J（1983） Genetic analysis of the symbiotic regions in *Rhizobium trifolii* and *Rhizobium parasponia.* In：Puhler A（ed.）Molecular genetics of the bacteria-plant interaction Berlin, Heidelberg, New York, Tokyo：Springer-Verlag, pp 188–203

[14]Rosenberg C, Huguet T（1984）The pATc58 plasmid of *Agrobacterium tumefaciens* is not essential for tumour induction. Mol Gen Genet 196：533–536

[15]Ruiz-Sainz JE, Chandler MR, Jimenez-Diaz R, Beringer JE（1984）Transfer of a host range plasmid from *Rhizobium leguminosarum* to fast-growing bacteria that nodulate soybeans. J Appl Microbio157：309–315

[16]Sambrook J, Fritsch EF, Maniatis T（1989）Molecular cloning：a laboratory manual, 2nd ed. Cold Spring Harbor, NY：Cold Spring Harbor Laboratory Press

[17]Vincent JM（1970）A manual for the practical study of root-nodule bacteria. Oxford：Blackwell Scientific Publications

[18]Wang CL, Beringer JE, Hirsch PR（1986）Host plant effects on hybrids of *Rhizobium leguminosarum* biovar *viceae* and *trifolii.* J Gen Microbiol 132：2063–2070

[19]Zhang ZM, Chen HK, Li FD, Fan YL（1991）Construction of gene library and isolation of pRaZi5 containing complete nodulation genes in *Rhizobium astragali.* Chin J Biotechnol 7：213–219

[20]Zhou JC, Zhang ZM, Huang CJ, Li FD, Chen HK（1987）Studies on *Rhizobium* plasmids II. Tests of plasmid elimination of *R. astragali.* J Huazhong Agri Univ 6：156–164

The Symbiotic Plasmid and Nodulation Genes in *Rhizobium* HUAKUII CHEN*

ZHANG, Z., NONG, G., HU, F., ZHANG, X., ZOU, X., ZHOU, J., LI, F., CHEN, H.

(Biotechnology Center, Huazhong Agricultural University, Wuhan, 430070, P.R. China)

INTRODUCTION

Rhizobium huakuii Chen(Chen et al., 1991)is the rhizobia isolated from nodules of *Astragalus sinicus L.*, which forms nodules mainly on *A. sinicus* and a few *Astragalus* spp. in a very narrow host-specific manner (Chen et al., 1944). *A. sinicus* is an important legume crop for green manure, traditionally grown in rice fields in Yangtze River Valley and southern China. The biology, physiology and ecology of the *A. sinicus-R. huakuii* symbiosis, as well as its application in agriculture have been extensively studied for several decades in China (Chen et al., 1992). Studies on its genetics have been conducted only lately. This paper focuses on the identification of symbiotic plasmid and cloning of early symbiotic genes in *R. huakuii.*

INDIGENOUS PLASMIDS AND SYMBIOTIC GENES IN *R. huakuii*

The plasmid profiles of 154 strains of *R. huakuii* were examined by a modified Eckhardt (1978) procedure of plasmid visualization. They contained 1 to 5 indigenous plasmids and the molecular weight of the plasmids varied from 30 to 600 MD(Zou et al., 1994). The Southern blots of plasmids of effective strains were hybridized with *nod*DABC (Downie et al., 1983) (from *R. leguminosarum*) and *nif*HDK (Cannon et al., 1979) (from *K. pneumoniae*) probes. The results showed that the symbiotic genes were located together on one of plasmids, the symbiotic plasmid, with only one exception, in which *nod* and *nif* genes were not detected on plasmid in a strain containing a single plasmid although proved to be Nod^+ Fix^+. The molecular weight of the symbiotic plasmid varied from 117 to 226 MD. *R. huakuii* strain 7653R is a Nod^+ Fix^+ rhizobia symbiotically specific to *A. sinicus*. It harbours two large plasmids, pRh7a(115 MD) and pRh7b(210 MD). Strain 7653R-1 is a Nod^- mutant derived from strain 7653R, which is devoid of the larger plasmid pRh7b (Zhou et al., 1987). In order to test the sym-plasmids response to early nodulation, experiment of root hairs infection was performed and the results are presented in Table 1. Root hairs on roots inoculated with strain 7653R show in average 58.32 percent of the elongated root hairs were deformed, and among them 24 "hepherd stick" were observed. Root hairs inoculated with strain 7653R-1 were not elongated, only 3.06 percent deformed and no "hepherd stick" observed. This indicated that the early nodulation genes *nod*DABC are located on the plasmid pRh7b, which is cured of in 7653R-1. The host-7653R-1 interaction stops at or before Hac stage. Hybridization with the probes of *nod*DABC and *nif*HDK showed that the tested genes were located on pRh7b.

---

*原载于 *Nitrogen Fixation: Fundamentals and Application.* 389～392, 1995.

Table 1 The test of root hairs infection

	Strain				Strain		
	CK	7653R	7653R-1		CK	7653R	7653R-1
No. of tested plants	30	30	30	No. of "shepherd stick"	0	24	0
No.of RT hairs observed	1290	1562	1860	RT hairs deformed (%)	0.93	58.32	3.06
No. of RT hairs deformed	12	911	57				

BEHAVIOR OF EXOGENOUS SYM-PLASMID IN *Rhizobium huakuii*

Some large rhizobia plasmids carrying symbiotic genes, such as pJB5JI, can be transferred between various *Rhizobium* species, but their behaviors are different in the new host strains, owing to the dissimilarities of genetic background(Beynon et al., 1980; Ruiz-Sainz et al., 1984; Wang et al., 1986). The aim of the experiment was to determine whether the *R. leguminosarum* host range plasmid pJB5JI could enable *R. huakuii* strain 7653R and its sym-plasmid-cured mutant 7653R-1 to form nodules on pea plants. The interaction of pJB5JI with the sym-plasmid of strain 7653R was studied. The sym-plasmid pJB5JI was transferred into *R. huakuii* strain 7653R and its sym-plasmid-cured mutant 7653R-1. Transconjugant 7653R-1 (pJB5JI) acquired the ability to form ineffective nodules on pea plants (Zhang et al., 1992), whereas transconjugant 7653R (pJB5JI) did not. This indicated that the indigenous sym-plasmid pRh7b restricted the functional expression of pJB5JI. Little is known about the mechanism of this. It is possible that the Nod factor produced by pJB5JI was modified by part of the *nod* genes on pRh7b, and could not be recognized by *Pisum sativum*. On the other hand, transconjugant 7653R(pJB5JI) showed higher nitrogenase activity on *A. sinicus* and higher dry weight(shoot) than those of the recipient strain 7653R(Zhang et al., 1995). The result is shown in the Table 2. The exogenous plasmid pJB5JI in *R. huakuii* strains 7653R and 7653R-1 remained stable during successive transfer on culture media. But in part of the isolates from nodules formed by them, the pJB5JI was not visualized on gel by Eckhardt procedure. Southern hybridization using Tn5 and *nod* gene probes showed that these isolates still reserved, at least part of DNA of pJB5JI, which was probably integrated onto the chromosome.

Table 2 Nitrogen fixation abilities of transconjuant

Inoculants	Acetylene reduced activity (μmol C_2H_4/g hour)	Shoot dry weight (g/plant)
7653R	15.8	0.1074
7653 (pJB5JI)	30.7[a]	0.1203
CK		0.0339

a: The difference compared to the parent strain 7653 is significant at level of p=0.01

CLONING AND ANALYSIS OF THE REGULATION ACTIVITY FRAGMENT OF THE *nod* GENES in *R. huakuii* STRAIN 7653R

It is known the *nod* promoter, NodD protein and flavonoid inducers are necessary for the expression of *nod*ABC (Mulligan, Long, 1985; Rossen et al., 1985). In order to clone early nodulation genes *nod*ABC promoter, DNA fragments of the sym-plasmid(pRh7b) from the *R. huakuii* strain 7653R were cloned into the transcriptional fusion vector pMP220 (Spaink et al., 1987) to construct the *lac*Z fusions. The *lac*Z fusions so constructed were transferred into strain 7653R and blue colonies were selected on SM plates containing *A. sinicus* seed extracts, x-gal(100 mg/ml), Sm(100 mg/ml) and Tc(10 mg/ml). A number of the blue colonies were chosen to assay β-galactosidase activity with or without *A. sinicus* seed extracts by Miller Assay (Miller, 1972). Transconjuant (blue colonies) were respectively designated HN11, HN12. As shown in Figure 1, the endogenous β-galactosidase activity of *R. huakuii* strain 7653R was very low. The β-galactosidase activity of transconjugant HN18 was higher in presence of seed extracts as compared with its absence. The β-galactosidase activity of transconjuants HN11，HN12，HN46 were lower in presence of seed extracts than without seed extracts. With or without seed extracts，the β-galactosidase activities of transconjuant HN19，HN112，HN113 were of the same level. So，the result indicated that the transconjugant HN18 might have contained the *nod*ABC promoter，or the regulating activity fragment of the *nod* genes.

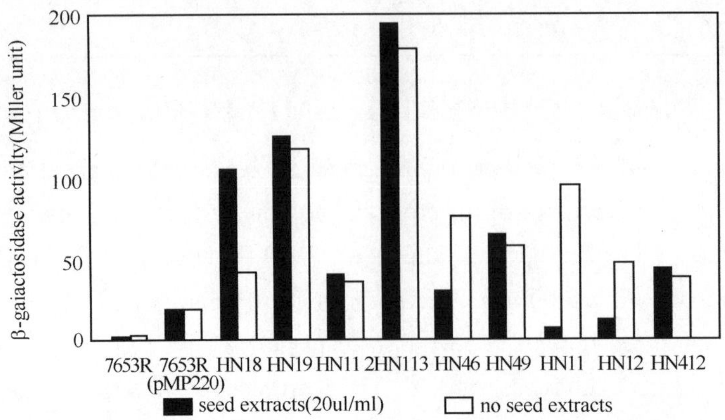

Fig.1 The β-galactosidase activity of transconjuants（blue colonies）

Subsequently，the plasmid（pHN18） isolated from transconjugant HN18 and retransferred into *E. coli* DH5a for genetic analysis. Hybridization of plasmid pHN18 with *nod* DABC（from *R. leguminosarm*） probe showed that there was a 1.7kb homogenous band. The plasmid pHN18 was transferred into sym-plasmid-cured Nod^- mutant strain 7653R-1. Transconjugant 7653R-1（PHN18） retained the induction activity in presence of seed extracts，which signified that the expression of plasmid pHN18 was independent of the symbiotic plasmid（pRh7b）. Three alternatives might confirm the experimental results：1） PHN18 contains *nod*D and *nod*ABC promoter from plasmid pRh7b. 2） pHN18 contains *nod* promoter only and nod is located somewhere else，say on chromosome，or 3） pHN18 contains *nod*D and *nod* promoter，and there is another *nod*D somewhere else.

PRELIMINARY IDENTIFICATION of *R. huakuii nod* GENE INDUCERS

To determine which of the flavonoids are able to induce *nod-lac*Z fusion of *R. huakuii*，responses to 14 different compounds were tested. The results are given in Table 3. A betain trigonelline and a flavonoid luteolin did induce *nod-lac*Z fusion of HN18 strain. Trigonelline and *A. sinicus* seed extracts were used respectively to culture *R. huakuii* strain 7653R on SM liquid medium. The extracts of bacteria cultures were prepared by centrifugation，filtration and concentration. The cell free concentrates were applied to the roots of cell-free *A. sinicus* seedlings. The results showed that cell-free extracts prepared from cultures with trigonelline or with *A. sinicus* seed extract were able to cause the elongation and deformation of the root hairs，which signified the presence of Nod factor produced by trigonelline induction in bacteria culture extracts.

Table 3 Reen for *nod-lacZ* inducibility by 14 compounds

Compounds（10μm）	β-galactosidase activity（Miller units）	Compounds（10μm）	β-galactosidase activity（Miller units）
Apigenin	315	Naringin	248
Biochanin A	194	Quercetin	10
Chalcone	361	Rutin（non-flavonoid）	275
Flavanon	367	Trigonellin（non-flavonoid）	925
Formononetin	36	Umbelliferon	392
Genistein	90	Zeatin（non-flavonoid）	292
Luteolin	871	Seed extract	1154
Naringenin	277	O（CK）	382

We are grateful for financial support from the Chinese Committee of Science and Technology and

National Natural Science Foundation of China（NSFC）.

REFERENCES

[1] Beynon JL et al（1980） J.Gen. Microbiol. 120，421-429.
[2] Cannon FC et al（1979） MGG，174，59-66.
[3] Chen HK et al（1944） Soli Sci. 51，291-193
[4] Chen HK et al（1992） In Hong GF，eds，Nitrogen Fixation in China，pp493-455，Berlin，Heidelberg，New York：Springer-Verlag.
[5] Chen WX et al（1991） Inter J. Syst. Bacteriol. 41，275-280.
[6] Downie JA et al（1983） MGG，190，359-365.
[7] Eckhardt T（1978） Plasmid，1，584-588.
[8] Miller JT（1972） In Experiments in Molecular Genetics. Cold Spring Harbor Laboratory，NY.
[9] Mulligan JT，Long SR（1985） Proc. Natl. Acad. Sci. USA 82，6609-6613.
[10] Rossen L et al（1985） EMBO J. 4，3369-3373.
[11] Ruiz-Sainz JE et al（1984） J. Appl. Microbiol. 57，309-315.
[12] Spaink HP et al（1987） Plant Mol. Biol. 9，27-39.
[13] Wang CL et al（1986） J. Gen. Microbiol. 132，2063-2070.
[14] Zhang XX et al（1992） J. Huazhong Agri. Univ. 11，175-181.
[15] Zhang XX et al（1995） Current Microbiol.（in press）
[16] Zhou JC et al（1987） J. Huazhong Agri. Univ. 6，156-164.
[17] Zou XH et al（1994） J. Huahong Agri. Univ. 13，325-331

紫云英根瘤菌共生基因分布及其共生效应的多样性*

邹向宏[1]　李阜棣[2]　曹燕珍[2]　陈文新[1]　陈华癸[2]

（1 中国农业大学生物学院；2 华中农业大学微生物系）

摘　要　从同一植株不同根瘤分离 40 株紫云英根瘤菌，所有菌株对 10 种抗生素的抗药性测定表明，该群体分为 22 个抗药类群。质粒检测显示所有分离株都含有质粒，质粒数 1～4 条。用快生型大豆根瘤菌 USDA205 质粒作参考，估测质粒 $M_τ$ 分布范围为 83～226Mu。根据图谱分析表明，该菌群可分为 6 个不同质粒型。各类型质粒通过与 Dig-*nod* ABC 和 Dig-*nif* HDK 杂交，结果显示带有 1 条质粒的菌株其共生基因定位在染色体上。带有 2 条或 2 条以上质粒的菌株各拥有 1 条共生质粒，共生质粒 $M_τ$ 范围有差异，大约为 117～226Mu。研究结果也显示，不同质粒型的菌株其共生效应存在明显差异，其中第 6 质粒型的菌株共生固氮率最强，第 1 质粒型菌株共生固氮率较低。共生固氮能力最强的第 6 质粒型菌株，只占总菌数 7.5%。

关键词　紫云英根瘤菌；共生基因；共生效应

近年来，根瘤菌多样性在很多根瘤菌群体中已报道[1,2]，紫云英根瘤菌是我国特有的生物固氮资源，它具有严格互接种族关系，只在紫云英根上结瘤，具有特异性。在分类上已被定为一个新种（*Rhizobium kuakuii*）。有关紫云英根瘤菌多样性也有较少结果报道，如从南方不同省内分离的紫云英根瘤菌，同一田间不同植株分离株，其质粒组成、抗药类群、生长代时、结瘤时间等存在较大差异[4]。在已有的文献报道中，人们一般以来自不同地理位置、同一地理位置不同植株的菌株为研究材料。事实上，在根瘤菌的生产实际应用中，人们发现施用具有高效固氮能力的根瘤菌并不都能使作物增产，这里可能存在着与土著根瘤菌竞争结瘤的问题，这就要求人们了解根瘤菌微生态学问题。同一田间不同植株的菌株多种多样，是否同一紫云英植株的分菌株也存差异，就这些问题我们对来自同一紫云英植株的分离株群体，初步研究它们的抗药类群、质粒数与大小、共生基因（*nod* 和 *nif*）的分布以及菌株共生效应，获得了一些有意义的结果。

1　材料和方法

1.1　菌株和质粒

紫云英根瘤菌从湖北省采集植株分离。质粒 pSA30 和 pJI1216 由华中农业大学生物固氮室提供。

1.2　根瘤菌天然抗药性测定

将溶化的 YMA 培养基冷却到 50℃左右。定量加入新配制的抗生素，混合均匀迅速制成平板，各抗生素的使用浓度（$p/μg.mL^{-1}$）分别为：庆大霉素（Gen）20，卡那霉素（Kan）20，壮观霉素（Spe）10，青霉素（Pen）40，红霉素（Ery）20，氯霉素（Chl）10，链霉素（Str）20，利福平（Rif）5，四环素（Tet）10，萘啶酮酸（Nal）20，将各菌株制成菌悬液，用多头接种器接种。每处理 5 次重复，28℃下培养，10d 后观察结果。

1.3　质粒检测

参考 Echhardt 的质粒快速方法[5]。

1.4　地高辛（Dig）非放射分子杂交

将紫云英根瘤菌快速检测的质粒通过 Southern 转移固定在杂交尼龙膜上，以肺炎克氏杆菌 *nif* HDK

*原载于《应用与环境生物学报》，3（1）：49～54，1997.

和豌豆根瘤菌 *nod* ABC 为探针，通过 Dig 标记，与根瘤菌质粒进行分子杂交。杂交方法及试剂参照 Boehringer Mannheimn 公司“Dig DNA Labelling and Detecting Kit”所附说明。

1.5　共生固氮效率测定

1.5.1　植物盆栽试验

紫云英种子常规表面消毒，萌发一致的种子播种于 Leonard 双层钵进行植株栽培（用 Jensen 无氮营养液）。每钵种 6 颗发芽种子，置光照室内（光照 12h，E=7000 lx，θ=20℃）条件下培养，待植株长出第 1 片真叶后，每钵接种 0.5mL 菌悬液（菌数约 10^8 / mL），每处理 5 个重复，同时用未接种根瘤菌盆栽苗作对照，7wk 后收获植株。

1.5.2　固氮酶活的测定（乙炔还原法）

剪取所有根瘤，每钵植株的根瘤分别置入 20mL 血清瓶内，用橡皮塞密封后，抽出 2mL 空气，再注入同体积的乙炔 28℃条件下反应 3h，用微量取样器取血清瓶内反应后的气体 50μL，用气相色谱仪（岛津 163 型）测定乙烯峰值。同样条件下测定标准乙烯并绘制标准曲线，由此计算样品的固氮酶活，用反应后单位鲜瘤量和单位时间的乙烯量[b（乙烯）• t^{-1} / μmol • g^{-1} • h^{-1}]表示。

1.5.3　植株干量和全氮量测定将收获的植株（包括根系与根瘤）

先 90℃杀青 30 min，然后 60℃、48 h 烘至恒重，称量每株植株干重，植株全氮量的测定采用凯氏定氮法。用根瘤的固定氮量表示根瘤的固氮能力。根瘤固定的氮量=植株的全氮量（接种根瘤菌）—植株全氮量（未接种根瘤菌）。

2　结果

2.1　紫云英根瘤菌的获得

从湖北省一块常年种植紫云英稻田（从未接种根瘤菌剂）采集多株紫云英，洗净根系后，从中选出 1 株结瘤数较多（40 个）的紫云英。将所选的根瘤按编号分别放入不同的无菌培养皿中，先用 95%的乙醇浸泡 5～10s，换用 0.1%的升汞处理 5min，接着用无菌水洗涤 7～8 次。在灭菌的血清管中加 1 滴无菌水，放入表面灭菌的根瘤，经灭菌的玻棒压碎后，用接种环取压碎汁在 YMA 平板上划线接种，28℃条件下培养，待长出菌落，取单菌落转接 YMA 斜面，同时进行革兰氏染色和镜检。紫云英根瘤菌为 G^-，小杆状，菌体含有折光性 β 羟基丁酸颗粒，在 YMA 平板上，菌体产黏液，菌落无色透明。淘汰不纯的菌株。选取优良的紫云英种子，常规表面灭菌后，催芽 2d 播种于 Jensen 无氮半固体培养基中，置光照室内培养待植株出现第 1 片真叶接种 0.5ml 菌悬液（约 10^8/mL），继续光照培养，6wk 内观察结瘤与否，确定纯化的结瘤菌株，每个根瘤获得 1 株分离株，共获 40 株的紫云英根瘤菌。

2.2　菌株的抗药性

对供试菌株采用 10 种抗生素，测定其天然抗药性，结果表明，群抗药性差别较大，供试的 40 个菌株可分为 22 个抗药类群（表 1）。其中第 3 和 11 抗药类群菌数较多。

表 1　紫云英根瘤菌天然抗药性类群

Table1　Instrintic antibiltic resistance patterns of *Rhizobium huakuii* strains

类群 Groups	抗生素 Antibiotics Gen	Spen	Kar	Pen	Ery	Chl	Str	Tet	Rif	Nal	菌株数 No. of isolates
1	−	−	−	−	−	−	−	−	−	−	2
2	−	+	−	−	−	−	−	−	−	−	2
3	−	−	−	−	+	−	−	−	−	−	4
4	−	−	+	−	−	−	−	−	+	−	1
5	−	+	+	−	−	−	−	−	−	−	1
6	−	−	−	−	+	−	−	−	+	−	1

续表

类群 Groups	抗生素 Antibiotics										菌株数 No. of isolates
	Gen	Spen	Kar	Pen	Ery	Chl	Str	Tet	Rif	Nal	
7	+	−	+	−	+	−	−	−	−	−	3
8	−	+	+	−	+	−	−	−	+	−	1
9	−	−	+	−	+	−	−	−	+	−	3
10	−	−	+	−	+	−	+	−	−	−	2
11	−	−	+	−	+	−	+	−	+	−	4
12	−	−	+	−	+	+	+	−	+	−	2
13	+	−	+	−	+	+	+	−	−	−	1
14	+	−	+	−	+	+	+	−	−	−	2
15	+	−	+	−	+	−	+	−	+	−	1
16	−	−	+	−	+	−	+	+	+	+	2
17	−	+	−	+	+	+	+	−	−	+	1
18	+	−	+	−	+	−	+	+	+	+	2
19	−	+	−	+	+	+	+	−	−	−	1
20	+	−	+	−	−	+	+	+	+	−	2
21	+	+	−	−	+	+	+	+	+	−	1
22	+	+	+	−	+	+	+	+	+	−	1

+示具有抗药性（indicating resistance），−示对抗生素敏感（indicating sensitivity）

2.3 菌株质粒

对所有分离株进行质粒快速检测，结果表明，它们都含有质粒，其质粒数为 1～4 条，因菌株不同而存在差异。已知大豆根瘤菌 USDA205 的 M_τ，分别为 83、117、226、600 Mu[6]，以此作参考，估计质粒分子量范围大约为 83～226 Mu 之间。根据质粒数目与大小，该菌群可分为 6 个不同质粒图谱类型（图 1a），各质粒类型菌株所占比例不同（表 2），其中第 2 质粒型菌株为优势菌群，占总菌数 4%。研究也表明，同一质粒型菌株在植株根系上不同位置都能分离到，说明同一质粒型菌株结瘤位点是随机的。

2.4 菌株共生基因分布

利用共同结瘤基因（*nod* ABC）和固氮酶结构基因（*nif* HDK）的高度保守性，探测紫云英根瘤菌共生基因分布。

2.4.1 *nod* ABC 基因定位

培养菌体后，碱抽提法获得的质粒 pJI1216 经 *Eco*R I 酶切后得到 6.6×10^3bp 的豌豆根瘤菌 *nod* ABC 片段，回收纯化后，进行 Dig 标记得到 Dig-*nod* ABC 探针。选择各质粒型代表菌，以大豆根瘤菌 USDA205 作参考。质粒检测得到供试菌株的质粒电泳图谱后，经处理转移到杂交尼龙膜上，采用地高辛（dig）标记的非放射性核酸杂交技术，与菌株的各质粒进行 Southern 杂交，杂交结果（图 1）表明，大豆根瘤菌 USDA205 的 *nod* 基因位于 M_τ 大约 117Mu 质粒上，与已有的报导一致。紫云英根瘤除第 1 质粒型菌株外，其余菌株均有质粒 DNA 与 *nod* ABC 基因探针有同源性杂交带。这些结果表明，第 1 粒型菌株的 *nod* 基因位于染色体上，其他质粒型的菌株均有 1 条大质粒载有 *nod* 基因，载有 *nod* 基因的质粒大小不同，其 M_τ 范围主要在 117～226Mu 之间。

2.4.2 *nif* HDK 基因定位

培养菌体后，碱抽提法将获得的质粒 pSA30 经 *Eco*RI 酶切后得到 6.2×10^3b *K. pneumonia* 的 *nif* HDK 片段，经回收纯化后制备成 Dig-*nif* HDK 探针，与供试菌株的质粒进行分子杂交，结果也表明，大豆根瘤菌 USDA205 的 *nif* 基因位于 M_τ，大约为 226Mu 质粒上，与已报道的结果一致。紫云英根瘤除第 1 质粒型菌株外，其余菌株均有 1 条质粒载有 *nif* 基因，载有 *nif* 基因的质粒大小不同，其 M_τ 范围大约为

117～226Mu。

2.4.3 共生质粒的确定

根据供试菌株质粒与 Dig-*nod* ABC 和 Dig-*nif* HDK 探针杂交带和质粒图谱相对位置，可确定第 1 质粒型菌株的共生基因（*nod* 和 *nif*）位于染色体上，带有 2 条或 2 条以上质粒的菌株，它们的 *nod* 和 *nif* 基因位于同一条大质粒上，这一质粒即为共生质粒，这些菌株只有 1 条共生质粒，其分子量大小存在差异。$M_τ$ 范围约为 117～226Mu。

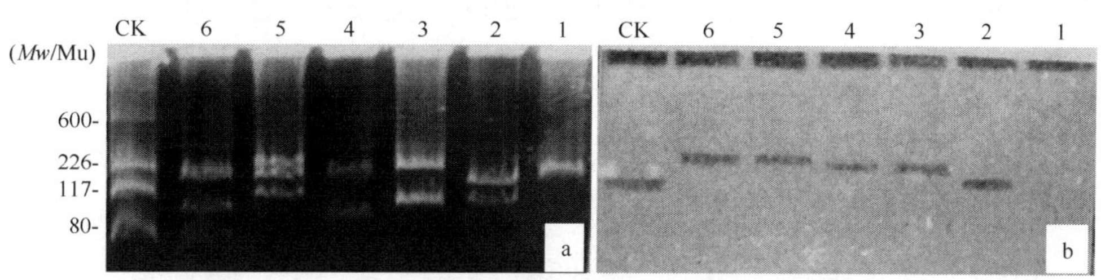

图 1　同一植株的紫云英根瘤不同质粒与 Dig-*nod*ABC 杂交

Fig.1　Various plasmids of *R. huakuii* strains from the same plant were Southern-hydridized with Dig-*nod*ABC.

a.质粒图谱，plasmid profiles；b. Southern 杂交图，Southern-hydridization. CK：USDA205；1-5：各质粒代表菌，typical strain for different plasmid types

2.5　菌株的共生效应

从同一植株分离株中，每个质粒型随机取了 3 个菌株，共选 18 个菌株，将这些菌株分别接种紫云英，每一处理 5 个重复，同时设未接种根瘤菌的植株作对照，测定各菌株的共生效应。接种 7wk 后收获植株，分别测定每株植物根瘤固氮酶活、植株干重、植株全氮。比较菌株的共生效应，各测定数据列于表 2。

表 2　同一植株不同质粒型菌株的固氮效应

Table 2　Effectiveness of nitrogen fixation of the strains with various plasmid types from the nodules on the same plant

质粒型 Plasmid types	菌株比例 Proportion of the strains with various plasmid types	代表菌株 Typical strains			植株增重 Added plant（m/mg）			固氮量（10^{-4}g/株） Amont of N fixed（10^{-4}g/plant）			固氮酶活 Acet.reduce [b（C_2H_2）$\cdot t^{-1}$/ $\mu mol \cdot g^{-1} \cdot h^{-1}$]		
CK						0			0			0	
1	15	AS31	AS32	AS33	3.6	4.7	7.7	2.46	4.31	4.52	9.8	5.5	6.0
2	40	AS34	AS35	AS36	13.8	14.5	15.0	8.58	7.87	8.20	11.3	7.2	7.6
3	12.5	AS37	AS38	AS39	12.3	12.1	11.5	8.33	8.27	8.07	2.7	6.2	4.8
4	15	AS40	AS41	AS42	5.7	7.9	8.8	4.52	4.44	4.80	6.1	7.1	5.6
5	10	AS43	AS44	AS45	13.7	13.0	16.4	9.46	8.94	10.24	7.4	6.5	9.1
6	7.5	AS46	AS47	AS48	28.7	21.3	22.0	15.94	11.81	11.70	9.2	8.4	10.1

表中数据是 30 株植株平均测定值，Values are the mean valules are the mean of thirty plants.

CK：未接种根瘤菌的植株，植株干重 13.2mg/株，全氮量为 2.40×10^{-4}g/株.

CK：Plant without inoculating strain；DW：13.2mg/plant and amount of total N；2.4×10^{-4}g/plant.

通常人们以固氮酶活、植株增重或植株含氮量表示菌株共生固氮能力。对于紫云英根瘤而言，哪一种指标表示其共生效应较合理？统计结果表明：植株增重与固氮量成正相关。植株增重与固氮量之间相关系数 r=0.97；而固氮酶活与固氮量（或植株增重）并不成对应关系。事实上，固氮酶活只能代表共生体系某一时刻的固氮能力，而固氮量则表示的是固氮的最终效果。某一时刻固氮酶活高并不一定表示最终固氮量高。各个时期固氮酶活存在差异。因此乙炔还原法只能测定植株生长发育某一阶段、某一时间的固氮酶活性，它对于紫云英根瘤菌共生体系只能作为共生效应测定的间接相对指标。在其他测定指标中，植株增重测定较简便，可直接用于比较紫云英根瘤菌的共生效应。

含有不同质粒数的根瘤菌，共生效应存在较大差异，而同一质粒型菌株共生效应差异不大（表 1）。第 1、4 质粒型的菌株固氮能力较低，第 6 质粒型的菌株固氮能力较高，在群体中比例较少，只占总菌数的 7.5%。而较低固氮能力的第 1、4 质粒型占菌数的 30%（表 2）。在群体中为优势群的第 2 质粒型菌株（占总菌数 40%），与第 6 质粒型菌株相比，其固氮能力也较低。可见，高效固氮能力的菌株在该植株上结瘤数并不占优势，反而较低。

3 讨论

本文以同一根系分离不同根瘤的所有分离株为材料，揭示了该菌群抗药类群非常广泛的特点。由于各抗生素作用位点不同，同时菌株抵抗各类抗生素机理复杂，揭示的抗生素类群不同反映出菌株的遗传背景及生理状态多样性。有关紫云英根瘤菌共生基因分布，金润之等（1993）也发现结瘤与固氮基因定位在同一质粒上[7]。魏辉等（1989）据不同紫云英根瘤菌为材料研究显示结瘤和固氮基因位于不同质粒上[8]。本研究表明，同一紫云英植物的分离株群体有的位于染色体上（如图 1 第 1 质粒类型），有的分布在同一质粒上，称为共生质粒，但共生质粒大小存在差异（图 1）。这些研究显示了紫云英根瘤菌结瘤和固氮基因分布的多样性.造成共生基因分布复杂的原因，可能与载有共生基因的质粒发生缺失和重组相关[9]，也可能与某些共生基因具有转座功能[10]有关。另外，Harrison（1988）指出，三叶草根瘤菌质粒数与植株干重呈现负相关[11]。本研究表明，紫云英根瘤菌其质粒数与其菌株的共生效应无相关性。根瘤菌的质粒与菌株的共生效应关系比较复杂，如某些非共生质粒存在是固氮根瘤形成必需的[12,13]，有些隐蔽质粒存在阻碍其菌株的共生固氮能力[14]，可见质粒间相互作用影响菌株的共生效应。同时，我们首次揭示：同一紫云英根系，共生固氮能力高的质粒型菌株在群体中所占比例少（即占瘤率少），而为优势菌群的质粒型菌株（即占瘤率较高的质粒型菌株）其共生固氮能力并不高。事实上，根瘤菌分布在土壤各层，同一植株根系被多种类型的菌株占有结瘤，如何使高效固氮能力的菌株成为优势菌群，具有非常重要的生态学和实际应用意义。

参 考 文 献

[1]Mozo T，Cabrera E，Ruiz-Agrueso T. Diversity of plasmid profiles and conservation of symbiotic nitrogen genes in newly isolated *Rhizobium* strains nodulating Sulla. *Appl Environ Microbiol*，1988，54：1262–1267.

[2]Laguerre G，Geniaux E，Amarger N. Conformity and diversity among field isolates of *Rhizobium leguminosarum* bv. *viacea*，bv. *trifolii* and bv. *phaseoli* revealed by DNA hybridization using chromosome and plasmid probes. *Can J Microbiol*. 1993，39：412–419.

[3]Chen WX，Li GS，Li Y *et a1*. *Rhizobium huakuii* sp. nov. isolated from the root nodules of *Asragnlus sinicus*. 1991，41：275–280.

[4]邹向宏，燕珍，李阜棣.紫云英根瘤菌天然抗药性和质粒研究。华中农业大学学报. 1994. 13：925–331.

[5]Echhardt T. A rapid method for the identification of plasmid deoxyribonucleic acid in bacteria. *Plasmid*. 1978，1：584–588.

[6]Masteron RV，Prakash PK，Atherly AG. Conversation of symbiotic nitrogen fixation gene sequences in *R. japonium* and *Bralyrhizobium japonicium*. 1985，165121–26.

[7]金润之.紫云英根瘤菌 Ra159 的巨大质粒上存在有 *nod* 和 *nif* 基因的证明。微生物学报。1993，33：170–176.

[8]魏辉，李阜棣.紫云英根瘤菌携带共生基因的物理证据。华中农业大学学报。1989，8：12–14.

[9]Sober-Chavez G，Najara R，Calva E *et al*. Partial deletion of the *Rhizobium phaseoli* CFN23 symbiotic plasmid implied an amplification of plasmid DNA sequence. *Mol Microbiol*. 1991，5：89–95.

[10]Galas DJ，Chandler M. Bacterial insertion sequences. In：Berg DE，Houe M ed. Mobile DNA，109～162. Washington DC：America Society for Microbiology.

[11]Harrison SP，Jones DG，Schunmann PHD *et al*. Variation in *Rhizobium leguminosarum* bv. *trifolii* sym plasmids and the association with effectiveness of nitrogen fixation. *J Gen Microbiol*. 1988，134：2721–2730.

[12]Hyne MF，McGregor NF. Two plasmids other than the nodulation plasmid are necessary for formation of nitrogen fixing nodules by *Rhizobium leguminosarum*. *Mol Microbiol*，1990，4：407–414.

[13]Susana Brom，Alejandro GS，Tomasz S *et al*. Different plasmids of *Rhizobium leguminosarum* bv. *phaseoli* are required for optimal symbiotic performance. *J Bacteriol*. 1992，174：5183–5189.

[14]叶政.增强快生型大豆根瘤菌共生固氮活性。武汉大学学报。1990，生物工程专刊：34-36

紫云英根瘤菌质粒功能研究*

邹向宏[1,2] 李阜棣[1] 陈华癸[1]

（1 华中农业大学微生物系；2 中国农业大学生物学院）

摘 要 紫云英根瘤菌 CH203 含有 3 条质粒（pRHa，97MD；pRHb，168MD；pRHc，251MD 为共生质粒），用带蔗糖敏感基因 Tn5-*sac*B 进行菌株质粒消除和质粒缺失突变株筛选，获得一系列突变株。与野生型菌相比，质粒 pRHa 的丢失导致菌株结无效根瘤，质粒 pRHb 的丢失使菌株失去共生能力，在 TY 培养基平板上菌落变得粗糙，失去了脂多糖（LPS1）。质粒 pRHc（共生质粒）的丢失显然失去其菌株的共生能力，同时使菌株抗酸性明显减弱。质粒回复能恢复突变株的表现特征和共生能力。此外，紫云英根瘤菌 CH205 含有 5 条大小不同的质粒（分子质量 42MD～230MD），该菌株某些质粒的消除能显著增强菌株的结瘤固氮能力。研究结果也表明除共生质粒外，紫云英根瘤菌其它质粒明显影响菌株的共生效应。

关键词 紫云英根瘤菌；质粒；共生效应

根瘤菌通过与豆科植物共生，将大气中的氮气转化为氨供植物生长利用，大多数根瘤结瘤菌的基因（*nod*）和固氮基因（*nif* 或 *fix*）定位在大质粒上（称为共生质粒）[1, 2]，除三叶草根瘤菌少数菌株含有三条共生质粒外，每个根瘤菌菌株一般只有一条共生质粒，但大多数根瘤菌含有至少一条有时高达 10 条其他不同质粒。对于苜蓿根瘤菌，共生质粒（1000～1500kb）至少带有 3 个基因簇涉及结瘤固氮，山羊豆根瘤菌含有 1700kb 巨型质粒[3, 4]，关于这些质粒的功能以及质粒间相互关系了解甚少。只是豌豆根瘤菌（*R. leguminosarum* bv. *viciae*）和菜豆根瘤菌（*R leguminosarum phaseoli*）个别菌株研究结果表明，除共生质粒外，隐蔽质粒也是有效共生行为必需的[5,6]。紫云英根瘤菌是我国特有的生物固氮资源，它具有严格的互接种族关系，在分类上被定为一个新种（*Rhizobium huakuii*）[7]，紫云英根瘤菌含有 1～5 条大质粒，分子量从 35MD 到 600MD 或更大[8]，人们很少了解这些质粒的生物学特性。本研究引用我国特有的紫云英根瘤菌为材料，运用 Tn5-*sac*B 转座子进行质粒消除和质粒缺失菌株筛选，初步探究质粒的多种功能及相互关系。

1 材料和方法

1.1 菌株和质粒

紫云英根瘤菌 CH203 和 CH205 由作者从湖北省武昌水稻田分离，农杆菌 GMI9023、质粒 pRK2013（含 *tra* 基因）由华中农业大学生物固氮室提供，质粒 pMH1701（含 Tn5-*sac*B）由加拿大构建者 Hynes MF 惠赠。

1.2 培养基

YMA 培养基[9]，TY 培养基[9]。所用抗生素浓度如下：新霉素（Nm），100μg/ml；萘碇酮酸（Nal），30μg/ml；利福平（Rif），50μg/ml；链霉素（Str），300μg/ml。

1.3 质粒检测和分子杂交

质粒快速检测参考修改的 Eckhardt 方法[10]，分子杂交参考 Dig-分子杂交手册（1993）。

1.4 脂多糖（LPS）测定

方法见参考文献[11]。

*原载于《微生物学报》，37（4）：245～251，1997.

1.5 质粒消除和菌株抗酸性测定

质粒消除见参考文献[5]，抗酸性测定见参考文献[12]。

1.6 植物试验

紫云英种子经表面消毒，将萌发一致的种子转接到经改进的 Leonard 双层钵，置光照室内培养（光照 12h，光强度 7000Lx，温度 20℃）。待长出第一片真叶时，接种 0.5ml 菌悬液（10^8/ml），每一处理 5 次重复，设未接种根瘤菌作对照，继续光照培养。7 周后收获植株，检查结瘤情况，测定固氮酶活，同时烘干植株，称量每株植株干重。

1.7 根瘤组织切片

将接种根 7 周后形成的根瘤，按 Johasen 描述的方法[13]，进行组织切片、固定、染色、在显微镜下观察拍照。

2 结果

2.1 菌株 CH203 被 Tn5-*sac*B 转座的质粒确定

菌株 CH203 含有 3 条大质粒，其分子量分别为 pRHa，97MD；pRHb，168MD；pRHc，251MD[8]。以豌豆根瘤菌的 *nod* ABC 和肺炎克氏杆菌 *nif* HDK 为探针，通过分子杂交，确定 pRHc 为共生质粒。将菌株 CH203 在 28℃纯培养条件下传代 100 次，以及与紫云英共结瘤七周后，各随机分离 100 个单菌落，质粒检测结果表明，它们的质粒组成不变，说明菌株 CH203 内源质粒较稳定。以该质粒稳定性菌株作为质粒消除出发菌株。

将菌株 CH203 与 *E. coli* 17-1（pMH1701）进行二亲本杂交，在 TY+Nm+Nal 平板上选择转移接合子 CH203 对 Nm 敏感，*E. coli* 对 Nal 敏感，只有转移接合子（被 pMH1701 上 Tn5-*sac*B 转座的菌株）才能在 TY+Nm+Nal 平板上生长。随机选择这些接合子单菌落，通过辅助菌株 MM294（含辅助质粒 pRK2013）与无质粒的农杆菌 GMI9023 进行三亲本杂交，将根瘤菌 CH203 被 Tn5-*sac*B 转座的质粒转移到无质粒的农杆菌中（选择平板为 TY+Nm+Rif+Str）。在该实验中，只有转座子 Tn5-*sac*B 插入到质粒上的菌株才能被转移，被 Tn5-*sac*B 转座质粒在辅助质粒 pRK20 13 协助下，从根瘤菌向农杆菌转移（其转移频率为 10^{-4}～10^{-6}），随机选取在 TY+Nm+Rif+Str 平板上生长的农杆菌转移接合子进行质粒检测，通过与出发菌株 CH203 质粒谱相比较，初步确定在菌株 CH203 中被 Tn5-*sac*B 转座的质粒。为了进一步确定质粒由 Tn5-*sac*B 转座，以 Dig-Tn5 为探针与农杆菌捕获的质粒进行杂交，农杆菌 GMI9023(pJB5JI∷Tn5)和野生型 CH203 的质粒作对照。图 1 显示，pJB5JI∷Tn5 质粒与 Dig-Tn5 探针反应呈阳性，CH203 的质粒与该探针反应呈阴性，农杆菌捕获的质粒与 Dig-Tn5 有明显的杂交信号。

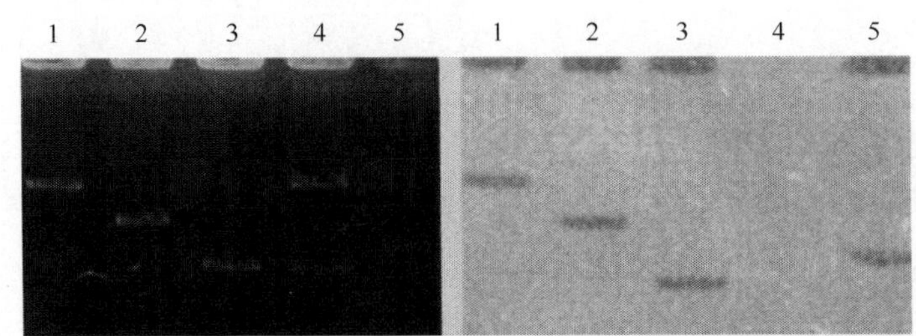

图 1 菌株 CH203 被 Tn5-*sac*B 转座质粒的鉴定

Fig.1 Identification of different plasmids in *R. huakuii* 203 labelled with Tn5-*sac*B

1. GMI9023（pRHa）；2. GMI9023（pRHb）；3. GMI9023（pRHc）；4. CH203；5. GMI9023（pJB5JI）

2.2 质粒缺失菌株的获得

将已确定被 Tn5-*sac*B 转座质粒的紫云英根瘤菌在 TY 液体中培养到一定浓度（10^7～10^8/m1），稀释（10^{-1}～10^{-3}）涂在 TY+7%蔗糖平板，首先置 39℃处理 2d，然后 28℃条件下继续培养 7d，若带有 Tn5-*sac*B 的质粒发生缺失或丢失，则该菌能在 TY+7%蔗糖平板上生长；否则，菌株由于蔗糖敏感基因（*sac*B）所编码的酶使蔗糖降解后的产物对革兰氏阴性细菌有毒害作用，导致菌体死亡。在 TY+7%蔗糖平板上出现的菌落再影印到 TY 与 TY+Nm 平板上，选取对新霉素（Nm）敏感的菌落进行质粒检测，确定质粒被消除的菌株。因为实验中可能 *sac*B 失活，而 Tn5 还存在，质粒并未消除，但能在 TY+7%蔗糖平板上生长，因此影印 TY 与 TY+Nm 可淘汰这些菌落，获得消除一条不同质粒的菌株（图 2）。以这些缺失一条质粒的菌株为出发菌株，重复利用转座子 Tn5-*sac*B 向筛选质粒缺失菌株可获得消除 2 条不同质粒的突变株，反复试验未能获得不含质粒或只含最小质粒的突变株（图 2），或许最大质粒或第二大质粒与菌株在纯培养条件下的生存能力相关。事实上，根瘤菌 CH203 的某些质粒显著影响菌株的生长速度。

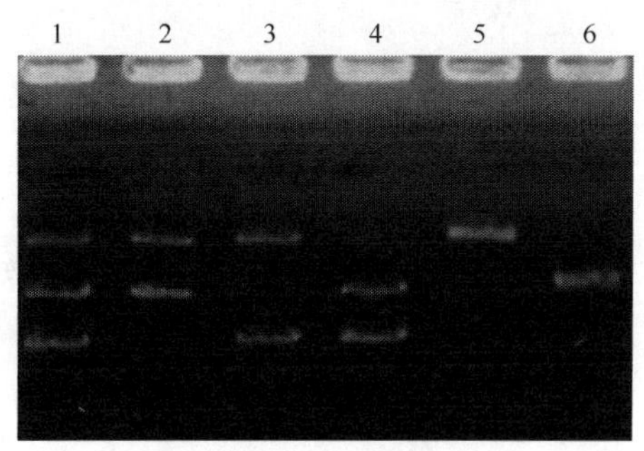

图 2　菌株 CH203 及其突变株质粒图谱

Fig.2　Plasmid Profiles of strain CH203 and its derivatives.

1.CH203; 2.CH203-1; 3.CH203-2; 4.CH203-3; 5.CH203-4; 6.Ch203-5.

2.3　质粒缺失菌株的表型特征

通过运用转座子 Tn5-*sac*B 正向筛选质粒缺失菌株。将这些突变株与野生型出发菌株进行表型特征比较，有些突变株的培养特征发生明显改变（表 1），如缺失了 pRHb 的菌株 CH203-2 和 CH203-4，在 TY 平板上菌落粗糙，在 PA 液体培养基中发生凝聚现象，出现絮状。在辅助菌株 MM294（含辅助质粒 pRK2013）协助下，以捕获了根瘤菌质粒（载 Tn5）的农杆菌为供体，缺失对应质粒的根瘤菌为受体进行反向三亲本杂交，在 TY+Nal+Nm 平板上筛选根瘤菌转移接合子（辅助菌和农杆菌对 Nal 敏感，受体根瘤菌对 Nm 敏感）。将这些根瘤菌转移接合子菌落进行质粒检测，获得一系列质粒回复菌株。CH203-6（pRHa$^-$/pRHa），CH203-7（pRHb$^-$/pRHb）和 CH203-8（pRHc$^-$/pRHc）。pRHb 质粒回复突变株 CH203-7（pRHb$^-$/pRHb）的菌落培养特征与野生型一致，如菌落表面光滑，在 PA 液体中菌体不发生凝聚等，该质粒回复能完全恢复菌株的培养特征，进一步证实了突变株 CH203-2（pRHb$^-$）表型变化是由质粒 pRHb 缺失引起的。

表 1　根瘤菌质粒缺失突变株的表型特征

Table 1　Characteristics of plasmid-cured derivatives of *R. huakuii* strain CH203

菌株 Strain	相关基因型 Gene type	菌落特征① Morphology	凝聚性② Flocculation	结瘤③ Nodulation	固氮酶活 Acetylene reduction
CH203	wild type	G	−	+	+
CH203-1	pRHa$^-$	G	−	+	−
CH203-2	pRHb$^-$	R	+	−	−
CH203-3	pRHc$^-$	G	−	−	−
CH203-4	pRHa$^-$ b$^-$	R	+	−	−
CH203-5	pRHa$^-$ c$^-$	G	−	−	−
CH203-6	pRHa$^-$/pRHa	G	−	+	+
CH203-7	pRHb$^-$/pRHb	G	−	+	+
CH203-8	pRHc$^-$/pRHc	G	−	+	+

①R：在 TY 平板上菌落粗糙；G：在 TY 平板上菌落光滑

R：Rough colony on TY medium；G：Glossy colony on TY medium.

②+：在 PA 液体培养基中菌体凝聚 +：Flocculation in PA broth.

③在接种根瘤菌 7 周内观察结果 Examined for 7 weeks after inoculation strains.

2.4　根瘤菌脂多糖

缺失质粒的突变菌株 CH203-2（pRHb$^-$）在 TY 平板上菌落粗糙，在 PA 液体中菌体凝聚，这种现象与 Cava（1989）[11]、Hynes（1990）[5]和 Sussan Brom（1992）[6]描述的 LPS 突变株表型特征相似。

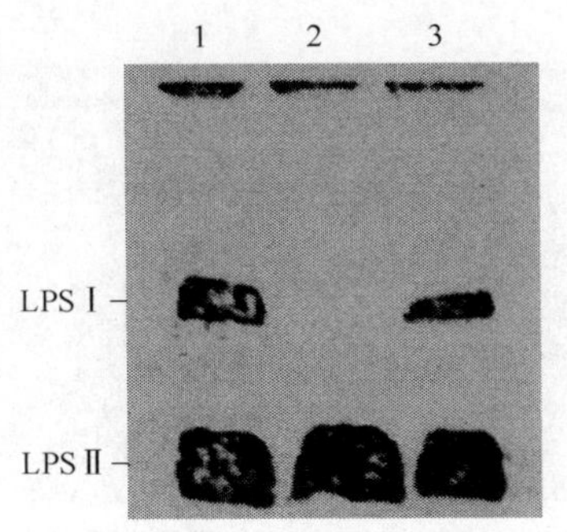

图3 紫云英根瘤菌脂多糖 SDS-PAGE 电泳图谱
Fig.3 LPS profiles of R. huakuii strains on SDS-PAGE.
1. CH203; 2. CH203-2（pRHb-）; 3. CH203-7（pRHb$^-$/pRHb）

用 SDS-PAGE 检测菌株的脂多糖（LPS）。结果（图 3）表明，质粒缺失菌株 CH203-2（pRHb$^-$）与野生型相比缺少脂多糖 I，而质粒恢复菌株 CH203-7（pRHb$^-$/pRHb）与亲本菌株 CH203 一样具有 LPS I 和 LPS II，进一步证实质粒 pRHb 与完整的 LPS 产生有关。

2.5 根瘤菌的抗酸性

根瘤菌 CH203 在 pH4.2 以上条件下能生长，低于 4.2 则不能生长。将各突变株转接于不同的 pH 缓冲培养液中，在温度为 28℃，转速为 150r/min 条件下培养一周，观察培养液是否变混浊，确定菌株生长与否。所有培养液接种培养一周后，缓冲液 pH 未变，结果表明（表 2）突变株的抗酸性能力有所下降，其中缺失了质粒 pRHb 的菌株，其抗酸性明显减弱。质粒回复试验表明。回复菌株 CH203-6（pRHa$^-$/pRHa）和 CH203-7（pRHb$^-$/pRHb）能完全恢复菌株的抗酸性，而质粒回复菌株 CH203-8（pRHc$^-$/pRHc）的抗酸性只能恢复到一定程度（pH=4.8），可能回复质粒 pRHc 上的 Tns 插入微弱影响其抗酸性功能，由此可见，菌株的质粒与菌株的抗酸性密切相关。

表 2 菌株在不同 pH 培养基条件下的生长情况
Table 2 Growth of the strains in the different pH TY broth

菌株 Strain	培养液的 pH 值 pH in TY broth							
	4.2	4.5	4.8	5.0	5.2	5.5	5.8	6.0
CH203	+	+	+	+	+	+	+	+
CH203-1（pRHa$^-$）	−	−	+	+	+	+	+	+
CH203-2（pRHb$^-$）	−	−	−	+	+	+	+	+
CH203-3（pRHc$^-$）	−	−	−	+	+	+	+	+
CH203-4（pRHa$^-$b$^-$）	−	−	−	−	+	+	+	+
CH203-5（pRHa$^-$c$^-$）	−	−	−	−	−	−	+	+
CH203-6（pRHa$^-$/pRHa）	+	+	+	+	+	+	+	+
CH203-7（pRHb$^-$/pRHb）	+	+	+	+	+	+	+	+
CH203-8（pRHc$^-$/pRHc）	−	−	+	+	+	+	+	+

+：表示在 TY 液体培养基中生长；−：表示在 TY 液体培养基中不生长。
+：Indicating growth in the TY broth；−，Indicating no growth in the TY broth.

2.6 质粒缺失菌株的共生效应

测定了菌株 CH203-1 的各质粒缺失菌株的结瘤和固氮能力。结果显示，只有质粒 pRHa 被消除的突变株 CH203-1（pRHa$^-$）能够在紫云英上结瘤，其它质粒缺失菌株均失去了它们的共生效应。根瘤呈白色，并且比野生型菌株 CH203 所结的根瘤小。菌株 CH203-1 所形成的根瘤无乙炔还原活性。这些白色根瘤的组织切片在光学显微镜下显示根瘤细胞量缺乏根瘤细菌，为无效根瘤。上述结果表明，紫云英根瘤菌其它隐性质粒也明显影响其共生效应。另外，关于紫云英根瘤菌 CH205 的质粒缺失菌株的共生效应结果也支持了这个观点，菌株 CH205 含有 5 条质粒，它们的大小分别为 pRGa，42MD；pRGb，72MD；pRGc，112MD；pRGd，161MD；pRGe，230MD[8]。质粒上未发现共生基因存在。实验结果（表 3）显示，质粒菌 CH205 中质粒的 pRGe 丢失明显降低菌株的共生效应（α=0.05），而质粒 pRGa 和 pRGd 同时丢失明显增强菌株的共生效应（α= 0.01）。

表 3　CH205 质粒缺失菌株的共生反应

Table3　Effectiveness of derivatives cured of plasmids in *R.hukuii* CH205

菌株 Strain	基因型 Gene type	结瘤数 No. of nodules	植株干重 Dry wt（mg/plant）	植株增重 Added Dry wt（mg/plant）
CK		0	18.10±0.30	0
CH205	Wild type	15±2	45.82±0.41	27.72±0.41
CH205-3	pRGe	14±1	32.16±0.39	14.06±0.39
CH205-7	pRGa d	25±3	65.75±0.62	47.65±0.62

CK：未接种紫云英根瘤菌，表中数据为 30 株植株平均测定值。

CK：Not inoculating R.huakuii strain valules are the mean　±SE of thirty plants

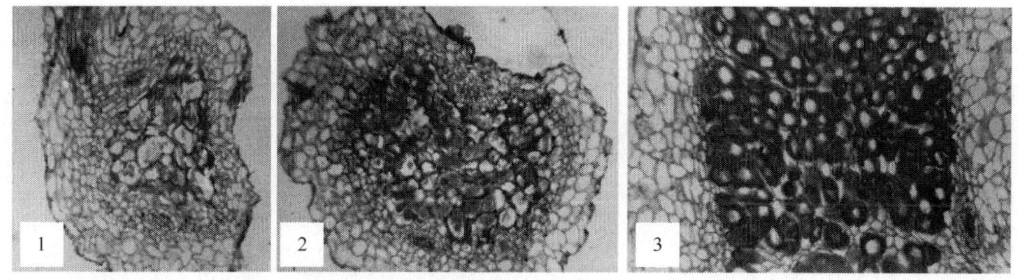

图 4　CH205 突变株根瘤切片（放大倍数 16×10）

Fig.4　Section of nodules formed on Astragalus sinicus L. by plasmid-cured derivatives of CH205（magnification 16×10）.

1. CH205-3（pRGe$^-$）；2. CH205；3. CH205-7（pRGa$^-$d$^-$）

根瘤切片光学显微图片（图 4）显示，与菌株 CH205 形成根瘤相比，突变株 CH205-3（pRGe$^-$）的根瘤小，含菌细胞少，胞内含菌量少；突变株 CH205-7（pRGa$^-$d$^-$）形成的根瘤大，含菌细胞多，胞内含菌量较多。

3　讨论

根瘤菌大都含有多个巨大质粒，人们通常将根瘤菌质粒转移到适当的受体，探测该质粒的功能，我们考虑到质粒的不相容性，质粒有可能形成多聚体，或者质粒在新的遗传背景下难以表达等因素，我们采用了消除质粒的方法探究质粒的功能，引用转座子 Tn5-*sac*B[14]快速检出质粒缺失菌株，为紫云英根瘤菌质粒功能研究奠定了基础。近年来，对个别豌豆根瘤菌和菜豆根瘤菌研究结果表明，除共生质粒外，其他非共生质粒是有效根瘤形成所必需的[5,6]。我们的结果表明，紫云英根瘤菌的隐性质粒显著影响菌株的共生行为，质粒间相互联系相互作用，共同决定根瘤菌的共生能力，也显示了涉及根瘤菌共生行为的基因复杂性与多样性。大多数情况下，质粒缺失使菌株失去共生能力或共生效率下降。但也有例外，我们研究发现消除紫云英根瘤菌 CH205 的某些质粒，能显著增强其共生效应，在百脉根根瘤菌和大豆根瘤菌 USDA194 也有类似现象[15]。消除与菌株某些功能无关的基因组，一部分基因组未能编码蛋白，但从总体上或许节省了能量，这有助于其他功能的表达；同时，质粒消除有可能排除一些限制性因子，使菌株的特定功能得以表达，如菜豆根瘤菌的共生质粒上存在抑制胞外多糖合成基因（*psi*），从而影响菌株结瘤[16]。迄今，除了对根瘤菌共生质粒上的共生基因外，人们对许多其他的质粒功能很少了解，事实上，根瘤菌许多其他重要性状与质粒紧密相关，如抗逆性、生长速率和胞外多糖产生等 [11]。随着研究方法改进以及研究性状范围的扩大，我们不但能揭示质粒新的生物学特性，获得一些特殊基因材料，而且会更多地揭示质粒间（基因间）的相互关系，为根瘤菌的研究提供新的材料和新的理论依据。

参 考 文 献

[1]Johnston AWB，Beynon JL，Beringer JE *et al. Nature*，1978，276：634-636.

[2]Rosenberg C，Boistard P，Denarie J *et al. Mol Gen Genet*，1981，184：320-333.

[3]Borthakur D，Lamb J W，Johnston AWB. *Mol Gen Genet*，1987，207：155-160.

[4]Selenska-Trajkowa S，Radewa G，Markov K. *Lett Appl Microbiol*，1990，10：123-126
[5]Hynes MF，McGregor NF. *Mol Microbiol*，1990，4：407-414.
[6]Susana Brom，Alejandro GS，Tomasz Stepkowsky *et al*，*J Bacteriol*，1992，174：5183-5189.
[7]Chen WX，Li GS，L Y *et al*，*Int J Syst Bacteriol*，1991，41：274-280.
[8]邹向宏，曹燕珍，李阜棣.华中农业大学学报，1994，13：325-331.
[9]Vincent MJ，上海植物生理研究所固氮室.根瘤菌研究实用手册.上海：人民出版社，1970. 3.
[10]Eckhardt T. *Plasmid*，1978.1：585-588.
[11]Crava J R，Elias PM，Neel KD *et al*. *J Bacteriol*，1989，171：8-15.
[12]Chen H，Cartner E，Rolfe BG. *Appl Environ Microbiol*，1993，59：1058-1064.
[13]Johansen DA，*Plant Microtechnique*. New York：McGraw Hill. 1940.
[14]Hyne MF，Quandt J，O'Conell MP *et al*，*Gene*. 1989，78：111-120.
[15]Pankhaurst CE. *J Gen Microbiol*，1986，132：2321-2328.
[16]Borhakor D，Johnston AWB. *M Gen Genet*，1987，207：149-155.

Characteristics of Plasmids in *Rhizobium huakuii*[*]

XIANGHONG ZOU, [1,2] FUDI LI, [1] HUAKUI CHEN[1]

(1Department of Microbiology, Huazhong Agricultural University, Wuhan 430070, P.R. China;

2College of Biological Sciences, Beijing Agricultural University, Beijing 100094, P.R. China)

ABSTRACT *Rhizobium huakuii* nodulates *Astragalus sinicus*, an important green manuring crop in Southern China, which can be used as forage. The plasmid profiles of 154 *R. huakuii* strains were examined with the Eckhardt procedure. The plasmid number of the strains varied from one to five, and their molecular weights were estimated from 42 to 600mDa or more. The plasmids were hybridized with probes *nod*ABC and *nif*HDK. The results showed that there was one plasmid carrying the *nod* and *nif* genes in the strains that harbor two or more plasmids, and the molecular weights of the symbiotic plasmids varied from 117 to 251mDa. Homology was not observed on plasmids in the strains having only one plasmid; presumably the symbiotic genes are on the chromosome. Plasmid curing was carried out with the *Bacillus subtilus sac*B to generate derivatives of *Rhizobium huakuii* strain CH203, which harbors three plasmids, pRHa (97MD), pRHb (168MD), and pRHc (251MD). The largest plasmid (pRHc) carried both nodulation and nitrogen fixation genes. When pRHc was cured, the strain lost its symbiotic ability. The other two plasmids were also related to symbiosis. The derivative cured of pRHb did not nodulate on the host plant, had an altered lipopolysaccharide, and grew much more slowly than the parent strain. Curing of the smallest plasmid (pRHa) resulted in delaying the strain nodulation and made it lose nitrogen fixation ability. Curing of each plasmid in strain CH203 reduced its acid tolerance. Complementation of plasmid-cured strains with appropriate plasmids restored their original phenotypes.

Bacteria of the genus *Rhizobium* are able to form nitrogen fixation nodules on roots of leguminous plants. In many *Rhizobium* species, the genes coding for nodulation (*nod*) and nitrogen fixation (*nif*) functions have been located on a plasmid called "symbiotic plasmid" or "pSYM" [1,19,29]. With the exception of *Rhizobium trifolii* strains (some strains harboring three *sym* plasmids), there is generally only one such plasmid per strain, but most of *Rhizobium* strains contain at least one, and sometimes as many as ten additional plasmids [32]. In *R. meliloti*, the plasmid (1000–1500 kb) carries at least three clusters of genes involved in nodulation and nitrogen fixation[5]. *R. galegae* strains harbor an extremely large (1700 kb) plasmid [30]. Very little is known about the function and interaction of plasmids in *Rhizobium*. In *R. leguminosarum* bv. *viciae* and *R. leguminosarum* bv. *phaseoli*, different plasmids are required for optimal symbiotic performances [4,15]. *Rhizobium huakuii* is a strictly host-specific species. It nodulates only on *Astragalus sinicus* [10], which is a winter green manuring crop for rice and a forage crop in China, Japan, and Korea. *R. huakuaii* was taxonomically defined by Chen and associates [11]. Little is known about plasmids of *R. huakuai*, but it is possible that some of the plasmids play a significant role in symbiosis and ecology. We isolated *R. huakuaii* strains from *Astragalus sinicus* in rice field in Southern China and studied their plasmid number, size, and location of symbiotic genes. By adopting an approach similar to that used by Hynes and McGregor [15], we obtained derivatives cured of each plasmid and identified some functions of different plasmids in *R. huakuii*

[*]原载于 *Current Microbiology*.35：215～220，1997.

strain CH203. The data showed variation and functional complexity of plasmids in *R. huakuii*.

MATERIALS AND METHODS

Strains and plasmids. The bacterial strains and plasmids used in the study are listed in Table 1.

Table 1 Bacterial strains and pladmids used in this study

Strain	Relevant characteristics	Reference
R. huakuii	wild type and their derivatives	This work
***A.tumdfaciens*:**		
GMI9023	Plasmid-free	[28]
GMI9023/pRHa:Th5-mob	Source of CH203 (pRHa::Th5-mob)	This work
GMI9023/pRHb:Th5-mob	Source of CH203 (pRHb::Th5-mob)	This work
GMI9023/pRHc:Th5-mob	Source of CH203 (pRHc::Th5-mob)	This work
Plasmid:		
pMH1701	Carries Tn5-mob-sac Derivatives Tn5B12	[16]
pRK2013	Conjugation helper	[12]
pSA30	Carries *nif*ABC	[6]
pJI1216	Carries *nds*HDK	[13]

Media and culture conditions. *Rhizobium* strains were grown at 28℃ on TY medium [2], minimal medium（MM）, or YMA medium [34]; *E.coli* strains at 37℃ on LB medium [22]; Agrobacterium strains at 28℃ on TY medium. Sucrose was added in medium at a final concentration of 7%（wt/vol）. Antibiotics were used at the following concentrations: neomycin, 100μg/ml; nalidixic acid, 30μg/ml; tetracycline, 50μg/ml; streptomycin, 300μg/ml; rifampin, 50μg/ml.

Plasmid profiles and hybridization analysis. Plasmid profiles were visualized by a modified Eckhardt technique [14] and blotted onto Hybon-Nylon membrane. Southern hybridization was carried out with the Dig DNA Labeling and Detecting Kit from Boehringer Mannheim, according to the manufacturer's instructions.

Plasmid curing and transferring. Mating was carried out as described previously [31]. By mating *E. coli* 17-1 harboring pMH1701 with *R. huakuii* strains, the *R. huakuii* strains were mutagenized with Tn5B12 on pMH1701. Through selection of TY plates with Nal and Nm, we obtained *R. huakuii* CH203 derivatives carrying Tn5B12. By randomly picking about 50 colonies from the mutagenized populations and crossing these in triparental mating with *Agrobacterium* strain GM9023（plasmid-free） and *E. coli* strain carrying pRK2013, insertions of Tn5*sac*B in different plasmids of *R. huakuii* strains were identified, since only the plasmid carrying the transposon could be mobilized into *Agrobacterium*. The transconjugants were examined for inherited plasmids on agarose gels. Plasmid curing was carried out by plating overnight cultures of the derivatives carrying Tn5B12 insertion plasmids on TY plates containing sucrose and incubating the plates for 1-2 days at 40℃ and then for one week at 28℃. All colonies were screened for loss in Nm resistance and then for change in plasmid contents by agarose gel electrophoresis. The plasmid derivatives were complemented with the corresponding Tn5B12-labeled plasmid by conjugation with *Agrobacterium* strains containing the plasmid.

SDS-PAGE for LPS. Washed cells of *R. huakuii* were examined for LPS production by SDS-polyacylamide gel electrophoresis（PAGE） by the method of Cava et al. [7]. Gels were fixed and stained by using a silver staining method as described by Tsai et al. [33].

Acid tolerance tests. The acid stress medium used for growth of *R. huakuii* strains at low pH was described by Richardson and Simpsin [26,27].

Plant tests. Seeds of *Astragalus sinicus* （Wuhan variety） were sterilized as described by Vincent [34], rinsed thoroughly in sterile distilled water, and germinated on sterile agar. They were then planted in big tubes with sterile, nitrogen-free nutrient solution [17] (five replicates for each treatment) and inoculated with 0.5 ml of rhizobial suspension （10^8 ml^{-1}） when cotyledons appeared. After growth in a light cabinet for 7 weeks at 20℃ and a 12-h light（7000 Lx） cycle, nodulation was scored every day, the plants were harvested, and acetylene reduction of the nodules was assayed as described previously [20].

Light microscopy of nodules. Nodules were excised from plants 7 weeks after inoculation with *R. huakuii*, fixed in formalin/acetic acid, dehydrated, and paraffin-embedded as described by Johansen [18]. About 5-μm sections were adhered to slides with a gelatin adhesive and stained with 0.3% methylene blue before microscopy.

RESULTS

Variation of plasmids in *R. huakuii*. We isolated 154 *R. huakuii* strains from *Astragalus sinicus* in a rice field of Southern China. Plasmids of all strains were examined; the number of the plasmids varied from 1 to 5, and their molecular weights were estimated from 42 to 600mDa or more, with the molecular weights of plasmids in USDA205 as reference [21]. According to plasmid number and size, there were eight different plasmid types among all 154 strains（Table 1）. Southern blots of different plasmids were hybridized with *nod*ABC and *nif*HDK. Results showed that the symbiotic genes were located on a plasmid（*sym* plasmid） in the strains harboring two or more plasmids, and the molecular weights of the *sym* plasmids varied from 117 to 251mDa (Table 2). But no homology was observed on the plasmids in the strains harboring only one plasmid; presumably the symbiotic genes are on the chromosome.

Table 2 Cryptic plasmids and symbiotic plasmids of *Rhizobium huakuii*

Plasmid type	Typical strain	Size of plasmid（MD）	Plasmid type	Typical strain	Size of plasmid（MD）
1	CH101	214	6	CH212	225, ^{ab}204, 104, 83
2	CH112	117, ^{ab}94	7	CH245	600, 239, ^{ab}112, 70
3	CH129	221, ^{ab}102	8	CH311	230, 161, ^{ab}112, 72, 42
4	CH164	198, ^{ab}147, 87	CK	USDA205	~600, 226, ^{b}117, ^{a}83
5	CH203	251, ^{ab}168, 97			

^{a} Indicating the plasmid carries *nod* genes.

^{b} Indicating the plasmid carries *nif* genes.

Isolation of the derivatives cured of different plasmids in *R. huakuii* CH203. The plasmids in *R. huakuii* CH203 showed stability after 50-100 generations in free-living in TY medium at room temperature. The plasmids labeled with transponson Tn5*sac*B were proved by transferring them into *A. tumefaciens*. The *A. tumefaciens* transconjugants containing each plasmid of *R. huakuii* CH203 were identified by hybridization with Tn5 and confirmed to be of the same strain as the original features in *R. huakuii* CH203（Fig. 1）. The derivatives, whose plasmids were confirmed labeled with Tn5*sac*B, were plated on TYmedium containing sucrose to select for loss of the plasmids (frequencies of loss: 10^{-5}–10^{-6}). The colonies, which grew on TY containing sucrose, were tested to make sure for loss of neomycin resistance (Tn5*sac*B) and then screened for loss of plasmid(s) by agarose gel electrophoresis. Derivatives lacking pRHa, pRHb, and pRHc were obtained (Fig. 2). A further curing for plasmids resulted in strains lacking pRHa/pRHb and pRHa/pRHc (Fig. 2). But the derivatives cured of both pRHb and pRHc and all three plasmids were not obtained. It might be because that loss of pRHb/pRHc caused death or the derivative grew much more slowly and therefore was difficult to be picked up on the medium.

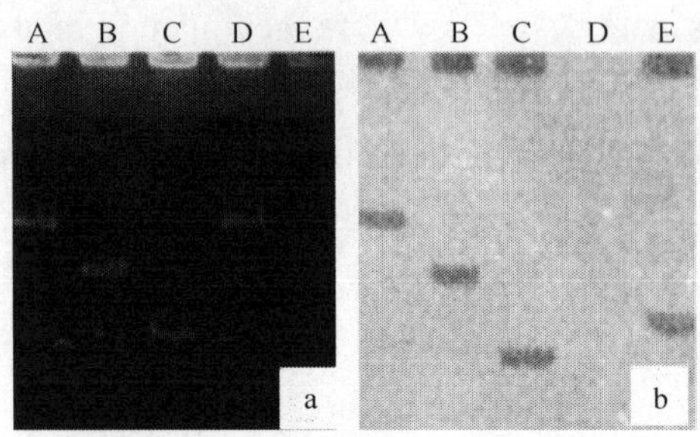

Fig. 1 Ind entification of different plasmids in *R. huakuii* strain CH203 labeled with Tn5*sac*B.

（a） Plasmid profile; （b） Southern hybridizationwith Dig–Tn5. Lane A, GMI9023（pRHa）; lane B, GMI9023（pRHb）; lane C, GMI9023（pRHc）; lane D, *R. huakuii* CH203（pRHabc）; lane E, GMI9023（pJB5JI）.

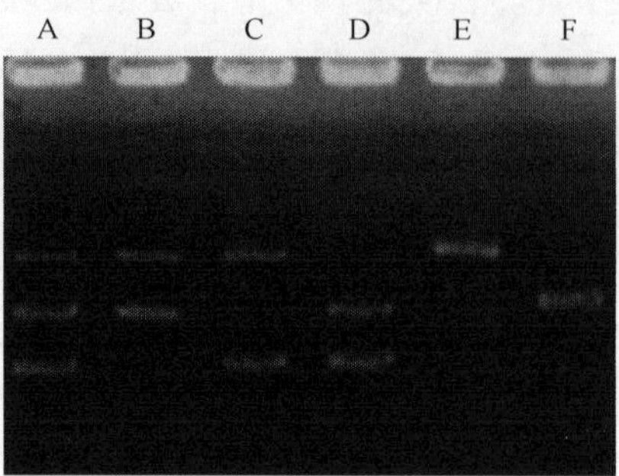

Fig. 2 Cured of plasmid in *Rhizobium huakuii* strain CH203.

Lane A, CH203; lane B, CH203–1（cured of pRHa）; lane C, CH203–2（cured of pRHb）; lane D, CH203–3（cured of pRHc）; lane E, CH203–4（cured of pRHa and pRHb）; lane F, CH203–5（cured of pRHa and pRHc）

Phenotypes associated with derivatives cured of different plasmids. The phenotypic traits of the plasmid-cured are presented in Table 3. It was obvious that the plasmid-cured derivatives were different from the parent strain. Strain CH203-2 lacking pRHb had a "rough" colony morphology on TY agar and flocculated when grown in MM broth. To ensure that the phenotypes of the derivatives are due only to plasmid loss, each derivative was complemented with the corresponding plasmid labeled by Tn5*sac*B. The transconjugants showed phenotypes identical to that of its parent strain（Table 3）.

Table 3 Characteristics of plasmid-cured derivatives of *R.huakuii* strain CH203

Strain	Gene type	Colony Morphology[a]	Flocculation[b]	Nodulation	Acetylene reduction
CH203	wild type	G	–	+	+
CH203-1	$pRHa^-$	G	–	+	–
CH203-2	$pRHb^-$	R	+	–	–
CH203-3	$pRHc^-$	G	–	–	–
CH203-4	$pRHa^- b^-$	R	+	–	–
CH203-5	$pRHa^- c^-$	G	–	–	–
CH203-6	$pRHa^-$/pRHa	G	–	+	+
CH203-7	$pRHb^-$/pRHb	G	–	+	+
CH203-8	$pRHc^-$/pRHc	G	–	+	+

[a]R, rough colony on TY medium; G, glossy colony on TY medium.
[b]+, flocculation in MM broth; examined 7 weeks after inoculation.

Tolerance to acid. The strain CH203 was tolerant to low pH and could grow in YMB at pH 4.2. The plasmid derivatives of CH203 were tested for growth in different pHs. Loss of different plasmids in *R. huakuii* CH203 reduced its capacity for acid tolerance. Strains lacking more than one plasmid showed less tolerance to acid（Table 4）. Plasmid-complemented strains（pRHa2/pRHa, pRHb2/pRHb） showed the same acid tolerance as the parent strain. The derivative CH203-8, inheriting the plasmid pRHc carrying Tn5*sac*B, grew only above pH 4.8（the parent strain grew at pH 4.2; Table 4）. It might be that inserting Tn5*sac*B in plasmid pRHc slightly reduced its tolerance to acid. The result showed that all plasmids in strain CH203 were associated with acid tolerance of the strain.

Table 4 Acid tolerance of derivatives of *R. huakuii* strain CH203

Strain	Gene type	pH in TY broth[a]							
		4.2	4.5	4.8	5.0	5.2	5.5	5.8	6.0
CH203	wild type	+	+	+	+	+	+	+	+
CH203-1	pRHa⁻	−	−	+	+	+	+	+	+
CH203-2	pRHb⁻	−	−	−	+	+	+	+	+
CH203-3	pRHc⁻	−	−	−	+	+	+	+	+
CH203-4	pRHa⁻ b⁻	−	−	−	−	+	+	+	+
CH203-5	pRHa⁻ c⁻	−	−	−	−	−	−	+	+
CH203-6	pRHa⁻/pRHa	+	+	+	+	+	+	+	+
CH203-7	pRHb⁻/pRHb	+	+	+	+	+	+	+	+
CH203-8	pRHc⁻/pRHc	−	−	+	+	+	+	+	+

+，indicating growth in the TY broth；−，indicating no growth in the TY broth.

Plasmid pRHb carries LPS genes. The pRHb$^-$ strain showed rough colony morphology on TY medium. It was flocculated in MM broth and did not move. The results are similar to LPS mutants previously described [4,7,15]. LPS in different strains was analyzed with silverstained SDS-PAGE gels. The LPS I band was completely missing from strain CH203-2（Fig. 3），and only the LPS II band corresponding to the LPS core was present [7]. Introduction of the pRHb restored production of LPS I（Fig. 3）.

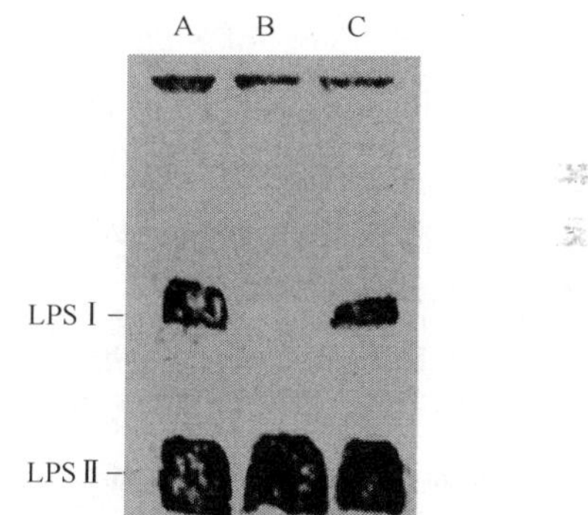

Fig. 3 LPS profile on SDS–PAGE gel.

Lane A，strain CH203；lane B，CH203–2（pRHb$^-$）；lane C，CH203–7（pRHb$^-$/pRHb）.

Symbiotic properties of plasmid-cured strains. Only CH203-1 lacking pRHa could nodulate on *Astragalus sinicus*, but the nodulation was delayed for 10 days. Other derivatives cured of different plasmids lost their nodulating ability. The nodules formed by strain CH203-1 were much smaller than those formed by its parent strain. They were white and had no acetylene-reducing activity. Sections of the white nodules were examined under the microscope. It appeared that the white nodules were largely devoid of bacteria（Fig. 4）.

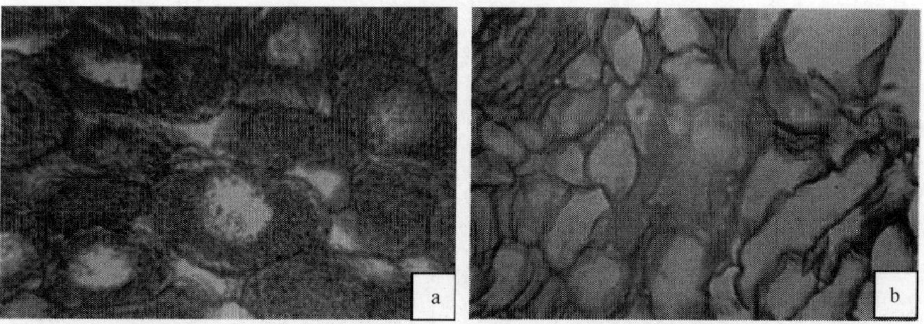

Fig. 4 Sections of nodules formed on *Astragalus sinicus* by plasmidcured-derivatives of *R. huakuii* strain CH203.

a：nodule formed by CH203（wild type）. b：nodule formed by CH203-1（pRHa2）. Magnification，400× for the nodules.

DISCUSSION

In this work，we have shown variation of plasmids in *Rhizobium huakuii* and discovered some functions of different plasmids in *R. huakuii* CH203. Hynes et al. [15] and Brom et al. [4] have reported the systematic curing of plasmids presented in *R. leuminosarum* VF39 and *R. leguminosarum* bv. *phaseoli* CFN42 respectively. Their studies showed that other plasmids in addition to the *sym* plasmid were necessary for development of efficient nodules. Our results showed that different plasmids in *R. huakuii* strain CH203 were also required for optimal symbiotic performance.

A previous report indicated that lacking the plasmid carrying the LPS gene（s） in *R. leguminosarum* VF39 resulted in inefficient nodules [15]. Our results showed that nodules were not completely formed when plasmid pRHb carrying LPS genes was cured. These suggest that the pRHb carries other genes involved in nodule formation in addition to LPS genes described by Noel et al. [23]. Some plasmids in *R. leguminosarum* bv. *trifolii* contain different genes responsible for the acid-tolerant phenotype [9]. Different plasmids in *R. huakuii* CH203 are associated with acid tolerance of the strain（Table 4）. Little is known about the physiological and genetic basis of acid tolerance in *Rhizobium*. At low pH，early stages in infection of leguminous plants by *Rhizobium* are restricted [27]. Some reports suggested that the ability of rhizobial acid tolerance was associated with the ability to control cytoplasmic pH [9,24]. The membrane permeability and proton pumping activity of *Rhizobium* are involved in regulation of the cytoplasmic pH at low pH. The acid-tolerant strain's cytoplasm may be more efficient in pumping proton out of the membrane and can maintain neutral internal pH in the cell， therefore surviving at low pH. For instance，compared with *R. leguminosaru* bv. *trifolii* AU1173，acid-sensitive mutants and native acid-sensitive strains showed more permeability to cell membrane and less proton extrusion activity [8,9].

All plasmids in *R. huakuii* strain CH203 are involved in symbiotic effectiveness. Curing of different plasmids resulted in the strain CH203 losing its effective symbiotic ability. Curing of some plasmids in other *R. huakuii* strains could significantly enhance their effectiveness in nitrogen fixation（data not showed）. Curing plasmids in *R. loti* could also enhance the symbiotic effectiveness of *R. loti* [25]. A polysaccharide inhibition gene（*psi*） was found in *R. phaseoli*，and the *psi* gene on the symbiotic plasmid inhibited exopolysaccaride synthesis and nodulation [3]. Presumably there may be some genetic factors on some plasmids of *Rhizobium* that restrict their symbiotic activity. This study suggests that plasmid preservation is also important for free-living cells. With expansion of the study of different functions of plasmids in *Rhizobium*，a number of interesting genes might be found.

ACKNOWLEDGMENTS

We thank Dr. M.F. Hynes for providing the plasmid pMH1701 and Prof. P.J.W. Young，Prof. W.X. Chen，and Dr. H. Ni for valuable help. Funding for the research was provided by the Chinese National Natural Sciences Foundation.

LITERATURE CITED

[1]Banfalvi Z，Sakanyan V，Koncz C（1981） Location of nodulation and nitrogen fixation genes on a high molecular weight plasmid of *R. meliloti*. Mol Gen Genet 184：318–325

[2]Beringer JE（1974） R factor transfer in *Rhizobium leguminosarum*. J Gen Microbiol 84：188–198

[3]Borthakur D，Johnston AWB（1987） Sequence of *psi*，a gene on the symbiotic plasmid of *Rhizobium phaseoli* which inhibits exopolysaccharide synthesis and nodulation and demonstration that its transcription is inhibited by *psr*，another gene on the symbiotic plasmid. Mol Gen Genet 207：149

[4]Brom S，Gacidelos A，Tomasz S，Floves M，Danle G，David R，Rafael P（1992） Different plasmids of *Rhizobium leguminosarum* bv. *phaseoli* are required for optimal symbiotic performance. J Bacteriol 174：5183–5189

[5]Burkardt B，Schillik D，Puhler A（1987） Physical characterization of *Rhizobium meliloti* megaplasmids. Plasmid 17：13–25

[6]Cannon FC，Riedel GE，Ausubel FM（1979） Overlapping sequences of *Klebsiella pneumonia nif* DNA cloned and characterized. Mol Gen Genet 174：1454–1456

[7]Cava JR，Elias PM，Turowshki DA，Noel KD（1989） *Rhizobium leguminosanum* CFN42 genetic region encoding lipopolysaccharide structures essential for complete nodule development on bean plants. J Bacteriol 171：8–15

[8]Chen H，Richardson AE，Gartner E（1991） Construction of an acid tolerant *Rhizobium leguminosarum* bv. *trifolii* strain with enhanced

capacity for nitrogen fixation. Appl Environ Microbiol 57：2005–2011

[9]Chen H，Cartner E，Rolfe BG（1993） Involvement of genes on a megaplasmid in the acid-tolerant phenotype of *Rhizobium leguminosarum* bv. *trifolii*. Appl Environ Microbiol 59：1058–1064

[10]Chen HK，Shu（1944） Note on the root-nodule bacteria of *Astragalus sinicus* L. Soil Sci 51：291–293

[11]Chen WX，Wang E，LiY（1991） *Rhizobium huakuii* sp. nov. isolated from the rootnodules of *Astragalus sinicus*. Int J Syst Bacteriol 41：275–280

[12]Datta N（1985） Plasmids as organisms. In Helinski DRl.（ed） Plasmid in bacteria. New York：Plenum Publishing Corp.，pp 3–16

[13]Downie JA，Knight CD，JohstonAWB，Rossen L（1985） Identification of genes and gene products involved in nodulation of peas by *Rhizobium leguminosarum*. Mol Gen Genet 198：255–262

[14]Eckhardt T（1978） A rapid method for the identification of plasmid deoxyribonuceic acid in bacteria. Plasmid 1：584–588

[15]Hynes MF，McGregor NF（1990） Two plasmids other than the nodulation plasmid are necessary for formation of nitrogen fixing nodules by *Rhizobium leguminosarum*. Mol Microbiol 4：407–414

[16]Hynes MF，Quandt J，O'connell MP，Puhler A（1989） Direct selection for the curing and deletion of Rhizobium plasmids using transposons carrying the *Bacillus subtilis sac*B. Gene 78：111–120

[17]Jensen HZ（1942） Nitrogen fixation in leguminous plants I. General characters of root nodule bacteria isolated from species of *Medicago* and *Trifolii* in Australia. Proc Linn Soc N S W 66：98–108

[18]Johanson DA（1940） Plant microtechnique. New York：McGraw Hill

[19]Johnston AWB，Beynon JL，Buchanan-Wollaston AV，Setchell SM，Hirsch PR，Beringer JE（1978） High frequency transfer of nodulating ability between strains and species of Rhizobium. Nature 176：634–636

[20]Martinez E，Pardo MA，Palacios R，Cevallos MA（1985） Reiteration of nitrogen fixation gene sequences and specificity of *Rhizobium* in nodulation and nitrogen fixation in *phaseolus vulgaris*. J Gen Microbiol 131：1779–1786

[21]Masteron RV，Prakash PK，Atherly AG（1985） Conversation of symbiotic nitrogen fixation gene sequences in *Rhizobium japonicum* and *Brady-Rhizobium japonicum*. J Bacteriol 5：89–95

[22]Miller JH（1972） Experiments in molecular genetics. Cold Spring Harbor，New York：Cold Spring Harbor Laboratory

[23]Noel KD，VandenBosch KA，Kulpaca B（1986） Mutations in *Rhizobium phaseoli* that lead to arrested development of infection threads. J Bacteriol 168：1392–1401

[24]O'Hara GW，Goss TL，Dilworth MJ，Glenn AR（1989） Maintenance of intracellular pH and acid tolerance in *Rhizobium meliloti*. Appl Environ Microbiol 55：1870–1876

[25]Pankhurst CE（1986） Enhanced nitrogen fixation and competitiveness for nodulation of *Lotus pedunculatntus* by a plasmid-cured derivative of *Rhizobium loti*. J Gen Microbiol 132：2321–2328

[26]Richardson AE，Simpson RJ（1989） Acid tolerance and symbiotic effectiveness of *Rhizobium trifolii* associated with a *Trifolium subterraneum* L. based pasture growing in an acid soil. Soil Biol Biochem 21：87–95

[27]Richardson AE，Simpson RJ，Djordjevic MA，Rolfe BG（1988） Expression of nodulation genes in *Rhizobium leguminosanum* biovar *trifolii* is affected by low pH and by Ca and Al ions. Appl Environ Microbiol 54：2541–2548

[28]Rosenberg C，Nigvet T（1984） The pATc58 plasmid of *Agrobacterium tumefaciens* not essential for tumour induction. Mol Gen Genet 196：533–536

[29]Rosenberg C，Boistard P，Denarie J，CassE-Delbart F（1981） Genes X. Zou et al.：Plasmids in *Rhizobium huakuii* 219 controlling early and late functions in symbiosis are located on a megaplasmid in *Rhizobium meliloti*. Mol Gen Genet 184：326–333

[30]Selenska-Trajkowa S，Radewa G，Markov K（1990） Comparison between *Rhizobium gaegae* and *Rhizobium meliloti* plasmid contents. Lett Appl Microbiol 10：123–126

[31]Simon R（1983） Abroad host-range mobilization system for in vivo genetic engineering transposon mutagenesis in gram negative bacteria. Biotechnology 1：784

[32]Thurman NP，Lewis DM，Jones GD（1985） The relationship of plasmid number to growth，acid tolerance and symbiotic efficiency in isolates of *Rhizobium trifolii*. J Appl Bacteriol 58：1–6

[33]Tisai CM，Frasch CE（1982） A sensitive silver stain for detectinglipopolysaccharides in polyacrylamide gels. Anal Biochem 119：115–119

[34]Vincent J（1970） A manual for the practical study of root nodule bacteria.（IBP Handbook 15） Oxford：Blackwell Scientific Publications

紫云英根瘤菌结瘤基因调控片段克隆和分析*

农　广[1]　张忠明[2]　胡福荣[2]　陈华癸[2]

（1 中山大学生物防治国家重点实验室；2 农业部华中农业大学农业微生物重点开放实验室）

摘　要　通过对紫云英根瘤菌 7653R 及其 Nod^- 突变株 7653R-1 侵染根毛试验证实，7653R 菌株 318×10^3bp 大质粒决定早期结瘤功能，分子杂交结果显示，其 *nod* DABC 定位于该质粒上。将 7653R 菌株的大质粒 DNA 进行酶切并克隆到表达载体 pMP220 上，构建 lacZ 重组子文库，将 lacZ 重组子文库导入 7653R 菌株，在加有 X-gal 和种子浸提物的根瘤菌培养基上选择蓝色菌落。通过 Miller 试验测定它们的 β-半乳糖苷酶活性，获得 1 株种子浸提物对其有诱导效应的菌株 HN18，对该菌株所含的重组质粒 pHN18 进行 *nod*DABC 探针杂交，发现有 1 条 1.7×10^3bp 的阳性杂交带，说明在 pHN18 中克隆有结瘤基因启动子。

关键词　紫云英根瘤菌；*nod* 基因；分子杂交；*lac*Z 基因

根瘤菌的共生结瘤作用涉及到其基因组中的数十个基因，早期的结瘤作用则与调节基因 *nod*D 和共生结瘤基因 *nod*ABC 有关，*nod*ABC 的表达需要 NodD 蛋白和植物根系分泌物的存在，根系分泌物中起诱导作用的是黄酮类物质。关于紫云英根瘤菌结瘤基因的研究报道不多，有报道提到 *nod* 基因定位于内源质粒上[1~3]，并且通过构建基因文库筛选到具有互补结瘤功能的重组质粒[4,5]，但都未涉及到结瘤基因的表达和调控。本研究利用表达型载体 pMP220 及结瘤基因诱导物（寄主植物种子浸提物）直接克隆和筛选具有调控结瘤基因表达活性的片段，以研究根瘤菌与寄主植物相互作用的分子机理。

1　材料和方法

1.1　供试菌株、质粒和培养基

供试菌株和质粒见表 1。培养大肠杆菌用 LB 培养基，培养根瘤菌用 SM、S40 培养基，培养紫云英植株用 Jensen 无氮培养基。

1.2　侵染根毛测试

紫云英种子经酒精和 *w*=0.1%升汞表面消毒，无菌水洗涤 6 次后，在 28℃萌发 2d 至胚根长出，移植于加有 Jensen 无氮培养基的软琼脂载玻片上，在 16～26℃培养，每天光照 12h，待 2 片子叶张开，感染根瘤菌，1d 后开始观察根毛形态的变化。对空白对照、感染 7653R 和 7653R-1 的植株进行根毛变形测试，连续观察 5d，每天 2 次，每次 3 个植株，统计所观察根毛的变形状况。

1.3　*nod* 探针杂交

使用宝灵曼公司的地高辛标记和检测盒进行杂交分析，并按所附说明书进行操作。

1.4　*lac*Z 重组子文库的构建和三亲转移接合子的筛选

按文献[6]方法提取根瘤菌大质粒，*Eco*RⅠ酶切 2h，将 DNA 片段与启动子缺失载体 pMP220 连接，连接产物转化到大肠杆菌 DH5α 中，在含 Tc（10μg/mL）的 LB 平板上筛选抗性菌落，从而获得 7653R 大质粒的 *lac*Z 重组子文库。以 *lac*Z 重组子文库为供体，根瘤菌 7653R 菌株为受体，在具有 tra^+功能的 pRK2013 质粒的协助下，通过三亲本杂交将 mob^+ tra^-的 *lac*Z 重组子导入到 7653R 菌株中。在加有 X-gal（100μg/mL）、植物种子浸提物（20μL/mL）、Sm（100μg/mL）和 Tc（10μg/mL）的 SM 上筛选蓝色菌落。

*原载于《中山大学学报（自然科学版）》，37（1）：73～77，1998.

表 1　实验菌株和质粒

Tab.1　Experimental bacterial strains and plasmids

菌株与质粒	相关性状
R. huakuii	
7653R[1)]	Nod^+ Fix^+ Sm^R
7653R-1[1)]	Nod^-，7653R 缺失 3.18×10^5 bp 大质粒
HN10[2)]	7653R 含有 *lacZ* 重组子
HN18[2)]	7653R 含有 *nod-lacZ* 重组子
HN46[2)]	7653R 含有 *lacZ* 重组子
HN49[2)]	7653R 含有 *lacZ* 重组子
HN114[2)]	7653R 含有 *lacZ* 重组子
R.leguminosarum	
T83K3[3)]	Nod^+ Fix^+ Km^R
R. fredii	
USDA205[3)]	Nod^+ Fix^+
E.coli	
DH5A	F^- λ^- Φ80*lacZ*M 15Δ（*lacZ*Y A *arg*F）
Plasmids	
pIJI216[3)]	*R.leg.* 6. 6×10^3bp *nod*DABC 克隆在 pKT230 上
pSA30[3)]	*K. p. nif* HDK 克隆在 pACYC184 上
pMP220[3)]	Inc P，无启动子的 *lacZ* 转录表达载体，Tc^R
pRK2013[3)]	Inc P. Km^+ tra^+ mob^-
pHN18[2)]	*R.huakuii* 7653R 的 *nod-lacZ* 重组子 克隆在转录表达载体 pMP220 上

文献及来源分别为：1）文献[3]；2）本工作；3）本室保存

1.5　种子浸提物的制备

取 3g 紫云英种子洗干净，用 φ=70%的酒精表面消毒，然后用 10mL 无菌水浸泡 2d，吸取棕黄色浸泡液，加无菌水至 10mL，过滤灭菌，于 4℃贮存。

1.6　β-半乳糖苷酶活性测定

将样品菌株活化，再转移至 S40 根瘤菌液体培养基培养 24h，在 600nm 测定 A 值，再将细菌稀释 A 至 0.1，每个样本各取 2 份，5mL，1 份加入种子浸提物 100μL，另一份加等量无菌水为对照，28℃振荡培养 10～24h，然后按文献[7]方法测定 β-半乳糖苷酶活性。

1.7　重组质粒的分离和酶切鉴定

从菌株 HN18 提取质粒，转化大肠杆菌 DH5α，在含有 Tc（10μg/mL）的 LB 平板上筛选抗性菌落，挑取菌落进行小量制备质粒 DNA，再进行内切酶分析。

1.8　重组质粒三亲转移接合子的筛选

以 pRK2013 为协助质粒，将重组质粒 pHN18 分别导入受体菌株 7653R 和 7653R-1，并在含 X-gal、种子浸提物、Sm 和 Tc 的 SM 平板上筛选蓝色菌落。

2　结果与讨论

2.1　紫云英根瘤菌 7653R 菌株的共生结瘤质粒遗传研究

2.1.1　7653R 菌株共生结瘤质粒的测定

以前对紫云英根瘤菌 7653R 菌株的质粒检测证实有 2 条内源质粒带，消除其中较大 1 条后产生不

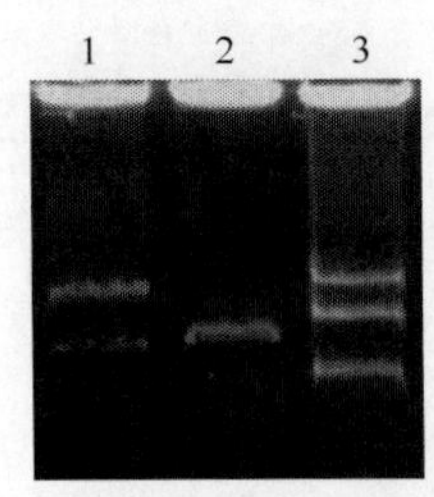

图 1 根瘤菌大质粒电泳快速检测

Fig.1 A rapid test of rhizobia megaplasmids.

1. *Rhizobium huakuii* 7653R 菌株；2. *Rhizobium huakuii* 7653R-1 菌株；3. *Rhizobium fredii* USDA205 菌株

能共生结瘤的菌株 7653R-1[3]。本文利用 DNAGEL[8]方法，以快生大豆根瘤菌 USDA205 菌株已知分子质量的内源质粒为标准，测得 7653R 菌株较大 1 条质粒带为 3.18×10^5bp，另 1 条为 1.74×10^5bp（图 1）。

2.1.2 侵染根毛测试

结果见表 2，统计数据显示，在空白对照中，紫云英植株根毛变形率低于 1%，它是由物理因素所引起；野生型菌株 7653R 所引起根毛变形的比率为 58.32%，显示了野生型菌株的侵染作用能引起极明显的根毛变形，变形根毛中有典型的“牧羊拐”状根毛；7653R-1 菌株所引起的根毛变形率为 3%，未发现“牧羊拐”状变形。观测结果表明 7653R-1 菌株没有侵染作用，丧失了共生结瘤能力。由此可判断，3.18×10^5 bp 大质粒决定 7653R 菌株的早期结瘤作用，此大质粒的缺失（7653R-1）使根瘤菌与植物根系的相互作用中止于 Hac 期或该期之前。

2.1.3 *nod*DABC 基因的物理鉴定

根毛实验显示早期结瘤基因定位于 3.18×10^5 bp 大质粒上，为了进一步获得直接证据，本文以豌豆根瘤菌 *nod*DABC 基因为探针，与紫云英根瘤菌 7653R 菌株大质粒进行分子杂交，杂交结果表明，正对照豌豆根瘤菌菌株 T83K3 呈现较强阳性杂交，另一正对照快生大豆根瘤菌菌株 USDA205 有相对弱的阳性杂交。7653R 的 3.18×10^5 bp 大质粒呈阳性杂交，7653R-1 没有阳性杂交，因此可判定 *nod*DABC 定位于 3.18×10^5 bp 大质粒上。根毛实验和探针杂交的结果是相一致的。

在初步了解内源质粒的遗传特性之后，通过共生结瘤的功能试验，发现大质粒与早期结瘤作用有关，亦即意味着由根系分泌物诱导表达的早期结瘤基因定位于大质粒上，分子杂交进一步证实早期结瘤基因的存在，并证实早期结瘤基因定位于 3.18×10^5 bp 大质粒上，从而为克隆结瘤基因提供了依据。

表 2 根毛感染测试

Tab.2 Test of root hair infection

项目	对照	7653R	7653R-1
供试植株数/株	30	30	30
观察根毛总数/条	1290	1562	1860
变形根毛总数/条	12	911	57
“牧羊拐”状根毛数/条	0	24	0
变形根毛百分率/%	0.93	58.32	3.06

2.2 *lac*Z 重组子酶活性诱导表达

2.2.1 根瘤菌三亲转移接合子蓝色菌落的筛选

以 *lac*Z 重组子库为供体、7653R 菌株为受体，进行三亲本杂交，抗性平板筛选蓝色菌落，通过平板计数，重组质粒向受体根瘤菌转移的频率约为 10^{-5}，其中诱导表达蓝色的菌落占三亲转移接合子的 1%左右。分别挑取蓝色和白色的菌落到同样的平板再培养，经传代培养证实颜色是稳定的。

2.2.2 蓝色菌落菌株的β-半乳糖苷酶活性测定

将分别命名为 HN10，HN18，HN46，HN49 和 HN114 的 5 个菌株，测定其β-半乳糖苷酶活性，结果表明（表 3）：在有或无诱导物时，HN49 和 HN114 的酶活没有显著差别，表现无诱导效应。有诱导物的 HN18 菌株比无诱导物的酶活高 2.5 倍，表现出诱导效应；无诱导物的 HN10 和 HN46 比有诱导物的酶活分别高 12.7 倍或 2.4 倍，说明诱导物对酶活有明显的抑制作用。各样本的酶活均显著高于对照菌株 7653R 的酶活。

实验结果显示，HN18 菌株的重组质粒上含有由黄酮类诱导物诱导表达的调控片段，它的诱导效应

与早期结瘤基因 *nod*ABC 基因受根系分泌物诱导所产生的效应相似，因此推测重组质粒可能含有 *nod*ABC 基因的启动子及调控序列。其他几个菌株中所克隆到的启动子则与 *nod* 基因无关。

表 3　蓝色菌落 β-半乳糖苷酶活性的诱导效应

Tab.3　Induction of β-galactosidase activity in blue colonies（10^{-9}mol s^{-1}）

菌株	β-半乳糖苷酶活性/U		菌株	β-半乳糖苷酶活性/U	
	加种子浸提物	无种子浸提物		加种子浸提物	无种子浸提物
7653R	50±5	67±17	HN46	517±55	1267±55
HN10	117±25	1584±457	HN49	1100±117	967±77
HN18	1750±307	700±28	HN114	1184±112	1084±190

1）实验设 3 次重复

2.2.3　重组质粒的酶切和 *nod* 探针杂交

图 2a 的电泳结果显示，从 HN18 菌株分离到的重组质粒 pHN18 含有外源片段。以豌豆根瘤菌 *nod*DABC 基因片段为探针进行分子杂交，结果如图 2b 所示。杂交结果表明，pHN18 有 1 条约 1.7×10^3 bp 的阳性杂交带，说明菌株 HN18 的质粒 pHN18 含有 *nod* 基因片段。

在 *lac*Z 重组子文库的构建中，所用表达载体 pMP220 只有 *lac*Z 基因及编码核糖体结合位点的序列而缺乏启动子和操作基因，因此没有表达 *lac*Z 的能力。只有与 7653R 菌株大质粒上具有启动子活性的片段融合到载体 pMP220 的 *lac*Z 基因上游，才能表达 β-半乳糖苷酶活性，*lac*Z 基因的表达水平由克隆到的调控序列的转录能力所决定。当克隆到早期结瘤基因时，黄酮类诱导物可激发 *nod* 基因的表达，从而带动 *lac*Z 的表达。

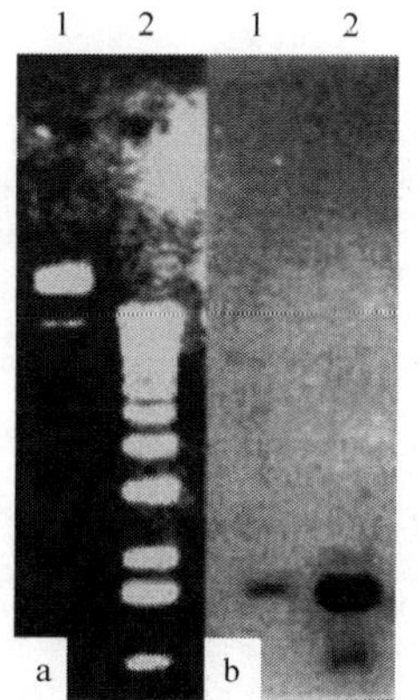

图 2　重组质粒 pHN18 的酶切分析和 *nod* 探针的分子杂交

Fig.2　Restriction enzyme analysis of recombinant plasmid pHN18 and its DIG-hybridization with *nod*DABC probe in *R. leg*.

a.重组质粒 pHN18 的酶切分析；b. *nod*DABC 探针杂交

1. pHN18+*Eco*R Ⅰ；2. 1000 bp 梯度分子质量标记

从所筛选到的菌株可看到，种子浸提物对 HN18 菌株有诱导效应，表现出黄酮类物质对早期结瘤基因表达的激发作用，同时也发现种子浸提物还有不同的诱导效应。由此可见，种子浸提物对早期结瘤有诱导作用，但其所含的成分并不局限于诱导早期结瘤作用，可能还有其他的诱导效应。进一步的探针杂交结果表明，从所筛选到的 HN18 菌株分离得到的重组质粒 pHN18 含有早期结瘤基因，诱导作用的结果和分子杂交的结论是相一致的。

2.3　重组质粒 pHN18 的 *lac*Z 诱导活性验证

将重组质粒 pHN18 导入受体菌株 7653R，在含诱导物的培养基上筛选到的三亲转移接合子均为蓝色，导入菌株 7653R-1 时，所获得的三亲转移接合子也全为蓝色。分别挑取菌落到同样的培养基（缺 Tc），菌落显示蓝色，而出发菌株 7653R 和 7653R-1 则是白色。

重组质粒 pHN18 再导入受体根瘤菌，所筛选到的三亲转移接合子均为蓝色，而对照菌株为白色，这一结果再次说明，pHN18 具有早期结瘤基因所表现的诱导表达活性，在功能上对前面的结果进行了验证。值得注意的是，pHN18 导入受体 7653R-1（缺 *nod*DABC）所得到的菌落也均为蓝色，其结果说明一个问题，pHN18 的诱导表达并不依赖于 3.18×10^5 bp 大质粒。进一步的研究正在进行之中。

参 考 文 献

[1]魏辉，李阜棣。紫云英根瘤菌质粒携带共生基因的物理证据。华中农业大学学报，1989，8：12.

[2]金润之，朱劲松，沈善炯，等。紫云英根瘤菌 *Ra*159 的巨大质粒上存在有 *nod* 和 *nif* 基因的证明。微生物学报，1993，33：170-176.

[3]周俊初，张忠明，陈华癸，等。紫云英根瘤菌质粒的研究（Ⅱ）紫云英根瘤菌经高温和吖啶橙处理后结瘤和固氮突变株的产生和质

粒消除。华中农业大学学报，1987，6：156-164.

[4]张忠明，陈华癸，范云六等。紫云英根瘤菌基因文库的构建及含完整结瘤基因的重组质粒 pRaZ15 的分离。生物工程学报，1991，7：213-219.

[5]周宗汉，江群益，沈善炯等。紫云英根瘤菌基因文库的构建和结瘤基因片段的分离。科学通报，1989，4：305-308.

[6]Hirsch P K，Von Montagu M，Johnston A W B，et al. Physical identification of bateriocinogenic，Nodulation and other plasmids in strain *Rhizobium leguminosarum*. J Gen Micr obio l，1980，120：403-412.

[7]Miller J T. Experiments in molecular genetics：β-galact osidase assay . New Yor k：Cold Spring Harbor Laboratory Press，1972. 352-355.

[8]Kieser T. DNAGEL：a computer program for determining DNA fragment sizes using a small computer equipped with a graphic tablet. Nucleic Acid Res，1984，12：679-688.

紫云英根瘤菌由种子浸提物诱导的基因调控片段的克隆和分析*

农　广　　张忠明　　胡福荣　　陈华癸

（华中农业大学农业微生物重点开放实验室）

关键词　紫云英根瘤菌；种子浸提物；DIG 杂交；*lac*Z 融合

在豆科植物中，种子浸提物和根系分泌物都能诱导根瘤菌的早期结瘤基因的表达，这因为它们都含有起诱导作用的黄酮类物质。作为植物的共生信号，黄酮类物质与根瘤菌的结瘤调节基因 *nod*D 的产物一起，诱导结瘤基因 *nod*ABC 等的表达，跟着形成根瘤菌结瘤因子-脂寡糖分子，结瘤因子分泌到菌体外并作用于根尖，诱导根瘤的形成。

紫云英根瘤菌的早期结瘤阶段，同样受黄酮类物质的诱导，并产生结瘤因子，引起根毛变形[1]。本研究通过在表达型载体 pMP220 对 7653R 菌株的共生质粒 DNA 片段进行克隆，利用结瘤基因诱导物（寄主植物种子浸提物）筛选具有调控表达活性的片段，获得一株种子浸提物对其有诱导效应的菌株 HN18，对该菌株所含的融合质粒 pHN18 进行 *nod*ABC 探针杂交，发现有一条 1.7kb 的阳性杂交带，表明在 pHN18 中克隆有结瘤基因启动子。

1　材料和方法

1.1　菌株和质粒

紫云英根瘤菌（*Rhizobium huakuii*）有效结瘤菌株 7653R 和大肠杆菌 DH5α 为本室保存；HN10、HN18、HN46、HNI14 为含有 *lac*Z 融合子的 7653R 菌株（本研究）；质粒 pHN18 是从 HN18 菌株分离到的 *lac*Z 融合质粒（本研究）。pMP220 为无启动子的 *lac*Z 转录表达载体，四环素抗性；pRK2013 为三亲结合转移的协助质粒；pU1216 含豌豆根瘤菌 *nodDABC* 基因片段于 pKT230 载体上，卡那霉素抗性。后三种质粒均为本室保存。

1.2　培养基

培养大肠杆菌用 LB 培养基[2]，培养根瘤菌用 SM[3]和 S40 培养基[4]。

1.3　*lac*Z 融合子库的构建和三亲转移接合子的筛选

按 Hirsch 法[5]提取根瘤菌大质粒，*Eco*RI 酶切 2h，将 DNA 片段与载体 pMP220 连接，酶连物转化到大肠杆菌 DH5α 中，在含 Tc（10μg/ml）的 LB 平板上筛选抗性菌落，从而获得 7653R 大质粒的 *lac*Z 融合子库。以 *lac*Z 融合子库为供体，根瘤菌 7653R 菌株为受体，在 pRK2013 质粒的协助下，将 *lac*Z 融合子导入到 7653R 菌株中。在加有 X-gal（100μg/ml）、植物种子浸提物（20μl/ml）、抗生素 Sm（100μg/ml）和 Tc（10μg/ml）的根瘤菌基本培养基（SM）上筛选蓝色菌落。

1.4　种子浸提物的制备

取 3g 紫云英种子洗干净，用 70%酒精表面消毒，然后用 10ml 无菌水浸泡两天，吸取棕黄色浸泡液，加无菌水至 10ml，过滤灭菌，于 4℃贮存。

1.5　β-半乳糖苷酶活性测定

将样品菌株活化，再转移至 S40 根瘤菌液体培养基培养 24h，在 600nm 波长测定 OD 值，再将细菌稀释至 OD 值为 0.1，每个样本取两份 5ml，一份加入种子浸提物 10μl，另一份加等量无菌水为对照，28℃振荡培养 10～24h，然后按 Miller[6]方法计算 β-半乳糖苷酶活性。

*原载于《微生物学报》，38（3）：225～228，1998.

1.6 重组质粒的分离，酶切和 *nod* 探针杂交

从菌株 HN18 提取质粒，转化大肠杆菌 DH5α，在含有 Tc（10μg/ml）的 LB 平板上筛选抗性菌落，挑取菌落进行小量制备质粒 DNA，再进行内切酶分析。使用宝灵曼公司的地高辛标记和检测盒进行杂交分析，并按所附说明书进行操作。

2 结果和讨论

2.1 根瘤菌三亲转移接合子中 *lac*Z 融合子蓝色菌落的筛选

以 *lac*Z 融合子库为供体、7653R 菌株为受体，筛选并得到了蓝色菌落（图略）。通过平板计数，融合子向根瘤菌转移的频率约为 10^{-5}，其中诱导表达所产生的蓝色菌落占三亲转移接合子的 1%。分别挑取蓝色和白色的菌落到同样的平板纯化，继代培养证实颜色是稳定的。

2.2 蓝色菌落菌株的β-半乳糖苷酶活性测定

将分别命名为 HN10、HN18、HN46、HN49 和 HN114 的五个菌株，测定其β-半乳糖苷酶活性，结果（表 1）表明，在有或无诱导物时，HN49 和 HN114 的酶活没有显著差别，表现无诱导效应。有诱导物的 HN18 菌株比无诱导物的酶活高 2.5 倍，表现出诱导效应；无诱导物的 HN10 和 HN46 比有诱导物的酶活高 12.7 倍或 2.4 倍，说明诱导物对酶活有明显的抑制作用。各样本的酶活均明显高于对照菌株 7653R 的酶活。

实验结果显示，HN18 菌株的融合质粒上会含有根瘤菌结瘤基因的启动子及 *nod* 基因无关。

表 1 蓝色菌落β-半乳糖苷酶活性的诱导效应*

菌 株	β-半乳糖苷酶活性(U)		菌 株	β-半乳糖苷酶活性(U)	
	加种子浸提物	无种子浸提物		加种子浸提物	无种子浸提物
7653R	3±0.3	4±1.0	HN46	31±3.3	76±3.3
HN10	7±1.5	95±27.4	HN49	66±7.0	58±4.6
HN18	105±18.4	42±1.7	HN114	71±6.7	65±11.4

*实验设三次重复

2.3 重组质粒的酶切和 *nod* 探针杂交

图 1-A 电泳结果显示，从 HN18 菌株分离到的重组质粒 pHN18 含有外源片段。以豌豆根瘤菌 *nod*DABC 基因片段为探针进行 DIG 杂交，图 1-B 杂交结果显示，pHN18 有一条约 1.7kb 的阳性带，说明菌株 HN18 的质粒 pHN18 含有 *nod* 基因片段。

在 *lac*Z 融合子库的构建中，所用表达载体 pMP220 只有 *lac*Z 基因及编码核糖体结合位点的序列而缺乏启动子和操作基因，因此没有表达 *lac*Z 的能力。只有在 7653R 菌株大质粒上具有启动子活性的片段融合到载体 pMP220 的 *lac*Z 基因上游，才能表达β-半乳糖苷酶活性，*lac*Z 基因的表达水平由克隆到的调控序列的转录能力所决定。当克隆到早期结瘤基因时，黄酮类诱导物可激发 *nod* 基因的表达，从而带动 *lac*Z 的表达。

从所筛选到的菌株可以看到，种子浸提物对 HN18 菌株有诱导效应，表现出黄酮类物质对早期结瘤基因表达的激发作用。同时也发现种子浸提物还有不同的诱导作用，但其所含的成分并不局限于诱导早期结瘤作用，可能还有其他的诱导效应。进一步的探针杂交结果表明，从所筛选到的 HN18 菌株分离得到的重组质粒 pHN18 含有早期结瘤基因调控片段。诱导作用的结果和分子杂交的结论是相一致的。

2.4 重组质粒 pHN18 的 *lac*Z 诱导活性验证

将重质粒 pHN18 经三亲结合再导入受体菌株 7653R，在含诱导物的培养基上筛选到的三亲转移接合子均为蓝色，挑取菌落到同样的培养基继代培养，菌落仍显示蓝色，而出发菌株 7653R 则是白色（图略）。

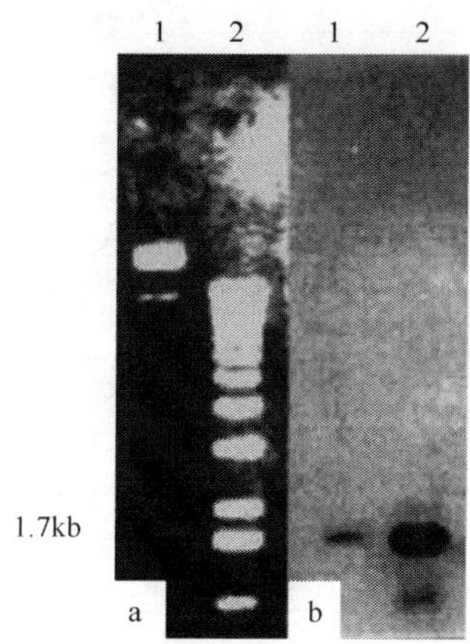

图 1 重组质粒 pHN18 的酶切分析和 nod 探针的分子杂交

Fig.1-a Digestion of fusion plasmid pHN18 with *Eco*RI：lane 1. pHN18+*Eco*RI；lane 2. 1kb ladder marker.

Fig.1-b Hybridization with *nod*DABC probe：lane 1. pHN18+*Eco*RI；lane 2. 1kb ladder marker.

重组质粒 pHN18 再导入受体根瘤菌的结果再次说明，pHN18 具有早期结瘤基因所表现的诱导表达活性，在功能上对前面的结果进行了复证。

参 考 文 献

[1]杨国平，朱军，娄无忌，微生物学报，1994，34（5）：406-408.

[2]Sambrook j，Fritsch E F，Maniatis T.，Molecular Cloning，2nd ed. New York；Cold Spring Harbor Laboratory Press，1989. A l.

[3]周俊初，张忠明，陈华癸等，华中农业大学学报，1987，6：156-164.

[4]Aguilar O M，Kapp D，Puhler A.，J Baceriol. 1985，164：245-254.

[5]Hirsch P K，von Monlagu M，Johnston A W B et al.，J Gen Micrabial. 1980，120：403-412.

[6]Miller J T.，Experiments in Molecular Cenetics，New York；Cold Spring Harbor Laboratory Press，1972. 325-355.

用 Tn5 标记带发光酶基因的华癸根瘤菌（*Mesorhizobium huakuii*）7653R 质粒*

郭先武　张忠明　胡福荣　莫才清　李阜棣　陈华癸

（华中农业大学农业部农业微生物重点实验室）

摘　要　在 pRK2073 的辅助作用下，通过三亲本接合转移，将 Tn5（*sac*B-*lux* AB）插入华癸根瘤菌 7653R 菌株基因组中。将 3200 个 Tn5 标记菌株影印到含有 8%蔗糖的 YMA 平板，检测这些菌株的发光酶活性和新霉素抗性。共获得 8 个菌株，其选择标记消失。质粒检测发现其共生质粒有不同程度的缺失甚至消除。结瘤实验证明，所有共生质粒部分缺失（有的缺失达 1/3）的菌株依然结瘤。经用 *lux* AB 探针的分子杂交证实，8 个菌株中有 3 个对应的标记菌株其 Tn5 插在共生质粒上。

关键词　华癸根瘤菌；共生质粒；分子标记；发光酶基因

生物固氮是豆科植物氮源的重要来源。豆科植物的固氮作用是在根瘤菌的参与下完成的，即根瘤菌对植物进行侵染等一系列过程，促使植物形成根瘤，根瘤菌在根瘤的特殊环境中进行固氮并将其产物输送给植物，也称之为共生固氮。紫云英（*Astragalus sinicus* L.）为我国南方大量种植的绿肥植物，与其共生的根瘤菌称为紫云英根瘤菌，40 年代，由陈华癸教授首次发现[1]，1991 年陈文新教授正式命名为华癸根瘤菌（*Rhizobium huakuii*）[2]，最近，归一新属 *Mesorhizobium*[3]。华癸根瘤菌的宿主范围极为狭窄，仅仅能与紫云英及 *Astragalus* 属少数种类[4]共生。目前为止，尚未发现与其他根瘤菌存在交叉宿主植物。本实验室对 7635R 菌株做了许多研究，认为它具有 2 个大质粒，分子质量较大的质粒为共生质粒（pSym）[5, 6]。共生质粒携带有大量与建立共生关系有关的遗传信息，因而也是根瘤菌遗传学研究的重要方面。本研究期望对其特定质粒进行分子标记，为观察其在土壤中的行为及研究在异源宿主中质粒的相互作用提供实验材料，同时，也期望获得缺失突变株。

1　材料与方法

1.1　材料

质粒 pHNC3[7]含转座子 Tn5（*sac*B-*lux*AB），其质粒载体来源于 pMH1701，在根瘤菌中不能复制而成为自杀质粒。含 *lux* AB 的细菌能在加入底物喹醛后发光，*sac*B 能使该细菌在含蔗糖的平板上不生长，因而 Tn5 上携带的这 2 种基因及本身的新霉素基因（*neo*）均可作为转移接合子的选择标记。

华癸根瘤菌（*M. huakuii* Chen）7653R 菌株由本实验室提供。

1.2　方法

1）质粒接合转移：将均处于对数生长期的根瘤菌 7653R、含 pHNC3 和 pRK2073（辅助质粒）的大肠杆菌以一定比例（大肠杆菌要比根瘤菌少）混合在离心管中，用无菌水洗涤后，再加无菌水悬浮后滴入预先放置的孔径为 0.22μm 的无菌微孔滤膜上。滤膜要提前 3h 左右放入不加任何抗生素的 TY 平板上。28℃培养 2d，滤膜取出放入盛有无菌水的小瓶中，摇荡洗下菌苔，稀释涂平板。

2）质粒的快速检测：采用修改的 Eckchadt 方法[8]。

3）质粒 DNA 的 Southern 转移：依常规方法转移到尼龙膜上。

4）分子杂交：应用同位素 α-32p-dCTP 分子杂交，见文献[9]；探针为随机引物标记。

*原载于《华中农业大学学报》，18（2）：147～150，1999.

5）结瘤试验：将获得的缺失菌株及其出发菌株感染紫云英植物，观察结瘤情况。结瘤实验在试管中进行，琼脂为 1.0%，使用 Fahraeus 无氮植物营养液。

2 结果与分析

2.1 含 Tn5（*lux* AB）质粒的接合转移

采用重组质粒 pHNC3-*lux* AB 作为 Tn5 供体。用大肠杆菌 HB101（pHNC3）和根瘤菌 7653R 在辅助质粒 pRK2013 的协助下进行三亲本接合转移。由于 pHNC3 在根瘤菌中为自杀质粒，其上的 Tn5 可跳入 7653R 基因组的任何位置，为了保证所得的接合子不是营养缺陷型，故以 SM+Neo+Str 为筛选平板，其中，SM 培养基不适于营养缺陷突变株生长，链霉素（Str）用于淘汰大肠杆菌，新霉素（Neo）用于淘汰受体菌 7653R。获得的转移接合子菌落在加底物后均能发光。共获得了 3200 个转移接合子。

2.2 含 Tn5- *lux*AB 质粒的菌株的筛选

将 3200 个转移接合子分别点种在 YMA 和 YMA+8%蔗糖平板上，将 2 种平板分别于 28℃和 37℃下培养，2d 后再将蔗糖平板转入正常培养 2d。总共获得 115 个在蔗糖平板上生长但不发光的菌落，进一步在 YMA+Neo 抗性平板上验证其抗性是否存在，最终获得 8 个没有新霉素抗性、能在 8%蔗糖平板上生长且不发光的突变菌株及其相应的出发菌株。在初选的 115 个菌株中，绝大部分依然表现新霉素抗性。

2.3 Tn5- *lux*AB 插在质粒上的菌株的确认

将上述 8 个菌株及其对应出发菌株做质粒快速检测，结果见图 1。可见，CIIL 菌株的大质粒缺失最多，达约 86kb，菌株 JIVL、M5L 的大质粒均有不同程度的缺失，P5L 和 O6L 菌株的质粒未检测出明显变化，菌株 W39L 的大质粒则完全消失。R4IVL 菌株在开始时能检测出明显部分缺失大质粒的存在，但重复检测时则出现弱带及完全消失的现象。

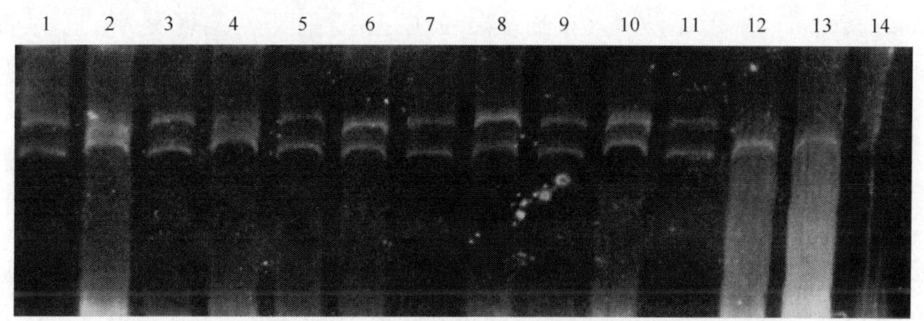

图 1 Tn5- *lux*AB 标记菌株及经处理后对应菌株的质粒图谱

Fig. 1 Plasmid profiles of the strains marked by Tn5- *lux*AB or the marked strains disposed with 8% sucrose.

泳道 Lanes：1.CIIR 2.CIIL 3.JIVR 4.JIVL 5.M5R 6.M5L 7.O6R 8.O6L 9.Q5R 10.Q5L 11.R4IVR 12.R4IVL 13.W39L 14.W39R

以 *lux*AB 为探针，并用 α-^{32}p-dCTP 标记它，然后在严谨条件下杂交（图 2）。可见，CIIR、R4IVR 和 JIVR 等含 Tn5-*lux*AB 但没有经过处理的相应出发菌株的共生质粒上均有明显的阳性杂交信号。所以 CIIR、R4IVR 和 JIVR 菌株 Tn5 插在共生大质粒上，其余含 Tn5-*lux*AB 未处理菌株的点样孔上有阳性信号，质粒上无信号，说明 Tn5-*lux* AB 插在染色体上；由此，本研究共获得了 Tn5-*lux*AB 插在质粒上的 3 个突变菌株和 pSym 不同缺失程度及完全消除的突变株。

2.4 突变株的共生能力

用上述突变株做结瘤实验。4 个星期后，除没有共生质粒的菌株未结瘤外，其余缺失菌株均结瘤，但瘤数少或较迟，重复一次的结果相同。本结果表明缺失未发生在共生基因上，因而不影响突变菌株与紫云英共生关系的建立，个别菌株缺失的片段甚至达共生质粒的 1/3。但缺失仅对其结瘤效率有一定影响。

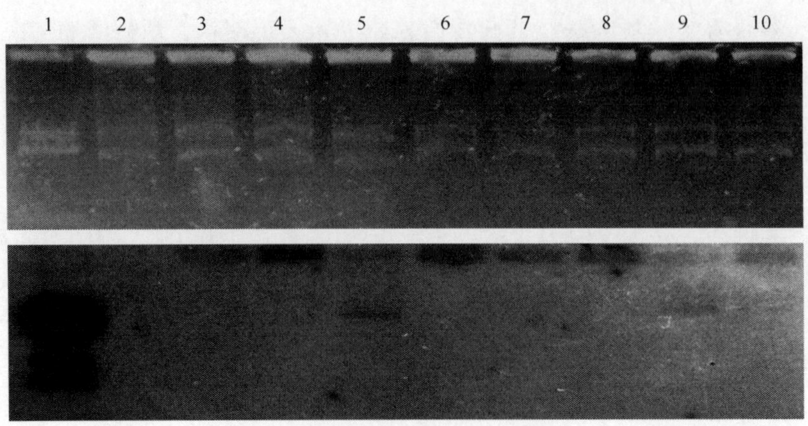

图 2 8 个 Tn5-*lux*AB 可能插在质粒上菌株的质粒图谱（上）及其用 *lux*AB 基因探针的分子杂交（下）

Fig. 2 The plasmid profiles of 8 strains which are probable to be inserted by Tn5- *lux*AB (up) and their hybridization map with the nodDBC probe (low).

泳道：1. pHNC3，2. 7653R，3. Z69R，4. W39R，5. R4IVR，6. Q5R，7. O6R，8. M5R，9. JIVR，10. CIIR

3 讨论

以 Tn5-*sac*B 为机制筛选消除质粒的菌株是由 Hyne 首先建立起来[10]，其原理是：如果 Tn5 插在染色体的必需基因上则将使突变株致死，但如果插在质粒上尤其是插在与质粒复制或稳定机制有关的基因上，则会使质粒丢失。而 *sac*B 基因可用作很好的正向选择标记（即消除选择标记者能生长），与用吖啶橙或高温处理不带选择标记的菌株的方法相比，其筛选效率要高得多。实际上 Hyne 在 27℃下于蔗糖平板上培养 Tn5 标记的根瘤菌株也可得到质粒消除突变株，但若同时提高培养温度则可加大其突变率。据报道在恶劣的环境下质粒的缺失、消除等会增加突变率[11]，高温、抗生素和 8%蔗糖的存在都应是不利因素的组成部分。许多研究者还发现，根瘤菌质粒在共生条件下常出现变异[12]，是否说明在共生条件下有些特殊因素使质粒不稳定，将是一个值得深入探讨的问题。自然界中根瘤菌质粒的多样性可能与此相关。

本研究所得到的质粒缺失或消除的突变菌株，其 Tn5 并非都插在缺失的质粒上，W39L 是质粒消除菌株，但其出发菌株 W39R 的分子杂交实验也未见其共生质粒有阳性信号，表明其他质粒的消除与 Tn5 是否插在其上并无必然关系，该类菌株是由其染色体上发生缺失而偶然发现的。但这并不说明该筛选法无效，因为染色体不仅稳定性高，而且 DNA 分子大得多，在这 3000 多个菌中也只筛选出 4 个是染色体缺失菌株。不管 Tn5 插在何处，质粒缺失菌株更容易获得，因此，该法用来获得质粒缺失菌株更为有效。

R4IVR 菌株被证明 Tn5-*lux* AB 插在共生质粒上，经处理（8%蔗糖和高温）的对应菌株 R4IVL 为 *lux*AB⁻ *sac*B⁻ *neo*⁻，说明其 Tn5 已丢失；质粒检测显示 2 个质粒未发生明显变化，但在随后的多次检测中发现共生质粒带由淡至消失，说明含 Tn5-*lux* AB 质粒片段丢失，影响了其质粒的稳定性。

本研究中未发现一个菌株在小质粒上发生明显的缺失或消除。周俊初等[6]用吖啶橙诱变 7653R 得到消除了共生质粒的菌株，也未能获得只有共生质粒的突变株，这一方面说明 Tn5 插入大质粒的几率较大，且插入片段影响了质粒的稳定性，另一方面也可能是 7653R 菌株中的非共生质粒更加稳定抑或是它含有什么特别重要的基因呢？有待深入研究。

pHNC3 质粒上的 Tn5 还含有能诱动质粒发生转移的 *mob* 基因片段，本人曾尝试多次、多种途径期望能将 Tn5 标记的共生质粒及非共生质粒转移到没有质粒的农杆菌 9023 菌株以及用 B30（另一个转座子）标记的 7653R-1 中，均未能获得结果。因而，对 pHNC3 质粒中 *mob* 片段的功能有待进一步证实。

参考文献

[1]Chen H K. Note on the root-nodule bacteria of *Astragalus sinicus* L，Soil Sci，1944，51：291-293

[2]Chen W X，Li G S，Qi Y I，et al. *Rhizobium huakuii* sp. nov. isolated from the root nodules of *Astragalus sinicus*. Int J Syst Bacterial. 1991，41：275-280

[3]Young J P W，Haukka K. Diversity and phylogeny of *Rhizobia*. New Phytol. 1996，133：87-94

[4]Chen H K，Li F D，Cao Y Z. Characteristics，Distribution，Ecology and Utilization of *Astragalus sinicus-Rhizobia* Symbiosis. In：Hong Guo-Fan ed. The nitrogen fixation and its research in China. Shanghai：Springer-verlag，1992，439-456

[5]农广，张忠明，胡福荣等. 紫云英根瘤菌 5 个互补结瘤菌株所分离到的重组质粒外源片段的酶切图谱.武汉大学学报（自然科学版），1995，41（4）：469-474

[6]周俊初，张忠明，黄诚金等. 紫云英根瘤菌的研究 II.紫云英根瘤菌经高温和吖啶橙处理后结瘤与固氮突变株的产生和质粒消除. 华中农业大学学报，1987，6（2）：156-164

[7]莫才清，周俊初，李阜棣. 含发光酶基因的转座子质粒载体的改造及向根瘤菌 HN01 的转座. 应用与环境生物学报，1997，3（1）：252-257

[8]Harrison S P，Jones D G，Schunmann P H D et al. Variation in *Rhizobium leguminosarum* biovar *trifolii* Symplasmid and the association with effectiveness of nitrogen fixation. J Gen Microbiol. 1988，134：2721-2734

[9]Sambrook J，Frit sch E F，Maniatis T. Molecular cloning：A laboratory Manual. 2nd ed. Cold Spring Harbor，NY：Cold Spring Harbor Laboratory，1989，474-488

[10]Hynes M F，Quandt J，Puhler A. A direct selection for the curing and deletion of *Rhizobium* plasmids using transposons carrying the *Bacillus subtilis* sacB gene. Gene 1989，78：111-120

[11]Woese C R. Bacterial Evolution. Microbiol Rev，1987，51（2）：221-271

[12]Chen W L，Zou J G，Hung Q Y et al. Diversity of Symplasmid in *Rhizobium fredii* strains and plasmid stability under free-living and symbiotic conditions. In：Li F D，Lie T A，Chen W X，et al. eds：Diversity and taxonomy of Rhizobia. Beijing：China Agricultural Scientech Press 1996. 16-125

紫云英根瘤菌分子遗传学研究进展*

张学贤　李阜棣　曹燕珍　陈华癸

（华中农业大学教育部农业微生物重点实验室）

摘　要　总结了紫云英根瘤菌分子遗传学研究进展。这类根瘤菌的寄主范围较窄，早先根据互接种族概念将其定名为 *Rhizobium astragali*，后来定名为 *Rhizobium huakuii*，最近并入 *Mesorhizobium* 属。对大量菌株进行的 16S 和 23S rDNA PCR-RFLP 分析和代表菌株的部分 16S rDNA 碱基测序，揭示了 7 个不同基因型，表明这类菌具有遗传多样性，可以认为存在不同的种。它们都有质粒，数量 1～5 个，因菌株而异，可分为不同质粒型，共生基因常定位在最大的一个质粒上，即共生质粒。它们的结瘤基因具有独特结构。采用包括 *nodA*、*nodB*、*nodC*、*nodD* 的苜蓿根瘤菌突变株进行基因互补，从紫云英根瘤菌基因文库中获得了相应的转移接合子。这类根瘤菌的 *nodA* 与 *nodBC* 分隔约 22kb 距离，不同于多数根瘤菌 *nodABC* 为一个操纵元的结构。紫云英根瘤菌结瘤基因的这种结构在 7 个基因型的菌株中都是一致的，说明结瘤基因的保守性。

关键词　紫云英；根瘤菌；中慢生根瘤菌；分子遗传学

紫云英（*Astragalus sinicus* L.）是能形成共生固氮体系的豆科植物，在长江中下游和华南地区广泛种植，主要用作水稻绿肥，也是很好的蜜源植物和饲料。日本、越南、朝鲜半岛等国家和地区也有生长。陈华癸和徐明光最先在我国分离获得紫云英根瘤菌有效菌株[1]。在初次栽种紫云英的土壤中，接种有效根瘤菌是种植成败的重要因素，多年种植紫云英的地区接种优良菌株也能促进地上部分增产。在许多情况下，接种并配合施用磷钾肥可提高青草产量和植株含氮量。半个世纪以来华中农业大学生物固氮领域的研究者们对紫云英根瘤菌进行了多方面深入研究，包括大面积接种应用、生物学、生态学、生理学与遗传学等[2]。遗传学方面的研究从物理和化学诱变开始，继而采用转座子诱变逐渐深入[3~5]，由于新技术的应用，研究工作在 90 年代取得重大进展。下面概述分子遗传学方面的主要研究成果，并结合国内外有关文献进行讨论。

1　紫云英根瘤菌的分类地位和遗传多样性

紫云英根瘤菌的感染性比较专一，寄主范围较窄。早先根据互接种族概念，将其定名为 *Rhizobium astragali*[6]。Chan 等研究了从中国南京和日本 Otaki 等地采集的 11 个菌株，将它归于 *Bradyrhizobium* 属[7]。陈文新等对来自湖北和江苏的 9 个菌株进行了数字分类和 DNA-DNA 杂交等研究，将其定名为 *Rhizobium huakuii*，在国际系统细菌学杂志发表[8]。后来一些学者根据根瘤菌系统发育研究，将这个种并入新建立的 *Mesorhizobium* 属[9]，种名是华癸中慢根瘤菌（*Mesorhizobium huakuii*），这一学名为研究者普遍采用。

上述 Chan 等和陈文新等研究结果的差别，可能主要是采用的菌株和方法不同，且数量不多，也与当时对根瘤菌分类的认识有关。我们近年开展了紫云英根瘤菌遗传多样性的研究，从中国南方 7 省 52 个不同地理和生态区域的紫云英根瘤中收集了 2000 多个分离体，筛选归并后选出 204 个菌株，运用 REP-PCR、16S 和 23S rDNA PCR-RFLP、碱基测序等方法研究了它们的遗传多样性。大多数菌株（78.4%）的 16S rDNA 基因型都与 *Mesorhizobium huakuii* 模式菌株 CCBAU2609 不同。对随机挑选的 136 菌株进行 REP-PCR 分析，它们在 60%相似性水平上可分为 27 个 REP 群。

---

* 原载于《华中农业大学学报》，22（1）：77～83，2003.

选取 12 个代表菌株和 16 个参比菌株用 9 种限制性酶进行 16S rDNA 和 23S rDNA 分析，全部 28 个供试菌株分为 16 种 16S rDNA 基因型。12 个代表菌株中只有 3 个菌株同百脉根根瘤菌模式菌株的酶切图谱类似，而与其他菌株均无相似性。这 12 个菌株包括 4 个 16S 和 5 个 23S rDNA PCR-RFLP 基因型[10]，当将 16S rDNA 和 23S rDNA PCR-RFLP 基因型结合分析时，可以将它们分为 7 个不同基因型（表 1）。对 7 个不同的 16S 和 23S rDNA 类型的代表菌株，还进一步作了 *recA* 基因和 2 个谷氨酰胺合成酶基因 *glnA* 与 *glnB* 的系统发育研究。

表 1 测试菌株的基因型

Table1 Genotypes of tested strains

菌株 Strain	寄主植物 Host plant	采集地 Site	16S rDNA^{D}	23S rDNA	16S&23S rDNA
Mesorhizobium sp.					
ZJ5B8	紫云英 *Astragalus sinicus* L.	浙江 Zhejiang	3（a）	1	4
GX8B23	紫云英 *Astragalus sinicus* L.	广西 Guangxi	3（a）	1	4
HB7B4	紫云英 *Astragalus sinicus* L.	湖北 Hubei	3（a）	1	4
JS6A16	紫云英 *Astragalus sinicus* L.	江苏 Jiangsu	3（a）	1	4
JS6A15	紫云英 *Astragalus sinicus* L.	江苏 Jiangsu	3（a）	3	5
JS4B2	紫云英 *Astragalus sinicus* L.	江苏 Jiangsu	3（a）	1	4
ZJ9A16	紫云英 *Astragalus sinicus* L.	浙江 Zhejiang	3（a）	3	5
JX2B5	紫云英 *Astragalus sinicus* L.	江西 Jiangxi	3（a）	5	6
HN14A16	紫云英 *Astragalus sinicus* L.	湖南 Hunan	8（c）	10	13
HB5A4	紫云英 *Astragalus sinicus* L.	湖北 Hubei	7（c）	10	11
HN15B23	紫云英 *Astragalus sinicus* L.	湖南 Hunan	7（c）	11	12
HN8A1	紫云英 *Astragalus sinicus* L.	湖南 Hunan	4（b）	1	2
M.huakuii CCBAU2609^{T}	紫云英 *Astragalus sinicus* L.	江苏 Jiangsu	4（b）	2	1
M.loti NZP2037	百脉根 *Lotus corniculatus* L.	新西兰 New Zealand	3	1	4
M.loti NZP2234	百脉根 *Lotus corniculatus* L.	美国 USA	4	4	3

续表

菌株 Strain	寄主植物 Host plant	采集地 Site	16S rDNA[D]	23S rDNA	16S&23S rDNA
M.loti NZP2213	百脉根 *Lotus corniculatus* L.	新西兰 New Zealand	9	7	15
M.plurifarium ORS1040	金合欢 *Albizzia julibrissin* L.	塞内加尔 Senegal	1	12	8
M.plurifarium ORS1001	金合欢 *Albizzia julibrissin* L.	塞内加尔 Senegal	2	13	9
M.ciceri UPM-Ca7[T]	鹰咀豆 *Cicer arvensis* L.	西班牙 Spain	10	9	14
M. sp. CIAM3502	芒柄花 *Omonis arvensis* L.	俄罗斯 USSR	3	1	4
M. sp. CIAM3402	小冠花 *Coronillia burifolia* Hance	俄罗斯 USSR	5	6	7
M.mediterranueum-UPM-Ca36[T]	鹰咀豆 *Cicer arvensis* L.	西班牙 Spain	6	8	10

1）数字是 9 种酶获得的 16S rDNA 基因型，括弧中字母为 4 种酶获得的 16S rDNA 基因型，上表中未包括测试的 3 个 *Rhizobium* 和 3 个 *Sinorhizobium* 参比菌株

Letters refer to the 16S rDNA genotype obtained by RFLP analysis with nine enzymes，and numbers in brackets refer to the 16S rDNA genotype obtained with four enzymes.

选用不同 16S rDNA 基因型的 4 个菌株进行部分 16S rDNA 碱基测序，验证了 REP-PCR 和 16S rDNA-RFLP 的分析结果[10,11]。能够在紫云英上结瘤的根瘤菌可以分为 2 个类群，如图 1 所示。96.6% 的菌株归入到优势群中，包括已经鉴定的华癸中慢根瘤菌，表明华癸根瘤菌是这一类群的真实代表。只有 7 株菌属于另一类群，它们与华癸中慢根瘤菌的遗传距离很远，很显然属于 *Mesorhizobium* 属中的一个新种（待命名）。中文名“紫云英根瘤菌”是指从该寄主植物获得的菌株，它不等同于“华癸根瘤菌”。

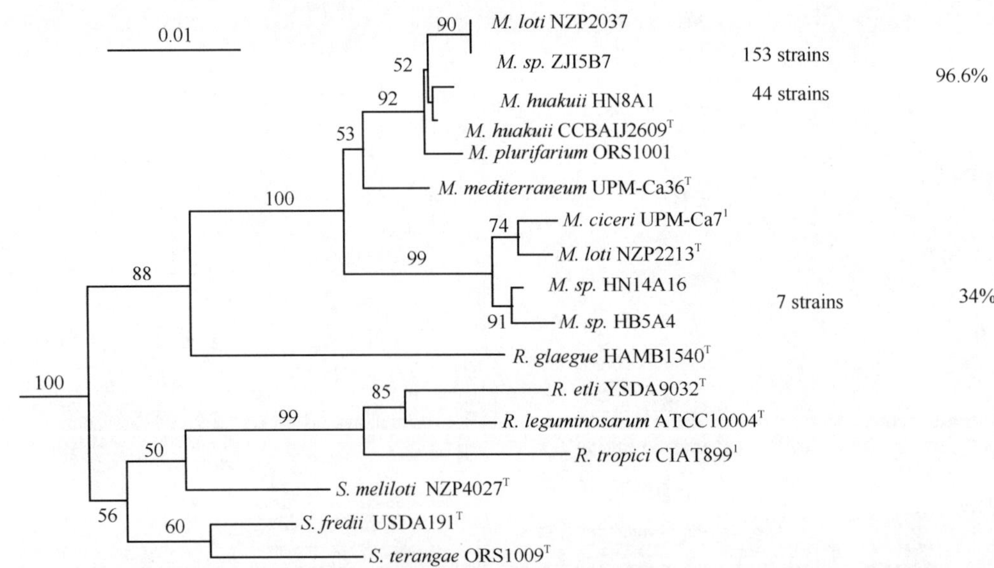

图 1 根据部分 16S rDNA 序列（600 bp）采用 neighbor-joining 法构建的系统发育树

Fig.1 Phylogenic tree constructed frompartial 16S rDNA sequences（600 bp）using the neighbor-joining method

对华癸根瘤菌和百脉根根瘤菌的比较研究表明，部分百脉根根瘤菌，包括全基因组序列已经测定了

的 MAFF303099 菌株，应属于 *M. huakuii*，而不是 *M.loti*。因此我们最近将这一百脉根根瘤菌作为一个生物型归入到华癸根瘤菌中，学名为 *Mesorhizobium huakuii* bv. loti[12]。这样一来，华癸根瘤菌就包括了能在紫云英和百脉根上结瘤的 2 类根瘤菌。

2　紫云英根瘤菌的质粒

紫云英根瘤菌的多样性也表现在质粒特征上，不仅在不同地域来源的菌株中显示出来，即使局部来源的菌株也显示出质粒多样性。

2.1　紫云英根瘤菌质粒的多样性

不同菌株的质粒数目和分子量大小可以在凝胶电泳上显示为不同图谱，据此可将根瘤菌分为不同质粒型。对前述不同地域来源的 100 个菌株进行了质粒检测，获得 16 种不同质粒图谱（图 2A）[13]，而不同地域的菌株也可以有相同的质粒图谱。我们曾在一块多年种植紫云英的稻田中采集植株，收集植物的全部根瘤来分离根瘤菌，纯化和回接试验后得到 154 个分离株。其中 100 株来自 10 株植物的根瘤（A 组），40 株来自同一植株的不同根瘤（B 组），14 株来自同一植株上的两个根瘤（C 组）。所有菌株都含有质粒，1～5 个，因菌株而异，分子质量为 35～600MD，或更大。根据凝胶电泳图谱，可将菌株分为不同的质粒类型，其中 A 组 8 个型（图 2B），B 组 6 个型，同一根瘤中来的菌株质粒数也非一致。从 A 组取 24 株菌（8 个质粒型各选 3 株）测定生长代时，小于 4h 的有 4 株，大于 6h 的有 7 株，其他 13 株在 4～6h 之间[14~16]。

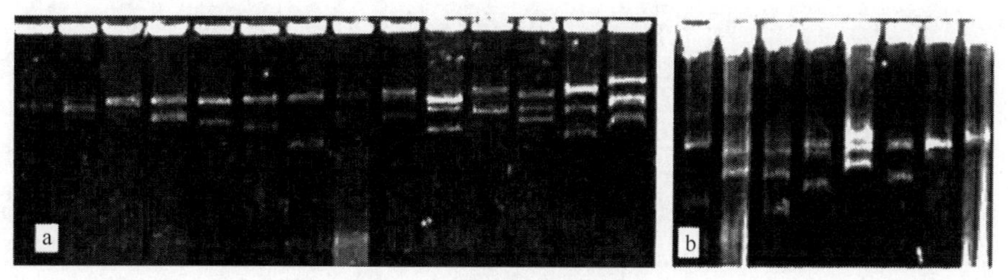

图 2　部分菌株的质粒图谱

Fig. 2　Plasmid profiles of part strains

a. 不同地域来的菌株 Strains from different geographic areas；b. 同一田块来的菌株 Strains from the same field

在这项研究之前我们曾测定从湖北、江苏、福建分离的 22 个菌株，它们也具有不同质粒特征[17]。

2.2　质粒功能

用高温和吖啶橙分别处理具有 2 条大质粒的紫云英根瘤菌 7653R 和 S52 菌株，获得了表型为不结瘤（Nod$^-$）和结瘤而不固氮（Nod$^+$，Nif$^-$）的突变株。同亲本株相比，突变株均丢失了一个较大质粒，可以推断菌株 7653R 的结瘤基因和菌株 S52 的固氮基因存在于被消除的大质粒上[18]。用含苜蓿根瘤菌 *nod*ABC 基因的 pSmSL52 对 7653R 菌株和 7653R-1 菌株（7653R 的 Nod$^-$突变株）进行分子杂交，显示结瘤基因存在于被消除的大质粒上[19]。这是紫云英根瘤菌质粒携带结瘤基因的直接证据。

用紫云英根瘤菌 CH203 菌株进行了质粒功能的进一步研究。该菌株含有 3 条大质粒，它们分别是 pRHa，97MD；pRHb，168MD；pRHc，251MD。经转管培养 100 次并与寄主结瘤后再分离获得培养体，检测其中 100 个单菌落，质粒组成都没有变化。采用 Hynes[20]的方法和他赠送的携带 Tn5-sacB 的质粒进行质粒消除。*sac*B 是蔗糖转葡聚糖基因，含此基因的菌株在提高蔗糖浓度（5%～7%）的培养基上不能生长，易于检测转移接合子，并利于高温（39℃）培养消除质粒。通过反复诱变和反复消除，获得了含 1 条和 2 条质粒的 5 种突变株（图 3），但未能消除这一菌株的全部质粒，也未获得仅含最小 1 条质粒（pRHa）的突变株[15]。

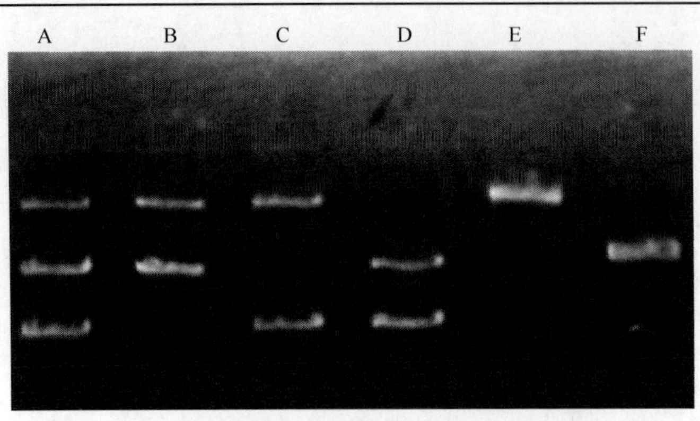

图 3 菌株 CH203 及其突变株的质粒图谱

Fig. 3 Plasmid profiles of strain CH203 and its mutants

A. CH203；B. CH203-1；C. CH203-2；D. CH203-3；E. CH203-4；F. CH203-5

用 *nod*DABC 作探针进行分子杂交，最大质粒 pRHc 有明显杂交信号，而且这一质粒缺失即丧失结瘤能力，可以肯定该质粒为共生质粒。缺失其他质粒的突变株在表型上有明显变化。如缺失了质粒 pRHb 的菌株 CH203-2 和 CH203-4 在 TY 平板上菌落粗糙，并在 PA 液中出现絮凝状。表面多糖分析揭示，粗糙型菌落缺少脂多糖 I。突变株的抗酸能力均有所下降。将缺失某质粒的菌株重新导入该质粒，则可以恢复原来的表型[16]。

研究了外源质粒对紫云英根瘤菌的影响。用紫云英根瘤菌 7653R 及其共生质粒缺失突变株 7853R-1 作受体，导入豌豆根瘤菌质粒 pJB5JI（插入了 Tn5 的共生质粒）。突变株 7653R-1（pJB5JI）获得了在豌豆上形成无效根瘤的能力，而突变株 7653R（pJB5JI）则不能。该结果表明，菌株的内源共生质粒对 pJB5JI 的功能有限制作用。随着 7653R 菌株结瘤因子（Nod factor）的化学结构测定之后，可以很好地解释上述现象。7653R 菌株具有 *nodHPQ* 基因，产生的结瘤因子含 1 个磺酸化的修饰基团[21]。因此，转移接合子 7653R 中 pJB5JI 所产生的豌豆根瘤菌结瘤因子就会发生磺酸化修饰，从而不能被豌豆识别。突变株 7653R（pJB5JI）在原寄主紫云英上形成的根瘤较野生型菌株表现较高固氮酶活性，而且植株干重也较高。这些结果表明，2 套不同来源的共生基因之间存在着较为复杂的相互作用[22,23]。

3 紫云英根瘤菌的结瘤基因

对紫云英根瘤菌 7653R 菌株的结瘤基因进行了遗传分析。根据这一菌株的结瘤因子的化学结构类似于苜蓿根瘤菌的结瘤因子，即它们都在还原端 C-6 的 O 位硫酸化，非还原端的 N 位被不饱和脂酰基链酰化[21]，于是设想它们的结瘤基因可能有较高 DNA 同源性，因而能采用苜蓿根瘤菌的各种 Nod^-突变株进行异源功能互补，可以从 7653R 菌株的基因文库中找到其结瘤基因。使用分别包括 *nodA*，*nodB*，*nodC*，*nodD* 等在内的 7 种突变株逐一杂交，每对杂交的转移接合子都用 200 株苜蓿苗来检测，除 *nodFL* 外（18%），有效结瘤的比例均高达 80%～97%。从根瘤中分离携带能够互补苜蓿根瘤菌突变株基因的培养体（也就是含特定基因的 cosmids），进行限制性分析。根据结果构建了菌株 7653R 结瘤区的物理和遗传图（图 4）[24]。

图中质粒 pHN50 是从苜蓿根瘤菌 *nodA*，*nodB*，*nodC* 和 *nodD1*，*D2*，*D3* 突变株的 Nod^+转移接合子分离获得的，说明 pHN50 含有这些基因。用 *nodD*、*nodA*、*nodC* 探针进行 Southern 杂交，得知 *nodC* 和 *nodD* 基因定位在 4.2kb *Eco*RI 片段上，*nodA* 是在 20kb *Eco*RI 片段和 10kb *Hind*III 片段上。于是可以认为 *nodA* 基因与 *nodBC* 基因是分离的。这种结构不同于大多数根瘤菌结瘤基因的排列（图 5）。对此我们还曾有一个间接的证据，用包括 *nodABC* 基因的质粒作探针从紫云英根瘤菌 7653R 菌株基因文库中选出同源的 25kb DNA 片段，在初步试验中它似乎有使 Nod^-菌株恢复结瘤的能力[25]。将携带这一片段的质粒称之为 pRaZ15，尽管经过保存后很稳定，用 *nodABC* 杂交仍为阳性，可是后来反复试验都未显示恢复突变株结瘤的能力[26]。因此，很可能这一质粒没有包括全部 *nodABC* 基因，因为它们是分隔的。

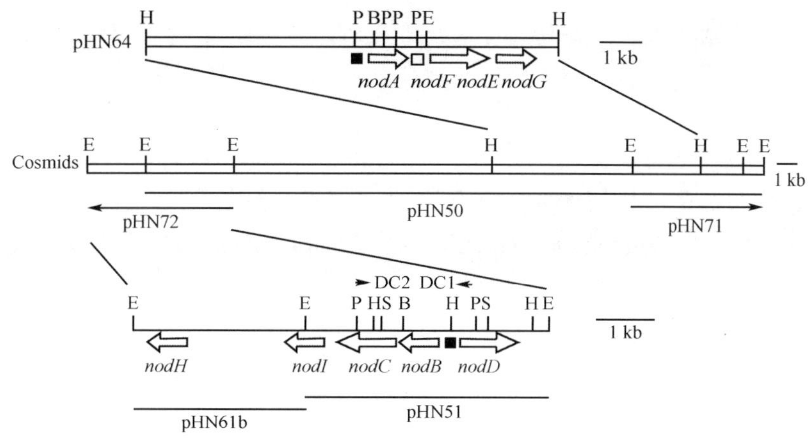

图 4　菌株 7653R 结瘤基因区的物理和遗传图

Fig. 4　Physical and genetic map of the nodulation gene region of *Mesorhizobium sp.*strain 7653R

质粒 pHN71 是从苜蓿根瘤菌 *nodF*、*nodL* 突变株的转移接合子分离的，它与携带 *nodA* 的 20kb*Eco*RI 片段共有 5.0 和 0.8kb *Eco*RI 长度。用 *nodE* 探针作 Southern 杂交，显示 5.0kb 片段含有 *nodE* 同源物。质粒 pHN72 是从 *nodH* 突变株的转移接合子分离的，它与携带 *nodDBC* 同源物的 pHN50 重叠 4.2kb *Eco*RI 片段。对和 pHN50 部分重叠的 pHN72 的 2.8kb *Eco*RI 片段进行部分测序，得知它含有 *nodH* 和部分 *nodI* 同源物。我们对紫云英根瘤菌 7653R 菌株的 *nodDBC* 和 *nodA* 基因的碱基进行了测序[24]，此处不作介绍。

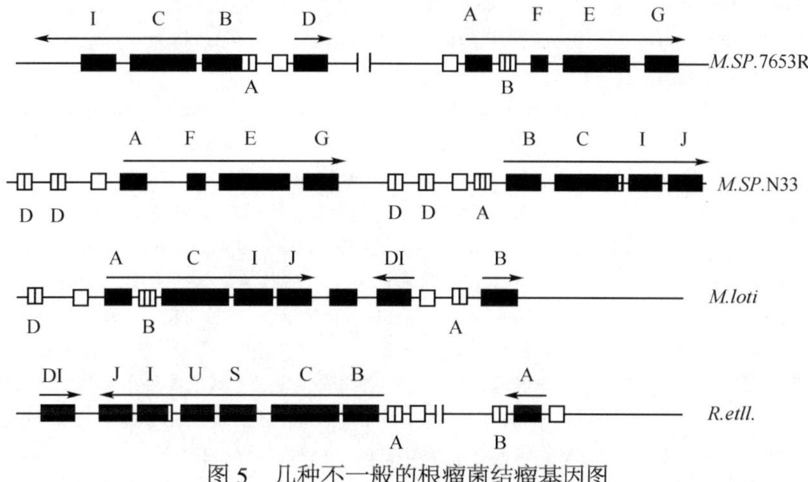

图 5　几种不一般的根瘤菌结瘤基因图

Fig. 5　Unusual nodulation gene maps of some rhizobial strains

我们揭示的紫云英根瘤菌结瘤基因的结构是较独特的。在研究过的多数根瘤菌中，*nodABC* 构成一个单独的操纵元，在其上游有一个转录方向与之相反的 *nodD* 拷贝[27]。已知只有 *Rhizobium etli*、*Mesorhizobium loti* 和 *Mesorhizobium* sp.N33 菌株 3 个例外。*Mesorhizobium loti* 的 *nodB* 与 *nodAC* 是分隔开的。在 *Rhizobium etli* 和 *Mesorhizobium* sp.N33 菌株中 *nodA* 同 *nodBC* 分隔开（图 5）。但是这些菌株 *nodA*、*B*、*C* 的转录方向都是相同的，而在 *Mesorhizobium* sp.（紫云英根瘤菌）7653R 菌株中，*nodA* 不但与 *nodBC* 基因分隔约 22kb，且 2 个操纵元的转录方向相反。

4　紫云英根瘤菌结瘤基因的保守性

紫云英根瘤菌结瘤基因 *nodA* 与 *nodBC* 相分隔的这种结构系 7653R 菌株所特有，还是有普遍性，抑或也存在于其他菌株中？我们设计了特定引物（DC1 和 DC2）来扩增 *nodDBC* 区段，它是 pHN51 的 *Pst*I 切割段（图 4）。从上述遗传多样性研究揭示的 7 种基因型菌株中选择 24 个代表，用引物进行 PCR 扩增的结果都是 1 条单一的 2.0kb 带，与从 7653R 菌株扩增的带相同，证明 *nodA* 与 *nodBC* 分隔是紫云英根瘤菌的一般特征。还对这 24 个菌株用 *nodA* 的扩增研究提出了旁证。

对 7 个不同基因型的紫云英根瘤菌进行了 *nodA* 的 PCR 扩增，并对扩增产物的 2 条链都测序。这 7 个基因型菌株 *nodA* 的碱基序列与 7653R 菌株 *nodA* 的序列相同，说明它们是保守的，尽管菌株的来源非常不同[24]。

不同染色体遗传背景的紫云英根瘤菌具有相同的结瘤基因结构和同样的 *nodA* 碱基序列，证明共生基因的侧向水平转移是根瘤菌多样性形成的重要原因。根据对结瘤基因的比较研究，我们认为这些基因的转移也可能发生在其他 *Mesorhizobiums* pp.之间。共生基因于自然条件下的横向转移已在百脉根根瘤菌中得到了充分证明[28,29]。

本文涉及的数据和讨论详见参考文献。紫云英根瘤菌是很好的研究材料，它在我国广泛栽种，植株小，易于操作。在已有的遗传研究基础上，可以进行许多有意义的研究，包括基因组和后基因组、同寄主相互作用的分子机理、分类研究和分子生态学等。

华中农业大学参与紫云英根瘤菌研究的前后有几十人，包括许多研究生。在遗传学方面除本文作者外，邹向宏、郭宪武、农广、周俊初、张忠明和胡福荣等是各个具体课题的主要工作者和参加者。

参 考 文 献

[1]Chen H K，Shu M K. Notes on the root nodule bacteria of *Astragalus sinicus* L. Soil Sci，1944，58：291-293.

[2]Chen H K，Li F D，Cao Y Z. Characteristic distribution ecology and utilization of *Astragalus sinicus-rhizobia* symbiosis. In：Hong K F ed. Nitrogen fixation research in China. New York：Springer-Verlag，1992. 39-455.

[3]曹燕珍，李阜棣. 紫云英根瘤菌的诱变. 微生物学通报，1979，6：7-9.

[4]李仲贤，张春生，李阜棣等. 紫云英根瘤菌的不结瘤（Nod⁻）突变. 华中农学院学报，1983，2：1-9.

[5]李阜棣，曹燕珍，黎若娅等. 易位子 Tn5 引入紫云英根瘤菌的研究.华中农学院学报，1984，3：52-54.

[6]陈华癸. 微生物学. 北京：高等教育出版社，1959.

[7]Chan C L，Lumpkin T A，Root C S. Characterization of *Bradyrhizobium* sp.（*Astragalus sinicus* L.） using serological agglutination，intrinsic antibiotic resistance，plasmid visualization and field performance. Plant and Soil，1988，109：58-59.

[8]Chen W X，Li G S，Qi Y L，et al. *Rhizobium huakuii* sp.nov. isolated from the root nodules of *Astragalus sinicu*. Int J Syst Bacteriol，1991，41：275-280.

[9]Jarvis B DW，van Berkum P，Chen W X，et al. Trnsfer of *Rhizobium loti*，*Rhizobium huakuii*，*Rhizobium ciceri*，*Rhizobium mediterraneum*，and *Rhizobium tianshannense* to *Mesorhizobium* gen. nov. Int J Syst Bacteriol，1997，47：895-898.

[10]Zhang X X，Guo X W，Terefework Z，et al. Genetic diversity among rhizobial isolates from field-grown *Astragalus sinicus* of southern China. Syst Appl Microbiol，1999，22：312-320.

[11]郭宪武. 华癸根瘤菌染色体基因和质粒基因群体遗传学比较研究：[博士学位论文] .武汉：华中农业大学，1998.

[12]Turner Sarah L，Zhang Xuexian，Li Fudi，et al. What does a bacterial genome sequence represent：reassignment of *Mesorhizobium loti* MAFF303099 to *Mesorhizobium huakuii* biovar loti. Microbiology，2002，148：3330-3331.

[13]Guo X W，Zhang X X，Zhang Z M，et al. Characterization of *Astragalus sinicus* rhizobia by restriction fragment length polymorphism analysis of chromosomal and nodulation gene regions. Curr Microbiol，1999，39：358-364.

[14]邹向宏，曹燕珍，李阜棣. 紫云英根瘤菌天然抗药性和质粒研究. 华中农业大学学报，1994，13（4）：325-331.

[15]邹向宏. 紫云英根瘤菌质粒及共生效应研究：[博士学位论文].武汉：华中农业大学，1995.

[16]Zou X，Li F，Chen H. Characteristics of plasmids in *Rhizobium huakuii*. CurrMicrobiol，1997，35：215-220.

[17]王常霖，陈金标. 紫云英根瘤菌的质粒特征. 华中农业大学学报，1988，7（1）：15-21.

[18]周俊初，张忠明，黄诚金等. 紫云英根瘤菌质粒的研究：II.紫云英根瘤菌经高温和吖啶橙处理后结瘤和固氮突变株的产生和质粒的消除. 华中农业大学学报，1987，6（2）：156-164.

[19]魏辉，李阜棣. 紫云英根瘤菌结瘤基因的定位研究. 微生物学报，1990，30：330-335.

[20]Hynes M F，Mcgragor N F. Two plasmids other than the nodulation plasmid are necessary for formation of nitrogen fixing nodules by *Rhizobium leguminosarum*. Mol Microbiol，1990，4：407-417.

[21]Yang G P，Debelle P，Savagnac A，et al. Structure of the *Mesorhizobium huakuii* and *Rhizobium galegae* Nod factors：a cluster of phylogenetically related legumes are nodulated by rhizobia producing Nod factors withα，β-unsaturated N-acyl substitutions. Mol Microbiol，1999，34：227-237.

[22]张学贤. 异源根瘤菌共生质粒相互作用的研究：[博士学位论文]. 武汉：华中农业大学，1993.

[23]Zhang X X，Zhou J C，Zhang Z M，et al. Behavior of plasmid pJB5JI in *Rhizobium huakuii* under free-living and symbiotic conditions. Curr Microbiol，1995，31：97-101.

[24]Zhang X X，Turner S，Guo X W，et al. The common nodulation genes of *Astragalus sinicus* rhizobia are conserved despite chromosomal

diversity. Appl Environ Microbiol，2000，66：2988-2995.

[25]张忠明，陈华癸，李阜棣等. 紫云英根瘤菌基因文库的构建及含完整结瘤基因的重组质粒 pRaZ15 的分离. 生物工程学报，1991，7：213-219.

[26]农广. 紫云英根瘤菌的共生质粒、共生基因克隆和早期共生关系的研究：[博士学位论文]. 武汉：华中农业大学，1994.

[27]Schlaman H R M，Philips D A，Knodorosi E. Genetic organization and transcriptional regulation of rhizobial nodulation genes. In：Spaink H P，Kondorosi A，Hooykaas P J J，eds. The Rhizobiaceae. The Nedlands：Kluwer Academic Publishers，1998. 361-386.

[28]Finan T M. Evolving insights：Symbiosis islands and horizontal gene transfer. J Bacteriol，2002，184：2855-2856.

[29]Sullivan J T，Trzebiatowski J C，Cruickshank R W，et al. Comparative sequence analysis of the symbiosis island of *Mesorhizobium loti* strainR7A. J Bacteriol，2002，184：3086-3095.

土壤中大豆根瘤菌之间竞争结瘤的研究 I、免疫荧光抗体技术在根瘤菌个体生态学研究中的应用*

王福生　陈华癸　李阜棣

（华中农学院生物固氮研究室）

本文详细介绍了免疫荧光技术，并报道了这一技应用于根瘤菌的个体生态学的研究结果。

前言

本世纪以来，根瘤菌剂的生产应用和研究结果基本上可概括为：①在无土著根瘤菌的新区，接种根瘤菌通常能使作物产量显著地高于未接种根瘤菌对照的产量，从而给人们带来经济效益；②在含有大量土著根瘤菌的老区田块，接种往往不能使作物增加产量。常规的接菌量无法使接种菌成为寄主豆科的主要占瘤者，与土著根瘤菌相比，处于竞争结瘤劣势地位[5,22]。面对这一现实问题，长期以来，研究者们试图考察自然土壤中，特定根瘤菌的数量变化与其竞争结瘤能力的关系，但又苦于缺乏相应的研究技术来区分供试菌株和土著根瘤菌，使这一有意义的研究工作——根瘤菌个体生态学的研究无法开展。尽管后来人们采用了试管凝集、凝胶扩散、抗性标记、菌株的固有抗药性、噬菌体分型、酶标（ELISA）以及植物感染测数（MPN）等方法，但仍不能满足研究者的要求。因为这些方法都不能追踪特定根瘤菌在自然土壤中的数量变化。最近，Bushby（1981）采用的选择性培养基直接测定土壤中特定根瘤菌数的方法太特殊以致无法广泛应用到实际研究工作中去（Brockwell，1982）[1]。

免疫荧光抗体（FA）方法具有不同于上述方法的独特优点，它除了能快速检测根瘤中根瘤菌的血清型（Trinick，1969）[20]外，更重要的是 FA 方法能广泛地用于测定自然土壤中特定根瘤菌的数量，研究根瘤菌的个体生态学（Schmidt，1974）[16]。这一方法的借鉴、革新和完善使之能够成功地应用于土壤微生物学研究领域，来自于 E.L.Schmidt 及他的同事们大量的探索性研究。1962 年，Schmidt 和 Bankole 首次将 FA 方法用于土壤微生物生态学的研究，他们用埋玻片法观察土壤中的黄曲霉（*Aspergillus flavus*）[17]。1968 年，Schmidt 等又将此方法应用子土壤中根瘤菌的研究，但发现由于土壤颗粒引起的非特异性荧光严重干扰荧光显微镜镜检[18]，效果并非理想。同年，Schnlidt 的学生 Bohlool 和他本人采用罗达明胶（RhITC） 染色技术，成功地排除了非特异性荧光的干扰[7]。通过大量的实践[8]，Schmidt（1974） [16]成功地摸索出了用 FA 测定自然土壤中根瘤菌数的方法，并通过与平板活菌测数对比，阐明了 FA 方法测定根瘤菌数量的可靠性。后来，有几位学者又继续作了 FA 测数方法的改进[12，13，24]，使这一方法更进一步完善。

免疫荧光抗体方法在生物固氮研究领域的成功应用，使研究者们长期以来所试图开展的根瘤菌个体生态学研究得以付诸实践。该方法的价值和重要性已愈来愈被广大的土壤微生物学者所认识。本文将以介绍荧光抗体技术为初旨，展示 FA 方法研究根瘤菌个体生态学的详细步骤，力图使对此方法感兴趣的学者能有一较全面的了解。

---

*原载于《华中农学院学报》，4（3）：38～47，1985.

1免疫荧光抗体技术用于根瘤菌生态学的研究是美国夏威夷大学斯蒂芬·都德(S.Dowdle)作为交流学者在我院学习和研究期间传授的。特此致谢。

材料与方法

一、试 验 材 料

试验菌株为大豆根瘤菌 PRC 005，由中国农业科学院土壤肥料研究所提供。用酵母汁甘露醇洋菜培养基（Bohlool 和 Schmidt，1970[8]培养。供试大豆栽培品种“矮脚早”由中国农业科学院油料作物研究所提供，供田间试验和田间盆栽试验之用；野大豆种子系 1983 年夏季采自本院试验农场，供植物感染测数（MPN）用。

二、试 验 方 法

1. 免疫荧光抗体方法

此方法包括：①抗原的准备；②抗血清的制备；③抗体提纯；④抗体标记；⑤标记抗体的提纯；⑥荧光抗体效价测定；⑦特异性荧光抗体测定自然土壤中的根瘤菌数量和根瘤中根瘤菌血清型的鉴定。

（1）抗原的制备[18]：大豆根瘤菌 PRC005 接种到酵母粉甘露醇培养液中，28℃振荡培养 3～4 天，10000*g* 离心 10 分钟，去上清液，用过滤了的 0.85%的生理盐水洗涤沉淀细胞，100℃煮 1 小时后离心使细胞沉淀，用生理盐水洗涤沉淀并再次离心，重复离心三次后用生理盐水悬浮沉淀细胞。用 721 型分光光度计调整菌悬液浓度，使菌液在波长等于 600 毫微米下的消光值为 0.45。加防腐剂，使防腐剂硫柳汞的终浓度为 1∶10000，4℃下保藏备用。

（2）兔子免疫——制备抗体：选用幼龄、健壮的 3 斤以上白兔进行免疫，注射抗原前从边缘耳静脉采 2～3ml 血作阴性血清对照。试验采用的免疫方案如下[18]：

天	注射途径	抗原（Ag）形式
1	静脉（IV）	0.5ml Ag
	皮下（SC）	1.0ml Ag（含 0.5ml 佐剂）
2	IV	1.0ml Ag
3	IV	1.0ml Ag
4-6	休息	——
7	IV	1.5ml Ag
8	IV	2.0ml Ag
9	IV	2.0ml Ag

兔子休息一周后，从边缘耳静脉采 2～3 ml 血，用试管凝集反应测定抗血清的效价[2]，效价低于 1280 的兔子加强注射一次，效价达 1280 以上的白兔禁食 24 小时，然后在颈动脉采血。除去血液中的红细胞，制成抗血清，将注射同一抗原的四个兔子血清混合后分装小瓶，加防腐剂，−20℃保藏备用。

（3）抗体的提纯[13,20]：取 15ml 抗血清装小烧杯后，缓慢加入等体积的 3.9mol/L（NH_4）$_2SO_4$ 同时用磁力搅拌器搅拌使两者均混合。4℃下静置 1 小时后 10000*g* 离心 30 分钟去上清液，用大约 13ml 的无菌水溶解沉淀，再加入等体积的 3.9mol/L（NH_4）$_2SO_4$ 不需静置直接离心，重复上述步骤离心三次后倒去上清液，加少量无菌水溶解球蛋白沉淀，用无菌吸管将抗体球蛋白转入透析袋对 0.85%的生理盐水透析，除去 NH_4^+ 和 $SO^{2-}{}_4$离子。透析过程中应每隔 4～6 小时换透析液生理盐水一次，透析 12～15 小时后，用 1% $BaCl_2$ 与透析液反应，如没有白色沉淀说明透析完全彻底。否则，需进一步透析。透析完全倒出透析袋内提纯的抗体球蛋白，供荧光染料标记之用。

（4）异硫氰酸荧光黄（FITC）标记抗体：用紫外分光光度计测定纯化抗体球蛋白浓度，用 0.85%

的生理盐水配制 10 毫升 1%的蛋白溶液。按每毫克蛋白需 0.05mg FITC 的比例，称好所需的 FITC，并将其倒入装有一定量 pH8.0、0.1mol/L 磷酸钠缓冲液的小瓶中，使之充分溶解。分别将 4ml pH9.0、0.1mol/L 磷酸钠缓冲液和 FITC 溶液先后加入 10 毫升 1%的抗体球蛋白液中，混合均匀，调 pH 8～9。盖好盖子，将溶液放在电磁搅拌器上搅拌，不要使溶液出泡，搅拌 6 小时标记过程结束。

（5）荧光抗体的纯化[3,4]：将标记的抗体通过葡聚糖凝胶 G-25 柱，除去其中游离荧光色素，从而得到精制的荧光抗体，其基本步骤如下：

按需要量称取葡聚糖凝胶 G-25，加到 1000 毫升蒸馏水中浸泡，使其充分膨胀，膨化完毕的凝胶在使用前一天用 pH7.2、0.01mol/L 磷酸盐缓冲液（PBS）平衡一夜。装柱时要求凝胶浓度在柱内上下一致，避免一段松一段紧而影响分离效果。待凝胶沉好，打开柱下端口用缓冲盐水继续平衡，等凝胶面洗脱液尚余 0.5cm 高度时，马上关闭下口停止流动，准备层析。

加样时吸管吸粗制的荧光抗体，沿管壁四周缓缓滴下，加样完了打开下口，并控制一定流速。样品进柱后，很快出现两种颜色不同的移动带，快速移动带是标记蛋白，慢速移动带是游离的荧光色素。两带之间是肉眼看不见的磷酸盐缓冲液带。因此，据颜色收集快速移动带部分，即是所求的荧光抗体液。将荧光抗体液放冰箱静置过夜，离心和过滤 FA 以达进一步纯化 FA 之目的。

（6）荧光抗体特性测定：用直接法检查所制备荧光抗体的免疫学特异性及染色效价。先将荧光抗体按二倍递减稀释法作好不同浓度的稀释液，与特异性抗原标本作系列染色，根据如下标准表示反应程度：①（-）无荧光；②（+） 荧光弱但清楚可见；③（++）荧光明亮；④（+++～++++）荧光闪亮。用出现（+++）荧光强度的最大稀释度，定为该荧光抗体的“染色使用单位”。

用所得的荧光抗体分别与其他大豆根瘤菌交叉进行非特异性的抗原-抗体染色，检查它们彼此间是否存在血清学的交叉反应。如有交叉反应，可采用交叉吸附反应除去它们的共同抗原，获得特异性较高的荧光抗体。本试验 PRC005 的荧光抗体与其他大豆根瘤菌无交叉反应。

（7）特异性荧光抗体在生物固氮研究中的应用

a. 根瘤中根瘤菌血清型的鉴定

根瘤放入“U”型孔的多孔塑料平板，加两滴无菌 0.85%的生理盐水，如是烘干的根瘤（70℃，48 小时）需放冰箱过夜让根瘤吸水，新鲜根瘤直接用牙签压碎涂片[19]，涂片后空气干燥，火焰固定，每个涂片样本加 2 滴罗达明胶（RhITc）染色，将片子置 55℃烘箱 1 小时后，用荧光抗体滴满样本面，放在湿盆中室温下染色，使荧光抗体与样本充分反应，20 分钟后取出样本，用 pH7.2，0.01mol/L PBS 滴洗，并将样本玻片放入该 PBS 缸内浸泡 20 分钟，取出样本使其通过蒸馏水一次，晾干、封片、镜检。

如根瘤中含有与荧光抗体相应的特异性根瘤菌，镜检时即可看到荧光闪亮的根瘤菌体。否则，整个视野黑暗，即负反应。

b. 自然土壤中根瘤菌数量的测定[13,16]

取 10 克含菌土样和 30 个玻珠，加入 250ml 的三角瓶中，向三角瓶加入 95ml 浸提液（1%明胶与 0.1mol/L（NH_4）$_2HPO_4$ 1∶9 混合），加 5 滴吐温 80 和 1 滴防沫剂 AF71，摇床振荡 15 分钟，加 10 毫升 5%偏磷酸钠或 0.7 克 Ca（OH）$_2$ 与 $MgCO_3$ 2∶5 的混合物，充分混匀后，静置 1 小时吸 1ml 上清液过滤。滤膜孔经 0.4μm。用碳素墨水或 Irgalan 黑染料将滤膜染成黑色。过滤完毕取下滤膜放于载玻片上，用过滤的罗达明胶（RhITC）染色滤膜，置 55℃烘箱 1 小时烘干后，又用过滤的荧光抗体（FA）染色滤膜，20 分钟后将滤膜放回过滤架上，用大约 100ml 过滤的生理盐水通过滤膜，以达到洗掉未结合，剩余 FA 之目的。过滤完毕，滤膜放回载玻片上封片，用落射荧光显微镜镜检计数，每个滤膜计数大约 20 个视野，按如下公式计算测数结果：

$$No/\text{g} = \frac{Nf \cdot A \cdot D}{a \cdot v}$$

公式中：*No*/g-细胞数 / 克土；*Nf*-平均每视野的菌数；*A*-滤膜的过滤面积；*D*-稀释倍数；*a*-荧光显微镜下每个视野所能观察到的滤膜面积，*v*-过滤时浸提液消耗的体积

2. 荧光抗体测数与平板测数的比较试验

称 10 克过 20 目筛的试验田土样，装入 18×180mm 的试管，121℃ 2 小时间歇灭菌二次，接入 3.5ml PRC005 振荡培养 3 天的稀释菌悬液，此时土样的含水量为 60%。置试管于 25℃培养箱无菌培养。定时取样测数，每次测数取三支试管样品。FA 测数与上述方法相同，平板测数按常规的无菌操作程序进行。

3. MPN 测数方法

称 20 克土倒入灭菌的 180ml 水和 30 个玻珠的三角瓶中，摇床振荡 15 分钟，制成土壤悬液，将其 10 倍系列稀释，每稀释度接 4 支无菌野大豆幼苗琼脂管（野大豆种子灭菌法：浓 H_2SO_4 中浸 20 分钟，用无菌水洗涤 8 次），每管接 1ml 土壤稀释液，接种完毕继续置灯光栽培室培养。灯光栽培室每天光照 14 小时，白天 24℃，晚上 22℃。植株培养一月后观察结果，根据各稀释度结瘤管数，查 Vincent 手册[1]算出最大可能数。

4. 田间试验和田间盆栽试验

试验田设在本院实验农场，该地三年前曾种植过大豆。预备试验结果表明：试验菌株 PRC005 与该地土著大豆根瘤菌没有血清学交叉反应。田间试验设置：不接种对照；不接种施（NH）$_4SO_4$；和其他 4 个不同接种量的处理，共 6 个处理，4 个重复（区组），每一重复（区组）内用随机对比法排列各个小区。用 PRC005 草炭菌剂接种。播种后定时从不接种 CK，不接种施硫铵，最高接种量小区和休闲小区采样。用 MPN 法测定大豆根际或休闲土壤大豆根瘤菌的最大可能数。并定时从同一最大接种量小区采样，用 FA 法测定大豆根际接种菌 PRC005 的数量。

田间盆栽试验设置的处理有：不接种对照和 5 个接种菌 PRC 005 不同接菌量处理。每处理八个重复。用 PRC005 的草炭菌肥接种，按由低接种量到高接种量处理顺序，每处理的接种菌肥先与无菌干草炭混合，再与定量的土混匀后分装 100×200mm 的无底管钵，管底用一有孔聚乙烯塑料布与田间土隔开。各处理管钵随机埋在试验田一侧的小区内。分别在播种后 35 天（苗期）和 55 天（花期）每钵随机采 30 个根瘤测定接种菌 PRC005 的占瘤率。

试验结果

一、荧光抗体测数法与稀释平板测数法比较

为了验证荧光抗体滤膜测数方法测数结果的可靠性，进行了荧光抗体测数方法与稀释平板测数方法两种测数方法所测定的结果比较试验（表 1）。从表 1 可以看出：荧光抗体（FA）方法测数比混菌平板法的低。两种方法的测数结果显著相关（r=0.968）。FA 法测数的计数效率为 30%。

表 1　两种测数方法测数结果的比较

采样时间（接种后天数）	根瘤菌数量（细胞数/克干土）		FA 测数效率（%）
	平板测数	FA 法测数	
1	1.45×10^6	4.36×10^5	30.1
2	1.28×10^6	3.95×10^5	30.9
4	1.49×10^6	4.47×10^5	30.0
8	1.60×10^6	4.87×10^5	30.4
16	1.41×10^6	4.45×10^5	31.6

二、大豆根瘤菌在根际和土壤中的消长

在田间试验中，我们用 MPN 测数方法测定了未接种 CK，未接种施硫铵和最高接种量小区大豆根际以及休闲小区土壤中大豆根瘤菌总数，用荧光抗体滤膜测数方法测定了最高接种量小区大豆根际接种

菌 PRC005 的菌数消长（图 1）。

从图 1 可以看出：休闲地土著大豆根瘤菌数量在整个田间试验期间变化较小。从未接种 CK 和未接种施硫铵处理中土著菌的数量变化可以看到大豆有明显的根际效应，使未接种 CK 处理小区菌数大大地超过休闲小区根瘤菌数，同时可以看出氮肥对根瘤菌的数量有轻微的影响。接种大豆根瘤菌并没有使接种小区大豆根瘤菌总数量明显增加。

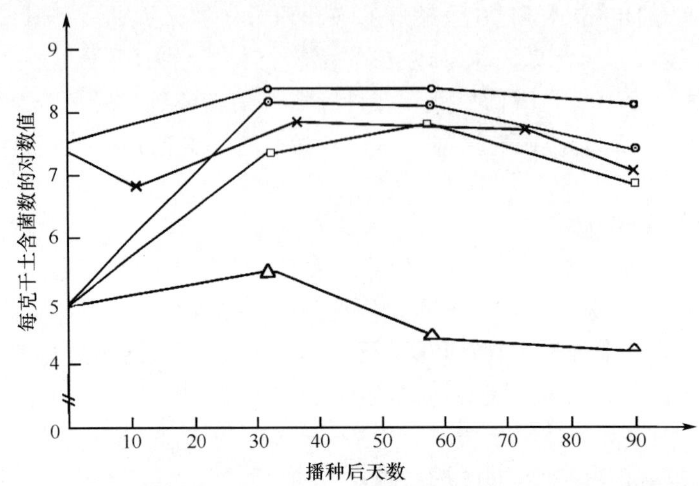

图 1　田间试验不同处理小区大豆根瘤菌总数和接种菌 PRC005 菌数的消长

图中符号表示：△：未接种休闲小区土样（MPN）；□：未接种施硫铵小区大豆根际土（MPN）；×：最高接种量处理小区大豆根际土 PRC005 菌数（FA 法）；⊙：未接种对照小区大豆根际土（MPN）；○：最高接种量小区大豆根际土（MPN）

三、 接种菌株 PRC005 与土著菌的竞争结瘤

图 2 是田间盆栽试验菌株 PRC005 不同接种量处理的占瘤率结果。试验结果表明：田间土壤中的土著根瘤菌有明显的竞争结瘤优势，但这种优势随接种菌数量增加而减弱，当接种菌数量与土著菌数的比例从 1.2 倍增长到 120 倍时，接种菌 PRC005 苗期（播种后 35 天）的占瘤率从 11.4%上升到 32.1%，花期（播种后 55 天）的占瘤率从 14.6%上升到 50.1%。

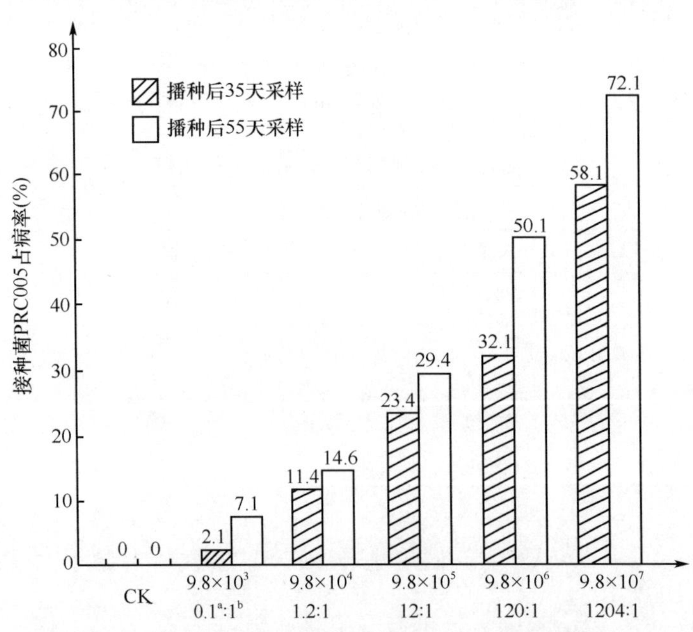

图 2　田间盆栽各处理接种菌 PRC005 占瘤率结果

a. 接种菌数量（个 / 克干土）；b. 接种菌数与土著根瘤菌数之比例

讨论

本试验 FA 测数方法与稀释平板测数方法两者结果显著相关（r=0.968），表明 FA 测数结果是可靠的，测数方法适合于土壤微生物学的研究。本试验 FA 测数方法的计数效率为 30%，这与已报道的结果一致[13, 15]。FA 测数方法较稀释平板测数低的主要原因是因为 FA 测数需要使浸提液澄清才能进行过滤，而在澄清的同时伴随着细菌和土壤颗粒同时下沉；稀释平板测数不需浸提液澄清，直接将浸提液系列稀释后平板测数。Kingsley 和 Bohlool（1981）[13]的试验指出，土壤质地影响 FA 方法计数效率，许多热带土壤与细菌结合紧，导致计数效率的下降，砂质土壤中根瘤菌的计数效率高于非砂质土壤。而且，FA 测数方法目前还不能区分死的和活的细胞，测数结果是活菌数和死菌数之和。FA 测数方法的另一不足之处，它要求所测的细菌每克土壤应含 10^4 个细胞以上，否则无法测定[16]。

在我们的试验中，未接种休闲小区土著菌数在 10^4～10^5 水平上波动，数量变化较小；加氮肥影响土壤中根瘤菌数量；大豆根际有明显刺激根瘤菌生长繁殖的作用；接种并没有使土壤总的根瘤菌数量明显增加。这些与已报道的结果一致[14,16]。Weaver 等（1972）[23]用 MPN 调查了美国衣阿华州 52 个不同田块的大豆根瘤菌数量，其结果表明土壤中大豆根瘤菌的数量与土壤结构、土壤 pH、土壤有机质和采样时土壤中是否种有大豆等特性之间没有显著的相关性。Elkins 等（1976）[11]也报道了类似的结果，表明在没有大豆寄主生长时，大豆根瘤菌也能在土壤中很好地生存。我们的试验地三年前种过大豆，后改种树苗，并且在我们进行大田试验的前一年，该地休闲未种任何作物。但田间试验时，土著大豆根瘤菌仍高于 10000 个 / 克干土，表明土著大豆根瘤菌在自然土壤环境中具有较强的生命力，即使没有大豆寄主存在也如此。“根际效应”早已被人们所认识，MPN 方法提供了根际根瘤菌数量明显高于非根际的证据。从我们的试验结果（图 1）也能看出这一点。近几年来，由于采用了 FA 方法测定根际特定根瘤菌的数量，对“根际效应”的问题有了更进一步的了解。一些研究结果表明：植物根际对根进菌的生长有刺激作用，而且豆科植物根际的刺激效应比非豆科作物更明显，但没有观察到前者对根瘤菌有“选择性的刺激作用”[6,14,15]。接种菌株 PRC005 接到没有灭菌的自然土壤后的 10 天内，菌数下降。这与菌株 PRC005 对一个新环境的适应性以及土壤生物因素的影响有关。因为接种后的 10 天内，大豆种子刚刚开始发芽，根际效应没有或很小，加之其他土壤微生物与接种菌竞争、拮抗或吞食等等方面都是使接种菌在接到土壤后短期内菌数下降的可能原因。Vidor 等（1980）[21]的试验结果也表明接种菌接到未灭菌土壤中第一周菌数下降 5～10 倍，而在相同的灭菌土壤中，接种后菌数迅速上升，他们认为大豆根瘤菌在未灭菌土壤中菌数下降是生物学因素所致。

在接种菌与土著根瘤菌的竞争结瘤中，往往是土著根瘤菌占优势，而增大接种菌的接种量常常可以提高接种菌的占瘤率[5,22]，我们的田间盆栽试验结果（图 2）也表明了这一点。土著根瘤菌的竞争结瘤优势可能与它所处的良好土壤条件或有利的生态分布有关。土著根瘤菌长期生活在土壤中，由外部因素造成的移动和土著菌自身的运功，会使土著根瘤菌均匀分布在土壤的各个微域之中，而按种菌虽在接种时与土壤进行了充分拌和，但仍不可能达到土箸根瘤菌那样的均匀分布程度，当豆科植物根系向土壤深处发展时，所到之处都可遇到土著菌而被感染结瘤，所以土著菌处于竞争结瘤的优势地位。Brockwell 等（1984）[10]的试验结果也说明了这一点：如果一个菌株在土壤中已生活了一年，而第二年用同一菌株再接种时，其竞争结瘤能力大大低于先进入土壤的根瘤菌。因为第一年进入土壤的根瘤菌相对于第二年刚接入的根瘤菌而言，前者可能处于较好的生态地位而有利于竞争结瘤。我们还可以用这一观点去分析和理解同一接种处理接种菌在花期的占瘤率比苗期的占瘤率高（图 2）。接种菌随着接入土壤的时间的延长，即接种菌在土壤生活的时间愈长，接种菌在大豆根际土壤中的分布就愈均匀，也就可能愈有利于接种菌的竞争结瘤。

参 考 文 献

[1]J. M.，芬森特著，1974，上海植物生理研究所固氛室译，根瘤菌实用研究手册.

[2]范秀容、沈萍，1980，徽生物学实验，P114-129，人民教育出版社，北京.

[3]中国人民解放军 59175 部队，1978，荧光显微术，上海科学技术情报研究所出版.

[4]Akiyoshi Kawamura，Jr.，1977，Fluorescent antibody techniques and their applications，2th Edition. Univ. of Tokyo press.

[5]Amarger，N. and J.P. Lobreau，1982，Quantitative study of nodulation competitivenes in rhizobium strains，Appl. Environ. Microbiol. 44：583-588.

[6]Bohlool，BB.，R. Kosslak and R.Woolfenden，1984，The ecology of rhizobium in the rhizosphere. In：Advances in Nitrogen Fixation Research（ed. C. Veeger and W.E. Newton），287-293.

[7]Bohlool，BB. and EL. Schmidt，1968，Nonspecific staining：its control in immunofluorescence examination of soil. Science. 162：1012-1014.

[8]Bohlool，BB. and EL. Schmidt，1970，Immunofluorescent detection of *Rhizobium japonicum* in soils. Soil Sci. 110：229-236.

[9]Brockwell，J.，1982，Plant infection counts of rhizobia in soils，In：Nitrogen Fixation in Legumes（ed. J. M. Vincent），Academic Press Australia，41-58.

[10]Brockwell，J.，R. J. Roughley and D. F. Herridger，1984，Impact of rhizobia Established in a high nitrate oil on a soybean inoculant of the same strain，In：Advances in Nitrogen Fixation Research（ed. C. Veeger and W. E. Newton） page. 328.

[11]Elkins，D. M.，Hamilton，G.，Chan，C. J. Y.，Briskovich，M. A. and Vandeventer，J. W.，1976，Effect of cropping history on soybean growth and nodulation aid soil rhizobia. Agron. J. 68：513-17.

[12]Hobbie，J. E.，R. J. Daley and S. Jasper，1977，Use of nuclepore filters for counting bacteria by fluorescence microscopy，Appl. Envlron. Microbiol. 33：1225-28.

[13]Kingiley，M. T. and B. B. Bohlool，1981，Release of *Rhizoblum* spp. from tropical soils and recovery for immunofluorescence enumeration，Appl. Environ. Microbiol. 42：241-248.

[14]Moawad，H. A.，W R. Ellis and E. L. Schmidt，1984，Rhizosphere response a s a factor in competition among three serogroups of indigenous *Rhizobium japonicum* for nodulation of field-grown Soybeans . Appl. Euiron. Mlcroblol. 47：607-612.

[15]Reyes. V. G. and E. L. Schmidt，1979，Population densities of *Rhizobium japonicum* Strain 123 estimated directly in Soil and rhizospherers. Appl. Earriron. Microbiol. 37：854-858.

[16]Schmidt，E. L.，1974，Quantitative autecological study of microorganism in Soil by immunofluorescence. Soil Sci. 118：141-149.

[17]Schmidt，E. L. and R. O. Bankole，1982，Detection of *Aspergilllus flavus* in Soil by immunofluorescent Staining. Science. 136：778-777.

[18]Schmidt，E. L.，R. O. Bankole and B.B. Bohlool，1968，Fluorescent antibody approach to Study of Rhizobia in soil，J. Bacterial，95：1987-1992.

[19]Somasegaran，P.，R. Woolfenden and J. Halliday，1983，Suitability of over-dried root nodules for Rhizobium Strain identification by immunofluorescent and agglutination，J. Appl. Bacteriol. 55：253-261.

[20]Trinick. M. J.，1969，Identification of legume nodule bacteria by the fluorescent antibody reaction，J. Appl. Bacteriol. 32：181-186.

[21]Vidor，C. and R. H. Miller，1980，Relative Saprophytic competence of *Rhizobium japonicum* Strain in Soils as determined by the quantitative fluorescent antibody technique（FA），Soil Biol. Biochem. 12：483-487.

[22]Weaver. R. W. and L. R. Frederick，1974，Effect of inoculum rate on competition nodulation of *Glycine max* L. Merrill，II Field studies，Agron. J，66：233-235.

[23]Weaver，R. W.，L. R. Frederick and L. C. Dumenil，1972，Effect of Soybean Cropping and Soil properties on numbers of *Rhizobium japonicum* in lowa Soil. Soil Sci，114：137-141.

[24]Wollum，A. G. and R. H. Miller，1980，Deuity centrifugation method for recovering *Rhizobium* Spp. from Soil for fluorescent - antibody Studies，Appl. Environ. Microbiol. 39：466-469.

快生型大豆根瘤菌的研究 I. 快生型大豆根瘤菌的分离及其生理生化性状*

曹燕珍　胡正嘉　黄诚金　石小岩　陈华癸

（华中农业大学生物固氮研究室）

从湖北省两种土壤中分离获得 11 株快生型大豆根瘤菌。在 YEM 培养基上产酸，世代时间低于 4 小时。这些菌株除对柠檬酸盐等五种碳源不能利用外。对供试的甘露糖等 16 种碳源均可利用。除一个菌株外，其余的 10 个菌株均能耐 1% NaCl，但不能超过 2%，用链霉素等 9 种抗素定量加入 YEM 培养基；培养分离出的菌株，根据其抗性反应，可将它们分为 5 个抗菌素抗性类型。

关键词　快生型大豆根瘤；世代时间

前言

中国是大豆的故乡。据有关专家考证，我国栽培大豆已有几千年的历史，栽培地域广，品种资源极其丰富。可以推测，与大豆共生的大豆根瘤菌的菌种资源在土壤中也是很丰富的。过去习惯把大豆根瘤菌看做是慢生型，生产上应用的优良菌种也都是从慢生型菌种中筛选出来的，对出现的快生型菌落，也未予以重视。自 1982 年 Keyser 等[7]从中国的大豆根瘤中分离出快生型菌株以来，在国内外引起了根瘤菌研究者们的浓厚兴趣，文献[2、3、4、5、6、8、9、10]先后报道了中、美、英、加等国的微生物学者从中国土壤和大豆根瘤中分离得到的菌株以及所进行的生理、生化、结瘤等习性、分类、遗传等方面的研究结果。

1983 年，在我室从事根瘤菌生态学研究的美国夏威夷大学的博士研究生[6]从湖北省洪湖县多年栽种大豆的土壤中分离出 41 株快生型大豆根瘤菌，经过抗菌素抗性的测定，将它们区分为 8 个不同的抗菌素抗性型；并且证明，在当地土壤的土著性大豆根瘤菌中，快生型大豆根瘤菌占有优势。

湖北省野生大豆资源丰富，栽培大豆亦较普遍。为开发我省快生型大豆根瘤菌这一新的固氮微生物资源，1984 年我们相继从本省长期栽培大豆的土壤中分离这类细菌，并对它们进行了生理、生化性状的研究。

材料和方法

一、土　　壤

采样地点及土壤列于下表 1。

表 1　土壤样本

采样地点	土类	质地	pH
钟祥（Z）*	黄棕壤	中壤	7.6
广济（G）	潮土	砂壤	8.0**
孝感（X）	黄棕壤	中壤	6.0
宜昌***（Y）	黄棕壤	中壤	5.8

*Z，G，X，Y 分别为钟祥、广济、孝感、宜昌、的代号；**无石灰反应；***自宜昌从未种过大豆的地区采取的土壤，以此为时照。

* 原载于《华中农业大学学报》，5（2）：149～156，1986.

二、大 豆 品 种

供试大豆品种为鄂豆二号（夏大豆），矮脚早（春大豆，代号为A），猴子毛（夏大豆，代号为H），1138-2（夏大豆），以上品种均为中国农科院油料所提供。野生大豆（*Glycine soji*）采自本校农场。

三、培 养 基

表2 培养基的种类和成分

	YEMA	无氮植物培养基		YEMA	无氮植物培养基
$K_2HPO_6 \cdot 3H_2O$	0.5g	0.2g	甘露醇	10.0g	——
$MgSO_4 \cdot 7H_2O$	0.2g	0.2g	微量元素（1）	4.0ml	——
NaCl	0.1g	0.2g	微量元素（2）	——	1.0ml
$CaHPO_4$	——	1.0g	水	900ml	1000ml
$FeCl_6$	——	0.1g	洋菜	18.0g	9.0g
酵母汁（10%）	100.0ml	——	pH	6.8	6.8

Jensen，1942。微量元素（1）：0.5% H_3BO_3 和 0.5% Na_3MoO_4 等量混合液；（2）：0.3% H_4BO3，0.2% $MnSO_4$，0.02% $ZnSO_4$，0.01% Na_3MoO_4，0.008% $CuSO_4$ 的混合液。

四、大豆根瘤菌的分离培养

将采自钟祥、广济、孝感、宜昌等地区的土壤风干后，分装入口径为9.0公分的塑料套钵中。每种土样分别播种前述五个大豆品种，播种前，种子均用常规0.1% $HgCl_2$ 进行表面灭菌，催芽后播种。每种土壤各重复播种3钵。置光照栽培室（光照强度为14000Lux，温度为25℃）培植，将无菌水加入塑料套钵底层，以保持土壤湿度。45天后，从大豆主根基部，取健壮根瘤，经表面灭菌，在YEMA上划线，保温28℃培养，在2～3天内，挑取平板上长出的菌落，移植于YEMA斜面（编号）培养（同时制片镜检），供进一步试验。

五、回 接 试 验

回接试验是在口径4公分，长35公分的大试管中进行的。试管盛无氮植物培养基40毫升，于121℃灭菌半小时，播入一粒经表面灭菌、并催芽的野生大豆种子，当出现第一片真叶时，接种所获菌株的悬液0.5毫升，置光照栽培室中培养，7～9天内（少数为6天）开始现瘤。根据结瘤结果，确定挑选出的菌株是否为根瘤菌。为了保证结果的准确性，回接试验重复了五次（与此同时菌株的纯度也检查了五次）。

六、世代时间的测定

用250毫升三角瓶盛装100毫升YEM培养液，pH调节至6.8，于121℃灭菌后，接入测试菌株，置保温28℃的摇床上振荡培养48小时后，以此培养液作种子菌，接入相同的YEM培养液中，使每毫升培养液含10^6个细胞为起始浓度，进行同样振荡培养。8小时后，每隔3小时用国产751型分光光度计测定光密度。4天后终止培养，并用国产PHS-Z型酸度计测定pH。

七、碳源的利用

以去掉甘露醇的YEM培养基为基础培养基，分别加入各种碳源，使终浓度为1%（*W/V*），调节pH

至 7.0，分装至小试管中，110℃灭菌 30 分钟。然后，接入各试验菌株，并设置不接种对照和不加碳源对照，保温 28℃培养，四天后检查结果。

八、生理生化特性的测定

各项生理、生化特性均按文献[11]的方法测定。

九、天然抗菌素抗性的测定

将新配制的各种抗菌素溶液分别加入定量、并冷却至 50℃的 YEMA 培养基中，混合均匀，立即制成平板。各抗菌素保持如下浓度（微克/毫升）：链霉素 10；卡那霉素素 20；利福平 5；氯霉素 50；青霉素 50；庆大霉素 30；四环素 15；红霉素 5；壮观霉素 40。将试验菌株在含不同抗菌素的 YEMA 平板上划线，各菌株均有三次重复，并接种未加抗霉素的培养基（YEMA）作对照，保温 28℃，6 天后观察结果。

结果与讨论

一、菌株的获得和回接试验

用广济等四个地区的土壤，经盆栽大豆结瘤，从根瘤中共分离得到 249 个纯菌株，其分布情况如表 3。

表 3　在四种土壤中获得的菌株

土壤	菌株数（株）	土壤	菌株数（株）
G（广济）	44	Y（宜昌）	5
X（孝感）	86	Z（钟祥）	114

从表 3 可见，多数菌株来自钟祥、孝感、广济栽培大豆的老区，也从未种过大豆的宜昌百里荒山地分离得到少数菌株。

我们对所获得的 249 个纯菌株进行了五次回接试验，在野生大豆上均结瘤良好。

二、菌株的生长速度和产酸反应

根据分离所得到 249 个纯菌株在培养平板和斜面上的生长快慢及菌落大小，初步判断其中 11 个菌落可能属于快生型。为了确证它们的生长类型，进行了世代时间和产酸的测定。结果见表 4。

表 4　菌株的世代时间和培养液 pH

菌株	世代时间（时间）	pH	菌株	世代时间（时间）	pH
ZA1	3.5	6.7	ZA11	4.0	5.8
ZA2	3.3	6.6	GA18	4.0	5.7
ZA3	3.5	6.5	GH35	3.5	6.2
ZA4	3.5	6.5	GH37	3.5	5.3
ZA5	4.0	5.3	USDA191	3.8	5.4
ZA6	3.5	6.5	113—2*	8.5	7.0
ZA7	3.2	6.4			

*为慢生型大豆根瘤菌

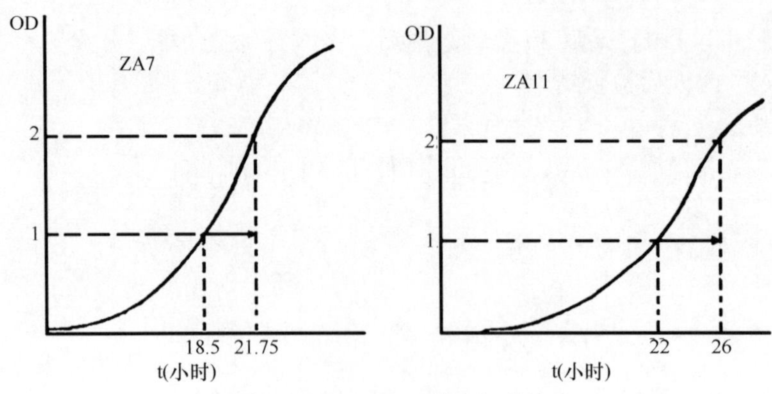

图 1 ZA7、ZA11 的生长曲线

试验证明，我们的分离的菌株确为快生型大豆根瘤菌。与对照菌株 USDA191 相较，在这两项特征上都相近。ZA7 号世代时间为 3.2 小时，是其中最快者，ZA11 世代时间为 4 小时，是其中的最慢者，它们的生长曲线见图 1。培养液 pH 值下降程度，不同菌株之间存在着差异，最高者为 6.7，而最低者仅 5.3。

这 11 株快生型大豆根瘤菌中，8 株来自钟祥县的黄棕壤，3 株来自广济县的潮土。

三、快生型大豆根瘤菌对碳源的利用

能利用多种碳源是快生型大豆根瘤菌不同于慢生型大豆根瘤菌的一项特征。为了鉴别我们所获得快生型菌株利用不同碳源的能力，试验中共使用了 21 种资源，结果列于表 5。

表 5 11 株菌株对碳源的利用能力

	ZA1	ZA2	ZA3	ZA4	ZA5	ZA6	ZA7	ZA11	GA18	GH35	GH37
甘露糖	+	+	−	+	+	+	+	−	−	+	−
甘露醇	+	+	+	+	+	+	+	+	+	+	+
葡萄糖	+	+	+	+	+	+	+	+	+	+	+
果糖	+	+	+	+	+	+	+	+	+	+	+
半乳糖	+	+	+	−	+	+	+	+	+	+	+
阿拉伯糖	+	−	−	−	−	−	−	−	−	−	−
木糖	−	−	−	−	−	−	−	−	−	−	−
鼠李糖	+	+	−	+	−	−	+	−	−	+	−
丁二酸钠	+	+	+	−	−	+	+	+	+	+	−
苹果酸	−	−	−	−	−	−	+	−	−	−	−
柠檬酸盐	−	−	−	−	−	−	−	−	−	−	−
丙酮酸盐	−	−	−	−	−	−	−	−	−	−	−
甘油	+	+	+	+	+	+	+	+	+	+	+
乳糖	+	+	−	+	−	−	+	−	−	+	+
蔗糖	+	+	+	+	+	+	+	+	+	+	+
麦芽糖	+	+	+	−	−	+	+	+	+	+	−
棉籽糖	+	+	+	+	+	+	+	+	+	+	+
萄糖	+	+	+	+	+	+	+	+	+	+	+
山梨醇	+	+	+	+	+	+	+	+	+	+	+
肌醇	+	+	+	+	−	+	+	−	−	+	−
糊精	+	+	+	+	+	+	+	−	−	+	−

Stowers 等曾作过快生型大豆根瘤菌利用不同碳源的试验，他们用了 20 种碳源，其中有 17 种与我们利用的碳源相同，这 17 种碳源中，在他们的试验中除对柠檬酸盐不能利用外，其他的均可被利用，而在我们的试验中，除柠檬酸盐外，还有阿拉伯糖、木糖、苹果酸盐、丙酮酸盐不能被利用，这说明同为快生型大豆根瘤菌，但在生理特征上仍有差异。

四、快生型大豆根瘤菌生理生化性状的检验

对新分离的菌株的其他生理生化性状也进行了测定，结果见表 6。

不能利用柠檬酸盐、不产生三酮基乳糖、在蛋白胨肉汤培养基中不生长或生长微弱，都是快生型根瘤菌的特点，我们的试验结果与经相符。在我们的试验中，除 GH37 外，其他菌株均能耐 1% NaCl，没有菌株能赖 2% NaCl。

表 6　快生型大豆根瘤菌的理化特征

		ZA1	ZA2	ZA3	ZA4	ZA5	ZA6	ZA7	ZA11	GA18	GH35	GH37	USDA191	113-2
蛋白胨肉汤		−	−	−	−	−	−	−	−	−	−	−	−	−
BTB 反应		产酸	产酸	产酸	产酸	产酸	产酸	产酸	产酸	产酸	产酸	产酸	产酸	产酸
3-酮基乳糖		−	−	−	−	−	−	−	−	−	−	−	−	−
柠檬酸盐		−	−	−	−	−	−	−	−	−	−	−	−	−
H_2S		−	−	−	−	−	−	−	−	−	−	−	−	−
耐盐性（1%NaCl）		+	+	+	+	+	+	+	+	+	+	−	−	+
石蕊牛奶	pH	中性	中性	中性	中性	中性	中性	酸性	酸性	酸性	酸性	酸性	△	△
	血清环	+	−	−	+	+	+	+	−	−	−	−	△	△

五、快生型大豆根瘤菌的天然抗菌素抗性类型

快生型大豆根瘤所具有的天然抗菌素敏感性高于慢生型大豆根瘤菌，因此，测定所获得菌株的固有抗性，也是一项重要的生理生化特征，而且也可以作为分群的一项指标，在我们的试验中，使用了九种抗菌素，测试所获菌株的抗菌素抗性，结果见表 7、表 8。

表 7　快生型大豆根瘤菌的天然抗菌素抗性

菌株	链霉素	卡那霉素	利福霉素	氯霉素	青霉素	庆大霉素	四环素	红霉素	壮观霉素
ZA1	−	+	+	+	+	+	+	+	−
ZA2	−	−	−	−	+	−	−	+	−
ZA3	−	+	−	−	+	−	−	+	−
ZA4	−	+	+	+	+	+	+	+	−
ZA5	−	−	−	−	+	−	−	+	−
ZA6	−	−	−	−	+	−	−	+	−
ZA7	−	−	−	−	+	−	−	+	−
ZA11	−	−	−	−	+	−	−	+	−
ZA18	−	−	−	+	+	−	−	+	−
GH35	−	+	+	+	+	+	+	+	−
GH37	−	−	−	−	+	−	+	+	−

+示有抗性；−示无抗性

表 8 快生型大豆根瘤菌对抗菌素的抗性型

抗菌素抗性类型	菌株数	抗性的抗菌素
1	5	青霉素、红霉素
2	3	卡那霉素、利福霉素、氯霉素、青霉素、庆大霉素、四环素、红霉素
3	1	卡那霉素、青霉素、红霉素
4	1	氯霉素、青霉素、红霉素
5	1	青霉素、四环素、红霉素

从以上结果可以看出，所有菌株对链霉素和壮观霉素均无抗性，对青霉素，尽管浓度较高，却均有抗性。同时，根据对九种抗菌素的反应，可将 11 个菌株分成 5 个抗菌素抗性类型。

从发现快生型大豆根瘤菌至现在仅四年时间，尽管国内外根瘤菌工作者在这方面已经做了不少工体，但对它们的各种性能，特别是生产性能的了解还是不够的。Keyser 等认为快生型大豆根瘤菌可能是根瘤菌属与慢生型根瘤菌属之间的重要链索，是否确系如此？快生型大豆根瘤菌与其他豆类植物的互接种族关系如何？快生型大豆根瘤菌的竞争结瘤能力如何？从快生型大豆根瘤菌中能否筛选出既有很强的竞争能力，又有很高固氮能力的菌株？这些都是既有一定理论意义又有一定生产意义的课题，是值得进一步研究和探索的。

我国土地面积辽阔，各地生态条件差异很大，从不同地区筛选这类细菌，既可以发现它们的共性，又可以找到各地特有的生态型，这将有利于加速和加深对快生型大豆根瘤菌的研究，是值得提倡的。

参 考 文 献

[1]Keyser，HH. et al.，1982，Fast-growing rhizobia isolated from root nodules of soybean，Science，215：1631-1632.

[2]Yelton，M. M. et al.，1983，Characterization of an effective salt-tolerant，fast-growing strain of *Rhizobium japonicum*，J. Gen. Microbiol，129：1537-1547.

[3]Stowers，M. D. and A. R. J. Eaglesham，1984，Physiological and Symbiotic Characteristics of fast-growing *Rhzobium japonicum* strains，Plant and Soil，77：3-14.

[4]Hatttori，J. et al.，1984，Fast-growing *Rhizobium japonicum* that Effectively Nodulates Several Commercial Glycine max，L. Merrill Cultivars，Appl. Environ. Microbial，48：234-235.

[5]Scholla，M. H. et al.，1984，Deoxyribonuleic acid Homology between fast-growing soybean nodulating bacteria and other rhizobia，Int. Syst. Bacteriol，34：283-286.

[6]Sudowsky. M. J. et al.，1983，Possible Involvement of a Megaplasmid in Nodulation of Soybeans by Fast-growing Rhizobia from China，Appl. Environ. Microbiol，46：906-911.

[7]DoWdle，S. F. and B. B. Bohlcol，1985，predominance of Fast-Growing Rhizobium in a Soybean Field in the People's Republic of china，Appl. Environ. Microbiol，50：1171-1179.

[8]徐珍玫等，1983，野生大豆根瘤菌分离初报，土壤肥料，（2）：7-8.

[9]徐珍玫等，1984，快生型大豆根瘤菌的理化特性和共生效应，大豆科学，（3）：102-108.

[10]张红樱等，1985，快生型大豆根瘤菌 QB113 的共生固氮及其对大豆的增产效果，大豆科学.（4）：267-277.

[11]中国科学院微生物所细菌分类组稿，1978，一般细菌常用鉴定方法.北京：科学出版社.

土壤中大豆根瘤菌之间竞争结瘤的研究Ⅲ.接种菌量对大豆生长的影响*

王福生　李阜棣　陈华癸

（华中农业大学土化系）

摘　要　本文研究了大豆根瘤菌PRC005的接菌量对大豆生长的影响。田间试验结果表明：接菌量在播种后40天和60天没有显著增加根瘤数、根瘤干重、地上部植株干重和植株含氮量。施氮肥处理和较高接菌量处理之间的大豆种子产量差异不显著，与不接菌对照处理相比，施氮肥处理和较高接种量处理之间的大豆种子产量差异不显著，与不接菌对照处理相比，施氮肥和较高接菌量两个处理的种子产量显著增加。施氮肥处理没有使植株含氮量增加，并且还妨碍了大豆的结瘤作用。室内盆栽试验结果表明：只有当接菌量高于土著菌数1200倍时，才能显著地提高大豆的结瘤数和植株干重。

众所周知，豆科作物接种根瘤菌是为了在缺乏相应根瘤菌的土壤中，建立一个有效的豆科-根瘤菌共生固氮体系[12]，或在含有土著根瘤菌的土壤中，用一个更有效的根瘤菌株取代土壤中的无效土著根瘤菌，从而达到使豆科作物增加产量，给人们带来经济效益之目的。然而，有时并不能达到增产的目的[6,8]。多年来的研究结果表明，种植豆科作物的老区的接种效果是一个复杂的问题，它受许多生态学和遗传学因子的制约[1,4]。其中土著根瘤菌与接种菌竞争寄主根系的结瘤部位表现得最为突出，常规的接菌量不能达到增瘤增产的预期效果[14,15]。一些研究者已经和正在采用多种途径试图克服土著菌和接种菌竞争结瘤的问题[9–11]。我们曾报道了在含大量土著菌的土壤中接菌量和接种位置对接种菌占瘤率的影响[1,2]。本文则主要讨论接种菌量对大豆的结瘤作用和种子产量的影响。

一、材料与方法

（一）试验材料

供试大豆根瘤菌PRC005由中国农科院土壤肥料研究所提供，用酵母汁甘露醇洋菜培养基培养[3]。供试大豆栽培品种“矮脚早”由中国农科院油料作物研究所提供。

（二）试验方法

1. 田间试验　在本校实验农场进行，试验田土壤为黄棕壤，pH5.9，有机质1.85%，全氮0.091%，全磷（P）0.055%，速效氮94ppm。试验前测定土壤含土著根瘤菌数量为每克干土8.14×10^4个（MPN法测数结果）。田间试验设置六个处理、四个重复，包括：不接种对照；不施$(NH_4)_2SO_4$；四个不同接种量的处理。每一个重复（区组）内用随机对比法排列各个小区。小区面积为5.5×2.4平方米，每小区播种4行，行距60厘米。播种前每小区沿播种行边缘施1040克过磷酸钙作底肥，加氮肥小区则沿播种行边缘施330克$(NH_4)_2SO_4$。菌肥接到播种行内，用锄头将菌肥与行内10厘米左右深的土混合，力求使接种菌均匀分布于土壤中。在播种后第40天和第60天，分别进行大豆苗期和盛花期采样，每次从每个小区的边行采10株，分析植株的根瘤数目、根瘤干重和植株地上部分干重，并将盛花期采集的植株地上部粉碎，用半微量K氏定氮法测定含氮量[3]。播种后90天测定每区中间两行4米长的种子产量。

2. 盆栽试验　用试验田土壤进行室内盆栽试验。试验设置不接种对照和4个不同接种菌量的处理。每处理重复6钵，盆栽共进行2次。用摇瓶培养3天的PRC005液体菌剂（10^9个/毫升）接种。大豆种

*原载于《土壤学报》，26（4）：1989.

子播种前进行了表面灭菌，播种后35天收获盆栽，测定大豆植株的根瘤数和植株地上部干重。

二、试 验 结 果

田间试验的测产结果表明（表 1）：随着接菌量的加大，大豆种子产量逐渐增加。当接菌量加大到土著大豆根瘤菌586倍时，其种子产量与不接菌CK处理的产量差异显著，即增产达显著水准。高接菌量处理与不接菌施氮肥处理之间的种子产量差异不显著。田间试验大豆苗期（表2）和花期（表3）的采样结果表明：加大接菌比例，仅仅使植株结瘤数、植株干重、地上部植物干重和植株含氮量稍微增加，四个接菌处理间的上述项目没有显著差异；无机氮肥影响大豆的结瘤能力，使植株的结瘤数和根瘤干重显著下降，而植株干重显著地高于四个接种处理。田间试验菌期（表 3）和花期（表 4）的不接菌 CK 处理的结果均未列入表中，是因为我们的田间试验采用了随机对比法排列，它比随机排列法需要更多不接菌 CK 小区，这将使其达到尽可能减少因试验田土壤肥力差异而造成的误差。

表1 接菌量对大豆种子产量的影响

Table 1 Effect of inoculation rate on soybean seed yield

接种比例 Inoculation rates	种子产量（公斤/亩） Seed yield（kg/mu）	接种比例 Inoculation rates	种子产量（公斤/亩） Seed yield（kg/mu）
CK	92.7 c*	173：1	108.3 abc
4.7：1	96.2 bc	586：1	115.1 ab
23.4：1	103 bc	+N	127.4 a

*英文字母表示5%统计学水平的差异性，表2、3、4类同。

表2 接菌量对大豆结瘤作用和地上部植株干重的影响

Table 2 Effect of inoculation rate on nodulation and plant top dry matter of soybean

接种比例 Inoculation rates	根瘤数/株 Nodule No./plant	根瘤干重/株（mg） Nodule dry weight/plant	地上部植株干重/株（mg） Plant tope dry weight/plant
47：1	3.8 ab*	3 a	0 a
23.4：1	4.2 ab	10 a	95 a
173：1	9.8 a	20 a	100 a
586：1	10.3 a	40 b	198 a
+N	−33 b	−34 c	390 b

表3 接菌量对大豆的结瘤作用、地上部植株干重的植株含氮量的影响

Table 3 Effect of inoculation rate on nodulation，dry matter and nitrogen status of plant tops of soybean

接种比例 Inoculation rates	根瘤数/株 Nodule No./plant	根瘤干重/株（mg） Nodule dry weight/plant	地上部植株 Plant tops	
			千克/株（g） Dry weight/plant	含氮（%） Total N
4.7：1	2.9 a*	48 a	1.19 a	0.17 a
23.4：1	5.0 a	58 a	1.29 a	0.20 a
173：1	5.1 a	60 a	1.39 a	0.29 a
586：1	6.4 a	63 a	2.28 ab	0.29 a
+N	−6.3 b	−19 a	2.98 b	0.24 a

表 4 不同接菌量的盆栽试验结果

Table 4 The pot experiment results of different inoculation rates

接种比例 Inoculation rates	根瘤数/株 Nodule No./plant	地上部根瘤干重/株（g） Plant tops dry weigth/plant
CK	21.2 b*	0.63 c
1.2：1	23.7 b	0.70 bc
12：1	24.9 ab	0.73 b
120：1	25 ab	0.73 b
1200：1	28.8 a	0.79 a

盆栽试验结果表明（表 4）：增大接菌比例能提高大豆的结瘤能力，增加地上部植株干重。但不接菌对照和几个接菌量低的处理之间，其根瘤数和植株干重无显著差异，仅在接菌量高于土著菌的 1200 倍时其大豆根瘤数和植株干重与其他各处理之间有显著差异。

三、讨 论

试验结果表明，在每克土含 10^4 个以上土著根瘤菌的土壤中，只有当接菌量高于土著菌 1200 倍时，才能显著地提高大豆的结瘤数和植物干重，否则没有明显效果。这与已报道的试验结果一致[7, 13]，他们用高于土著菌几十或几百倍的菌量接种，通常只是增加大豆生长早期土壤中的根瘤菌数量，使大豆的结瘤作用提前，但不能使植株的根瘤干重和植株干重显著地增加。

我们曾报道[2]了增大接菌量使一定的大豆根瘤数中高效的接种菌成瘤比例增加，相对地减少了无效土著菌的占瘤比例，从而增强了作物的固氮效率，使种子产量提高，但采用过高的接菌量企图使豆科作物增产的方法在生产实践中是不现实的，因为在现有的生产技术条件下，还不可能使大田的接菌量高于土著菌 1000 倍以上，即使做到也是没有经济效益。因此，我们必须对大豆种植老区影响根瘤菌接种效果的诸因子进行全面深入地研究，寻求解决该问题的切实可行途径，以期达到提高根瘤菌的应用效果，给实际生产带来经济效益。

参 考 文 献

[1]王福生、陈华癸、李阜棣，1985，土壤中大豆根瘤菌之间竞争结瘤的研究 I. 免疫荧光抗体技术任根瘤菌个体生态学研究中的应用. 华中农学院学报，第 4 卷 3 期，38-47 页.

[2]李阜棣、陈华癸、王福生，1986，土壤中大豆根瘤菌之间竞争结瘤的研究 II.土壤中根瘤菌的数量分布与占瘤率之关系.中国土壤研究的当代进展（英文），中国土壤学会编，第 117-123 页，江苏科技出版社.

[3]史瑞和、鲍士旦主编，1980，土壤农化分析.农业出版社.

[4]姚惠琴、朱增炎、曹景勤、许月蓉、陈碧云，1983，某些生物因子对苕子根瘤菌剂和土著根瘤菌竞争结瘤的影响.土壤学报，第 20 卷 1 期，85-91 页.

[5]Bohlool BB and Schmidt EL，1970，Immuofluorescent detection of *Rhizobium japonicum* in soils，Soil Sci.，110：229-236.

[6]Caldwell BE and Grant Vest，1970，Effect of *Rhizobium japonicum* strains on soybean yields，Crop Sci.，10：19-21.

[7]Bllis WR，Ham GE and Schmidt EL，1984，Persistence and recovery of *Rhizobium japoniaum* inoculm in a field soil，Agron J，76：573-576.

[8]Ham E，Cardwell VB and Johnson HW，1971，Evaluation of *Rhizonbium japomicum* inoculants in soils containing naturalized populations of rhizobia，Agron. J，63：301-303.

[9]Hunt PG，Matheny TA and Wollum AG，1985，*Rhizobium japonicum* nodular occupancy，nitrogen accumulation，and yield for determinate soybean under conservation and conventional tillage，Agron J，77：579-584.

[10]Hunt PG，Wollum AG and Matheny TA，1981，Effect of soil water on *Rhizobium japonicum* infection，nitrogen accumulation，and yield in bragg soybeans，Agron J，73：501-505.

[11]Kapusta G and Rouwenhorat DL，1973，Influence of inoculum size on *Rhizobium japonicum* serogroup distribution frequency in soybean nodules，Agran J，65：916-919.

[12]Smith RS，Ellis MA and Smith RE，1981，Effect of *Rhizobium japonicum* inoculant rates on soybean nodulation in a tropical soil，Agron J，73：505-508.

[13]Weaver RW and Frederick LR，1972，Effect of inoculum size on nodulation of *Clycine max* L. Merrill. variety ford，Agron J，64：597-599.

[14]Weaver RW and Frederick LR，1974，Effect of inoculum rate on competitive nodulation of *Clycine max* L. Merrill. I：Greenhouse studies，Agron J，66：229-232.

[15]Weaver RW and Frederick LR，1974，Effect of inoculum rate on competitive nodulation of *Clycine max* L. Merrill. II：Field studies，Agron J，66：233

慢生型大豆根瘤菌基因文库的构建和共同结瘤基因的分离*

彭文涛　周俊初　陈华癸

（华中农业大学土化系）

摘　要　pLAFR1 为载体构建了慢生型大豆根瘤菌 22-10 的总 DNA 基因文库。利用种间 DNA 分子杂交和功能互补的植物筛选法，从基因文库中检出三个含有共同结瘤基因的阳性克隆。

关键词　慢生性型大豆根瘤菌；基因文库；共同结瘤基因

根瘤菌和豆科植物的共生固氮是一个由双方基因调控的复杂过程。已知参与这一过程的根瘤菌基因有固氮酶基因（*nif*）、结瘤基因（*nod*）和固氮基因（*fix*）等。研究表明：多数快生型根瘤菌的上述共生固氮基因定位在称为共生质粒（pSym）的大质粒上[1−7]，而慢型根瘤菌则未见共生质粒的报道[8, 9]。

有关慢生型根瘤菌共生固氮基因的研究难度较大。Hennecke 等[10]已在慢生型大豆根瘤菌（*Bradyrhizobium japonicum*）USDA110 的染色体 DNA 上鉴定出两个基因簇：第一簇至少有 *nif* H.D.K.E.N.B.S 和 *fix* A.B.C.X 等 11 个基因，第二簇包括有 *common nod* A.B.C.D.I.J 和 *nif*A 及 *fix*A.R 等基因。本研究所采用的慢生型大豆根瘤菌 22-10，是自黑龙江分离并已大面积推广应用的优良生产菌株，为了在分子水平上研究其共生固氮基因定位和调控机制，我们选择 Friedman 等[11]构建的柯斯质粒（Cosmid）pLAFR1 为载体，先构建成总 DNA 基因文库，然后从库中分离出含有共同结瘤基因的三个阳性克隆。

1　材料和方法

1.1　材料

（1）细菌菌株和质粒（见表 1）

（2）培养基

a.常规培养基：YMA[12]，TY，LB，PA[13]。

b.TB 培养基：每升含 Tryptone 10 克，NaCl 8 克，麦芽糖 2 克，维生素 B_1 5 毫克。pH 自然，110℃，30′ 灭菌。

c.NZ-Amine 培养基：每升含 NZ-Amine 20 克，NaCl 5 克，$MgCl_2$ 2 克，葡萄糖 2 克，pH7.0。

d.SM 培养基：每升含甘露醇 10 克，KNO_3 0.5 克，K_2HPO_4 0.5 克，NaCl 0.1 克，$CaCl_2 \cdot 2H_2O$ 0.1 克，Rh 微量元素液[12] 4ml，混合维生素液 1ml，pH6.8-7.0。

（3）主要酶与抗菌素

限制性内切酶 *Eco*R I、*Bam*H I；T_4 DNA 连接酶；DNase 和碱性磷酸酶等来自华美公司；Pronase E 和 RNase 来自 Merck 公司，Lysozyme、四环素、卡那霉素和壮观霉素来自 Sigma 公司。同位素 α-^{32}p-dCTP 来自英国 Amersham 公司。

（4）其他

a.预杂交液：3×SSC，1%SDS，1×Denhardt's solution[14]，100 μg/ml 鲑鱼精子（salmon sperm）DNA。

b.Fahraeus 无氮植物营液见参考文献[15]。

---

*原载于《华中农业大学学报》，9（3）：230～239，1990.

表 1 细菌菌株和质粒

菌株或质粒	特性	来源
1.*Rhizobium*		
B. jnponicum 22-10	Nod^+.Fix^+（wild type）	From X. T. Dot
B. jnponicum USDA 110		From H. Hennccke
SPC4 NODA 1	Nod^-（nod A）	
SPC4 168	Nod^-（nod C）	
R. leguminosarum bio	pRL1jI（nod D：：Tn5）	From Q.S.Ma
Phascoli JI8400		
2. *E. coli*		
LE392	F^-.hsdR supE lacY	Our Lab.
	galk metB tryR	
HB101	F^-.hsdS recA ara pro	Our Lab.
	lac gal xyl mtl supE	
ED8767	recA supE hsd met	From P.R..Hirsch
BHB2688	recA（λimm CIts Eam Sam）	From T.R.Diman
BHB2690	recA（λimm CIts Dam Sam）	From T.R.Diman
3.Plasmid		
pLAFR1	T_c^r mob^+tra^-	From Q.S.Ma
pRK2013	K_m^r mob^-tra^+	From Q.S.Ma
pRmSL26	pLA FR1：：common nod genes	From Q.S.Ma
pRmSL42	pBR325：：common nod genes	From Q.S.Ma
pRmZ1	PBR325：：8.7kb common	Our Lab.
	Nod gcnes from *R.meliloti*	
pBjP1	PLA FR1：：common nod genes	This study
	From B.japonicum 22-10	
pBjP 2	Same as above	This study
pBjP3	Same as above	This study

1.2 *方法*

（1）供体 DNA 的制备

慢生型大豆根瘤菌 22-10 经 PA 培养基培养至对数期，用 Pronase E 和 SDS 在 37℃下溶菌，以苯酚和氯仿先后多次抽提，除尽蛋白等杂质后，用等量异丙醇沉淀，离心洗涤后溶于少量 TE，于 4℃下保存。按 Maniatis 等 [14]的方法对提取的总 DNA 用 *Eco*R I 作部分酶切（每微克 DNA 用 0.5 单位酶，37℃，30′），然后按 Downie[13]的方法于 1.25～5mol/L 氯化钠密度梯度离心（Beckman，SW41 转头，140000*g*，4hr），分部收集并经电泳检测回收 20～30kb 的 DNA 片段。

（2）载体 $PLAFR_1$ 的制备

按 Maniatis 等[14]的碱变性法分离质粒 pLAFR1，并经 CsCl 梯度分离心纯化，再用 *Eco*R I 完全酶切后用碱性磷酸酶处理黏性末端，以防止自连现象。

（3）包装蛋白的制备及其试包装效价测定：

参照 Priefer 等[16]的方法，从 *E.coli* 2688 和 2690 菌株中提取包装蛋白，分装小管后于液氮保存备用，要求试包装入 DNA 的效价不得低于 1×10^7pfu/1μg λDNA。

（4）载体 pLAFR1 与外源 DNA 的 *Eco*R I 酶切片段的连接，按文献〔14〕的方法，用 T_4 连接酶在

12℃下经 16hr，获得酶连 DNA。

（5）酶连 DNA 的体外包装和转染受体菌

按 Hohn[17]的方法对酶连 DNA 进行体外包装，并以包装混合物感染受体菌 ED8767，在 LB+Tc 抗性夹板上长出的克隆即为 22～10 的基因文库，将克隆全部洗出，用甘油法在-20℃下保存备用。

（6）菌落原位杂交，质粒的分离、转化，Southern 印迹杂交及 DNA 酶切片段的回收

菌落原位杂交除杂交膜采用点种法[18]和预杂交液采用本文配方外，其余照 Maniatis 等[14]的方法。根瘤菌质粒的分离照 Hirsch[19]的方法进行，大肠杆菌质粒的分离，用质粒 DNA 转化受体菌，探针的缺口转译，Southern 印迹杂交等照 Maniatis 等[14]的方法进行。从琼脂糖凝胶用冻融法回收 DNA 酶切片段[12]。

（7）三亲本杂交和植物结瘤试验

克隆在基因文库 $PLAFR_1$ 上的外源基因，能在协助质粒 pRK2013 的帮助下，向根瘤菌或其他 G-细菌转移，并能复制传代，因而有可能通过三亲本杂交，以不结瘤突变株为受体从库中钓出含有结瘤基因的克隆。其操作方法是：供体菌 22-10 基因库在 LB+TC，协助菌 pRK2013 在 LB+Km 培养基中于 37℃振荡过夜，受体不结瘤根瘤菌在 TY 斜面上长好后制成菌悬液。按菌量取样，并混合上述三菌培养液，离心后再悬浮于 2ml 无菌水中，过滤到微孔滤膜上，并转至 TY 平板上，于 28℃培养 1～2 天，用无菌水洗下菌苔，将菌悬液涂 SM+TC（15μg/ml）选择平板上，由于合成培养基 SM 不适于供体菌 22-10 基因库和协助菌生长，受体菌不抗 TC，因此，只有转移了重组或自连 pLAFR1 的受体菌才能长成菌落。

除野大豆种子的表面灭菌用浓硫酸处理外，其他种子的表面灭菌、催芽及砂培盆栽结瘤试验均照 Vincent[15]的方法进行。植物在光照室内按需要的温度和光照条件培养。一般播种一月后检查结瘤状况，并按常规方法[15]从根瘤中分离根瘤菌。

2　结果

2.1　*B. japonicum* 22-10 基因文库的构建

（1）根瘤菌总 DNA 的提取和 *Eco*R I 部分酶切

总 DNA 提取的关键是要在充分溶菌的基础上，反复用苯酚和氯仿抽提，以除去蛋白等杂质。并在全部操作过程中避免剧烈振荡，以使提取的总 DNA 低于 70kb。取少量总 DNA 进行 *Eco*R Ⅰ 部分酶切试验的目的在于选择最佳酶量。结果见图 1。

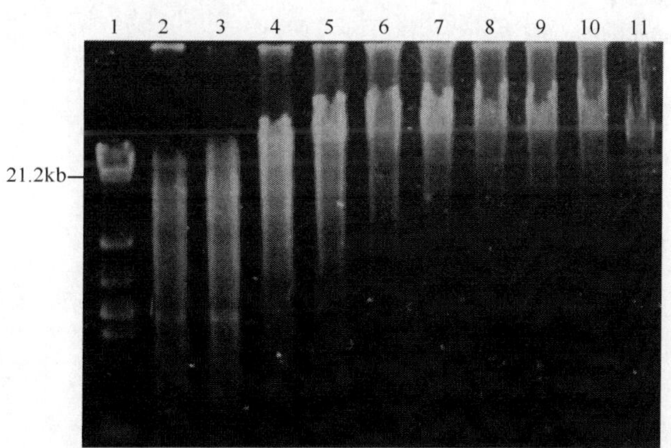

图 1　*B. japonicum* 22-10 总 DNA 的部分酶切图谱

1. λ DNA+*Eco*R Ⅰ+*Hind* Ⅲ Marher；2. Total DNA+4u *Eco*R Ⅰ；3. Total DNA+2u *Eco*R Ⅰ；4. Total DNA+1u *Eco*R Ⅰ；5. Total DNA+0.5u *Eco*R Ⅰ；6. Total DNA+0.25u *Eco*R Ⅰ；7. Total DNA+0.125u *Eco*R Ⅰ；8. Total DNA+0.0625u *Eco*R Ⅰ；9. Total DNA+0.312u *Eco*R Ⅰ；10. Total DNA；11. λ DNA

由图 1 可见：第 5 号的酶量（0.5μl/μg 总 DNA）可用于大量总 DNA 的部分酶切，结果也令人满意（图 2）。

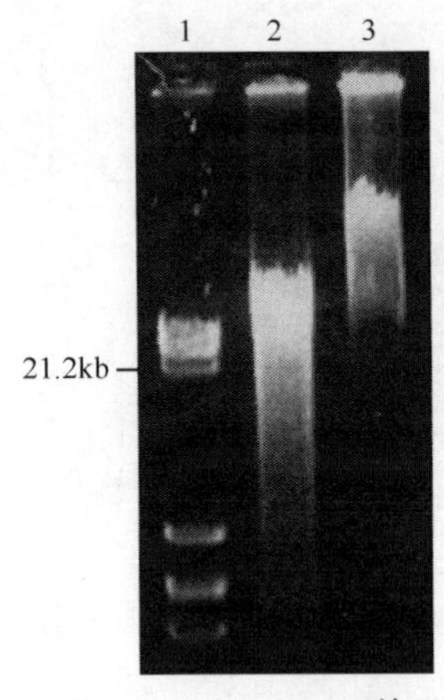

图 2 *B. japonicum* 22-10 总 DNA 的大量部分酶切图谱

1. λ+*Eco*R I marker；2.Partial digested total DNA；3.Total DNA

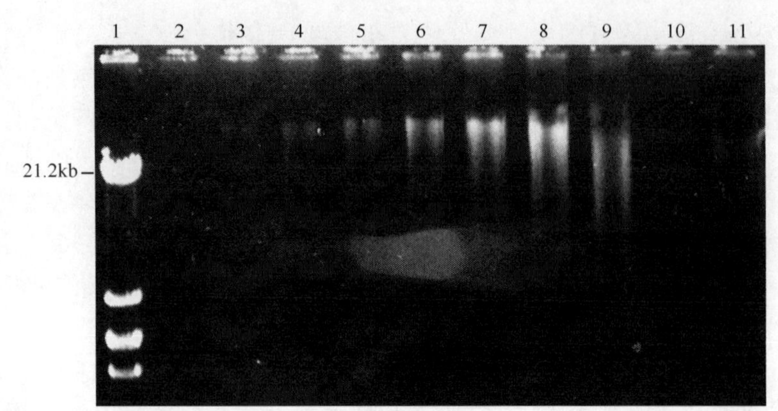

图 3 氯化钠密度梯度离心后部分收集的 *B. japonicum* 22-10 *Eco*R I 酶切的总 DNA 图谱

1.λ+*Eco*R I marker；2～11. Fractionated *Eco*R I digested total DNA

（2）氯化钠密度梯度离心分离与回收 20～30kb 总 DNA 酶切片段

氯化钠梯度由分层冻结的下层 5M 和上层 1.25M 溶液在 4℃融化时自然形成。加入酶切总 DNA 后超速离心 4hr，从底部穿孔分部收集，图 3 是收集各管的电泳结果。由图 3 可见，第 8 管中较大量 20～30kb 片段，故取该管透析去盐并沉淀回收 DNA。

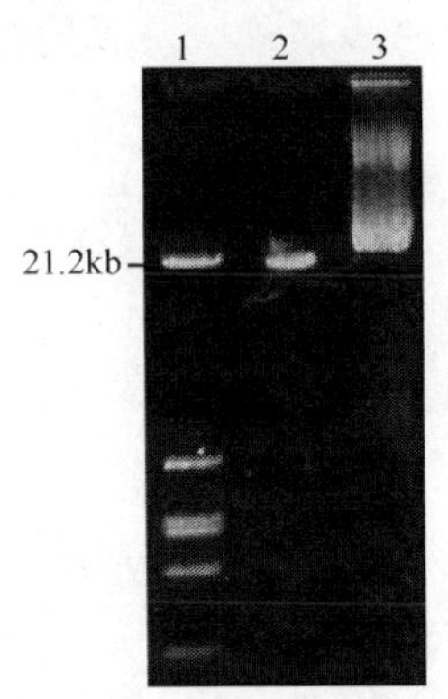

图 4 载体 pLAFR1 的 *Eco*R I 的酶切图谱

1.λ+*Eco*R I marker；2. pLAFR1+*Eco*R I；3. pLAFR1

（3）载体 pLAFR1 的提取、酶切和碱性磷酸酶处理：

用碱变性法提取的 $PLAFR_1$ 经 EB-CsCl 密度梯度离心纯化后，用 *Eco*R I 完全酶切，获得一条 21.6kb 的酶切带（图 4）。酶切后的 pLAFR1 进一步用碱性磷酸酯酶处理，弃除 5′ 末端磷酸基后已不能自我连接。

（4）包装蛋白的提取和试包装：

用热诱导法提取的包装蛋白试包装 λNA 的效价平均为 1.64×10^8pfu/1μg λ DNA（表 2），符合建库要求。

（5）酶连、包装和转染受体菌

取约 2μg 总 DNA 的 *Eco*R I 酶切片段与约 1μg $pLARF_1$ 的 *Eco*R I 酶切片段，由 T_4 连接酶连接后，经包装蛋白体外包装的混合物立即用于转染受体菌 ED8767，在 LB+TC 平板上获得的抗性菌落数超过 2.0×10^4 个，它们就是：*B. japonicum* 22-10 的总 DNA 基因文库。

表 2 λ DNA 的试包装效价

试验次数	噬菌斑数/1μg λ DNA	对照（无 λ DNA）
No.1	1.23×10^8	0
No.2	1.38×10^8	0
No.3	2.3×10^8	0
平均	1.64×10^8	0

（6）基因文库中抗性菌落的质粒和外源 DNA 片段检测

从文库的 Tc 抗性菌落中随机推选 22 个，用碱变性法[14]提取质粒 DNA，经 0.5%琼脂糖凝胶电泳检测，发现其中 15 株的质粒带谱不同于载体 pLAFR1（图 5）。由图可见：克隆 2、3、5、7 和 8 的质粒带谱不同于 pLAFR1。

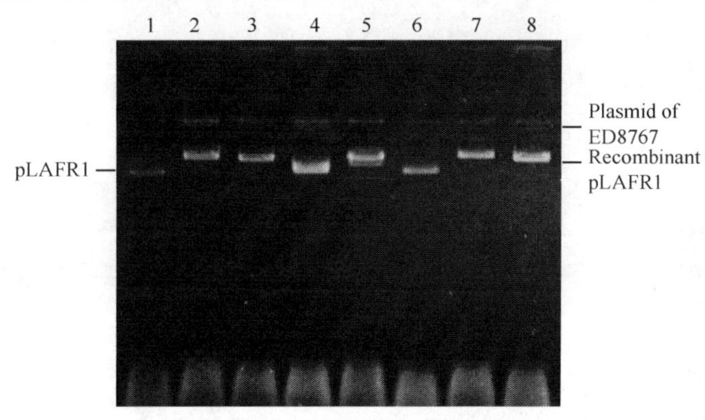

图 5　Tc 抗性菌落的质粒图谱

1. pLAFR1；2～8. Tc^r clones

进一步将这 15 株的质粒 DNA 用 *Eco*R I 完全酶切，部分结果见图 6。由图可见：克隆 2、3、4 和 5 的酶切带谱除第一带（为 pLAFR1 载体）外，其余片段均不相同，为外源 DNA 酶切片段，其大小总和约为 19kb。

照 Maniatis 等[14]的公式，构建基因文库需要的重组子克隆数 $N=\dfrac{\mathrm{In}(1-\mathrm{p})}{\mathrm{In}(1-\mathrm{f/g})}$，若几率按 99%，插入外源 DNA 片段大小 f 为 19kb，根瘤菌的总 DNA g 为 1×10^4kb 计算，N=2.4×10^3。在本研究中获得的总抗性克隆数>2.0×10^4，重组子比例达 68.2%，估算的总重组子克隆数>1.63×10^4，为上述理论值的 6.39 倍，因而符合建库要求。

2.2　用 DNA 分子杂交法分离含有共同结瘤基因的阳性克隆

根据不同根瘤菌之间共同结瘤基因的同源性，可以用已知的其他根瘤菌的 *nod* 基因为探针，从 22-10 基因文库中钓取含有共同结瘤基因的阳性克隆。

（1）菌落原位杂交

随机从基因文库中取 2000 个菌落点种在两张硝酸纤维滤膜上，以 α-^{32}P 标记的 pRmSL42 为探针，pRmSL26 为阳性对照，通过菌落原位杂交和放射性自显影，筛选出二个阳性克隆并命名为 pBjP1 和 pBjP2（图 7）。

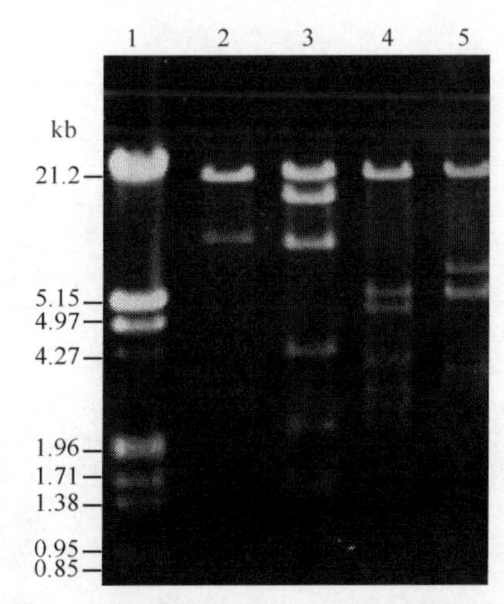

图 6　重组质粒的 *Eco*R I 完全酶切图谱

1.λ+*Eco*R I + *Hin*dⅢ marker；2～5：*Eco*R I digestion of recombinant plasmid

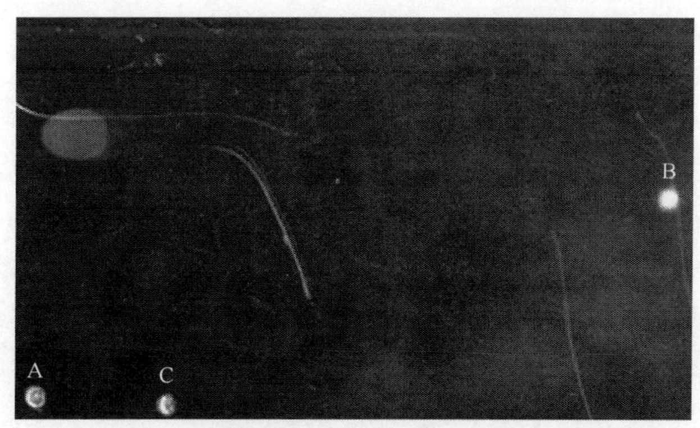

图 7　菌落原位杂交的放射性自显影

A. RmSL2b（阳性对照）；B. pBjP1；C. pBjP2

（2）阳性克隆的复证

从上述两个阳性克隆分离出质粒 DNA，经 *Eco*R I 完全酶切后进行凝胶电泳和 Southern 转移，用从

$pRmZ_1$上回收的α-^{32}P标记的3.5kb的*Eco*RI-*Bam*H I 片段（含有 *common nod* DABC）为探针进行分子杂交，结果见图8和图9。由此图可见：pBjP1上的约3kb和5kb两个片段和$pBjP_2$上的5kb片段上含有共同结瘤基因。

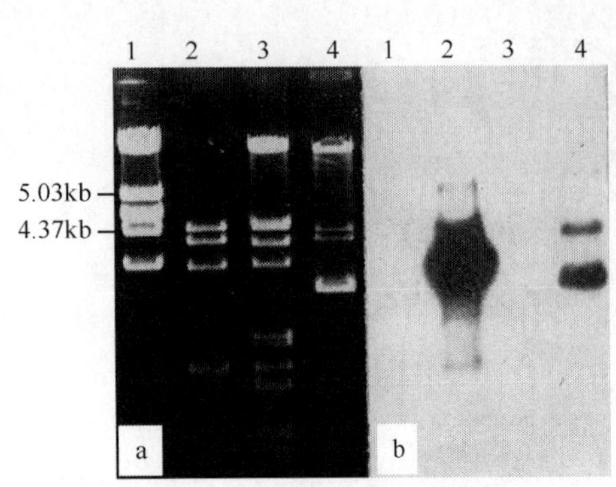

图8 pBjP1的Southern印迹杂交图谱

a.凝胶电泳；b.用Common nod DABC为探针杂交的放射自显影

1. λ+*Eco*R I marker; 2. pRmZ1+ *Eco*R I+*Bam*H I; 3. λ+*Eco*R I+*Bam*H I marker；4. pBjP1+*Eco*R I

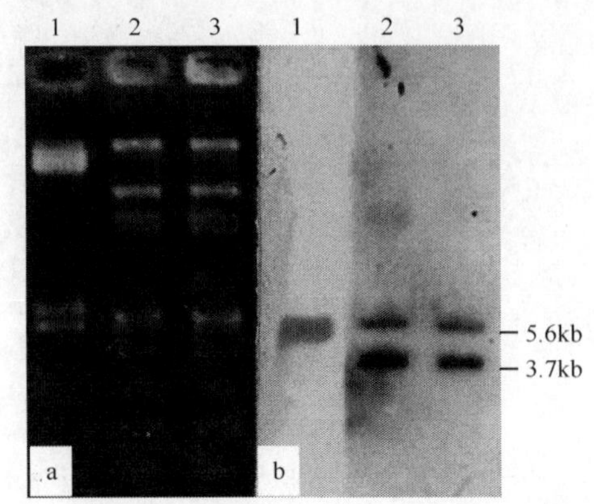

图9 pBjP2和pBjP3的Southern印迹杂交图谱

a.凝胶电泳；b.用共同结瘤基因DABC为探针杂交的放射自显影

（3）$pBjP_1$和$pBjP_2$上共同结瘤基因的功能互补试验

由于DNA分子杂交只能说明序列的同源性，尚难确定杂交同源DNA区段的功能。为此，我们以*B. japonicum* USDA110、SPC4 $NODA_1$和168及*R. leguminosarum* bio. *Phaseoli* 8400等三个不结瘤突变株为受体，以$pBjP_1$和$pBjP_2$为供体作三亲本杂交。结果表明只有$pBjP_1$含有功能性片段，能使上述突变株恢复在相应宿主植物上的结瘤能力。

2.3 从基因文库中直接分离带功能性共同结瘤基因的重组克隆

（1）功能互补结瘤试验

在上述试验的基础上，我们用上述三个不结瘤突变株为受体，通过三亲本杂交和植物结瘤实验，直接从基因文库中钓取带有功能性共同结瘤基因的阳性克隆。表3是用转移接合子接种相应豆科植物恢复结瘤的情况。

表3 转移杂合子的植物结瘤实验[1)]

供试菌株	$NodA_1$		168		8400	
	对照	杂合子	对照	杂合子	对照	杂合子
野大豆	0/5	10/40	0/5	14/40		
菜豆					0/5	3/20

1）结瘤株数/供试总株数

（2）根瘤分离株中重组质粒的检测

从上述杂合子结瘤植株的根瘤中分离并纯化出根瘤菌，然后按Hirsch[19]法分离质粒，电泳结果表明上述根瘤分离株均比原出发受体菌多出一条50kb左右的质粒带（图略）。

（3）重组质粒上共同结瘤基因的鉴定

用互补$NodA_1$突变株的接合子根瘤分离株的重组质粒转化HB101[14]，在LB+Tc（15μg/ml）平板上筛选转化子。从转化子中分离质粒，经*Eco*R I完全酶切，Southern转移后以pRmZ1的3.5kb的*Eco*R I-*Bam*H I 片段为探针作为分子杂交，结果见图9。由图可见该重组质粒（命名为$pBjP_3$）的3.7kb和5.0kb *Eco*R I片段上也含有 *common nod* 基因。

3　讨论

目前，从根瘤菌基因文库中分离带有结瘤基因的重组 DNA 克隆主要有三种方法。即功能互补的植物筛选法[20]；Tn5 插入法[13]；和利用共同结瘤基因的保守性，以已克隆的结瘤基因为探针，通过 DNA 分子杂交，直接从基因文库中筛选出阳性克隆的方法，为了准确地分离到目的基因，我们认为，先通过功能互补法从基因文库中钓到功能性目的基因，然后用基因探针复证并作进一步的功能互补试验较好。其优点是：①对整个基因文库作了检测，②分离到有功能的 DNA 片段，③用基因探针作分子杂交，以进一步复证。

Friedman 等[11]在用 pLAFR1 为载体构建根瘤菌基因文库时直接采用 pLAFR1 的 *Eco*R I 片段，为减少自连而加大外源 DNA 片段的用量，因而使重组克隆在文库中的百分率降低。本试验对载体 pLAFR1 的酶切片段进行了碱性磷酸酯酶处理，从而提高了重组克隆的百分率。

本研究钓出的含有慢生型大豆根瘤菌 22-10 共同结瘤基因的阳性克隆可供进一步深入研究结瘤基因的调控机制和向其他根瘤菌或 G^-细菌转移之用。

参 考 文 献

[1]Ruvkun GB et al.，Directed transposon Tn5 mutagenesis and complementation analysis of Rhizobium meliloti symbiotic nitrogen fixation genes. Cell，1982，29：551-553.

[2]Krol A J M et al.，On the operon structure of the nitrogenase genes of *Rhizobium Leguminsarum* and *Azotobacter vinelandii*. Nucleic Acid Research，1982，10：4147-4157.

[3]Rolfe B G et al.，Nitrogen Fixation Research Progress（Evans，Bottommly and Newton eds.），1985，79-85.

[4]Kondorosi E et al.，Physical and genetic analysis of a symbiotic of *Rhizobium meliloti*：Identification of nodulation genes. Mol. Gen. Genet. 1984，193：415-452.

[5]Broughton W J et al.，Plasmid-linked nif and "nod" genes in fast-growing rhizobia that nodulate *Glycine max*，*Psophocarpus tetragonolbus* and *vigna umguiculata*. Proc. Natl. Acad. Sci. USA 1984：81：3093-3097.

[6]Masterson R V et al.，Nitrogen fixation（nif）genes and large plasmids of *Rhizoium japonicum*. J. bacteriol. 1986，152：928-931.

[7]Stanly J et al.，Genetic locus in *Rhizobium japonicum*（fredii）affecting soybean root nodule differentiation. J. Bacterial. 1986，166：628-634.

[8]Fischer H M and Hennecke H. Linkage map of the *Rhizobium japonicum* nif H and nif DK operons encoding the polypeptides of the nitrogenase enzyme complex. Mol. Gen. Genet. 1984，196：537-540.

[9]Jagadish M N et al.，Advances in Molecular Genetics of Bacteria-plant Interaction（Szalay，A，A. and Legocki，R. P. eds.）1985，27-31.

[10]Hennecke H et al.，Regulation of nif and fix genes in *Bradyrhizobium japonicum* occurs by a cascade of two consecutive gene activation steps of which the second one is oxygen sensitive. Nitrogen：Fixation Hundred years after. Proceedings of The Inter. Congress on Ntirogen Fixation，（Bothe et al. eds）Gustar Fischer Press. 1988，339-344.

[11]Friedman A M et al.，Construction of a broad host range cosmid cloning vector and its use in the genetic analysis of *Rhizobium* mutants. Gene，1982，289-296.

[12]Smith H O.，Recovery of DNA from gels，Methods Enzymol. 1980，65：371-380.

[13]Dowine J A et al.，Cloned nodulation genes of *Rhizobium leguminosarum* determine host-range specificity. Mol. Gen. Genet. 1983，190：359-365.

[14]Manistis T et al.，Molecular Cloning：a laboratory manual. Cold Spring Harbor Laboratory，Cold Spring Harbor. 1983.

[15]Vincent J M. A manual for the practical study of the root nodule bacteria. Blackwell scientific publications，Oxford，1970.

[16]Priefer U et al.，Clone with cosmid. Advanced Molecular Genetics.（Puhler. A. and Timmis，K. N. eds.）Springer-Verlag. 1984，190-201.

[17]Hohn B. In vitro packaging of landa and cosmid DNA. Methods Enzymol. 1979，68：299-309.

[18]Grunstein M and Hogness D. Colony Hybridization. A method for the isolation of cloned DNAs that contain a specific gene. Proc. Natl. Acad. Sci. 1975，72：3961.

[19]Hirsch P R et al.，Physical identification of bacteriocinogenic，nodular and other plasmids in strains of *Rhizobium leguminosarum*，J. Gen. Microbiol.1980，120：403-412.

[20]Long S R et al.，Cloning of *Rhizobium meliloti* nodulation genes by direct complementation of Nod mutants. Nature，1982，298：485-488.

快生型大豆根瘤菌（*R. fredii*）和慢生型大豆根瘤菌（*B. japonicum*）基因文库的构建*

周俊初　沈　辉　胡志浩　陈华癸

（华中农业大学土化系）

摘　要　以能在 G-细菌中复制和转移的广谱质粒 pLAFR1 为载体构建成快、慢生大豆根瘤菌总 DNA 的基因文库，获得 Tc^r 克隆的总数分别为 3.9×10^4 和 2.1×10^4；其中含外源片段的重组克隆百分率达 40%和 42%，符合建库要求。

早期克隆生物 DNA 上特定区段的方法是先将 DNA 提出并酶切后通过制备凝胶电泳或反向色谱柱分离，再用探针鉴定出需要的片段后克隆到相应的载体上[1,2]；1997 年，Hohn 和 Murray 报道了 λ 蛋白体外包装技术[3]，Beton 和 Davis 研究出 DNA 分子分子原位杂交技术[4]，从而有可能先将生物 DNA 的全部片段克隆到适当载体上构建成“基因文库”之后再从中调取需要的阳性克隆。目前用于建立基因文库的载体主要是 λDNA，它能克隆 20-30kb 的外源 DNA 片段，便于体外包装和感染。但由于它不能在根瘤菌中复制而不适于构建根瘤菌的基因文库。Fridman 等（1982）在广谱质粒上加上 λ 的 cos 位点构建出能为 λ 包装蛋白进行体外包装的质粒 pLAFR1，并以它为载体构建出根瘤菌的基因文库[5]。

研究表明：苜蓿、三叶草、豌豆等快生型根瘤菌的共生固氮基因定位在大质粒上[6,7,8]；慢生型的大豆根瘤菌定位在染色体上[9]；快生型大豆根瘤菌则两者皆有并随菌株而异[11,12]。因此研究大豆根瘤菌的共生固氮基因的最佳选择是先构建其总 DNA 的基因文库。

材料和方法

一、供试菌株和质粒

见表 1。

表 1　供试菌株和质粒

菌株或质粒	特性	菌株或质粒	特性
Bradyrhizobium japonicum 22-10	$Nod^{\pm}$，$Fix^{\pm}$	*EcoR* Ⅰ　BHB 2690	Dam，$Temp^s$
Rhizobium fridii B_{32}	$Nod^{\pm}$，$Fix^{\pm}$	*EcoR* Ⅰ　ED 8767	$RecA^-$
EcoR Ⅰ BHB 2688	Eam，$Temp^s$	pLAFR1	Tc^r，$mob^{\pm}$，tra^-

二、方　　法

参照 Maniatis[12]介绍的方法，用 E.coli2688 和 2690 提取的包装蛋白，经试包装 λ DNA 的效价高于 10^7 个噬菌斑/1μg λ DNA 之后，于液氮或−80℃保存备用。用 Birnboim 等[13]的碱变性法提取 pLAFR1，然后通过 CsCl 超离心纯化，再用 *Eco*R Ⅰ酶切后备用。用链丝蛋白酶 E 和 SDS 在 37℃下溶菌及苯酚-氯仿反复抽提法和快慢生大豆根瘤菌中提取总 DNA，经 EcoR1 部分酶切后用 NaCl 梯度离心分离并回收 20-30kb 的片段。

取上述总 DNA 的 20～30kb 片段与 pLAR1 和 *Eco*R Ⅰ片段按 8：1 的量混合，经 T4-ligase 连接后，用上述 λ 包装蛋白作体外包装并感染受体菌 ED8767，将感染液涂抹在 LB+Tc 平板上的硝酸纤维素膜上，

*原载于《农业生物技术》，西安：陕西省科技出版社：66～69，1990.

经 37℃培养后对长出的克隆抽样作质粒和酶切检测，当含有外源 DNA 片段的重组克隆超过建库最低要求理论值后，即构建成该根瘤菌的基因文库。

结果

一、根瘤菌总 DNA 的提取和 *Eco*RI 部分酶切

总 DNA 提取的关键是充分溶菌和彻底去除蛋白质等杂质，全部提取过程中应避免剧烈振荡以保证获得的总 DNA 大小不低于 70kb。

部分酶切的第一步是在固定其他条件的情况下选择最佳 *Eco*R Ⅰ 酶量，结果见图版 XII-1。

由图版 XII-1 可见，适宜的酶浓度在 7（1.625）与 8（3.25）之间。选择 2.5u 酶量作大量酶切试验，结果见图版 XII-2。

由图版 XII-2 可见，未经酶切的总 DNA 大小超过 λ DNA（50kb），总 DNA 的 *Eco*RI 酶切片段大都在 20～50kb 范围内。

二、氯化钠梯度离心分离与回收 20～30kb 总 DNA 片段

氯化钠梯度由分别冷冻的下层 5mol/L 和上层 1.25mol/L 溶液在 4℃融化时自然形成的，酶切总 DNA 加入后用 BeckmanL8 离心机 35000r/min，4℃离心 2 小时后从底部穿孔部收集，图版 XIII-1 是分部收集液的电泳结果。

由图版 XIII-1 可见：20～30kb 片段主要存在于第 5 管中，故将该管的 DNA 透析回收。

三、pLAFR1 载体的提取和酶切

pLAFR1 为 21.6kb，有 *Eco*R Ⅰ 等单一酶切位点和四环素抗性，克隆 15-35kb 的外源片段后能为 λ 蛋白体外包装并感染受体大肠杆菌。

用碱变性法提取的 pLAFR1 经 EB-CsCl 梯度离心（40000r/min，40hr 20℃后获得清晰的 pLAFR1 质粒带）（图版 XIII-2）。

pLAFR1 的 *Eco*R Ⅰ 完全酶切结果见图版 XIV-1。由图可见未经酶切的 pLAFR1 由于产生二聚体等原因而变现出二个质粒带，但经 *Eco*RI 酶切后仅有一个 21.6kb 的带。

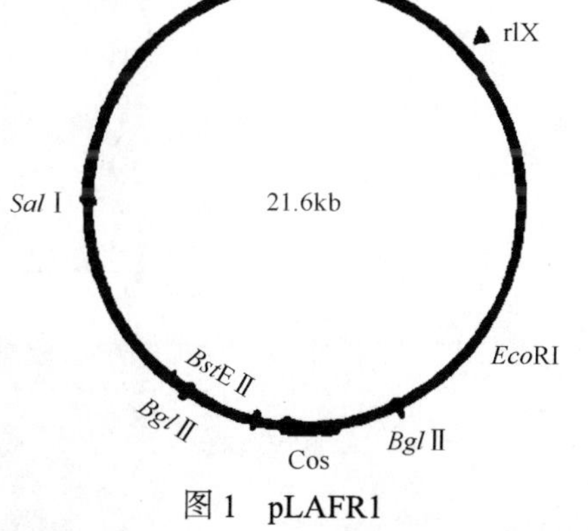

图 1 pLAFR1

四、包装蛋白的提取

包装蛋白的质量是建库成功的重要条件，提取时应注意：①菌株需在使用前作温度和紫外检测试验；②在培养时控制使两菌株的生长速度相当；③经 42℃诱导后要随时检查控制好提取时间。

在试包装 λ DNA 时，要注意受体菌的选择和处理。我们试包装的效价达 1.38×10^8（图版 XIV-2），符合建库要求。

五、酶连和体外包装

大豆根瘤菌总 DNA 的 20～30kb 片段与 pLAFR1 的酶切载体片段按 5～8∶1 的用量混合，在 T_4 连

接酶作用下连接，其结果见图版 XV-1。由图可见，连接 DNA 的大小与 λ DNA 相近，符合建库要求。

体外包装是构建基因文库的又一个重要环节，酶连 DNA 应在包装蛋白融化前加入，在包装中途添加第二管包装蛋白能提高包装效率，选用新鲜处理的受体菌并选择好涂抹平板的浓度，为便于基因库的保存，每张 Φ13cm 的硝酸纤维滤膜上的克隆数应不低于 1×10^8；并同时用甘油法保存另一份基因文库。

六、基因文库中克隆的质粒检测

从基因文库的 Tc 克隆中随机挑选 50 个，用 Bimboim 法提取质粒，并进一步用 *Eco*R1 酶切，以 pLAFR1 为对照来判断带有外源片段的重组克隆百分比。图版 XV-2 是部分抗性克隆质粒的酶切图谱。由图可见：2、3、5、7、8 等克隆均含有外源 DNA 片段。

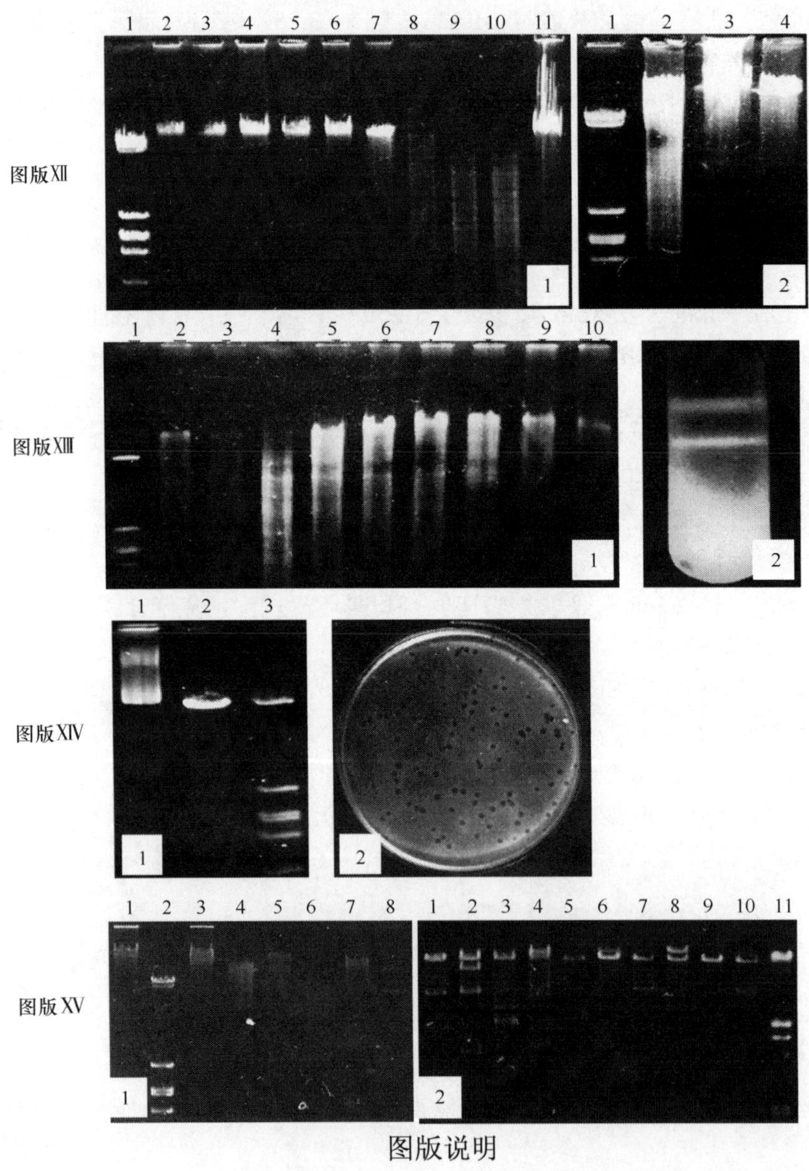

图版说明

图版 XII-1. *B. japonicum* 22-10 总 DNA 的部分酶切试验，1. λ DNA+*Eco*RI；2-10. 为总 DNA 的 EcoR1 酶切浓度试验：其中 10 为 13u、9 为 6.5u、8 为 3.25u、……按二倍稀释法向前类推；11. λ DNA。

图版 XII-2. *B. japonicum* 22-10 总 DNA 的大量酶切，1. λ DNA+*Eco*RI；2. 22-10 总 DNA+EcoRI；3. 22-10 总 DNA；4. λ DNA。

图版 XIII-1. 22-10 总 DNA 部分酶切样品的 NaCl 梯度离心，1. λ DNA+*Eco*RI；2-10. 为部分收集管：其中 10 为第 1 管，9 为第 2 管……依次类推。

图版 XIII-2. pLAFR1 的 EB-CsCl 梯度离心。

图版 XIV-1. pLAFR1 的 *Eco*RI 酶切：1. pLAFR1；2. pLAFR1+*Eco*RI；3. λ DNA+*Eco*RI.

图版 XIV-2. 试包装形成的噬菌斑，受体菌：ED8767；稀释度：10^{-5}。

图版 XV-1. 大豆根瘤菌DNA片段与载体DNA的连接：1. λ DNA；2. λ DNA+*Eco*RI；3. B52片段+pLAFR1；4. B52片段；5. 22-10片段+pLAFR1；6. 22-10片段；7. λ DNA；8. pLAFR1+*Eco*RI。

图版 XV-2. *R. fredii* B52 基因文库中 Tc^r 克隆质粒的 EcoRI 酶切图谱：1. pLAFR1；2-11. Tc^r 克隆。

在 B52 基因库供试的 30 个克隆中，有 12 个带有外源 DNA 片段，达 40%；在 22-10 基因库中则为 42%。按 Maniatis[12]的公式，构建基因文库需要的重组克隆数 N=In（1-P）/In（1-f/g），若按机率 P 为 99%；插入外源 DNA 片段大小为 2.5×10^4 bp；根瘤菌总 DNA 为 1×10^7 bp 计算，N=1.8×10^3 个。考虑到本研究的重组克隆数小于 50%，将 N 再增加一倍为 3.6×10^3 个。我们在本研究中获得的总克隆数值分别是 B52：3.9×10^4；22-10：2.1×10^4，它们均大大超过上述理论值，因而符合建库要求。

本研究中构建的基因文库中含有外源 DNA 片段的重组克隆百分率不高的主要原因在于 pLAFR1 酶切片段的自连（如图版 XV-2 中的 4、6、9、10、11）。避免载体自连的有效方法是对酶切载体进行碱性磷酸酯酶处理，去掉 5’ 磷酸后的载体不能自连。我们曾做过多次处理尝试但均未获得理想的结果，于是只好选择加大外源 DNA 用量的办法。

用硝酸纤维滤膜保存基因文库的优点是操作方便和易于下一步从库中调取目的基因；缺点是需要控制好抗性克隆的数量和成本较高。

参 考 文 献

[1]Tilghman，S.M. et al，1977，Proc. Natl. Acad. Sci.，74：4406.

[2]Hardies，S.C. and R.D. Wells，1976，Proc. Natl. Acad. Sci.，73：3117.

[3]Hohn，B. and K. Murray，1977，Proc. Natl. Acad. Sci.，74：3259.

[4]Beton，W.d. and R.W. Davis，1977，Science，196：180.

[5]Friedman，A.M. et al，1982，Gene.，18：289-296.

[6]Kondorosi，E. et al，1984，Mol. Gee. Genet.，193：445-452.

[7]Rolfe，B.G. et al，1985，“Nitrogen Fixation Research Progerss” pp：79-85.

[8]Johnston，AWB. et al，1978，Natute，276：634-636.

[9]Hennecke，H. et al，1978，“Nitrogen Fixation Hundred Years After”（Bothe，Bruijn.

[10]Robert，V. et al，1985，J.Bacteriol. 163（3）：21-26.

[11]Barbour，W.M. et al，1985，Appl. Environ. Microbiol. 50（1）：41-44.

[12]Maniatis，T. et al，1982，“Molecular Clonning”，Cold Spring Harbor Lab.，270-274.

[13]Birnboim，H.C. and J. Doly，1979，Nucleic Acids Res.，7：1513.

高效结瘤固氮大豆根瘤菌基因工程菌株的构建及应用*

周俊初　沈　辉　张忠明　胡福荣　莫才清　陈华癸

（华中农业大学生物技术中心）

窦新田　李晓鸣　李新民　李树蕃

（黑龙江省农业科学院土肥所）

摘　要　利用大型柯斯质粒 pLAFR1 为载体先构建快生型大豆根瘤菌 B52 总 DNA 的基因文库。在协助质粒 pRK2013 帮助下用基因文库同慢生型大豆根瘤菌 22−10 作三亲本杂交，从获得的转移接合子中检测到一个转入的大质粒。经大豆盆栽结瘤试验，从中筛选出 32、43、B−2−6 和 B−3−8 等 4 个工程菌株，它们的产量依次比不接种的增产 16.8%、12.0%、14.3%和 5.0%；比出发菌株 22−10 增产 7.8%、3.6%、5.6%和−2.6%。用 32 与 *E. coli* HB101 作三亲本回交，已从其 Tc 抗性转移结合子中检测到一个外源大质粒，DNA 分子杂交试验证明它是含有外源 DNA 的重组 pLAFR1。

关键词　大豆根瘤菌；基因文库；基因工程；田间应用

根瘤菌与豆科植物的共生固氮作用在自然、氮素循环和农业生产中有着重要的作用。长期以来，高效根瘤菌株的筛选一直沿用自然筛选或人工诱变的传统方法。研究表明，根瘤菌和豆科植物的共生固氮是一个由双方基因调控的复杂过程。直接参与的根瘤菌基因有固氮酶（*nif*）基因、结瘤（*nod* 和 *hsn*）基因和共生固氮（*fix*）基因等，在快生型根瘤菌中，它们定位在大的共生质粒（pSym）上[1 2 3]；在慢生型大豆根瘤菌中则定位在染色体上[4]；在快生型大豆根瘤菌中，多数在大质粒上，少数在质粒和染色体上均有[5 6]。Dejong 等（1982）曾将豌豆根瘤菌 128C53 的可转移共生质粒 PIJ1008 引入另一菌株 300，增加了结瘤植物的干重和固氮量。鉴于多数共生质粒没有自我转移能力，Simon 等（1984）构建出一套 Tn5-Mob 诱动转移系统，并成功地将苜蓿根瘤菌的共生质粒转移到根癌农杆菌 C58 中，使苜蓿形成无效根瘤[8]。韩素贞等（1990）试图用这一系统诱动紫云英根瘤菌共生质粒的努力未获成功[9]。此外，王常霖等（1987）证明宿主植物对根瘤菌共生质粒有选择性[10]。

Friedman（1982）在质粒 pRK290 上加入噬菌体的 Cos 位点构建成质粒 pLAFR1（图 1）。它不仅能在根瘤菌和大肠杆菌细胞内复制，而且能在质粒 pRK2013 的协助下通过接合作用在 G^-细菌之间转移[11]。考虑到大豆根瘤菌存在快生和慢生两种类型和共生固氮基因定位的多样性，本文选择了一条新路线来构建和筛选高效结瘤的大豆基因工程菌，取得了较好的结果。

1　材料和方法

1.1　供试菌株和质粒（见表 1）

表 1　供试菌株和质粒

菌株或质粒	表型或基因型	来源
Rhizobium fredii B52	Nod^+，Fix^+	本室分离
Bradyrhizobium japonicum 22-10	Nod^+，Fix^+	黑龙江农业科学院土肥所
E.coli BHB2688	Eam，$Temp^s$	
E.coli BHB2690	Dam，$Temp^s$	英国 John Innes 研究所
E.coli ED8767	$recA^-$，met^-	
E.coli HB101	met^-，recA13，$proA_2$	
pLAFR1	Tc^r，mob^+，tra^-	广西农学院马庆生
pRK2013	Kan^r，mob^-tra^+	

*原载《武汉大学学报》生物工程专刊，11~18，1990.

1.2 基因文库的构建

参照 Maniatis 等[12]的方法，在 *E.coli* BHB2688 和 2690 提取包装蛋白。用 Birnboim 等[13]的碱变性法提取 pLAFR1 并经 *Eco*RI 完全酶切。*R.fredil* B52 经链丝蛋白酶 E（Pronase E，Merck）和 1%SDS 溶菌后按常规方法提取总 DNA，经 *Eco*RI 部分酶切后用 NaCl 梯度离心法分离并回收 20~30kb DNA 片段。

将上述片段与酶切的 pLAFR1 按 8:1 混合、酶连和体外包装后转染受体菌 *E.coli* ED8767，在 LB+Tc（15μg/ml）平板上筛选抗性克隆。

1.3 三亲本杂交

分别用 TY 培养 *B.japonicum*22-10、LB+Tc 培养 *R.fredii* B52 基因库和 LB+Kan 培养 *E.coli* LE392（pRK2013）同步至对数期，按 1:1:1 的比例混合后取 0.5~1.0ml 用 0.22μm 微孔滤膜过滤，将滤膜移放在 TY 平板表面于 28~30℃下培养过夜。次日将膜上的细菌洗于无菌水中用于涂 SM+Tc 平板以筛选转移接合子。

工程菌株复证时的三亲本回交以 32 为供体，*E.coli* HB101 为受体，在 pRK2013 协助下进行，转移接合子的筛选平板用 LB+Tc。

1.4 大豆盆栽结瘤试验

大豆（*Glycine max*（L.）Merr.）用黑龙 29 品种，催芽后播入双层砂瓶中，同时接种根瘤菌或转移接合子菌悬液。用 Fahraeus 无氮植物营养液[14]，每年的 5 至 10 月在自然光照棚室培养，11 月至次年 4 月在人工光照培养室培养，每日光照 12h。30~40 天后取出观察植株长势和结瘤状况，并进行根瘤固氮酶活性测定。

1.5 根瘤固氮酶活性测定

用日立 163 型气相色谱仪按乙炔还原法进行[15]。

1.6 田间小区比较试验

小区试验设不接种、22-10 和 32、43、B-2-6 和 B-3-8 等 6 个处理，三次重复，随机区组排列。小区面积为 $14m^2$，起垄条播的同时施根瘤菌剂和少量氮磷肥。于分枝期、初花期和初荚期进行了三次田间取样调查，最后在收获期分小区计产和考种。

1.7 DNA 分子杂交

工程菌株经三亲回交后分离的质粒经 *Eco*RI 酶切后作 Southern 转移，以缺口转移法制备的标记 pLAFR1 为探针进行 DNA 分子杂交和放射自显影，操作方法参见文献[12]。

2 结果

2.1 *Rhizobium fredii* B52 基因文库的构建

总 DNA 经部分酶切后进行大量酶切和 NaCl 梯度离心，以收集 20~30kb DNA 酶切片段。

pLAFR1（图 1）的大小为 21.6kb，有 *Eco*RI 单一酶切位点和四环素（Tc）抗性。经 *Eco*RI 酶切后仅产生一个 21.6kb 的片段。

包装蛋白提取时应注意，在经 42℃诱导后随时检查裂解情况。包装蛋白试包装 λDNA 的效价达到 1.38×10^8pfu/1μg λDNA，符合建库要求。

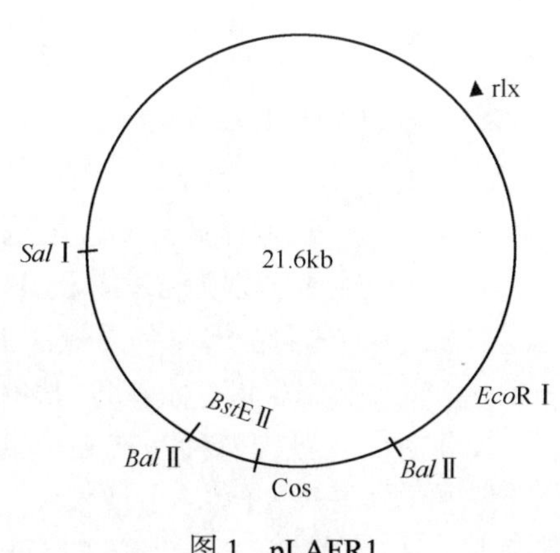

图 1 pLAFR1

B52 总 DNA 的 20~30kb 片段与 pLAFR1 用 T_4 连接酶连接，结果表明，酶连后的 DNA 大小与 λDNA，酶连物经体外包装后转染 ED8767 菌株，共获得 3.9×10^4 个 Tc 抗性克隆，它们构成了 *R. fredii* B52 的总 DNA 基因文库。

随机选取 30 个 Tc^r 克隆提取质粒，经 *Eco*RI 酶切。从图 2 可见，2、3、5、7、8 等 5 个克隆含有外源 DNA 片段。在 30 个克隆中，12 个带有外源 DNA，重组率为 40%。按 Maniatis 等[12]的公式，构建基因文库需要的重组克隆数，式中 p 为基因文库中含有全部基因的几率，一般取 99%；f 为载体携带的外源 DNA 片段大小，在本研究中为 2.5×10^4 bp，根瘤菌总 DNA 大小 g 可取 1×10^7 bp，由此代入公式后计算出的 $N=1.8\times10^3$。获得的 Tc^r 克隆数为 3.9×10^4，按 40%计算的重组克隆数亦达 1.56×10^4，已大大超过了上述理论值。

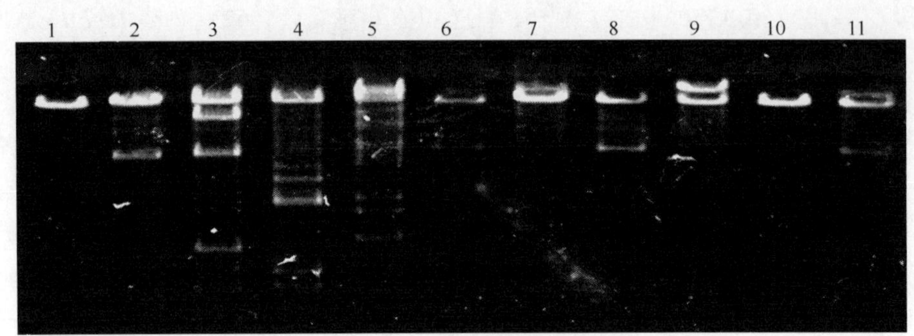

图 2 部分基因文库中 Tc^r 克隆质粒的 *Eco*R Ⅰ酶切

1. pLAFR1；2~11. Tc^r clones

2.2 三亲本杂交和植物盆栽结瘤试验

三亲本杂交的 Tc 抗性转移频率为 3.0×10^{-5}，在相同条件下 22-10 的 Tc 抗性自发突变频率却小于 10^{-8}。

部分转移接合子的质粒快速检测结果见图 3。由图可见，除个别样品外，多数转移接合子均比受体菌 22-10 多了一条大质粒带。

用转移接合子进行了三批大豆盆栽结瘤试验，每批至少接种 50 钵，同时用 22-10 接种与不接种的对照。试验表明，其接种大豆植株的长势明显地优于对照和 22-10。其根瘤的平均乙炔还原活性为 22-10 的三倍。

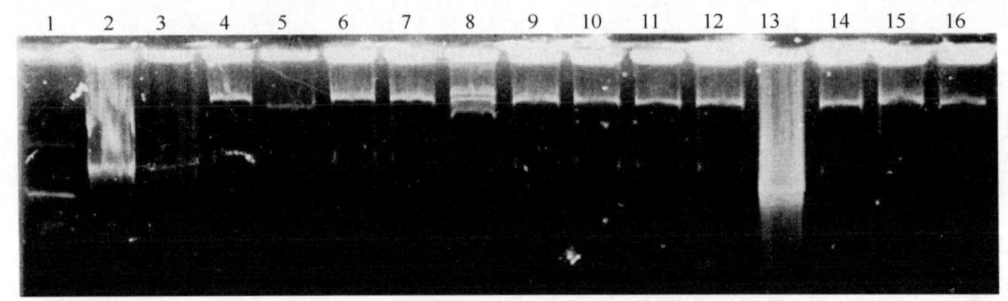

图 3 转移接合子的质粒快速检测

1. pLAFR1；2~7 和 9~16. 转移接合子；8. *R. fredii* USDA205

2.3 小区田间比较试验

田间小区试验中，由于有土著大豆根瘤菌，故不接种仍结瘤。在分枝期（V_2）、初花期（R_1）和结荚始期（R_3）三次取样调查结果表明，所有接种处理的结瘤和固氮水平均优于不接种对照，在 V_2 期，32 和 43 的根瘤数分别为 22-10 的 115%和 125%。在 R_1 期，32、43 和 B-2-6 在瘤数和固氮酶活等方面也优于 22-10，它们在结瘤和固氮能力上的优势继续保持到初荚期。B-3-8 在各次调查中均与 22-10 的水平接近，32、22-10 和对照在各时期瘤数和固氮酶活性的比较见图 4。

收获时的各处理考种与产量结果见表 2。可见，32 较不接种增产 16.8%，亩增收大豆 28.2kg；较出发菌株 22-10 增产 7.8%，亩增收大豆 14.25%。小区产量统计的方差分析表明，试验各处理间的 F 值为 11.342，大于 $F_{0.01}$（5.64）。处理间差异达到极显著标准。用新复极差法（LSR 法）进一步对各处理产量进行的差异显著测定（表 3）结果表明，工程菌株 32 与出发菌株 22-10 间的差异已达到显著标准；各接种处理与不接种处理间的差异均达到显著标准；32 与 B-3-8 和不接种两个处理间的差异还达到极显著标准。

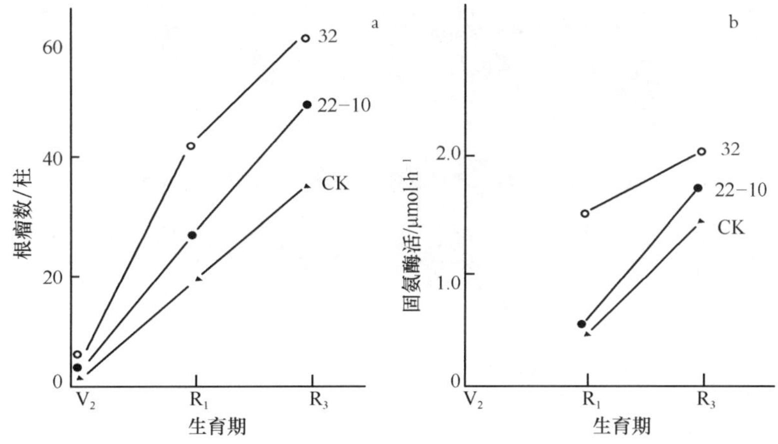

图 4 32 与 22-10 和对照在大豆不同生育期中瘤数和固氮酶活性的变化

a. 瘤数；b. 固氮酶活性

2.4 基因工程菌株 32 的复证试验

为了在分子水平上比较工程菌株与出发菌株 22-10 的差异，以 32 为供体菌在 pRK2013 协助下与 *E.coli* HB101 进行了三亲本回交，在 LB+Tc 平板上获得 485 个转移接合子，转移频率为 6.5×10^{-7}。选取其中 8 个代表性接合子检测质粒，均发现有一个比 pLAFR1 大的外源大质粒（图 5）。

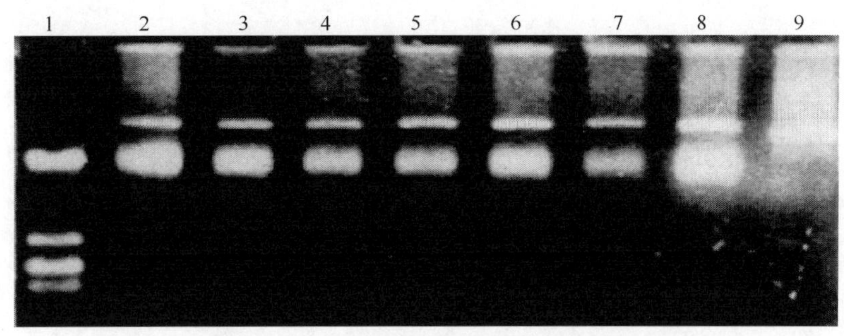

图 5 32 与 HB101 转移接合子的质粒检测

1. λ+*Eco*R1；2~8. 转移接合子（32H1-32H7）；9. pLAFR1

选接合子 $32H_3$ 和 $32H_7$ 的质粒电泳凝胶经 Southern 转移后以 pLAFR1 为探针的检测结果见图 6。由图可见它们均呈明显的正反应，在大小上也高于 pLAFR1。

取 $32H_3$ 以 pLAFR1 为对照经 *Eco*RI 酶切的电泳结果见图 7A，Southern 转移后与 pLAFR1 探针的杂交结果见图 7B。由图可见，$32H_3$ 含有一个能与 pLAFR1 杂交但比它更大的酶切片段，这就是含有外源 DNA 的重组 pLAFR1，$32H_3$ 的另一个酶切小片段不能与 pLAFR1 杂交，它也是插入外源 DNA。

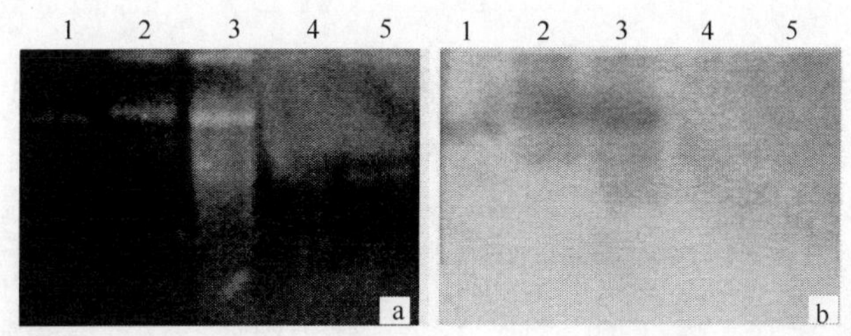

图 6 转移接合子 $32H_3$ 和 $32H_7$ 的电泳和 DNA 杂交图谱，探针为 pLAFR1

a（凝胶电泳），b（放射自显影）：1. pLAFR1；2. $32H_3$；3. $32H_7$；4. $43H_2$；5. pRK013

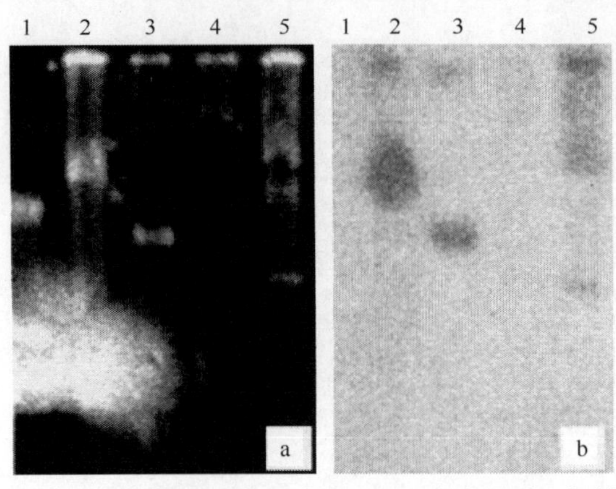

图 7 转移接合子 $32H_3$ 的 *Eco*RI 酶切和 DNA 杂交图谱，探针为 pLAFRI

a（凝胶电泳），b（放射自显影）：1. λ+*Eco*RI；2. $32H_3$+*Eco*RI；3. pLAFR1+*Eco*RI；4. $32H_3$；5. pLAFR1

表 2 大豆根瘤菌基因工程菌田间小区试验收获期考种与产量*

处理菌株	株高 cm	分指数 个/株	一粒荚 个/株	二粒荚 个/株	三粒荚 个/株	四粒荚 个/株	秕荚 个/株	总荚数 个/株	百粒重 g	折合亩产（斤）	为22-10的%	为CK的%
22-10	91.5	0.77	4.83	8.67	8.93	3.60	0.70	26.0	18.9	364.2	100.0	108.3
32	100.0	0.57	4.77	8.57	9.87	5.57	0.60	28.8	18.7	392.7	107.8	116.8
43	103.0	0.33	5.30	9.60	9.47	3.87	0.83	28.2	19.1	377.4	103.6	112.0
B-2-6	102.5	0.23	5.10	8.40	10.13	4.53	0.80	28.2	19.8	384.0	105.6	114.3
B-3-8	101.1	0.30	4.43	8.63	7.97	4.03	0.93	25.1	19.3	354.6	97.4	105.0
CK	94.6	0.40	4.90	7.93	9.23	3.70	0.70	25.9	18.9	336.3	92.3	110.0

*除亩产外，均系每小区取 10 株调查的平均值

表 3 小区产量及 LSR 分析

处理	平均数（$\bar{x}$）	0.05	0.01
32	392.7	a	A
B-2-6	384.5	ab	AB
43	377.4	ab	AB
22-10	364.2	b	AB
B-3-8	354.6	b	B
不接种	336.3	c	B

3 讨论

本文提出的选育高效结瘤固氮菌株的新路线与传统方法相比，其优点是：①通过构建基因文库和用整个库与受体菌作三亲本杂交，可以为供体菌的全部基因提供分区段进入受体菌的机会。②用广谱性质粒作载体，能避免供体菌共生质粒与受体菌固有质粒存在互不相容性的困难。③若供体菌与受体菌皆为根瘤菌，可能通过三亲本杂交形成共生固氮基因的局部异源二聚体，为共生固氮能力的提高创造了条件。④以植物结瘤试验初筛，避免逐一进行单个接合子纯化和结瘤试验，可从结瘤植株长势淘汰接合子，从长势好的根瘤中回收高效接合子菌株。⑤构建的基因文库可用于共生固氮基因的分离鉴定及其他研究。

影响根瘤菌在“老区”接种效果的另一个问题是，土生性根瘤菌的竞争和它们在土壤中均匀分布的位置效应[16]，Triplett 等（1990）从三叶草根瘤菌无效菌株 T24 中分离到产生三叶草根瘤菌素（Trifolitoxin）的基因 *tfx*，将其克隆到载体质粒 pLAFR3 之后用于同有效三叶草根瘤菌 TA_1 作三亲本杂交，获得的转移接合子菌体 TA_1::10-15 在同对三叶草根瘤菌素敏感菌株的竞争结瘤试验中表现出很强的优势[17]。对我们构建高效结瘤固氮的大豆根瘤菌工程菌株亦有较大的启迪。

基因工程菌株 32 经三亲本回交从转移接合子中检测到一个分子量比 pLAFR1 大的外源质粒，经酶切和 Southern 转移，用 pLAFR1 为探针进行的 DNA 杂交结果证明该质粒为含有外源 DNA 的重组 pLAFR1，进一步的研究正在进行之中。

参 考 文 献

[1]Kondorosi E et al. Mol Gen Genet，1984，193，445-452.

[2]Rolfe B G et al. Nitrogen fixation Research Progress. （Evans B and Newton eds），1985，79-85.

[3]Johnson A W B et al. Nature，1978，276，634~636.

[4]Hennecke H et al. Nitrogen Fixation Hundred Years After. （Bothe B and Newton eds），1988，339-344.

[5]Masterson R V et al. J Bacteriol，1985，163（3），21-26.

[6]Stanly J et al. J Bacteriol，1986，166，628-634.

[7]DeJong T M et al. J Gen Microbiol，1982，128，1829-1838.

[8]Simon R et al. Mol Gen Genet，1984，196，413-420.

[9]韩素贞等. 中国微生物学杂志，1990，2（1），56-63.

[10]Wang et al. J Gen Microbiol，1986，132，2063-2070.

[11]Friedman A M et al. Gene，1982，18，289-296.

[12]Maniatis T et al. Molecular cloning. Cold Spring Harbour Lab，1982，270-274.

[13]Birnboim H C and Doly J. Nucleic Acids Res，1979，7，1513.

[14]Vincent J M. Amanual for the practical study of the root nodule bacteria. Black-well Scientific Publications，oxforcl，1970.

[15]Bergersen F J. Methods for Evaluation Biological Nitrogen Fixation. John Wiley and Scientific Publications， oxford，1970.

[16]王福生等. 华中农业大学学报，1985，4（3），38-47.

[17]Triplett W E. Appl and Envir Microbiol，1990，56（1），98-103.

高效结瘤固氮大豆基因工程菌株的构建及小区田间试验结果*

周俊初　沈　辉　张忠明　胡志浩　彭文涛　胡福荣　陈华癸

（华中农业大学土壤化学系）

窦新田　李晓鸣　李新民

（黑龙江省农业科学院土壤肥料研究所）

根瘤菌与豆科植物的共生固氮作用在自然界氮素循环和农业生产中有着重要的作用，根瘤菌剂接种的增产效果也早已为农业生产的实践所公认。但是长期以来高效根瘤菌株的筛选一直沿用自然分离和人工诱变的传统方法，仅仅依靠个别基因的突变往往难于获得共生固氮效率大幅度提高的新菌株，以致使菌剂的接种效果在"老区"逐年下降并最终失去对接种的需要性。

研究表明根瘤菌和豆科植物的共生固氮是一个由双方基因控制的复杂过程。已知直接参与这一过程的根瘤菌基因有固氮酶基因（*nif*）、结瘤基因（*nod*）和共生固氮基因（*fix*）等三种类型。在苜蓿、三叶草和豌豆等快生型根瘤菌中，它们均定位在称为共生质粒（pSym）的大质粒上；在慢生型大豆根瘤菌总则定位在染色体上；而快生型大豆根瘤菌的定位情况则随菌株而异，多数在质粒上，少数则质粒和染色体上均有。按照提高基因拷贝数有可能提高基因表达能力的设想，DeJong 等（1982）曾将豌豆根瘤菌 128C53 的可能转移共生质粒 pJJ1008 引入另一菌株 300，并明显地增加了其结瘤植株的干重和固氮量。但是已知多数根瘤菌的共生质粒缺少自我转移能力。Simon 等（1984）曾构建出一套 Tn5-Mob 诱变转移系统并成功地将苜蓿根瘤菌 2011 的共生质粒诱动转移到根癌农杆菌 C58 中并使之能在苜蓿上形成无效根瘤。但韩素贞等（1987）曾试图用这一系统来诱动紫云英根瘤菌共生质粒的努力却未获成功。此外，王常霖等（1987）的试验还证明宿主根物对根瘤菌的共生质粒具有选择性，它们在共生过程中会排斥导入受体根瘤菌的异源共生质粒，从而限制了它对共生固氮的作用。

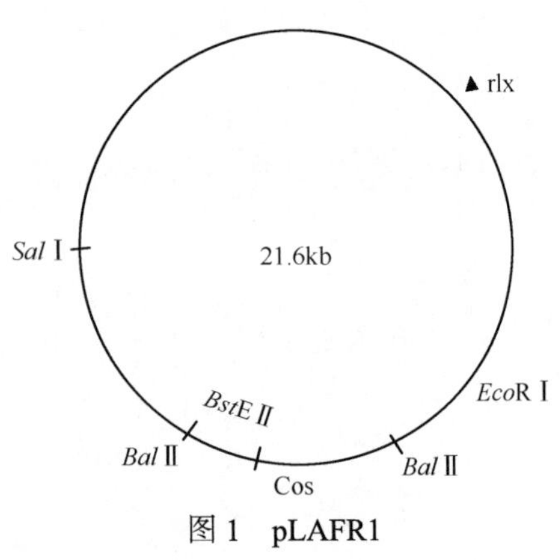

图 1　pLAFR1

Fredman 等（1982）在广谱质粒 pRK290 上加入噬菌体的 Cos 位点构建出能插入 20~30kb 外源基因并可为入包装蛋白进行体外包装的质粒——pLAFR1（图 1），它不仅能在根瘤菌和大肠杆菌等 G⁻ 细菌细胞内复制，而且能在 pRK2013 的协助下通过接合作用在 G⁻ 细菌之间转移。考虑到大豆根瘤菌存在快生和慢生两种不同的类型和它们的共生固氮基因定位的多样性，我们在本研究中选择了一条新的路线来构建和筛选高效结瘤固氮大豆基因工程菌（图 2），并取得了成功。这一路线的特点是先以 pLAFR1 为载体构建快生型大豆根瘤菌的总 DNA 基因文库，然后用基因文库同目前已经在推广应用的慢生型大豆根瘤菌作三亲本杂交，并通过植物盆栽结瘤试验对转移接合子进行初筛，再从入选植株的根瘤中分离回收接合子，再进入田间小区比较试验和分子生物学的复证工作。

*原载于《高技术新技术农业应用研究》，北京：中国科技出版社:189~194，1991.

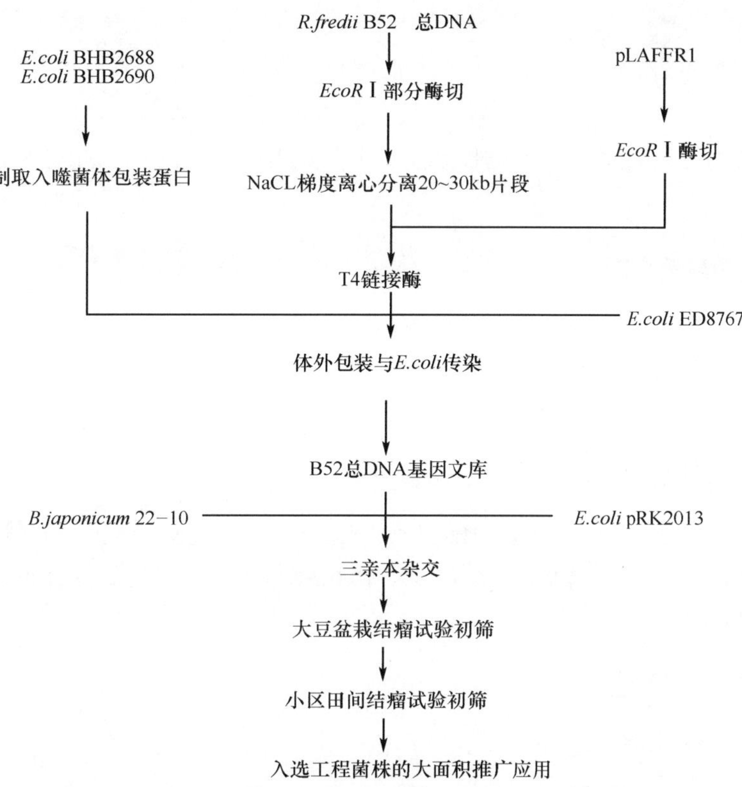

图 2　高效结瘤固氮大豆基因工程菌的构建和筛选路线

一、材料和方法

（一）供试菌株和质粒

见表 1。

表 1　供试菌株和质粒

菌株或质粒	表型或基因型	来源
Rhizobium fredii B_{52}	Nod^+，Fix^+	本室分离
Bradyrhizobium japonicum 22-10	Nod^+，Fix^+	黑龙江农科院土肥所
E.coli　BHB 2688	Eam，$Temp^s$，	英国 John Inners 研究所
E.coli　BHB 2690	Dam，$Temp^s$	英国 John Inners 研究所
E.coli　ED 8767	$RecA^-$，met^-	英国 John Inners 研究所
E.coli　HB 101	F^-，recA13，$proA_2$	英国 John Inners 研究所
pLAFR1	Tc^r，mob^+，tra^-	广西农学院马庆生
pRK2013	Km^r，mob^-，tra^+	广西农学院马庆生

（二）基因文库的构建

参照 Maniatis 等的方法，用 2688 和 2690 提取包装蛋白，试包装入 DNA 的效价高于 10^7pfu/1μg DNA 之后，与液氮中保存备用。用 Birnboim 等的碱变性法提取 pLAFR1，经 CsCl 梯度离心纯化后用 *Eco*R Ⅰ部分酶切后备用。用链丝蛋白酶 E 和 SDS 在 37℃下溶菌及苯酚、氯仿抽提法从 B52 中提取总 DNA，经 *Eco*R Ⅰ部分酶切后用 NaCl 梯度离心法分离并回收 20~30kb 片段。

将上述 20~30kb 片段与 pLAFR1 的酶切片段按 8∶1 的比例混合，经 T_4 连接酶连接后，用上述装蛋白体外包装并感染受体菌 ED8767，在 LB+Tc 平板上获得抗性克隆，对克隆随机抽样作质粒分离和酶切分析，当带有外源 DNA 片段的重组克隆数超过建库理论值后，即构建成该根瘤菌总 DNA 的基因文库。

（三）三亲本杂交

用微孔滤膜法进行 B52 基因库、22-10 和 pRK2013 的三亲本杂交，在 SM+Tc 平板上筛选转移接合子，经影印和点接三次纯化后用于植物结瘤实验。复证时用 32、HB101 和 pRK2013 进行三亲本杂交，筛选平板为 LB+Tc。

（四）大豆盆栽结瘤试验

大豆用黑龙 29 品种，催芽后播入自制的双层砂培瓶中，同时接种供试根瘤菌或转移接合子菌悬液，每年 11 月至次年 4 月送人工光照室培养，5~10 月在自然光照棚室培养，30~40 天后取出测定。

（五）田间小区比较试验

田间试验在黑龙江省农业科学院试验场进行。6 个处理三次重复，随机区间排列，小区面积为 $14m^2$，起垄条播后同时施菌剂和少量氮磷肥，于分枝期、初花期和初荚期进行三次田间取样调查，最后在收获期分小区测产和考种。

（六）根瘤固氮酶活性测定

用气相色谱仪和乙炔还原法进行。

二、结　　果

（一）*Rhizobium fredii* B52 基因文库的构建

1.提取的包装蛋白试包装效价为 1.38×108pfu/1μg DNA。

2.提取的 B52 总 DNA 大小超过 70kb，部分酶切的最适 *Eco*R Ⅰ酶用量为 2.5u/μg 总 DNA，经大量酶切和 NaCl 梯度离心后收到足够量的 20~30kb 片段。

3.提取的 pLAFR1 经 CsCl 梯度离心纯化后能被 *Eco*R Ⅰ酶切。

4.总 DNA 20~30kb 片段与 pLAFR1 酶切片段的酶连混合物感染 ED8767 后，在 LB+Tc 平板上获得 3.9×104 个克隆随机选取 50 个克隆进行的质粒分类和酶切分析，证明带有外源 DNA 片段的重组克隆数占 23%。由此而计算出本基因库中的重组克隆总数为 9×10^3 个，已大大超过建库所需的理论值 1.8×10^3。

（二）三亲本杂交和植物盆栽结瘤试验

用快生型大豆根瘤菌 B52 基因文库在 pRK2013 协助下与慢生型大豆根瘤菌 22-10 作三亲本杂交，在 SM+Tc 上测得 Tc 抗性转移频率为 3.0×10^{-5}，而同样条件下 22-10 的 Tc 自发突变频率 $<10^{-8}$。

用于每批转移接合子的盆栽为 50 钵，同时有 22-10 接种和不接种对照，经 30~40 天光照培养后按批取出，先观察记录叶色、株高、每株瘤数和瘤重，然后用气相色谱仪测定每株根瘤的乙炔还原活性，最后用 YMA+Tc 培养基从高活性根瘤中分离转移接合子菌株，表 2 是三批盆栽结瘤试验的汇总表。共从表中选出 32、43、B-2-6 和 B-3-8 四个接合子菌株进入田间小区比较试验。

表 2 *R. fredii* B_{52} 基因文库与 *B. japonicum* 22-10 三亲本杂交转移接合子的盆栽结瘤试验结果*

观测项目	株高（cm）				根瘤重（g/株）				乙炔还原活性（μMol C_2H_2/g 根瘤 /小时）							
试验次数	1	2	3	平均	1	2	3	平均	1	为 20-10(%)	2	为 20-10(%)	3	为 20-10(%)	总平均	为 20-10(%)
CK	28.3	44.9	65.5	46.2	0	0	0	0	0	0	0	0	0	0	0	0
22-10	36.8	57.7	63.2	52.6	1.00	0.72	0.85	0.86	5.08	100	11.00	100	24.94	100	13.67	100
转移接合子	38.0	51.2	92.9	60.7	0.67	1.18	0.72	0.86	42.73	841	30.99	274	59.19	237	41.27	302

*均为 10 株以上的平均值。

由表 2 可见转移接合子在平均水平上株高为 22-10 的 115.4%，瘤重持平，但固氮酶活性高 2 倍以上。

（三）小区田间比较试验

小区田间试验设 22-10、32、43、B-2-6、B-3-8 和 CK6 个处理，三次重复，小区随机区组排列并预留有取样区。在大豆分枝期 V_2、初花期（R_1）和结荚始期 R_3 进行的三次田间取样调查结果见表 3。由表 3 可见所有接种处理在结瘤和固氮水平均优于对照。在分枝期，32 和 43 的结瘤能力分别为 22-10 的 115%和 125%；在初花期，32、43 和 B-2-6 在瘤数和固氮酶活等方面也优于 22-10；他们在结瘤和固氮梦里上的优势继续保持到初荚期。B-3-8 菌株在各次测定中均与 22-10 的水平接近，未见其明显的优势。

表 3 大豆根瘤菌基因工程菌株的田间小区试验调查结果表*

生育期	处理菌株	株高（cm）	地上部鲜重（g/株）	地上部干重（g/株）	地下部鲜重（g/株）	地下部干重（g/株）	根瘤总数（个/株）	为 22-10 的（%）	瘤重（g/株）	为 22-10 的（%）	固氮酶活性（μMol $C_2H_4g^{-1}hr^{-1}$）	为 22-10 的（%）
分枝期（v_2）	22-10	11.2	3.85	0.88	1.07	0.24	4.60	100	0.015	100		
	32	12.5	4.38	1.03	1.10	0.25	5.27	115	0.015	100		
	43	10.9	4.03	0.91	1.24	0.25	5.73	125	0.016	107		
	B-2-6	10.4	4.68	1.08	1.11	0.24	4.57	99	0.016	107		
	B-3-8	11.5	4.08	0.93	1.28	0.26	4.57	99	0.016	107		
	CK	11.2	3.77	0.88	1.06	0.23	2.63	57	0.014	93		
初花期（R_1）	22-10	35.2	31.2	5.38	6.4	1.05	27.2	100	0.35	100	0.55	100
	32	37.8	38.0	6.36	6.7	1.33	42.5	156	0.51	146	1.50	273
	43	29.6	37.1	5.95	6.0	1.07	29.1	107	0.34	97	0.60	109
	B-2-6	36.5	39.8	5.48	6.4	1.15	36.6	135	0.37	106	1.33	242
	B-3-8	36.3	32.8	5.30	6.2	1.02	26.0	96	0.35	100	0.49	89
	CK	33.3	31.7	5.10	6.4	1.00	19.1	70	0.30	86	0.45	82
结荚期（R_3）	22-10	84.1	98.3	17.4	10.9	2.9	48.8	100	0.66	100	1.72	100
	32	90.1	108.6	23.5	11.6	2.9	60.2	123	0.94	142	2.00	116
	43	89.0	104.2	20.5	12.8	3.0	57.5	118	0.99	150	2.24	130
	B-2-6	88.0	114.8	23.0	11.3	2.7	63.2	130	1.01	153	2.18	127
	B-3-8	88.7	86.9	20.2	11.7	2.5	45.7	94	0.72	109	1.94	113
	CK	82.9	82.9	22.3	10.7	2.4	35.4	73	0.64	97	1.46	85

*系每小区取 10 株调查的平均值。

秋收的小区产量比较和考种结果见表 4。由表 4 可见：与对照相比，32 增产 16.8%，亩增收大豆

28.2kg；B-2-6 增产 14.3%，亩增收大豆 24.1kg；43 增产 12.2%，亩增收大豆 20.6kg。以上三个菌株分别比出发菌株 22-10 增产 7.8%，增收 14.25kg（32）；5.6%与 9.9kg（B-2-6），3.6%与 6.6%（43）。统计分析的结果表明基因工程菌株 32 与出发菌株 22-10 的产量差异达到显著标准。经小区田间试验选出的基因工程菌株 32 于 1990 年在黑龙江作大面积推广应用试验。

表 4 大豆根瘤菌基因工程苗田间小区试验收获期考种与产量表*

处理菌株	株高（cm）	分枝数（个/株）	一粒荚（个/株）	二粒荚（个/株）	三粒荚（个/株）	四粒荚（个/株）	秕荚（个/株）	总荚数（个/株）	百粒重（g）	折合亩产（斤）	为 22-10 的（%）	为 CK 的（%）	显著性水平（5%）
22-10	91.5	0.77	4.83	8.67	8.93	3.60	0.70	26.0	18.9	364.2	100.0	108.3	B
32	100.0	0.57	4.77	8.57	9.87	5.57	0.60	28.8	18.7	392.7	107.8	116.3	A
43	103.0	0.33	5.30	9.60	9.47	3.87	0.83	28.2	19.1	377.4	103.6	112.2	AB
B-2-6	102.5	0.23	5.10	8.40	10.13	4.53	0.80	28.2	19.8	384.0	105.6	114.3	AB
B-3-8	101.1	0.30	4.43	8.63	7.97	4.03	0.93	25.1	19.3	354.6	97.4	105.0	B
CK	94.6	0.40	4.90	7.93	9.23	3.70	0.70	25.9	18.9	336.3	92.3	100.0	C

*除亩产外，均系每小区取 10 株调查的平均值。

（四）基因工程菌株 32 的复证试验

为了取得工程菌株 32 与出发菌株 22-10 在分子水平差异的直接证据，我们用 32 为供体菌在 pRK2013 协助下与 *E. coli* HB101 进行了反向的三亲本杂交，已获得数十个转移接合子，电泳结果表明，接合子有一条转入的外源大质粒带。进一步的研究工作尚在进行中。

三、讨 论

在近年来的固氮分子遗传学研究中，国内外的重点都放在固氮基因的定位和调控等方面，考虑到生物固氮过程的复杂性和多样性，要彻底揭开这一过程的奥秘，进一步扩大生物固氮的范围和规模无疑还需要作大量的工作和较长的时间。从应用方面来考虑，我们应当进一步充分发挥现有固氮微生物，特别是根瘤菌的作用，为此急需根据我们现有的关于固氮作用的分子遗传学知识提出一个有效快捷的筛选高效结瘤固氮菌株的新路线。本文提出并成功地按新路线在两年内筛选获得了一个高效大豆基因工程菌株。我们认为本路线的优点是：

1. 通过构建基因文库和三亲本杂交，让根瘤菌的全部基因分段进入受体菌比较；

2. 让植物作为转移接合子的初筛工具，省去了大量纯化单个接合子的工作，并大大减少了试验工作量和扩大了筛选范围；

3. 利用了真核生物多倍体的生长优势原理，在原核生物中通过构建部分二倍体或提高基因的拷贝数来加强固氮基因的表达水平并筛选获得基因工程菌株。

大豆基因工程根瘤菌田间结瘤和共生固氮效应*

窦新田　李晓鸣　李新民　周继松　李树藩

（黑龙江省农科院土肥所微生物室）

周俊初　沈　辉　张忠明　胡志浩　彭文涛　胡福荣　陈华癸

（华中农业大学土化系生物固氮室）

利用三亲本接合转移方法，将快生型大豆根病菌 B52 的基因文库克隆转移到受体菌慢生型大豆根瘤菌 2210 中，获得 4 株大豆基因工程根瘤菌（32、43、B−2−6、B−3−8）。田间接种试验表明：在大豆分枝期（V2），43 和 32 结瘤数分别比不接种对照增加 117.9%和 100.4%；在初花期（R_1），32 和 B−2−6 结瘤数分别比对照增加 122.5%和 91.6%，根瘤固氮酶活性增加 233.3%和 195.6%；在结荚始期（R3），B−3−8 和 32 结瘤数比对照增加 78.5%和 70.1%，根瘤固氮酶活性 43 和 B−2−6 分别比对照增加 53.4%和 49.3%。产量统计表明：32、B−2−6 和 43 比对照分别增产 16.8%、14.3%和 12.2%，亩增收大豆分别为 28.2、24.1 和 20.6 公斤。以上三个菌株均比受体菌株 2210 增产率高。

关键词　大豆根瘤菌；基因工程；结瘤；固氮

大豆接种根瘤菌能提高大豆的产量和蛋白质的含量，已被国内外大量的试验和研究所证实[1, 2]。黑龙江省是我国大豆的主产区，年播种面积、总产量和出口量均居全国首位。1983~1989 年全省大豆根瘤菌接种面积累计达 850 万亩。为了提高大豆根瘤菌的接种效果，不断向农业生产提供高效固氮的优良菌种则是一项重要的研究任务。

目前，在大豆生产上应用的主要是慢生型大豆根瘤菌（*Bradyrhizobium japonicum*）。1982 年，H.H.keyser 等从中国大豆上发现了快生型大豆根瘤菌（*Rhizobium fredii*），它具有生长速度快和对大豆品种选择性较强等特点[3]。近年，随着分子生物学和生物技术的发展，已对根瘤菌固氮和结瘤基因进行了定位和转移研究。周俊初等[4]在 1988 年完成了快、慢生型大豆根瘤菌基因文库的基础上，用三亲本接合转移方法将快生大豆根瘤 B_{52} 的基因文库克隆转移到慢生型大豆根瘤菌 2210 中，经植物结瘤试验选出 4 株基因工程菌。1989 年在黑龙江省大豆田间进行了结瘤和共生固氮试验，本文即是此试验结果的报导。

材料和方法

（一）根瘤菌菌种

1. 2210，黑龙江省农科院土肥所选育出的慢生型大豆根瘤菌，已在全省应用 100 万亩。
2. 32、43、B-2-6、B-3-8，华中农业大学土化系选育的大豆基因工程菌。

（二）试验方法

1. 田间小区：采取完全随机区组设计；6 个处理分别为：CK（不接种）、2210、32、43、B-2-6、B-3-8；每小区 4 垅，5 米长，面积为 14m^2。重复三次。

*原载于《生物技术》，1（3）:26~29，1991.

2. 接种方法：所有菌种进行液体振荡培养，菌数达 20 亿/ml 时用草炭吸附成菌剂。菌剂用量，每小区 120g；种肥磷酸二铵每小区 158g。

3. 种植方法：人工开沟，深 15-20cm，施种肥后复土；人工点籽，粒距 5cm；施菌剂，复土，压实。大豆品种为黑农 33 号。

（三）调查和测定

1. 结瘤调查：在大豆生育期挖取 10 株大豆，洗根，摘瘤，调查纪录单株瘤数、瘤重。

2. 固氮酶活性：用气相色谱仪应用乙炔还原法，按乙炔乙烯峰高比计算根瘤固氮酶活性。

结果与讨论

（一）大豆基因工程根瘤菌选育过程

1. 大豆根瘤菌基因文库的构建

（1）根瘤菌总 DNA 的提取：用 PA 培养基培养的大豆根瘤菌 2210 和 B52 以链丝蛋白酶 E 和 SDS 在 37℃下溶菌，用苯酚和氯仿多次反复抽提以除去蛋白等杂质，最后用异丙醇在－20℃下沉淀总 DNA。

（2）总 DNA 片断提取：用 *Eco*RI 对总 DNA 作部分酶切。然后对酶切物用 NaCl 梯度离心分离，分小管逐层回收后上样电泳，挑选并沉淀回收 20~30kb 的总 DNA 片段。

（3）用碱变性法从 *E.coli* 提取质粒 pLAFR1。并经 CsCl 超速离心进一步纯化，回收的 pLAFR1 经 *Eco*RI 完全酶切后于-20℃下保存备用。

（4）从 *E.coli* 2688 和 2690 中提取包装蛋白，于液氮保存备用。该蛋白试包装 λDNA 的效价达到 $1.38\times10^{8}/1\mu g$ λDNA。

（5）将大豆根瘤菌的 20~30kb 片段与 pLAFR1 酶切片段按 5-8:1 的量混合，在 T4 连接酶作用下连接，并随之用包装蛋白进行体外包装，让包装物感染受体菌 *E.coli* ED8767，在 Tc 平板上获得的 Tc^{r} 克隆数分别是 3.0×10^{4}（B52）和 2.1×10^{4}（2210）。

2. 共生固氮基因的转移

（1）用快生型大豆根瘤菌 B_{52} 的基因文库在协助质粒 pRK2013 的帮助下与受体菌慢生型大豆根瘤菌 2210 作三亲本杂交，在 SM 合成培养基（淘汰未与基因库杂交的受体菌 2210）平板上筛选 2210 接合转移子，转移频率为 3×10^{-5}。

（2）将转移重组子接种在水培条件下生长的大豆植株，通过结瘤数和根瘤固氮酶活性等指标，选出优于受体菌 2210 的工程菌株 32、43、B-2-6 和 B-3-8。

（二）大豆基因工程根瘤菌田间结瘤和固氮效果

在大豆分枝期（V2）、初花期（R1）、结荚始期（R3）三次调查表明，所有接种处理在结瘤和固氮效果上均好于不接种对照。其中在分枝期，以 43 和 32 结瘤数最多，比不接种增加 117.9%和 100.4%；在初花期，以 32 和 B-2-6 结瘤数最多，比不接种增加 122.5% 和 91.6%，固氮酶活性也以 32 和 B-2-6 最高，比不接种增加 233.3%和 195.6%；在结荚始期，结瘤数最高的为 B-2-6 和 32，分别比不接种增加 78.5%和 70.1%，固氮酶活性 43 和 B-2-6 最高，分别比不接种增加 53.4%和 49.3%（表 1）。

表 1　大豆基因工程根瘤菌田间结瘤和固氮效果

生育期	处理	株高 cm	地上部鲜重 g/株	地举部干重 g/株	地下部鲜重 g/株	地下部干重 g/株	瘤数 个/株	瘤鲜重 g/株	固氮活性 $\mu mol \cdot g^{-1} \cdot h^{-1}$
分枝期（V2）	2210	11.2	3.85	0.88	1.07	0.24	4.60	0.015	
	32	12.5	4.83	1.03	1.10	0.25	5.27	0.015	
	43	10.9	4.03	0.91	1.24	0.23	5.73	0.016	
	B-2-6	10.4	4.68	1.08	1.11	0.24	4.57	0.016	
	B-3-8	11.6	4.08	0.93	1.28	0.26	4.57	0.016	
	ck	11.2	3.77	0.88	1.06	0.23	2.63	0.014	
初花期（R1）	2210	35.2	31.2	5.38	6.4	1.05	27.2	0.35	0.55
	32	37.8	38.0	6.36	6.7	1.33	42.5	0.51	1.50
	43	29.6	37.1	5.95	6.0	1.07	29.1	0.34	0.60
	B-2-6	36.5	39.8	5.48	6.4	1.15	36.6	0.37	1.33
	B-3-8	36.3	32.8	5.30	6.2	1.02	26.0	0.35	0.49
	ck	33.2	31.7	5.10	6.4	1.00	19.1	0.30	0.45
结荚始期（R3）	2210	84.1	98.3	17.4	10.9	2.9	48.8	0.66	1.72
	32	90.1	108.6	23.5	11.6	2.9	60.2	0.94	2.00
	43	89.0	104.2	20.5	12.8	3.0	57.5	0.99	2.24
	B-2-6	88.0	114.8	23.0	11.3	2.7	63.2	1.01	2.18
	B-3-8	88.7	86.9	20.2	11.7	2.5	45.7	0.72	1.94
	ck	82.9	92.8	22.3	10.7	2.4	35.4	0.64	1.46

与受体菌株 2210 相比，32、43、B-2-6 三个菌株对大豆的结瘤和共生固氮影响均高于 2210，只有 B-3-8 低于 2210，但它优于不接种对照。

（三）大豆基因工程根瘤菌对大豆产量的影响

秋收考种和测产说明：大豆接种基因工程根瘤菌均提高了大豆产量。其中 32 比不接种对增产 16.8%，亩增收大豆 28.2 公斤；B-2-6 增产 14.3%，亩增收大豆 24.1 公斤；43 增产 12.2%，亩增收大豆 20.6 公斤；B-3-8 增产 5.4%，亩增收大豆 9.2 公斤；受体菌株 2210 增产 8.3%，亩增收大豆 14.0 公斤（表 2）。与受体菌株 2210 相比，基因工程根瘤菌 32、B-2-6、43 均比 2210 增产率高，它们比 2210 分别增产 8.5%，6%和 3.9%；亩多增收大豆 14.2、10.1 和 6.6 公斤。只有 B-3-8 增产率低于 2210。

从表 2 中可看出，基因工程根瘤菌增产的原因是因为产量构成因子的单株荚数和单株粒数得到增加，因而使大豆产量得到提高。应用 Dunca's 新复极差多重比较说明：所有接种处理增产数均达到了 0.05 显著水平。

表 2　大豆基因工程根瘤菌对大豆产量的影响

处理	单株荚数	单株粒数	亩产 kg	增产 %	增产 kg/亩	$LSR_{0.05}$
32	28.8	73.8	196.4	16.8	28.2	a
B-2-6	28.2	70.4	192.3	14.3	24.1	ab
43	28.2	68.4	188.7	12.2	20.5	ab
2210	26.0	63.4	182.1	8.3	13.9	b
B-3-8	25.1	61.7	177.3	5.4	9.1	b
ck	25.8	63.3	168.2	—	—	c

综上所述，应用三亲本接合转移方法获得的大豆基因工程根瘤菌能提高大豆田间结瘤和共生固氮作用，可选育出优于受体菌株的高效固氮大豆根瘤菌。有关大豆基因工程根瘤菌在不同土壤上对不同大豆品种的增产效果已在全省进行了多点试验，其试验结果另报。

参 考 文 献

[1]窦新田，生物固氮，农业出版社，北京，156-161，1989.

[2]窦新田主编，大豆根瘤菌剂的研究与应用，黑龙江科技出版社，哈尔滨，85-88，1988.

[3]Keyser，H.H.et al.，World soybeen Research Confevence II proceedings，926-930 1985.

[4]周俊初等，大豆根瘤菌基因文库的构建和共生固氮基因的转移，华中农业大学生物技术研究论文集，武汉，35-36 1989.

高效固氮大豆根瘤菌中重组质粒 $pBj32H_2$ 的物理图谱*

张忠明 莫才清 沈 辉 周俊初 陈华癸

（华中农业大学生物技术中心）

关键词 物理图谱；质粒；大豆根瘤菌

KEY WORDS Restriction map，plasmid，*Bradyrhzobum japomcum*

以在我国黑龙江地区广泛应用的慢生型大豆根瘤菌 22-10[1]为出发菌株，通过导入克隆有快生型大豆根瘤菌菌株 B52[2]的 DNA 片段的重组质粒 $pBj32H_2$，构建了高效固氮大豆基因工程根瘤菌 32[3]。该菌株在黑龙江地区进行田间中试和大面积（15 万亩）推广应用，获得比出发菌株 22-10 增产大豆 7.4% 和 8.5%的效果。为了研究 $pBj32H_2$ 中所克隆基因的结构和功能，本工作对 $pBj32H_2$ 进行了限制性内切酶图谱分析，确定了 $pBj32H_2$ 中外源 DNA 片段的大小及 *Eco*R I、*Bgl* II、*Xho* I 的酶切位点，建立了该片段的物理图谱。

1 材料和方法

采用三亲交配法将高效固氮大豆基因工程根瘤菌 32 中的重组质粒转移到大肠杆菌 HB101 中，经在含四环素的 LB 培养基上分离和纯化后，通过碱性-SDS 法[4]提取质粒并用苯酚/氯仿反复抽提而得到纯化的质粒 DNA。质粒 DNA 的酶切分析和琼脂糖凝胶电泳按常规方法进行。

2 结果与分析

2.1 $pBj32H_2$ 质粒 DNA 单酶和双酶消化片段的分析

用 *Eco*R I、*Bgl* II 和 *Xho* I 单酶消化 $pBj32H_2$ DNA，分别得到 3 个 *Eco*R I 片段（其中 21.6kb 为载体质粒 pLAFR1，3.2kb 和 0.5kb 为克隆的外源片段），3 个 *Bgl* II 片段和 1 个 *Xho* I 片段。根据双酶消化环状 DNA 的片段总数，必然是两个酶分别单独消化片段数之和[5]。因此，*Eco*R I 和 *Bgl* II 双酶消化 $pBj32H_2$DNA 时，得到 6 个片段；*Eco*R I 和 *Xho* I 双酶消化时，得到 4 个片段；*Bgl* II 和 *Xho* I 双酶消化时得到 4 个片段。各片段相加的分子量总和为 25.3kb。详见表 1 和图 1。

表 1 单酶和双酶消化 $pBj32H_2$DNA 的片段数及其大小（kb）

片段	*Eco*R I	*Eco*R I/*Bgl* II	*Bgl* II	*Xho* I	*Eco*R I/*Xho* I	*Bgl* II/*Xho* I
A	21.6	18.8	21.7	25.3	21.6	21.7
B	3.2	2.9	2.0		2.5	2.0
C	0.5	1.6	1.6		0.7	1.6
D		1.2			0.5	0.4
E		0.5				
F		0.3				
总分子质量	25.3	25.3	25.3	25.3	25.3	25.3

本文 1991-06-21 收到

* 国家高技术“863”项目资助

原载于《华中农业大学学报》，11（1）:94~96，1992.

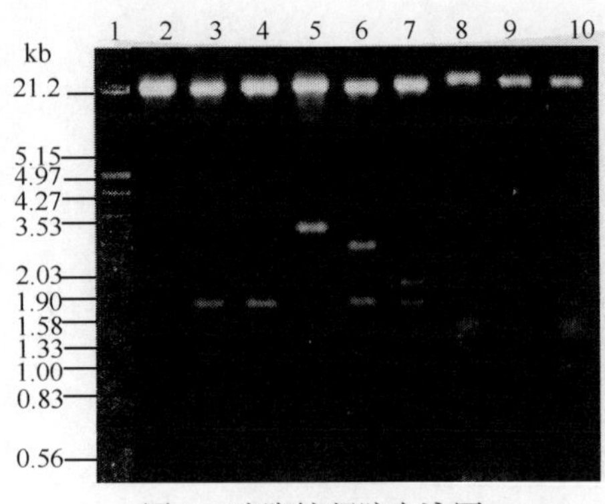

图 1 琼脂糖凝胶电泳图

1.λDNA+*Eco*R Ⅰ + *Hind*Ⅲ分子量标准；2.pLAFR1+*Eco*R Ⅰ； 3. pLAFR1+*Eco*R Ⅰ + *Bgl*Ⅱ；4. pLAFR1+ *Bgl*Ⅱ；5.pBj32H2+*Eco*R Ⅰ；6. pBj32H2+*Eco*R Ⅰ + *Bgl*Ⅱ；7. pBj32H2+ *Bgl*Ⅱ；8. pBj32H2+*Xho* Ⅰ；9. pBj32H2+*Eco*R Ⅰ + *Xho* Ⅰ；10. pBj32H2+ *Bgl*Ⅱ+ *Xho* Ⅰ

2.2 pBj32H_2质粒的物理图谱

由于 pBj32H_2 的载体质粒是 pLAFR1，其物理图谱也很清楚 （图 2）[6]，因此从上述结果很容易得到 pBj32H_2 的图谱，结果见图 3。

上述研究结果表明，pBj32H_2 的分子质量为 25.3kb，其中载体为 21.6kb，克隆的外源 DNA 片段为 3.7kb，在 3.7kb 中包含 2 个 *Eco*R I 片段，1 个为 3.2kb，另一个为 0.5kb；同时含有 1 个 *Bgl* II 和 1 个 *Xho* I 的酶切位点，结果见图 4。这一结果的获得，不仅证实了导入到慢生型大豆根瘤菌 22-10 中的外源 DNA 片段为 3.7kb，而且为该片段的亚克隆，基因结构和功能的分析打下了基础。

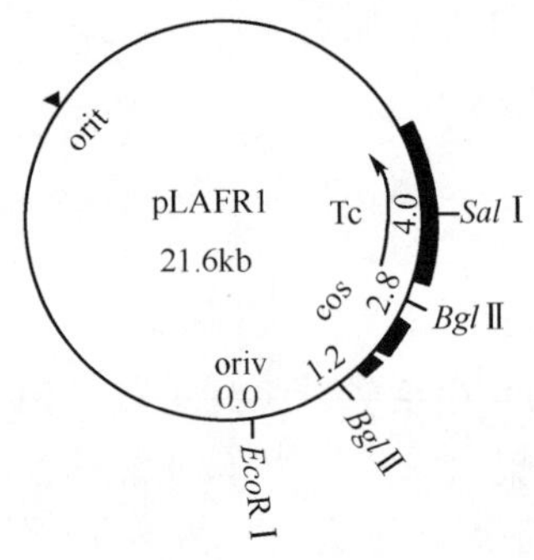

图 2 载体质粒 pLAFR1 的物理图谱

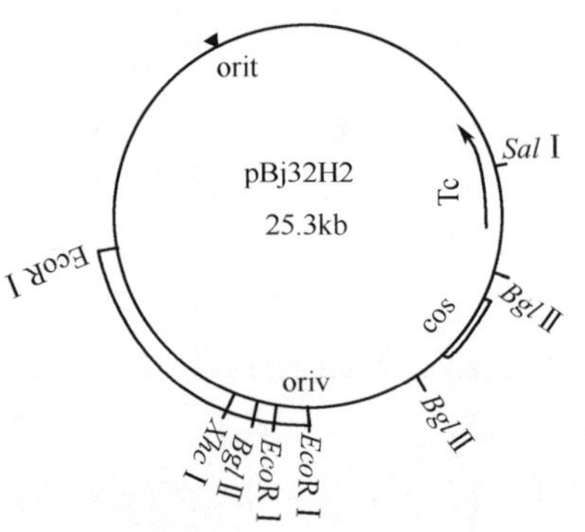

图 3 重组质粒 pBj32H2 的物理图谱

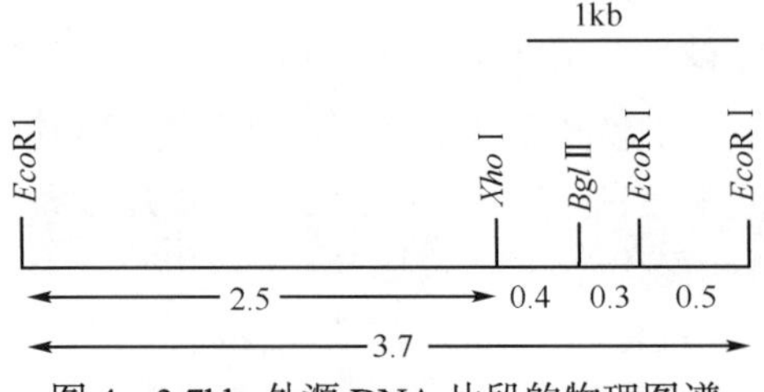

图 4 3.7kb 外源 DNA 片段的物理图谱

参 考 文 献

[1]窦新田等，大豆根瘤菌 22-10 和 27-50 新菌种选育和接种效果. 黑龙江农业科学，1990，6:14-18.

[2]沈辉等，快生型大豆根瘤菌基因库的构建. 中国微生态学杂志，1990，2（4）:51-56.

[3]Zhou Junchu （周俊初） et al，Constructing and Screening high effective strains of *Bradyrhizobium japonicum* by the method of genetic engineering. Nitrogen Fixation: Achivements and Objectives，Chapman and Hall，1990，p545.

[4] Birnbolm H C et al.，A rapid alkaline extraction for Screening recombinant plasmid DNA. J. Nucleic Acids Res. 1979，7:1513-1523.

[5]齐义鹏等. 基因工程原理和方法. 成都，四川大学出版社，1989，p115.

[6]Friedmann A et al.，Construction of a broad host range cosmid cloning vector and its use in the genetic analysis of *Rhizobium*，Gene，1982，18:289-296.

快生型大豆根瘤菌（*Rhizobium fredii*）3.7kb 增效片段的亚克隆序列分析*

张学贤　马立新　周俊初　陈华癸

（华中农业大学微生物系）

摘　要　利用柯氏质粒 pLAFR1 为载体构建了快生型大豆根瘤菌 B52 菌株的基因文库，并通过植物结瘤试验筛选到一个 3.7Kb 增效片段，将其导入慢生型大豆根瘤菌 2210，能提高其共生固氮效率。所得高效菌株 HN32 在黑龙江、四川和广西大面积应用，平均增产 11.4%[5]。本文测定了 3.7Kb 片段的核苷酸序列，计算机分析发现含有两个可读框（ORFs）。ORF1 编码含 190 个氨基酸的多肽，与已报道基因和多肽的同源性很低，但其 N 端 19 个氨基酸与发根农杆菌的乳糖转移酶高度同源。ORF2 编码含 235 个氨基酸的多肽，与豌豆根瘤菌的 hupE 及苜蓿根瘤菌的糖基转移酶基因显示 54%的同源性。

关键词　大豆；根瘤菌；共生；基因

一、引　　言

根瘤菌能感染豆科植物形成具有共生固氮作用的根瘤，将大气中的氮气转化为植物可以利用的氨。根瘤菌作为接种剂应用于豆科植物生产也有近百年的历史。Maier 和 Brill 于 1978 年[1]首次证明通过较为简单的遗传操作可以提高根瘤菌的共生固氮能力。此后遗传学手段被广泛地应用于根瘤菌高效菌株的选育，方法包括自发突变株的选育、化学或辐射诱变、转座子诱变以及重组 DNA 技术[2]。大豆是重要的经济作物之一，是人类植物蛋白和油料的主要来源。“七五”以来，我们开展了大豆根瘤菌高效菌株的构建及其应用研究。众所周知，根瘤菌与豆科植物之间共生固氮体系的建立是一个非常复杂的过程，涉及的基因很多，各类基因的定位、表达、调控及其相互作用研究仍在不断深入。但目前针对某一菌株，我们往往不知道哪一个基因是关键。在这种情况下，我们设计了一条新的技术路线，即通过植物结瘤实验直接从基因文库中筛选起增效作用的克隆，并在“八五”期间成功地从快生型大豆根瘤菌 B52 中分离得到了一个 3.7Kb 的增效片段[5]。本文将报道增效片段的核苷酸序列及其中一个启动子的表达验证。

二、材料和方法

快生型大豆根瘤菌 B52 及慢生型大豆根瘤菌 2210 分别分离自湖北及黑龙江，皆为野生型菌株。克隆载体 pUC19、pUC118、M13mp18 以及大肠杆菌受体菌株 DH5α、JM109 皆来自 New England Biolabs 公司。根瘤菌培养在 YEM 培养基[3]，大肠杆菌培养在 LB 或 M9 培养基[4]。抗生素使用浓度：四环素（Tc，20μg/ml），卡那霉素（Km，100μg/ml），氨苄青霉素（Ap，50μg/ml）。

绝大多数序列分析采用的是 Promega 公司的 TaqTrack Sequencing Kit，在 Pharmacia LKB-Macrophar 序列分析仪上进行。部分序列在 ABI370A 型全自动序列分析仪上完成。有关基因克隆等方法皆参照《分子克隆实验指南》上的方法进行[4]。

* 原载于《高技术通讯》，4：4~7，1996.

三、结果与讨论

1．37Kb 增效片段的序列分析

利用 10 种不同的限制性内切酶建立了 pHN32 的物理图谱（图 1）。其中 0.5Kb 的 *Eco*R Ⅰ片段以两个方向克隆到载体 M13mp18 上，然后作为模板直接测序。与此同时也将 3.2Kb 的 *Eco*R Ⅰ片段克隆到 M13mp18 上，结果发现，由于外源片段较大，所得重组质粒极不稳定，表现出不同程度的缺失和重组。进而将 3.2Kb 的 *Eco*R Ⅰ片段转而克隆到另一常用载体 pUC19 上，然后利用 Promega 公司的 Erase-a-Base 系统生成一个嵌套顺序缺失克隆，再对每一克隆进行测序。各个亚克隆核苷酸序列的拼接及开放阅读框架的识别皆在 Berckman 公司的 Microgenic 软件上进行，结果表明在全长为 3667bp 的序列内含有两个可能的开放阅读框架（图 3）。

2．ORF2 启动子活性的验证

为了验证找到的ORFs的正确性，我们对ORF2的启动子活性进行了检测（图2）。首先将3.7Kb *Eco*R Ⅰ片段克隆到载体 pUC118 上，转化子在含 IPTG 和 X-gal 的平板上呈白色；然后经 *Bam*HⅠ、*Bgl*Ⅱ双酶切及 Klenow 酶补平后连接，可以将 ORF2 启动子与 LacZ 结构基因以正确的方式连接在一起（图 2）。转化 *E. coli* 后发现转化子能在仅含 X-gal 的 LB 平板上呈蓝色。该结果表明 ORF2 启动子是正确的，而且能组成型地启动 *LacZ* 基因的表达。

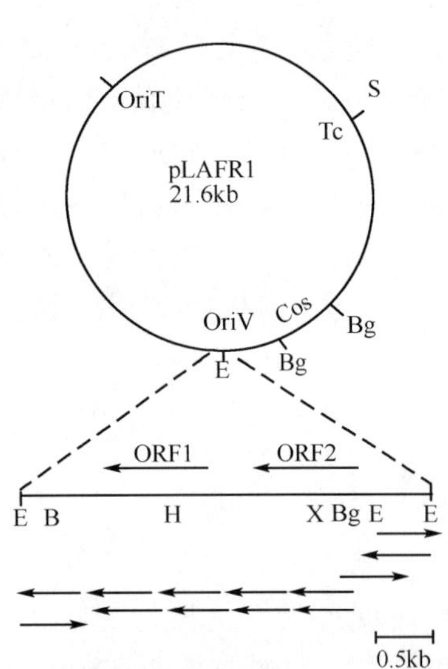

图 1　3.7kb 增效片段的限制性酶切图谱

E-*Eco*RⅠ，B-*Bam*HⅠ，Bg-*Bgl*Ⅱ，X-*Xho*1，H-*Hind*Ⅲ，S-*Sal*Ⅰ，粗箭头表示蛋白质编码区及方向；细箭头表示各亚克隆的长度及序列分析的方向

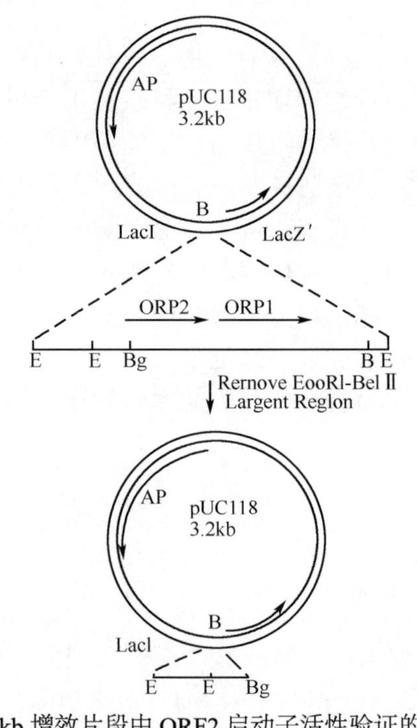

图 2　3.7kb 增效片段中 ORF2 启动子活性验证的策略

E-*Eco*RⅠ，Bg-*Bgl*Ⅱ，B-*Bam*HⅠ

3．同源性分析

利用 EMBL 数据库对两个 ORFs 相应的氨基酸序列进行了同源性分析。ORF1 编码 190 个氨基酸，其中含 27 个精氨酸，没有从数据库中找到同源基因，仅 N 端的 19 个氨基酸与发根农杆菌的乳糖转移酶高度同源。ORF2 编码 235 个氨基酸，与豌豆根瘤菌的 hupE 及苜蓿根瘤菌的糖基转移酶基因显示出 54%的同源性。对这两个基因的生化功能及其增效作用机理尚有待作进一步研究。

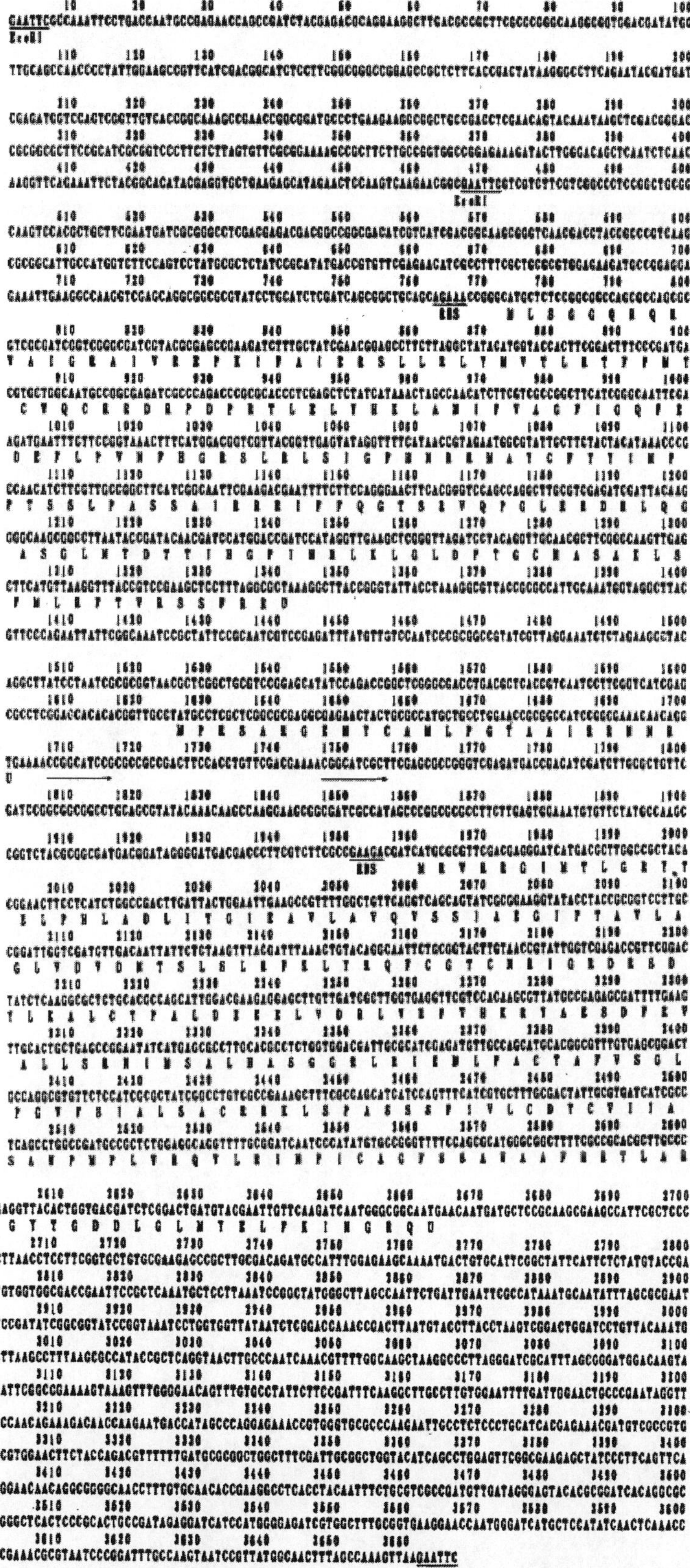

图 3 3.7kb 增效片段的 DNA 序列及两个可能编码区相应的氨基酸序列

RBS 表示可能的核糖体结合位点，箭头提示存在的重复序列

参 考 文 献

[1] Maier R *et al. Science*，1978，201（4），448.

[2]Pau AS *et al*. Nitrogen Fixation: Achievements and Objectives，Gresshoff PM *et al.*（eds），Chapman and Hall，1990:617.

[3]Vincent JM. A Manual for the Practical Study of Root-nodule Bacteria，Blackwell Scientific Publications，Oxford，1970.

[4]Sambrook J *et a1*. Molecular Cloning: a Laboratory Manual，2nd eds，New York: Cold Spring Harbor Laboratory press，1989.

[5]Zhou JC *et al*. Agricultural Biotechnology，You C B *et a1*.（eds）China Science and Technology press. 1992:739.

不同化合态氮浓度对大豆根瘤菌结瘤和固氮作用的影响*

刘　莉　周俊初　陈华癸

（华中农业大学农业部农业微生物重点实验室）

ABSTRACT A comparative study was conducted on the effect of different concentrations of NH_4NO_3，KNO_3 and （NH_4）$_2SO_4$ on the symbiosis between soybean and rhizobium strains of HN32，HN01，12-2 and 9-1-9 under sand culture and pot experiment. It was found that the symbiosis was completely inhibited by 10mmol/L of NH_4NO_3 or （NH_4）$_2SO_4$ which was named as zero-nodulation concentration. But low levels （0.5～1.0mmol/L） of combined nitrogen improved their nitrogen fixation efficiency.The mechanism revealed by specially designed micro-root-box equipment was that high levels of combined nitrogen inhibited root hair infection by blocking the symbiosis at the early stage. Their inhibitory effects ranked （NH_4）$_2SO_4$> NH_4NO_3> KNO_3.

KEY WORDS Soybean；Rhizobia；Combined nitrogen；Root hair infection；Zero-nodulation concentration

关键词　大豆；根瘤菌；化合态氮；根毛侵染；O−结瘤浓度

已有的研究表明，根瘤菌与豆科植物的共生固氮作用是一个非常复杂的相互作用过程。Vincent[2]提出共生过程受根瘤菌基因组、植物基因组和环境因子等三方面因素的影响。已知影响共生固氮作用的环境因子很多，其中土壤化合态氮素含量对共生效应的影响早已为人们所关注。本文比较研究了不同化合态氮浓度对大豆根瘤菌结瘤和固氮作用的影响，并探讨了产生O-结瘤现象的作用机理。

1　材料与方法

1.1　材料

供试植物、菌株与质粒　大豆品种是来自窦新田的黑农37；慢生型大豆根瘤菌是本室构建的HN32[1]；快生型大豆根瘤菌用本室保存的HN01、9-1-9、12-2。

1.2　方法

1.2.1　根瘤菌共生效应的测定　盆栽采用本室改进的双层钵和无氮植物营养液[2]进行。在各化合态氮浓度下设接种12-2、9-1-9、HN01、HN32及不接菌的对照处理，每个处理重复5次，收获测定每株根瘤数、根瘤重和植株干重。干重增量按接菌植株干重减去同一浓度下未接菌对照植株干重计算。全氮增量按接菌植株全氮量减去同一浓度下对照植株全氮量计算。

1.2.2　根毛侵染试验　采用本室研制的微根盒装置进行。该装置由双层有机玻璃制成的微根盒和烧杯组成。微根盒底板大小为5cm×7cm，四周用厚2mm的有机玻璃条密封，上端留一个宽约10mm的播种口，下端留5～6个宽4mm的培养液连通口，盖板尺寸与底板相同。试验前微根盒用95%乙醇浸泡消毒，吹干后用透明胶条封住底板上、下端，用1%琼脂加供试根瘤菌倒于混菌平板，取下透明胶条盖上盖板后用橡皮筋固定。将已催至根长3cm左右的大豆根从播种口插入后将微根盒转入盛有灭菌无氮或加氮植物营养液的塑料烧杯中进行光照培养，分别于5d和10d后在倒置显微镜下观察根毛形态并照相。

* 原载于《中国农业科学》，31（4）：87~89，1998.

2 结果与分析

2.1 大豆根瘤菌在不同（NH_4）$_2SO_4$ 和NH_4NO_3 浓度下的共生效应

2.1.1 不同$(NH_4)_2SO_4$浓度对大豆根瘤菌结瘤能力的影响

在砂培条件下，不同浓度的（NH_4）$_2SO_4$对HN32和12-2在大豆黑农37上结瘤能力随浓度增大而下降，到10mmol/L时，供试植株均不结瘤，从而确定10mmol/L浓度的（NH_4）$_2SO_4$为12-2和HN32两个供试大豆根瘤菌与大豆共生体系的O-结瘤浓度。

2.1.2 不同NH_4NO_3浓度对大豆根瘤菌共生效应的影响

供试快生型大豆根瘤菌12-2、HN01和9-1-9在砂培条件下于10mmol/L NH_4NO_3浓度下均未结瘤。0～5mmol/L的NH_4NO_3浓度对瘤数、瘤重、干重和干重增量的影响，随化合态氮浓度的增加，无论接菌与否，植株干重都随之而增加，表明在无氮砂培条件下化学氮肥对植株生长有明显的促进作用。在0.5mmol/L NH_4NO_3浓度下的接菌植株干重增量均比其他浓度的增量高，大于2.5mmol/L的NH_4NO_3浓度对供试大豆根瘤菌的结瘤有明显的抑制作用。接种根瘤菌最适宜化合态氮浓度还因菌种不同而异。

2.1.3 不同NH_4NO_3浓度对HN32结瘤固氮的影响 在盆栽条件下不同NH_4NO_3浓度对HN32共生效率的影响中，在0.5mmol/L的NH_4NO_3浓度下，HN32在大豆上的瘤数、瘤重、干重与全氮增量明显高于其他浓度，说明该浓度能明显地促进其共生固氮作用。随着NH_4NO_3浓度的增加虽然大豆植株的干重和全氮量不断提高，但其瘤数、瘤重、干重和全氮增量均逐渐下降。

2.2 化合态氮对大豆根瘤菌侵染过程的影响

在显微镜下观察并记录了在微根盒培养装置中大豆根瘤菌感染大豆根毛的全过程。观察到在较低的KNO_3、NH_4NO_3和（NH_4）$_2SO_4$浓度下接种根瘤菌的大豆植株根毛发生变形、卷曲、呈分叉状、螺旋状或顶部膨大等形态，并记录到典型的"牧羊拐"状卷曲。但在较高化合态氮浓度下，则很少观察到类似的根毛变形现象。不同浓度的化合态氮素对大豆根瘤菌感染根毛的影响见下图。

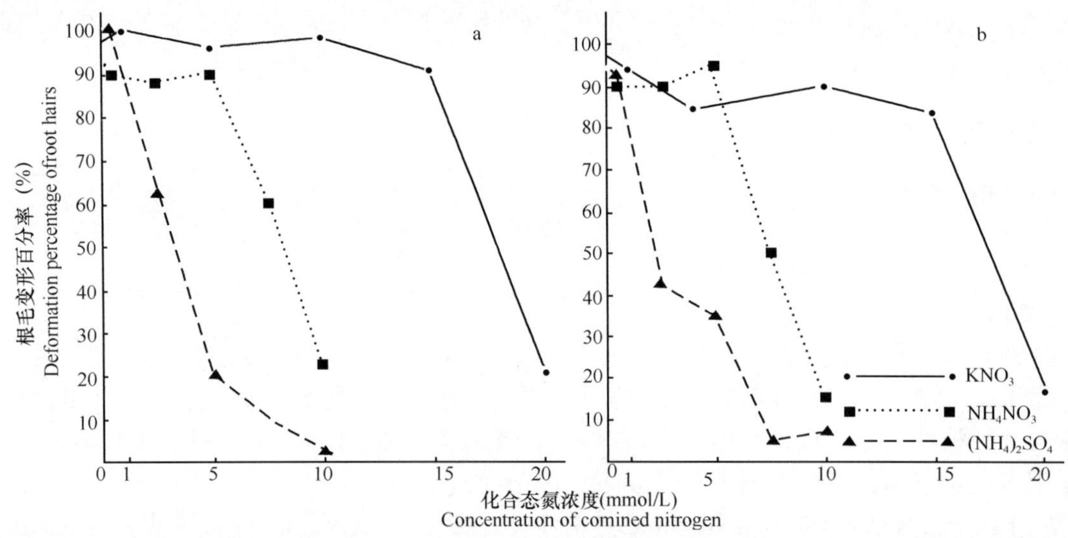

图 不同化合态氮对快生型大豆根瘤菌 HN01（a）和慢生型大豆根瘤菌 HN32（b）感染产生的根毛变形的影响

Fig.The effect of different combined nitrogens on root hair deformation of *S. fredii* HN01（a） and *B. japonicum* HN32（b）

由图（A）可见，随3种供试化合态氮浓度的增加，HN01产生的根毛变形率下降的快慢依次为:（NH_4）$_2SO_4$ > NH_4NO_3 > KNO_3。当KNO_3浓度为15mmol/L时，根毛变形率仍可达90%，至20mmol/L时的根毛变形率才下降到20%。而对于NH_4NO_3，根毛变形率为90%的浓度是5mmol/L，当浓度增加到7.5mmol/L时的根毛变形率就只有60%。对（NH_4）$_2SO_4$来讲，浓度为2.5mmol时的根毛变形率就只有62.5%。图（B）中化合态氮对HN32产生的根毛变形影响的变化趋势与HN01大致相同。本结果表明，高浓度的化合态氮对大豆-根瘤菌共生体系的抑制作用表现为在早期形成阶段抑制根瘤菌对大豆根毛的侵染，不同种类的化合态氮产生抑制作用所需要的浓度不同。

3 讨论

本研究结果发现抑制结瘤作用的NO_3^-浓度高于NH_4^+浓度；在研究方法上，使用本室设计的微根盒培养装置与Fahraeus设计的载玻片培养装置[3]相比较，其特点是既可满足大种子植株生长的要求，又便于在显微镜下进行早期根毛侵染的定位研究。

参 考 文 献

[1]周俊初，等.慢生型大豆根瘤菌（*Bradyrhizobium japonicum*）工程菌株 HN32 中异源增效因子的鉴定，农业生物技术进展与展望，合肥: 中国科技大学出版社，1993:142-162.

[2]Vincent J M.，In Nitrogen Fixation Vol. III，（eds Newton WE and Oreme-Johns on. W. H.），University Park Press. USA. 1980.

[3]Fahraeus G. The infection of clover root hairs by nodule bacteria studied by a simple glass slide technique. J. Gen. Microbol.，1957，16:374-381.

导入外源DNA片段的花生根瘤菌高效基因工程菌株的构建*

李立家　周俊初　陈华癸

（华中农业大学生物技术中心）

摘　要　以柯斯质粒pLAFR1为载体先构建快生型花生根瘤菌85−7菌株总DNA的基因文库。在协助质粒pRK2013帮助下用此基因文库分别同菌株85−7和另一株慢生型花生根瘤菌菌株CO2−5作三亲本杂交。转移接合子接种沙培条件下生长的花生，通过结瘤试验初筛选和复筛选，共筛选出3株工程菌HN11、HN12（以85−7为受体）和HN13（CO2−5为受体）。其每植株根瘤的固氮酶活分别比出发菌株的提高302%（HN11）、356%（HN12）和284%（HN13），每植株干重分别比出发菌株的提高85%（HN11）、83%（HN12）和28%（HN13），工程菌株的重组质粒酶切分析表明，除载体pLAFR1外还分别含有25kb（pHN11）、14kb（pHN12）和18kb（pHN13）的外源片段。

关键词　花生根瘤菌；基因文库；工程菌；外源DNA

根瘤菌高效菌株的筛选一直沿用自然选育或人工诱变育种的传统方法[1)]。随着固氮分子生物学和遗传操作技术的发展，许多学者已利用遗传工程技术来构建和筛选优良高效菌株[1, 2, 3]。周俊初等[4]考虑到共生固氮基因的复杂性，提出了一条选育高效结瘤固氮菌株的新路线。即通过转移大豆根瘤菌基因文库到另一大豆根瘤菌菌株中，然后通过大豆盆栽及田间结瘤试验而选育高效大豆根瘤菌菌株，结果筛选到一株高效工程菌株32。其田间试验增产16.8%。本研究采用同样的技术路线，试图构建选育花生根瘤菌高效工程菌。结果筛选到3株盆栽条件下高效的工程菌HN11、HN12和HN13，并对它们的重组质粒进行了初步的酶切分析。现作如下报道。

1 材料和方法

1.1　供试菌株、质粒和培养基

供试菌株和质粒见表1。培养大肠杆菌用LB培养基[5]，培养根瘤菌用TY、PA与SM培养基[6]。

表1 供试菌株和质粒

Table 1　Bacteria strains and plasmids used in this study

菌株与质粒 Strains and plasmid	有关性状 Relevant characteristics	来源 Soruces
花生根瘤菌85-7	野生快生型 Nod^+，Fix^+	四川农业大学提供[1)]
Groundnut Rhizobium 85-7	Wild type Nod^+，Fix^+	
花生根瘤菌CO2-5	G诱变菌株，慢生型 Nod^+，Fix^+	四川农业大学提供[1)]
Groundnut Rhizobium CO2-5	Slow growth type Nod^+，Fix^+	
E.coli HB101	F^-，recA13，proA2	英国John Innes研究所提供[2)]
E.coli LE392	F^-，hadR514	英国John Innes研究所提供[2)]
E.coli BHB2688	Eam. Temp	英国John Innes研究所提供[2)]
E.coli BHB2690	Dam，Temp	英国John Innes研究所提供[2)]
pLAFR1	Tc，mob+，tra^-	广西农学院马庆生提供[3)]
pRK2013	Kan^r，mob^-，tra^+	广西农学院马庆生提供[3)]

1）Sichuan Agriculture University supplied. 2）England John Innes research institute supplied. 3）GuangXi Agricultural College Ma Qingsheng supplied.

1.2　基因文库的构建

参考周俊初等人构建大豆根瘤菌基因文库方法[4]。用PA培养液培养85-7菌株到对数期，然后溶菌

原载于《华中农业大学学报》，13（3）：219~225，1994.

1）陈华癸等，《中国共生固氮研究五十年》。1987年，南京农业大学内部出版

提总 DNA，*Eco*R I 酶部分酶切，蔗糖梯度离心分离收集 20～30kb 大小的 DNA 片段；碱变性法提取 pLAFR1，*Eco*R I 酶完全酶切；用 *E.coli* BHB2688 和 *E.coli* 2690 提取包装蛋白。最后将 DNA 片段与载体 pLAFR1 酶连，体外包装、侵染 *E.coli* LE392，在 LB+Tc（15μg・ml^{-1}）上选择抗性菌落。

1.3 三亲本杂交

将 85-7 基因文库菌（用 LB+Tc 培养基），MM294（pRK2013）[用 LB+Km（50μg・ml^{-1}）培养基]、85-7 或 CO2-5（用 TY 培养基）分别培养至对数期。然后按 1:1:1 比例混匀过滤到微孔滤膜上，于 TY 平板上 28℃培养 10h（85-7）或 24h（CO2-5）后用无菌水洗下膜上的菌体，作系列稀释涂布 SM+Tc（对 85-7 为 25μg・ml^{-1}，对 CO2-5 为 35μg・ml^{-1}）选择 T_c^r 转移接合子。

1.4 花生结瘤试验

选择籽粒饱满大小一致的花生（品种为天府 3 号）种子用 0.1% $HgCl_2$ 灭菌，用无菌水洗涤催芽。然后选择发芽一致的种子播于灭菌砂钵中，待长出第一片真叶时挑选生长一致的砂钵接种根瘤菌，以不接种作对照。生长 40 多天后即在花开时进行收获。

1.5 根瘤固氮酶活性测定

用日立 163 型气相色谱仪按乙炔还原法进行。

1.6 质粒转化，提取和酶切参见文献[5]

2 结果

2.1 快生型花生根瘤菌 85−7 基因文库的构建

*Eco*R I 酶完全酶切的 pLAFRI 和 *Eco*R I 酶部分酶切的 85-7 总 DNA 的 20~30kb 片段，经酶连、包装和侵染 *E.coli* 392 后，在 LB+Tc 上共筛选到 T_c^r 克隆 1.4×10^4 个，此即 85-7 菌株总 DNA 基因文库。

随机挑取 21 个抗性菌落抽提质粒，用 *Eco*R I 酶酶切分析表明，质粒含有外源片段的克隆数占总克隆数的 57%（见图 1）。构建此基因文库理论上需要的克隆数 N=ln（1-p）/ln（1-f/g）[5]。在本研究中 p=99%，g=4.1×10^6bp[7]，f=1.8×10^4 bp，即 N=1047 个。本试验构建的基因文库的实际克隆数为 14000×57%=7980 个。大大超过其理论值，满足建库要求。

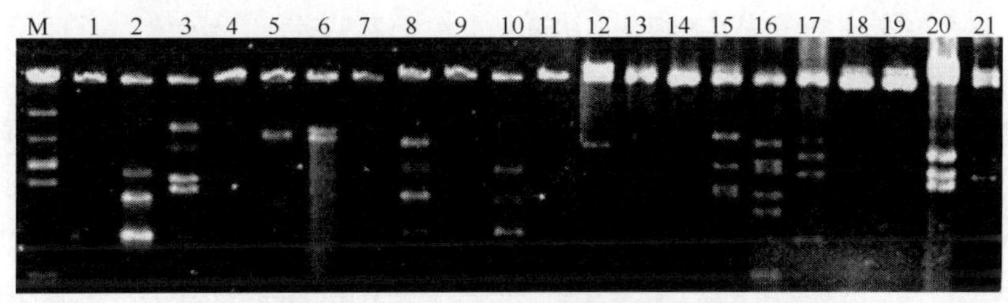

图 1 基因文库中部分 Tc^r 克隆质粒的 *Eco*R I 酶切

Fig 1 M. λDNA/*Hind* III；1~21. Tc^r 抗性克隆质粒+*Eco*RI 酶切。1、4、5、6、9、10 为有外源片段的重组质粒；2、3、7、8 为无外源片段的载体 pLAFR1

2.2 三亲本杂交及花生盆栽试验结果

通过三亲本杂交将 85-7 菌株总 DNA 基因文库的重组质粒。分别引入到 85-7 和 CO2-5 菌株中，为了考查载体 pLAFR1 对根瘤菌结瘤固氮作用的影响，以 pLAFR1 为对照同时进行转移。转移接合子未发现对 Kan 有抗性。这表明协助质粒 pRK2013 未转移进入受体菌中。T_c^r 相关联的质粒转移频率平均约为 2.3×10^{-4}。在相同条件下受体 Tc 抗性自发突变率小于 10^{-8}。部分转移接合子质粒检测进一步证实了 Tc^r 菌落含有质粒。

将 SM+Tc 上生长的转移接合子在同样平板上再纯化一次后，用无菌水洗下接种砂培条件下生长的花生。收获花生植株后测定固氮酶活、植株地上部分干重、叶片大小和叶色等指标。根据测定结果，从 85-7 转移接合子接种的花生植株中初步筛选出几株长势较好的植株。从其主根根瘤中共分离出 6 个 T_c^r 菌株，分别编为 85-7 的 1、2、3、4、5、6 号以用于复筛选试验。同样从 CO2-5 转移接合子结瘤试验中获得 8 个 T_c^r 菌株，分别编为 CO2-5 的 1、2……8 号，也用于下轮复筛选结瘤试验。

将上述初筛选获得的T_c^r菌株进一步进行复筛选盆栽结瘤试验。对照有3个，它们是出发受体菌株85-7（或CO2-5），含有载体pLAFR1的出发菌株85-7（或CO2-5）以及不接种。收获后同样测定固氮酶活、植株干重等指标，并对结果进行统计方差分析。结果表明复筛选中各供试菌株在固氮酶活和植株干重方面都存在极显著的差异。这说明从整体上看我们的处理是有效的。为检验各个处理和对照之间有否差异，我们采用最小显著差数法（LSD法）进一步进行多重比较，结果见表2和表3。由表中结果可见，不论从酶活还是植株干重上，只引入载体pLAFR1，对受体85-7和CO2-5的结瘤固氮都没有显著影响。表2表明1号和2号菌株作为进一步试验对象，并分别定名为HN11和HN12。表3表明只有4号菌株在固氮酶活和植株干重上都比CO2-5的有极显著的提高，将其命名为HN13。HN11和HN12分别比受体菌株85-7酶活提高302%和356%，植株干重分别提高85%和83%；HN13比受体菌株CO2-5酶活提高284%，植株干重提高28%。

表2 各转移接合子与出发菌株85-7盆栽试验结果的比较

Table 2 Diffferent analysis of nitrogenase activity and plant dry weight between transfermants and strain 85-7

处理 Treatment	固氮酶活平均值 Average nitrogenase activity	显著水平[1)] Significance level	植株干重平均值 Average plant dry weight	显著水平[1)] Significance level
85-7	1.571	-	1.15	-
85-7（pLAFR1）	1.668	C	1.17	C
1	6.310	A	2.13	A
2	7.160	A	2.11	A
3	1.815	C	1.27	C
4	2.345	C	1.18	C
5	2.963	B	1.13	C
6	2.879	B	1.21	C

1）A，极显著 B，显著 C，不显著。1）A，The most significant level B，The significant level C，The unsignificant level. 酶活单位（nitrogenase activity unite）：μmol·（plant）$^{-1}$·h^{-1}，植株干重单位（Plant dry weight unti）：g 表中数均为5株花生植株测得的平均数。The numbers in table 2 are the average calculated from five groundnut plants.

2.3 复筛选出的工程菌株的重组质粒酶切分析

由于根瘤菌中可能含有遗传背景不清楚的内源质粒，因此不便于直接从工程菌株中提取重组质粒作酶切分析。本试验采用小样抽提法先从根瘤菌中提取重组质粒直接用于转化大肠杆菌，转化频率平均为10^{-6}，这样可以很方便地得到转化子。受体85-7和CO2-5提取物作转化对照无抗性菌落。然后从大肠杆菌转化子中提取重组质粒，用*Eco*R Ⅰ酶切，结果见图2。由图2可见复筛选工程菌株HN11、HN12和HN13的重组质粒（分别命名为pHN11、pHN12 、pHN13）除有载体pLAFR1外，还分别含有约25kb、14kb和18kb的外源片段。

表3 各转移接合子与出发菌株CO2-5盆栽试验结果的比较

Table 3 Different analysis of nitrogensae activity and plant dry weight between transformants and strain CO2-5

处理 Treatment	固氮酶活平均值 Average nitrogenase activity	显著水平[1)] Significance level	植株干重平均值 Average plant dry weight	显著水平[1)] Significance level
CO2-5	1.240	-	1.208	-
CO2-5 （pLAFR1）	1.120	C	1.206	C
1	1.710	C	1.200	C
2	0.992	C	1.120	C
3	1.107	C	1.230	C
4	4.762	A	1.548	A
5	1.201	C	1.220	C
6	1.422	C	1.252	C
7	0.580	C	1.148	C
8	2.236	B	1.334	C

1）A，极显著 B，显著 C，不显著。1）A，The most signifrcant level B，The significant level C，The unsignificant level. 酶活单位（nitrogenase activity unit）：μ mol·（plant）-1·h-1，植株干重单位（Plant dry weight unti）：g 表中数均为5株花生植株测得的平均数。The numbers in table 3 are the average calculated from five groundnut plants.

3 讨论

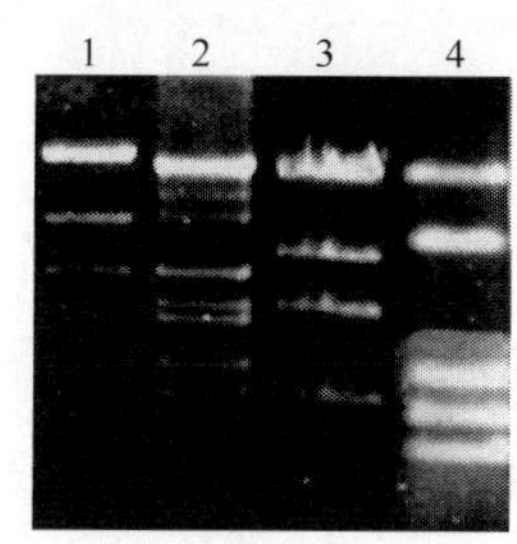

图 2 重组质粒酶切分析

Fig 2 1. λDNA/*Hind* III；2. pHN11/*Eco*RI；3. pHN12/*Eco*RI；4. pHN13/*Eco*RI

考虑到根瘤菌固氮基因的多样性和复杂性，本研究选以广谱性质粒 pLAFR1[7]为载体构建了快生型花生根瘤菌 85-7 的基因文库，将根瘤菌 85-7 的全部基因片段分存于文库的不同克隆 DNA 片段上，然后转移此文库 DNA 片段到一定的受体中，通过结瘤试验以筛选可能含有增效重组质粒的工程菌株。结果获得 3 株花生根瘤菌高效工程菌株。自从周俊初等人[4]建立此方法以来，本实验室用此方法又构建获得 3 株大豆根瘤菌高效工程菌[8]。根据这些结果，可以得出如下共识，受体菌获得了一个外源质粒，在外源质粒中含有一段给体根瘤菌菌株的 DNA 片段，能够产生提高共生固氮作用的外源 DNA 片段的几率是比较大的。这作为一种基因工程育种手段是可行的。

本工作中获得了有增效作用的转移接合子菌株，即由 85-7 菌株与 85-7 菌株基因文库的重组质粒 pHN11 或 pHN12 构成的菌株。本实验已证实空载体 pLAFR1 对受体菌 85-7 和 CO2-5 结瘤固氮作用没有影响。由于这两个转移接合子基因背景相同，所以它们的功能只能是局部同源双倍体所产生的优势。已经获得的异源转移接合子 HN13（CO2-5 为受体）的功能则可能是局部 DNA 加倍的优势作用，也可能是基因背景不同的二基因片段的杂种优势作用。这些优势的产生只能是随机构成的。

我们进一步的工作将重复验证这 3 个工程菌株的结瘤固氮作用，以及进行田间试验。另外在试验或田间试验中必须保证重组质粒的稳定性和菌株的高竞争力。

致谢 本工作得到李阜棣教授、张忠明、沈辉和莫才清老师以及张学贤博士等的指教，特此致谢。

参 考 文 献

[1]Triplett E W. Construction of a symbiotically effective strain of *Rhizobium leguminosarum bv. Trifolii* with increased nodulation competitiveness. Appl and Envir Microbiol. 1990. 56（1）. 98-103

[2]Dejong T M. Brewin N J. Johnston AWB et al. Improvement of symbiotic properties in *Rhizobium legumnosarum* by plasmid transfer. Joural of General Microbiology. 1982. 128:1829-1838

[3]Birkenhead K. Manian S S. Q'Gara F. dicarboxylic acid transport in *Bradyrhizobium japonicum* use of *Rhizobium melilotidct* gene（s） to enhance nitrogen fixation. J. Bacteriol. 1988. 170:184-189

[4]周俊初. 沈辉. 张忠明等. 高效结瘤固氮大豆根瘤菌基因工程菌株的构建及应用. 武汉大学学报（自然科学版），1990，（生物工程专刊）:11-17

[5]Maruatis T. Fritsch E F. Sambrook J. Molecalar cloning: a laboratory manual. New York: Cold Spring Harbor Laboratory. 1982. 168-175

[6]张忠明. 陈华癸. 李阜棣等. 紫云英根瘤菌基因文库的构建及含完整结瘤基因的重组质粒 pRaZ15 的分离. 生物工程学报. 1991，7（3）:213-219

[7]Friedman A M. Long S R. Brown S E et al. construction of a broad host range coamid cloning vector and its use in the genetic analysis of *Rhizobium* mutants. Gene. 1982，18. 289-292

[8]谬礼红. 导入外源片段大豆根瘤菌基因工程菌株的构建与遗传分析. [硕士学位论文]. 武汉华中农业大学土化系. 1991

重组质粒在花生根瘤菌中的稳定性*

李立家 周俊初 陈化癸

（华中农业大学生物技术中心）

摘　要　以花生根瘤菌高效工程菌株HN11、HN12和HN13各自的增效重组质粒为材料，在共生条件下和非共生人工液体培养基中，研究其在各自宿主快生型花生根瘤菌85–7和慢生型化生根瘤菌CO2–5中的稳定性。结果表明：在根瘤内，85–7宿主中重组质粒与空载体pLAFR1丢失程度相似，与外源片段存在无关；而在CO2–5宿主中HN13的增效重组质粒比无效重组质粒及空载体pLAFR1丢失程度较轻，在人工液体培养基中，pLAFR1及其重组质粒在同一宿主中丢失程度相似。并讨论了以pLAFR1为载体的重组质粒丢失较严重的原因，提出了保持工程菌中重组质粒稳定性的相应措施。

关键词　共生；人工液体培养基；pLAFR1；增效重组质粒

研究导入工程菌株中的重组质粒的稳定性，对于基础研究和生产应用有着重要意义。只有构建高表达、高稳定的重组质粒才能达到高产目的[1]。HN11，HN12和HN13是在协助质粒pRK2013帮助下，用快生型花生根瘤菌菌株85-7基因文库分别与85-7及另一株慢生型花生根瘤菌菌株CO2-5作三亲本杂交，再通过盆栽结瘤试验筛选到的3个花生根瘤菌高效工程菌株，遗传分析表明它们分别含有不同酶切图谱的重组质粒pHN11，pHN12和pHN13[2]。为了进一步有效地将它们应用于生产之中，我们对其增效重组质粒在各自宿主菌中通过结瘤和人工液体培养之后的稳定性进行了比较研究，并详细探讨了各种丢失的原因。

1　材料和方法

1.1　供试菌株、质粒和培养基

表1　供试菌株和质粒

菌株（号）和质粒	特征	来源
花生根瘤菌		
85-7	野生快生型 Nod^{+} Fix^{+}	四川农业大学
CO2-5	C诱变菌株，慢生型 $Nod^{+}Fix^{+}$	四川农业大学
HN11	85-7中引入增效质粒pHN11，Tc^{r}	本室
HN12	85-7中引入增效质粒pHN12，Tc	本室
HN13	CO2-5中引入增效质粒pHN13，Tc^{r}	本室
1#	CO2-5中引入无效重组质粒，Tc^{r}	本室
2#	CO2-5中引入无效重组质粒，Tc^{r}	本室
3#	85-7中引入无效重组质粒，Tc^{r}	本室
质粒		
pLAFR1	Tc^{r}，mod^{+}, tra^{-}	广西农学院
pHN11	Tc^{r}pLAFR1上含有外源片段	本室
pHN12	Tc^{r}pLAFR1上含有外源片段	本室
pHN13	Tc^{r}pLAFR1上含有外源片段	本室

供试菌株、质粒见表1. 其中1#，2#和3#菌株中的无效重组质粒是将85-7总DNA上的某个片段随机克隆到载体pLAFR1上而构建的，对各自宿主菌的固氮作用没有影响[2]。根瘤菌培养用YMA培养基

*原载于《武汉大学学报（自然科学版）》，4：115~119，1994.

[3]，花生盆栽营养液用 Fahraeus 无氮植物营养液[2]。

四环素（Tc）在培养基中使用浓度是：以 85-7 为宿主的含质粒菌为：25μg • ml^{-1}；以 CO2-5 为宿主的含质粒菌株为：35μg • ml^{-1}。

1.2 花生结瘤试验

花生去壳后挑选籽饱满、大小一致的种子，用 0.1%升汞溶液进行表面灭菌，然后用无菌水洗涤多次后保持适当水分让种子充分吸胀，28℃催芽，最后选择发芽一致的种子播于灭菌砂钵中，待长出第一片真叶时挑选生长一致的花生砂钵接种根瘤菌。生长约 40d 后即花生开花时收获。

1.3 根瘤中重组质粒在花生根瘤菌中稳定性测定

在 YNA+Tc 培养基上培养供试花生根瘤菌，然后接种盆栽花生进行结瘤试验，收获长好的花生植株，从其根部剪下洗净的根瘤并进行表面灭菌，然后压破根瘤挤出浆液，用少量无菌水制成菌悬液，经系列稀释后涂布于 YMA+Tc 平板，28℃培养至长出菌落后计数。

1.4 人工液体培养基中重组质粒在花生根瘤菌中稳定生测定

从 YMA+Tc 平板上分别挑取供试菌株菌落各 2 个，接种于 5mL YMA 液体培养基中（含 Tc），28℃振荡培养 12h（对 85-7 为宿主的菌株）或 24h（对 CO2-5 为宿主的菌株），然后分别取 1mL 菌液转接于 5ml YMA 液体培养基中（不加 Tc），28℃振荡 24h（以 85-7 为宿主的菌株）或 72h（以 CO2-5 为宿主的菌株）后经系列稀释并分别涂布 YMA 和 YMA+Tc 平板，28℃培养至长出菌落后计数。

2 结果

2.1 共生根瘤根瘤菌中重组质粒的稳定性

从花生的根部剪下根瘤，分别随机测定 2 个根瘤中 Tc 抗性菌数所占的比例，结果见表 2 和表 3。由表可见，和 Tc 抗性相关联的重组质粒在根瘤共生条件下丢失较严重。各根瘤中保持有 Tc^r 的菌数占的百分率小于 50%。表 2 统计分析的 F 值小于 1，这表明各个平均数之间没有显著差异，亦即根瘤 85-7 宿主中，pLAFR1 及其重组质粒丢失程度的相似。表 3 统计分析的 F 值大于 $F_{0.05}$ 值，表明这些数值在 α=0.05 水平上有显著差异。进一步的多重比较结果（表 4）表明：HN13 菌株在根瘤 CO2-5 中 Tc^r 保持百分率与另三个菌株 Tc^r 保持百分率之间有显著差异性（α=0.05 水平）；而 1#，2#菌株与 CO2-5（pLAFR1）菌株 Tc^r 保持百分率之间差异不显著。

表 2 在根瘤中重组质粒（Tc^r）在 85-7 受中的稳定性

菌株或菌株号	85-7	85-7（pLAFR1）		HN11		HN12		3#	
根瘤号		1	2	1	2	1	2	1	2
菌落数（YNA+ Tc）×10^2	0	133.3	7.7	13.0	12.7	13.0	10.7	11.0	9.7
菌落数（YNA）×10^2	23.0	31.3	24.7	31.0	24.0	23.0	25.3	24.0	32.3
根瘤内 Tc 抗性（%）	0	42.5	31.2	42.0	52.9	56.5	42.3	45.8	29.7
平均值*（%）		3.6		47.5		49.4		37.8	
F 值	0.95<1								

*菌数是两个平板上的菌落数平均值

表 3 在根瘤中重组质粒（Tc^r）在 CO2-5 受体中的稳定性

菌株或菌株号	CO2-5	CO2-5（pLAFR1）		HN13		1#		2#	
根瘤号		1	2	1	2	1	2	1	2
菌落数（YMA+ Tc）×10^5	0	29.0	58.7	56.0	84.3	88.013	130.3	136.0	64.7
菌落数（YM）×10^6	8.7	14.7	18.7	11.0	19.3	31.3	49.3	42.3	19.7
根瘤内 Tc 抗性（%）	0	19.7	31.4	50.9	43.7	28.1	26.4	32.0	32.8
平均值*（%）	0	25.6		47.3		27.3		32.4	
F 值	8.1								
$F_{0.05}$	6.59								
$F_{0.01}$	16.69								

*菌数是两个平板上的菌落数平均值

表 4 在根瘤中重组质粒在 CO2-5 受体中稳定性结果的多重比较（LSR 法）

菌株或菌株号	根瘤内 Tc 抗性（%）	差异显著性	
HN13	47.3	a	A
2#	32.4	b	A
1#	27.3	b	A
CO2-5（pLAFR1）	25.6	B	A

2.2 人工液体培养基中重组质粒在花生根瘤菌内的稳定性

重组质粒（Tc^r）在宿主根瘤 85-7 或 CO2-5 中通过人工液体培养后的稳定性测定结果见表 5 和表 6。我们把丢失百分率定义为在非选择性培养条件下生长的某一特定质粒所丢失的菌数在总菌数中所占的数百分率。对于某一特定质粒及受体菌来讲，在一定的培养条件下和时间内，丢失百分率可以是一个比较稳定的数值，它间接从一个侧面反映该质粒的稳定性。由表 5 和 6 可知，pLAFR1 及其重组质粒在同一宿主中丢失情况相似。在 85-7 中约 58.0%，而在 CO2-5 中约为 50.0%。丢失百分率都大于或接近于 50%，说明丢失比较严重。

表 5 人工液体培养基中重组质粒（Tc^r）在 85-7 受体中的稳定性

菌株	85-7		85-7（pLAFR1）		HN11		HN12	
菌落号	1	2	1	2	1	2	1	2
菌数*（YMA+Tc）/10^{10}.L^{-1}	0	0	12.5	13.0	14.0	15.5	13.5	14.5
菌数*（YMA）/10^{10}.L^{-1}	22.5	39	29.5	26.5	36.5	42.0	33.5	32.5
丢失百分率（%）	–		57.6	50.8	61.6	63.1	59.7	54.7
平均值（%）	–		54.3		62.4		57.2	

*表中菌数是两平板上的菌落数平均值而算得

表 6 在人工液体培养基中重组质粒（Tc^r）在 CO2-5 受体中的稳定性

菌株	CO2-5		CO2-5（pLAFR1）		HN13	
菌落号	1	2	1	2	1	2
菌数*（YMA+Tc）/10^{10}.L^{-1}	0	0	82.5	91.5	80.5	73.5
菌数*（YMA）/10^{10}.L^{-1}	181.5	164.5	170.5	179.5	143.5	164.5
丢失百分率（%）	–		51.5	49.0	43.9	55.3
平均值（%）	–		50.3		49.6	

*表中菌数是由两平板上的菌落数的平均值而算得

3 讨论

在共生根瘤中，以 85-7 为宿主，pLAFR1 及其重组质粒丢失程度近似，这表明外源片段的存在不影响重组质粒的丢失。除了 85-7 宿主的影响外，重组质粒的丢失主要由 pLAFR1 载体决定。可是以 CO2-5 为宿主时，增效重组质粒 pHN13 与空载 pLAFR1 及以 pLAFR1 为载体，克隆有外源 DNA 的无效重组质粒相比较丢失程度较轻，这暗示增效质粒的功能片段影响它在 CO2-5 菌株中的稳定性。Long 等人也曾观察到在共生状态下重组质粒大量丢失的情况[4, 5]，他们推测这可能是由于编码早期结瘤基因的质粒在结瘤早期为其所必须，而一旦侵染成功，就失去选择压力，在进一步的分裂过程中可能被丢失。他们以此来解释重组质粒在根瘤内大量丢失的原因。由此我们推测 pHN13 重组质粒具有固氮增效作用，很可能是由于它编码与固氮作用有关的基因在结瘤后为固氮作用所必须，并赋予选择压力，所以相对于空载体 pLAFR1 丢失较轻。对增效重组质粒外源片段的功能进一步的分析将给此推测提供直接而充分的证据。

众所周知，一般构建的质粒其分配基因（partion gene）缺失，分配便成为完全随机，因而分裂过程中的丢失情况由其在受体菌中的拷贝数决定[1]。pLAFR1 质粒拷贝数较高，应该在根瘤菌中是比较稳定的，因此它本不应该有如此高的丢失率。但结合着不论从大肠杆菌还是根瘤菌中提取的 pLAFR1 或其重组质粒往往呈多联体现象[6]，我们认为这是由于 pLAFR1 含有 COS 位点[3]，使许多质粒连接成为一个分配单位，降低随机分配的拷贝数，表现出丢失较为严重。

为了使工程菌应用于生产试验，必须保持其质粒的稳定性。笔者认为可以在生产菌剂中加抗生素，不过往往成本高；也可以克隆到另一根瘤菌载体质粒上，且此载体拷贝，其间不应相互连接；最有效的方法是将外源片段插入到根瘤菌染色体上。

参 考 文 献

[1]吴卫星，细菌质粒的稳定性，生物工程进展，1988，2:27~30

[2]李立家，花生根瘤菌基因工程育种的研究：[硕士论文] 。武汉：华中农业大学生物技术中心，1992

[3]Friedman AM，Long SR，Brown SE，et al. Constructiom of a broad host rang cosmid cloning vector and its use in genetic analysis of *Rhizobium* mutants. Gene，1982，18:289~296

[4]Long S R，Buikema WJ，Ausubel FM. Cloning of *Rhizobium meliloti* nodulation genes by direct complementation of Nod mutants. Nature，1982，298:485~488

[5]Lambert G R. Symbiotic expression of cosmid-borne *Bradyrhirobium japonicum* hydrogenase gene. Appl Environ Microbiol，1987，53:422~428

[6]沈辉，张忠明，曹燕珍，快生型大豆根瘤菌基因文库的构建。中国微生态学杂志，1990，（4）：51~56

重组质粒 pHN32 对花生根瘤菌固氮的影响*

李立家　周俊初　陈华癸

（华中农业大学生物技术中心）

摘　要　通过三亲本杂交将大豆根瘤菌高效基因工程菌株 HN32 所携带的增效重组质粒 pHN32 引入一株快生型花生根瘤菌菌株 85−7 和另一株慢生型花生根瘤菌菌株 CO2−5 中。花生盆栽结瘤试验结果统计分析表明：pHN32 载体质粒 pLAFR1 对根瘤菌 85−7 和 CO2−5 的固氮功能没有影响；pHN32 对 85−7 固氮影响不论在植株干重方面还是在固氮酶活方面统计学上都不显著；而对 CO2−5 则植株干重提高了 25.7%，存在极显著差异，固氮酶活尽管提高了 51.8%，但统计学上不显著。根瘤中分离的 Tc^r 根瘤菌 CO2−5（pHN32）中的 pHN32 *Eco*R I 酶切分析表明：pHN32 没有发生结构上的变化且固氮增强作用可能由 3.7kb 外源片段引起。

关键词　pHN32；快生型花生根瘤菌；慢生型花生根瘤菌

将 *S. fredii* B52 菌株基因文库与 *B. japonicum* 22-10 菌株三亲本杂交，然后通过结瘤试验筛选到到一株高效固氮大豆根瘤菌菌株 pHN32 [1]。该菌株抽携带的增效重组质粒命名为 pHN32，其酶切图谱表明它除了载体 pLAFR1 外还含有 3.2kb 和 0.5kb 两个 *Eco*RI 酶切片段[2]。菌株 85-7[3]和 CO2-5[4]是两株已在生产上推广应用的花生根瘤菌。本研究将 pHN32 引入这两株花生根瘤菌菌株中，研究了 pHN32 对它们固氮作用的影响。

1 材料和方法

1.1　供试菌株、质粒和培养基

供试菌株、质粒见表 1。培养大肠杆菌用 LB 培养基，培养根瘤菌用 TY、YMA、SM 等培养基[5]。

1.2　三亲本杂交

用 TY 液体培养基培养花生根瘤菌 85-7 或 CO2-5、LB+Tc（15μg/ml）培养 HB101（pHN32、LB+Kan（50μg/ml）培养 MM294（pRK2013）同步生长至对数期，然后按 1:1:1 比例混合均匀，取 2ml 用 0.22μm 微孔滤膜过滤，将滤膜移放在 TY 平板表面于 28℃下培养 10hr（对 85-7 受体）或 20hr（对 CO2-5 受体）后，用无菌水洗下细菌并用于涂 SM+Tc(20μg/ml 或 35μg/ml)选择平板以筛选转移接合子 85-7(pHN32)、或 CO2-5 （pHN32），再在同样的 SM+Tc 平板上纯化一次，最后用无菌水洗下菌体制成悬液用作花生盆栽试验。

表 1　供试菌株及质粒

菌株或质粒	表型或基因型	来源
快生型花生根瘤菌 85-7	Nod^+Fix^+	四川农业大学
慢生型花生根瘤菌 CO2-5	Nod^+Fix^+	四川农业大学
E. coli HB101	F^-，recA13，proA-2	英国 John Innes 研究所
E. coli MM294	F^-，hsdR17	英国 John Innes 研究所
pRK2013	Kan^r，mob^-，tra^+	广西农学院
pLAFR1	Tc^r，mob^+，tra^-	广西农学院
pHN32	Tc^c，mob^+，tra^-	本室

*原载于《生物技术通报》，1:9~12，1994.

pHN32 由转移接合子 CO2-5（pHN32）向 HB101 转移时，转移接合子 HB101（pHN32）用 LB +Tc（15μg/ml） 选择。

1.3　花生结瘤试验

花生去壳后挑选籽粒饱满大小一致的种子置培养皿中用 95%乙醇浸润 3 分钟，弃去乙醇后向皿中加入 0.1%升汞溶液表面灭菌 8 分钟。倒掉升汞，无菌水洗多次后保持适当水分于 28℃下催芽 3~4 天。选择发芽一日的种子播于砂钵中，待长出第一片真叶时挑选生长一致的植株接种根瘤菌，并以不接种为对照。

1.4　重组质粒提取及酶切

使用碱变性法[6]小样抽提纯化然后用 *Eco*RI 酶酶切电泳检测分析。

1.5　根瘤固氮酶活性测定

用日立 163 型气相色谱仪乙炔还原法进行。

2 结果

2.1　花生结瘤试验结果及统计分析

花生种子使用四川天府 3 号，经灭菌催芽后播种，选出苗整齐一致的接种根瘤菌。40 多天后即花生开花时进行收获并测定固氮酶活、植株地上部分干重等指标，结果见表 2、表 3。由表知，85-7（pHN32）植株地上部分平均干重是受体 85-7 的 91.6%，略低，而固氮酶活则为 105.6%，略高，但表 4 统计分析都不显著，这提示 pHN32 对 85-7 的固氮作用没有影响。pLAFR1 对受体 85-7（表 2、表 3）和 CO2-5（表 6）的固氮作用影响统计学上都不显著。而 CO2-5（pHN32）的植株地上部分干重平均值比受体 CO2-5 的提高了 25.7%，表 5 统计分析表明植株干重方面存在极显著差异，但在固氮酶活方面不显著，尽管也提高了 51.8%。表 6 多重比较进一步表明 CO2-5（pHN32）与受体 CO2-5 对花生结瘤固氮作用（在植株干重上）存在极显著差异，提示 pHN32 对受体 CO2-5 固氮有增强作用。

表 2　85-7 为受体转移接合子的花生结瘤试验*

菌株	植株干重（g）	是 85-7 结瘤植株干重的百分比%	固氮酶活（μmol/小时/植株）	是 85-7 结瘤植株固氮酶活的百分比%
85-7	1.31	100.0	0.220	100.0
85-7（pLAFR1）	1.36	103.8	0.196	89.1
85-7（pHN32）	1.20	91.6	0.232	105.6
CK	0.81	61.8	—	—

*每个样品重复 5 次

表 3　CO2-5 为受体转移接合子的花生结瘤试验*

菌株	植株干重（g）	是 85-7 结瘤植株干重的百分比%	固氮酶活（μmol/小时/植株）	是 85-7 结瘤植株固氮酶活的百分比%
CO2-5	2.30	100.0	2.55	100.0
CO2-5（pLAFR1）	2.28	100.9	2.88	112.9
CO2-5（pHN32）	2.89	125.7	3.87	151.8
CK	1.38	60.0	—	

*每个样品重复 5 次

表 4 85-7 为受体转移接合子结瘤试验固氮酶活及植株干重方差分析

	变异来源	DF	SS	MF	F	F0.05	F0.01
固氮酶活方差分析	处理间误差	2	0.00336	0.00168	0.282	3.89	6.93
		12	0.07136	0.00595			
	总变异	14	0.07472				
	变异来源	DF	SS	MF	F	F0.05	F0.01
植株干重方差分析	处理间误差	2	0.0631	0.0316	1.133	3.89	6.93
		12	0.3343	0.0279			
	总变异	14	0.3974				

表 5 CO2-5 为受体转移接合子结瘤试验固氮酶活及植株干重方差分析

	变异来源	DF	SS	MF	F	F0.05	F0.01
固氮酶活方差分析	处理间误差	2	4.723	2.364	3.60	3.89	0.93
		12	7.873	0.656			
	总变异	14	12.60				
	变异来源	DF	SS	MF	F	F0.05	F0.01
植株千重方差分析	处理间误差	2	1.189	0.595	8.62	3.89	0.93
		12	0.828	0.069			
	总变异	14	2.017				

表 6 CO2-5 为受体转移接合子结瘤试验结果的比较

处理	植株干重平均值	显著水平
CO2-5	2.30	
CO2-5（pLAFR1）	2.28	C
CO2-5（pHN32）	2.89	A

A.极显著；C.不显著

2.2 CO2-5（pHN32）质粒的确证

CO2-5（pHN32）菌株对花生植株干重有明显的增效作用，为了进一步确证 pHN32 的存在作用，我们从根瘤中分离出 Tc 抗性菌株，然后将其重组质粒转入大肠杆菌 HB101 中，从中提取质粒用 *Eco*RI 酶切，结果表明该质粒除载体 pLAFR1 外还含 3.7kb（3.2kb+0.5kb）的外源片段，这与最初 pHN32 质粒相一致，说明没有发生结构上的变化，初步提示 CO2-5（pHN32）固氮增效作用可能由重组质粒 pHN32 所引起。

3 讨论

pHN32 是由载体 pLAFR1 和 3.7kb 的外源片段组成，在大豆根瘤 HN32 菌株中，已初步证实其固氮增效作用。主要由 3.7kb 外源片段引起，pLAFR1 无增效作用[1,2,7]。本试验结果证实 pLAFR1 对 85-7 和 CO2-5 两株花生根瘤菌固氮没有影响，这也暗示在 CO2-5 中 pHN32 的增效作用主要由此段引起。进一步对此片段的功能研究和在花生大豆根瘤菌中的作用机理研究正在进行中。

将 pHN32 引入花生根瘤菌菌株 85-7 中没有显著影响，而引入菌株 CO2-5 中对花生植株干重有明显的增效作用，这说明 pHN32 的增效作用随不同背景的根瘤菌菌株而异。

参 考 文 献

[1]周俊初等：武汉大学学报（自然科学版），1990，11-17（生物工程专刊）
[2]张忠明等：华中农业大学学报，1991，11（2）：94-96
[3]黄怀琼：四川农业大学学报，1990，8（3）：188-193
[4]陈华癸等：中国共生固氮研究五十年，1987，南京农业大学出版社
[5]Wang，C.L. Acta Genetic Sinica，1988，15（1）：25-33
[6]Sambrook，J. et al. Molecular Cloning，Cold Spring Harbor Laboratory，Press.1989
[7]沈辉等：华中农业大学学报，1993，12（1）：40-44

外源质粒（基因）导入花生根瘤菌的行为分析*

朱光富　周俊初　陈华癸

（农业部华中农业大学农业微生物重点开放实验室）

摘　要　利用二亲本或三亲本杂交的方法，将携带有共生固氮基因的外源重组质粒或外源载体质粒导人慢生型花生根瘤菌[*Bradyrhzobium* sp.（*Arachis*）] 147−3 和快生型花生根瘤菌[*Rhizobium* sp.（*Arachis*）] 85−7 中。探讨了转移接合子中外源质粒在人工培养条件下和共生条件下的稳定性，发现外源质粒在花生根瘤菌中的稳定性与质粒的类型、受体菌的特性和环境条件有关。同时还探讨了外源质粒上的共生基因对受体菌 147−3 共生固氮效率的影响。结果表明，外源共生基因对共生固氮能力的影响是复杂的，既可以产生正效应，也可以产生负效应。

关键词　花生根瘤菌；转移接合子；外源质粒的稳定性；共生固氮效率

花生是一种重要的食、油两用经济作物：全世界花生种植面积为 3~4 亿亩，以印度、中国、尼日利亚、美国和塞内加尔面积最大，占全世界播种面积的 75%。我国每年的播种面积为 3~4 千万亩，占世界播种面积的 10%。花生根瘤菌是一类能在花生根部结瘤固氮的土壤细菌，它们能通过有效的共生固氮作用，提供花生生育期中所需氮素的 1/3~2/3 [1]。构建和选育固氮效率高、竞争结瘤能力强的花生根瘤菌对减少化学氮肥用量，提高花生产量和减少环境污染有着十分重要的意义。

将外源共生质粒或重组质粒通过接合转移导入花生根瘤菌以提高竞争结瘤能力和共生固氮效率的研究少见报道。Pankhurst [15]、Trinick [19]和 Kondorosi [12]等曾将苜蓿根瘤菌（*Rhizobium meliloti*）的共生质粒 pRme4lb∷pAK11 和 pRme41b∷pAK12 导入花生根瘤菌 PN4003 和 MPIK3030（即 NGR234）中，获得的转移接合子虽然能够在苜蓿根上形成大小不一的根瘤，但只能发育至一定的阶段，而且没有固氮酶活性。李立家等[2]将快生型花生根瘤菌[*Rhizobium* sp.（*Arachis*）]85-7 的基因文库分别引入到 85-7 及慢生型花生根瘤菌[*Bradyrhizobium* sp.（*Arachis*）]CO2-5 中筛选获得了 3 株在固氮酶活性和植株干重上比受体菌有极显著提高的转移接合子，同时他又将大豆根瘤菌 HN32 含有增效因子的重组质粒 pHN32 分别导入 85-7 和 CO2-5 中，发现 pHN32 虽然对 85-7 的结瘤固氮能力没有明显促进作用，但却使 CO2-5 接种植株的干重提高 25.7%，固氮酶活性提高 51.8%。

本研究利用滤膜二亲本或三亲本杂交的方法将外源质粒（基因）导人野生型花生根瘤菌，考查了外源质粒（基因）在受体菌中的稳定性，并通过植物盆栽实验比较研究了转移接合子与亲本的共生固氮效率，进而探讨了通过接合转移的方法构建和选育优良花生根瘤菌的可能性和潜在的问题。

1 材料和方法

1.1　供试菌株和培养条件

1.1.1　供试菌株与质粒（表 1）

*原载于《遗传学报》，23（2）：131~141，1996.

表1 供试菌株与质粒

Table 1 Strains and Plasmids tested

菌株或质粒 Strain or Plasmd	表型或基因型 Genotype or phenotype	来源或参考文献 Source or reference
大肠杆菌 （*E. coli* ）		
HB101	F^-，*sup*E44，*rec*A13，*pro*A2，*lac*Y，*gal*K2	[14]
MM294	F^-, hsdR17	本室保存 Our laboratory
花生根瘤菌 （*Peanut rhizobium*）		
147-3	慢生型，Nod^+，Fix^+， Cm^5（slow-growing strain）	中国农业科学院油料所 Oil Crop Research Institute，CAAS
85-7	快生型，Nod^+，Fix^+ （fast-growing strain）	四川农业大学 Sichuan Agricultural Uniyersity
质粒（Plasmid）		
pLAFR1	Mob^+，tra^-，21.6kb Cosmid，Tc^r	本室保存 Our laboratory
pRK2013	Mob^-，tra^+，Km^r	[8]
pRaZ15	载体 pLAFR1 上克隆有 *R.huakuii7653R* 的共生基因，Tc^r，46 .6kb （ containing symbioticgenes of *R.huakuii* 7653R on pLAFR1）	本室张忠明构建 Construeted by Zhang Zhongming
pRmSL26	载体 pLAFR1 上克隆有 *R.melilori* 8.7 kb 的 *Eco*R I 片段，含 *nod*DABC，Tc^r，41.2kb （containing 8.7kb *Eco*R I fragment of *R.mililot nod* DABC genes on pLAFR 1）	本室保存 Our laboratory
pHN32	载体 pLAFR1 上克隆有 *R.fredii* B52 3.7kb *Eco*R I 增效基因片段，Tc^r，25.3kb， （containing 3.7kb *Eco*RI fragment of *R.fredii* B52 enhancement genes on pLAFR 1）	本室构建 Construeted by ou r laborator y
pMN53	载体 pMNZ 上携带有 *R.legumiosarum* bv. *Phaseoli* RCR362 的共生基因 Tc^r，AP^r 166kb （ containing symbiotic genes of *R.leguminosarum* bv. *Phaseoli* RCR3522 on pMN2）	本室莫才清构建 Construeted by Mo Cai qing
pCK3	载体 pRK290 上克隆有 *K.pneumoniae nif* A 基因，mob^+，Tc^r （pRK290 derivative carrying the *K.pneumoniaen nif* A gene）	北京农业大学 Beijing Agricultural University
pKT230	广谱性载体质粒 mob^+，Sm^r，11 .9kb （ wide host-range Plasmid ）	本室保存 our laboratory

1.1.2 培养基与培养条件 大肠杆菌采用 LB 培养基，于 37℃下培养。花生根瘤菌采用 YGA（Yeast extract Glycerol Agar）或 SM 培养基，于 28℃下培养。

1.1.3 抗生素使用浓度 四环素（Tc）：对 147-3 为 100μg/ml；对 85-7 为 20μg/ml。卡那霉素（Km）：50μg/ml。氨苄青霉素（Ap）：50μg/ml。氯霉素（Cm）：50μg/ml。链霉素（Sm）：300μg/ml。

1.2 接合转移

采用滤膜二亲本或三亲本杂交法[5]。

1.3 质粒检测

大肠杆菌采用碱变性法[3]提取，根瘤菌采用 Hirsch 修改的碱变性法[11]提取。TE 溶解后，一起点样电泳，琼脂糖凝胶浓度为 0.65%，电泳缓冲液为 TBE。

1.4 花生植物结瘤实验

采用周平贞等的水培法[6]。

1.5 根瘤固氮酶活性测定

采用乙炔还原法。

1.6 植株地上部分含氮量的测定

采用半微量凯氏定氮法。

2 结果

2.1 外源质粒（基因）向花生根瘤菌的转移

重组质粒 pRaZl5、pRmSL26 和 pHN32 的载体为 pLAFRl，pCK3 的载体为 pRK290，本身均不能自我转移，必须在辅助质粒 pRK2013 的帮助下，通过三亲本杂交才能进行接合转移。pMN53 和 pKT230 有自我转移能力，通过二亲本杂交即可实现接合转移。由于以上这些质粒的宿主均是营养缺陷型大肠杆菌，利用加相应抗生素的根瘤菌合成培养基 SM 即可将供体菌、受体菌和辅助菌菌株淘汰并筛选得到转移接合子。实验结果见表 2 和表 3。

表 2 外源共生质粒（基因）向 147-3 的转移

Table 2 Transference of exogenous Plasmids （genes） into 147-3

质粒 Plasmid	选择平板 Selective Plate	计数平板 Calculation Plate	选择平板菌落数 Colonies on Selective Plate	计数平板菌落数 Colonies on Calculation Plate	转移频率 Frequeneyof transference	自发突变频率 Frequeney of spontaneous mutation
pRaZ15	SM + Tc	SM	2.64×10^4	7.50×10^7	3.52×10^{-4}	$<10^{-8}$
pRmSL26	SM + Tc	SM	108×10^4	1.07×10^6	1.01×10^{-3}	$<10^{-8}$
pHN32	SM + Tc	SM	3.21×10^4	2.73×10^8	1.17×10^{-4}	$<10^{-8}$
pMN53	SM+AP+ Cm	SM+Cm	1.82×10^3	5.12×10^6	3.56×10^{-4}	$<10^{-8}$
pCK3	SM + Tc	SM	4.12×10^4	2.53×10^8	1.63×10^{-4}	$<10^{-8}$
pKT230	SM +Sm	SM	2.38×10^3	4.05×10^6	5.88×10^{-4}	$<10^{-8}$

注:表中菌落数均为 3 个平板上菌落数的平均值

Note: The colony number in the Table is means of the colony numbers on three plates

表 3 外源共生质粒（基因）向 85-7 的转移

Table3 Transference of exogenous Plasmids （genes） into 85-7

质粒 Plasmid	选择平板 Selective Plate	计数平板 Calculation Plate	选择平板菌落数 Colonies on Selective Plate	计数平板菌落数 Colonies on Calculation	转移频率 Frequeney of transference	自发突变频率 Frequeney of spontaneous mutation
pRaZ15	SM + Tc	SM	2.647×10^5	9.00×10^7	2.97×10^{-4}	$<10^{-8}$
pRmSL26	SM + Tc	SM	4.05×10^3	4.75×10^6	8.53×10^{-4}	$<10^{-8}$
pHN32	SM + Tc	SM	2.41×10^6	3.86×10^8	6.24×10^{-4}	$<10^{-8}$
pMN53	SM +Tc	SM	2.75×10^3	2.55×10^7	1.08×10^{-4}	$<10^{-8}$
pCK3	SM + Tc	SM	1.28×10^5	3.61×10^8	3.55×10^{-4}	$<10^{-8}$
pKT230	SM +Sm	SM	1.38×10^5	2.41×10^8	5.37×10^{-4}	$<10^{-8}$

注:表中菌落数均为 3 个平板上菌落数的平均值

Note :T he colony number in the Table is means of the colony numbers on three plates

由表 2 和表 3 可见，受体菌 147-3 和 85-7 的自发突变率均小于 10^{-8}，而二亲本或三亲本杂交的接合转移频率均大于 10^{-4}，说明质粒转移是成功的。

2.2 转移接合子中外源质粒的检测

用碱变性法从供体大肠杆菌和纯化的转移接合子提取质粒。部分质粒的琼脂糖电泳结果见图 1。

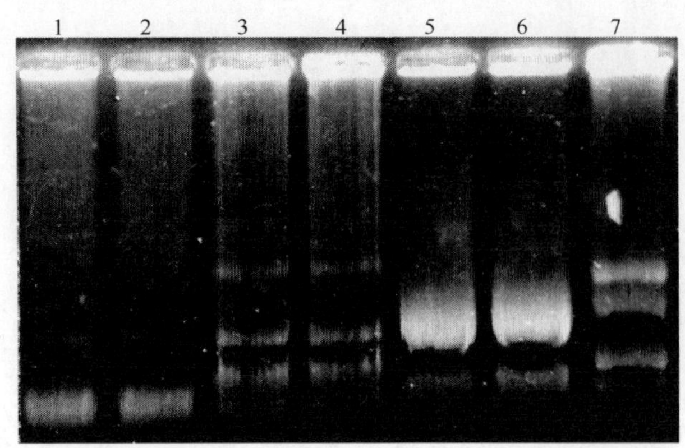

图 1 转移接合子 147-3（pHN32）的质粒图谱

Fig.1 Plasmid electrophoresis pattern of transconjugant 147-3 (#1N32)

1、2. 147-3；3、4．147-3（pHN32）；5、6. HB101（pHN32）； 7. HB101（pLAFR1）

电泳结果表明，外源质粒已成功地导入了花生根瘤菌。图l中pLAFR1及以它为载体的重组质粒在电泳时均出现多条质粒带，这是由于它们容易形成多聚体的缘故。由于pMN53不稳定，容易丢失所携部分外源DNA而出现多条质粒带，实际检测样品为含有不同大小重组质粒细菌的混合物。

2.3 外源质粒在人工继代培养条件下的稳定性

将在选择平板上长出的花生根瘤菌转移接合子再在相应的选择平板上纯化后，随机挑取5个单菌落，在不加抗生素的YGA培养基上转接2次，稀释培养后各挑取100个单菌落分别点种YGA加相应抗生素的平板和YGA平板，于28℃培养至长出菌落，计数并依以下公式计算出外源质粒所携抗性的丢失百分率，结果见表4和表5。

$$\text{抗性丢失百分率（\%）} = \frac{\text{YGA+抗生素平板上的菌落数}}{\text{YGA平板上的菌落数}} \times 100$$

表 4 外源质粒导入 147-3 后再人工继代培养条件下的稳定性

Table 4 Stablility of exogenous plasmids in 147-3 under artificial culture condition

转移接合子 Transconjuants	抗性平板菌落数 Colonies on resistant plant	YGA 平板菌落数 Colonies on YGA plant	丢失百分数 Loss percentage of resistance
147-3（pRaZ15）	279	500	44.2
147-3（pRmSL26）	264	500	46.2
147-3（pHN32）	400	500	20.0
147-3（pMN53）	443	500	11.5
147-3（pCK3）	400	500	20.0
147-3（pKT230）	500	500	0

表 5 外源质粒导入 85-7 后再人工继代培养条件下的稳定性

Table 5 Stablility of exogenous plasmids in 85-7under artificial culture condition

转移接合子 Transconjugants	抗性平板菌落数 Colonies on resistant plant	YGA 平板菌落数 Colonies on YGA plant	丢失百分数 Loss percentage of resistance
85-7（pRaZ15）	239	500	52.2
85-7（pRmSL26）	378	500	24.4
85-7（pHN32）	434	500	13.2
85-7（pMN53）	500	500	0
85-7（pCK3）	344	500	31.2
85-7（pKT230）	500	500	0

从表4可以看出，转移接合子l47-3（pKT230）全部具有链霉素抗性，随机进行的质粒检测结果表明均含有pKT230质粒。表明在人工条件下，不含外源共生基因的质粒pKT230在转移接合子147-3（pKT230）中能稳定存在。其他转移接合子的抗性则出现或多或少的丢失现象，对丢失抗性的菌株进行质粒检测，没有发现相应的外源质粒，表明除pKT230以外的其他外源质粒在转移接合子中存在程度不等的不稳定性，其中尤以pLAFR1系列的重组质粒更为突出。

从表5可见，转移接合子85-7（pMN53）租85-7（pKT230）全部保持了相应抗性，进一步作质粒检测，结果证明均含有导人的外源质粒，表明在人工条件下，plaiN53和pKT230在转移接合子中能稳定存在。其他转移接合子的抗性在不同程度上发生丢失，对抗性丢失的菌株进行质粒检测，发现它们已不含有导人的外源质粒，表明pRaZ15、pRmSL26、pHN32、和pCK3在转移接合子中也有程度不等的不稳定性。

图 2 147-3 及其转移接合子接种植株的生长势
Fig.2　Plant growth potential inoculated with 147-3 and its transconjugants

2.4　花生根瘤菌147−3及其转移接合子共生固氨效率的测定

将纯化的转移接合子制成菌悬液，接种花生植株，并分别以出发菌株和不接种处理作对照。自然光照下培养生长至40天左右于花生植株处于盛花期时收获植株。部分实验植株的生长势见图2。从图2可以看出，接种l47-3（pCK3）和147-3（pHN32）的植株生长势比接种出发菌株147-3的植株好，而接种147-3（pRaZl5）、147-3（pRmSL26）和147-3　（pMN53）　的植株生长势比接种147-3的植株差。

表 6　147-3 及其转移结合子的共生效应

Table 6　Symbiotic efficiency of 147-3 and its transconjugants

菌株 Strain	叶色 Colour	株高（cm） Plant highness	地上部分干重（克/株） Shoot dry weight (g/plant)	根瘤数（个/株） Nodule number (per plant)	根瘤重（克/株） Nodule weight (g/plant)	固氮酶活性（μmol 乙烯/株·小时$^{-1}$） Nitrogenase activity（μ mol ethylene/plant·hr^{-1}）	植株含氮量（%） Nitrogen percentage of plant	植株总氮量（克/株） Total nitrogen contant (g/plant)
147-3	绿（green）	17.4	1.39	60.6	0.57	3.82	4.213	0.060
147-3（pRaZ15）	黄（yellow）	14.8	1.22	44.4	0.34	2.15^{c}	2.476d	0.029d
147-3（pRLS26）	黄（yellow）	15.6	1.07	48.4	0.41	5.30	3.014	0.035a
147-3（pHN32）	绿（green）	18.7	1.46	87.8	0.89^{a}	3.57	3.840	0.057
147-3（pMN53）	黄（yellow）	13.7^{e}	1.16	98.6	0.37	1.16^{d}	1.316^{d}	0.016d
147-3（pCK3）	绿（green）	19.6	2.01^{c}	87.2	0.91^{c}	4.62	5.062^{c}	0.010
147-3（pKT230）	绿（green）	13.7	1.59	74.0	0.44	4.52	2.972	0.046
CK	黄（yellow）	12.3^{d}	0.77^{d}	-	-	-	1.478^{d}	0.011d

注：接合子与 147-3 相比，147-3 与 CK 相比，a:在 α=0.01 水平上差异显著；c：在 α=0.05 水平上差异显著；d：在 α=0.01 水平上差异显著.

Note: Transconjguants compared with 147-3；　147-3 compared with CK；　a: significant at the α=0.01 level ；　c: significant at the α=0.01 level ；　d: significant at the α=0.01 level .

综合考查植株叶色、株高、地上部分干重、根瘤数、根瘤重、固氮酶乙炔还原活性、地上部分含氮量和地上部分总氮量等指标，并进行数理统计，结果见表 6。从表 6 可见，转移接合子 147-3（pRaZl5）所结根瘤瘤数、瘤重、固氮酶乙炔还原活性分别比出发菌株 147-3 减少 26.7%、40.4%和 43.7%，其中固氮酶乙炔还原活性在 α=0.05 水平上达到显著降低。植物高度、地上部分干重、含氮量和总氮量也分别比 147-3 所接种植株减少 14.9%、12.2%、41.2%和 51.7%，其中地上部分含氮量和植物总氮量 α=0.01 水平上达到极显著减小。

转移接合子 147-3（pRmSL26）除固氮酶乙炔还原活性比 147-3 增加 38.7％外，其他各项指标均比 147-3 降低。其中根瘤数和根瘤重分别减少 20.1％和 28.2%。植株高度、地上部分干重、含氮量和总氮量分别减少 10.3％、23.0％、28.4％和 41.7％。

转移接合子 147-3（pHN32）所结根瘤瘤重比 147-3 增加 56.1％，且在 α =0.10 水平上达到显著水平。根瘤数、植株高度和地上部分干重分别比 147-3 增加 44.9％、7.5％和 5.1％，但均未达到 α=0．05 的显著水平。在固氮酶活性、植株地上部分含氮量和总氮量等方面比 147-3 略有减少。

转移接合子 147-3（pMN53）所结根瘤数比 147-3 增加 62.7％，但从观察根瘤发生过程来看，其根瘤形成比 147-3 迟、细小且多为白色。根瘤重、固氮酶活性、植株高度、地上部分干重、地上部分含氮量和植株总氮量分别比 147-3 减少 35.1％、69.6％、16.6％、68.8％和 73.3％，其中，植株高度在 α=0.05 水平上达到显著降低，表现出明显的负效应。

转移接合子 147-3（pCK3）的根瘤重、地上部分干重、含氮量和总氮量分别比 147-3 增加 44.6%、60.0%、20.2％和 68.3％，且均在 α=0.05 水平上达到显著水平。植株高度、根瘤数和固氮酶活性亦分别比 147-3 增加 12.6%、43.9％和 20.9％，但未达到 α=0.05 的显著水平。

2.5 外源质粒在共生条件下的稳定性

从转移接合子接种植株的根瘤中分离出单菌落，随机挑取 500 个分别点种 YGA 加相应抗生素的平板和 YGA 平板，培养后，计数并计算出丢失百分率，结果见表 7。

表 7 转移接合子中的外源质粒在共生条件下的稳定性

Table 7 Stablility of exogenous plasmids in 147-3 under symbiotic condition

转移接合子 Transcojuants	抗性平板菌落数 Colonies on resistant plant	YGA 平板菌落数 Colonies on YGA plant	丢失百分数 Loss percentage of resistance
147-3（pRaZ15）	155	500	69.0
147-3（pRmSL26）	201	500	59.8
147-3（pHN32）	398	500	20.4
147-3（pMN53）	460	500	8.0
147-3（pCK3）	456	500	8.8
147-3（pKT230）	500	500	0

可见，以 pLAFR1 为载体的外源重组质粒，经过与植物共生后的不稳定性加剧；pMN53 和 pCK3 经过共生后，虽然也存在一定程度的抗性丢失现象，但在共生条件下比在人工培养条件下相对稳定；pKT230 在共生条件下也能够稳定地存在于转移接合子 147-3（pKT230）中。

对抗性丢失的菌株进行质粒检测结果表明均不再含有相应的外源质粒，证明抗性的丢失是由于外源质粒的丢失所致。

3 讨论

本研究结果表明，外源质粒在根瘤菌中的稳定性除受质粒基因的自我控制以外，还与受体根瘤菌的特性和环境条件有关。在人工培养条件下，以 pLAFR1 为载体的外源重组质粒在 147-3 和 85-7 中均表现出较高的丢失率，pMN53 虽在 85-7 中能稳定存在，但在 147-3 中却表现出不稳定现象，pCK3 无论在 147-3 还是 85-7 中都有程度不等的丢失率，唯不含外源基因的广谱质粒 pKT230 能在 147-3 和 85-7 中稳定存在。外源质粒导人 147-3 后得到的转移接合子经过与植物共生后，pRaZ15 和 pRmSL26 的不稳定性加剧，但 pMN53 和 pCK3 比人工培养条件下稳定性相对提高，唯有 pKT230 仍能够稳定存在。研究表明，细菌质粒在宿主中的稳定遗传必须具备以下两个条件：一是能在宿主细胞内复制并具有一定的拷贝数；二是在细胞分裂时能实现质粒在子细胞中的均等分配。国内外已有不少学者也观察到在人工培养条件下和共生状态下外源质粒丢失的现象[4, 7, 13, 18]，他们认为在人工培养条件下能够稳定存在的外源质粒经与植物共生后，出现全部或部分丢失现象的原因主要是由于宿主植物的功能不相容性所致，即宿主植物会影响非互接种族根瘤菌共生质粒的稳定性。作者认为本研究中以 pLAFR1 为载体的外源。

质粒在转移接合子中的不稳定性比其他外源质粒高的原因很可能是由于 pLAFR1 含有 COS 位点，质粒间易于连接形成多聚体的缘故，在细胞分裂时若以多聚体为分配单位，自然会降低质粒分配的随机性，以至出现较高的丢失百分率。为使外源质粒能在转移接合子中稳定地遗传下去，我们认为可以从以下两方面着手，一是借助于抗生素的选择压力，但在实际生产应用中常由于成本高而难于采用；二是用转座、同源双交换或其他方法使质粒整合到受体根瘤菌染色体上，使其随染色体一道复制和分配。ROSS 等[16]报道了一种在不用选择压力的条件下使外源质粒稳定存在的方法。他们将外源共生基因克隆到有控制胸腺嘧啶合成基因的质粒载体中，然后再将这一重组质粒引入胸腺嘧啶缺陷型的受体根瘤菌中，这样，外源质粒丢失的细胞因不能合成胸腺嘧啶而被淘汰，生存下来的细胞均含有外源质粒。另外，我们在本研究中还发现广谱性载体质粒 pKT230 无论在人工培养条件下还是在共生条件下，均能稳定地在 147-3 中遗传，因而在花生根瘤菌的遗传工程和育种研究中，可以改用 pKT230 作为外源基因的载体，以克服外源共生基因的丢失现象。

本研究结果表明，外源质粒（基因）对花生根瘤菌共生固氮效率的影响是复杂的。pCK3 导人 147-3 后使转移接合子所接种植株的根瘤重、地上部分干重、含氮量和总氮量都明显高于出发菌株，植株高度、根瘤数和固氮酶活性也比出发菌株高，对共生固氮作用表现出明显的正效应。pHN32 对受体菌 147-3 的共生固氮效率也有一定的正效应。但 pRaZl5、pRmSL26 和 pMN53 对受体菌 147-3 共生固氮的影响却表现出负效应，原因可能与多拷贝基因对共生固氮作用的正调节基因产物（如 Nod D 和 Nif A 等）的竞争性抑制作用有关，由于结构基因拷贝数的增加，使每一基因分配到的正调节基因产物减少，从而导致负效应的产生。而不带共生基因的 pKT230 的影响则不显著。近年的一些研究结果还表明，*nif* A 基因不仅可以控制固氮酶基因，而且可以控制与根瘤成熟和类菌体存活期有关的基因[9 10]。我们认为 147-3（pCK3）工程菌株固氮效率的提高可能是由于导人的 *nif*A 基因在共生固氮过程中加强了正调节因子 NifA 的作用，从而提高了共生固氮效率。Sanjuan 等[17]将 pCK3 和 pCKl(pKT230 上携带有 *K .pneumoniae nif*A 基因）引入苜蓿根瘤菌 2011、L5、30、GR013 和 41 中，也发现 *nif* A 能明显提高受体菌的竞争结瘤能力。本研究得到的共生固氮效率提高的工程菌株 147-3（pCK3）在实际生产应用中能否表现出高竞争结瘤能力和高效固氮能力，还有待作进一步的田间实验。

参 考 文 献

[1]湖北油料研究所植物系微生物组，1976. 油料作物科技，（4）:30-36

[2]李立家等，1994.华中农业大学学报，13（3）:219-225

[3]徐建国等，1991.微生物学通报，18（1）:44-47

[4]张学贤等，1993.农业生物技术进展与展望，中国科学技术大学出版社，133-141

[5]张忠明等，1989.华中农业大学学报，8:285-290

[6]周平贞等，1979.中国油料，（2）:60-62

[7]邹向宏，1994.华中农业大学学报，13（4）:325-331

[8]Ditta G *et al*，1980 Proc.Natl.Acad Sci，USA，77:7347-7351

[9]Fisher HM *et al*，1986. EMBO J.5:1165-1173

[10]Hirsch A M et al，1987．J.Bacteri01.169:1137-1146

[11]Hirsch P R et al，1980．J.Gen.Microbi01.120:403-412

[12]Kondorosi A et al，1982.M01.Gen.Genet，88:433-439

[13]Long S R et al，1982.Nature，298:485-488

[14]Maniatis T et al，1982. In: Molecular Cloning :a laboratory manual，Cold Spring Harbor laboratory，Cold Spring Harbor，N.Y

[15]Pankhurst C E，1977. Can. J. Microbiol. 23:P1026-1033

[16]Ross P et al，1990. Appl. Environ. Microbiol .56 :2164-2169

[17]Sanjuan J et al，1991. Mol. Plant-Microbe Interactions，4[4 :365-369

[18]Szeto WW et al，1984.Cell，36 :1035-1043

[19]Trinick M J，1980. J.Appl.Bacteriol. 49:39-53

菜豆根瘤菌 *sym* 和 *mel* 基因在农杆菌中的转移和表达*

王常霖 陈华癸

（华中农业大学土化系）

菜豆根瘤菌（*Rhizobium phaseoli*）共生质粒 pSYM3622 具有诱导宿主植物（*Phaseolus vulgaris* cv. "jamapa"）结瘤固氮的有效基因，以及菜豆根瘤菌特征性的产黑素基因（melanin production gene）在诱动因子 RP4 的存在下，共生质粒能够有效地向三叶草根瘤菌和农杆菌转移。种间和属间杂交子都能诱导菜豆形成无效根瘤，并且具备在特定培养基上产生黑素的能力。pSYM3622 在三叶草根瘤菌菌株 RCR226 中具有明显的不亲和性，但是在农杆菌杂交子中，这些结瘤和产生黑素的特征在植物根瘤分离菌中能够稳定地保持下去。试验同时研究了 pSYM3622 向假单胞菌和大肠杆菌的诱动转移。

关键词 共生质粒转移；农杆菌；结瘤；产生黑素

根瘤菌与豆科植物共生有着较为严格的宿主专一性，但是，通过对共生质粒在根瘤菌中间杂交转移能够打破这种专性关系，使杂交子具备有诱导非种族内宿主植物形成根瘤的能力[1-3]。比如，在结瘤缺陷型（Nod^-）菜豆根瘤菌 JI8400 中导入豌豆根瘤菌（*R. leguminosarum*）共生质粒 pRL1JI，杂交子能够在豌豆植物上形成无效根瘤[2]；而在野生型（Nod^+）三叶草根瘤菌（*R. trifolii*）中导入豌豆根瘤菌和共生质粒 pJB5JI，杂交子在三叶草和豌豆两种植物上都能诱导产生根瘤[4, 5]。但是，更有效的共生基因表达范围的研究是把共生质粒转到根瘤菌属以外的一些其他细菌中，以消除与共生过程相关的一切遗传背景。比如，将苜蓿根瘤菌（*R. meliloti*）共生基因导入农杆菌（*Agrobaterium tumefaciens*）以及大肠杆菌（*E.coli*）中并给予表达[3, 6, 7]。最近，Plazinaki 和 Rolfe[8]将三叶草根瘤菌的共生基因转入到 *Lignobacter* 和 *Pseudomonas* 中，用这些杂交子接种三叶草植物，只能诱导根毛畸形，但不能形成正常的根瘤形状。

本文报道菜豆根瘤菌共生大质粒 pSYM3622（MW 285kb）在三叶草根瘤菌以及三种表型不同的农杆菌菌株中的表达范围，同时对 pSYM3622 向假单胞菌和大肠杆菌的诱动转移也做了初步研究。

材料与方法

（一）细菌菌株和杂交子的获得

细菌菌株和质粒见表 1。根瘤菌培养基采用 TY[1 2]，农杆菌和其他细菌的培养基 ONA[Oxoid Nutrient Broth Powder 1.3%（*W*/*V*），Agar 1.5%（*W*/*V*）]。细菌的生物膜杂交是在 TY 平板上进行的。由无菌水制取的受体和供体菌悬浮液中各取 0.1ml 在生物膜上混匀并置于无菌操作柜中风干，然后倒置于 28℃中培养过夜，菌体用无菌水从生物膜上洗下来，用适当稀释度在 ONA 添加卡那霉素（50μg/ml）的平板上涂皿。杂交子经过对形成菌落的分离纯化，并通过质粒电泳胶验证后获得。

*原载于《生物工程学报》，4（4）：271~277，1988.

表 1 细菌菌株和质粒

Table 1 Bacterial strains and plasmid

菌株 Strains	有关特征及质粒 Relevant characteristics and plasmid	来源 Reference
Rhizobium phaseoli 3622-15	Sym plasmid pSYM3622 marked with Tn5-Mob，Km，mobilizing plasmid RP4-4Te	(1)
R. trijolii RCR226	wild type	Rothamsted collection
Agrobacterium tumefaciens GMI9023	Derivative of C58C1 cured of both $pAtC_{58}$ and $pTiC_{58}$，a plasmid-free strain，Sm^r，Rif^r	(9)
UBAPF2	Derivative of LBA_{275} cured of both $pAtC_{58}$ and $pTiC_{58}$，Rif^r	(10)
UBATi1	Derivative of LBA_{275} cured of $pAtC_{58}$	(10)
E. coli $J_{5\text{-}3}$	Pro，met，nal^r	(11)
Pseudomonas putida NCIB10007		Stanier et al. (1966)
P. fluoroscens NCIB 10527		Ibid

（二）细菌的生物学检测技术

快速琼脂糖凝胶电泳，分子原位杂交以及产生黑素的鉴定均按文献[1]的方法。

（三）植物接种方法

接种三叶草植物的无菌培养同以前描述的方法[5]。接种菜豆植物的无菌培养是采用双层塑料沙钵和蛭石培养钵，在 Fahreaus[1,3]培养液中培养。开口的塑料沙钵用双层牛皮纸封闭，并留下对剪约 1cm 的十字小孔让植物幼苗伸出，然后用胶带封闭，置于 22℃灯光室中生长（16h/8h：明/暗）。根瘤在 4~5 周内形成。从根瘤中分离根瘤菌按常规方法在 YMA 刚果红平板上划线获得。

（四）光学显微镜检查

根瘤形态使用植物解剖镜，根毛检查用相差光学显微镜按常规方法拍摄。

结　果

（一）菜豆根瘤菌共生质粒向三叶草根瘤菌转移

在菜豆根瘤菌菌株 RCR3622 中至少含有 4 个分子质量大小不等的质粒。采用转座子 Tn5-Mob 诱变技术，两个分子量较小的质粒受到标记并转出该诱变菌株，将其中的一个（pSYM3622）导入结瘤缺陷型（Nod^-）的菜豆根瘤菌菌株 JI8400 中，能够恢复其本菌株的 Nod^+ Fix^+ Mel^+特征。对菌株 3622-15 作 37℃热处理诱变，发现该质粒的消除或缺失与诱导寄主植物结瘤能力的丧失是一致的[1]。在 DNA 分子同源杂交试验中，^{32}P 标记的 *nif* HDK（*Klebsiella*）能够对该质粒进行标记（王常霖，私人通讯）。可以认为，质粒 pSYM3622 载有结瘤基因（*nod*）：固氮酶结构基因（*nif*）以及产黑素（*mel*）的全部有效基因。同时，由于转座子 Tn5-Mob 插入所赋予质粒的卡那霉素抗性和诱动转移特征，使菌株 3622-15 成为共生基因研究中有效的供体菌株之一。

在向三叶草根瘤菌菌珠 RCR226 诱动转移试验中，pSYM3622 以 10^{-5} 频率导入受体菌株。经杂交子 RCR226（pSYM3622）琼脂糖凝胶电泳检测中，可见到相应于 pSYM3622 质粒带的存在，把这些分离菌同时用来接种三叶草（*Trifolium repens*，*T. pratens*，*T. subterraneans*）和菜豆四种植物，杂交子不但保持诱导三叶草寄主植物结瘤固氮的原有特性，同时诱导菜豆植物形成无效根瘤。在含有卡那霉素平板上检查来自不同植物的根瘤分离菌的抗药性时，发现有 90%的三叶草根瘤分离菌都丧失了卡那霉素抗性能力（Km^s）。其中白三叶，红三叶和地三叶的 Km^s 分别为 95%，95%

和 57%（Km^s 落数/实际测试菌落数各为 203/214，57/60 和 27/47，每个根瘤分离平板中挑两个菌落）。琼脂糖凝胶电泳进一步证实，在这些菌株中，菜豆根瘤菌共生质粒 pSYM3622 已经丢失（见图版 I-A）。从菜豆根瘤中获得的所有分离菌依然保持卡那霉素抗性。将这些依旧载有 pSYM3622 质粒的根瘤分离菌按同样的方法在 4 种植物上做两次接种试验，得到与第一次回接试验类似的结果。

为了检查载有 pSYM3622 杂交子在实验室培养条件下的稳定性，将 6 株杂交子接连 4 周在不含 Km 抗性的富养平板上培养。把这些培养菌制成菌悬液的涂皿得到的单菌落，在含卡那霉素的平板上复制，发现约 10%的菌落在实验室培养条件下也丧失 Km 抗性。虽然在试验中还没有发现三叶草根瘤菌内源质粒的丢失现象（由于选择是以卡那霉素抗性为基础的），但可以认为，菜豆根瘤菌的共生质粒 pSYM3622 与三叶草根瘤菌 RCR266 中的某内源质粒同属于一个质粒不相容组。有关与三叶草根瘤菌质粒不相容性的详细材料将另文讨论。

（二）菜豆根瘤菌共生质粒向农杆菌的转移

为了排除 Ti-DNA 诱导双子叶植物形成的肿瘤对共生基因诱导根瘤形成结果的干扰，在这个试验中，受体菌株选用了三株具有不同表型的农杆菌。GMI9023 是消除了 pAtC58 和 pTiC58 的无质粒菌株；UBAPF2 也是无质粒菌株，但是与 Hynes 等[10]报道有所不同的是在本试验中，总能看到一条分子质量很大的巨型质粒带（>1000kb）的存在；另一个菌株是保留有 Ti 质粒的菌株 UBATi1。采用与根瘤菌类似的杂交转移方法，pSYM3622 能够很容易转入这些农杆菌菌株中，诱动转移的频率均在 10^{-4} 左右。从每一杂交组合中挑选 5 个菌落，在相同培养条件下分离纯化，然后用琼脂糖凝胶电泳检测。导入质粒 pSYM3622 在所有的杂交子中清晰可辨（见图版 I-B），初步证明共生基因已经转入农杆菌中。

将上述杂交子分别用来接种菜豆寄主植物。在 4~5 周后检查菜豆结瘤情况，发现含有 pSYM3622 的全部农杆菌杂交子均能诱导菜豆形成灰白色无效根瘤。这些根瘤均小于由三叶草根瘤菌 RCR226（pSYM3622）形成的无效根瘤，但具有典型的根瘤形状（见图 1），与单纯农杆菌诱导的根部肿瘤明显不同。有趣的是尽管对照菌株 GMI9023 和 UBAPF2（消除 Ti 质粒的农杆菌）不能诱导菜豆产生根瘤，通过光学显微镜对这些处理中植物根系的检查，发现根毛的形状呈“7 形”（见图版 I-C），既不像未接种处理的正常根毛，也不像根瘤菌侵染所引起的根毛弯曲。

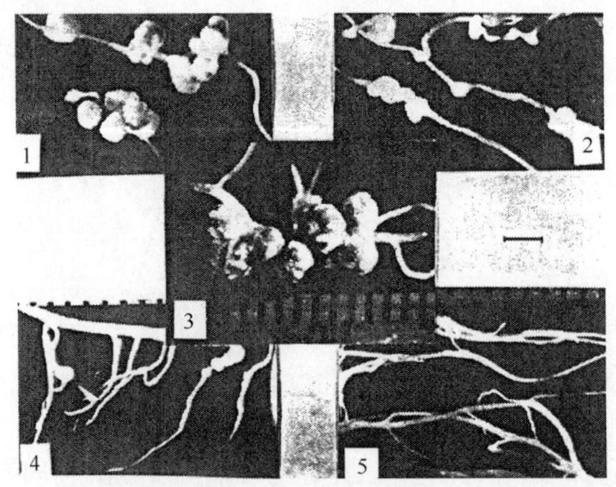

图 1 由三叶草根瘤菌杂交子和农杆菌杂交子侵染菜豆形成的根瘤

Fig.1 Bean nodules induced by transconjugants of RCR226（pSYM3622）and GMI9023（pSYM3622）.

1，RCR226（pSYM3622）；2，GMI9023（pSYM3622）；3，control of *R. phaseoli* 3622-15；4，control of *R. trifolii* RCR226；5，control of *A. tumefaciens* GMI9023.

为了证实这些根瘤的形成是由农杆菌杂交子诱导而不是由于根瘤菌污染造成，通过对根瘤分离菌进行生长特征和抗药性标记的鉴定表明，这些分离菌都具有在 ONA 平板上形成粗糙菌落和具有卡那霉素抗性的能力。并且，用这些分离菌对菜豆植物做二次接种试验，依然产生形状类似的根瘤。

比较三株不同质粒内容的农杆菌杂交子的植物结瘤试验结果，在植物生长和根瘤形态等方面，三者之间没有明显差异。本实验没有用电镜对根瘤内部显微结构作分析。

（三）菜豆根瘤菌共生质粒 pSYM3622 向假单胞菌和大肠杆菌的转移试验

在诱动共生质粒 pSYM3622 向假单胞菌菌株 *P. putida* NCIB10007 和 *P. fluorescens* NCIB10527 以及大肠杆菌 *E. coli* $J_{5\text{-}3}$ 的转移试验中，卡那霉素抗性能够以 10^{-6} 左右向这些受体菌株转移，这一频率比卡那霉素抗性的自发突变频率高 10~100 倍。但是在质粒电泳检测时，除了在菌株 NCIB10007 杂交子中检测到 pSYM3622 的存在外，其余 2 个菌株没有发现相应于 pSYM3622 质粒带的存在。对这些杂交的表型分析结果表明，它们既不能在植物回接试验中诱导菜豆结瘤，也不具备产生黑素的能力（见表 2）。

表 2　产生黑素基因在不同寄主细胞内的表达范围

Table 2　Host range of melanin genes expresslon by introduction of plasmid $pSYM_{3622}$

菌株 Straias	在含酪氨酸的 TY 培养基上产生黑素的特征 Melanln charact erfstice on TY-lyrostne	
	单本菌株 Parental straias	杂交子 Transconjugants
Rhlioblum phaseoll Control RCR	−	+
3622	+	
JI_{8400}	−	+
R.Ieguminosarum B_{151}	−	+
R.trifolii RCR_{226}	−	+
Agrobacterium iumefaciens		
GMI_{9023}	−	+
$UBAPE_2$	−	+
$UBATi_1$	−	+
Pseudomonas putida $NCIB_{10007}$	−	−
P.jluor escens $NCIB_{10527}$	−	−
E.coil J5-3	−	−

（四）*mel* 基因的表达范围

在根瘤菌科（*Rhizobiaceae*）中，绝大多数菜豆根瘤菌具备产生黑素的特征。对于黑素生成的起因和生化研究还不清楚，但遗传研究已经证实了产生黑素基因（*mel*）存在于 *nod* 和 *nif* 基因附近，而它的产物生成独立于共生基因的表达。尽管如此，利用 *mel*、*nod*、*nif* 基因的连锁可用作鉴别共生基因存在的识别标记。因此，研究 *mel* 基因的表达范围，在根瘤菌分子生物学研究中具有一定的实用价值。

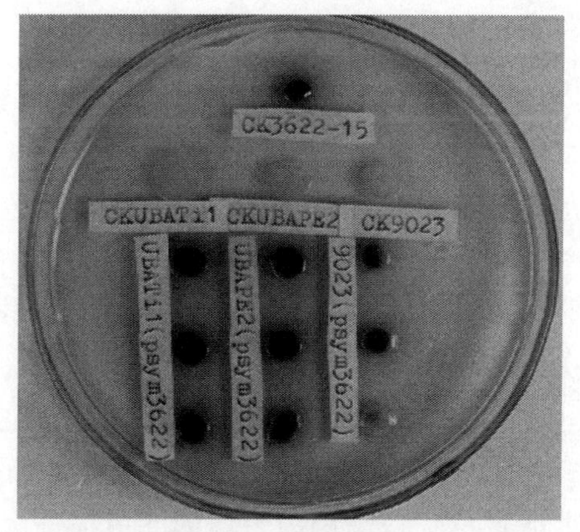

图 2　农杆菌杂交子在平板上合成黑素

Fig.2　Melanin production of *A. tumefaciens* introduced pSYM3622

在本实验中，丢失共生质粒 pSYM3622 的三叶草根瘤菌分离菌株全部不具备产黑素的能力（Mel^-）。对 Mel^-突变株筛选获得少数 Km^r，Mel^-突变株仍保留诱导三叶草和菜豆宿主结瘤的能力（Nod^+）。可以推断，这些 Mel^-突变株的形成可能是自发突变的结果。在琼脂糖凝胶电泳中，质粒 pSYM3622 在这些自发突变的 Km^r Nod^+ Mel^-突变菌中遭受大片段 DNA 缺失，导致该质粒的前移（见图版 I-A，第 10 孔）。这一结果进一步证实，在 pSYM3622 质粒上的 *mel* 基因与 *nod* 基因毗邻，但独立于共生基因（*sym*）的表达。比较共生质粒 pSYM3622 在不同细菌寄主中的表达范围（见表 2），发现除了根瘤菌以外，*mel* 基因还能够在不同的农杆菌菌株之间表达（见图 2）。

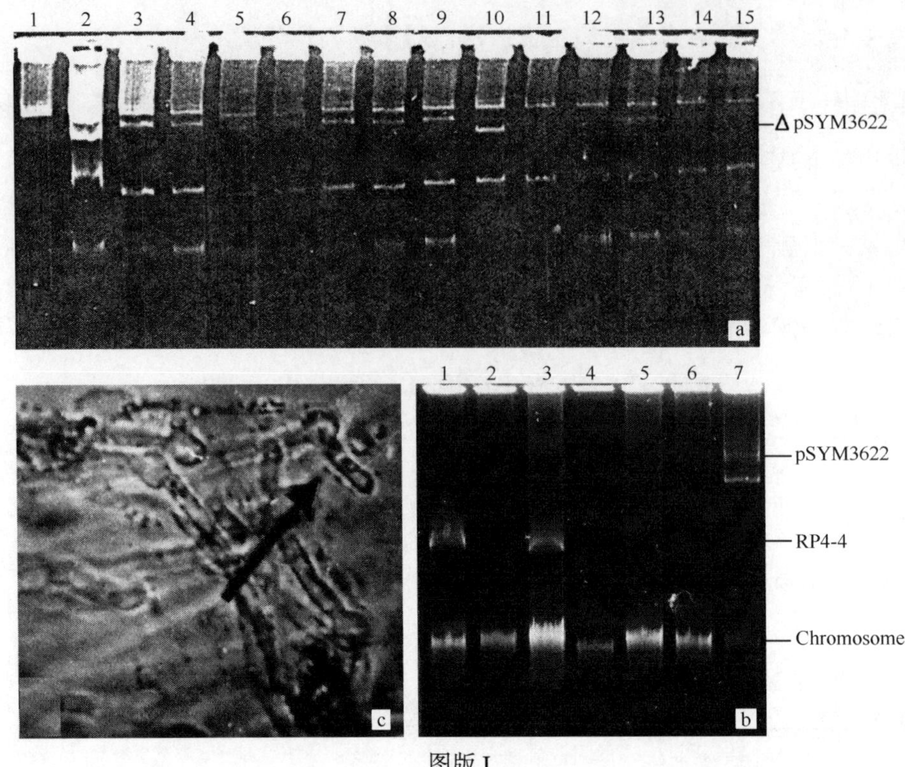

图版 I

A.由杂交子 RCR226 （pSYM3622）接种的根瘤分离菌质粒图谱。Plasmid pattern of nodule isolates inoculated with transconjugants RCR226 （pSYM3622）. 1，2，3. 分别为对照菌株三叶草根瘤菌 RCR226，菜豆根瘤菌 RCR3622 和杂交子 RCR226 （pSYM3622）；4-9. 来自菜豆植物的根瘤分离菌株；10~12. 来自白三叶草的根瘤分离菌株，其中 10 号菌株质粒 pSYM3622 遭受大片段 DNA 缺失，导致 Mel-的自发突变产洼；13~16. 来自红三叶草的根瘤分离菌，其中 13 号表型为 Nod+ Mel+ Km^r 菌株。

Lane 1，2 and 3. Control *R. trifolii* RCR226，*R. phaseoli* RCR3622 and transconjugants RCR226 （pSYM622）； Lanes 4-9. Nodule isolates from beans; Lanes 10~12. Nodule isolates from white clover，derivative in lane 10 sufferred a large deletion of pSYM3622 causing Km^r Mel^- spontaneous mutation; Lanes 13-15. Nodule isolates from red clover，derivative in lane 13 was Nod^+ Mel^+ Km^r.

B.导入质粒 pSYM3622 的农杆菌杂交子的质粒图谱。Plasmid pattern of transconjugants of *Agrobacterium tumfaciens*. Lanes 1 and 2，UBAPF2 （pSYM3622）； Lane 3 UBAPF2; Lane 4 and 5，GMI9023 （pSYM3622）； Lane 6，GMI9023; Lane 7，control *R. phaseoli* 3622-15. C.农杆菌 GMI9023 诱发的根毛不正常卷曲。Root hair deformation induced by *A. tumfaciens* GMI9023.

讨 论

Beynon 等[14]曾经报道豌豆根瘤菌质粒在菜豆根瘤菌寄主细胞中的不亲和性。最近，Hooykeas 等[15]对三叶草根瘤菌质粒在菜豆根瘤菌中的不稳定性也作了报道。这种外源质粒的不稳定性一般认为是由于寄主细胞中存在同一不相容组质粒所致。在本实验中，10%的菜豆根瘤菌共生质粒在三叶草根瘤 RCR226 中自动丢失，这种不稳定性在通过非种族内寄主植物的接种后加剧发生（90%丢失）。根据笔者早先[5]的研究，寄主植物能赋予非种族内的共生质粒在功能上的不相容性。因此，这种现象可能也是植物寄主在结瘤功能上的选择压力所致。

pSYM3622 同时能够导入农杆菌中并诱导菜豆宿主植物产生大量的无效根瘤。由根瘤中获得的分离菌仍保持农杆菌杂交子所具有的特征，并重新侵染菜豆，诱导形成根瘤组织。这一结果排除了根瘤的形成是由于根瘤菌污染造成的可能性。因此，可认为质粒 pSYM3622 带有诱导菜豆形成根瘤的全部有效因子（包括 *nod* 和 *hsn* 基因），这些因子全部能够在农杆菌寄主细胞中表达。

除了大多数菜豆根瘤菌菌株产生黑素外，Cubo 等[16]报道在少数豌豆，三叶草和快生型大豆根瘤菌中也发现有产黑素基因，并且这些基因都是由质粒携带。最近，有些研究者[17, 18]报道，在菜豆根瘤菌中，有两组基因控制着共生质粒 pRP2JI 的黑素合成。一组是酪氨酸酶合成基因 *mel*A，$MelA^-$突变子不合成黑素，但不影响根瘤的有效固氮能力。对另一组突变株的研究表明，*mel*A 的合成受 *nif*A 基因调控，$NifA^-$突变子既不能合成黑素，也不能固氮。并且在大肠杆菌中，*mel*A 基因受肺炎克氏杆菌（*Klebsiella*）*nif*A 激活，但必须有 *ntr*C 基因的存在。在我们的试验中，*mel* 基因在农杆菌中表达，但是在假单胞菌和大肠杆菌中不能表达。Hawkins 等[17]研究表明 *mel*A 基因在大肠杆菌中的表达是温度敏感型的。因此，这些假单胞菌和大肠杆菌杂交子中的黑素不能合成，可能也是受温度所制约。

在农杆菌杂交子中，*mel* 基因表达的时间通常比对照的菜豆根瘤菌要快得多。在相同条件下，黑素在 3-4 天则明显可见，而菜豆根瘤菌产生的黑素通常要在一周以后才能出现。我们认为这可能与农杆菌的生长周期短，代谢旺盛有关。*mel* 基因的表达有助于共生基因在转移试验中对杂交子的检出，设想将整接有目标基因的重组 *mel* 基因次级克隆接到 Ti 质粒上，用构建新质粒来感染植物细胞，采用黑素分析方法，可以简单快速地检查细胞中目标基因的存在与否。

参 考 文 献

[1]Wang CL and Hirsch PR，Acta Genet Sinica，15:25-33. 1988

[2] Johnston AWB et al.，Nature，276:634-636. 1981

[3]Wong CH et al.，J. Cell Biol，97:787-794. 1983

[4]Christensen A H et al.，J. Bacteriol.，156:592-599. 1983

[5]Wang，C.L.et al.，J. Gen Microbiol.，132:2063-2070. 1986

[6]Hirsch Am et al.，J. Bacteriol，162:1261-1269. 1984

[7]Truchet GC et al.，J. Bacteriol，157:134-142. 1984

[8]Plszinski J and Rofe BG，J. Bacteryol，162:1261-1269. 1985

[9]Rosenberg C and Huguet T，Mol Gen Genet，196:533-536. 1984

[10]Hynes MF et al.，Plasmid，13:95-105. 1985

[11]Meynell GG，Bactrial peasmids，Macmillan，London，and Basingstoke. 1972

[12]Beringer JE，J. Gen Microbiol，84:188-198. 1974

[13]Fahraeus G，J. Gen Microbiol，16:374-381. 1957

[14]Beynan JL et al.，J. Gen Microboil，120:421-429. 1980

[15]Hooykaas PJJ et al.，Plasmid，14:47-52. 1985

[16]Cubo MT et al.，Abstracts of the 7th International Symposium on Nitrogen Fixation，FRG. pp.10-42. 1988

[17]Hawkins FKL and Johnston AWB，Abstracts of the 7th International Symposium on Nitrogen Fixation，FRG. pp.10-23. 1988

[18]Borthakar D et al.，Mol Gen Genet，207:158-160. 1987

三叶草素基因在快生型大豆根瘤菌中的表达*

万　钧[1]　覃筱婷[2]　周俊初[2]　陈华癸[2]

（1 北京农业大学生物学院微生物系）

（2 华中农业大学农业部农业微生物学重点实验室）

摘　要　研究了三叶草根瘤菌的三叶草素基因在快生型大豆根瘤菌中的表达情况，发现导入的三叶草素基因得到了表达，使快生型大豆根瘤菌获得了产生三叶草素的能力，产生的三叶草素对指示菌产生了抑制作用。并且发现导入的三叶草素基因对受体的共生能力没有影响。

关键词　三叶草素基因；基因表达；快生型大豆根瘤菌

一、引　　言

根瘤菌与豆科植物形成的共生固氮体系是生物界效率最高的固氮体系，根瘤菌与豆科植物共生关系的建立和维持是一个复杂的过程，有宿主和根瘤菌双方的许多基因参加。人们为了提高共生固氮的效率，在根瘤菌育种方面做了大量的工作，但根瘤菌在大田应用时增产效果很不稳定，原因主要是在长年种植相应豆科植物的地区，土壤中存在大量的土著根瘤菌，人工接种的优良根瘤菌由于对当地环境的适应性、对栽培品种的亲合性等原因难于与土著菌竞争，占瘤率很低，增产效果也就不明显[1]。

最近，美国的 Triplett 等人在三叶草根瘤菌（*Rhizobium legunninosuram* bv *trifolii*）上的研究表明三叶草素的产生对三叶草根瘤菌的竞争结瘤能力有很大促进作用。用产生三叶草素的三叶草根瘤菌和对三叶草素敏感的三叶草根瘤菌以 1:1 的比例混合接种三叶草时，产三叶草素的根瘤菌占瘤率在 90%以上。用 Tn5 诱变该产三叶草素菌株，使之失去产三叶草素的能力后，与同样的对三叶草素敏感的菌株以 1:1 的比例一起接种，突变株的占瘤率只有 40%。而且他们的研究还发现三叶草素对于快生型大豆根瘤菌（*Rhizobium fredii*）有抑制作用[2]。因此我们设想将三叶草素基因导入具有良好共生性状的快生型大豆根瘤菌中，观察其表达情况，用以进一步考察其对受体菌竞争结瘤性的影响。

二、实验材料与方法

本实验所使用的受体菌 HN01 是本实验室分离自湖北洪湖长年种植大豆的灰潮土中的快生型大豆根瘤菌，具有良好的共生性能[5]。质粒 pT2TX3K 和 pT2TFXK 是 Triplett 博士惠赠的重组质粒，pT2TX3K 上带有不完整的三叶草素基因簇，部分编码基因及其启动子缺失，对三草素的抗性基因和它们的启动子是完整的。pT2TFXK 上带有完整的三叶草素基因簇，既有产素基因也有抗性基因。两个质粒的载体为 pTR102，上面有 *mob* 基因没有 *tra* 基因，可以在帮助质粒 pRK2013（$mob^{-}tra^{+}$）的帮助下通过“三亲本杂交”法在 G^{-}细菌中接合转移[3, 4]。

质粒 pT2TX3K 和 pT2TFXK 以及质粒 pRK2013 均存在于营养缺陷型的大肠杆菌中，这些大肠杆菌均不能在根瘤菌基本培养基 SM 上生长[6]。质粒 pT2TX3K 和 pT2TFXK 上分别带有四环素抗性（Tc^{r}），卡那霉素抗性（Km^{r}）和氨苄青霉素抗性（Amp^{r}）作为标记，受体菌 HN01 对 Tc（20μg/ml）、Km（50μg/ml）和 Amp（100μg/ml）均表现出敏感[5]。因此可以在加有相应抗生素的 SM 平板上得到导入质粒的转移接

*原载于《高技术通讯》，10：50~52，1996.

合子。转移接合子在大肠杆菌和根瘤菌都可以生长的 YMA 抗性培养基上反复划线分离，通过生长速度和菌落形态的不同淘汰大肠杆菌以后，就得到了导入质粒的根瘤菌转移接合子。将转移接合子命名为 HN01（pT2TX3K）和 HN01（pT2TFXK）。

将转移接合子点接种于没有加抗生素 YMA 平板上培养至菌苔长出后用打孔器打下，移至混有指示菌的 YMA 平板上，继续培养，观察混菌平板中指示菌的生长情况，以检测是否产生了三叶草素。

本实验中所使用的指示菌是快生型大豆根瘤菌 USDA205 和三叶草根瘤菌 TA1。

三、实 验 结 果

将带有质粒 pT2TFXK 和 pT2TX3K 的转移接合子 HN01（pT2TFXK）和 HN01（pT2TX3K）纯化点接种于 YMA 平板上长成菌苔，用打孔器将菌苔连同培养基一起打成菌块放于混有 USDA205 和 TA1 的 YMA 平板上，继续培养。发现 HN01 （pT2TFXK）对 USDA205 和 TA1 都有抑制作用而 HN01（pT2TX3K）则没有（图 1、图 2）。

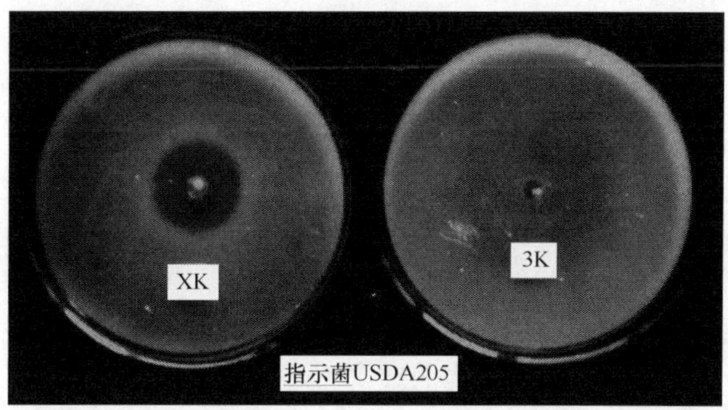

图 1　转移接合子 HN01（pT2TFXK）和 HN01（pT2TX3K）对 *R. fredii* USDA205 的抑制作用

左图 XK 为 HN01（pT2TFXK）右图 3K 为 HN01（pT2TX3K）

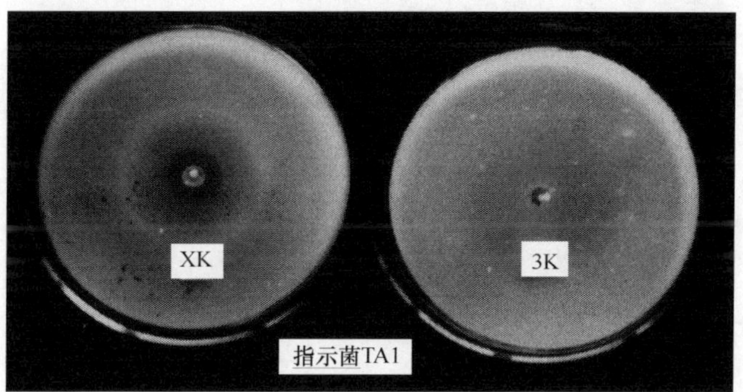

图 2　转移接合子 HN01（pT2TFXK）和 HN01（pT2TX3K）对 *R.leguminosarum* bv. *trifolii* TA1 的抑制作用

左图 XK 为 HN01（pT2TFXK）右图 3K 为 HN01（PT2TX3K）

将带有 HN01（pT2TFXK）的菌块放于混有 HN01 和 HN01（pT2TX3K）的混菌平板上时，只在 HN01 的混菌平板上产生抑菌圈（图 3、图 4）。

可见，质粒 pT2TFXK 上完整的三叶草素基因簇在快生型大豆根瘤菌中得到了表达，使受体获得了产生三叶草素的能力。质粒 pT2TX3K 上的不完整的产素基因没有表达，而完整的抗性基因得到了表达，HN01（pT2TX3K）对 HN01（pT2TFXK）产生的三叶草素有抗性，生长不被抑制。

我们将 HN01（pT2TFXK）和 HN01（pT2TX3K）分别接种大豆植株，发现两者均能在大豆上结瘤固氮，其结瘤固氮能力与出发菌株 HN01 没有差异。从两者形成的根瘤中又分离出了带有与质粒的抗性标记一致的快生型大豆根瘤菌。

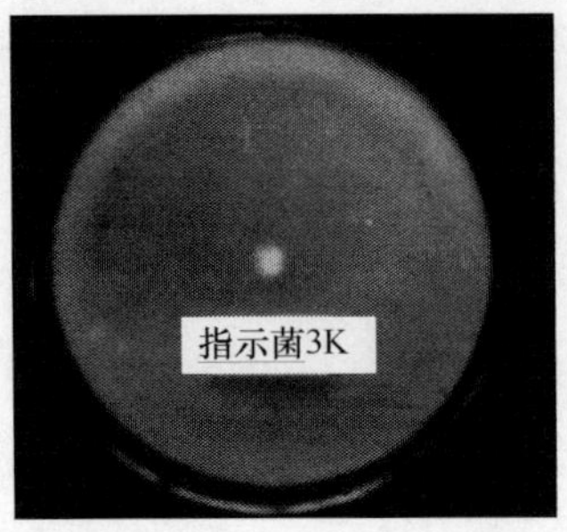

图3 转移接合子 HN01 (pT2TX3K) 对三叶草素的抗性

图中菌块上的菌为转移接合子 HN01（pT2TFXK），混菌平板中的指示菌为 HN01（pT2TX3K）

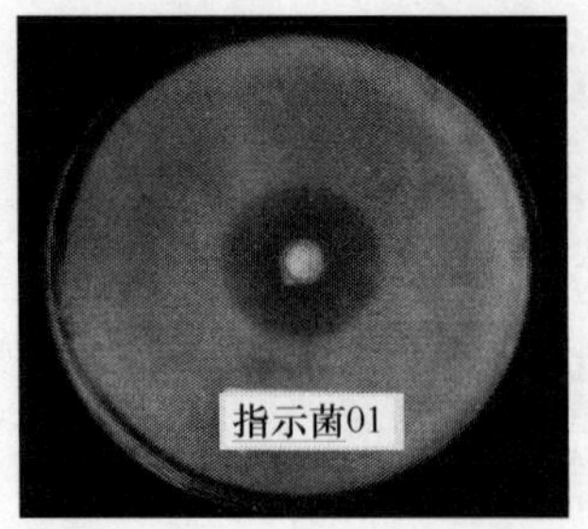

图4 受体 HN0I 对三叶草素的敏感性

图中菌块上的菌为转移接合子 HN01（pT2TFXK），混菌平板中的指示菌为 HN01

四、讨　　论

以上实验表明三叶草根瘤菌的产三叶草素基因在快生型大豆根瘤菌中得到了表达，使快生型大豆根瘤菌获得了产三叶草素的能力，而且这种能力对快生型大豆根瘤菌的共生固氮作用没有影响。下步就是要测定产生三叶草素对快生型大豆根瘤菌的竞争结瘤能力的影响。

在中国的许多大豆产区都发现了作为土著菌的快生型大豆根瘤菌，而且在有的地方，快生型大豆根瘤菌是大豆的主要共生体。如果三叶草素的产生可以增强快生型大豆根瘤菌竞争结瘤能力。在实践上将有巨大的应用前途。同时这些研究也提示我们可否在大豆根瘤菌中寻找产生细菌素的菌株，并加以利用，这对提高大豆根瘤菌的竞争能力一定更有作用。细菌素对共生行为的作用对于微生物生态学的理论和实践也有着重要的意义。可以使我们对于微生物的拮抗和这种拮抗在微生物之间、微生物与环境之间的作用有更多的认识。

有关三叶草素基因对快生型大豆根瘤菌竞争结瘤能力的影响的研究我们正在深入进行中。

参 考 文 献

[1]华中农业大学，南京农业大学主编．微生物学（第4版）．农业出版社，1993，ppl37．

[2] Triplett E W et al.，Plant Physiol，1987，85:335

[3] Triplett E W et al.，Appl Environ Microbiol．1994，60:4136

[4] Weinstein M et al. J. Bacterial. l992. 174: 7486

[5]陈文莉，洪湖灰潮上中快型号大豆根瘤菌多样性的研究，华中农业大学博士论文．1994

[6]李立家等，武汉大学学报（自然科学版）．1994，86:115

以绿荧光蛋白基因为报告基因的广宿主启动子探针载体的构建和应用*

马立新　史巧娟　周俊初　陈华癸

（农业部农业微生物重点实验室　华中农业大学　武汉 430070）

摘　要　将以绿荧光蛋白基因（*gfp*）的cDNA为模板，用人工合成引物经PCR扩增获得的0.9kb DNA片段克隆到表达载体pET−11C上构建成gfp表达载体pHN115。从pHN115上切下的不含启动子，但保留了SD序列的gfp基因经克隆载体SK（+）和pIJ2925亚克隆后再克隆到广谱稳定性质粒pTR102上构建成广谱、稳定、可视的启动子探针载体pHN127。并用它从费氏中华根瘤菌HN01的总DNA中成功地钓出组成型和诱导型表达的启动子。

关键词　绿荧光蛋白基因（*gfp*）；费氏中华根瘤菌；启动子探针载体

绿色荧光蛋白（GFP）是Morse等[1]在研究水母生物荧光现象时首先报道的。其后Shimomura[2]和Ward等[3]相继对GFP的结构和功能进行了较详尽的研究，证实它是分子量为30kD的单体发光蛋白，受395nm近紫外光或（470nm）蓝色光激发后，在Ca^{2+}催化下，能发出波长为510nm的绿色荧光。该作用不需要酶参与，也不需作用底物，发光蛋白对细胞没有毒性、结构稳定、作用持久，可以在活体组织、细胞内部以及经甲醛等固定后的组织切片标本上直接检测。

Prasher等[4]从多管水母（*Aequorea vicotoria* ）的cDNA文库中用人工合成的绿色荧光蛋白基因*gfp*的5'寡核苷酸为探针筛选出编号为*gfp*10的阳性克隆。1994年2月，美国的Chalfie等[5]首次报道分别用T_7启动子和mec-7启动子成功地使*gfp*作为报告基因在*E. coli* 和线虫等两个异源宿主表达。

启动子探针载体是分离和测定启动子效率的主要工具，通常借助于易检测的报告基因来完成。在细菌研究中目前常用的报告基因主要有抗性基因和非抗性基因。但不论抗性基因或非抗性基因标记方法本身都存在一定局限性。而且，传统细菌启动子探针载体的不足之处还表现在宿主范围窄和稳定性不高。为此，以*gfp*为报告基因并选择适用于G^-细菌的广谱稳定性质粒pTR102为载体来构建广宿主、稳定和可视的启动子探针载体，并将其应用于大豆根瘤菌启动子的分离与克隆。

1 材料和方法

1.1　菌株和培养条件

1.1.1 菌株与质粒；　实验用菌株和质粒见表1。

表 1　供试菌株与质粒

Table 1　trains and plasmids tested

供试菌株或质粒 Strain or plasmid	基因型或表型 Genotype or phenotype	来源或参考文献 Source or reference
大肠杆菌 *Escherichia coli*		
DH5α	*sup E*44，*hsdR* 17，Δ*lacU*169，*rea*A 1，*endA* 1，*gryA* 96，*thi*-1 *relA* 1	本室保存 This laboratory
BL 21（ DE3）	*hsd* 5，*gal*λ*cIts*857，*ind* 1，*sam* 7，*nin*5，*lacUV s-T7 gene* 1	[6]

*原载于《微生物学报》，39（5）：408~416，1999.

续表

供试菌株或质粒 Strain or plasmid	基因型或表型 Genotype or phenotype	来源或参考文献 Source or reference
费氏中华根瘤菌 *Sinorhizobium fredii* HN01	Nod^+，Fix^+	本室保存 This laboratory
质粒 plasmid		
pAC-*gfp*	克隆有 *gfp* cDNA 的杆状病毒转移载体，Ap^r	武汉大学齐义鹏教授 Wuhan University
SK（+）		本室保存 This laboratory
pET-11C	表达载体，Ap^r	[6]
pIJ2925	衍生于 pUC19 的克隆载体，Ap^r	本室保存 This laboratory
pTR102	广谱性稳定载体，mob^+，tra^-，Ap^r，Tc^r	[7]
pBR322	克隆载体，Ap^r，Tc^r	本室保存 This laboratory
pRK2013	协助质粒 mob^-，tra^+，Km^r	[8]
pHNI14	SK（+）上克隆有 *gfp* 的 cDNA	本研究 This work
pHNI15	pET-11C 克隆有 *gfp* 的表达载体	本研究 This work
pHNI16	SK（+）上携带有 pHNI15 上的 *gfp* （去掉了 Φ10 启动子）	本研究 This work
pHNI17	pIJ2925 携带有 pHNI16 上的 *gfp*	本研究 This work
pHNI27、pHNI28	pTR102 上携带有来源于 pHNI17 的 *gfp*，两者的插入方向相反	本研究 This work
pHNI29	pHNI27 上克隆有来源于 pBR322 的具双向启动子活性的 670bp 的 *Sau*3AI DNA 片段	本研究 This work

1.1.2 培养基与培养条件：大肠杆菌采用 LB 培养基，于 37℃下培养。根瘤菌采用 TY 或 SM 培养基，于 28℃下培养。BL21（DE3）（pHN115）的诱导表达见文献[6]。

1.1.3 抗生素及使用浓度：四环素（Tc）为20μg/ml；卡那霉素（Km）为50μg/ml；氨苄青霉素（Ap）为100μg/ml。

1.2 试剂和酶

DNA片段回收试剂盒购自Clontech公司。IPTG和X-gal购自Sigma公司。M13正反向引物购自Promega公司。限制性酶、碱性磷酸酶和Taq酶等购自Bio-Rad、Promega、Gibco/BRL等公司。

1.3 PCR扩增

引物序列Pgfp （GAATAAAAGCTAGCAAAGATGAGTAAAG）由Sangon公司合成，分别与M13的正向、反向引物配对。PCR扩增程序为95℃ 2min、40℃ 1 min、65℃ 1min 后，94℃ 1 min、52℃ 1 min、72℃ 1 min 共30个循环后于72℃ 维持8 min。PCR反应体积50μl，内含模板DNA 5μl、引物各25pmol和*Taq*DNA聚合酶2μl。

1.4 DNA的分离及操作

根瘤菌总DNA的分离参见Long等[9]的方法。质粒抽提、酶切、去磷酸化、连接和转化等一系列克隆操作参照Sambrook等[10]的方法。

1.5 接合转移

采用三亲本滤膜杂交法[11]。

1.6 大豆种子提取物的制备

参照Mulligan等[12]的方法。

1.7　荧光检测与照相

用Fotodyne在360nm的长波紫外光照射下检测菌落和菌液荧光。用日立RF-5000型荧光分光光度计测定其激发光谱、发射光谱与不同启动子的相对强度。用彩色摄影记录菌体绿色荧光时，选用绿色滤光片。

2　结果

2.1　绿荧光蛋白基因在大肠杆菌中的表达

2.1.1　*gfp*表达载体的构建

用*Eco*RI酶切载体pAC-*gfp*，回收约0.9kb的片段，插入到SK（+）的*Eco*RI位点，得到重组质粒pHN114并经酶切鉴定（图1）。然后以合成引物Pgfp与M13的正向、反向引物分别组成引物对，以pHN114为模板进行PCR扩增，结果只有Pgfp与M13的反向引物对扩增出了0.9kb的片段。PCR产物经*Eco*RI和*Nhe*I双酶切后，插入到经相同双酶切的表达载体pET−11C，得到的重组质粒pHN115，经*Eco*RI和*Nhe*I消化证实含有0.9kb的片段（图2）。*gfp*表达载体pHN115的构建过程见图3。

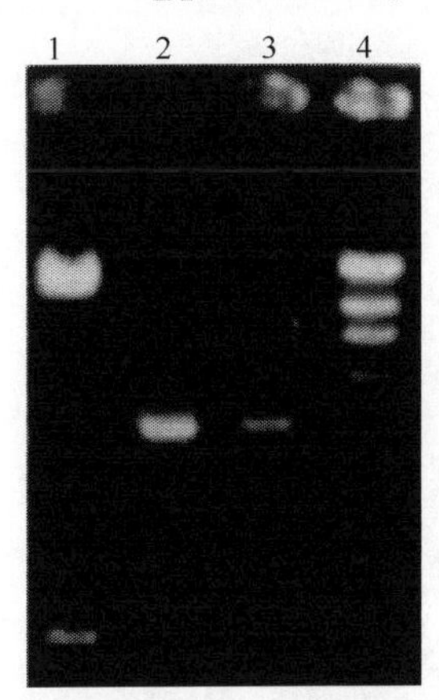

图 1　重组质粒 pHN114 的酶切鉴定

Fig. 1　Electropphoresis identification of HN114 by the method of enzyme digestion

1.pAC-*gfp*/*Eco*RI；2.SK（+）*Eco*RI；3.pHN114/*Eco*RI；4.λDNA/*Hind* III.

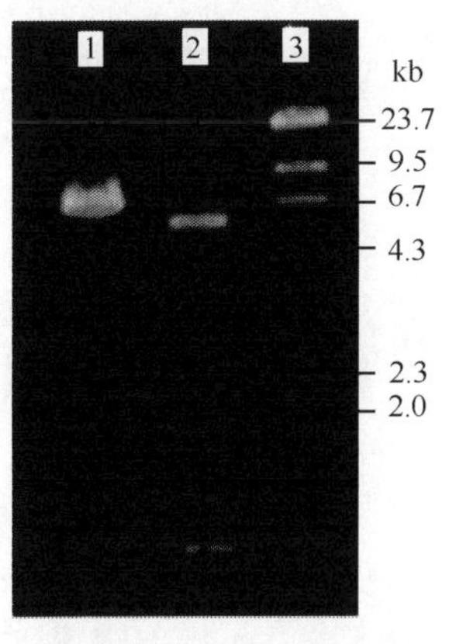

图 2　重组质粒 pHN115 的酶切鉴定

Fig.2　Identification of pHN115 by the method of enzyme digestion

1.pET-11C/*Eco*RI+*Xba* I；2.pH115/*Eco*RI+*Xba* I；3.λDNA/*Hind* III.

2.1.2　*gfp*在BL21（DE3）中的表达

将用pHN115转化BL21（DE3）得到的转化子接LB+Ap液体培养基，经37℃摇瓶培养至对数中期，用终浓度为1mmol/L的IPTG诱导3~4h，经长波紫外光照射检测未见到绿色荧光。改将转化子在LB+ Ap平板上于37℃培养形成的单菌落也未检测到绿色荧光。但将培养温度改为28℃后重复上述过程，均能检测出荧光。说明*gfp*只有在28℃培养时才能在大肠杆菌中正确表达（图版I-1）。

2. 2　广宿主、可视启动子探针载体pHN127的构建

用*Eco*RI和*Xba*I双酶切质粒pHN115，回收约1.0kb的片段（该片段保留了*gfp*表达单元的SD序列，但不含φ10启动子），插入到经同样双酶切的克隆载体SK（+）得到重组质粒pHN116，用*Eco*RI和*Xba*I双酶切复证，产生1.0和2.9kb两个片段。用*Hin*d III和*Xba*I双酶切质粒pHN116，回收1.0kb的片段，插入到经同样双酶切的表达载体pIJ2925上，得到重组质粒pHN117，经*Hind* III和*Xba*I双酶切复证产生1.0和2.6kb两个片段（图4）。用*Bgl* II酶切pHN117，回收1.0kb的*Bgl* II片段所含*gfp*基因的上游SD序列和单一*Bam*HI位点，克隆到pTR102的*Bam*HI位点；得到的重组质粒用*Kpn*I酶切可鉴定出*gfp*的两种插入方向。将能切出约4.0kb片段的重组质粒命名为pHN127，切出约3.0kb片段的重组质粒，命名为pHN128（图5）。启动子探针载体pHN127的构建过程见图6。

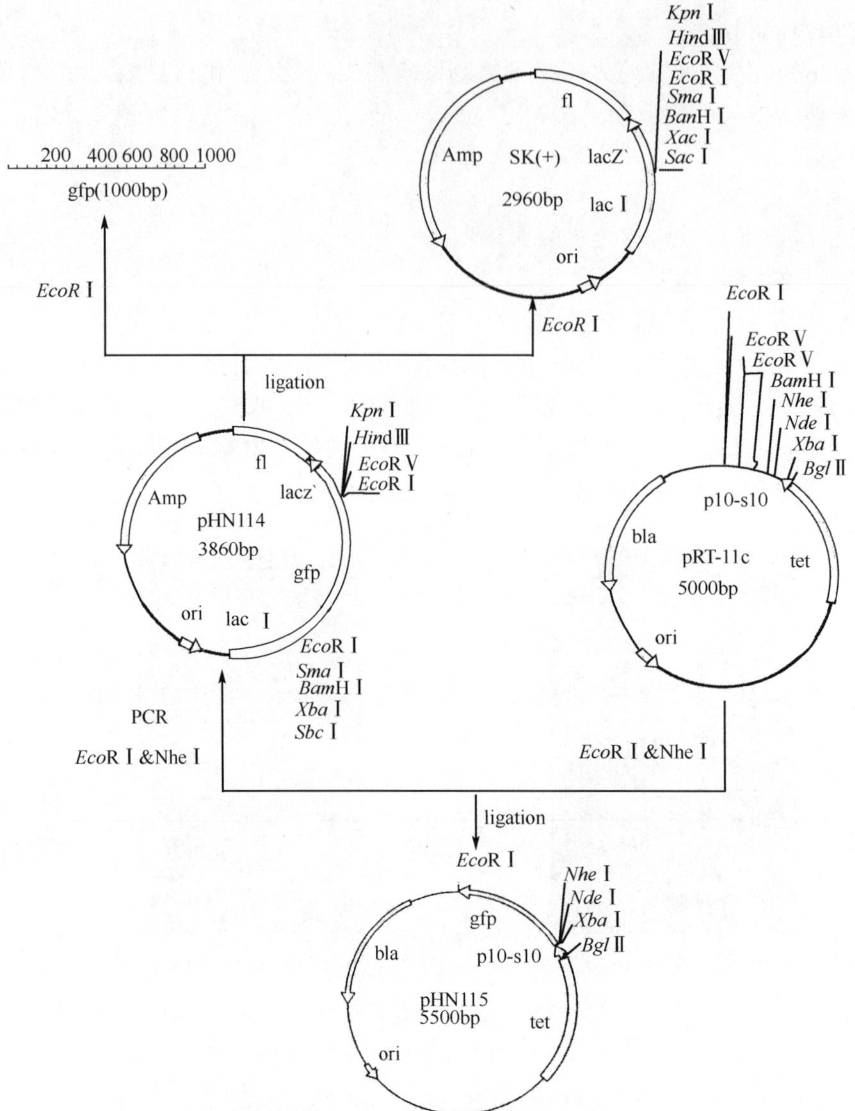

图 3 *gfp* 表达载体 pHN115 的构建

Fig.3 Construction of the *gfp* expression vector pHN115

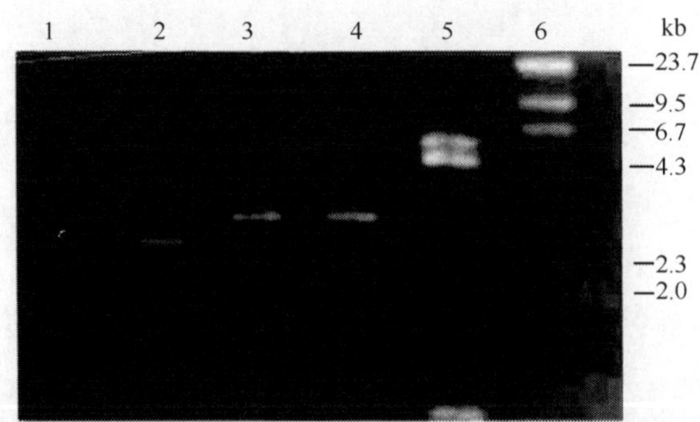

图 4 重组质粒 pHN116 和 pHN117 的酶切鉴定

Fig. 4 Identification of pHN116 and pHN117 by the method of enzyme digestion

1. pIJ2925/ *Hind* III+ *Xba* I；2. pHN117/ *Hind* III+ *Xba* I；3. KS（ + ） / *EcoRI*+ *Xba* I；4. pHN116/ *EcoRI*+ *Xba* I；5. pHN115/ *EcoRI*+ *Xba* I； 6. λ DNA/ *Hind* III

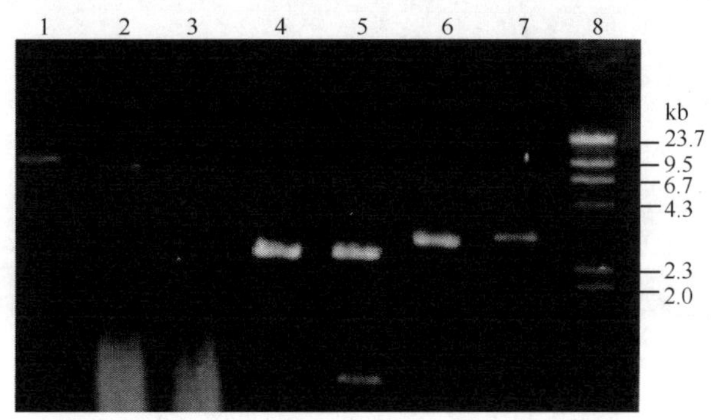

图 5 重组质粒 pHN127 和 pHN128 的酶切鉴定

Fig. 5 Identification of pHN127 and pHN128 by the method of enzyme digestion

1. pHN127/ *Bam*HI; 2. pHN128/ *Bam*HI; 3. pHN127/ *Kpn* I; 4. pIJ2925/ *Bgl* II; 5. pHN117/ *Bgl* II; 6.KS（+ ） / *Xba*I+ *Hind* III; 7. pHN116/ *Xba*I+ *Hind* III; 8. λ DNA/ *Hind* III

2.3 pHN127启动子探针特性的证实

用*Sau*3AI酶切pBR322，回收具有双向启动子活性的约670bp的片段[13]，插入到pHN127的*Bam*HI位点得到质粒pHN129，转化DH5α，得到的转化子用长波紫外光照射可发绿色荧光。挑取发光菌落提取质粒，用*Sau*3AI酶切检测到670bp的片段，从而证实pHN127的启动子探针特性。通过三亲本接合转移法将质粒pHN129导入费氏中华根瘤菌HN01，检测转移接合子均能发绿光（图版I-2），说明gfp也能在根瘤菌中表达。

2.4 用"鸟枪克隆"法钓取大豆根瘤菌的启动子

用*Sau*3AI部分酶切费氏中华根瘤菌HN01的总DNA，与经*Bam*HI酶切并经脱磷酸化处理的pHN127连接，转化大肠杆菌DH5α后，将在LB+Tc平板上长出的转化子混合培养后通过三亲本接合转移导入HN01中，用长波紫外光检查转移结合子，有3%的菌落发光，且发光有强有弱。表明已克隆到不同强度组成型表达的根瘤菌启动子。将所有不发光的HN01转移结合子转接到加大豆种子抽提物的TY培养基上，筛选到3个受种子抽提物诱导发光的启动子（图版I-3）。

3 讨论

迄今尚未见到启动子探针载体同时具备稳定性、广宿主性和报告基因极易检测等特性的报道。本研究构建的启动子探针载体pHN127的上述优点是由其载体pTR102和报告基因*gfp*决定的。pTR102衍生于革兰氏阴性细菌的广宿主质粒RK2，由miniRK2和RK2的*par*区域构成。后者由两个反向转录的操纵子*par* CBA和*par* DE组成[14]，通过*par* CBA和*par* DE的协调作用，能以一种不依赖于复制子、广宿主的方式稳定质粒。研究表明*par* DE通过编码能杀伤无质粒细胞的蛋白来维持质粒稳定[15]。*par* CBA则可能编码某种有效分配系统来维持质粒稳定[16]。由于*gfp*基因在任何细菌中表达都不会有本底荧光干扰并易于定性和定量检测，连同pTR102的广宿主性和稳定性，从而赋予启动子探针载体pHN127以广宿主性、稳定性和可视的优点。研究结果表明*gfp*在大肠杆菌中正常表达的关键是培养温度。据文献资料分析我们未能在37℃下检测到转化子的诱导发光的原因可能与表达出的GFP蛋白在37℃下的错误折叠并形成包涵体有关[17, 18]，转化子在28℃培养条件下表达的GFP由于能有效折叠，因而发出较强的绿色荧光。

已知根瘤菌与豆科植物共生关系的建立涉及根瘤菌的早期结瘤基因与植物基因在分子水平上的对话：组成型表达的根瘤菌调节基因*nod*D产生的NodD蛋白经植物根分泌的类黄酮化合物诱导活化后，能激活根瘤菌的早期结瘤基因（如*nod*ABC等）表达产生特殊的寡糖类结瘤因子（Nod factor）。后者又反过来诱发植物根的皮层细胞分裂产生根瘤[19]。快生型大豆根瘤菌虽然首先在我国发现并广为分布，但目前对它们参与早期结瘤过程的基因定位、结构和功能的了解几乎处于空白状态。用启动探针载体pHN127去钓取它们的早期结瘤基因，必将进一步促进快生型大豆根瘤菌分子生物学研究的进展。

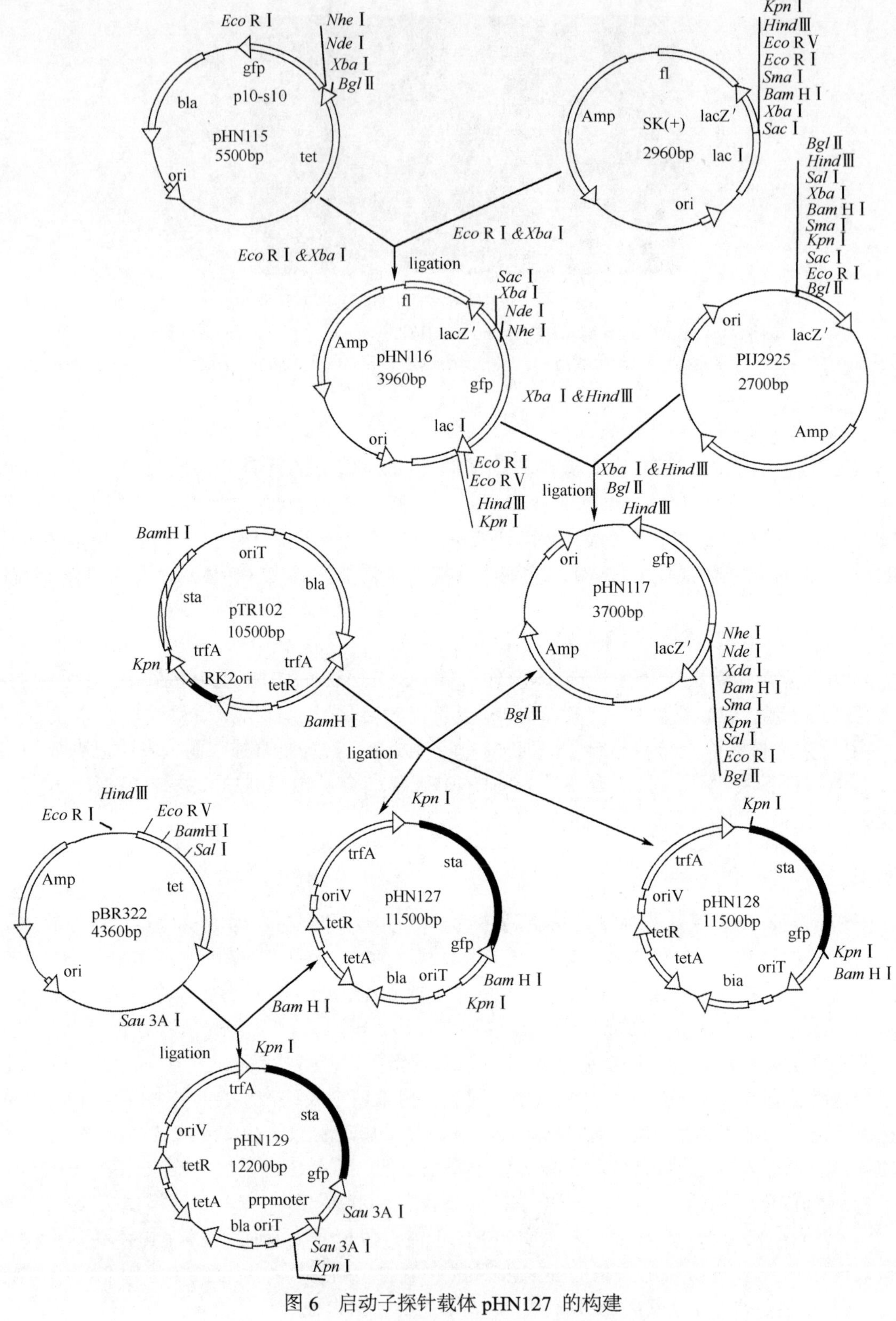

图 6 启动子探针载体 pHN127 的构建

Fig. 6 Construction of promoter-probe vector pHN127

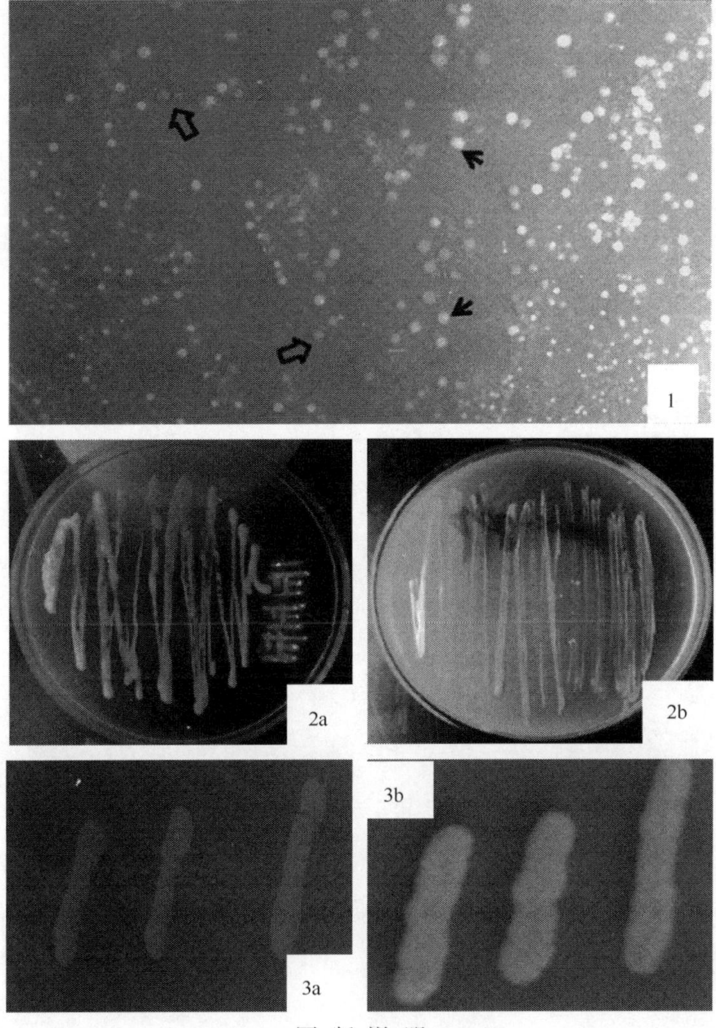

图 版 说 明

1. *E. coli* BL21(DE3)受 IPT G 诱导发绿光(→ 示发荧光的转化子菌落 ⇨ 示受体菌菌落)。2. gfp 在快生型大豆根瘤菌 *Sinorhizobium fredii* HN01 中的表达（a. 发绿光的转移接合子 HN01（pHN129）；b. 受体菌 HN01）。3. 三个诱导发绿光的转移接合子（a.未经种子抽提物诱导，b.经种子抽提物诱导）。

1. *E. coli* BL21（DE3） emitting green light after IPTG induction （Black arrow points the colony of transformant emitting green fluorescent light and empty arrow points the colony of recipient strain）. 2. Expression of gfp in *Sinorhizobium fredii* HN01（a. the conjugant HN01（pHN129）emitting green light； b. recipient srain HN01）. 3. Three induced green light emitting transcon- jugants （a. without soybean seed exudat einduction； b. with soybean seed exudate induction）.

参 考 文 献

[1]Morise S G，Shimomura O，Johnson F H，et al. Biochemistry，1974，13:2656-2662.

[2]Shimomura O，Johnson F H，Saiga Y.J. Cell Comp Physiol，1962，59:223-240.

[3]Ward W W，Cody C W，IIart R. et al.，Photochem. photobiol，1980，31:611-615.

[4]Prasher D C，Eckenrode V K，Ward W W. et al.，Gene，1992，111:229-233.

[5]Chalfie M，T u Y，Euskirchen G. et al.，Science，1994，263:802- 805.

[6]William S，Alan H R，John J D. et al.，Methods in Enzymol，1990，185:61-89.

[7]Weinstein M，Roberts R C，Helinski D R.，J Bacteriol，1992，174:7486-7489.

[8]Ditta G，Stanfield S，Corbin D. et al.，Proc Natl Acad Sci USA，1980，77:7347-7351.

[9] Long S R，Baikema W J，Ausubel F M.，Nature，1982，298:485-488.

[10] Sambrook J，Fritsch E F，Maniatis T，Molecular cloring. A Laboratory Manual，2nd ed. New York: Cold Spring Harbor Laboratory 1989.

[11]朱光富，周俊初，陈华癸. 遗传学报，1996，23（2）: 131-141.

[12]Mulligan J T ，Long S R. Proc Natl Acad Sci USA，1985，82: 6609-6613.

[13]Bolivar F，Rodriguez R L，Greene P J. et al. Gene，1977，2: 95

[14]Eberl L，Giskov M，Schwab H.，Mol Microbiol，1992，6:1969-1979.

[15]Roberts R C，Strom A，Helinski D R. J Mol Biol，1994，237:35- 51.
[16]Davis T L，Helinski D R，Roberts R C. Mol Microbiol，1992，6: 1981-1994.
[17]Cormack B，Valdivia R H，Falkows. Gene，1996，173:33-38.
[18]Heim R，Prasher D C，Tsien R Y. Proc Natl Acad Sci USA，1994，91:12501-12504.
[19] Long S. Annu Rev Gen，1989，23: 483-506.

第二项　非豆科-弗兰克菌(Frankia)

马桑共生固氮根瘤及其内生放线菌*

周平贞

（中国农业科学院油料作物研究所）

吴　捷　陈华癸

（华中农业大学）

马桑型非豆科内生放线菌的共生固氮根瘤的含菌组织与桤木型不同。前者为一马蹄形整体，围绕着中柱维管束；后者完全包围中柱，其中含菌细胞和不含菌细胞交错存在或形成分散的含菌细胞组织团。尼泊尔马桑（*Coriaria nepalensis* Wall.）根瘤的内生放线菌菌丝形态与桤木根瘤的内生菌形态相同，具有一层电子密度很高的单层壁和具有单层壁的横隔膜。菌丝分枝，具有中间体。成熟的含菌细胞内有一围绕着中央液泡的栅栏排列的柱形泡囊，它们是菌丝的顶端细胞，其电子密度比菌丝高。含菌组织的衰老部分有颗粒体，颗粒体有双层壁，内层壁与菌丝的相同，外层壁比内层壁厚数倍，电子密度低。

关键词　马桑根瘤；内生放线菌；非豆科共生固氮

非豆科植物和内生放线菌的共生固氮根瘤已在桤木等 21 属植物中发现[1, 2]。1978 年以来已报道从 9 个属的植物根瘤中分离出纯培养物并成功地回接寄主植物，形成新的共生根瘤。从桤木等属分离的放线菌已定为 *Frankia* 属[3-5]。

非豆科植物和内生菌的共生固氮根瘤是变形的支根。按照根瘤内含菌组织和内生菌的形态特征，可以分为两个类型：①桤木型，代表除马桑属（*Coriaria*）和 *Datisca* 属外已发现的各属植物。在横断面上，桤木型根瘤的含菌组织完全包围中柱，含菌细胞和不含菌细胞交错存在或形成分散的含菌细胞团。含菌细胞中部充满内生放线菌的分枝菌丝。在寄主细胞壁附近，菌丝顶端形成球形或其他形状的泡囊。衰老的含菌组织中含有颗粒体，Dijk 和 Merkus[6]指出他们来自孢子囊。②马桑型，Bond[7]最先报道马桑属及其内生菌的共生固氮组含菌组织在根瘤内为一整体，横断面为马蹄形。其后 Newcomb 等报道 *Datisca* 属及其内生菌的共生固氮组织与马桑同型[2,8,9]。但至今都没有从马桑或 *Datisca* 根瘤中分离出内生菌的纯培养物和回接成功的研究报告。

本文报道了尼泊尔马桑（*Coriaria nepalensis* Wall.）的共生固氮根瘤的含菌组织及其内生菌的形态研究。尼泊尔马桑广泛分布在长江中上游流域的丘陵山地上。

材料和方法

（一）根瘤来源

1.试验根瘤采自中国农业科学院油料作物研究所（武汉）种植的马桑苗圃园内。

---

*原载于《微生物学报》，26（3）：277~281，1986.

2.广口瓶中装土壤和蛭石混合物，灭菌。播种表面灭菌的马桑种子，接种马桑根瘤的压碎汁。植株生长，形成根瘤后采取根瘤样品。未接种马桑根瘤压碎汁的对照处理没有结根瘤。

（二）光学显微镜观察

用 FAA 固定根瘤，按常规进行石蜡包埋。用旋转式切片机切片（厚度 5~8μm），在玻片上脱蜡后用番红-固绿双重染色，高碘酸席夫试剂染色，和碘液染色。在光学显微镜下观察、摄影。

（三）扫描电镜观察

最初，用常规方法制样：用 2.5%戊二醛固定根瘤指形小枝，经系列酒精脱水，用双面刀片切开，暴露含菌组织，将组织块黏在样品台上，临界点干燥，喷金后观察，结果含菌细胞中的内含物丢失。后改为石蜡包埋块，切成 20~24μm 厚片，黏在盖玻片上，脱蜡后固定在样品台上，用 Eiko IB-5 离子溅射仪喷金 8 分钟，用 Hitach^{i} Ts-50 型扫描电镜观察，摄影。

（四）透射电镜观察

取根瘤的含菌组织，切成 1mm^3 大小，2.5%戊二醛前固定，磷酸缓冲液清洗，乙醇脱水，以上处理均在 4℃进行。转入乙醇-丙酮（V/V 1:1）液和纯丙酮脱水。Epon 812 或$^{\#}$618 树脂包埋，用玻璃刀超薄切片。醋酸双氧轴-柠檬酸铅双染色，各染 10 分钟。Jeol CX-100 型透射电镜观察、摄影。

结　　果

（一）根瘤形态和含菌组织

马桑根瘤为变形的分枝侧根，形成一丛指形物（图版 I-1），每一指形物中间为中轴维管束组织。横断面的皮层组织感染后形成马蹄形的含菌组织，不完整地包围着中轴维管束（图版 I-2）。指形物的纵切片可分为五段：①顶端分生组织；②顶端后的细胞分化为维管束、皮层和表皮层；③新感染的含菌组织；④成熟的含菌组织；⑤衰老的含菌组织。

（二）含菌组织中的内生放线菌及其泡囊

含菌组织中的内生菌呈现为分枝有隔膜的菌丝（图版 I-3），宽 0.4~0.8μm，细胞壁为电子密度很高的单层壁，细胞质内有明显的单位膜中间体（图版 I-4），菌丝纵剖面中可观察到中间体与原生质膜连接。这些特征表明内生菌是放线菌。

在双重染色的光学显微镜下，含菌细胞的中央为液泡，周围为一整圈栅栏状排列的柱状体，番红着色。柱状体圈外被固绿着色（图版 I-5）。在透射电镜下，与番红着色的栅栏结构相应的是电子密度高的丝状细胞，长 3.0~8.0μm，宽 0.7~1.0μm，Newcomb[10]和 Mirza[11]称它们为泡囊；而与固绿着色相应的是电子密度低的菌丝。电镜观察表明泡囊是菌丝的连续，比菌丝稍粗并与菌丝以横隔膜分开（图版 I-6）。还不能解释同一菌丝的两段在染色反应上有这样明显的区别。

（三）含菌组织的外围细胞和菌丝侵入细胞的初期

靠近中柱的几层皮层细胞充满淀粉粒，碘液染成蓝色，无菌丝感染。含菌组织前端新感染的皮层细胞碘液染色除有大颗粒的淀粉粒外，还含有大量未能鉴定的无色或浅蓝色小颗粒，用高碘酸席夫试剂染色，淀粉粒（大颗粒）紫红色，小颗粒深红色。扫描电镜观察，新入侵的分枝菌丝附着在淀粉粒上生长，还有球形体或球杆状体，表面光滑，直径与菌丝相当（图版 I-7）。

（四）衰老的含菌细胞

在含菌组织底部的衰老区，栅栏状泡囊的原生质浓缩，质壁之间形成空隙。同时出现聚集在一起的颗粒体群，颗粒状内部电子密度较高，具有电子通透的类脂颗粒，颗粒体壁两层，内层电子密度高，厚度与菌丝和泡囊的胞壁相同，外层电子密度较低，但比内层厚数倍（图版 I-8）。

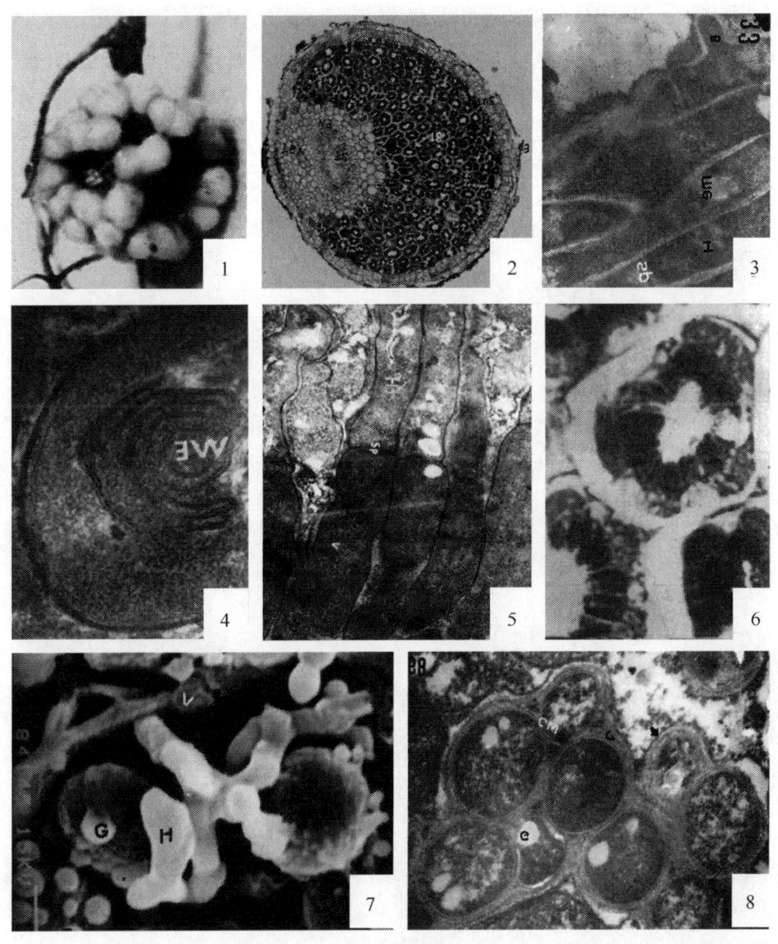

图版说明

1.马桑的共生固氮根瘤。Symbiotic nitrogen-fixing nodule of *Coriaria*. ； 2.马桑根瘤横切面，马蹄形含菌组织不完整地包围中柱。（光学显微镜照片）. Transverse section of *Coriaria* nodule lobe. Horseshoe-shaped tissue containing endophyte，incompletely surrounding the stele. （Micrograph）；3.分枝的内生放线菌菌丝，有隔膜，中间体与原生质膜连接。（透射电镜照片×17000）. Branched hyphae of actinomycetal endophyte with septa and mesosomes attached to plasma membrane. （TEM）；4.菌丝的中间体，显示单位膜结构。（透射电镜照片×116000）. Mesosome of hypha，showing unit membrane.（TEM）；5.内生菌泡囊（电子密度高）和菌丝，注意单层隔膜将泡囊和营养菌丝分开。（透射电镜照片×14000）. Endophytic vesicles （electron-dense）and hyphae. The endophytic vesicles are separated from the vegetative hyphae by septa.（TEM）；6.含菌细胞中栅栏状排列的泡囊，泡囊番红着色，其它内含物固绿着色。（双重染色，光学显微镜照片）. Palisade of vesicle in endophyte-infected host cell，vesicles stained by safranin 0 and other contents stained by fast green.（Double staining micrograph）；7.新感染的寄主细胞，含淀粉粒、分枝菌丝、球状、球杆状体。（扫描电镜照片）. Freshly infected host cell containing starch granules，branched hyphae，cocci and coccids.（SEM）；8.颗粒体群，显示双层壁。（透射电镜照片×19000）. Cluster of granular bodies，showing double-layered cell wall.（TEM）

讨 论

1.马桑型根瘤的含菌组织与桤木型的含菌组织形态有明显区别。本文报道的尼泊尔马桑含菌组织的形态与 Bond[7]、Newcomb 等[10]和 Mirza 等[11]的相同，并增添了一些新内容。

2.马桑根瘤的内生放线菌菌丝的超薄切片电镜图像与 Becking[12]的相同，为分枝的丝状体。胞壁单层，电子密度高。菌丝内有单位膜中间体，与原生质膜相连。这与桤木型的内生菌丝相同。

3.在含菌细胞内大量菌丝的顶端形成泡囊，泡囊紧密排列，形成栅栏状结构，围绕着寄主细胞的中央液泡。泡囊的电子密度比菌丝的高，石蜡切片经番红-固绿双重染色，在光学显微镜下泡囊被番红着色，菌丝和寄主细胞物质被固绿着色。这些与桤木型含菌细胞内的泡囊不同。周平贞等曾认为栅栏状菌体是细菌，电镜观察证明这是不正确的。

4.靠近中柱的几层皮层细胞含淀粉粒。在含菌组织的前端，含淀粉粒的皮层细胞被菌丝侵入形成新的含菌细胞，其中含有分枝菌丝和与菌丝宽度相等的表面光滑的球状或球杆状体。

5.在衰老的含菌细胞中，泡囊内容物与胞壁有明显的质壁分离现象。同时也形成一些颗粒体，颗粒体有两层壁，一层为电子密度高的内壁，一层为电子密度低且比内壁厚几倍的外壁，颗粒体的形态与Becking[6, 14, 15]等报道的桤木型内生放线菌的颗粒体（granule）或孢囊孢子（spore）相似，但尚不能证明它们来自孢子囊。

6.由于马桑型的共生放线菌和含菌组织与桤木型的共生放线菌和含菌组织有以上异同，在肯定其为内生放线菌的前提下，是否与从桤木等植物根瘤分离得到的*Frankia*相同，有待于从马桑共生根瘤中分离出纯培养物和回接结瘤成功后解决。

参考文献

[1]周平贞等：《中国油料》，2:1-3，1980.

[2]Akkermans，A.D.L.et al.，In Advances in Nitrogen Fixation Research，Martinus Nifhoff/Dr W.Junk Publisher，pp.311-319，1984.

[3]Becking，J.H. et al.，*Antonie van Leeuwehoek*，30:343-376，1964.

[4]Becking，J.H.，*Plant and Soil*，32:611-654，1970.

[5]Becking，J.H.，*Int.Syst.Bact.*，20:2，1970.

[6]Bond，G.，In Symbiotic Nitrogen Fixation in Plants，Cambridge University Press，Cambridge，London，New York，Melbourne，pp.443-475，1976.

[7]Bond，G.，In The Biology of Nitrogen Fixation，North-Holland Publishing Company，Amsterdam，Oxford American Elsevier Publishing Company，Inc.New York，pp.342-378，1974.

[8]Callaham，D. et al.，*Science*，199:899-902，1978.

[9]Gardner，I.C.，In Symbiotic Nitrogen Fixation in Plants，Cambridge University Press. Cambridge. London，New York，Melbourne，pp.485-496，1976.

[10]Hafeez，F. et al.，*Plant and Soil*，79:383-420，1984.

[11]Lechevalier，M.P.，*Plant and Soil*，78:1-6，1984.

[12]Mirza，M.S. et al.，In Advances in Nitrogen Fixation Research，Martinus Nifhoff/Dr W.Junk Publisher，1984.

[13]Newcomb，W. et al.，*New Zealand J.Botany*，20:93-103，1982.

[14]Newcomb，W. et al.，*New Zealand*，*J.Botany*，20:105-113，1982.

[15]van Dijk，C.and E.Merkus，*New Phytol*，77:73-91，1976.

马 桑 简 介*

周平贞　胡传炯

（中国农科院油料所）

陈华癸

（华中农业大学）

形态学特征　尼泊尔马桑在中国常用名为马桑（*Coriaria rcepalensis*，Wall，1932 年）又名中华马桑（*C.sinica*，Maxim 1988 年）后者为前者的异名。属于马桑科、马桑属、尼泊尔马桑种。是多年生共生固氮薄叶灌木，通常高 2~3 米，有时可高达 6 米。枝条斜展，幼枝有棱，无毛，每株有枝条 10~20 根。叶对生；叶柄粗短，长约 1~3 毫米，常带紫色；叶片椭圆形或阔椭圆形，长 3~13 厘米，宽 2~8 厘米，先端急尖，基部近圆形，全缘，两面有粗大明显的基出脉三条，绿色，无毛或仅沿下面叶脉有细毛，纸质或薄纸质。花杂性，集结成总状花序并腋生于前一年的枝条上；雄花序通常先叶开放，花柄长不及 5 毫米，花托小，花被下位；萼片 5 枚，分离呈复瓦状排列；花瓣 5 枚，略较萼片为小；雄蕊 10 枚，紫蕊 5 枚，离生；各子房内有一倒生胚珠；花柱丝状，柱头不明显，整个表面被乳头状突起，瘦果 5 枚，外包有宿存的肉质花瓣，初红色，成熟后渐变紫黑色，直径 6~8 毫米（照片 1，2）。

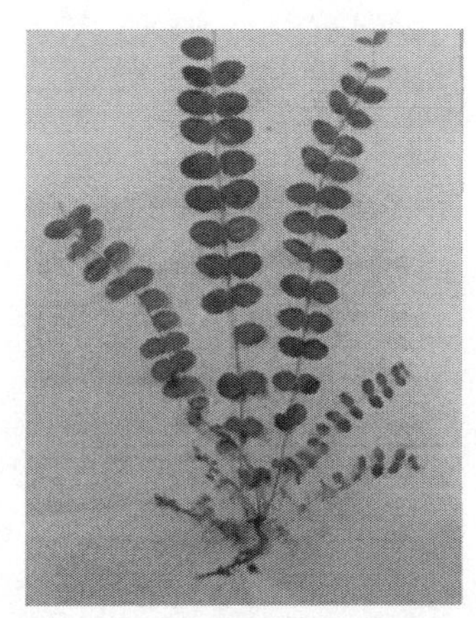

图 1　枝条上有大量马桑果实

图 2　植物根部有根瘤

化学成分　马桑茎叶，种子及马桑寄生子（指寄生在马桑树上的桑寄生科植物的通称）含有马桑内酯、蛋白质、油酯、没食子酸、山萘酚、叶绿素、鞣质、多糖、树酯、黄酮、色素等。马桑内酯类含量在马桑叶中为 0.115%，马桑子中为 0.193%，马桑寄生子中为 0.173%~0.216%。内酯成分含有马桑毒素（A）、羟基马桑毒素（B）、马桑亭（C）及未知内酯（D）。马桑叶中含 B 和 D 内酯，马桑子中含 B、C、D 三种内酯，马桑寄生子中含有 A、B、C、D 四种内酯。马桑茎叶含蛋白质 11.63%，种子含蛋白质 21.83%，马桑子含油脂 19.33-20.3%，油主要为 13-羟基顺-9-反 11 十八碳二烯酸达 58.96%。马桑果的肉质花瓣含可溶性糖 8.69-9.81%，并含有较浓的水溶性紫色素。

用途　马桑具有生长快，耐阴、耐瘠薄适应性强的特点，并且每对叶腋处又可生长二次分枝，枝叶繁茂，枝条斜展，是水土保持的优良树种。马桑根部具有共生固氮根瘤，内有共生固氮放线菌，系 *Frankia* 属，每年每亩固氮 45 公斤，产干茎叶 2448 公斤，茎叶肥嫩，易腐烂，是理想的优质绿肥。叶子可养蚕，

---

*原载于《中国油料》，（2）：92-93，1990.

每 4 千头蚕可产丝 1 公斤。茎叶、种子及寄生子中的内酯是治疗精神分裂症的药物，它具有疗效显著、稳定、费用低廉，药源广的特点。肉质花瓣可制酒和提炼色素。

分布及繁殖 主要分布在喜马拉雅山脉的尼泊尔和中国。我国大多分布在西南和华中各省的海拔 400~2100 米的丘陵山坡地带，适宜范围，温带及亚热带地区均能正常生长。它主要与黄荆、黄栌、胡枝子等灌木丛生，或与栎树、松树、柁木等乔木混种。土壤 pH5.5~7.5 的砂质壤土及壤土均生长正常。可用种子或插条繁殖，春秋雨季时繁殖为佳，要求温度 15~25℃昼夜温差发芽生根最快。第一年生长较慢，一般仅 40~80 厘米高。第二年以后生长迅速，可达 2~6 米高。

尼泊尔马桑放线菌共生固氮根瘤的感染和发育*

吴 捷 陈华癸

（华中农业大学土化系）

周平贞

（中国农业科学院油料作物研究所）

摘 要 弗兰克氏放线菌通过感染尼泊尔马桑的根毛侵入根的皮层细胞。由于内生菌侵入的刺激，部分皮层细胞分裂和增大，产生前根瘤原基。根瘤原基分裂、分化，形成初生根瘤瘤片。瘤片顶端分生组织不断双叉分枝，发育，并伴随着内生菌感染的寄主细胞，产生多次双叉分枝的珊瑚状根瘤。观察瘤片的横切面，含菌是一个马蹄形的致密整体，不完全地包围着稍偏的中柱。观察瘤片的纵切面，可将其划分为6个区域，即顶端分生组织，未感染皮层细胞组织、新感染含菌组织、成熟含菌组织、衰老含菌组织和中柱及其外围数层富含淀粉粒的皮层细胞。

关键词 马桑根瘤；生物固氮；弗兰克氏菌；非豆科共生体系

在豆科植物以外，能与固氮微生物共生、结瘤的植物中，只有榆科的 *Parasponia* 一属是和根瘤菌（*Rhizobium*）形成共生固氮根瘤的[7-8]，其他都是和一属称为弗兰克氏菌（*Frankia*）的放线菌形成共生固氮根瘤。搜集现有的报道，和弗兰克氏放线菌形成共生固氮根瘤的植物有200种以上，分布在7目，8科，23属中[2, 5, 10]。我国已报道的有5目，5科，6属42种[1]。

关于弗兰克氏放线菌感染寄主以及共生根瘤的组织发育过程，前人的报道都是片面的，只有综合各家的报道，才能取得弗兰克氏放线菌共生根瘤组织形态发育的基本模式。本文报道了弗兰克氏放线菌与尼泊尔马桑（*Coriaria nepalensis* Wall.）形成共生根瘤的组织形态发育的全过程。

材料和方法

1 植物和根瘤来源

（1）尼泊尔马桑和四川桤木（*Alnus cremastogyne*）的种子和根瘤，最初采集于云南、四川和湖北宜昌，在中国农业科学院油料作物研究所（武汉）的苗圃内种植，从中采取种子和根瘤。

（2）塘瓷杯和广口瓶中装蛭石或土壤蛭石混合物，灭菌。播种表面灭菌的马桑种子，接种马桑根瘤的压碎汁，暨温室中培养。植株生长，形成根瘤后，采取根瘤和根的样品。

2 光学显微镜观察

（1）用相差显微镜观察根毛感染。

（2）石蜡切片，用FAA液固定根瘤，按常规进行石蜡包埋。用旋转式切片机切片（厚度5~8μm）。在玻片上脱蜡后用番红-固绿、高碘酸席夫试剂、苏丹VI和碘液染色。在光学显微镜下、摄影。

（3）半薄切片，取受感染的根段和根瘤组织，切成$1mm^3$大小，2.5%戊二醛前固定，磷酸缓冲液清洗，1%锇酸后固定，乙醇和丙酮系列脱水，Epon 812树脂包埋，用玻璃刀在超薄切片机上切成半薄

*原载于《武汉植物学研究》，9（1）：1~6，1991

切片（厚度 1~2μm），片切置玻片上，在温台（70℃左右）上展开，用 10% KOH 溶液溶去树脂，复红染色，用光学显微镜观察、摄影。

3 扫描电镜观察

将石蜡包埋块切成 20~24μm 厚片，黏在盖玻片上，脱蜡后固定在样品台上，用 Eiko IB-5 离子溅射代喷金 8 分钟，用 Hitachi TS-50 型扫描电镜观察、摄影。

结果和讨论

1 菌丝的入侵

关于弗兰克氏菌侵入寄主的途径，前人报道有两种。一种是毛区的新鲜根毛受弗兰克氏菌的刺激而伸长和弯曲，弗兰克氏菌菌丝侵入变形根毛，进入皮层细胞[3, 4, 6, 9]。另一种途径是弗兰克氏菌菌丝直接在根表皮细胞间穿透进入寄主体内。我们观察弗兰克氏菌菌丝侵入尼泊尔马桑属前一途径（图片 I：1，2）。这里应该指出，这种根毛感染和入侵与根瘤菌感染和入侵豆科寄主根毛不同，虽然都刺激根毛伸长、弯曲，并在弯曲处入侵，根瘤菌在入侵处刺激根毛内陷，内陷逐渐深入形成侵入线，根瘤菌在侵入线内随着侵入线的延伸而大量增殖，在弗兰克氏菌的根毛感染过程中，我们通过光镜观察到，菌丝穿入根毛，并向根毛基部发展，但未观察到菌丝在根毛内部大量增殖。

2 前根瘤的形成，内生菌感染部分皮层细胞，根瘤原基的发生

尼泊尔马桑和其他一些寄主一样，在被弗兰克氏菌感染后，首先形成前根瘤。前根瘤是在入侵菌刺激之下，寄主的皮层细胞分裂增殖而成的（图版 I：3，4）。在前根瘤的顶端，一小部分细胞先被菌丝感染，成为含菌细胞。在寄主细胞内的菌丝称为内生菌，内生菌是穿壁进入寄主细胞的。在前根瘤的底部，原根中柱鞘位置的细胞恢复分生能力，产生分生组织，称为根瘤原基（图版 I：3）。

3 真根瘤的形成

由于内生菌的逐渐穿壁入侵，含菌细胞逐渐增多，前根瘤中的根瘤原基不断发展分化，产生维管组织和皮层细胞。当部分皮层细胞被来自前根瘤中的内生菌侵染后，就形成初生根瘤（图版 I：5）。初生根瘤的顶端分生组织随后双叉分枝、发展，形成两个生长点，位于双叉连接处的含菌细胞中的内生菌分别向两面穿壁入侵，形成两个含菌瘤片（图版 I：6，7）。由于两个瘤片的分生组织是从一个共同的分生组织分叉产生的，随着分生组织的向前发展，细胞的伸长和分化，形成底部相连，分叉的中轴组织，并在内皮层外面形成含菌组织。由于分生组织的多次快速分叉，形成几十个生长点，发育成为几十个瘤片，在外形上幼嫩根瘤的许多紧密团聚在一起（图版 I：8）；成长的根瘤表现为多次分叉的珊瑚状根瘤（图版 I：9），每个分叉的前端有一个具有含菌组织的瘤片。含有几十个瘤片的根瘤的自然直径可达 20~40mm，固氮活性最旺盛的根瘤其直径为 8~10mm。Becking 将放线菌根瘤的外形分为两类：桤木型和杨梅/木麻黄型。杨梅/木麻黄型根瘤的每一瘤片顶端都有一根背地性生长的“根瘤根”。桤木型的瘤片顶端没有这种根瘤根。按这种分型法，尼泊尔马桑属于桤木型。

4 瘤片的构造

一个瘤片的横切面在显微镜下显示出图版 II：10 的形态，马蹄形的含菌组织不完全地我围着偏心的中柱。在内皮层和含菌组织之间有 4-5 层不含菌，但在量淀粉粒的皮层细胞。在皮层细胞（包括含菌和不含菌的）和表皮细胞之间，有 6-7 层栓质化的细胞，用苏丹 VI 染色均呈红褐色，靠外 2~3 层细胞栓质化程度高，靠内 3-4 层栓质化程度低。观察瘤片的纵切面，从顶端向后，依次为顶端分生组织、未感染皮层细胞组织、新感染含菌细胞组织、成熟含菌细胞组织和衰老含菌细胞组织，纵切面中间为中柱，

中柱外围有数层不含菌但富含淀粉粒的皮层细胞（图版 II：11，12）。在未感染皮层细胞组织中，细胞含有大量淀粉粒和一种尚鉴定的小颗粒（图版 II：13）。在新感染含菌细胞组织中，细胞中除含淀粉粒和小颗粒外，还含有刚侵入的、分枝的放线菌菌丝（图版 II：14）。

5 尼泊尔马桑和四川桤木根瘤含菌组织的比较

图版 II：10 和图版 II：15（四川桤木瘤片的横切面）相比，显示出明显的区别。尼泊尔马桑根瘤的含菌组织是一个致密的整体，其中每一个细胞都是含菌细胞，而不含菌皮层细胞位于含菌组织的另一边；四川桤木的含菌组织则是少量含菌细胞成簇地分散在不含菌细胞内，这种成簇分散的含菌细胞在木麻黄（*Casuarana*）、杨梅（*Myrica*）、胡颓子（*Eleaegus*）、仙女木（*Dryas*）和悬钩子（*Rubus*）等属的根瘤中都有报道[9]。

关于尼泊尔马桑根瘤含菌细胞内生菌的形态在文献中已做初步报道，关于尼泊尔马桑根瘤含菌细胞和内生菌的超微构造将另文报道。

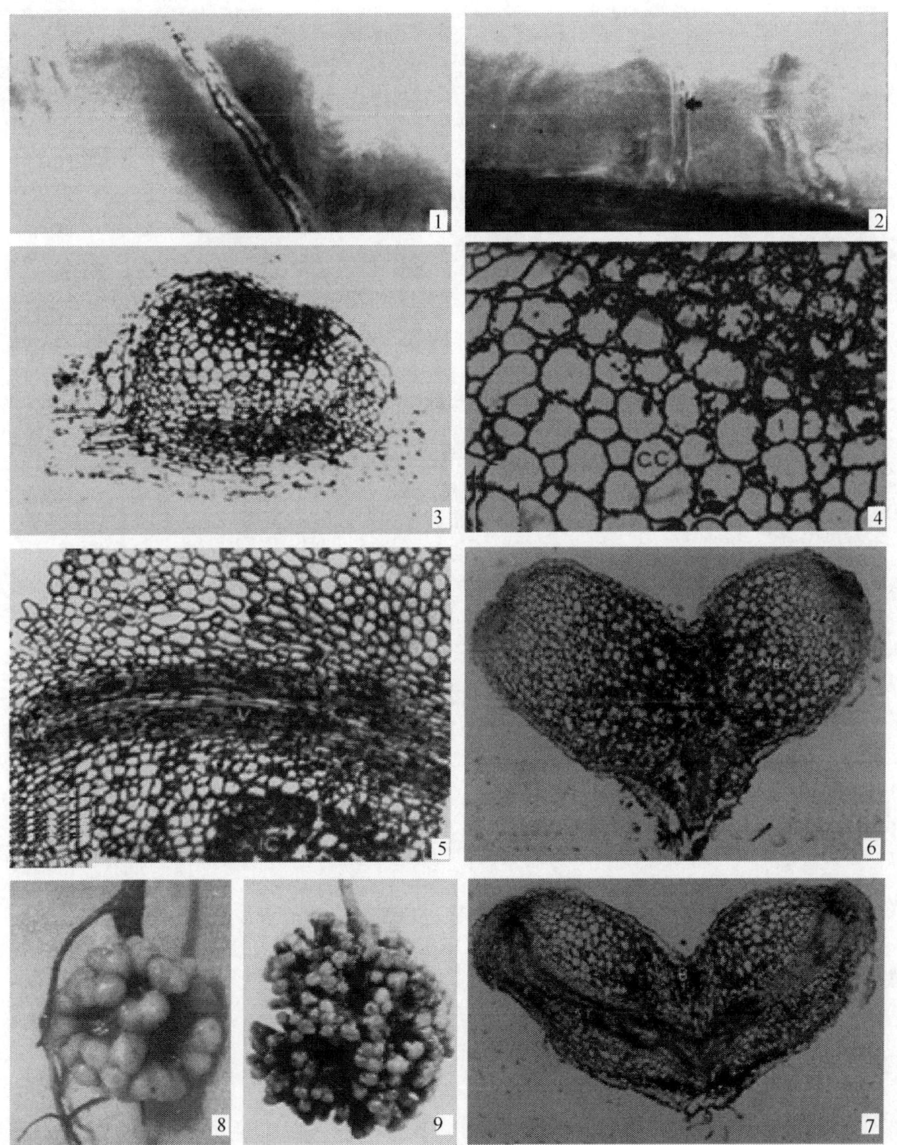

图版 I：1. 未感染根毛；2. 被感染根毛，示侵入线（箭头）；3，4. 示根瘤，示含菌细胞（IC），未感染皮层细胞（CC）和分生组织（MT）；5. 示主根瘤，示含菌细胞（IC），维管组织（V）和顶端分生组织（AM）；6，7. 瘤片分叉，示含菌细胞（IC），维管 （V）；8. 健旺幼嫩根瘤；9. 成熟的大根瘤

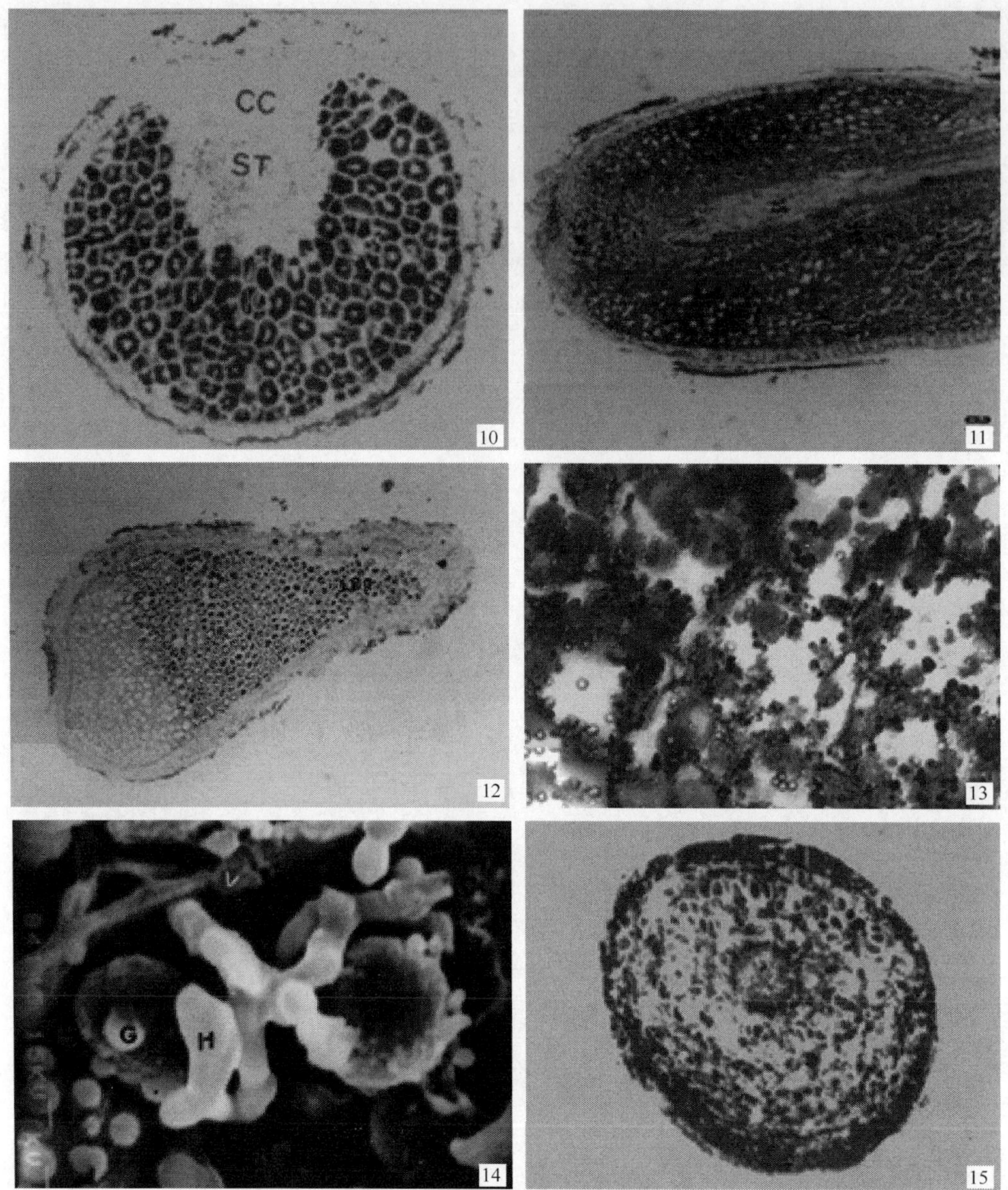

图版 II：10. 成熟瘤片含菌区域的横切面，示马蹄形含菌组织（IC），稍偏的中柱（ST）和未感染的皮层细胞（CC）；11，12. 成熟瘤片的纵切面，示顶端分生组织（A），未感染的皮层组织（B），新感染含菌组织（C），成熟含菌组织（D）和衰老含菌组织（E）；13. 未感染细胞，富含淀粉粒和一种未鉴定的小颗粒；14. 新感细胞，含菌丝（H）、淀粉粒和一种未鉴定的小颗粒（G）；15. 成熟的四川桤木瘤片的横切面

参 考 文 献

[1]黄家彬等.第五届国际固氮会议论文集.荷兰.1983

[2]Bond G Ed . P S Nutman.1976:443-474

[3]Berry A M et al . Can J Bot.1986；64:292-305

[4]Callaham D et al . Bot Gaz.140（Suppl.）1979；S1-S9

[5]Lechevalier MP . Can J Bot.1983；61:2964-2967

[6]Lalond M Eba . Newton W，L R Postgate and Rodriguer-Barruc.Academic Press . London .1977；569-589

[7]Trinick M J .Nature（London）.1973；2224；459-460

[8]Trinicik M J .Can J Microbion.1979；25:565-578

[9]Torrey J G . Amr J Bot . 1976；63:335-344

[10] Zhongze Z et al . Plant and Soil，1985；87:1-16

尼泊尔马桑根瘤维管束形态发育研究*

胡传炯 周平贞

（中国农业科学院油料作物研究所）

陈华癸

（华中农业大学）

摘　要　以生物制片和光学显微技术研究了尼泊尔马桑根瘤维管束的形态结构和发育特点。尼泊尔马桑根瘤维管束呈多重二叉分枝的树状结构，这是由根瘤顶端分生组织不断分裂形成的。它在基部与植物维管系统联成一体，在本质上是由植物侧根原基衍生而来。马桑根瘤维管束的这种形态结构和发育特点不同于豆科植物根瘤，也不同于杨梅属、木麻黄属等非豆科植物的根瘤。

关键词　尼泊尔马桑；根瘤维管束

在研究根瘤形态发生和发育过程中，必须研究根瘤内部维管束的形态民结构和发育特征，以揭示根瘤的发育演化类型和规律。过去，研究者们常以外形来确定根瘤类型。在非豆科植物根瘤中，杨梅型根瘤是由很多簇状基生的瘤瓣组成，每个瘤瓣顶端都长有负地性生长的根瘤根；而桤木型根瘤则是由很多二重分枝瘤片组成的珊瑚状结构[1]。但这两种外形有差异的根瘤在形成早期却较难区分，且两种根瘤外形在一定条件下可以转变[1, 2]。因此只有通过对根瘤内部维管束形态和结构的观察，才能从本质上确定其根瘤类型。

因根瘤结构复杂，外皮坚韧，维管组织被紧密地包围在根瘤内部，很难完整、准确地观察和描述其形态发育。作者在研究尼泊尔马桑根瘤形态发育过程中，逐渐摸索出一套剥离根瘤维管束的方法，能有效显示出维管束形态发育特征。

1　材料与方法

1.1　根瘤切片观察材料为尼泊尔马桑根瘤。取不同发育时期的马桑根瘤，经FAA固定，以常规石蜡切片制作方法制片，显微镜下观察。

1.2　根瘤维管束形态观察综合离析法、透明法和整体封固法[3]。具体做法简述如下。

取不同大小和发育阶段的马桑根瘤，小心洗净后，转入35%Na_2CO_2中沸水加热20分钟，然后以1mol HCl于60℃下水解。水解后用解剖器具小心剥离，最后以5mol NaOH透明。材料经染色封固后便可显微观察。

2　结果与讨论

2.1　根瘤维管束的形态结构

研究表明，尼泊尔马桑根瘤维管束呈多重二分枝的树状结构（图版I：1，2）。以其基部与主根的维管系统相连。基部维管组织发育最为成熟，是早期初生根瘤维管束遗留下来的。维管束由近端及远端维管组织渐次幼嫩，最前端为尚未分化完全的初生维管组织和少量残余的顶端分生组织，维管束近端导管密集，染色呈半透明的深色。

2.2　根瘤维管束的发生及发育特征

马桑根瘤维管束在本质上起源于植物侧根原基。根瘤形成须具备两个先决条件，即内生菌侵染和分

*原载于《武汉植物学研究》，11（3）：211~213，1993

生组织的形成。单有内生菌侵染只能引起植物根部膨大，形成所谓“前根瘤”（pre-nodule）[5]；单有来源于植物的分生组织只能形成侧根。发生于前根瘤形成区域的侧根原基向外分裂形成突起。当它进入前根瘤的膨大区域内并感染了来自前提瘤的内生菌时，便逐渐衍化为根瘤原基，后者逐渐分化出根瘤的顶端分生组织和维管组织，并和根系维管束相连。至此根瘤才成为植物根系上完整的一部分（图版I：3-5）。

根瘤维管束的发育与根瘤的发育相联系。当根瘤发育到一定程度，根瘤含菌组织就不断增加，根瘤顶端的分生组织一分为二（图版I：6），形成两个分离的顶端分生组织，后者不断分裂形成新生的皮层和维管组织，皮层受根瘤含菌组织内生菌感染，不断扩大形成裂片；维管组织向后渐次成熟并与基部初生根瘤的维管束相连。从而形成顶端分离而底部相连的二裂瘤片（图版I：7）。当这两个根瘤裂片发育到一定程度时，每个顶端分生组织重又分裂产生新的分生组织，这些分生组织又逐渐发育形成新的皮层和维管组织，继而根瘤裂片数目成倍增加，维管组织依次相连。由初生根瘤逐级分裂形成的裂片和维管束珊瑚状排列。

尼泊尔马桑根瘤及其维管束这种多重二分叉结构和形成机制，与杨梅/木麻黄型根瘤明显不同，后者所有根瘤裂片呈簇状丛生在一起，而不像马桑根瘤那样所有裂片由根部同一位点最初的一次有效感染形成。这是由于杨梅/木麻黄型根瘤分生组织多点起源于侧根的中柱鞘细胞[4, 5]。且每一起源只形成单一的裂片，每个裂片顶端不再分叉，而非限制性生长形成负地性生长的“根瘤根”。马桑根瘤及其维管束的结构和形成机制与Moiroud（1984）关于桤木根瘤多重二分叉形成机制的假说相印证。

马桑根瘤在外形上属桤木型，即裂片呈多重二分叉结构。但两者在根瘤组织形态上有极大差异。马桑根瘤因含菌组织呈马蹄形不完全包围中柱，内生菌泡囊以栅栏状紧密排列在细胞周围从而被定为独特的马桑型[7]。

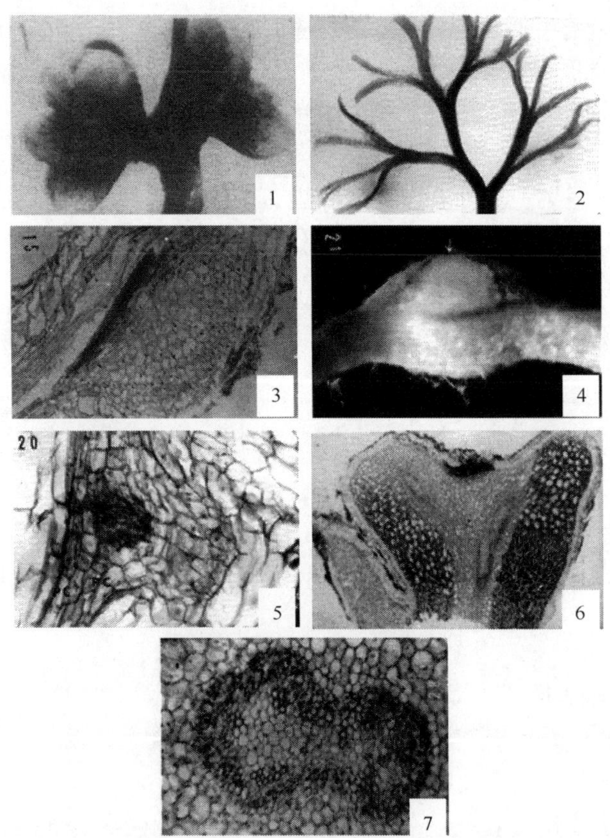

图版 I：1. 示典型的根瘤外形；2. 根瘤维管束，示其多重二分枝结构；3. 前根瘤的膨大结构，没有维管组织分化，箭头示植物维管束；4. 示发生于前根瘤处的侧根原基（箭头），以后转化为根瘤原基，向外突出形成初生根瘤；5. 初生根瘤切片，示其内起源于侧根原基；6. 根瘤纵切面，示顶端分生组织分裂，形成两个根瘤裂片，每个顶端分生组织向后分化形成的维管组织在基部相连；7. 二裂根瘤横切面，示根瘤分生组织最初分裂的部位，图中维管组织已分化成熟

参 考 文 献

[1]李正理.植物制片技术.北京：科学出版社，1978，86-140

[2]周平贞，吴捷，陈华癸.马桑共生固氮根瘤及其内生放线菌.微生物学报.1986，26（3）：277-281

[3]Becking J H. Root nodules in non-legumes.In:The development and function of roots，Eds:Torrey J G and Clarkson D T.Academic Press London，1975.507-560

[4]Bond G.The results of the IBP survey of root nodule formation in non-leguminous angisperms.In:Symbiotic nitrogen fixation in plants，Ed:Nutman P S.1978.443-474

[5]Bowes B et al. Time-lapse photographic observations of morpho-genesis in root nodules of Comptonia peregrina （Myricaceac）.Amer *J*Bor，1977，64（5）:516-525

[6]Flet cher W W.The developments and structure of the root nodules of *M*yrica gale L.with special reference to the nature of the endophyte.*A*nn Bot（London），N.S.1955，19（76）:143-156

[7]Moiroud A et al.Symbiotic relationshop in Actinorhizse.in:Genes involved in microplant interaction，Eds: D P S Vema and Th Hohn Springer-Verlag.Vien New York，1984.205-223

Nodulation and Molecular Characterization of Pure Cultures Isolated from Root Nodules of *Coriaria nepalensis*[*]

HU CHUANJIONG[1], ZHOU PINGZHEN[2], ZHOU QI[1], CHEN HUAKUI[2] AND AKKERMANS A.D.L.[3]

(1. College of Life Science and Technology, Huazhong Agricultural University, Wuhan 430062, China;

2. Institute of Oil Crops, Chinese Academy of Agricultural Sciences, Wuhan 430062, China. 3. Department of Microbiology, Wageningen Agricultural University, 6703 CT. Wageningen, The Netherlands)

ABSTRACT Four strains with typical morphology of actinomycete genus of *Frankia* were isolated from root nodules of *Coriaria nepalensis*. They were shown to nodulate the seedlings of host plant and hybridize with *Frankia* 16S rRNA targeted oligonucleotide probes, indicating that they did belong to the genus *Frankia*. Furthermore, by *nif* HDK probe hybridizations, the homologous fragments of *nif*HDK genes were detected among the bacteria, and they were located in various sizes of restriction fragments of total DNA, showing diverse patterns of restriction fragment length polymorphisms of *nif*HDK gens (*nif*HDK-RFLP^s) The PCRbased amplification and cloning of *nif*H gene throw light on the molecular phylogeny of *Coriaria*-infective *Frankia*.

KEY WORDS *Frankia Coriaia nepalensis*, nodulation, molecular, characterizatlon

Members of actinomycete genus *Frankia* are root symbionts that nodulate a wide range of woody dicotyledonous plants belonging to more than 200 species, 24 genera and 8 families [1]. After the first isolation of *Frankia* pure culture in 1978, several hundreds of typical *Frankia* strains have been successfully isolated from root nodules of 9 species belonging to *Alnus*, *Casuarinaceae*, *Elaeagnaceae*, *Myricaceae*, However, so far there has been no report on infective and effective *Frankia* strain isolated from *Coriaria*[2]. For several years of attempts, we have isolated some actinomycete strains with typical morphology of *Frankia* from root nodules of *Coriaria nepalensis*, i.e. formation of sporangia and vesicles brone on hyphae.

The aim of this study was to characterize these nodular isolates on the basis of nodulation test, *nif*HDK probe hybridization and *nif*H amplification and cloning.

1 MATERIALS AND METHODS

The pure cultures of Cs030, Cs103, Cs150 and Cs641 were isolated from root nodules of *Coriaria nepalensis* sampled from various localities in China[3]. The confirmed *Frankia* strains ArI_3 (from *Alnus rubra*), Ccol (from *Casuarina cunninghamiana*), Hr18 (from *Hippophae rhamnoides*) and Mg^+ (from *Myrica gale*), gifts from Prof. Ding Jian, were grown on Bap liquid medium[4] at 28℃ for 2~4 weeks.

The nodulation tests were carried out in greenhouse. Seeds of *Coriaria nepalensis* were surface-sterilized and germinated as described in ref.[3]. The axenic seedlings were then transplanted to steril containers containing autoclaved soil/ vermiculite (2:1, v/v), perlite and semi-solid agar, respectively. The semi-solid agar was made of 0.8% agar and 1/4-strength of Hoagland's nutrient solution. Six-week-old seedlings were inoculated with 2 ml of cell suspension per plant of 2-week-old *Frankia*, cultured in Bap or Bap nitrogen free

*原载于 *Chinese Science Bulletin*, 43 (8) : 695~698, 1998.

liquid medium. Three to four replications were conducted for each strain, and the same number of replications designed as controls, which were separated from the treatment pots or tubes. Three inoculations were completed in total during plant growth phase at intervals of 2 weeks. The steril nitrogen-free 1/8-strength of Hoagland's solution was supplemented once a week.

Total DNA and RNA of *Frankia* strains were extracted according to Simonet et al. [5] and Hahn et al.[6], respectively. DNA fragments, used for labelling and transformation, were isolated and purified by the method of low-melting-temperature gel electrophoresis.

Radioactive labelling of DNA, dot and slot hybridization and Southern blotting were manipulated following Maniatis et al.[7]. DIG-labelling and hybridizations were carried out following the manufacturer's specification. Cool *nif*HDK homologous fragment harbored in plasmid pC-clGX[8] was isolated and used as *nif*HDK probe.

Oligonucleotide probes specific for *Frankia* genus and *Coriaria/Datisca* nodular endophytes were designed targeting specific sequences of 16S rRNA[2]. They were used to characterize *Coriaria* isolates in this study. Primers *nif*HF（5'CACGGATCCGCAAG GGTGGTATTGG3'）and *nif*HR（5' GTGAAGCTTCTCGA TGACCGTCATCCGGC） [9] were used perform specific PCR amplification. The conditions for PCR reaction were the same as those described by Mirza et al.[3] The *nif*H amplification was purified prior to transforming it into *E. Coli* JM109 by using pGEM-T vector system（see *PromegaTechnical Bulletin*, 1995）.

2 RESULTS AND DISCUSSION

2.1 Re-inoculation of *Frankia* isolates from *Coriaria nepalensis*

The results of *Coriaria* isolates nodulating the host plants are shown in table 1.

Table 1 Results of re-inoculation tests with *Coriaria* isolates

Strains	*Nodulated No. Of seedlings living on*		
	soil/vliteermicu	Perlite	Semi-soid agar
Cs030	1/17a)	1/12b)	1/9c)
Cs103	1/20	0/12	1/8
Cs150	1/19	1/11	0/9
Cs641	1/20	0/12	0/0d)
Controls	0/20	0/12	0/9

a) The average values from 4 growing pots with 5 seedlings each; b) the average values from 4 growing pots with 3 seedlings each; c) the average values from 3 growing tubes with 3 seedling each; d) the seedlings on the tubes were dead completely when recorded.

The nodulation experiments in the greenhouse indicated that all the 4 isolates of *Coriaria* nodular endophytes were able to nodulate the plants grown on soil/vermiculite. Strain Cs030 can infect and form nodules on plants grown on soil/vermiculite, perlite and agar. Other 3 strains showed little variety in nodulating plants grown on soil or perlite and agar. However, in all 3 cases, the nodulation rates of these strains were quite low, ranging from 5.0% to 12.5%. Previous work demonstrated that the nodulation rates on *Coriaria* seedlings were quite low even inoculated with the suspensions of crashed nodules and host rhizosphere soils.

Various authors showed that there were many factors such as temperature, pH, nutrient and growth of plants influencing the ability and effectivity of nodulation by *Frankia* [1]. Our previous work indicated that soil condition was also an important factor to plant growth and nodulation.

2.2 16S rRNA-targeted oligonucleotide probe hybridization

By comparing the sequence differences of 16S rRNA among several *Frankia* strains and 23 strains from other actinomycete genera, Hahn et al. designed a *Frankia* genus-specific oligonucleotide probe （Fp）, which is complementary to partial sequence of 16S rRNA. By the same strategy, Mirza et al. developed a 16S rRNA targeted oligonucleotide probe, namely Cor/dat probe which specifically reacts with nodule endophytes of *Coriaria* and *Datisca*. The nodules of *Datisca* are anatomically similar to that of *Coriaria*, so far no nodular endophyte has been successfully isolated from the two genera. Hybridizations of probes Fp and Cor/dat with RNA extracts from *Coriaria* isolates are shown in fig.1. All the four isolates reacted well with both Fp and Cor/dat probe, whereas stronger positive signals were detected with Cor/dat probe hybridization. From this result these pure cultures proved to be nodular endophytes of *Coriaria nepalensis* and members of genus *Frankia*.

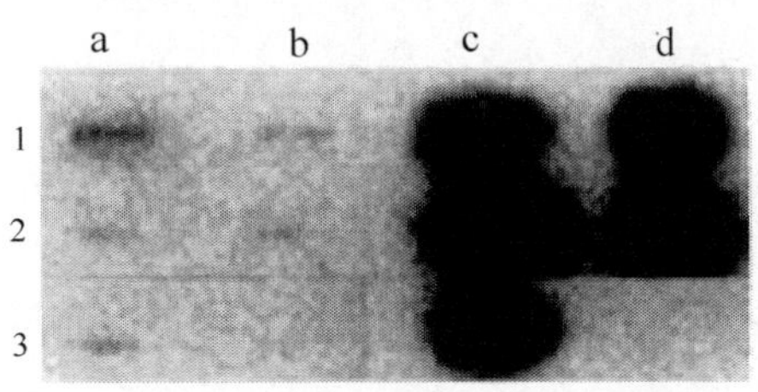

Fig.1 Slot hybridization of Fp and Cor/dat probes with nucleic acids preparations from pure cultures.

al, c1: ARI3; b1, d1:Cs030; a2, c2:Cs103; b2, d2:Cs150; a3, c3:Cs641; b3, d3: *E.coli*.

2.3 Identification of *nif*HDK homologous fragments and their diversity

*Nif*HDK genes encoding complex of nitrogenase are highly conserved and constant so that they were used to probe the homologous fragments of the parallel genes in *Frankia*[8-10]. We previously revealed that the homologous fragments of *nif*HDK were detected among six *Coriaria* isolates by hybridization with pSA 30, and one strain was shown to have 3 sizes of nifHDK hybridizing bands[10]. In this study, a *Frankia* homologous *nif*HDK fragment （about 5.5 kb in size） from Ccol was used as probe. Four *Coriaria* isolates were shown to give positive hybridizations with pCc1GX confirming the results of nodulation experiment （fig.2 （a））. Moreover, the hybridizing bands were located in different sizes of BamHI-digests, indicating that these *Frankia* isolates can be differentiated on the basis of restriction fragment length polymorphisms of *nif*HDK genes （*nif*HDK-RFLPs）. Cs150 and Cs030 were shown to contain more than two *nif*HDK hybridizing bands, and one of these bands was shown to be in similar size of 5.5kb, same as Ccol, whereas the other bands were of different sizes. Both Cs103 and Cs641 only have one hybridizing band at approximate size of 11 kb （fig.2（b））. The hybridization results obviously revealed the genetic diversity among *Coriaria* isolates on the basis of RFLP of *nif*HDK sequences. Up to now, only one promising *Frankia* isolate has been obtained from *Coriaria nepalensis*[2], this isolate, however, neither forms vesicles nor nodulates the host plant. This study, for the first time, provides the confirmation of *nif* gene being involved in *Coriaria*-isolates.

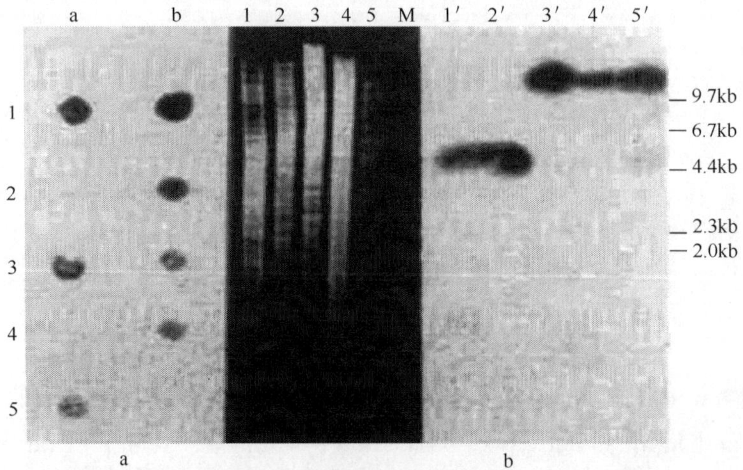

Fig.2 DIG-hybridization of *nif* HDK （from pCc1GX） -probe with total DNAs of pure cultures （dot blotting, （a） ） and their *Bam*HI digests （b） . （a） al,a3,a5 were MI,Hr18,and Mg^{++} respectively;b1-b4 were Cs641,Cs150,Cs103 and Cs030 respectively;b5,*E.coli*.(b)Lanes 1-5 and 1′-5′ were Cs150,Cc01,Cs641,Cs103,Cs030;M, λ-*Hind*Ⅲ.

2.4 Amplification and cloning of *nif*H gene

Due to the similarities in morphology, cell chemistry and ability to nodulate host plants, these *Coriaria* isolates were characterized previously on the basis of various biological characteristics. The finding of *nif*HDK homologous sequences for *Coriaria* isolates was also confirmed by amplification of *nif*H gene. As shown in fig.3, *nif*H gene of Cs030 has already been amplified and cloned into *E.coli* JM109 *via* pGEM-T vector system. The amplifications and their clones were tested further to hybridize with *nif*HDK fragment from *Rhizobium* sp. indicating that the amplification could be the *nif*H gene fragment of *Coriaria* isolate. However, this needs to be confirmed directly by sequencing the *nif*H amplification. This sequence data will provide a precise means to study the molecular phylogeny of *Coriaria*-infective *Frankia*.

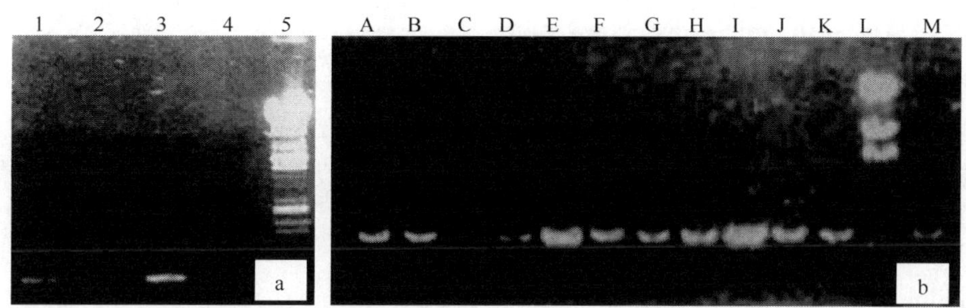

Fig.3 *nif*H amplified (a) from pure cultures and its clones (b).

(a) 1, Cs030; 2, blank; 3, ArI3; 4, negative control; 5, λ-*Eco*RI/*Hin*dIII; (b) lanes A-K and M are clones of *nif*H amplified (about 3.86kb) from Cs030; L, λ-*Hind* III.

Acknowledgement This work was supported by the International Foundation for Science (IFS, D2678-1), Hubei Provincial Science Foundation (Grant No.97J057) and partly supported by the Foundation of Wageningen Agricultural University.

REFERENCES

[1]Benson, D.R., Silvester, W.B., Biology of *Frankia* strains, actinomycete symbionts of actinorhizal plants, *Microbiol. Rev.*, 1993, 57:293.

[2]Cui, Y.H., Wang, Y.L., Qin, M.et al., Isolation and comparing of fragments of *nif* genes from *Frankia*, *Chinese Science Bulletin* (in Chinese), 1990, 35 (18) :1419.

[3]Hahn, D., Starrenbury, M.J.C., Akkermans, A.D.L., Oligonucleotide probes that hybridize with rRNA as a tool to study *Frankia* strains in root nodules, *Appl. Environ. Microbiol.*, 1990, 56:1342.

[4]Hu, C.J., Zhou.P.Z., Zhou, Q., Biological characterization of nodular isolates from *Coriaria nepalensis*, *Acta Microbiologica Sinica* (in Chinese), 1996, 36 (2) :132.

[5]Hu.C.J., Zhou, P.Z., Zhou, Q., DNA restriction pattern and *nif*HDK probe hybridization of *Frankia* isolates from *Coriaria nepalensis* in China, in *Abstracts for 7th International Symposium on Nitrogen Fixation With Nonlegumes*Faisalabad 1996 (eds. Malik, K.A., Mirza, M.), Sheikh Ghulam: Ali & Sons Ltd., 1996.189.

[6]Maniatis, T., Fritsch, E.F., Sambrook, J., *Molecular Cloning, a Laboratory Manual*, 1st ed., New York: Cold Spring Harbor, 1982, 538.

[7]Mirza, M.S., Characterization of uncultured *Frankia* strains by 16S rRNA sequence analysis, *The Ph D thesis*, Den Haag:Koninklijke Bibliotheek, 1993, 49.

[8]Murry, M.A., Fontaine, M.S., Torrey, J.G., Growth kinetics and nitrogenase induction in *Frankia* sp. HFP ArI3 grown in batch culture, *Plant Soil*, 1984, 78:61.

[9]Normand, P., Simonet, P., Bartin, R., Conservation of *nif* sequences in *Frankia*, *Mol.Gen. Genet.*, 1988. 213:238.

[10]Simonet, P., Normand. P., Moiroud, A. et al., Restriction enzyme digestion patterns of *Frankia* plasmids, *Plant Soil*, 1985, 87:49.

Isolation, Nodulation and Characterization of Nodular Endophytes of *Coriaria nepalensis*[*]

HU CHUANJIONG ZHOU PINGZHEN[2] ZHOU QI[1] CHEN HUAKUI[1]

(1.College of life Science and Technology, Huazhong Agricultural University, Wuhan 430070;

2.Oil Crops Institute, Chinese Academy of Agricultural Sciences)

ABSTRACT Using the micro-dissection method, 273 actinomycete isolates were obtained from root nodules of *Coriaria nepalensis*. Twenty four *Frankia*-like isolates were shown to nodulate the host plants through pot experiment. Phenotypic characterization of ten isolates demonstrated that they were of the typical morphological and cultural characteristics of the genus *Frankia*. These strains were divided into physiological group A, B and A/B where the latter was in intermediate group between the A and B strains. It was evident that group B but not A strains were able to nodulate the host. However, one strain of the group A/B unexpectedly nodulated the host. The utilization of carbon and nitrogen source for the strains was similar to other *Frankia*. Some strains were of type II call wall; others were of type II. Whole cell sugar patterns were diverse. Besides the common non-diagnostic sugars, most strains were shown to contain xylose and galactose, some contain xylose, galactose and arabinose and few contain galactose or xylose only. The G+C mol % of strains were ranging from 69 to 74.

KEY WORDS *Coriaria nepalensis* Endophyes Isolation Back-inoculation Phenotypic characterization

Members of the actinomycete genus *Frankia* are root symbionts nodulating more than 200 species of plants belonging to 24 genera and eight families that are called actinorhizal[1, 2]. After the first successful isolation of *Frankia* from *Comptonia peregrina* in 1978, hundreds of *Frankia* strains have been isolated from about ten actinorhizal genera. However, there is no report of successful isolation of both effective and infective *Frankia* strains from the plant genus *Coriaria*[2]. Zhou *et al.* (1986) initiated the study in China in early 1980's on symbiosis of *Frankia* and *Coriaria nepalensis*. Since then more than six hundred actinomycete isolates were obtained from the root nodules[1]. Wu Jie (1988) performed the first back-inoculation tests using totally 538 strains, three isolates were shown to nodulate the host plant grown in water jars (PhD. thesis). Mirza *et al.* (1992) reported a *Frankia* isolate from *Coriaria nepalensis* in Pakistan, but this pure culture neither formed vesicles nor nodulated the host [3]. Based on the previous exploration on nodulation techniques, we repeated the nodulation trials on modified conditions where host plant seedlings were grown on sterile soil, perlite and semi-solid agar. In this paper, we are reporting the promising results of isolation, back-nodulation and biological characterization of *Coriaria* nodular endophytes.

1 ISOLATION AND BACK-NODULATION OF NODULAR ENDOPHYTES

Repeated attempts of isolating nodular endophytes direct from the field nodules of *Coriaria nepalensis* have failed. But actinomycete symbionts were successfully isolated from the nodules that were induced in greenhouse by using field nodules as inocula. The fresh and compact nodules were sliced aseptically into the tiny thin slides and then transferred into the screwed test tubes containing liquid or solid isolation media. The microscopic observations were undertaken to check the morphology of isolates after 4-8 weeks. Totally 650

*原载于 *Agricultural Science in China*, 1:114~119, 2000.

actinomycete strains were obtained. Back-inoculation trials with 273 strains were carried out at greenhouse through pot experiment. Four months later 24 strains were proven able to nodulate the host plant on sterile soil（Table 1）. Among them strain Cs030，Cs103 and Cs150 were further confirmed by nodulating the seedlings grown on sterile perlite and semi-agar slants watered by 1/4-strength Hoagland nutrition solution.

Table 1 Results of nodulation trials on pot experiment

Strains	*Nodulation rate*	*Nodules per nodulated plant*	*Lobes per nodule*	*Strains*	*Nodulation rate*	*Nodules per nodulated plant*	*Lobes per nodule*
Cs013	14.2[1)]	1	3	Cs431	6.3	1	7
Cs030	5.9	1	5	Cs459	14.3	1	4
Cs44-2	5.9	1	5	Cs472	6.7	1	4
Cs103	5.0	1	6	Cs481	5.6	1	4
Cs108	3.1	1	4	Cs501	13.3	1	5
Cs136	4.8	1	1	Cs505	7.7	1	6
Cs150	5.3	2	4.5	Cs521	7.8	1	1
Cs157	4.8	1	3	Cs537	14.3	1	8
Cs213	11.1	3	3.3	Cs613	2.9	1	6
Cs383	14.3	1	7.0	Cs620	2.9	2	4.5
Cs396	20.0	1	2	Cs641	8.0	1	3
Cs414	5.9	1	1	CK	0[2)]	0	0
Cs421	5.3	1	4	Average	8.3	1.2	4.2

1）Average values from 2 replications；2） Average values from 20 replications as total controls.

2 MORPHOLOGICAL AND CULTURAL CHARACTERISTICS

Nodular isolates were microscopically observed after cultivation for more than four weeks on liquid media S [4]，Qmod[5]，Bap[6] and nitrogen-free Bap（Bap-N）[6]. Most strains were shown to have typical morphology of multi-lamella sporangia，some of them were also shown to have vesicles depending on type of medium used. It has been demonstrated that nitrogen-free medium Bap-N likely generated the highest frequency of formation of vesicles for most strains. Strain Cs103 was observed to form Reproductive Toulouse Hyphae （RTH） on both Bap and Bap-N liquid media.

All the *Coriaria* isolates and reference strains grew much better on liquid media than on solid media. On liquid media under static condition，hyphae were submerged in the lower layer of the broth or adhere to the bottom of container in the form of cotton or granule. The amount of biomass，coloration of hyphae and soluble pigments varied with strains，media components and cultural conditions. Hu et al. revealed that most strains grew better under shaking condition [7]. On solid media S，Bap and Qmod，usually it took more than 4 weeks to form small pale or colored colonies for most strains. The growth rate，colonial properties and production of soluble pigment varied with strains.

Based on colonial morphology 24 *Frankia*-like strains were evidently divided into four groups，namely，group IH with diffusive and hard colonies，group IIH with raised and hard colonies，IS with diffusive and soft colonies and IIS with raised and soft colonies. It was evident that all the soft-textured colonies produced soluble pigment whereas all the hard colonies did not.

3 PHYSIOLOGICAL GROUP

Ten *Coriaria* isolates were determined by the methods described by Lechevalier*et al.* （1983）[4]. The results were shown in Table 2. Four *Coriaria* isolates and two reference strains had demonstrated increase in the dry weight of biomass on Tween-80 containing media than that of Tween-80 free media, indicating the synergic effect on growth with increment of glucose. In contrast, the growth rate and glucose utilization of other five strains were inhibited in Tween-80 containing media, indicating that they did belong to physiological group A. However, two strains showed no difference in growth rate in both TW-80 containing and free media. The dry weights of biomass under the two media varied insignificantly. These two strains showed the intermediate properties between A and B strains which was defined A/B type （Fig.）.

Table 2 Physiological groups of pure cultures

Effective strains		*Ineffective strains*	
Strains	Physiological group	Strains	Physiological group
Cs150	B	Cs120	A
Cs641	B	Cs205	A
Cs521	B	Cs470	A
Cs103	B	Cs642	A
Cs146	A, B	Cs155	A, B
Reference strains			
Cc01	B	$Airl_2$	A

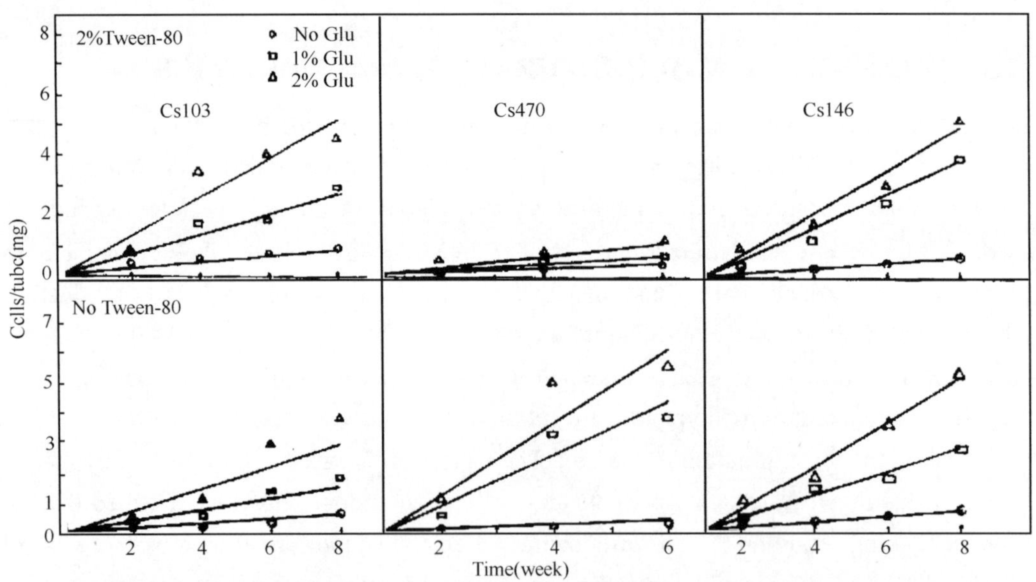

Fig. Influence of Tween-80 and glucose on growth of *Coriaria*Isotates Cs146, Cs470 and Cs103 in S medium

4 UTILIZATION PATTERN OF CARBON AND NITROGEN SOURCES

To investigate the utilization pattern of *Coriaria* isolates, four organic acids or their salts, eight sugars and five nitrogen sources were applied as sole carbon or nitrogen sources. All the strains were shown not use or use poorly of arabinose, galactose, xylose and mannitol. Most strains grew well on medium containing formic acid, sodium acetate, sodium pyruvate, and sodium propionate. There were only two strains utilizing sodium citrate. Obviously most effective strains did not utilize the monosaccharides or oligosaccharides whereas al the ineffective strains did use most sugars. For nitrogen utilization, all the strains grew well on medium containing NH_4Cl or casein, but did not utilize the KNO_3 as the sole nitrogen source. All the strains utilize the two amino acids tested.

5 CALL CHEMICAL COMPOSITION

Pure cell walls of *Coriaria* isolates and reference strains were prepared by the modified alkali method [8]. The cell wall preparations were hydrolyzed and analyzed by thin layer chromatography （TLC） and amino acid analyzer to determine the cell wall types. All the strains were detected to contain glutamic acid, glycine, alanine, and diaminopimelic acid （DAP） in the cell wall hydrolysates. Using amino acid analyzer the contents of the abovementioned amino acids were determined quantitatively and the molar ratios were calculated. Eight strains were shown to share the same range of characteristic value of molar ratios with known members of the genus *Micromonospora*, which is of cell wall type II. Two other strains shared the same range of characteristic molar ratios with cell wall type III strains like *Streptosporangium* sp. [9]. By TLC all the DAP constituents were determined as meso-DAP. Table 3 shows the results of diagnostic sugars of whole cell hydrolysates. Five strains are shown to have galactose plus xylose, two have galactose, xylose plus arabinose, two have galactose or xylose only.

Table 3 Cell chemical composition of pure cultures

Effective strains			Ineffective strains		
Strains	CWT1)	WCSP2)	Strains	CWT1)	WCSP2)
Cs150	II	g+x+a	Cs642	I	g
Cs641	II	g+x	Cs205	I	g+x
Cs521	II	g+x	Cs120	I	g+x
Cs103	III	g	Cs470	I	x
Cs146	II	g+x+a	Cs155	III	g+x

1）CWT: Cell wall type; 2）WCSP: Whole cell sugar pattern; g: Galactose; x: Xylose; a: Arabinose

6 DEOXYRIBONUCLEIC ACID BASE COMPOSITION

The percent contents of Guanine plus Cytosine （G+C mol %） in DNA of *Coriaria* isolates and *Frankia* reference strains were determined by using High Proficient Liquid Chromatography （HPLC）. The percent G+C contents for ten isolates were assayed ranging between 65% and 74%. The values for two reference strains were 69% and 75%, respectively. These figures are within the ranges of genus *Frankia* reported by other authors, namely 66%-72%[11], and 66%-76%[12].

7 DISCUSSION

Actinorhizal endophytes normally do not need very strict nutrient requirements *in vitro*. However, because of the very slow growth rate (normally the generation time is 2-5 d or longer), it is difficult to isolate *Frankia* strains directly from field nodules which are old and incompact harboring other fast growing microorganisms. It was postulated that *Frankia* strains could be killed by the toxic compounds produced by host cells of *Coriaria* during isolation [1, 13]. In this study, we only obtained isolates from induced nodules where the frequency of successful isolation was largely increased compared to that from field nodules. One explanation is that *Coriaria*endophytes were activated and concentrated in the newly induced nodules. Furthermore, the induced nodules were normally more compact and cleaner than the field ones. Thus complete surface sterilization was easier to perform during isolation.

Nodulation trials on pot experiment indicated that at least 24 strains were capable of infecting the host plant and forming nodules. Some stains were confirmed by further nodulation test on various modified conditions. These results showed the first evidence of successful isolation and back-nodulation of *Coriaria*-compatible *Frankia*. However, the nodulation rate and the averaged number of nodules formed per plant were just 8.3% and 1.2, respectively. The apt condition and techniques for the re-nodulation trials are yet to be optimized.

Coriaria isolates were shown to have structures of sporangia and vesicles that are typical of genus *Frankia*. Physiological group B strains were able to infect the host plant but group A stains could not. Most group B stains did not utilize the monosaccharides and oligosaccharides whereas all the A strains did. All these phenotypic features are very similar to other known *Frankia* strains. However, some strains displayed differences with other frankiae in the following aspects: （1） Besides the physiological group A and B, there was a presence of an intermediate group A/B.（2） Two cell wall types, namely II and III were grouped among the isolates. So far there is only one kind of cell wall type, namely III reported for the genus *Frankia*[2, 11]. （3） Whole cell sugar patterns of the isolates were proven very diverse, new patterns （diagnostic sugar compositions） like xylose plus galactose, xylose plus galactose plus arabinose, and xylose or galactose were found. All these results indicate the diversity of *Frankia* and will provide useful data for further typing of genus *Frankia*.

ACKNOWLEDGEMENT This study was supported by International Foundation for Science （D/2678） and National Natural Science Foundation of China （39870037）. Thanks are given to Professor Ding Jian and Ruan Jisheng for provision of *Frankia* reference strains, and to Professor Liang Rongfang, Yuan Dejun and Dr. Bai Lingquan for their assistance in cell wall chemical analysis.

REFERENCES

[1]An C S, *et al.* Deoxyribonucleic acid base composition of 12 *Frankia* isolates. Can.J.Bot.1983, 61:2859-2862

[2]Bai Linquan, et al. Identification of actinomycete cell wall types using amino acid molecular ratio. Microbiology （China）. 1997, 24（1）:9-14

[3]Benson D. Biology of *Frankia* strains, actinomycete symbionts of actinorhizal plants. Microbiol. Rev.1993. 57:293-319

[4]Fernandez M P, *et al.* Deoxyribonucleic acid relatedness among members of the genus *Frankia*. Int.J.Syst. Bacteriol. 1989, 30:424-429

[5]Hu Chuanjiong. *et al.* Biological characterization of nodular endophytes of *Coriaria nepalensis*. Acta Microbiologica Sinica. 1996, 32 （2） :132-137

[6]Lalonde M.Confirmation of the activity of a free living actinomycete isolated from *Comptoniaperegrina* root nodules by immunological and ultrastructural studies. Can. J. Bot. 1978, 56:2621-2635

[7]Lechevalier M P, *et al.* Genus Frankia Brunchorst 1886, 174AL. In: Bergey's manual of Systematic Bacteriology, Vol. 4. Baltimore: The Williams Wilkins Co. 1989:2410-2417

[8]Lechevalier M P, *et al.* Physiology, chemistry, serology and infectivity of two *Frankia* isolates from *Alnus incana* subsp. *rugosa*. Can. J. Bot. 1983, 61:2826-2833

[9]Liang Rongfang, *et al.* A rapid method for cell wall chemical analysis of actinomycetes. Microbiology. 1990, 17（4）:247-249

[10]Mirza M S, *et al.* Isolation and characterization of *Frankia* strains from *Coriaria nepalensis*. Syst. Appl. Microbiol.1992, 5:289-295

[11]Murry M A, *et al.* Growth kinetics and nitrogenase induction in *Frankia* sp. HFP ArI3 grown in batch culture. Plant Soil. 1984 , 78:61-78

[12]Wu Jie, *et al.* Infection and development of symbiotic nitrogen-fixing actinorhizae of *Coriaria nepalensis*. J.Whuan Bot. Res, 1991, 9(1):1-4

[13]Zhou Pingzhen, *et al.* Studies on nitrogen-fixing actinorhizae and the actinomycete symbionts of *Coriaria nepalensis*. Acta Microbiologica Sinica. 1986, 26（3）:277-281

第三项　固氮螺菌

固氮螺菌耐高铵突变株的选育*

罗孝扬　蒋亚平　蔡金芝　陈华癸

（中国科学院武汉病毒研究所）

严家骐

（武汉大学病毒系）

应用亚硝基胍（N-nitrosoguanidine，NTG）诱变剂对固氮螺菌菌株Ma241、Ma99、Sp7和G14进行诱变处理后，在添加了铵的类似物乙撑二胺（ethylene diamine）的Dobereiner无氮培养基中进行筛选，反复纯化，获得了在45mol/L NH_4^+ 浓度以上，保持固氮酶活性的耐铵突变株共9株。突变株22的耐锭固氮酶活性最强，在75m mol/L NH_4^+ 浓度下，固氮酶活性达到464nmol乙烯/mg蛋白小时，在200mol/L NH_4^+ 浓度下，固氮酶活性仍有32nmol/（mg蛋白·小时）。

关键词　固氮螺菌；耐铵突变菌株

当环境中有化合态氮（NH_4^+ 或 HO_3^-）存在时，固氮细菌和蓝绿藻就会失去固氮酶活性[1-3]，只有在化合态氮被耗尽后，它们才表现固氮酶活性[4-6]。因此，研究固氮酶产生的调节机制并寻找在有 NH_4^+ 存在下固氮酶活性不受抑制的耐铵菌株，便成为人们重视的课题。

1974年，Brill[7]发现蛋氨酸亚砜亚胺（谷酰胺合成酶的一种抑制剂）和 NH_4^+（28m M）同时存在的条件下，*Azotobactervinelandii*固氮酶活性不完全受抑制。此后，探讨在较高 NH_4^+ 浓度下产生固氮酶活性又有报道[8, 9]。Kennedy等人[8]和Brill[9]总结了克氏肺炎杆菌（*Klebsilla pneumoniae*）固氮基因的调节机制，结果表明：nifLA操纵子的产物调节所有其他*nif*基因的表达；*nifA*基因的表达仅受 NH_4^+ 的阻抑。朱家璧等将编码半乳糖苷酶基因（*LacZ*）组装在*K.pneumoniac*的nifLA的启动子上。以LacZ表达的快速变色反应作为检测手段，筛选nifLA不受 NH_4^+ 抑制的突变株，获得了在15mmol/L NH_4^+ 浓度下仍有固氮酶活性的突变株。

Maiicyprh等[9]用NTG对固氮螺菌Sp7进行诱变处理，用乙撑二胺筛选出一株在15mmol/L NH_4^+ 浓度下保持固氮酶活性的突变株（No.42）。

作者参照Maiicyprh等人的方法，加以修改，获得了*Azozpirillumbrasilense*的一些耐铵突变株，它们在45~200mmol/L NH_4^+ 浓度下具有强弱不等的固氮酶活性。

材料和方法

（一）供试菌株

采用菌株Sp7（ATCC29145）、Ma241和Ma99及来源于Ma99的突变株G14为出发菌株。（菌株Sp7来自Dobereiner，其余均为本实验室分离或诱变所得。均属*Azospirillumbrasilensc*[10]）。

*原载于《微生物学报》，26（1）：47~52，1986.

（二）培养基成分和培养条件

采用参考文献[11]所报道的含不同 NH_4^+ 浓度的半固体培养基的配制方法：准确称取一定量的 NH_4Ac，加入半固体Dobereiner无氮培养基中（采用高压蒸汽灭菌，自然放气，经试验证明：NH_4^+ 浓度在灭菌前后保持不变）。灭菌后，分装到25ml的血清瓶中。

（三）亚硝基胍（NTG）处理和突变株的筛选

参照Maiicyprh等人的方法，加以修改。将上述供试菌株，分别接种Dobereiner液体培养基，32℃振荡培养（摇床转速：4000r/min）48小时，离心（3000r/min）30分钟后，收集沉积菌体。用柠檬酸缓冲液（pH5.5）稀释至菌数 1×10^{12} 细胞/ml。加NTG诱变剂（浓度为50~100μg/ml，死亡率在99%以上）。随即用同样的缓冲液稀释，终止NTG的作用。离心（3000r/min）30分钟，收集沉积菌体，并将该菌体悬浮于0.05%乙撑二胺的Dobereiner无氮液体培养基中。再将全部悬浮液接入该固体培养基平皿中，32℃培养5~6天。挑出单个菌落。

（四）菌体蛋白测定

采用Lowry等人[12]所报道的Folin-phenol试剂法。用牛血清蛋白作标准曲线，72型分光光度计比色（波长660nm）。

（五）固氮酶活性测定

活细胞的固氮酶活性测定采用乙炔还原法。用容积为25ml的血清瓶，内装6ml不含 NH_4^+ 或含不同浓度 NH_4^+ 的Dobereiner半固体培养基，接入对数生长期的菌液0.6ml，菌数不低于 5×10^5 细胞/ml。置32℃培养10~82小时（视细菌生长情况而定）。注入乙炔，转化8小时，用102G型气相色谱仪测定固氮酶活性。

结　　果

（一）耐铵突变株的来源及其固氮酶活性

经反复选育后，共获得突变株141株。其中耐7-45mmol/L NH_4^+ 的有55株，耐高于45mmol/L NH_4^+ 的共有9株，它们分别为四个出发菌株的后代（表1）。在无 NH_4^+ 条件下，出发菌株与突变株都有较高的固氮酶活性。突变株22、47、48、136-1和24的固氮酶活性低于出发菌株，其余突变株的酶活性均高于出发菌体。在15mmol/L NH_4^+ 存在下，出发菌株无酶活性，而突变株22、23、19-1和24在 NH_4^+ 浓度达到75mmol/L时，仍有较高的酶活性（表2）。

表 1　在无条件 NH_4^+ 下耐铵突变株及其出发菌株的固氮酶活性比较（单位：nmol 乙烯/mg 蛋白·小时）

Table 1 Comparison of nitrogenase activities of mutant strains in the presence of NH_4^+ and their correspouding starting sirains in the absence of NH_4^+ （unit:nmol ethylent/mg protein/hr）

出发菌体 Starting strain		由出发菌株诱变后所获的耐 NH_4^+ 突变株 Mutant strains fixing nitrogen in the presence of ammonium ions derived from starting strains after mutagenesis	在无 NH_4^+ 条件下出发菌株的固氮酶活性（a） Nitrofenase activity of starting strains in the absence of NH_4^+ （a）	在无 NH_4^+ 条件下耐 NH_4^+ 突变株的固氮酶活性（b） Nitrogenose activity of mutant strains in the absence of NH_4^+ （b）	h/a
菌号 Strain	来源地点及其联合植物 Source and associated plants				
Sp7	巴西俯仰马唐 Brazil Digiteria decumbens	22	669.95	513.70	0.77
		47		231.04	0.32
		48		179.83	0.27
		136-1		401.55	0.60
Ma241	广西玉米研究所太189×柳干-4-2-1 GuangXi Institute of maize Tail89×Liu7-4-2-1	19-1	330.81	434.94	1.31
		23		525.36	1.58
Ma99	湖北省农科院川农7号 Hubei Academy of Agricultural Sciences Chuannong No.7	14	417.60	623.79	1.32
G14	由Ma99诱变的突变株（本实验室） Mutant strain derived from Ma 99 by mutagenesis	24	649.67	585.11	0.90
		79		845.21	1.30

注：上表数字系 4 个重复的平均值。

The figures listed in this table are the mean values of quadraplicalapecimens.

表 2　在不同 NH_4^+ 浓度下出发菌株与耐铵突变株固氮酶活性比较（单位：nmol 乙烯/mg 蛋白·小时）

Table 2　Camparison of nitrogebase activities produced by starting strains and mutant strains in the presence of different concentrations of NH_4^+ （units：nmol ethylene/mg protein/hr）

菌株 Strain	NH_4^+ /（mmol/L）			
	0	15	45	75
Sp7	669.75	0*	0*	0*
22	513.70	297.18	223.83	464.30
47	35.24	4.92	2.35	0.56
48	179.83	147.40	7.58	0*
136-1	401.55	151.55	24.26	0*
Ma241	294.87	0*	0*	0*
19-1	439.94	273.57	243.69	80.49
23	525.36	52.18	335.32	197.95
Ma99	417.10	0*	0*	0*
14	623.79	144.23	12.18	24.5
G14	613.82	0*	0*	0*
24	585.01	57.86	80.51	72.75
79	845.21	39.54	2.02	1.91

注：表内数字系三个重复的平均值。

*表示生长良好，但无酶活性。

The figures listed in this table are the mean values of triplicate specimens.

*Denotes no detectable nitrogenase activity，but those strains grow well.

表 3 在 5mmol/L NH_4^+ 存在下出发菌株的固氮酶活性和残余 NH_4^+ 的含量（nmol 乙烯/mg 蛋白·小时）

Table 3 The nitrogenase activities of strains starting in the presence of 5mmol/L NH_4^+ and the residual NH_4^+ in growth medium after different period of cultivation （nmol ethylene/mg protein/hr）

出发菌体 Starting strain	培养10小时 After cultivation for 10hrs		培养16小时 After cultivation for 16hrs	
	固氮酶活性* Nitrogenase activity	残余 NH_4^+ ** Residual NH_4^+	固氮酶活性* Nitrogenase activity	残余 NH_4^+ ** Residual NH_4^+
Sp7	0	+	748.21	–
Ma99	0	+	800.17	–
Ma241	0	+	184.16	–
G14	0	+	240.96	–

*表中数字系 4 个重复的平均值。

**用 Nesslar 试剂检测半固体培养基中的残余 $NH^+{}_4$（肉眼观察）。

*The figures listed in this table are the mean values from quadriplicafe specimens.

** Residual $NH^+{}_4$ in the Semi-solid growth medium as detacted by Nessler's reagent with naked eyes.

（二）不同 NH_4^+ 浓度对出发菌株及其突变株的固氮酶活性的影响

1、出发菌株在5mmol/L NH_4^+ 的半固体培养基中培养10小时，固氮酶活性为零，这时培养基中有 NH_4^+ 残留（表3）；培养16小时，固氮酶活性表现出来，这时培养基中的 NH_4^+ 已被耗尽（用Nesslar试剂检测，肉眼观察）。

2、在本实验条件下，在半固体培养基中，当 NH_4^+ 浓度在50mmol/L以上时，各突变株的生长速度随 NH_4^+ 浓度的提高而减慢。因此，对不同 NH_4^+ 浓度，耐铵突变株酶活的测定采用不同的培养时间和相同的转化时间。即0-50mmol/L NH_4^+ 培养10小时，转化8小时；75~125mmol/L NH_4^+ 培养27小时，转化8小时；150~175mmol/L NH_4^+ 培养38小时，转化8小时；200mmol/L NH_4^+ 培养72-82小时，转化8小时。当 NH_4^+ 浓度达到225mmol/L时，突变株22、23、24、19-1生长良好，但无固氮酶活性。

3、耐铵突变株的固氮酶活性随 NH_4^+ 的增加而逐渐下降（图1）。突变株22的表现最为突出，在 NH_4^+ 浓度达到100-125mmol/L时，其酶活性仍在250nmol乙烯/mg蛋白·小时以上，并且在75mmol/L NH_4^+ 浓度下表现的酶活性几乎与无 NH_4^+ 条件下的酶活性相接近。根据各突变株在高 NH_4^+ 浓度下的固氮酶活性大小，可将它们排成以下顺序：22>23>19-1>24。

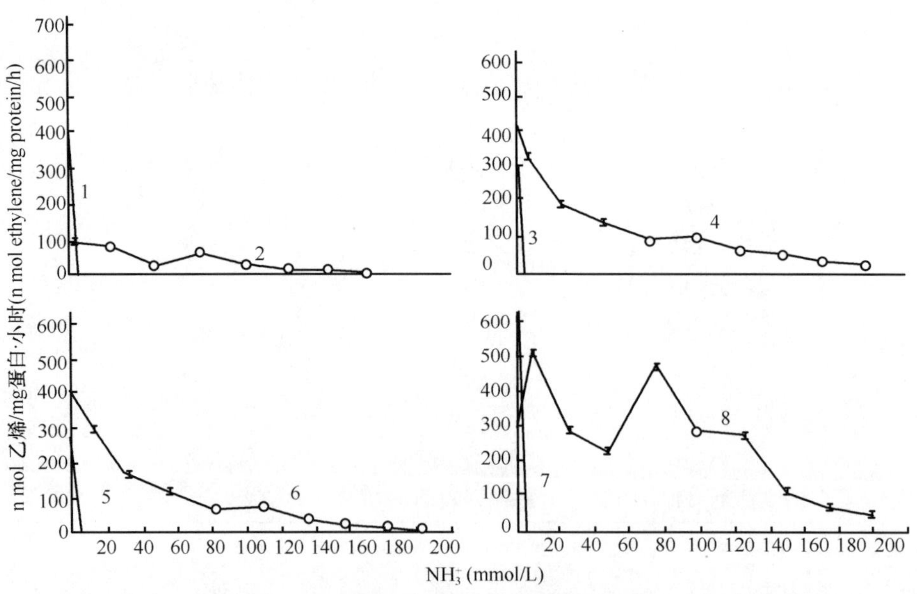

图 1 不同 NH_4^+ 浓度对出发菌株和突变菌株固氮酶活性的影响

Fig.1 The effect of different concentrations of NH_4^+ on the nitrogenase activities of starting strains and their mutant strains. 1.出发株（starting strain）G14 2.突变株（mutant strain）24 3.出发株（starting strain）Ma241 4.突变株（mutant strain）19-1 5.出发株（starting strain）Ma241 6.突变株（mutant strain）23 7.出发株（starting strain）Sp7 8.突变株（mutant strain）22

（三）不同突变株多次传代后固氮酶活性的持续性

试验结果表明（图2），突变株22在75mmol/L NH_4^+ 浓度下，经过30次转接，固氮酶活性仍然保持较高水平。此菌株在同样 NH_4^+ 浓度下，培养27小时，再经不同时间转化后测定固氮酶活性，结果（图3）表明，转化持续150小时，其固氮酶活性仍然有12nmol乙烯/mg蛋白·小时。

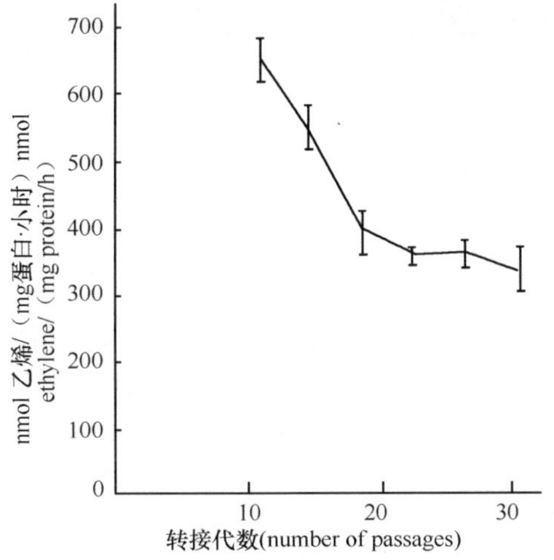

图2 22号耐铵突变株的不同转接代数在75mmol/L NH_4^+ 下的酶活性

Fig.2 The nitrogenase activities of different number of passages of mutant strain No.22 in the presence of 75mmol/L NH_4^+

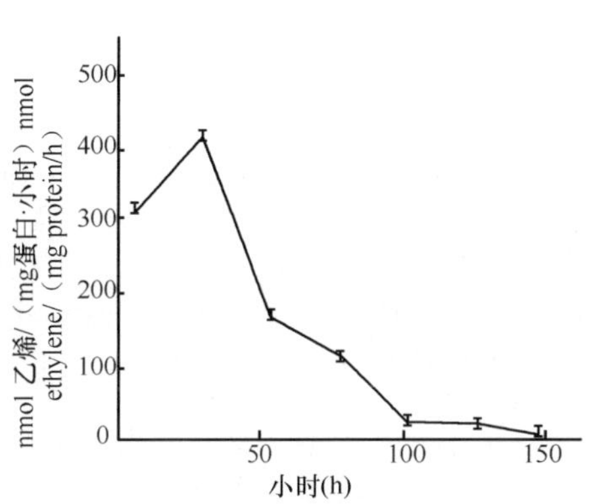

图3 22号耐铵突变株经不同时间转化的酶活性变化培养液中 NH_4^+ 浓度为75mmol/L

Fig.3 Nitrogenase activities of mutant strain No.22 at different transformation time

The concentration of NH_4^+ in growth medium was 75mmol/L

以苹果酸钠为唯一碳源时，突变株24的培养物pH由7.0上升到9.0，虽然培养物中的菌体很浓，菌膜也厚，但测不出固氮酶活性。在半固体培养基中添加麦芽糖（dl-苹果酸钠：麦芽糖=3:2），突变株24在培养过程中pH就保持在7.5左右，可以测出固氮酶活性。

讨 论

Maiicyprh等人报道，Sp7 在5mmol/L NH_4^+ 条件下培养48小时，转化18小时，固氮酶活性已下降至零。这与我们对Sp7菌株的试验结果不一致。我们的试验结果（表3）表明：培养10小时，培养基中还残留有 NH_4^+，测不出固氮酶活性。这表明残留的 NH_4^+ 量对固氮酶活性仍然起着抑制作用。而培养16小时，培养基中已无残留 NH_4^+，固氮酶活性就表现出来了。此外，在谷酰胺对固氮酶活性的影响的试验中，Maiicyprh等人的试验表明：Sp7能耐35mmol/L的谷酰胺，而我们的实验菌株Sp7则不耐谷酰胺。Maiicyprh等人和我们虽然都采用Dobereiner Sp7菌株，但所得的结果不同，这可能是由于在不同条件下长期传代，菌株的属性已经不完全相同的缘故。

采用乙撑二胺作筛选剂，其特点在于它和 NH_4^+ 不同，不能被固氮螺菌利用作为氮素养料。但它又与 NH_4^+ 相同，具有阻遏固氮调节基因的功能，因而阻遏了固氮酶活性的表达。在含有乙撑二胺的条件下，不耐 NH_4^+ 的菌株无固氮酶活性，它由于缺乏氮源而不生长，而耐 NH_4^+ 的突变株不受乙撑二胺的阻遏，仍具有固氮酶活性，能利用空气中的氮而正常生长。

我们筛选所得的固氮螺菌的突变株22，耐 NH_4^+ 能力比Maiicyprh等人所得突变株No.42高出许多倍。

突变株22，在15mmol/L NH_4^+浓度下的固氮酶活性为297nmol乙烯/ mg蛋白·小时；在45mmol/L NH_4^+浓度下为223nmol乙烯/ mg蛋白·小时；在75mmol/L NH_4^+浓度下出现一高峰，达464nmol乙烯/mg蛋白·小时；在220mmol/L NH_4^+浓度下仍达32nmol乙烯/mg蛋白·小时。该菌株在75 mmol/L NH_4^+浓度下出现固氮酶活性高峰的特性，有待深入研究。

参 考 文 献

[1] Fred，E.B.et al.: University of Wisconsion Studies in Science，Madison Sisc.No.5，1932.

[2]Wilson，P.W.:Proc. Nat. Acad.Sci.USA.29:289-294，1943.

[3]Zelitch，I.:ibid，37:559-565，1951.

[4]Goldbery，R.B.et al.: J.Bacteriol.，118:810-814，1974.

[5]Streicher，S.L.et al.:ibid，120:815-821，1974.

[6]Gordon，J.K.and W.J.Brill: Biochem. Biophys. Res. Commun，59:967-971，1974.

[7]Brill，W.J.et al.:J.Becteriol，145:348-357，1981.

[8]Kennedy，C.et al.:In “Current Perspective in Nitrogen Fixation Proceedings of the Fouth International Symposium on Nitrogen Fixation”，ed. Gilbson and Newton，Australian Academy of Science，PP.146-156，1981.

[9]Maiicyprh，A.H.:Renetu，4:575-597，1982.

[10]罗孝扬等：微生物学报，23（1）：68-72，1983。

[11]湖北省微生物研究所生物固氮组：微生物学报，19（2）：160-165，1979。

[12]Lowry，O.G.et al.: J.Biol. Chem，193:265-275，1951.

固氮螺菌CWV-22突变株与玉米、小麦联合体的固氮作用*

罗孝扬　蒋亚平　蔡金芝　杨宝玉

（中国科学院武汉病毒研究所）

严家骐

（武汉大学）

陈华癸

（华中农学院）

应用乙炔还原法和同位素^{15}N示踪技术证明固氮螺菌CWV-22是具有耐安铵、泌铵能力的突变株。它在含有50mM NH^{+}_4浓度的纯培养固氮试验中，吸收^{15}N为135μg/mg蛋白，不接种的对照吸收^{15}N量为零，接种Sp7的只有1μg/mg蛋白，在测试允许误差之内，证明Sp7的固氮作用是不耐铵的。

在密闭培育装置中，用^{15}N示踪植物试验：①种植玉米或小麦，接种不耐铵的Sp7或耐铵的CWV-22菌株，不论在有或无NH_4 Ac的条件下，植物吸收的生物固定^{15}N量均远远高于不接种细菌的对照植物，达4.5~9.0倍；②接种耐铵CWV-22菌株不受NH_4Ac存在的抑制，接种不耐铵Sp7菌株则受到显著的抑制，但仍有一定的固氮作用，并将固定的^{15}N输送到植物根、茎、叶里。经检测，在有NH^{+}_4Ac条件下，根际砂中含NH^{+}_4量降低到1~2mM（远离根际的砂中含NH^{+}_4为29.1mM），低于对Sp7菌株的抑制浓度，可能是Sp7仍有固氮活性。

关键词　固氮螺菌；玉米；小麦

固氮螺菌CWV-22突变株是由固氮螺菌Sp7经诱变、筛选获得的耐高铵固氮螺菌突变株[1]。它在有固定态氮的环境中能否泌铵，在和玉米、小麦形成联合体时，植物能否从它在耐铵条件下固定的氮得到氮素营养，本文报道了这方面的试验结果。Stephen和Brill等[2]用^{15}N示踪试验指出，从玉米根际分离的固氮菌能固定$^{15}N_2$气体被转化的^{15}N量只能滞留在玉米植株的根系上。我们采用同位素^{15}N示踪法考查了CWV-22与接种的植物玉米和小麦——细菌联合体的固氮作用和植物吸收、运输作用。

材料和方法

（一）供试菌株和植物

1. 菌株：（1）固氮螺菌Sp7菌株（ATCC29145），不耐铵，在>3mM NH^{+}_4条件下无固氮能力；（2）中国科学院武汉病毒研究所固氮螺菌耐铵泌铵突变株22（简称CWV-22）由本实验室从Sp7菌株诱变筛选获得[1]。

2. 植物：玉米品种为予农704，小麦品种为苏麦2号。种子表面灭菌处理，按常规方法进行。

（二）$^{15}N_2$混合气体的制备

$^{15}N_2$气体的制备参照尤崇杓等的制$^{15}N_2$方法[3]，在制备$^{15}N_2$气体前，将装置的各个磨口部位用真空脂

*原载于《微生物学报》，27（3）：277~283，1987.

或真空泥密封。装置的各个部分抽至760mmHg柱。所用$^{15}(NH_4)_2SO_4$的^{15}N原子丰度为11.59%。制备的$^{15}N_2$气体，再按比例加入O_2和CO_2，制成$^{15}N_2$混合气体，体积比为$^{15}N_2$ 80%，O_2 19.8%，CO_2 0.2%，用排水集气法搜集于贮气瓶中，备用。

（三）纯培养物 ^{15}N 示踪试验

1. 在Dobereiner半固体培养基[3]中，分别进行如下处理：

（1）无NH_4Ac+Sp7菌株；

（2）无NH_4Ac+CWV-22菌株；

（3）50mM NH_4Ac+CWV-22菌株；

（4）50mM NH_4Ac+Sp7菌体；

（5）对照：无NH_4Ac，不接种；

（6）对照：50mM NH_4Ac，不接种。

每个处理重复三次。

2. 纯培养物^{15}N示踪固氮试验方法：

在容积为130-135ml的刻度血清瓶中，装入不同处理的45ml Dobereiner半固体培养基，接种对数生长期的菌液5ml，培养18~20小时，将棉塞换成翻口橡皮塞，抽真空，导入贮气瓶中的^{15}N混合气体，继续培养72~96小时后，加入2ml浓H_2SO_4待蒸馏。

（四）密闭培养装置和 ^{15}N 示踪植物试验

1. 密闭培育装置构造，见图1。

图 1 密闭培育装置

Fig.1 The bermitic cultivation-equipment

2. 密闭培育装置中的植物-细菌联合体固氮试验处理：

砂培试验营养液成分为（单位g/L）：K_2HPO_4 0.5，$MgSO_4$ 0.5，NaCl 0.6，$Ca_3(PO_4)_2$ 0.26，$MnSO_4$ 0.01，$NaMoO_4$ 0.01，$FeCl_3$ 0.01，H_3BO_3 0.01，苹果酸钠1.0，pH7.0，H_2O 1000ml。

在上述营养液中分别进行五种不同处理：

（1）无NH_4Ac+$^{15}N_2$混合气体+植物+Sp7；

（2）无NH_4Ac+$^{15}N_2$混合气体+植物+CWV-22菌株；

（3）NH_4Ac+$^{15}N_2$混合气体+植物+Sp7；

（4）NH_4Ac+$^{15}N_2$混合气体+植物+CWV-22；

（5）无NH_4Ac+$^{15}N_2$混合气体+植物，不接种。

营养液NH_4Ac处理浓度，玉米试验为40mM，小麦试验为30mM。

密闭培育装置试验中的器皿、砂、营养液均经灭菌和除氮处理。每个装置装入砂300g，营养液62~65ml。种子经表面灭菌后，催芽，露白后播种，每个装置中放玉米种4粒，小麦种8~10粒，细菌接种两次：一次在播种前将露白种子浸于菌液中；一次在幼苗长出了3~4片真叶时，加入菌液5ml。追施菌液后24小时，将密闭培育装置抽真空处理（用高频火花测验器测真空度，达到实验要求）。在温室中培养，自然光照，温室温度为26~28℃，用黑布包裹装置的装砂部位，并在夹层中通自来水。

5月27日播种，6月6日进行第二次接种菌液，7日导入$^{15}N_2$混合气体，保持1个大气压。再经4~5天后收获。取出植株，洗净，烘干（80℃），称重，保存待测试。

（五）用乙炔还原法测定培养物和在密闭培育装置中联合体固氮活性的原位测定

1. 纯培养物的固氮酶活性测定参考文献[3]。处理见表1，每个处理5次重复。

2. 密闭培育装置中，固氮活性的原位测定培育方法同（四），处理见表2。

当玉米植株高达13~15cm时，抽出装置中部分气体，同时注入等体积的乙炔，经24小时后，取气体样，测乙炔生成量。测定仪器为102G型气相色谱仪。

（六）植物中 ^{15}N 原子%的测定

将浓缩到2-3ml的液体样品，送中国科学院南京土壤研究所质谱组（ZHT-1301质谱计，参比标准0.003^{15}N%）进行^{15}N原子%的测定。

（七）总氮量测定

采用克氏法[7]。

（八）菌体蛋白的测定

采用Lowry等人[4]的Folin-phenol试剂法测定。

表1　培养物固氮酶活性比较*

Table 1　Comparison of the nitrogenase activities of *Azospirillum* Sp7 and its mutant CWV-22 in the presence or absence of ammonium ions

菌号 Strain	处理 Treatment	固氮酶活性（nmol C_2H_4/mg 蛋白·小时） Nitrogenase activity（nmol C_2H_4/mg protein/h）
Sp7	—NH_4Ac	669.75±0.19
CWV-22	—NH_4Ac	503.77±0.89
Sp7	50mM NH_4Ac	0
CWV-22	50mM NH_4Ac	204.53±0.26

* 新发生的 C_2H_2 测不出 C_2H_4 含量。

Freshly prepared acetylene，no ethylene was detected.

表2　密闭培育装置中玉米植株分别同 Sp7 菌株及 CWV-22 菌株联合体固氮酶活性原位测定

Table 2　Determination in situ of the nitrogenase activities of combinants of maize plant with *Azospirillum* Sp7 and its mutant CWV-22 respectively in hermetic cultivation apparatuses

菌号 Strain	处理 Treatment	固氮酶活性（nmol C_2H_4/mg 蛋白·小时） Nitrogenase activity（nmol C_2H_4/mg protein/h）
Sp7	玉米+35mM NH_4Ac maize+35mM NH_4Ac	0
CWV-22	玉米+35mM NH_4Ac maize+35mM NH_4Ac	331.68
不接种对照 Control	玉米+35mM NH_4Ac maize+35mM NH_4Ac	0

结　　果

（一）菌株 Sp7 和 CWV-22 的固氮酶活性比较

1. 培养物的固氨酶活性比较

试验结果（表1）表明，在无NH^+_4存在时，CWV-22菌株的固氨酶活性是Sp7菌株的75%；有50mM NH^+_4存在时，由于CWV-22菌株固氨酶活性不受抑制，其固氮酶活性为204.53nmol乙烯/mg蛋白·小时；而Sp7菌株的固氨酶活性被NH^+_4抑制，数值为零。

2. 密闭培育装置中联合体的乙炔还原活性的原位测定

结果见表2。在含35mM NH^+_4的营养液中，Sp7菌株无乙炔还原活性，而CWV-22菌株的乙炔还原活性达到331.68nmolC_2H_4/mg蛋白·小时。以上培养试验和玉米、细菌联合体试验，结果均表明：CWV-22菌株的固氮作用是耐铵的，Sp7菌株无耐铵性。

（二）有或无 NH^+_4 对纯培养物固氮作用的 ^{15}N 示踪测定

试验结果（表3）表明，当培养基中无NH^+_4存在时，Sp7菌株吸收固定^{15}N量为CWV-22菌株的1.5倍；50mM NH^+_4存在时，Sp7菌株吸收固定^{15}N量只有1μg/mg蛋白，处于质谱分析的允许误差范围内，而CWV-22菌株则为135μg/mg蛋白。^{15}N示踪证明，CWV-22菌株在有或无NH^+_4存在时，均有固氮作用，而Sp7菌株只在无NH^+_4或微量NH^+_4存在时才有固氮作用。

表 3 有或无 NH^+_4 对纯培养物吸收固定 ^{15}N 的影响

Table 3 The effect of the presence or absence of ammonium ions on the absorption and fixation of ^{15}N by pure cultures

菌号 Strain	菌体细胞 mg 蛋白 Protein of bacterial cells	总 N 量（mg） Total amount of nitrogen	培养物过量 ^{15}N 原子% ^{15}N atom% excess of pure culture	吸收固定 ^{15}N 量（μg） Amount of absorption and fixation of ^{15}N
Sp7	30.0	8.784	1.94	55
CWV-22	26.2	8.296	1.18	36
Sp7	32.5	68.530	0.05	1
CWV-22	25.0	70.488	0.49	135
不接种对照 Control	0	0	0	0
不接种对照 Control	0	70.000	0	0

注：1.贮气瓶中 $^{15}N_2$ 丰度%为 10.64。2.上表数字系三个重复的平均值。

1.The percentage of $^{15}N_2$ enrichment in gas reservoir was 10.64. 2.The figures listed in this table were the mean values of triplicate specimens.

表 4 有或无 NH^+_4 对玉米植株分别同 SP7 菌株及 CWV-22 菌株联合体的固 ^{15}N 的影响

Table 4 The effect of the presence or absence of ammonium ions on the ^{15}N fixation by the combinants of maize plants with Sp7 and CWV-22 respectively

菌号 Strain	烘干重（g） Dry weight		总 N 量（mg） Total amount of nitrogen		样品中过量 ^{15}N 原子%** ^{15}N atom% excess of samples		吸收生物固 ^{15}N 量（μg） Amount of absorption and fixation of ^{15}N		
	根 Root	茎、叶 Stem，leaf	根 Root	茎、叶 Stem，leaf	根 Root	茎、叶 Stem，leaf	根 Root	茎、叶 Stem，leaf	根、茎叶 Root+Stem，leaf
Sp7	0.1809	0.3112	21.1592	38.4062	0.007	0.008	15	30	45
CWV-22*	0.1276	0.1432	2.3041	60.6172	0.016	0.009	-	53	>53
Sp7	0.1391	0.3808	48.3321	57.6576	0.009	0.005	42	28	70
CWV-22	0.1731	0.3593	53.2236	42.2736	0.049	0.022	254	91	345
不接种对照 Control	0.0993	0.2042	9.8691	11.0382	0.003	0.002	6.2	4.6	10.8

* 生长发育受阻碍，其原因不明。

** 贮气瓶中 $^{15}N_2$ 丰度%与表 3 相同。

* The growth of roots has been retarded for unknown reason.

** The percentage of $^{15}N_2$ enrichment in gas reservoir was 10.64.

（三）有或无 NH^+_4 在密闭培育装置中联合体固氮作用的 ^{15}N 示踪测定

表4与表5的结果表明，当营养液中无或有NH^+_4存在时，与不接种对照相比，联合体植株的过量^{15}N原子%均提高，表明均有一定的固氮作用，并能将固定的^{15}N运送到茎叶部位。在30mM 或40nM NH^+_4存在条件下，接种CWV-22菌株的小麦或玉米植株联合体比较种Sp7菌株的联合体植株吸收生物固定^{15}N量大5~8倍；在无NH^+_4存在时，前者为后者的1~2倍，但接种CWV-22菌株的玉米，由于生长发育不够正常，无法加以讨论。从装置中取出的小麦干重和总氮量。在无NH^+_4时，接种CWV-22菌株的比较种Sp7菌株的差别小，和不接种的对照相比，干重差别也不大，但总氮量都比对照高4~5倍。在30mM NH^+_4存在时，接种CWV-22菌株与接种Sp7菌株的植株干重和总氮量差别也不大，但植株吸收生物固氮量，CWV-22菌株比Sp7菌株大8倍。

表 5　有或无 NH^+_4 对小麦植株分别同 SP7 及 CWV-22 联合体的固 ^{15}N 影响

Table 5　The effect of the presence or absence of ammonium ions on the ^{15}N fixation by the combinants of wheat plants with Sp7 and CWV-22 respectively

菌号 Strain	烘干重（g） Dry weight		总 N 量（mg） Total amount of nitrogen		样品中过量 ^{15}N 原子%* ^{15}N atom% excess of samples		吸收生物固 ^{15}N 量（μg） Amount of absorption and fixation of ^{15}N		
	根 Root	茎、叶 Stem，leaf	根 Root	茎、叶 Stem，leaf	根 Root	茎、叶 Stem，leaf	根 Root	茎、叶 Stem，leaf	根、茎叶 Root+Stem，leaf
Sp7	0.0390	0.1330	25.8462	28.5188	0.046	0.050	116	138	254
CWV-22	0.0484	0.1071	19.8350	32.6080	0.197	0.037	380	111	491
Sp7	0.0680	0.1388	51.9941	59.0058	0.01	0.006	51	36	87
CWV-22	0.1056	0.2068	36.9318	87.3114	1.29	0.026	464	221	685
不接种对照 Control	0.0613	0.0993	12.6656	8.7028	0.004	0.003	4.93	2.55	7.48

* 贮气瓶中 $^{15}N_2$ 丰度%与表 3 相同。

* The percentage of $^{15}N_2$ enrichment in gas reservoir was 10.64.

结论与讨论

1. CWV-22是一株耐高铵、泌铵菌株，它所分泌的铵能被植株吸收利用，并能运送到茎和叶中。当试验环境中存在50mM NH^+_4时（本实验室试验结果表明，在200mM NH^+_4条件下，CWV-22菌株仍有固氮酶活性[1]），CWV-22菌株纯培养物的固氮酶的固氮作用不受NH^+_4抑制，而Sp7菌株纯培养物的固氮作用则因NH^+_4的存在而不能进行。说明前者能耐高铵，而后者无耐铵能力。

2. 密闭试验中，接种Sp7菌株或CWV-22菌株的过量^{15}N原子%均高于不接种对照，表明植株在有或无NH_4Ac条件下，或多或少都吸收了细菌固定的氮。玉米试验由于无NH_4Ac，接种CWV-22菌株的植株生长和发育不正常，无法加以讨论。小麦试验，在无NH_4Ac条件下，CWV-22菌株供给植物的氮为Sp7菌株的一倍，在有NH_4Ac条件下，则为8倍。这表明，尽管Sp7菌株在有NH_4Ac条件下，仍有一些固氮作用，但显著地被抑制，而CWV-22菌株则不受抑制。

3. Sp7菌株不耐铵，在联合体中也有一定的固^{15}N能力，也能被植物吸收利用，并转运到植株的茎叶部位。在纯培养物中，Sp7菌株在含有50mM NH^+_4的培养基中不能固氮，但植物-Sp7菌株联合体在30~40mM NH^+_4培养液中都表现能固氮，这可能是因为植物吸收培养液中的NH^+_4，降低了根际培养液中NH^+_4的浓度，短期和局部地消除了NH^+_4的抑制作用，因而表现出固氮作用。收获时，用奈氏法测定根系周围的砂中的NH^+_4含量只有1~2mM（低于抑制浓度），而根系未达到部位的砂中NH^+_4含量仍有29.1mM表明NH^+_4在砂中的扩散速度低于植物根系吸收速度，造成局部的低NH^+_4环境，而这正是固氮螺菌能大量生存的场所（固氮螺菌主要生活在根表），因而Sp7菌株也有固氮酶活性，能进行固^{15}N作用。

4. 联合体吸收固定的^{15}N量，能被植物吸收利用，并且也能够转移到植物的茎、叶中。在密闭培养装置中，$^{15}N_2$的试验结果（表4、5）表明：在有或无NH^+_4条件下，联合体-CWV-22菌株和联合体-Sp7菌株所固定的^{15}N量，均被植物的根系吸收固定，并且能运送到植物的茎、叶中。Stephen[2]等用^{15}N示踪法研究玉米植株根系的固氮菌的固氮作用，其固^{15}N只能滞留在植株的根部。我们的结果表明，^{15}N不但能被植株根系吸收固定，而且还能运送到植株的茎、叶中。

参 考 文 献

[1]罗孝杨等：微生物学报，26（1）：47-52，1986.

[2]Stephen，W. & W. J. Brill：Plant Physiol.，70：1564-1569，1982.

[3]湖北省微生物研究所生物固氮组：微生物学报，19（2）：160-165，1979.

[4]Lowry，O. G. et al.：J. Biol Chem. 193：265-275，　1951.

[5]尤崇杓等：原子能，第 11 期，第 995-1000 页，1965.

[6]尤崇杓等：原子能，第 6 期，第 535-541 页，1965.

[7]Burris，R. H. Methodology，in "The Biology of Nitrogen fixation"，Quispel. A. Ed. North Holland Pub. Co.，9-13，1974.

固氮螺菌突变株CWV-22对提高小麦固氮量的初步研究*

罗孝扬　蒋亚平　蔡金芝　杨宝玉

（中国科学院武汉病毒研究所）

严家骐　陈华癸

（武汉大学）（华中农业大学）

摘　要　固氮螺菌突变株CWV-22与玉米、小麦联合体的固氮作用经采用纯培养物和密闭培育装置试验已被证实。在此基础上进行了盆栽（砂培）试验，结果表明：在无NH^{+}_{4}存在下，接菌CWV-22和接菌Sp7的小麦联合体对比，小麦苗期每克干物质中的含氮量，前者比后者高2.1倍；在30mmol/L NH^{+}_{4}存在下，前者比后者高46mg；而在30mmol/L $^{15}NH^{+}_{4}$存在下，小麦苗期植株中吸收的^{15}N量，接菌CWV-22的比接菌Sp7的小麦联合体高13μg。

关键词　固氮菌突变株CWV-22；NH^{+}_{4}或$^{15}NH^{+}_{4}$浓度的影响；小麦

固氮螺菌突变株CWV-22[1]与玉米、小麦联合体的固氮作用已被证实[2]。本文首次报道在环境中含有30mmol/L NH^{+}_{4}或30mmol/L $^{15}NH^{+}_{4}$浓度下，接种固氮螺菌突变株CWV-22对提高小麦苗期固氮量的一些结果。

材料和方法

（一）材料

1. 供试小麦品种及其处理：品种为苏麦2号。处理按常规方法进行[2]。每缸定植小麦幼苗12株。

2. 播种及生长天数：第一批于1985年5月13日~5月27日。生长期为14d。第二批于1985年9月20日~10月23日，生长期为34d。

3. 菌种培养及使用原始菌株为Sp7（ATCC29145），突变株为CWV-22。其培养条件和使用方法按参考文献[2]所描述的程序进行。

4. 河砂及砂培的培养液成分：河砂用2%的HCl浸泡，用无菌水洗到无NH^{+}_{4}。每缸装处理过的砂子1000g。砂与营养液的比例按22%配制。营养成分见参考文献[2]。

（二）方法

1. 同位素过量^{15}N原子%的测定：混合样品送中国科学院南京土壤研所质谱组进行检测。

2. 总氮量测定：采用凯氏比色法[3]。

3. 吸收^{15}N量的计算：本试验采用（$^{15}NH_4$）SO_4中^{15}N的丰度为11.59%。

固^{15}N量（mg）=（C_1-C_0）/（C_2- C_0）×（W_0+W_1）

W_0——小麦植株吸收^{15}N的量（mg）；

W_1——小麦植株吸收^{14}N的量（mg）；

（W_0+W_1）=小麦植株中测得的总氮量（mg）；

*原载于《微生物学报》，29（2）：137~140，1989.

C_0——^{15}N的自然丰度；

C_1——收获时小麦植株中^{15}N的丰度；

C_2——施入缸中^{15}N的丰度。

4. 试验处理：

（1）不接菌加30mmol/L $^{15}NH^+_4$（对照）[$^{15}NH^+_4$为（$^{15}NH_4$）$_2SO_4$，下同]；

（2）接菌（CWV-22）加30mmol/L $^{15}NH^+_4$；

（3）接菌（Sp7）加30mmol/L $^{15}NH^+_4$；以上为第一批试验。

（4）不接种，不加NH^+_4（NH^+_4为NH_4Ac，下同）。

（5）接菌（CWV-22），不加NH^+_4；

（6）接菌（Sp7），不加NH^+_4；

（7）不接菌，加30mmol/L NH^+_4（对照）；

（8）接菌（CWV-22），加30mmol/L NH^+_4；

（9）接菌（Sp7），加30mmol/L NH^+_4；以上是第二批试验。

上述两批试验，均采用随机排列，每个处理四个重复。

结　　果

（一）在不加 NH^+_4 的条件下，各处理含氮量的变化

在不加NH^+_4的试验中（表1），接菌CWV-22的小麦植株每克干物质的含氮量比接菌Sp7的小麦植株高于2.1倍，接菌Sp7的小麦植株含氮量均低于CWV-22-小麦联合体和不接菌的对照。这一结果表明：接菌Sp7的小麦植株，在苗期显示固氮量很低，可能是由于Sp7的固氮量甚微，在其繁殖过程中本身消耗了部分氮，因而小麦植株中的氮量减少了。

表1　无 NH^+_4 对小麦植株分别接菌 CWV-22 和 Sp7 联合体固氮的影响

Tabe 1　The effect ot the absence of ammonium ions on the N_2-fixation by the combinants of wheat plants with Sp7 and CWV-22 respectively

处理 Treatment	含氮量（mg-氮/g 干物质） Ammount of nitrogn content（mg-N/g·dry matter）
不接种对照 Without inoculum	30.99±2.75
Sp7	18.37±1.20
CWV-22	39.22±2.34

（二）在有 NH^+_4 存在下，接菌 CWV-22-小麦联合体和接菌 Sp7-小麦联合体的固氮量变化

由表2结果可知：接菌CWV-22的小麦植株的含氮量比接菌Sp7的小麦植株含氮量约高46mg氮/g干物质。比不接菌的对照约高29mg氮/g干物质。小麦植株中茎叶部位的含量比根部含氮量高3~7倍。各处理间的含氮量变化次序是：CWV-22-小麦联合体>不接种的对照>Sp7-小麦联合体。

表2　在有 NH^+_4 存在下，接菌 CWV-22-小麦联合体和接菌 Sp7-小麦联合体的固氮量变化

Table 2　The effect ot ammonium ions on the ammount of N_2 fixed by the combinants of wheat plants with Sp7 and CWV-22 respectively

处理 Treatment	含氮量（mg-氮/g 干物质） Amount of nitrogen content （mg-N/g·dry matter）		
	茎叶 Stem and leaf	根 Root	茎叶+根 Stem leaf and root
CK+30 mmol/L NH^+_4	119.53±2.3	34.78±2.30	154.31±4.6
Sp7+30mmol/L NH^+_4	121.5±3.81	14.52±3.20	137.03±7.01
CWV-22+30mmol/L NH^+_4	159.98±3.11	23.71±2.08	183.59±5.19

（三）用同位素示踪技术测定各处理间吸收固定氮量的变化及其在植株中的分配

表3的结果表明：接菌CWV-22的小麦植株中的含氮量比接菌Sp7的小麦植株高38.79%，比不接菌的对照的小麦植株含氮量高24.12%。在植株含氮量的分配上，根部的含氮量比叶部高出1~1.5倍。Sp7-小麦联合体的含氮量仍低于其他处理。根据同位素^{15}N稀释法[3，4]的原理，联合体中样品^{15}N富集度%均低于不接菌的对照，表明对照只能吸收以^{15}N的化合态氮作为氮素来源。CWV-22-小麦联合体与Sp7-小麦联合体的叶和根的^{15}N富集度%尽管没有差异，但由于Sp7不耐铵，其固氮酶的固氮作用受NH^+_4所抑制，因而表明在叶和根吸收固定的氮量比不接菌的对照低26.38%，比CWV-22-小麦联合体低62.18%。联合体中固定的^{15}N量不仅根内存在，而且根内存在的部分^{15}N量能输送到植株的茎叶部位。这一结果与参考文献[2]的结果一致。

表 3 有 $^{15}NH_4$ 对小麦植株分别接菌 CWV-22 和 Sp7 联合体的固氮影响*

Table 3 The effect of ammouniuminos on the ^{15}N-fixation by the combinants of wheat plants with Sp7 and CWV-22 respectively

处理 Treatment	烘干重（g） Dry weight		总 N 量（mg） Total amount of nitrogen		样品中过量 ^{15}N 原子%* ^{15}N atom% excess of samples		吸收生物固 ^{15}N 量（μg） Amount of absorption and fixation of ^{15}N		
	叶 Leaf	根 Root	叶 Leaf	根 Root	叶 Leaf	根 Root	叶 Leaf	根 Root	叶+根 Leaf and root
CK+30mmol/L（$^{15}NH_4$）$_2SO_4$	0.6835	0.2386	10.2838	20.746	9.06	9.64	8.3040	17.9170	26.2210
Sp7+30mmol/L（$^{15}NH_4$）$_2SO_4$	0.5364	0.2102	5.7215	19.3102	8.96	9.40	4.5690	16.1779	20.7469
CWV-22+30mmol/L（$^{15}NH_4$）$_2SO_4$	0.5668	0.2048	16.2438	24.6533	8.96	9.41	12.9719	20.6763	33.6482

* 表中数据为四缸样品等量混合。

结论与讨论

Sp7-小麦联合体无论在无NH^+_4或有NH^+_4（$^{15}NH^+_4$）存在条件下，其固氮量均低于CWV-22小麦联合体与不接菌的对照。其原因在无NH^+_4存在时，Sp7-小麦联合体固定的氮量不能满足小麦苗期对氮素的需要，同时Sp7在其本身繁殖过程中需要消耗部分的氮素营养，此时小麦植株的缺氮现象就显得更为突出，致使氮的含量低。在有30mmol/L NH^+_4或30mmol/L $^{15}NH^+_4$存在的环境中，由于Sp7不耐铵，其固氮酶的固氮作用受NH^+_4或$^{15}NH^+_4$所抑制，固氮作用不能进行，而其本身又要消耗部分氮，因此，其含氮量既低于不接菌的对照，又低于CWV-22-小麦联合体。

在含有30mmol/L NH^+_4或30mmol/L $^{15}NH^+_4$的环境中，接菌CWV-22与小麦形成的联合体所吸收固氮（^{15}N）量与原始菌株Sp7或不接种的对照相比，均有明显提高。因此，CWV-22小麦联合体的固氮体系不仅已经建成，而且在较高的NH^+_4或$^{15}NH^+_4$浓度下仍然能够固定空气中的氮以供给禾本科植株的氮素来源。这一结果将给禾本科植物的氮素营养提供了依据。

参 考 文 献

[1]罗孝扬等. 微生物学报. 26（1）：47-52. 1980

[2]罗孝扬等. 微生物学报. 27（3）：277-283. 1987

[3]林沧等. 中国农业科学. 4：53-58. 1984

[4]Boddey，R. M. et al.：International Symposium on N_2-fixation with Non-leyumes （Program Abstract）p. 19.1982.

[5]Burris，R. H.：Methodology，in：The Biology of Nitrogen Fixation，Ed. Ouispel A.North Holland Publ. Co.，pp. 9-13，1974.

固氮螺菌突变株CWV-22对作物物质生产和氮素循环的影响
II.CWV-22对小麦生长发育产量及吸氮量的影响*

刘　强[1] 苏国栋[1] 罗孝扬[2] 蒋来平[2] 蔡金芝[2] 杨宝玉[2] 严家骐[3] 陈华癸[4]

（1 湖南农学院农学系）

（2 中国科学院武汉病毒研究所）

（3 武汉大学病毒系）

（4 华中农业大学土化系）

摘　要　施用固氮螺菌突变株CWV-22能使小麦株高，干物重，产量，吸氮量和土壤全氮含量明显增加；在产量构成因子中，CWV-22影响最大的是每穗实粒数，其次是千粒重，但不能增加每亩有效穗数；CWV-22的固氮作用缓慢而持久，施用该菌后仍应在小麦生长中，后期补充适量化学氮肥；CWV-22的施用方式以一次拌种为好。

关键词　固氮细菌；固氮作用；小麦；氮素营养；产量

禾本科作物的微生物固氮，一直是世界各国农学界和微生物学界密切关注并努力探索的重大研究课题，在已发现的几种能与禾本科作物根系形成联合体固氮的微生物中，固氮螺菌*Azospirillumbrasilense*是研究较多且较有希望的一种。1975年中国科学院武汉病毒研究的在我国率先开展对该菌的研究，1983年罗孝扬等人对该菌进行诱变筛选，获得一批耐高铵泌铵突变菌株[1, 2]，其中选出的CWV-22能与小麦、玉米根系形成密切的联合体，固定空气中的N_2，转化为$NH^+{}_4$后泌出体外，供作物吸收利用[3]，特别是它在含$NH^+{}_4$较高的环境中仍有较强的固氮活性[2]，能增加小麦玉米的氮素积累[4]，促进小麦玉米的生长发育，从而提高其生物产量，因此具有诱人的利用前景，CWV-22在砂培条件下的上述作用已得到证实，但在土培及田间条件下表现如何尚未见报道，本研究即是在1989年砂培试验的基础上，进行土培及田间试验，以进一步探讨CWV-22对小麦的增产增氮作用规律，为大田推广应用提供可靠依据。

1　材料与方法

1.1　供试土壤

供试土壤为湖南农学院教学实验场第四纪红土红壤，前作为辣椒间作甘蓝，肥力中上，土壤主要理化性状见表1。

表1　供试土壤主要理化性状

Table 1　Maina physicochemical properttes of tested soils

项目	有机质（%）	质地	pH（水浸）	全氮（%）	全磷（%）	全钾（%）	碱解氮（mg/100g 土）	有效磷（ppm）	速效钾（ppm）
含量	2.71	壤土	5.8	0.187	0.09112	1.59	15.2	15.9	82.6
测试方法	重铬酸钾法	中国制	酸度计法	高氯酸硫酸消化扩散法	NaOH 碱融钼锑抗比色法	NaOH 碱融火焰光度法	扩散法	$NaHCO_3$ 法	NH_4AC 浸提火焰光度法

1.2　供试菌液

出发菌株*Azospirillumbrasilense*（ATCC29145，代号Sp7）和突变株CWV-22均由中国科学院武汉病

---

*原载于《湖南农学院学报》，17：419~426，1991.

毒研究所提供，菌液浓度：活菌$1.5*10^8$个/ml。

1.3 供试作物

小麦11101品系，中熟，春性。

1.4 土培试验

设5个处理，I.Sp7拌种（代号$Sp7^1$）；II.CWV-22拌种（代号22^1）；III.Sp7拌种十分蘖期追施（代号$Sp7^2$）；IV.CWV-22拌种十分蘖期追施（代号22^2）；V.对照（代号CK）。

重复4次，完全随机排列，采用20cm×25cm下口瓷盆，每盆装土7.5kg，肥底为N0.75，P_2O_5 0.75，K_2O 1.125g/盆，形态分别为尿素、过磷酸钙、氯化钾，作基肥1次施下，1989年11月13日播种，1990年5月14日收获，全生育期182d，每盆定苗6株。

1.5 田间试验

5个处理：I. CWV-22拌种；II. Sp7拌种；III. 播种40d每亩施分蘖氮肥3kg（形态为尿素，代号分蘖肥）；IV. 播后40d每亩施分蘖氮肥3kg+3月份每亩施拔节氮肥1kg（形态为尿素，代号拔分肥）；V. 对照（代号CK）。重复5次，拉丁方设计，每小区0.03亩，肥底为N 3， P_2O_5 3，K_2O 8kg/亩，形态同上，做基肥一次施下。1989年11月18日播种，1990年5月27日收获，全生育期190d。

1.6 观测与分析项目

观测株高、分蘖动态：测定各时期干物质，收获期产量；采用过氧化氢硫酸消化，蒸馏法测定植株含氮量，统计分析时采用邓肯氏新复极差法（SSR法）进行多重比较。

2 结果与分析

2.1 CWV-22对小麦株高的影响

土培和田间试验结果均表明：分蘖期、拔节期各处理株高差异不显著，抽穗期则株高差异逐渐出现，到成熟期各处理间株高差异达显著水准，收获前土培试验的株高（cm）顺序为：22^2（93.69）>22^1（93.28）>$Sp7^2$（92.60）>$Sp7^1$（92.40）>CK（91.11），方差分析达显著水准，其中22^2、22^1与CK间差异分别达极显著，显著水准，而$Sp7^2$、$Sp7^1$和CK间差异不显著，22^2与22^1间差异也不显著，收获前田间试验的株高（cm）顺序为：拔分肥（97.5）>分蘖肥（96.9）>22^1（95.7）>$Sp7^1$（95.3）>CK（93.5），方差分析达极显著水准。其中，拔分肥、分蘖肥的株高与对照组间差异极显著，而其他各处理间差异不明显，这表明，施用Sp7对小麦株高影响不大，而施用CWV-22对促进小麦高度增加有一定作用，这种作用在土培试验中表现明显，在田间试验中则不明显，这是因为土培试验容易控制试验条件，实现单一差异原则，而田间试验误较大的缘故，CWV-22施用2次（即拌种加追施）与施用1次（拌种）相比，两处理间无本质差异，表明CWV-22拌种的作用是主要的。CWV-22虽有增高植株作用，但不如分蘖期每亩施3kg和分蘖期施3kg加拔节期施1kg氮处理的那样明显，其原因可能与CWV-22固氮作用缓慢而持久有关。

2.2 施用CWV-22对小麦分蘖的影响

施用CWV-22能促进小麦的前期分蘖，田间试验动态观测表明，施用CWV-22后的最高分蘖数可达4.4个/株，明显高于Sp7，分蘖肥，拔分肥（均为3.6个/株）和对照（3.4个/株），由于小麦中，后期生长旺盛，CWV-22固定的氮不能充分满足分蘖转化为有效穗的需求，所以，施用该菌后最终形成的有效分蘖数没有增加，田间试验收获前调查的各处理有效穗数（穗/亩）为：CWV-22，196933；Sp7，199033；对照，216067；分蘖肥，210467；拔分肥，210000，经方差分析，各处理间差异不显著。土培试验的趋势也与田间试验一致。

2.3 施用CWV-22对小麦干物重的影响

在本研究中，小麦的干物重指地上部小麦植株体烘干重，收获时土培试验的茎叶干重列于表2，方差分析表明，各处理间差异极显著（F=8.67**），多重比较显示，施用2次CWV-22的处理与对照相差极显著，但与施用1次CWV-22的处理差异不显著；同时还可看出，施用CWV-22的与施用Sp7的差异显著，而施用Sp7的比对照也有明显差异，这说明Sp7有增加小麦茎叶干重的作用，但CWV-22的这种作用又明显大于Sp7，而施用2次CWV-22与施用1次CWV-22相比，没有本质差异。

表 2　土培试验收获时小麦茎叶干重（g/盆）及多重比较结果

Table 2　Wheat plant dry weights and multiple comparisons in various treatments in soil culture（g/pot）

处理	重复				$\overline{X}$	多重比较		比对照增加（%）
	1	2	3	4		5%水准	1%水准	
22^2	32.8	32.3	31.9	32.2	32.30	a	A	6.43
22^1	32.3	31.4	31.2	31.5	31.60	ab	AB	4.12
$Sp7^2$	32.1	32.1	30.9	31.8	31.50	b	AB	3.79
$Sp7^1$	31.4	30.4	31.6	31.2	31.15	b	BC	2.64
CK	29.8	30.2	30.5	30.9	30.35	c	C	0

表3为田间试验不同时期各处理的干物重，按双向分组资料进行方差分析，各处理间差异达显著水准（F=4.08*），其中拔分肥、分蘖肥比对照干物质明显增加；拔分肥、分蘖肥与CWV-22间无显著差异；CWV-22与Sp7和对照间差异也不明显，各处理不同时期平均干物重顺序为：拔分肥>分蘖肥>CWV-22>Sp7>对照（表3），成熟期CWV-22比对照增重25.63%，说明CWV-22仍有一定的增加小麦干物重的效果。

表 3　田间试验各处理不同时期干物重（g/50 株）及多重比较结果

Table 3　Wheat plant dry weights and multiple comparisons in various treatments at different stages in field experiment（g/50 plant）

处理	分蘖期	拔节期	抽穗期	成熟期	$\overline{X}$	多重比较		成熟期比对照增加（%）
						5%水准	1%水准	
拔分肥	4.82	47.44	152.20	199.30	100.94	a	A	32.69
分蘖肥	4.96	46.24	136.70	185.30	93.30	ab	AB	23.37
CWV-22	5.58	47.74	120.00	188.70	89.00	abc	AB	25.63
Sp7	4.97	36.46	114.30	159.50	78.81	bc	AB	6.19
CK	5.29	35.37	109.90	150.20	75.19	c	B	0

2.4　施用CWV-22对小麦产量的影响

小麦产量是衡量CWV-22综合作用的决定性指标，土培试验的结果（表4）表明，各处理间差异达极显著水准（F=9.48**），其中，凡是施菌的处理（包括CWV-22和Sp7）比对照增产均达极显著水准，施用2次CWV-22比施用1次Sp7的显著增产，而施用2次CWV-22与施用1次CWV-22的差异不显著。田间试验，各小区产量列于表5，方差分析表明，各处理间差异极显著（F=30.39**），多重比较显示（表6），拔分肥、分蘖肥，CWV-22和Sp7均能明显提高产量，比对照分别增产30.16%，19.58%，9.97%和8.79%，拔分肥、分蘖肥比CWV-22和Sp7明显增产，CWV-22与Sp7间无明显差异。

表 4　土培小麦产量（g/盆）及多重比较

Table 4　Wheat grain weights and multiple comparisons in various treatments in soil culture（g/pot）

处理	重复				$\overline{X}$	多重比较		比对照增加（%）
	1	2	3	4		5%水准	1%水准	
22^2	25.7	26.4	25.8	23.8	25.43	a	A	14.91
22^1	24.6	24.1	25.6	24.6	24.73	ab	A	11.75
$Sp7^2$	24.4	25.2	23.8	24.2	24.40	ab	A	10.26
$Sp7^1$	24.2	23.9	24.5	22.8	23.85	b	A	7.7
CK	21.5	21.3	22.8	22.0	22.13	c	B	0

表 5 田间拉丁方设计各小区小麦产量（kg/小区）

Table 5 Wheat grain weights of various plots in latin square design（kg/plot）

CWV-22	对照	Sp7	拔分肥	分蘖肥
5.50	5.58	5.26	6.82	6.52
分蘖肥	拔分肥	CWV-22	Sp7	对照
5.74	6.50	4.64	5.72	5.44
Sp7	分蘖肥	对照	CWV-22	拔分肥
4.95	5.55	3.75	5.76	6.93
对照	Sp7	拔分肥	分蘖肥	CWV-22
4.58	5.21	5.58	6.31	5.53
拔分肥	CWV-22	分蘖肥	对照	Sp7
6.15	5.59	5.26	5.22	5.57

表 6 田间各处理平均小麦产量（kg/小区）及多重比较

Table6 Average wheat grain weights and multiple comparisons in various treatments in field experiment（kg/plpt）

处理		拔分肥	分蘖肥	CWV-22	Sp7	CK
$\overline{X}$		6.396	5.876	5.404	5.346	4.914
比对照增加（%）		30.160	19.580	9.970	8.790	0
多重比较	5%水准	a	b	c	c	d
	1%水准	A	B	C	CD	D

表 7 田间收获时小麦产量性状调查与实际产量

Table 7 Forming factors of wheat grain production in the field investigation and practical production

处理	有效穗数（穗/亩）	每穗实粒数（粒/穗）	千粒重（g/千粒）	估测产量（kg/亩）	实际产量（kg/亩）
CWV-22	196933	34.0	36.2	242.39	180.13
Sp7	199033	32.7	34.7	225.84	178.20
分蘖肥	210467	32.2	37.1	251.43	195.87
拔分肥	210000	33.9	38.5	274.08	213.20
CK	216067	30.0	34.5	223.63	163.80

综合土培及田间试验结果可以看出：a，施用CWV-22能明显增加小麦产量，增产幅度为9.97%~14.91%；施用Sp7也有明显的增产作用，增产幅度为8.79%~10.26%；CWV-22的作用大于Sp7，但土培中两者相差较明显，而田间则不显著，这可能与田间试验条件不易控制，误差较大有关。a，施用分蘖肥，拔节肥的小麦产量明显高于CWV-22处理，说明小麦中，后期氮素营养是十分重要的，施用CWV-22虽能使小麦吸氮量增加，但在小麦生长旺盛期，CWV-22固定的氮素却难以满足小麦急剧生长发育的需要，因此在产量上表现出拔分肥>分蘖肥>CWV-22的趋势。

CWV-22虽不能使小麦有效分蘖增加，但它可使每穗实粒数增多，使千粒重加重，这可能是无效分蘖的养分转移集中到有效穗中所致，表7是收获前田间小麦产量性状调查数据及实际亩产，5种处理中，CWV-22的每穗实粒数居于首位（34.0粒），千粒重也明显高于Sp7和对照处理，调查数据估测的产量与实际收获的产量趋势是一致的，两者间的绝对差异主要是收获时脱粒未净及风、雨、鸟造成的浪费引起，估测产量与实际产量的换算公式为：实际产量= −5.910+0.801*估测产量，相关系数为0.95，达显著水准。

2.5 施用CWV-22对小麦氮素吸收的影响

CWV-22的主要特点就是在高NH^{+}_{4}环境中仍具有较强的固氮活性，它与小麦根系形成的联合体的固氮作用可通过小麦吸氮量反映出来，小麦吸氮量指小麦地上部吸收氮素的总量。表8为土培收获时小麦吸氮量，方差分析表明各处理差异极显著（F=6.42**），其中，22^2与$Sp7^2$、CK差异达显著或极显著水准，$Sp7^1$与CK差异也显著，表明施菌确能增加小麦植株体吸氮量，CWV-22的作用又大于Sp7，而2次施菌与1次施菌并无本质差异，这与收获时土培茎叶干重和产量趋势大体一致。

表 8 收获时土培小麦吸氮量（N.g/盆）及多重比较

Table 8 N uptake and multiple comparisons of wheat in various treatments in soil culture（N.g/pot）

处理	重复				$\overline{X}$	多重比较		比对照增加（%）
	1	2	3	4		5%水准	1%水准	
22^2	1.137	1.356	1.202	1.062	1.189	a	A	25.95
22^1	1.084	1.072	1.181	1.074	1.103	ab	AB	16.84
$Sp7^2$	1.043	1.006	1.170	1.018	1.059	b	ABC	12.18
$Sp7^1$	1.054	1.059	1.020	1.031	1.041	bc	BC	10.28
CK	0.924	0.918	0.952	0.981	0.944	c	C	0

表 9 田间小麦不同时期各处理吸氮量（N.g/盆）及多重比较

Table 9 Nuptake and multiple comparisons of wheat in various treatments at different stages in field experiment（N.g/pot）

处理	时期				$\overline{X}$	多重比较		成熟期比对照增加（%）
	分蘖期	拔节期	抽穗期	成熟期		5%水准	1%水准	
拔分肥	0.289	1.541	2.265	3.315	1.808	a	A	66.22
分蘖肥	0.296	1.523	1.974	2.881	1.669	a	AB	52.76
CWV-22	0.343	1.332	1.712	2.773	1.541	ab	ABC	47.03
Sp7	0.304	1.020	1.532	2.114	1.243	bc	BC	12.09
CK	0.319	0.769	1.447	1.886	1.105	c	C	0

对田间试验不同时期各处理吸氮量进行了双向分组资料的方差分析，结果表明（表9），各处理间差异也达极显著水准（F=6.23**），其中拔分肥、分蘖肥、CWV-22三者间无本质差异，Sp7与对照间差异也不显著，但CWV-22与对照间差异显著，说明施用CWV-22能明显增加小麦地上部吸氮量，而Sp7的这种作用不大，分蘖期、拔节期施用氮肥能有效地促进小麦氮素积累，就成熟期小麦吸氮量而言，拔分肥，分蘖肥，CWV-22和Sp7比对照分别增加66.22%、52.76%、47.03%、12.09%。

2.6 施用CWV-22对小麦土壤全氮的影响

由于田间试验有施用化学氮肥处理，故只对土培试验的土壤进行了试验后土壤全氮量的测定，取样深度为20cm得到的土壤全氮量如图所示，可以看出，土壤全氮含量可分为3个等级：原始土壤因未种作物，全氮含量最高，施用2次CWV-22和施用一次CWV-22的土壤全氮量居中，$Sp7^1$和$Sp7^2$与CK属于最低的一级，施用CWV-22的处理因该菌的固氮能力强，除供小麦吸收利用外，尚有部分固定的氮残留于土壤，Sp7在环境含有NH^+_4时固氮能力较弱，基本上没有残留氮，故与对照相差无几。

3 讨论与小结

a. CWV-22是从Sp7中筛选出来的耐高铵突变株，在含NH^+_4较高的土壤中，CWV-22的固氮活性大于Sp7，因此施用CWV-22能较多地固定空气中的氮，使小麦吸氮量、株高、生物产量、经济产量和土壤全氮含量不同程度地增加，土培和田间试验结果表明，同对照相比，上述5个指标增加的幅度依次为16.84%~47.03%，2.35%~2.83%，4.12%~25.62%，9.97%~14.91%，15.56%~17.04%；而Sp7不耐铵，故它的上述作用较弱。

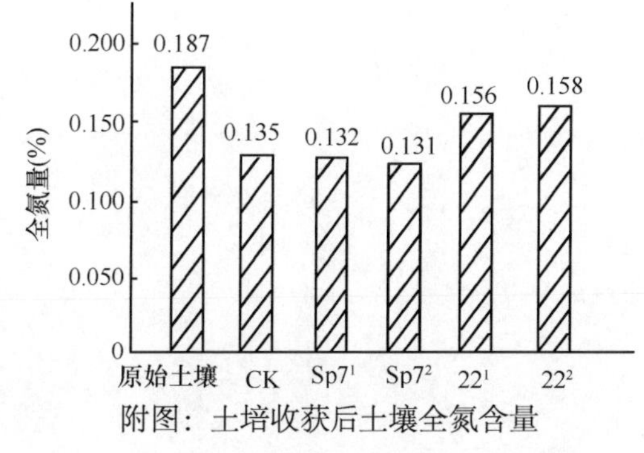

附图：土培收获后土壤全氮含量

Fig. Soil total N after harvesting in soil culture

b. 就施用次数而言，CWV-22拌种加追施的效果与CWV-22拌种的效果没有本质差异，因此CWV-22拌种是行之有效的施用方法，为省工省时计，分蘖期追施可不必进行。

c. 在小麦产量构成因子中，CWV-22影响最大的是每穗实粒数，其次为千粒重，施用CWV-22虽能促进前期小麦分蘖，但对每亩有效穗数没有明显影响，其原因有待于进一步研究。

d. CWV-22有明显的固氮作用，且其作用缓慢而持久，小麦是一种需氮量较大的作物，特别是生长中，后期对氮的需求量较高，CWV-22固定的氮尚难完全满足其需要，因此，小麦即使施用CWV-22，也应考虑到在中、后期适当补施少量化学氮肥。

e. 已发表的资料只涉及CWV-22的固氮作用，但CWV-22是否还有其他增产机理，亦有待于进一步研究。

参考文献

[1]罗孝扬，曾宽容，蒋亚平，蔡昌建，周亿闺，杨宝玉，王子方. 禾谷类作物根表固氮螺菌的分类鉴定和分布. 微生物学报. 1983，23（1）：68-72

[2]罗孝扬，蒋亚平，蔡金芝，陈华癸，严家骐. 固氮螺菌耐高铵突变株的选育. 微生物学报. 1986. 26（1）：47-52

[3]罗孝扬，蒋亚平，蔡金芝，杨宝玉，严家骐，陈华癸. 固氮螺菌 CWV-22 突变株与玉米、小麦联合体的固氮作用. 微生物学报. 1987，23（3）：277-283

[4]罗孝扬，蒋亚平，蔡金芝，杨宝玉，严家骐，陈华癸. 固氮螺菌突变株 CWV-22 对提高小麦固氮量的初步研究. 微生物学报. 1989，29（2）：137-140

[5]Stephen，W. W J Brill. Screening and selection of maize to enhance associative bacterial nitrogen fixation plant physiology. 1982，70：1564-1568

第二部分　水稻田土壤微生物

提　要　这是一项 1948 年开始研究的新课题。陈华癸教授针对我国水稻生产的重要性和水稻土肥力的特殊性，率先开拓我国水稻田土壤微生物的研究。研究内容有两项，发表论文 13 篇[注]。

第一项　水稻土营养元素的生物循环

1953 年陈化癸教授领导 4 名研究生对绿肥耕翻蓄水种水稻后微生物种群消长的研究。论文在 1956 年第六届国际土壤学大会（巴黎）上专题讨论。1957 年他领导另一位研究生研究烤田对土壤微生物区系的影响。结果为这项对水稻生长有利的技术措施提供了理论依据。1981 年这项研究结果又编入《土壤微生物学》专著。由陈华癸、李阜棣、陈文新、曹燕珍编著，上海科学技术出版社出版。

第二项　硝化作用和硝化细菌

这是一项具有突出贡献的项目。首先发现水稻田土壤中的硝化细菌是兼厌气性的，在好气和嫌气条件下均能进行硝化作用，而旱地土壤（对照）的硝化细菌只能在好气条件下进行，表明硝化细菌的多样性。否定了自 1890 年以来一直认为硝化作用和硝化细菌是绝对需氧的普遍性规律。

有关硝化细菌在嫌气条件下进行硝化作用的研究已编入陈华癸主编的《微生物学》第二版，1962：一篇题为 *Facultatively Anaerobic Nitrification and Nitrite-forming Organisms* 的论文在 1964 年第八届国际土壤学大会（布加勒斯特）上宣读，和一篇论文 *The Activity Nitrifying and Denitrifying Bacteria in Paddy Soil* 作为中国土壤学研究成果的代表作之一在美国 *Soil Science* 上发表后引起强烈反响。十几个国家的学者发来 20 多封信件索取单印本；国内同行自 60 年代以来一直在引用这些研究结果。

研究还获得亚硝酸细菌及其伴生菌的纯培养。伴生菌很特殊，不仅在好气和嫌气条件都能进行反硝化作用，而且还能在完全没有有机碳化合物的亚硝酸培养基中正常生长和繁殖，此外，还证明亚硝酸细菌和伴生菌生活在一起是造成土壤中无机氮素营养大量流失的主要原因。由此发展和修订了日本学者 Shingo Mitsui 1956 年提出的水田土壤的分层及无机氮的转化和损失的经典图式。

[注]包括已在《微生物与农业》上刊登的论文，见附件 1。

第一项　水稻土营养元素的生物循环

中国科学院土壤微生物学座谈会和我国土壤微生物学发展概况*

陈华癸

（华中农学院）

土壤微生物学在我国已有二十多年的历史，但在解放前是处于断断续续、自生自灭的状态中。解放后，科学工作得到共产党和人民政府的关怀，土壤微生物学几年来也获得一些成绩，但是发展进度和成绩远不能满足国家建设的需要。因此，中国科学院邀请了院内外有代表性的土壤微生物学工作者，在7月30日至8月2日举行座谈会，交流工作经验，讨论土壤微生物学研究工作的发展方向和今后几年内的任务，并初步交换建立经常联系和组织的意见。这些都是很必要而且适时的。

解放前二十多年中，在土壤微生物学方面也进行了一些科学研究工作，但是研究范围很狭窄。主要是被看成为肥料学中的一些个别问题。在根瘤菌方面会进行过一些形态和生理的研究、接种关系的研究和品种比较试验，这些工作虽然有它的实践意义，但一直没有推广，使其在生产中发挥作用。在堆肥方面，曾进行过堆制方法的研究、高温性细菌的分离培养以及堆制过程中的利用等工作，这方面的工作虽在一些地区零星地推广过，但都没有在农业实践中生根。此外，在粪尿处理方面，土壤中氮素转化方面，土壤微生物的种类和数量方面也进行过一些研究工作。这些科学研究工作在解放前是得不到支持的，研究者经常处于生活上和工作上的不安定状态中。因而二十多年的断断续续自生自灭的科学研究工作，并没有建立起一个比较安定的工作据点，不能够有系统、连续地进行科学研究工作。

解放后，土壤微生物学工作有了很大的发展，目前大致进行着以下几方面的研究工作。

1　细菌肥料的研究

这方面包括根瘤菌、固氮菌和磷细菌的工作。根瘤菌方面由于生产上的需要和苏联先进经验介绍得较早，同时我国土壤微生物学在这方面较有基础，因此首先得到发展。华北农业科学研究所关于花生根瘤菌的工作，科学院应用化学研究所(即前长春综合研究所)农业化学室和东北农业科学研究所关于大豆根瘤菌的工作，都已经从选择有效菌种发展至大量制造接种剂和大面积推广应用，取得了大豆和花生的增产效果。华北农业科学研究所于1953年推广花生根瘤菌接种62万亩，增产15%；1954年推广200万亩。在东北，大豆根瘤菌1953年推广150万垧。对绿肥作物的根瘤菌接种问题，许多农业科学研究机构也都在试验研究或试行推广中。在祖国社会主义建设中，豆类的食粮作物和技术作物. 绿肥和牧草等的栽培面积都要扩大，因而根瘤菌接种工作一定要大规模发展。今后工作重点将放在提高接种剂质量和生产效率，选育根瘤菌和进一步了解和利用这共生现象上。

在固氮菌方面。科学院应用化学研究所农业化学室已进行了两年的工作，在马铃薯、甜菜、玉米、黍上试验，获得一些成绩。华北农业科学研究所和东北农学院也曾做了一些接种试验。关于固氮菌的工

---

*原载于《科学通报》，9：61~62，1954.

作现在只是方才开始，工作重点将放在菌种的有效性，接种剂的制造法和接种方法，固氮菌在植物根际的生存发育条件和固氮条件等问题上。

关于磷细菌，科学院应用化学研究所农业化学室和东北农业科学研究所都已开始研究，从黑土中分离得一些菌种，并初步试验了它们的矿物质化能力。科学院土壤研究所研究磷灰石转化的微生物学过程，从土壤中分离得几种类似硅酸盐细菌，并试验了它们利用磷灰石中磷的能力。北京农业大学研究堆肥堆制过程中磷灰石的转化，初步知道土壤中磷细菌的种类可能很多。今后的重点工作为研究磷细菌的种类和有效性，最合适的活动条件，磷转化过程中的化学分析等，以便在生产实践上应用。

2　根际微生物的研究

这方面的范围很大，工作复杂。目前关于微生物群的研究方法以及主要种类的分类和鉴定都遗存着很大困难，因此还停留在探索工作方法的阶段上。科学院土壤研究所、科学院应用化学研究所农业化学室、东北农业利用研究所和北京农业大学在这问题上正做着多方面努力。科学院土壤研究所正进行着小麦和棉花根际微生物的研究，测定微生物的数量，辨识主要种类。华中农学院在研究水稻田土壤的微生物学过程的基础上，计划开展水稻根际、耕层的微生物群的研究。

3　土壤微生物的分类和鉴定问题

过去在这方面的基础很差，对土壤微生物学的科学研究有着很大的障碍。在这方面，科学院菌种保藏委员会正在放线菌部分展开有系统的工作，并在霉菌、酵母和丙酮丁酸细菌等方面做了些准备工作。关于纤维素分解微生物，山东大学已从中温性类型开始研究，并准备扩充到其他类型上去。在微生物范围内，对“种”的概念的具体应用还很模糊，现有的各种分类系统都还不能反映微生物的系统发展规律，因此分类和鉴定的工作还存在很多困难。

4　其他研究工作

华中农学院在研究水稻田氮素转化的微生物学过程的基础上，将展开水稻田绿肥耕作制的微生物学过程的研究。东北农业科学研究所正在研究草田轮作制中的微生物学过程。科学院应用化学研究所农业化学室关于菌根的研究，科学院菌种保藏委员会和福建师范学院关于放线菌的分离方法和抗生性的研究等。都在土壤微生物学的各个据点上展开工作。

解放五年来，土壤微生物学的科学研究工作是有发展的，有些工作已在农业生产中起着作用，有些工作正在开始，有些工作还在准备阶段。这些工作使土壤微生物学在祖国的科学事业中露出了一枝幼苗。

这次的土壤微生物学座谈会上，到会的人都报告了自己的工作，提出存在的问题，供大家讨论，使自己的工作得到大家的帮助，各人的工作方向更为明确，工作信心也大为加强。

在座谈会的第一天，会上介绍了苏联土壤微生物学的发展情况，大家认识到由于苏联学者们的努力，土壤微生物学已经成长为一门独立的科学。关于这门科学的发展，苏联一直占据全世界的第一位。大家也认识到，苏联土壤微生物学是在深入生产实际、不断地解决社会主义建设所提出的问题中成长的。苏联土壤微生物学的发展过程，指出了我国土壤微生物学应走的道路。我们应该在解决生产实践所提出的问题中发展土壤微生物学，在这基础上深入研究土壤中微生物发展的规律和微生物对土壤肥沃性与植物土壤营养的作用。在祖国几千年来农业生产实践中，累积了十分丰富的经验，在目前的农业建设中，也提出了一系列必须解决的问题，总结这些经验，提高到理论水平，再用来解决目前存在的问题，是必要的。在生产实践中，含蓄着土壤微生物学工作者们无限的智慧的源泉、丰富的科学生活。

土壤微生物学的发展和其他有关科学的发展是密切联系着的。土壤微生物学的发展使我们能够更深刻地了解土壤的发展规律，同时它也能够丰富和发展对于生命现象的一般规律的知识。土壤微生物学、工业微生物学和医学微生物学之间的关系在日趋密切中。对于这些，土壤微生物学工作者也都应有充分的认识。

会上也讨论了今后发展中的一系列具体问题，如统一研究计划问题，各单位的分工问题，组织和领导问题，培养干部问题，加强联络问题等等，并且针对实际情况，提出了一系列具体建议和办法，对开展我国土壤微生物学的工作都有积极的帮助。

与会的人一致希望中国科学院能具体领导全国土壤微生物学的研究工作。

土壤肥力的微生物学方面*

陈华癸

（华中农学院土壤农化系、中国科学院武汉微生物研究所）

序

可以这样说，土壤微生物学有狭义的和广义的两个范畴。狭义的土壤微生物学研究土壤肥力的微生物学方面。广义的土壤微生物学有两个方面的任务，一个方面是研究土壤肥力的微生物学方面，另一个方面是从发掘国民经济多种资源的角度研究土壤微生物学。

本文试图从对于土壤的生物活性的认识出发，多方面地提出土壤微生物的生命活动对于土壤肥力的作用。写这篇文章的目的，是想指出，即使是狭义的土壤微生物学，所包含的内容和科学命题也是十分丰富的。发展我国的土壤微生物学，对于一个工作单位来说，自然要集中力量深入一个或少数问题，而对于全国许多工作单位来说，就有必要分别钻研这门学科的多方面的问题，因为，只有从多方面开展土壤微生物学研究工作，才能和土壤微生物对于土壤肥力的多方面作用相适应。

土壤的生物活性　我在华中农学院主编的《土壤学》（农业出版社，1962，4~6 页）的绪论中提出了这个术语。当然，这并不是什么新的概念，而是在上世纪末已提出的，物质的生物循环的概念中就含有了这种意思。十九世纪末期的农业化学家和土壤学家就已经指出，生物活动是土壤形成的主要因素，微生物的生命活动是土壤中植物养料的主要制造者（参考 Russell，1961，18-22 页；Pochon et Barjac，1958，2 页）。Вильямс（1949，6 版，57-82 页）的土壤形成学说和 Лысенко（1953）的植物土壤营养的生物学理念进一步地强调了这个概念。概括地说，土壤的生物活性就是土壤中的生命活动（植物根系、小动物和微生物）所引起的（或决定的）土壤性质。

土壤的“代谢性”和“可塑性”，土壤具有“代谢性”，也就是说，它具有不断地同化外条件为土壤本身、同时也不断地异化自己为外界物质的性能。土壤的“代谢性”是它的最本质的性质之一，在土壤形成的整个过程中，无休止地进行着。在土壤形成的特定阶段中，具有与这个阶段相应的生命活动，进行着特定的代谢过程。土壤从一个阶段发展到另一个阶段也是通过不断的代谢作用来实现的。由于外界条件的变化，土壤的生命活动及其所引起的土壤代谢过程也引起了相应地变化，从而表现出新的土壤肥力特性和形态特征。这样变化体现出土壤的“可塑性”（或土壤性质的可改造性），因而，土壤的“可塑性”也主要是土壤生命活动的体现。

没有土壤生命活动及其所引起的代谢过程，就没有土壤的形成和发展过程，就没有土壤的量和质的变化过程。

土壤中的活泼因素　土壤是一种十分活泼的自然体系。土壤作为一种由无机、有机、生物复合体（Лазарев Н М.1949；Lycn & Buckman，1943，15-16 页；Pochon et Barjac，1958，38 页），它的最活泼的因素是：土壤的生物活性、土壤的胶体活性和土壤的水、气动态。

土壤的无机部分包括：①土壤的骨骼物质；②土壤的无机胶体；③胶体表面吸附的离子；④土壤溶液中的无机成分。从土壤中物质变化的相对速度来说，土壤中骨骼物质的变化是很缓慢的，土壤无机胶体的变化也是比较缓慢的。从农业利用的角度，可以把这两部分的性质看成为土壤的不变的因素（个别情况除外）。胶体表面吸附的离子和土壤溶液中的无机成分的变化是密切相连的，而且变化的速度较快，

*原载于《中国土壤学会、中国微生物学会 1964 年土壤微生物学专业会议专题报告及研究报告摘要集》，专 1~19，1964.

从农业利用的角度，它们是可以通过一定的农业利用方式和农业技术措施而变化和改善的。

土壤的有机成分包括：①土壤腐殖质（它是胶态物质，和土壤无机胶体结合而成为土壤的吸收性复合体）和②生物的残体、分泌物、溶细胞产物和分解产物。两者都是变化得较快的，它们的质和量都可以通过一定的农业利用方式和农业技术措施而变化。土壤有机成分的来源和变化都是生物学的。

土壤生物是十分活泼的因素。土壤生物主要包括高等绿色植物、土壤微生物和在土壤中生活的小动物。

植被的演化和更替是土壤类型的演化和更替的重要因素，按照 Вильямс 的学说，则是主导的因素。

在农业生产中，不同的轮作制度产生不同的人工植被，对土壤肥力起着不同的影响。

如果我们以土壤腐殖质含量的变化代表农业利用和农业措施对土壤肥力的影响，可以看出，这种影响确实是很活泼的（Tenny，1941，252-257 页；Кононова，1951，中译 238-314 页）。Tenny（1941，252-253 页）引用的数据指出，美国中部生荒地开垦 60 年（生荒植被改为农业植被），含氮量（可以作为腐殖质含量的指标，矿质土的腐殖质含量大致为含氮量的 20 倍）从原来的 100%下降到 60%左右。同一土壤上五种不同的轮作制度所产生的影响是很不一样的。以 1894 年不同轮作制度开始时的土壤含氮量为 100%，30 年后，三区轮作地的土壤含氮量为 80%强，连作玉米地的土壤含量只剩下不到 40%了。

至于土壤中微生物和小动物的生命活动就更为活泼了。上述植被的影响实际上是高等绿色植物和土壤微生物和小动物的协同作用。土壤微生物的生长、繁殖和死亡不仅仅有季节性的变化，而且有每日的变化，因而，由于农业技术措施而引起的土壤环境条件的短暂变化（耕作、施肥、灌排等）能导致土壤微生物生命活动的重大变化（Pochon et Barjac 1958，563-575 页；陈华癸，1962，309-320 页）。关于土壤微生物的活泼性质及其对土壤肥力的多方面影响是本文探讨的主题。

土壤中小动物的活泼性及其对土壤肥力的影响，早在达尔文的著作中就认真地探讨过了（Darwin，1881）。可是关于土壤小动物的研究工作却又是土壤生物学中的最薄弱的部门。土壤小动物既是土壤有机物质分解和腐殖质形成的积极参加者，又是土壤物理机械状态的改造者（Russell，1961，169-196 页；Jacks，1963）。本文没有条件讨论这方面的问题，只能在这里着重指出，土壤生物学的这个分科在我国还是空白的。

本文主要从三个方面讨论土壤微生物对土壤肥力所起的作用：

（1）土壤中植物营养元素循环的微生物学；

（2）土壤有机质转化的微生物学；

（3）土壤微生物对于土壤水、气条件和其他植物居住环境的影响。

土壤中植物营养元素循环的微生物学

从土壤微生物学的开创时期起，土壤中植物营养元素循环的微生物学就是一项主题，几十年来，取得了不少重大的科学成果。然而，以现有的成绩和这项问题对于土壤肥力和植物营养的重要性来比，和农业生产实践对于这项问题所提出的要求来比，现有的成绩实在是太微小了。即使是对氮、磷这两种主要植物营养元素来说，研究工作也是做得不深、不透的，因而也还不能在指导农业生产实践中充分发挥其应有的作用。在这里提出这几句话是因为，近来有些研究者们认为，土壤中植物营养元素循环的微生物学已经过时了（例如，Красильников（1958）的书——《土壤微生物和高等植物》——就没有讨论这个方面的问题），这个问题已经是土壤微生物学的历史问题了，研究植物土壤营养的微生物学应该是研究土壤微生物所产生的生物活性物质的问题了。当然不能否认研究生物活性物质的重要性（后面要提出这个问题），但是，不能说这种把土壤中植物营养元素循环的微生物学看成为过时的问题的见解是符合实际的，甚至于可以说，这种见解是有害的。

氮 可以说，土壤中氮素变化的主要环节都是生物学的。这是一向受到足够的重视的。这里提出其中的几个问题。

关于固氮微生物的种类及其作用的知识，近二十年来大大地增加了（Jensen 1950；Mortensen，1962；Каленская，1963；Moore，1963）。这就需要对于自生微生物的固氮作用对土壤氮素积累和植物氮营养的重要性进行新的估价（Mortensen，1962），还没有较明确的答案。蓝藻在水稻田中的固氮量是很可观的（De & Mandel，1956；黎尚豪，1964）。有些文献指出，Clostridium 的固氮作用可能大于 Azotobacter（Parker，1954，1955）。还没有条件估计其他自生固氮微生物在土壤中的作用强度。至于“微嗜氮性”究竟是什么性质（Pochon et Barjac，1958，85-86 页）？在时常遇到的微嗜氮性土壤微生物中究竟有没有能固氮的？种类多少？用现代方法研究这个问题还是处于开始阶段，有待于系统深入。

关于固氮菌（Azotobacter）肥料的实效以及其作用的实质，仍然是在争论中（Cooper，1959；Рубенчмк，1960，229-276 页；Мишустин и Наумова，1962；Мишустин，1963）。樊庆笙（1963）总结了我国的研究工作。鉴于我们正在对土壤中自生固氮微生物的种类、数量和固氮强度进行着新的估价，自然也值得探讨，用其他固氮微生物做细菌肥料的可能性。除蓝藻以外，这方面的工作基本上还没有开展。

Azotobacter 作为耕作土壤的熟化指标问题，看来是有比较广泛的意义的（Рубенчик，1960，160-165 页），但是，还没有人提出 Azotobacter 的发展和耕作土壤熟化相联系的原因的见解。

固氮蓝藻的固氮作用在自然界中的重要性是可以肯定下来了（Штина，1962；黎尚豪 1964）。在水稻田中应用固氮蓝藻作为生物肥料的技术正在发展阶段（黎尚豪，1964）。除水稻田以外，独立生活的固氮蓝藻和含有固氮蓝藻的地衣及其他植物共生而产生的固氮作用也是不可忽视的（Штина，1962）。

关于豆科植物的共生固氮作用的生物学、生物化学和农业技术问题一向是受到重视的。我国的研究单位不少，绝大部分的力量是致力于解决应用技术和与应用技术紧密关联的某些应用基础问题，这自然是恰当的（陈华癸，1964），当前的问题是如何把有关的研究力量组织起来，把这项应用技术问题解决得更快更好。

应该对非豆科植物的共生固氮作用投入一定的研究力量（参考 Allen & Allen，1958；Mckee，1962，71-80 页，98-102 页）。

可喜的是，我国也已经开展了生物固氮作用机制的研究工作，而且是在无细胞提取液的水平上迎头赶上的（张宪武，1963）。

关于硝化细菌的研究工作，长期以来都着重于理论方面，这自然是必要的。关于土壤中硝化作用的利弊问题，应该将硝化作用和反硝化作用或硝酸盐的淋洗作用结合起来看。鉴于植物既能吸收铵态氮，又能吸收硝态氮，两者的有效性差别不大（McKee，1962，9-18 页），似乎硝化作用的有利方面不及它和反硝化作用或硝酸盐淋洗作用结合所起的有害作用大（指土壤含氮量的损失），特别是在水稻土中（三井进午，1956，中译，58-61 页；陈华癸，周啟，1963）。土壤淋溶采集器的研究结果指出，硝酸盐淋洗损失的量是很大的（Russel 1961，314-317 页，Alison，1955）。土壤中由于反硝化作用而损失的氮量也很强（Loewenstein et al，1957）。

水稻田中有兼嫌气性的硝化细菌和硝化作用的发现（陈华癸，周啟，1963）更加指出了这个问题对土壤肥力和农业生产的重要性。

美国正在发展防止硝化作用的农用药剂（Goring，1962；Turner，Warren & Andreisson，1962）。

关于氨化作用，将在后面结合着有机物质的分解问题讨论。

磷　虽然在早期的土壤学和土壤微生物学研究工作中也一般地指出了土壤中生命活动对于土壤磷素的转化起着重要的作用（Stoklasa，1911；Waksman & Starkey，1930，173-176 页），可是，在 40 年代以前，研究者们并没有认真地研究这项微生物学问题。人们主要是把决定土壤磷素营养状况的因素归结为非生物学的化学变化，亦即不可给性含磷化合物的化学风化作用和可给性含磷化合物的化学固结作用。近二十年来的研究成果逐渐地指出了土壤中生命活动（亦即土壤磷素的生物循环）对于土壤中磷素营养状况的重要性（Peirre，1948；Thompson，1957，242-266 页；陈华癸，1963）。

土壤含有机磷量约为全磷量的 1/4-1/2（表层土壤）。土壤有机磷的矿化（生物作用）对于植物磷素营养的重要性是不可忽视的（Thompson，1957，244 页），有时它还是土壤中可给性磷含量消长的决定因素（Mcdonnell & Walsh，1957）。对有机磷细菌肥料的研究和推广所做的努力，也说明了近些年来人们对于这些问题的重要性的认识（Верезове пр.，1959，119-137 页）。然而，由于有关土壤磷素的生物循环的研究工作开展得比较迟，现有的知识还是比较贫乏的（陈华癸，1963）。对于有机磷细菌肥料的实践效果和作用本质也还不很清楚（Cooper，1959；Мишустин и Наумова，1962；Мишустин，1963）。

前前后后的研究工作指出了，即使是含磷矿物的可给化作用，也在很大的程度上受到土壤中微生物活动的促进作用（Gerretson，1948；Краснльников и Котелев，1956；Березова пр.1956；Муромцов，1957），包括：①产酸性强的微生物对磷酸三钙和磷灰石的水解作用；②并非由于产酸性强而起的作用，例如硅酸盐细菌（陈廷伟和陈华癸，1959）以及其他微生物的作用（Муромцев，1958；Кубаицкая，1963）。

硫 在非钙质土中，除水溶液里含有少量无机硫酸盐以外，几乎全部的硫素都含蓄在有机硫化物中（Williams et al，1960）。可以说，土壤硫素循环的生物学性质是和氮素循环一样强烈的。在这个过程中，硫酸盐的有机质化和有机硫化物的分解是高等植物和土壤微生物共有的生命活动。蛋白质分解和矿化的终产物既有氨，也有硫化氢。可是土壤中硫化物氧化为硫磺并进一步氧化为硫酸的作用则完全是土壤微生物（主要是硫化细菌）的作用（Starkey，1950；Walker，1957）。土壤中的硫酸盐在缺氧条件下还能被反硫化细菌还原而产生硫化氢。水田土壤中的反硫化细菌数量可以很大（Takai，Koyama & Kamura，1956）。硫化氢的毒害作用是日本老朽田水稻烂根的一项重要原因（三井进午，1956 中译 80-93 页）。至于在我国水稻烂根问题中，硫化氢毒害究竟起多大的作用，还是一项有待澄清的问题。

铁 微生物多方面地参与铁素转化作用（Alaxander，1961，387-388 页）：①早在十九世纪末叶，Виноградский[1888，1952（中译）39-41 页]就指出了铁细菌氧化亚铁为高铁的作用，近来，Glesn（1951）用现代研究方法证实了这个作用。②土壤中有机质的分解能显著地降低周围氧化还原电位，因而使得高铁还原为低铁。Bromfield（1954a，1954b）指出，在有适宜的给氢体条件下，许多土壤微生物能还原高铁化合物为低铁化合物。③不少有机营养型的微生物能分解含铁的有机物质，产生无机铁化物。Crowford（1956）认为，微生物分解溶解性高的含铁有机化合物，将铁质沉淀下来，是灰壤的铁质淀积层形成的原因。④微生物分解有机物质，产生有机酸，后者又能和铁结合而成为有机酸的铁盐。⑤在嫌气条件下，由于反硫化作用而产生的硫化氢和溶解性的亚铁结合而形成硫化铁沉淀。三井进午[1956（中译）20-22 页，80-93 页]认为这种作用是防止水稻田硫化氢毒害的土壤化学机制。⑥很多低产冷侵田富含亚铁。锈水田实际上是土壤含亚铁特高的表现（亚铁在水面氧化为高铁）。冷侵田低产和亚铁过多之间的关系还有待于深入研究。于天仁、刘畹兰（1957）指出，土壤含亚铁量高于 100ppm 对水稻有毒害作用。

其他元素 显然，凡是植物和微生物生活所需的元素都随着生物的合成和分解而进行着相应的生物循环。同时，植物和微生物的代谢产物也或多或少地对于各种元素的变化发生影响。这方面的研究工作比较零散，这里就不逐一提出了。

土壤有机质转化的微生物学

Внльямс 的土壤形成学说指出，有机质的合成和分解是土壤形成的本质。对于土壤形成、土壤肥力和农业生产来说，碳素的生物循环的重要意义是多方面的：①二氧化碳是植物的必需养料，而土壤空气和大气中的二氧化碳只是自然界碳素生物循环中的一个组成环节，而且是一个薄弱环节；②土壤水中的碳酸根对于成土母质的风化、多种矿质养料的可给性和土壤 pH 值都有十分深刻的影响；③随着有机物质的合成和分解（亦即碳素的有机质化和无机质化），伴随着各种植物营养元素的不可给化和可给化。④有机物质的分解作用是土壤微生物生活的主要能源；⑤土壤腐殖质的形成和分解是自然界碳素循环的一个特殊方面，它对于土壤形成、土壤肥力和农业生产的意义又是多方面的。腐殖质是土壤胶体的组成成分，腐殖质是土壤结构形成的重要因素，腐殖质的形成和分解以及它的吸着作用对于植物养料的供给情况起着重要的调节作用。⑥最后，在土壤中还不断地产生着许多种有机物质，它们是复杂有机物质的分解产物，溶细胞产物和代谢分泌物，这些物质对于土壤肥力、植物营养以及土壤微生物的生活也起着多方面的影响，其中，特别值得重视的是维生素类物质和生理活性物质的作用。

生物物质（主要是植物物质和有机肥料）在土壤中的分解作用 这是十分复杂的微生物学过程。作为土壤微生物学问题，主要是研究：①土壤中有哪些微生物能分解生物物质中的哪些成分；②在特定的土壤条件下，哪些种类的作用最为突出；③在特定的土壤条件和农业技术措施下，生物物质分解的动态及其对土壤肥力和农业生产的影响。

土壤中有哪些微生物能分解生物物质中的哪些成分 千方百计地从土壤中分离出能够分解各种生物物质的土壤微生物种类来，确定它们的分解能力，研究它们的生活条件和作用条件，在这方面，尽管已经积累了一些知识，可是和面对着的问题相比，仍然是很残缺、笼统的。

关于能分解木质素和复杂的脂类和芳香族化合物的微生物种类和其作用条件的知识，仍旧差不多是空白的（Pochon et Barjac，1958，237-245 页；Henderson，1960）。这不仅是由于木质素分解得很缓慢，更主要的是由于：①我们还不能制备出比较单一的、没有起什么变化的木质素样本来；②我们还不很清楚木质素的化学结构。就目前的知识来说，木质素的分解作用可以分为两个阶段（Hendsrson，1960）。第一个阶段是木质素分解为简单的环状化合物（丁香醛、香草素等）的阶段。在木材腐朽的过程中，这阶段主要是“白色腐朽菌”（高等真菌）的作用。还不知道它们是否也在土壤（特别是耕作土壤）中起作用。第二个阶段是简单的环状化合物的进一步分解的作用，似乎能起作用的种类较多。

对于分解其他生物物质成分的微生物种类来说，我们遇到的困难是，各种成分并不是只由一种或少数种类所分解的，而是，土壤中有很多种类都能够分解同一种成分，要把能分解各种生物物质成分的土壤微生物种类查清楚，几乎是做不完的工作。

由于这种困难，使得有些研究者们满足于把鉴定工作做到形态类群或生理类型。这样做，自然是深入不下去的。

由于这种困难，有必要将上述第①项和第②项任务结合起来做，即研究特定的土壤和农业措施条件下，哪些种类的作用最为突出，或优势种的研究。着力于优势种的研究，就有可能，集中力量深入地研究少数种类。

优势种的研究 生物物质在土壤中分解，本身就起了加富的作用，刺激了相应的微生物种类的发展。Виноградский（1924-1945，见（中译）345-442 页）从正面和反面提出了，当动植物有机残体进入土壤中后，发酵性微生物种类，由于旺盛地分解这些有机物质而大量发展，在特定的土壤条件下，由于参与了特定有机成分的分解而旺盛发展的微生物种类就是相应的优势种类。

研究工作指出，就是在特定的土壤条件下，进行特定的有机物质的分解作用，占优势的种类并不是前后一致的，而是分为几个接替的阶段的，在各个阶段中，优势种类各有不同（Красильников и Ничпкина，1945；Рыбалкина и Кониненко，1959；Garret，1956，33-34 页）。这种优势种类更替的情况，不仅出现于成分复杂的植物物质的分解过程中，也出现在单一有机成分的分解过程中，例如，比较纯的纤维素材料（Tribe，1957，1960；Maiejowska & Williams，1963）。这样就进一步地增加了这项问题的复杂性。

关于上述第③项事务，即在特定的土壤和农业技术措施条件下，生物物质分解的动态及其对土壤肥力和农业生产的影响，似乎，土壤学家和农业化学家的兴趣比土壤微生物学家的兴趣要大些，尽管这个过程是微生物学的。

有机物质的分解不单纯是碳素转化问题，也是氮素、磷素以及其他有关元素的转化问题。

几十年来，研究者们已经对于厩肥、堆肥、植物物质分解的动态提供了很有价值的基础知识[Waksman，1932，588-642 页；Harmsen & Van Schreven，1955（氮）；陈华癸，1963（磷）]。把这些基础知识和特定的自然条件和农业利用结合起来，将是很有生产实践意义的。

特别值得重视的是绿肥的分解问题。以现有的科学资料和这个问题在农业上的重要性相比，实在是太不够了。至于水田绿肥的，以及结合我国生产实际情况的科学资料就更少了。

生产经验一般地指出，绿肥是既能提高后茬作物的产量，也能培育土壤肥力的。可是，这些年来，科学研究者们的见解却是分歧的。如果说，研究者们一般都肯定了施用厩肥对于提高和保持土壤腐殖质含量的积极意义，不少研究者们却怀疑绿肥在这方面的作用。

Russell（1961，255-263 页）认为，绿肥通常只能改善土壤的氮素营养状况，对于提高土壤腐殖质含量没有什么作用。Broadbent（1948）用同位素 C^{13} 和 N^{15} 追踪幼嫩的苏丹草的分解过程，发现它不但没有提高土壤中的腐殖质含量，还加速了腐殖质的矿化。Mortensem（1963）用 C^{14} 和 N^{t5} 追踪苜蓿绿肥的分解作用，他指出，绿肥分解是否产生加速土壤有机质矿化的作用是决定于多种因素的。

当然，这是一个十分复杂的问题。评价绿肥的肥效，不能孤立地研究它的分解作用，需要结合着实际的农业生产情况，逐年地考查有关的动态平衡（包括农作物产量的增减、土壤有机质含量的增减、养料元素水平的变化等），这是一项急待开展的研究项目。

腐殖质的形成和分解　对比同一本书的前后三版[指 *Russell：Soil Condition and Plant Growth* 7 版（1937），8 版（1950）和 9 版（1961）]，可以鲜明地看出，关于腐殖质的本性的见解，二十多年来有很大的变化。目前的主导见解认为，腐殖质的主要成分是胡敏酸，后者是一种高分子体系，它是由一些简单的有机成分聚合而成的，主要的成分有：简单糖类、糖醛酸、氨基酸、氨基糖、多元酚以及它们的多缩物等（Кононова，1963，33-38 页；Russell，1961，268-276 页）。因此，腐殖质的合成作用可以分为两个阶段：①这些有机成分的形成作用，和②各种有机成分缩合成为高分子的，以胡敏酸为主的腐殖质物质的过程。

上述第一阶段无疑是和土壤微生物的生物活动密切关系的，可能包括两个方面的微生物活动：①微生物分泌酶（主要是水解酶），将植物物质（以及其他生物物质）直接地分解为上述各种简单有机成分；②微生物的代谢产物、分泌物和溶细胞产物。

我们还不能判断第二个阶段是生物学的还是非生物学的变化。Кононова（1951，中译 132-143 页，1963；78-84 页）认为，微生物分泌的氧化酶（过氧化酶和多元酚氧化酶）参与腐殖质形成的酶促作用。然而，我们还没有研究资料证明，在土壤生物活动基本停止的情况下，有没有酶促的或非酶促的腐殖质物质的合成作用。

腐殖质是在不断地形成着和消失着的。腐殖质的消失过程无疑地是微生物的分解过程，但是，对于分解腐殖质的微生物的研究工作，却是很残缺的。Виноградский（1924-1945）的腐殖质分解试验表明，土壤中的土生性微生物区系有分解腐殖质的能力。最近，Мишустин и Накитин（1961）报道，分离出了能分解胡敏酸的细菌纯培养（属于 *PS.fluorescens*）。然而，Burges 和 Latter（1960）的研究却指出，用胡敏酸加富培养所获得的真菌种类并没有分解胡敏酸的明显能力。

关于土壤有机质或腐殖质的形成、积累和消失的规律，参考 Jenny（1941）和 Кононова（1951，1963）的书。Jenny 系统地探讨了美国气候、植被和农业利用对土壤有机质的动态平衡的影响。Кононова 系统地探讨了苏联主要土类的腐殖质的质、量特征，以及各土类的水、热条件对于土壤生物活性的影响，将后者看作为土壤腐殖质的形成、积累和消失的主要原因。

总括而言，耕作土壤中的腐殖质含量低于相应的生荒土壤（Jenny，1941，251-253 页；Кононова，1963，208 页），可是，按每克土（或每克腐殖质、每克氮）的微生物数量计算，耕作土壤又高于相应的生荒土壤（Мишустин，1956，中译 125 页）。这就提出了这样一个问题，究竟耕作利用（照大多数实际利用的情况而言）是提高了土壤肥力，还是降低了土壤肥力？

如果说，每克土的微生物数量代表土壤中生物循环的强度，因而在一定意义上代表土壤的有效肥力；而土壤的腐殖质含量则代表腐殖质形成和分解的库存量，因而在一定意义上代表土壤的潜在肥力；那么，土壤的熟化过程（耕作利用过程）通常是从潜在肥力较高而有效肥力较低的生荒土壤转变为潜在肥力较低而有效肥力较高的农业土壤的过程。

从这个概念推导，能不能说，农业土壤的培肥土壤和提高产量的矛盾和统一，在一定程度上，意味着，保持或提高腐殖质含量和调节及提高生物循环的强盛度的统一。

可以几个主要的水稻土为例来说明这个见解。例如，不少湖泥土，腐殖质含量较高，但是排水困难，或有些冷浸，微生物的生命活动较弱，不能及时地提供充足的有效养料。培肥这类土壤，提高产量的主要问题就在于如何加强微生物的生命活动（排水、改善土壤的通气情况），以提高有效肥力。又例如，不少白善土，水、气条件好，土壤燥性，土壤微生物的生命活动旺盛，有效肥力高。但是正因为如此，腐殖质形成和分解的库存量低，潜在肥力低，对作物生长而言，比较容易脱肥，对土壤肥力而言，比较容易衰退。培肥这类土壤，提高产量的主要问题，就是大量施用厩肥或其他有机肥料，种植绿肥，提高腐殖质含量，提高潜在肥力（普通白善土变为油白善土）。

生物物质分解和微生物代谢所产生的有机物质和生理活性物质　土壤中含有各种各样的、生物来源的有机物质。

土壤中含有多种简单有机化合物。土壤中所含的简单有机酸主要是甲酸、乙酸、乳酸和丁酸，而以乙酸为主（Schwartz，Verner & Martin，1954；Takai，Koyama & Kamnda，1956；泷鸣康夫，1960，1961）。有机酸和其他挥发性有机物的累积可能对高等植物有毒害作用（泷鸣康夫等，1960；Самцевич и Вогисова，1963），土壤翻耕和排水有清除这些有害物质的作用。

土壤中有很多种多缩糖和多缩糖醛酸，它们既是植物物质的成分，又是土壤微生物的代谢产物。这类物质或者自由存在，或者与胡敏酸合在一起（因而两者都包含在腐殖质的含意之内），或者是更大的有机-无机复合物的组成成分（Forsyth，1947，1950；Graveland & Lynch，1961）。

土壤中有各种氨基酸，它们既包含在高分子的腐殖质和生物物质的蛋白质分子的水解物中，也以自由状态存在于土壤的溶提物中（Dadd，Fowden & Pearsall，1953；Payne，Rouett & Katznelson，1956；Parl & Schmidt，1960）。

土壤中含有氨基酸，不仅是有机氮矿化的原料，而且可以为植物和微生物直接吸收利用。它们是氨基酸异养型的微生物的必要养料，它们在土壤中的存在丰富了土壤微生物的种类，并调节了各种氮素营养类型的微生物种类的相对强度（Lochhead，1952；Starhey，1958）。

土壤中还有氨基糖及其他的简单有机氮化物（Bremner，1958；Sowden，1959）。

含氮基是土壤核酸的组成成分。土壤中还有核酸、磷脂和植酸的盐类。植酸可能是土壤中最主要的有机磷化物（Bremner，1956；Ulrich & Benzler，1956；陈华，1963）。

土壤中的有机硫可能存在于动植物残体和腐殖质（Williams，Williams & Scott，1960）中，以及含硫的有机代谢分泌物中（含硫氨基酸、乙族维生素等），在嫌气条件下，硫醇和硫化氢一同产生（Takai & Asami，1962）。

土壤微生物是许多种维生素类物质*的产生者（Schmidt，1951；Schmidt & Starkey，1951；Lochhead，1958；Красильников，1958，中译，298-262 页）。

土壤微生物还产生赤霉素和类似赤霉素的物质，能刺激植物生长，同时，土壤微生物也产生一些能抑制植物生长的物质（Красильникоа，1958 中译 333-361 页；Красильников，1963）。

土壤微生物产生多种抗菌素。现在已经有充足的证据证明，抗生性微生物不仅在试验室和其他特殊的培养条件下有强盛的抗菌作用，在土壤中也在产生着，并且起着显著的作用，它是土壤中群体关系的一项重要调节因素（Красильников，1958，中译 382-387 页；Brian，1957，1960）。

土壤有很强的酶活性，这些酶或者是微生物和其他生物的自溶产物，或者是它们的分泌物（郑洪元，1963）。

提出以上这类土壤有机成分，主要是因为，一方面它们都是生物起源的，另一方面它们又是土壤肥力的一个不可忽视的方面，或者对植物起着直接的作用，或者通过它们对土壤微生物的影响而对植物起着间接的作用。

土壤微生物对于土壤水、气条件和植物居住环境的影响

土壤结构和土壤孔隙　这是影响土壤水、气的重要因素。改善土壤的结构和孔隙状况是土壤耕作和土壤改良的主要任务之一。

苏联和其他国家的研究工作充分地说明了土壤生物活性对于土壤结构形成和水、气动态的重要意义（Внльямс，1948，中译 177 页；Мальцев，1954，中译 10-46 页；Russell，1961，423-446 页）。Jacks（1963）总结了有关的研究工作指出，土壤的物理肥力，与其说是一种物理化学现象，不如说是一种生物物理现象。他的比喻是：成土母质是砖，生活在土壤中和地面上的整个作物群是泥灰，而生活在土壤中的生物则是城市（指土壤结构）的建设者。

决定土壤结构和孔隙状况的因素包括：质地、无机质胶体、腐殖质胶体、代换离子、根系的机械作用和分泌物、微生物的分泌物和机械作用、土壤动物的机械作用和排泄物、土壤的耕作、改良措施等。其中，比较活泼的因素主要是生物学性质。

土壤微生物的生命活动对于土壤结构和孔隙状况的影响是多方面的（Jacks，1963）。首先，有机质分解和腐殖质的形成和分解是土壤稳定性结构（微团粒和较大的团粒）的重要形成因素之一（McCalla，1945；Iimura & Egawa，1956；Egawa & Seklya，1956）。其次，微生物的黏液物质的胶黏作用也是土壤团粒结构形成的重要因素（Гелцер，1940；Martin，1946，Рудаков，1953；Swaby，1949）。微生物菌丝的缠绕作用对于土壤团粒结构的形成也起着一定的作用（Swaby，1949；Iimura & Egawa，1956；Martin，et al 1959）。值得指出的是，改善稳定性和不稳定性土壤结构的各种作用，对于改良土壤、提高产量都是有价值的，过分地强调稳定性结构，因而过分地贬低不稳定性结构的意义是有片面性的。

除了由于改变了土壤结构和孔隙状况而影响了土壤水的运动之外，土壤微生物的旺盛发展对于土壤的渗漏性还起着另外的阻碍作用。Allison（1947）认为，微生物的代谢产物能阻碍土壤孔隙，阻碍水的运动。这种代谢产物的影响也是两方面的。一方面是微生物产生的黏液物质溶解于水中，提高了水的滞性（渗漏率与水溶液的滞性成反比例，Russell，1961，378 页）；另一方面，微生物代谢所产生的气体阻塞了土壤孔隙，隔断了水柱，从而降低了渗漏作用（Christennin，1944，1947）。

土壤的气体成分　土壤气体成分也受着土壤生物（主要是植物根系和微生物）的强烈影响。土壤气体成分不同于大气成分主要是土壤中生命活动的结果。在粘性重的旱地土壤中，土壤气体中二氧化碳气的含量可以达到 11.5%，而氧气只剩下 10%（见 Russell，1961，366 页）。在淹水或其他嫌气条件下，土壤气体中可以完全不含氧气，CO_2 成为主要的气体成分，同时还产生很多可燃性气体（Takai，Koyoma & Kamura，1956）。在严格嫌气条件下，稻草发酵发生等量的甲烷和二氧化碳（Acharya，1935），在粪坑和草塘里发生的沼气可以达到能够燃点和利用的程度。

Takai，Koyoma & Kamura （1956）指出，在淹水条件下，土壤微生物对于土壤气体的影响可以分为三个交替的阶段：①好气性代谢消耗氧气的阶段；②嫌气性代谢阶段，产生二氧化碳、氢和有机酸；③甲烷发酵和硫酸还原阶段，在这阶段中，以二氧化碳和硫酸盐为受氢体，氢和有机酸为给氢体，产生甲烷和硫化物。

土壤反应　决定土壤反应的因素包括：无机胶体、腐殖质胶体、离子、二氧化碳、根系和微生物的分泌物、有机质分解的产物等。

对于土壤反应的影响因素，研究者们一向侧重于探讨土壤胶体性质和代换离子（主要指阳离子）之间的相互作用，而往往忽视土壤中不断产生着的碳酸和其他酸类（有机酸和无机酸）的影响。Russell（1961，109 页）指出，在石灰性土壤和盐碱土中，植物生活所接触的实际 pH 要比通常测定的数值低得多，因为根际的 CO_2 量很高，对 pH 有很显著的影响。Whitney 和 Gardner（1943）的试验表明，在通常条件下测定 pH 为 9.5 的土壤，如果放在含 1%CO_2 的空气中测定就下降到 8.0 以下。如果空气中的 CO_2 提高到 0.77 大气压，土壤 pH 就从 9.2 下降到 6.4。然而，在水稻田土壤中，CO_2 占土壤气体含量一半左右，却是平常的现象（Takai，Koyoma & Kamura，1956）。

前面提到，土壤中有一些有机酸存在，在嫌气条件下，有机酸（以乙酸为主）含量可以达到 2m.e./100g 土，对于阳离子代换量在 10-30m.e./100g 土壤，特别是<10m.e./100g 的土壤来说，对于土壤 pH 的影响不会是很小的。

土壤中矿酸也必然对土壤 pH 有一定的影响，对于特别酸的土壤（pH<3.0）来说，矿酸是造成这样低的 pH 的主要原因。Jensen（1927）报道了含硫铁矿的土壤的 pH 为 2.2，它是由于 Thiobacilius thiooxydans 氧化硫铁矿产生硫酸而造成的。

土壤中有机酸和矿酸的产生都是土壤微生物的作用。

在一丘田地中，土壤 pH 值随部位和时间而有很明显的变化。Мишустин（1956 中译 56 页）引用 Ремезов 的资料，同一丘田的不同部位的 pH 值的变化为 6.0~8.2，同一部位隔两个月的变化为 6.7~8.2。没有对这丘田施石灰。无疑地，生命活动的季节变化和不一致性是造成这种变化的主要因素之一。

土壤的氧化还原电位　决定土壤氧化还原电位的因素包括：大气氧气的透入（决定于孔隙状况和含水情况），无机物氧化还原体系，有机物氧化还原体系，根系和微生物的生物氧化过程，此外，水稻根系还向土壤分泌氧气。

作为一个总的氧化还原体系来看，土壤的氧化还原电位主要决定于大气氧气的透入（氧化因素）和植物根系和微生物的生物氧化作用（还原因素）（于天仁等，1959）。

氧气进入土壤的情况随土壤孔隙状况和水分状况而变化。在排水良好的砂性土壤的表土中，土壤空气中的氧气含量大致和大气差不多；而在粉砂性的黏土中，土壤空气中的氧气含量就比大气少得多了，即使在表土中，有时也会不到10%，在3-5尺深入，有时可以下降到1%左右（见Russell，1961，366-367页引用的资料）。在泡水条件下，土壤气体中的氧气完全消失（Takai， Koyoma & Kamura，1956，1957）。

加易分解有机物质于土壤中，土壤的氧化还原电位迅速下降（Кононова，1951，中译300~302页；Takia，Koyoma & Kamura，1957）。这些变化，都是由于根系和微生物的代谢作用所引起的。土壤不是一种匀质的体系，对于氧化还原电位来说，更为突出。即使在透气性优良的土壤中也包含有高度缺氧的微区，因而也有严格嫌气性的微生物生活。

植物根系（实际上是根系加根际微生物）对土壤的氧化还原电位有很强的影响，对一般旱作物来说，由于根系和根际微生物的强盛呼吸作用，根际内的氧化还原电位比根际外要低些（于天仁、李松华，1957）。但是，在淹水的水稻田中，水稻根际的影响却是相反的，根际的氧化还原电位比根际外高些（于天仁、李松华，1957），这主要是水稻根系分泌氧气的缘故（Неуныпов，1948）。

问题和任务

在以上几节中，粗放地摊开了土壤微生物的生命活动对于土壤肥力的多方面作用。无疑地，人们对于这些方面的知识愈丰富、愈深透，就愈加能够能动地维护、改造和提高土壤肥力，从而提高农业生产水平。因此，关于土壤肥力的微生物研究应该认为是，为农业生产服务的应用基础研究，其中也有不少问题是从土壤培肥和改良的特定问题和技术措施提出的，因而就属于应用技术研究的范畴了。

做好这些应用基础研究，为相应的应用技术提出充分的科学根据，从而科学地解决一系列有关的应用技术问题，并且把比较成熟的应用技术成就推广到生产实践中去，是为我国社会主义建设服务的土壤微生物学工作者们的主要职责。

去年夏天，我在中国土壤学会1963年学术年会的分组会上提出了发展土壤微生物学的三项指标的意见，即，以十五年到二十年的努力做到：①凡是已经能够应用到农业生产上去的土壤微生物学专业科学技术，都用上去；②全面地掌握土壤微生物学的课题，研究方法和研究成果；③在某些理论和实际问题上做出些突出的成绩来。我认为，这不是什么不现实的要求，我国是地大、物博、人口众多的社会主义国家，在中国共产党的正确领导之下，有信心、有条件，在不很长的时间内，在每一个重要的科学领域中，都赶上世界先进水平。

就土壤微生物学的目前发展情况来说，已经能够应用到农业生产中去的土壤微生物学专业科学技术的项目还不很多，主要的问题已经在国家的科学技术规划中落实下来了。认真地做好这些工作是土壤微生物学研究的首要任务。

其次是分别地占领土壤微生物学的每一个领域问题。当然，这并不是单纯从学科范围出发的问题，而是如何将土壤微生物学的每一个小领域和农业生产中实际存在的问题联系起来的问题。

从上面几节的内容中可以看出，提出的土壤微生物学的每一个小领域都是和土壤肥力的特定问题相联系的，在不同的自然环境和农业实践中，不同的土壤微生物学问题成为培肥和改良土壤肥力、提高农业生产水平的关键。

因此，我们就有可能、有必要将全面占领土壤微生物学领域的问题和研究特定的农业科学应用技术和应用基础的问题统一起来，在解决一系列应用技术和应用基础问题中占领学科领域。

如果我们把土壤微生物学（狭义的）分为若干个小领域，譬如说，把氮素循环分为5~10个小领域，磷素循环分为 2~4 个小领域，分别地与特定的土壤肥力和农业生产问题结合起来。譬如说，把硝化作用的微生物学作为一个小领域，把反硝化作用的微生物学作为另一个小领域，针对着土壤（特别是水稻土）的氮素损失问题进行研究，在阐明无机氮素转化的土壤微生物学规律中谋求减少土壤氮素损失的科学技术及技术原则。

氮素循环	5-10
磷素循环	2-4
硫素循环	2-4
其他元素循环	1-2
碳素循环和生物物质的分解	5-10
腐殖质的微生物学	3-6
土壤微生物的代谢产物	5-10
土壤微生物对土壤的水、气条件和其他植物居住环境的影响	3-6
其他问题	10-20
共计	36-72

这样，36~72 个小领域，分别地落实在科学院、农业科学研究单位和学校的土壤微生物学的教学和科研单位，没有理由不能在 5~10 年内，全面地掌握每一个小领域，包括这小领域的主要知识、研究方法、发展历史和当前的研究情况，并且针对着生产实践中的需要，在这个小领域中的某些具体问题上开展研究工作。

至于第三项指标，只要我们认真地达到了前两项指标，在党的正确领导下，在毛主席思想的指导之下，理论联系实际地开展科学研究，那么，经过 10 年、15 年的工夫，在某些问题上作出突出的成绩来，应该是统计学的必然性。

关键在于，面对生产，适当分工，做好规划，定下来，坚持下去。

最后，还需要指出，本文的讨论范围只是狭义的土壤微生物学，但这并不意味着，对于土壤微生物学应该给以狭义的理解。广义的土壤微生物学的重要意义是不容忽视的。只是限于知识面，限于水平，限于篇幅，在这篇文章中，不能不把学习和写作的范围缩小一些。

参考文献

[1]于天仁，谢建昌，杨国治. 1959. 水稻土中决定氧化还原电位的体系问题.科学通报. 6：205-206

[2]于天仁，李松华. 1957. 水稻土中氧化还原过程的研究，II 土壤与植物的相互影响.土壤学报. 5（2）：166-174

[3]于天仁，刘畹兰. 1957. 水稻土中氧化还原过程的研究. 土壤学报. 5（4）：292-304

[4]华中农学院（主编）. 1962. 土壤学（下册）. 北京：农业出版社

[5]陈华癸（主编）. 1962. 微生物学（2 版）. 北京：农业出版社

[6]陈华癸. 1963. 土壤生物活性和土壤有机态磷的来源、积累的矿化——主要研究成果和问题，微生物学进展（高尚荫主编）. 北京：科学出版社

[7]陈华癸. 1964. 做好发展根瘤菌肥料的科学技术工作.人民日报

[8]陈华癸，周啟. 1963. 水稻田土壤中的硝化作用和硝化微生物的研究 II，水稻田土壤的亚硝酸细菌加富培养体在培养条件下的繁殖和亚硝酸化强度. 中国土壤学会 1963 年学术年会论文摘要. 46 页

[9]陈廷伟，陈华癸. 1959. 钾细菌的形态生理及其对磷钾矿物的分解能力.微生物. 2（3）：104-112

[10]张宪武. 1963. 第三届全国微生物学会（北京）分组报告

[11]郑洪元. 1963. 土壤活体. 手稿

[12]黎尚豪. 1964. 蓝藻的固氮作用及其在农业上应用的前途，湖北省土壤学会、微生物学会 1964 年 6 月学术讨论会报告

[13]樊庆笙. 1963. 我国土壤中固氮细菌的研究及其展望. 土壤学报. 11：220-228

[14]三井进午. 1959. 水稻无机营养、施肥和土壤改良（中译）. 上海：上海科技出版社

[15]泷鸣康夫. 1960. 水稻土壤と有机酸代谢の水稻生育阻害性に关する研究，第一报，国产シソカグルとる为有机酸の chromatographyよるの应用. 日本土肥志. 31：435-440

[16]泷鸣康夫，盐岛光洲，有田裕. 1960. 土壤と有机酸代谢の水稻生育阻害性に关する研究，第二报，有机酸の根生长并に养分吸收阻害，日本土肥志. 31. 441-446

[17]泷鸣康夫. 1961. 土壤と有机酸代谢の水稻生育阻害性に关する研究，第三报，土壤による有机酸の吸收よ土壤有机酸定量法，日本土肥志. 32：130-134

[18]Березова Е.Ф.，Доросинский А. М.，Лонатина Г. В.，Менкина Р. А.，и Лазарев Н. М. 1955. Примепение батериальных удобрений Сельхозгиз，Москва

[19]Березова Е. Ф.，Уарога В. Н.，Нестерова Г. Н. и Нестерова Е.И. 1956. Иепользование бактерий，расщепляющих трехкальценый фосфаг，для улучшения фосфорного питанкя растений и повышенкя эффективнести фосфоритной муки Бюл. Н. -Т. Инф. по с. -х.микробиология ио. 1，18-

[20]Вильямс В. Р. 1948. 耕作学原理[中译].北京：中华书局.1953

[21]Вильямс В. Р. 1949. Почвоведенме，6 изд．Гос-изд-во，Москва

[22]Виноградский С. Н. 1888. 于铁细菌，土壤微生物学问题[中译]. 北京：科学出版社. 1962

[23]Виноградский С. Н. 1924-1945. 土壤微生物学中的方法[中译]，土壤生物学问题. 科学出版社. 1962

[24]Гельцер Ф.Ю. 1940. Значение миероорганязмов в образованми перегноя п прочности сгруктуры поягы сельхозгиз，Москва

[25]Калепиская Т.А. 1963. Азогофиксирующие мнкобактерня，выделенные из дерновподзолистых чочв Почвенная и с. -х. микробиология изд．АНУССР，Ташкент

[26]Кононова М.М. 1951. 土壤腐殖质问题及其研究工作的当前任务[中译]. 北京：科学出版社. 1956

[27]Кононова М.М. 1963. Органическое пещество почвы изд．АНССОР，Москва

[28]Краснльников Н.А. 1958. 土壤微生物和高等植物[中译]. 北京：科学出版社. 1962

[29]Краснльников Н.А. и Котелев В.В. 1956. Влияыко нечвепиых бактормн на усвоенве растениями соедмнений фосфора доклады АНСССР，110，5，858-861

[30]Краснльников Н.А. 1963. Осьювные пребхемы и задачи севременной с.-х. и почвенной микробиологик Микроорганизмы в сельскомхозяйстве 9-34 изд．Московского университета，Москва

[31]Краснльников Н. А. и Нитикяна Н.М. 1945. Влияиме разлагающихея иерией на состав микрофлэры и печве Печвеведение 1945.（2）：131-135

[32]Кублицкая М.А. 1963. Бактерни，разрущаюшие соедимемие фосфора，в почвах крыма в связн с культурой винограда Михрооргамизмы в сельскомхозяйстве изд．Московского университета，Москва

[33]Лысеико Т.Д. 1954. 在提高农作物产量中植物土壤营养的科学任务[中译]. 苏联农业科学. 3：1-9

[34]Мальцев Т.С. 1956. 促使农作物获得连年丰收的土壤耕作法及种植法采夫耕作法全苏会议讨论录[中译]. 10-46，北京：财经出版社

[35]Машустнн Е.Н. 1956. 微生物和土壤肥力[中译]. 北京：科学出版社，1958

[36]Машустнн Е.Н. 1958. Бактериальные удобреиия и их эффективнесть Микробиология 32：911-917

[37]Машустнн Е.Н. и Наумова А.Н. 1962. Бактериальные удобреиия，их эффективнесть и механизм действия Минробиолегия，31：543-555

[38]Муромцев Г. О. 1957. О роли продуктов жизнедеятельнести почвениых микроорганизмов в мобилизации P_2O_5 фосфоритов Агробиолотия. 103：96-103

[39]Нёунылов Б.А. 1948. Окислительно—восстановительные процессы в почвах рисовых полей и методы управления ими с целью повышения урожайности Приморокое краевое управление сельского хозяйства сб.ваучвых работ，вып.1，Владивосток

[40]Рубенчик Л.И. 1960. Азотобактер и его применение в сельскомхозяйстье изд．АНУССП，Киев

[41]Рудаков К.И. 1953. Почвенная структура и деятельный перегной Труд конференции по вопресы почвой микробиология изд．АНССОР，Москва

[42]Рыбалкина А. В. й Кононенко Е. В. 1959. Микрофлора разлагаюшпхся растельных остатков Почвоведеиие. 5：21-34

[43]Самцевач С. А. и Борисова Б.Н. 1963. О токснчности летучих веществ образуеиых микроорганизмами в почве Микробкологая.32：484-492

[44]Штена З. А. 1962. 蓝藻的固氮作用[中译].生物学动态. 北京：科学出版社

[45]Acharya，C.N. 1935. Studies on the anaerobic decomposition of plant materials III.comparison of the course of decomposition of rice straw under anaerobic，aerobic and partially aerobic conditions，Biochem. J.，29，1111-1120

[46]Alexander，M. 1961. Introduction to soil microbiology John Wiley & Sons，New York

[47]Allen，E.K. Allen，O.N. 1958. Biological aspects of symbiotic nitrogen fixation Eandbuch dor pflanzonphysiologie，B. 8，48-118

[48]Allison，L.E. 1947. Effect of microorganisms on permeability of soil under prolonged submergence.Soil Sci.，63，439-450

[49]Allison，F.E. 1955. The enigma of soil nitrogen balance sheets.Adv Agron.，7，213-250

[50]Bremner，J.M. 1956. A reveiew of recent work on soil organic matter

[51]Bremner，J. 1958. Amino-sugars in soil J.Sci.Food & Anric.，8：528-532

[52]Brian，P. W. 1957. The ecological significanco of antibiotic production Microbial Ecology，The Univ.press，Cambridgc，168-188

[53]Brien，P. W. 1960. Antagenistic and competitive mechanisms limiting survival and activity of fungi in soil The Ecology of Soil Fungi，Liverpool Univ. Press，Liverpool

[54]Broadbent，F. E. 1948. Nitrogen release and carbon loss from soil organic matter during decomposition of added plant residues Proc.Soil Sci. Soc. Amer.，12：246-249

[55]Bromfield，S. M. 1954a. The reduction of iron oxide by bacteria J. Soil Sci.，5:129-139

[56]Bromfield，S. M. 1954b. Roduction of ferric compounds by scil bacteria J. gen. Microbiol.，11，1-6

[57]Burges，N.A. Latter，P. M. 1960. Microbio'ogical problems associated with the decompositlon of humic acid The Ecology of Soil Fungi，Liverpool Univ. Press，Liverpool，239-245

[58]Christlansen，J. E. 1944. Effect of entrapped air on the permeability of soils Soll Sci.，58:355-365

[59]Christiansen，J. E. 1947 Some permeability characteristics of saline and alkali soil Agr. Engin.，28：147-150

[60]Cooper，R.，1959 Bacterial fertilizer in the Soviet Union.Soil & Fert.，22：327-333

[61]Crawfeld，D. V. 1956 Microbiological aspects of podzolization Rapports，be I. C. S. S.，paris，C，197-202

[62]Dadd，C.，Fowden，L.，Pearsall，W. 1953 An investigation of the free amino-acids in organic soil types using paper partition chromatography. J. Soil. Sci.，4：69-71

[63]Darwin，C. 1881 Darwin on Humus and the Earthworms.Faber & Faber，London，1945

[64]De，P. K. Mandel，L.N. 1956 Fixation of nitrogen by algae in rice soils. Soil Sci.，81：453-558

[65]Egawa，T. Sekiya，K. 1956 Studies on humus and aggrega te formation.Soil & Plant Food，2：75-82

[66]Forsyth，W. 1947 Studies on the more soluble complexes of soil organic matter I. Method os fractionation，Biochem. J.，41：176-181

[67]Forsyth，W. 1950 Studies on the soluble complexes of soil organic matter II. The composition of the soluble polysaccharide fraction Biochem.J.，46：141-146

[68]Garret，S. D. 1956 Biology of Root-infecting Fungi The Univ. Press，Cambridge

[69]Gerretsen，F. C. 1948 The influence of microorganisms on the phosphate intake by the plant.Plant and Soil，1：51-81

[70]Gleen，H. 1951 Some aspects of the metabolism of a new group of iron-oxidizing microorganisms，J. Gen. Microbiol.，5，xv

[71]Goring，C. A. I. 1962 Contral of nitrification by 2-chloro-2（trichloromethyl） pyridine.Soil Sci.，93：211-218

[72]Graveland，D. Lynch，D.1961 The distribution of uronides and polysaccharides in the profiles of a soil.Soil sci.，91：162-165

[73]Harmsen，G. M. Van Schreven，D. A. 1958 Mineralization of organic N in soil. Adv. Agron.，7：299-398

[74]Henderson，M. E. 1960 Studies on the physiology of lignin decomposition by soil fungi The Ecology of Soil Fungi，Liverpool Univ. press，Liverpool，286-296

[75]Iimura，K，Egawa，T.，1956 A study of decomposition of organic matter and aggregate formation. Soil & Plant Food，2：83-88

[76]Jacks，C. V. 1963 The biological nature of soil productivity.Soils & Fert.，25：147-150

[77]Jenny，H. 1941 Factors of Soil Formation，McCrwa-Hill Book Comp.，New York

[78]Jensen，H. L. 1927 Vorkomen von Thiobacillus thiooxydans im daenischen Boden Centbl. f. Bakt.，pasasitenk. u. Jnfektionskr.，72（Abt II），242-246

[79]Jensen，H. L. 1950 A survey of biological nitrogen fixation in relation to the world supply of nitrogen Trans.，4th I. C. S. S.，Amsterdam，1：165-172

[80]Lochhead，A. 1952 Soil microbiology，Ann. Rev. Microbiol.，6：185-206

[81]Lochhead，A. G. 1956 Soil bacteria and growth-promoting substances，Bact. Rev.，22：145-153

[82]Loewenstein，H.，Engelbert，I. E.，Attoe，O. J. & Allen，O. N. 1957 Nitrogen loss in gaseous form from soils as influenced by fertilizers and management Soil Sci. Soc. Proc.，21：397-400

[83]Lyoms，T. L. Buckman，H. O. 1943 The Nature and properties of Soils（4th ed.）

[84]Maiejowska，Z. Williams，E. B. 1963. The effect of cellulose additions and moisture level on mycoflora of soil，Canad. J. Microbiol.，9：155-162

[85]Marthin，J. P. 1946. Microorganisms and soil aggregation II.Influence of bacterial polysaccharides on soil structure，Soil Sci.，61：157-166

[86]Martin，J. P.，Ervin，J. O. Shepherd，R. A. 1959. Decomposition and aggregating effect of fungus cell material in soil，Soil Sci. Soc. Amer. proc.，23：217-220

[87]McCalla，T. M. 1945. Influence of microorganisms and some organic substances on soil structure，Soil Sci.，59：287-297

[88]McDonnell，P. M. & Wasch，T. 1957. The phosphate status of irish soils with particular references to farming systems，J. Soil Sci.，8：97-112

[89]McKee，H. S. 1962. Nitrogen metabolism in plants，Clerendon press，Oxford

[90]Moore，A. W. 1963. Occurrence of non-symbiotic nitrogen-fixing microorganisms in Nigerian soils，Plant and Soil，19：385-395

[91]Mortensen，L. E. 1962. Inorganic nitrogen assimilation and ammonia incorporation The Bacteria，Acd.press，New York & Londen，119-166

[92]Mortensen，J. L. 1963. Decomposition of organic matter and mineralization of nitrogen in Brookston silt loam and alfalfa green manure，Plant and Soil，19：374-384

[93]Parker，C. A. 1954. Non-symbiotic nitrogen-fixing bacteria in soil I. Studies on *Clostridium butyricum*，Aust. J. Agr Res.，5：90-95

[94]Parker，C. A. 1955. Non-symbiotic nitrogen-fixing bacteria in soil II. Studies of Azotobacter Aust. J. Agr. Res.，6：388-397

[95]Paul，E. D. Schmidt，E. 1960. Extraction of free amino-acids from soil，Soil Scil. Soc. Amer. proc.，24：195-198

[96]Payne，T.，Rouatt，J，Katznelson，H. 1956. Dectectection of free amino-acids in doil，Soil Sci.，82：521-524

[97]Pierre，W. H. 1948. The phosphorus cycle and soil fertility J1/2，Amer. soc. Agron.，40：1-14

[98]Pochon，J. et Ed Barjac，H. 1958 Traited Microbiologie des Sols，Dunnod，Paris

[99]Russell，E. M. 1961. Soil Condition and plant Growth（9 ed），Longmans，London

[100]Schmidt，E. L. 1951. Soil microorganisms and plant growth substances I. Historical Soil Sci.，71：129-140

[101]Schmidt，E. L. Starkey，R. L. 1951. Soil microorganisms and plant growth substances II. Transfomation of certain B-vitamins in soil，Soil Sci.，71：221-231

[102]Schwartz，S. M. Vernr，J. E. Martin，W. P. 1954. Separation of organic acids from several dormant and incubated Ohio soils，Soil Sci. Soc. Amer. Proc.，18：174-177

[103]Sowden，E. 1959. Investigation on the amounts of hezasamines found in various soils and methods of their determination，Soil Sci.，88：138-143

[104]Starkey，R. L. 1950. Relations of microorganisms to transformation of sulfur in soils，Soil Sci.，70：55-65

[105]Starkey，R. L. 1958. Interrelations between microorganisms and plant roots in the rhizosphere Bact. Bev.，22：154-172

[106]Stoklasa，J. 1911. Biochemscher Kreilouf des phosphateions in Boden Centbl. Bakt.，29（Abt.II），358-519

[107]Swaby，R. J. 1949. The relationship between microorganisms and soil aggregntion，J. gen. Microbiol.，3：236-254

[108]Takai，Y.，Koyoma，T. & Kamuda，T. 1956. Microbiogical studies on the reduction process of paddy soils，Rapports，6e I.C.S.S.，paris，C，527-531

[109]Takai，Y.，Koyama，T. & Kamuda，T. ??? Microbial metabolism of paddy soils III-IV，Effect of iron and organic matter on the reduction process，J. Agri. chem. Soc. Japan，31：211-220

[110]Takai，Y.，Asmi，T. 1962. Eormation of methyl mereaptan in paddy soils，Soils Sci. & plant Nutr.，8：40-41

[111]Thompson，L.M. 1957. Soils and Soil Eertility，MsCraw-Hill Book Comp.，New York

[112]Tribe，H. T. 1957. Ecology of microorganisms in soils as observed during their development upon buried cellulose film，Bacterial Ecology，The Univ. press，Cambrige，287-298

[113]Tribe，H. T. 1960. Decomposition of buriod cellulose films with special reference to the ecology of soil fungi，The Ecology of Soil Fungi，Liverpool Univ. press，Livcrpool，246-256

[114]Turner，C. O.，Warren，I. E. & Andriessem，E. G. 1962. Effect of 2-chloro-6（trichloromethyl）pyridine on the nitrification of ammonium fertilisers in field soils，Soil Sci.，94：270-273

[115]Ulrich，Von B. U. Benzler，J. H. 1955. Der organische gebundene in Boden Zeits. f. Pflanzenernahr.，Dung. u. Bodendund.，70，220-249.

[116]Waksman，S. A. 1932. principles of Soil Microbiology（2 ed.），Balliere，Tinda. & Cox，London

[117]Waksman，S. A. & Starkey，R.L.，1928.The Soil and the Microbe，J. Wiley & Sons，New York

[118]Walker，T.W. 1957. The sulfur cycle in glassland soils，J. Brit. Crassland Soc.，12：10-18

[119]Whitney，R.S. & Gardner，R. 1943. The effect of carbon dioxide on soil reaction，Soil Sci.，55：127-141

[120]Whitney，R. S. & Gardner，R.，1943. The effect of carbon dioxide on soil reaction II. An apparatus for the electrometric titration of soil suspension with carbonated water，Soil Sci.，56：63-65

[121]Williams，C. H.，Williams，E. G. & Scott，N. M. 1960. Carbon，nitrogen，sulphur and phosphorus in Scottish soils，J. Soil Sci.，11：339-346

根际微生物群对植物土壤营养的作用*

陈华癸

（华中农学院）

（一）

自从 1953 年 9 月李森科在全苏列宁农业科学院大会上做了关于“植物土壤营养科学在提高农作物收获量中的任务”的报告[1]以后，对于李森科在该报告中所提出的植物土壤营养的微生物学路线问题，在苏联引起了学者们的重视，研究和争论，有好几篇有关的文献已经译成中文，在《苏联农业科学》杂志上发表。同时，在中央农业部的直接领导下，我国也展开了全国范围性质的混合施肥法的试验研究工作。李森科所提出的混合施肥法的理论基础就是植物土壤营养的微生物学理论。因此，无论在关于植物土壤营养的理论方面，或者是对农业生产中的实践意义方面，都可以说是第一等重要的问题。然而，学者们对这问题的意见是不一致的，正在热烈地争论着。

目前争论的焦点，并不在于是否承认微生物在土壤肥力和农业生产上的重要性，这是一致承认了的；也不在于是否承认微生物的生命活动对植物土壤营养起着重要的作用，这也是一致承认了的。问题在于这重要性应该如何正确的体会；在这方面，意见是很不一致的。

道拉辛斯基等[2]认为：“施入土壤的肥料，想必不能直接进入植物，而乃是一种原料，从这些原料在土壤中形成本身组成非常复杂的可给态的养分，这些养分基本上是土壤微生物生命活动的产物。”李森科在某种程度上是支持这概念的，土尔钦和康诺诺娃等是反对这概念的。[3]虽然他们承认微生物的生命活动的产物对植物对植物土壤营养的重要性，但是不能承认施入土壤中的肥料如可溶性的化学肥料（例如：过磷酸石灰等）不能直接进入植物，而只是一种原料，要基本上经由微生物的生命活动才转变为可给态的养分，土尔钦在他的报告中所提到的，用示踪原子研究植物吸收磷质的过程，完全可信服的证明植物可以直接自土壤中吸收可溶性磷质化学肥料，而无需先由微生物转化。

百多年来累积的科学资料，尤其是近 20 年来的精确的试验研究，足以充分证明，植物可以从水中或土壤中直接吸收可溶性的无机养料，而不是一定要作为原料，再由于土壤微生物的生命活动，将这些原料改造为可给态的植物养料。

但是，这并不等于我们就可以否定土壤微生物的生命活动对于植物土壤营养的直接的意义，在植物土壤营养的实际情况之下，在农业生产的实际情况之下，我们施入的肥料，即使是可溶性的无机肥料，究竟有多大比重是植物直接吸收的？多大比重是先经由微生物的生命活动再行供给植物的？两项是同等的重要，还是那一项是主要的？这是问题的焦点。

（二）

植物生长在土壤中，每天不断的吸收养分，我们施用肥料，即使是可溶性的无机肥料，是以每一季作物一次或最多 3、5 次施用的。不可能设想，可溶性的无机肥料在土壤中维持 30、50 天乃至于 3、5 个月不变，静候植物吸收，土壤中微生物强盛的生命活动，不容许这样，根据我们研究水稻田中的氮质养料的变化[4]，硝酸态氮在水田中一两天内就转化为非硝酸态氮，尿素和石灰氮等肥料也在几天之内转化为别种氮化物，因此，即使植物可以直接吸收硝酸和尿素，而在植物土壤营养的实际情况下，在全部

*原载于《湖北农业科学》，1：1~4，1955.

植物生长过程中，所吸收的养料，主要是微生物生命活动的产物，我们所施用的肥料，不论是迟效性的，或者是直接可给性的，主要都是先作为微生物生命活动的原料，经微生物转化后再被植物徐徐吸收。关于这一点，我们应该接受道拉辛斯基的意见。

然而，道拉辛斯基所提出的，由于微生物生命活动所产生的“非常复杂的可给态养分”究竟是什么东西？对于这一点，可以说是一种没有起码的事实根据的假设，这种假设和康诺诺娃所提出的微生物生命活动的少量特殊产物（胡敏酸，维生素类的物质等）的作用的学说是不相侔的，康诺诺娃是作为某些生长因素而提出的，道拉辛斯基则认为，即使是N、P、K质也要先成为这种“非常复杂的可给态养分”，才被植物吸收。

（三）

究竟微生物对植物土壤营养起那些具体的作用？就目前所知，微生物的作用是多方面的。由于微生物的生命活动，一方面改变了土壤中的养料状态，提高或者降低了土壤中植物养料的含量和可给性，另一方面改变了土壤的物理肥沃性，提高或者降低了植物吸收养料的能力。对于后一方面，这里不拟讨论，对于前一方面，其作用也是多种多样的，概括起来可以分为下列四类。

（1）由于微生物的生命活动，增加或者减少了土壤中植物养料元素的总含量。在增加方面，可以固氮作用为典范。在减少方面，可以反硝化作用为例。

（2）由于微生物的生命活动，使得植物养料元素不断的有机质化和无机质化。无机质化强盛，则可给态的养料加多。有机质化强盛，则相反。虽然土壤中植物养料元素的总量并无增减。

（3）由于微生物的生命活动使可给态的养料加多，并不仅限制于有机物质的无机质化。近来的研究工作证明，岩石矿物的化学风化的强度，在很大的程度上决定于微生物的生命活动（生物化学风化）。在这方面，矽酸细菌是一个很突出的例子，它可以把不溶性的，几乎没有肥效的矽酸盐矿物中的磷素和钾素转化为可溶性的植物养料成分。在我们的实验室中，从施用过包含磷矿石粉的混合肥料的土壤（植物根际土壤）中分离出很多矽酸细菌，能够转化磷矿石粉。

（4）由于微生物的生命活动，制造了一些比较复杂的有机化合物，对植物起生长素类物质的作用。在这方面，除微生物能产生一些已知的维生素和植物生长素类物质外，胡敏酸也起着显著作用，这里还值得提出一个问题：即植物营养是否和动物或某些微生物一样，需要一些必需的，不可代替的氨基酸类物质，而这类物质是由微生物的生命活动可以制造的？

但是，无论在上述四类之中，或在四类之外，我们已知的某些微生物的作用，肯定的，只是极少的一部分。正如李森科所说：“植物生活需要的，不只是已阐明的土壤微生物中的一种有相互联系，而且还和许多其他科尚未阐明的土壤微生物有相互联系，在微生物的酶的作用下，不适于吸收的有机质，矿物质和大气环境中的物质和元素转化成适于吸收的状态。”因此，“农业科学应当进一步揭发，植物与土壤环境间，植物营养必需的，自然的，生物学的相互联系，并在这基础上制定出农业技术的方法与措施，以便改善植物营养，增加收获量，减轻劳动并提高劳动生产率。”

（四）

关于上述的争论的问题，我们也有一些实验资料，可以报道出来作为参考。这项实验是在1944年、1945年和1946年在四川北碚，进行的盆栽实验，研究土壤（微生物）接种对望江南（*Cassia occidentalis* 一种不生根瘤的豆颊植物，在西南有些地区作为水稻田绿肥）的营养的作用。由于当时对于微生物在植物土壤营养中的作用的认识十分狭窄，所得结果，无法总结，因此一直没有报道。

这项工作的目的是：研究望江南根际土壤中的微生物的生命活动对于望江南生长的影响。

1944年的试验

土壤　每盆用河沙7kg，掺和土壤（重庆系黏壤质中和性紫色土）750g，用蒸气灭菌（每盆蒸20分钟，并非完全灭菌，但经实验证明可以杀死如根瘤菌等无孢子细菌）。

试验处理

	第1组	第2组	第3组	第4组	第5组	第6组
K_2HPO_4（g）	0.60	0.60	0.60	0.60	0.60	0.60
KNO_3（g）	—	0.95	—	0.95	—	0.95
土壤接种 I（g）	—	—	20.00	20.00	—	—
土壤接种 II（g）	—	—	—	—	20.00	20.00

所加 K_2HPO_4 和 KNO_3 都经加压灭菌，土壤接种 I 为采自泸县的多年种植望江南的土壤。土壤接种 II 为采自内江多年种植望江南的土壤。

播种和管理　种子表面灭菌（用 0.1%$HgCl_2$ 杀死种子表面细菌）。4 月 13 日播种，9 月 16 日收获，在玻璃室内栽培，用冷开水浇灌。

试验结果

组号	一	二	三	四	五	六
每株平均风干重量（g）	3.03	2.81	5.75	6.45	5.18	4.00
株数	15	16	8	10	14	22
每盆平均风干重量（g）	5.69	5.00	7.67	10.65	8.06	11.00
盆数	8	9	6	6	9	8

接种土壤比不接种土壤平均产量高近一倍。施氮肥与否无显著效果。

1945 年试验

土壤　每盆用河沙 5.5kg，掺和土壤（同前）625g，用蒸气灭菌，蒸 30 分钟，土壤含氮量 0.015%。

试验处理

组号	一	二	三	四
K_2HPO_4（g）	0.075	0.075	0.075	0.075
KNO_3（g）	—	0.250	—	0.250
土壤接种 I（g）	—	—	325	325
对照灭菌土壤（g）	325	325	—	—

所加 K_2HPO_4 和 KNO_3 均经加压灭菌。土壤接种来自 1944 年的盆栽实验植物根际。对照灭菌土壤来源相同，但经加压灭菌。

播种与管理　于 4 月 28 日播种，8 月 7 日收获，方法同去年。

图片 I

上	左	1945 年	第一组	（不接种）
上	右	1945 年	第二组	（不接种）
下	左	1945 年	第三组	（接种）
下	右	1945 年	第四组	（接种）

试验结果 （参照图片 I）每盆 4 株，平均风干株重（g）

组号重复	一	二	三	四
1	0.27	0.41	5.80	5.60
2	0.33	0.28	5.80	6.80
3	0.43	0.45	4.30	6.20
4	0.40	0.26	3.70	4.70
平均	0.358	0.350	4.90	5.60

土壤接种比不接种，增产约 12 倍，不施氮肥组每盆中土壤和株的含氮量如下：

组号重复	一（不接种）		三（接种）	
	植株	土壤	植株	土壤
1	4.6mg	869mg	83.3	991
2	5.6	919	85.2	862
3	7.2	987	60.2	899
4	6.7	1105	55.3	874
平均	6.025	970.0	71.0	906.5

这结果指出接种后，土壤中氮素营养料的吸收性增加，但是与第二组（不接种而加施 250mg $NaNO_3$）相比，则氮素养料的吸收性的增加并不是仅只因为微生物将有机氮质无机质化的原因。第二组施用了足够的无机养料（三要素）但仍不能满足植物的需要，这指出所施无机养料（虽然是水溶性）的可给性很低，而接种以后，由于微生物的活动，可给性大为增加了（第二、四组）。

1946 年的试验

土壤 除接土壤以接种剂外，完全用河沙，含氮量很低（见后），每盆装沙 6kg，另加 200cc 无氮培养液，含下列成分：

K_2SO_4	0.9g	Fed_3	0.02
KH_2PO_4	0.5	硼砂	0.02
CaH_2PO_4	0.5	$MnSO_4$	0.02
$MgSO_4 \cdot 7H_2O$	0.5		
NaCl	0.5		

灭菌法如前。

试验处理

	第一组（对照）	第二组（接种）
土壤接种	—	400g
灭菌对照土壤	400g	—

土壤接种来自 1945 年盆栽试验用土。

播种与管理 方法同前。于 4 月 24 日播种，8 月 26 日收获。

图片 II

上	1946 年	第二组	（接种）
下	1946 年	第一组	（不接种）

试验结果（参照图片 II）

每盆植株风干重		每盆植株含氮量		每盆土壤含氮量收获后	
对照	接种	对照	接种	对照	接种
0.971g	4.885	10.8mg	33.4	256.0mg	249.6
0.814	5.880	11.9	35.1	448.0	377.6
0.760	4.830	8.9	34.2	493.0	326.4
0.580	4.225	8.0	26.6	546.0	320.0
平均 0.7812g	4.9550	9.82mg	32.32	435.80mg	324.53

本年试验结果，土壤接种增加植株重量约 6 倍。

（五）三年试验结果总结

（1）接种来自望江南根际的土壤，对望江南生长有重要作用，这种作用是由于微生物群的活动。如将细菌杀死，虽用土壤接种，也没有效果。

（2）微生物的作用是复杂的，其中也包括对于所施用的肥料的影响，如果不接种，虽然施用各种无机肥料，它们的可给性还是很差，接了种，不论是在肥料充分或肥料不足（氮少或不少）的情况下，都能使植物生长得更好。而且，施用了氮肥，如果没有接种，氮肥无效果。

（3）这试验所指出的微生物的作用并不能用已知的任何具体的微生物生命活动的规律来理解。

（4）对于道拉辛斯基的学说，这试验有参考价值，即指出肥料（无机的，可溶性的）施入土壤后，也要求微生物的生命活动，使之转化为植物能吸收的状态。但是，对于道拉辛斯基所假设的“非常复杂的可给态的养分”，本实验不能提供进一步探讨的资料。

参考文献

[1]李森科. 植物土壤营养科学在提高农作物收获量中的任务（中译）. 苏联农业科学. 1954，（3），1-8

[2]道拉辛斯基. 关于微生物在植物根部营素中作用问题（中译）. 苏联农业科学. 1955，（2），76-77

[3]土尔钦. 植物的营养和肥料的使用（中译）. 苏联农业科学. 1954，（12），7-12

[4]陈华癸，萧泽宏. 水稻田土壤中的无机氮素化合物，武大理科季刊. 1948. 第 9 卷（1），79-88

[5]亚历山大雷夫. 扎克. 分解铝矽酸盐的细菌——矽酸盐的细菌（中译）. 苏联农业科学.1953，（11），8-11

[6]克里斯吉娃. 关于胡敏酸和其他有机物质参与高等植物的营养作用的研究（中译）. 在翻译中，将在《土壤译报》中发表

水稻土特性的发展和水稻田的绿肥耕作制*

陈华癸

（华中农学院）

序　言

过渡时期农业建设对农业科学提出的要求是：建立一套先进的科学的耕作制度。这个制度必须能够不为提高土壤的肥力，这个制度要有实践基础，要有不断发展的前途，要与农业生产合作化的发展相适应，这样才能保证不断地提高农业生产，获得高额而稳定的产量。威廉斯的草田耕作制对苏联社会主义农业建设的伟大贡献就在于此。它就是这样一种能够满足上述各项要求的耕作制度。但问题在于今天我们如何创造性地学习草田耕作制，使它适应于我们具体的自然和社会条件。

威廉斯在其经典著作中指出：阐明土壤生物循环的发展方向和速度是土壤科学的任务；掌握和控制生物循环的发展方向和速度是耕作学的任务。本文想根据杜库查耶夫-柯斯特切夫-威廉斯的土壤学说来分析讨论水稻田土壤特性的发展；并根据威廉斯的土壤肥沃性学说来分析和讨论水稻田的绿肥耕作制。

所引用的资料，一部分是本人过去的一些试验研究结果，一部分是已发表的各专家的科学报告，另一部分是 1954 年 7 月在农业部土壤肥料技术会议上交流的十分丰富的生产经验和研究结果。

一、水稻土特性的发展

由于种植水稻的缘故，使得水稻田土壤具有与别种土壤不同的特性。各种不同的母质和土壤在长期种植水稻的影响下具有了一系列共同的特性。如果说在各种不同类型的植被下所引起的土壤肥力发展的质的特性是土壤分类的主要根据，那么水稻可以看做是一种特殊的植被。水稻植被之下可能发生特殊的植物群社特征，因此进行着特殊的生物循环过程。这就使得水稻土与非水稻土有显著不同，即当非水稻田土壤改变为水稻田土壤时，就引起了一系列的土壤特性的改变。同时，各地水稻土之间都有许多共同的特性，尤其表现在植物群社和物质的生物循环过程方面。

关于水稻和水稻田的土壤，有许多偏见和错觉流传着，需要事先予以澄清：

有人认为水稻是热带和亚热带的植物，水稻和水稻田局限于热带和亚热带，这是完全不正确的。在东北已有广大面积的水稻田，而且尚在不断扩大中。甘肃、河北、陕西等省凡水源充足的地方，就能够种植水稻，而且能够得到高额产量。

有人认为水稻要求酸性土壤，这也是完全不正确的。水稻对于酸碱度的适应范围很广，pH4~9（5.5~8.5）都能生长[3]。实际上，在盐碱地的利用和改良上，水稻占有重要地位。

有人认为水稻田表土浸在水中，土壤颗粒完全分散，这也是不正确的。水稻土还是有小团粒的形成，根据团粒分析结果，珞珈山农场水稻土含>2 毫米的团粒约 20%，2~0.02 毫米团粒 79.6%，机械分析结果没有>2 毫米的颗粒，2~0.02 毫米的颗粒约 34.6%，但这些小团粒结构并不能改变土壤的孔隙性，仍以毛管孔隙为主。

也有人认为水稻根系不需要氧气，这也是不正确的。水稻可以在通气情况差（泡水）的条件下生长，但这只是忍耐，并非理想。当水的条件和通气条件矛盾时，水稻比其他作物更需要水，更能适应于低气压。然而，在满足水的条件下，增加土壤空气供应则是丰产因素之一。在水源有保证的条件下，反复灌

*原载于《土壤学报》，3（2）：97~111，1955.

水和放水晒田是最理想的生长条件，即所谓“浅水勤灌”、“干干湿湿”。

种植水稻的条件和旱作有所不同。种植水稻的地形要平坦，要能修田埂保水。因此，水稻田中一般很少径流和冲蚀现象。另外，在水稻生长时期要有足够的水分（水稻需水较一般作物更多，苏联灌溉水稻田需水率约为 6000 公方/公顷，我国缺少这方面的研究，但估计一般水田当比前述数值要小得多），在整个或绝大部分水稻生长时期，一般的水稻田中表面积水。至于终年积水，对水稻田而言，可以说是不得已，而不是理想的情况。

1. 水稻田的地形，水文和气候条件：地形和水文对水稻的种植和水稻土的分布起决定性的作用。水稻土多发育在起伏地形的中部和下部，很少种植在分水岭上，像这些地方是完全靠雨水的。一般水稻田都在半山腰以下，直至谷地和浅湖，在潮湿多雨的地带，梯田可以接近于山顶，但不能达到山顶（如贵州）。气候干旱的地带，水稻田的位置低限于坡麓。换言之，即水稻田的水源依靠着降雨和山坡流下的水。从这两个来源所得的水量减去水分的蒸发量（包括蒸腾量）和渗漏量，要能经常满足水稻生长的需水要求（土壤水分要接近饱和或更多）。

在干旱地带，除上述水源外，主要依靠人工灌溉补充给水的不足，也就是利用广大面积流来的水量，供给较小面积的水稻田的需要。

（1）湖泊：因湖水太深，不能种水稻。

（2）冬水田：实际是浅湖，终年浸水，可种植水稻。

（3）两作田：比冬水田地势略高，但基本上还是谷地（垅田或畈田）；夏季蓄水种水稻，冬季放水种冬作，但有些地方由于排水困难，冬季休闲。

（4）塝田：在垅田或畈田的上坡，水旱两作。

（5）在水旱两作的塝田之上，冬天蓄水以保障来年插秧的水源，这种冬水田与上述冬水田正相反，不是排水困难，而是冬天不敢排水。

（6）更上则不能种水稻，在丘陵地带，梯田高度视气候与水源而定。在两湖的低丘地带，很少有高于 1/3 坡的，在广州可高到 2/3 坡，有如下列图解所示。

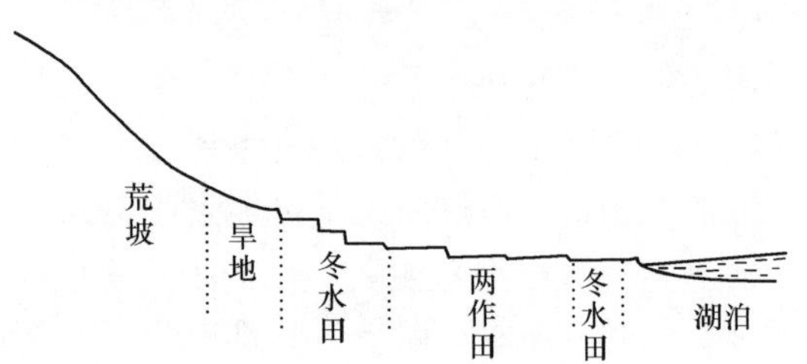

按照水位和地面的相对关系，水稻土可分为下列四种：

（1）浅湖冬水田：大多数冬水田实际是一种浅湖，也就是说，水面在土面之上。

（2）潜育性水稻土：地下水位高，自由地下水面或地下水面的毛细管运动范围直接影响水稻土的性质。

（3）潴育性水稻土：地下水位低，但在地下水位以上有一层排水困难的土层（称为潴育层）。由于潴育层漏水慢，在这层上的土壤经常或大部时间为水分所饱和，而影响水稻土的性质。

（4）渗育性的水稻土：地下排水无阻碍，水稻田中的水分完全依靠充沛的天雨或灌溉，给水量（在水稻生长期内）大于渗漏。当然，就是在这样的条件下，也需要相当黏重的土层。在地下水位低、排水畅顺的砂地上，既灌溉亦很难满足水稻对水的要求，故不能种水稻，也自然没有水稻土发育。

母质对水稻土的影响主要表现在质地和保水层方面，其他母质的性质只能影响各种水稻土的多样性，而对水稻土的发育与否，则不起决定性的作用。

2. 水稻土的植物群社和生物循环：研究水稻土的植物群社特性，须先从较简单的冬水田着手，再进到较复杂的两作田。

冬水田终年积水，水稻是唯一的高等绿色植物（少数杂草除外）。水稻田中的藻类植物和浮萍对水稻土的生物循环也起一定的作用，但属于次要的。

由于每年收获，冬水田中每年的有机质供给量不多（主要是根兜、根系和有机质肥料）。

由于终年积水，水稻土中的主要微生物是嫌气性细菌。有机质的腐解作用主要是嫌气性的。好气性的微生物作用微弱（但不是完全停止的），非强酸性的土壤中硝酸与铵的相对含量是土壤中好气性微生物作用和嫌气性微生物作用的相对强弱的可靠指南。在通气良好的条件下，硝化作用比氨化作用强盛。硝酸是含氮有机物磷化的最终产物，铵则是一种能积累的中间产物。在一般的旱作土壤中，硝酸和氨的相对含量就是这样。在多年生牧草地中，好气性微生物和嫌气性微生物的作用强度相若，硝酸和铵的相对含量也大致相近。

在冬水田中正相反，铵成为无机氮素的主要状态。下面是北碚天生路农场冬水田土壤含无机氮量的比较[6]：

表 1　冬水田中无机氮含量（%）比较

测定时期	冬水田的土壤		冬水田的水	
	NH_3-M	（NO_2+NO_3）-N	NH_3-M	（NO_2+NO_3）-N
43′-12-24	10.6	微量	—	—
44′-1-10	12.7	″	0.3	微量
44′-2-16	5.9	″	—	—
44′-3-1	3.6	″	0.2	—

上述结果指出：在冬水田土壤中，硝化作用极弱。土壤中主要微生物作用是嫌气性的；但这并不是说，冬水田土壤中没有好气性微生物作用。在根孔壁上，一般可见到许多铁锈条痕，这是二价铁化物氧化为三价铁化物的证明。根孔是通气比较好的地方，好气性作用比较强盛。与根孔的铁锈条痕相对比的铁青色的背景，也正好证明冬水田土壤嫌气性作用为主的特性。

水稻土含有机质和其他嫌气性湿土很不相同，表现在它的有机质累积不起来。我国各地水稻土的有机质的含量均在 1%~2%，土壤有机质含量的多少决定于下列因素的相对强弱：

冬水田种植水稻的每年收获给予土壤的有机物质不多。虽说水稻田土壤中以嫌气性作用为主，有利于腐殖化作用，但毕竟由于每年有机物质供给量不多，不能累积腐殖质。在水稻田中，有机物质的分解作用是很强盛的，主要表现于：①压绿肥后，有机氮素化合物矿化作用很强；②腐殖质腐熟程度很高，C/N 比率在 8~12 比 1 之间。

水稻土的生物循环的特殊性还表现在铁质的运动规律上，即所谓“潜水离铁作用”，这在后面将讨论到。

其次，讨论两作田土壤中的生物循环：两作田夏季蓄水种植水稻，冬季排水种谷类或技术作物或绿肥作物。这样，植被情况就有改变，蓄水、排水使空气情况也有了改变。然而，根据土壤中微生物作用过程的研究，它们仍保留水稻土的生物循环的基本特性。首先，在蓄水时期（水稻生长时期），微生物作用仍以嫌气性作用为主，主要的无机态氮素是铵，硝酸含量甚微。在排水种冬作物时期，硝化作用虽然增强，但仍不能与氨化作用相比拟[5]。当田地再度灌水时，表现很强烈的反硝化作用，硝酸在下雨天内完全消失。上述情况说明在两作田土壤中，微生物作用仍以嫌气性作用为主，只是好气性作用在一定时期中比较冬水田土壤旺盛些。

关于水稻田中众多的动物群对水稻土特性发展的影响，还没有人研究过。

3. 水稻土剖面特征的发育：水稻田土壤可以在各种土壤和母质上形成，因此经的特征中必然保有它前一时期土壤特性的某些残余。研究水稻土剖面特征的发育，必须将各种残余特征和新生特征区别开来。各地水稻土的新生剖面特征有本质上的共同性，这就是水稻土的质的特征。

由于不断的种植水稻而产生的剖面特征，主要表现在下列几项：

（1）有一经常潮湿、接近水分饱和的土层，这层或在地面（冬水田），或在地面以下的地下水面或潴育层。一方面是由于有一层比较黏重的土层；另一方面也是水稻土剖面发育的结果，即铁质胶体的淋溶和淀积的结果。

（2）在泡水情形下，由于嫌气性微生物作用的结果，产生了三价铁化物还原为二价铁化物的作用，因二价铁化物的溶解度较大而向下移动。另外，由于季节性的泡水和干燥以及根孔的通气作用，向下移动的二价铁化物氧化为溶解度小的三价铁化物。因此，铁质在一定的层次中淀积出来，这就出现了水稻土剖面的铁锈条痕层和铁质淀积层（即“潜水离铁作用”）。

（3）水稻土的“潜水离铁作用”不同于灰壤化过程中的铁铝移动现象[2, 6]。潜水离铁作用是由于嫌气性细菌还原三价铁的作用，因此只有铁质的移动而没有铝质的移动。在灰壤化过程中，由于克连酸的影响，克连酸的铝盐和铁盐都向下移动，因此表现为铁铝的移动而不单是铁的移动。

（4）关于水稻土腐殖质的特性，除前述累积情况外，还没有作更深入的研究。

4. 水稻土保水特性的发展：旱地改变为水稻田，一般均增强了土壤的保水性。这说明水稻土的保水性是在不断的种植水稻的作用下发展的，主要是由于：

（1）铁质胶体的淀积作用，增加了底土层的黏重性，因而增加了保水性，降低了渗漏性。

（2）犁底的镇压作用，暂时地增加了犁底层的保水性，降低了渗漏性。

（3）土壤微生物所分泌的复糖类胶体物质（荚膜物质），增加了土壤保水性，降低了渗漏性。

（4）土壤微生物呼吸产生的大量CO^2微气泡，割断水流，降低渗漏性。

四项中除第（1）项有持久性的作用外，其他三项都只是在种植水稻的时期中起作用，是没有持久性的。因此，水稻土的保水性需要通过正确的耕作方法，不断地发展和提高。

以上是对水稻田土壤的肥力发展规律的初步分析和讨论，希望能作为今后深入研究水稻土的一个基础和开端。

二、水稻田的绿肥耕作制

水稻和绿肥作物轮作在我国普遍的实行着，而且有极悠久的历史，其中包括极为丰富的生产经验，但却亟待整理。

解放以后，水稻田施用绿肥的科学技术研究和推广工作，成为农业生产技术中一个突出的重点。长江区域各农业科学机构都在努力钻研这个问题。但由于我们对它的基本特征了解不够，工作上或多或少的陷入片面性和狭隘性，把它孤立地看成为一种特殊的施肥制度，而没有把它看成为一种优良的极有发展前途的特殊的水稻生产制度中的一个主要环节，因而不能从整个制度的观点来研究绿肥施用问题。

由于绿肥作用的多样性，并由于各地施用绿肥的多种多样性，如果不能从整个的水稻生产制度上去了解绿肥的多方面的作用，以及不能正确地了解这种特殊的水稻生产制度的基本特性，科学工作者就无法清晰地总结这项十分丰富的农民生产经验，把它提高到现代科学水平的高度，因而也就无法将它发展成为适应于社会主义建设要求的先进生产制度。

本节的目的在于讨论水稻田绿肥耕作制的基本特性，分析这制度各个主要环节的意义，并指出深入了解这制度所应进行的调查研究工作，和实施这一制度的一些关键性的问题。无疑的，这是一个大胆的尝试，但为着推动这方面的科学工作，这种尝试还是必要的。

1. 水稻田的绿肥耕作制：这是一种特殊的农业生产制度，它的特征有下列诸项：①以水稻为主要作物，每年生产一季（或两季）；②以冬季绿肥作物为培育土壤肥沃性的主要作物；③按照各种不同情况，找出一套完整的轮作制，其中包括其他粮食作物或技术作物；④将一个完整的轮作制当做一个生产制度的整体，考虑它的生产总量、劳动生产率和对于土壤肥力的影响；⑤在一个完整的轮作制基础上，考虑和设计土壤的耕作法、施肥法，以及灌溉和排水、品种和播种量等等农业技术措施，要在这个基础上来了解水稻田的绿肥耕作制和别种水稻作制的主要差异。同时也要在这个基础上来分析了解各种水稻田的绿肥耕作制之间的共同性和特异性。

下面是水稻田绿肥耕作制的一种类型：每年在夏季灌水种一季中熟水稻，水稻收种后种一季紫云英，来年春季将紫云英压入田内作绿肥，再种水稻，这样周而复始的轮作。在这样轮作制中应考虑作物品种和种子的选育，土壤的耕作方法，施肥法，灌溉和排水法。在这个制度中以水稻为主要作物，以绿肥为主要的培肥措施，这是一种极简单的水稻田绿肥耕作制。

下面是另一种类型：分三区（区数不拘）轮作，每年夏季各区都种水稻，每年冬季三区轮换，一区种冬季绿肥作物，另两区种麦类、油菜或其他技术作物。待冬季作物收获后，于春天将绿肥分别施用于三区中，然后春耕灌水，进行插秧。 在这样的轮作制度中，也应考虑作物品种和种子的选育、土壤的耕作法……。这制度中同样的是以水稻为主要作物，以绿肥为主要的培肥措施，但同时还包括其他作物，这是一种比较高级的水稻田绿肥耕作制度，也是今后发展绿肥耕作制度的主要方向。

下面又是另一种类型：和前述第一种类型相似，夏季种植水稻，冬季种植绿肥作物，因为绿肥作物同时也是饲料作物，故可收割地上部分的全部或一部分作饲料，发展一定比例的畜牧业（主要是养猪喂牛）。在这样耕作制度上也应考虑各种农业技术措施。

下面是一种极特殊的类型：上述各种类型都应用于两作田中（夏水田、冬旱地）。水田的养萍法则是一种应用于冬水田的水稻田绿肥耕作制。水稻田终年蓄水，夏季种植水稻，冬季在水面养萍，作为绿肥。显然这里水稻是主要作物，而浮萍则是绿肥作物。

当然，还可以举出更多的类型和各种过渡的形式。但是用以上几个类型做例子已经足够说明水稻田绿肥耕作制和其他水稻耕作制度的不同点，以及各种水稻田绿肥耕作制之间的共同点和不同点。

总的说来，水稻田的绿肥耕作制和其他水稻耕作制度的不同点，就在于它包括有对土壤培肥具有重大意义的绿肥作物。同时也就是它比其他水稻耕作制度优越的地方，适合于社会主义农业制度的耕作制度必须能够培育土壤的肥力，并且在土壤肥力不断提高的基础上获得高额而稳定的产量，这是正确的运用水稻田绿肥耕作制所能达到而其他的水稻作制所不能够达到的任务。

培育土壤肥力的关键在于经常的供给植物以大量的有机质。通过土壤微生物的腐解作用，一方面不断地产生新鲜的活性腐殖质，来改良土壤的物理性质；另一方面使有机物质矿化来供给充足的植物营养料。除了绿肥耕作制外，其他水稻田的耕作制度不能普遍而且经常的供给这样大量的有机物质，来促进土壤微生物活动。

各种水稻田的绿肥耕作制，适用于不同的环境条件和不同的耕作技术发展阶段。以下将以长江区域种植一季中熟稻的两作水稻田为主要对象，涉及其他情况，讨论水稻田绿肥耕作制的各个主要环节和它们的关键问题及发展前途。

2. 水稻田绿肥耕作制中的绿肥作物：首先讨论冬季绿肥对土壤肥力和对主要作物（水稻）的作用。种植一季冬季豆类绿肥作物如紫云英或苕子，在长江区域可获得1000斤到8000斤的地上部分（青重）和相应的地下部分，这些有机物 在春天翻入土中，在泡水的情况下进行腐解作用。参与这腐解作用的主要微生物群社是嫌气性的，在这种特定的腐解作用下，土壤的物理性质和化学性质有了很显著的改善。物理性质方面，最显著的表现是压绿肥的田比不压绿肥的田松软易耕，减轻板结现象；同时由于土质松软，盛夏时放水晒田，不至于造成过大过深的裂隙，引起再灌时严重的漏水现象。在天气干旱时，压绿肥的田比不压绿肥的田保水性强，裂隙性弱，因而可经受更长期的旱期，以上这些是保证高额而稳定的水稻产量的重要条件。至于更深刻的了解水稻土的物理性质对水稻生产的作用和压绿肥对水稻土物理性质的影响，则需要深入的有系统的科学研究工作。

绿肥压青以后， 矿质化很快。根据在武汉的测定，压绿肥后的两个星期，土壤中的无机氮素养料就有显著增加[5]。无机氮素养料的显著增加，不仅说明土壤中氮素养料水平的提高，而且说明在有机物质矿质化作用进行中一系列的微生物活动的加强，和各种有关植物养料水平的提高。这方面进一步的深刻研究也是十分需要的。

以一季中熟稻为主要作物，在长江流域，有充足的时期种植紫云英或苕子，秋后播种不成问题。开春后一般都在盛花时压绿肥，但也可以稍晚，譬如说在四月压绿肥，五月上、中旬插秧，可以取得很好的肥效。在四月中旬压绿肥，有些品种还在盛花期，有些品种则已结实，可以将需要的种子收下后再行压青。一般来说，绿肥种子成熟较晚，故不能在压绿肥的田中收种子。在成都平原有一种好的办法，即在种子半熟时就将需要的种子收下，在屋顶上晒熟，这样既可以不致压青太晚，也不需另留种子地。当然，种子问题也可以用其他方法解决，限于篇幅，此处从略。

根据农民的经验和我们的实验结果，也可以压绿肥以培育秧田；三月中、下旬压绿肥，以后灌水整地，四月上、中旬播种，肥效很好[4]。

目前绿肥的产量普遍很低，每亩地上青重只 1000 斤左右，只能勉强满足本区水稻的肥料需要。因此，前述的第一种类型目前是最普遍的形式，这种情况就限制了水稻田绿肥耕作制度的进一步扩张，因为因为它使得土地的种植指数不能太于“1”。如果每亩绿肥产量能提高到 5000~6000 斤的地上青重，则一亩绿肥作物可以施三亩地。每年夏季种水稻，冬季三区中有二区种谷类或技术作物，则土地的种植指数可以提高到“$1\frac{2}{3}$”。这样，几乎所有两作田都可以推行水稻绿肥耕作制。

上述情况是完全办得到的，成都平原的农民、各地试验研究机关和先进农民也都证实了每亩收获 5000 斤以上的紫云英或苕子（地上青重）是有把握的。种植得法，在入冬以前就能得 1000 斤以上青重，开春后生长得更旺盛，到四月中旬可达 5000 斤以上。

分析各地紫云英或苕子产量提高的原因，其关键在于：①提高种子的发芽率和增加播种量；②注意开沟排水；③绿肥作物施肥。以上三项，各地都有很好的先进经验，目前有足够的条件提出这样的初步要求：即每亩绿肥只有 1000 斤左右的提高到 2000 斤左右，使得能完全满足本区土壤的要求；已达 2000 斤的，提高到 4000~6000 斤，使得一亩绿肥可供二三亩田地用，从而提高粮食和技术作物的种植指数到“$1\frac{1}{2}$~$1\frac{2}{3}$”。达到这个水平后，就有条件更普遍的推广水稻田绿肥耕作制。

上面是指以一季中熟稻为主要作物。在许多种植双季稻的地区，问题就不在于设法增加冬季谷类或技术作物的面积，而在于如何使得在较短的生长期中获得较多绿肥青重。一般都是在双季稻晚作尚未收获时，将紫云英或苕子撒播在田中（生板法），既不整地，又是撒播，绿肥生长难得健旺。同时，如果不整地、不压播稻根，则可加重螟害，对来年水稻将有很不好的影响。另外，撒播绿肥固然因为有机质的增加可以改良土壤耕作，但却牺牲了秋后耕犁，所以就是对土壤耕性来讲，也是有利有弊。因此，双季稻推行绿肥耕作制，在于双季稻晚作收获后，争取能够先翻耕整地，再行播种绿肥种子。如果选育晚播的品种，播种前要用催芽处理，播种后要注意排水及其他田间管理工作，以提高绿肥产量。如果在这些方面能做出成绩，对于推广双季稻可起更大的作用。双季稻要求土壤肥沃，肥料充足，因而也就更要求能实行绿肥耕作制。种一季晚熟稻中（如粳稻）秋天的问题和双季稻相同，春天将不像种双季稻那样时间紧迫。

以上是指单播紫云英或苕子绿肥。根据各地先进经验，混播绿肥效果更好，湖南多将紫云英和肥田萝卜混播，四川有将苕子和黑麦混播的，据说都比单播好。这些生产经验都值得详细地加以总结，并应配合土壤物理和化学性质的分析研究工作。创造优良的混播绿肥方法，可以大大地提高水稻绿肥耕作制对培育土壤肥力和提高作物产量的效能。

目前，有些地方在推广绿肥作物（紫云英或苕子）和冬麦或油菜间行条播法。实际上这相当于以后行为单位的分区轮作法，作物指数均等于“$1\frac{1}{2}$”。推广者指出这方法的主要好处，在于可以保证绿肥的田间管理和施肥。如果分区种绿肥和冬麦，则一般农民对绿肥田的管理和施肥就不够重视了。在这种间行条播法中，两种植物间的相互关系，应该加以研究。

水稻田养萍是目前所知的唯一有实践基础的适宜于冬水田的水稻绿肥耕作制，值得深入的研究。

3. 水稻田绿肥耕作制的田间管理、灌溉排水、土壤耕作、施肥：首先应该指出，研究水稻田绿肥耕作制的田间管理问题，必须从耕作制的整体着眼和着手。如果孤立的讨论其中某一季作物的田间管理问题，必然的会造成研究问题上的片面性，并且引起许多难以克服的矛盾。

灌溉排水是实行水稻田绿肥耕作制的最重要的环节之一，除水稻田养萍以外，主要的绿肥耕作制必然是应用于两作田。夏季种水稻，在水稻生长的全部时间或绝大部分时间内稻田灌水，土壤成分在饱和情况下，通气情况很差；而冬季绿肥作物或其他冬季作物是旱生作物，冬季作物生长期间，水田变为旱地，土壤水分不可过多，通气情况必需大大改善。实际上，各地生产经验充分证明：正确的开沟排水是提高绿肥作物产量，并从而提高水稻田绿肥耕作制水平的关键问题。目前，大多数绿肥每亩只有 1000 斤左右的青重，其主要原因是由于排水措施不好。

同时，在坡地的冬水田，不敢放水种冬作的主要原因之一，是怕春天灌不上水，耽误了水稻插秧。因此，在大范围内（一个自然水系单位）或在水小范围内（一个自然村或一个农业生产合作社）改进灌溉排水系统是减少冬水田面积，扩大和提高水稻田绿肥耕作制的关键问题。在长江区域，很多地方上坡的水顺坡穿田而过，首先应改为修建池塘蓄水并另开辟沟渠，引水宣泄。

关于土壤耕作法问题，首先要提出的是应尽一切要能避免采用“生板”法。所谓“生板”法，是指在水稻收获后，不经适当的土壤耕作，就将绿肥种子撒播在地面上。“生板”法有许多坏处：首先，将种子播在由于水稻根兜所造成的凸凹不平的地面上，发芽很不一致，出苗率很低，撒在凸起地方的种子因干旱而发芽迟缓，甚至初发芽后幼苗被烈日晒死，较低地方的种子则因水分太多而不能发芽，只有在高低适中的一部分种子得到优良的发芽条件，因此“生板”田的出苗情况一般是斑斑点点的，这就显著地影响了地面覆盖率和青重产量。各地试验结果证明，提高播种量和发芽率是提高绿肥产量的重要关键点之一。提高发芽率一方面要靠种子播种前的处理，一方面要靠改进土壤的耕作质量。

“生板”法的另一个主要缺点是妨碍水稻除螟。杀死螟虫卵块是治螟的主要方法。但用“生板”法种绿肥，水稻根兜仍留在田中，加以有绿肥枝叶的荫蔽，就为虫卵越冬创造了良好的条件。所以过去甚至有反对水稻田种绿肥的建议，这虽是因噎废食的看法，但这矛盾毕竟是应该解决的。

因此，废除“生板”法，提倡水稻收获后绿肥播种前的土壤耕作，是推广和提高水稻田绿肥耕作制的关键问题。水稻收获后即行翻耕，将根兜完全压入土中，耙平，必要时开排水沟，用条播法播种绿肥种子，是结合稻田治螟提高绿肥的重要措施。

必须指出，废除“生板”法是有相当困难的：①劳动力缺乏；②绿肥的生长期间题晚稻或双季稻田和③农民思想上不重视绿肥作物。要克服第一项困难，在组织起来的基础上是完全有条件的。对于克服第三项困难应该指明，通过提高每亩绿肥的产量不仅可以肥田和增加水稻产量，并且可以增加冬季谷类和技术作物的面积。如果一亩地生长 5000~6000 斤绿肥青重，可肥三亩地，则可推行三区轮作法；如果一亩地只生长 1000~2000 斤绿肥青重，则只能实行最简单的水稻绿肥轮作法。前者种植指数为“$1\frac{2}{3}$”，后者只有“1”。克服第二项困难，应从选育绿肥品种和用播种前催芽法解决。选育适宜于各种情况的绿肥品种是十分重要的工作，用搓伤种皮和浸种相结合的催芽法可以提早种子出芽（约一星期左右）。争取一星期的时间，对上述第二项困难的克服往往是成败关键。

以上是针对绿肥作物的栽培而论土壤耕作，至于如何就整个水稻田绿肥耕作制而论的完整的土壤耕作法，目前尚无条件讨论。在各个较先进的地区，总结农民生产经验，分析整理出一套基本耕作法是十分重要的工作。

4. 关于施肥法问题：我们的工作一向陷于把绿肥作为一种肥料，并将它与别种肥料对比，考究它们间的优劣利弊，这是一种很不正确的研究方法。首先绿肥的效果与其他肥料的效果有质的不同，不能就其一点两点做片面的对比，例如，比较其含氮量、含三要素量，或在等氮量的基础上比较其对主要作物的肥效等。绿肥的肥效不能只用它所含的三要素量的多少来评价的，它有更重大的培育土壤肥力的意义，而且，即使我们在等氮量的基础上或在三要素含量的基础上比较其对主要作物的肥效，得到了比较数字，这并不能回答我们所要解决的问题。我们的问题是如何在水稻田绿肥耕作制的基础上进行合理施肥，争取更高的产量和更高的劳动生产率。因此，不是绿肥与其他肥料相互竞争、相互淘汰的问题，而是相互配合、相得益彰的问题。不纠正这观点，我们的研究工作是找不到出路，得不到实际结果的。

因此，应该首先肯定绿肥作物是水稻田绿肥耕作制的基本环节之一，同时也要肯定水稻田绿肥耕作制比其他水稻耕作制在农业建设中的先进性，它是目前所知的有生产实践基础的（不同于在我国尚在想象阶段的水稻草田轮作制），能广泛实行的，并在不断提高土壤肥力的基础上获得高额而稳定的产量的耕作制（不同于现有的其他水稻耕作制）。然后，才有条件研究水稻田绿肥耕作制的施肥法问题。

在施用氮肥方面，各地的经验和试验结果证明以豆科植物为基本绿肥作物，其所含的氮素，基本上可以满足整个轮作制中各作物对于氮素肥料的需要。一亩地生产2000斤绿肥（青重），连地下部分共约有3000斤，含氮15斤，其中10斤以上是从空气中来的，可供给比较高额产量的水稻的基本要求。如果一亩地生产4000斤绿肥（青重），在其他条件配合不上的情况下，水稻会产生施氮肥太多的不良现象，故应该分施二、三亩田中。因此，在水稻田绿肥耕作制中施用其他肥料，重点应放在争取进一步的提高产量的施肥法上，而不是与绿肥比短长。在绿肥播种时施用少量（每亩1~2斤N）氮素肥料（与其他肥料配合），对于绿肥作物的早期发育十分有利，能提高整个绿肥耕作制的水平。 在水稻及其他冬季作物的生长期中，按照生长情况和当时的其他条件，施用适量的追肥是进一步提高产量的措施。除作为追肥的有机肥料（如粪水等）外，其他农家有机肥料应该主要用在不适合于实行水稻田绿肥耕作制的田地上，或者配合秋耕，施用于冬季作物。

根瘤细菌接种，在水稻田绿肥作物的新推广区有很显著的作用，不进行人工接种，依靠天然感染，在新种绿肥的头几年往往生长不好。进行人工接种配合其他优良栽培方法，第一年都可以获得较高的产量。

顺便讨论到秧田绿肥问题。根据农民经验及试验研究，秧田绿肥（秧田泡青）对于以一季中稻为主要的水稻地区是完全有条件的、现实的。根据各地反映，问题不在于水稻播种期早，得不到足够的青重或来不及腐解生效。而在于如果泡青太多，氮素太多，其他条件配合不上，如天阴多雨、阳光和温度不够，将造成烂秧或其他秧田病害等现象。显然，这是利用绿肥的技术问题，不是用得用不得的问题。秧田泡青不可太多，在适量的基础上，按照秧苗生长情况，配合施用追肥，是秧田的正确施肥法。用绿肥长得很好的田做秧田，应在压绿肥之前割去一部分。

这里还可以指出，根据秧田泡青的成功经验，在推广水稻直播的地带，水稻田绿肥耕作制仍旧是可行的基本耕作制。

在磷肥方面，应该指出，绿肥中含有相当丰富的磷素，压绿肥后，在土壤中转化为植物能够吸收的有效性磷。因为绿肥是完全肥料，突出的问题在于绿肥中磷肥本身是要有来源的。因此，强调对绿肥作物施用磷质肥料，在水稻田绿肥耕作制中应该是一个重点。目前，对绿肥作物重视不够，一般不愿对它施肥。但各地试验证明，对绿肥施用是提高绿肥产量，从而将水稻田绿肥耕作制推向更高水平的关键措施之一，对绿肥作物施用磷质肥料，是间接地对主要作物施肥，这点在苏联的草田轮作制和其他绿肥耕作制的实践中早就充分的证明了。肥效迟缓的磷质肥料，如磷矿石粉等更适合于施用于绿肥作物。

钾肥应该和磷肥同样看待，但水稻对钾肥的需要性不很显著，有关的试验研究资料又少，这里暂不讨论。

关于磷钾肥料在水稻土中的生物循环过程，过去还没有研究过，今后应该重视，必须了解其生物循环过程，才能够掌握科学的施肥方法。

根据现有的极少资料，十字花科植物，如肥田萝卜等同化土壤中磷化物的能力比较强，这指出紫云英（或苕子）和肥田萝卜混播的优越性。农民经验也证明混播绿肥比单播好，这点需要更深入的总结农民经验和试验研究。

最后，关于供给有机质方面。水稻田绿肥耕作制是目前有实践基础的、唯一能够普遍大量供给土壤以有机质的水稻耕作制。在讨论这问题之前，首先要指出，杜库查耶夫在他的经典著作中早就指出，黑钙土含腐殖质丰富的原因是和多年生草本植被和它特定的自然条件分不开的。后来的学者们往往疏忽了特定自然条件，而希望通过种植多年生混合牧草，普遍的、机械的追求增加土壤腐殖质含量。这点，在近年来苏联的草田轮作制实践中得到了充分的批判。土壤腐殖质含量的增减决定于：①有机物质的供给量；②腐殖化过程的强弱和③矿物质化过程的强弱。在实行水稻田绿肥耕作制的特定地带和特定条件之下，我们是否应当追求土壤腐殖质含量的不断增长？土壤腐殖质含量应达到什么要求才能保证土壤高度的肥力？这一点是还没有解决的问题。根据在热带和亚热带、半湿润和湿润气候区域土壤的有机质含量、腐殖质化过程等的一般规律来看，初步的意见认为不应该强调土壤腐殖质总量的不断增加，重点应该放在不断的供给新生的、活性腐殖质的问题上。

威廉斯及他的继承者们指出：对土壤肥力有关的首要因素是新生的活性腐殖质，因此，如果每年不断的供给新生的活性腐殖质，而不强求腐殖质总量的不断增加，乃是我们努力的方向。水稻田的绿肥耕作制能够满足这项要求，每年能供给新鲜的有机质材料，经过嫌气性为主的腐殖化作用而产生新生的活性腐殖质。当然，关于水稻田中腐殖质的形成和变化的规律还要求我们更深入的研究。

以上是对于水稻田绿肥耕作制的概括介绍。本文的目的，与其说是说明水稻田绿肥耕作制的基本特性，不如说是希望对这问题引起注意和争论。研究这个有实践基础、有发展前途的水稻耕作制的基本特性的重要性和迫切性是无可怀疑的。只有这样，才能够建立科学的水稻耕作制度，适应于我国农业建设的要求。

总　结

本文主要分两部分：

1. 讨论水稻田土壤的基本特性（各地水稻土壤所共有的而且是水稻田土壤特有的）的发展。首先阐述水稻田土壤的外界环境条件，着重于地形和水文。其次讨论到水稻田土壤生物循环的发展，指出水稻田土壤的植物群社特性为：水稻是主要植被，嫌气性微生物是主要的非绿色植被；指出水稻田土壤有机物质来源的贫乏，和有机质矿化过程的相对旺盛。也讨论了水稻田土壤剖面特征的发展，着重指出铁质的移动和淀积（由于生物化学的还原和氧化作用）。最后讨论水稻田土壤保水性的发展。上述主要特性的发展都指出了水稻田土壤中生物过程的主导作用。由于水稻田土壤有共同的特性的发展，而且这些特性的发展是由于特殊的植物群社的作用，水稻田土壤应看作为一个特殊的土壤类型——水稻土。至于它应作为一个土类或亚类。则是一个尚待解决的问题。

2. 讨论水稻田的绿肥耕作制，分析组成这制度的各个主要环节，指出这制度的先进性，它是有实践基础的、有发展前途的一种水稻田的耕作制度。它以水稻为主要作物，绿肥为培育土壤肥力的基本措施，也可以包括其他冬季作物。按照不同情况制定的水稻田绿肥耕作制具有广泛的实践性，这制度能够在提高土壤肥沃性的基础上获得高额而稳定的产量。指出以一季中熟稻为主要作物的水稻田绿肥以耕作制的进一步提高，首先在于提高绿肥作物的单位面积产量，争取每亩绿肥生产 5000~6000 斤的地上部分青重，因而使水稻田绿肥耕作制的种植指数提高到$1\frac{1}{3}$。指出以双季稻为主要作物的水稻田绿肥耕作制的关键问题，首先在于解决播种绿肥和土壤耕作之间的矛盾。对于上两问题都指出了解决问题的途径和有关的先进生产经验，讨论了争取绿肥作物高产量的一系列的关键问题，包括：轮作制、品种和种子的选择和处理、单播绿肥和混播绿肥、灌溉和排水、土壤耕作法、根瘤细菌接种和施肥方法。并且强调指出，必须就整个水稻田绿肥以耕作制来研究这制度的全面意义，不可孤立地将绿肥看作为一种施肥方法来和别种施肥方法比较某一方面的短长优劣。

参 考 文 献

[1]朱莲青，马溶之，李庆逵. 1941. 中国之土壤概述. 土壤季刊，2（1），4-95

[2]侯光炯. 1941，对于吾国土壤分类法之建议. 土壤季刊，2（1），113-143

[3]侯学煜. 1950.川黔境内水稻土与地形气候和土壤的关系. 中国土壤学会会志.1（3-4），131-140

[4]陈华癸. 1952.秧田泡青的试验报告及讨论。土壤学报，2（1），46-54

[5]陈华癸，萧泽宏.1948.水稻田土壤中的无机氮质化合物。武汉大学理学季刊，9（1），79-89

[6]熊毅. 1941. 水稻土的化学性质. 土壤特刊，甲种第四号

[7]1954 年 7 月农业部土壤肥料技术会议上交流经验的笔记，本文摘引很多，又大都是综合意见，未能详引出处

秧田泡青的试验报告和讨论*

陈华癸

（武汉大学农学院）

一、序　　言

将青嫩的植物茎叶（以冬作豆类为主）泡在水稻秧田里，让它腐解产生肥效，称为秧田泡青。这方法，在长江中游，有不少地方实施着，但是，由于过去农村间相互闭塞，这些点的经验没有广为推行。我们认为这是一种优良的农业技术，经过适当的实验研究，长江流域和长江以南可以普遍推广。1944 年，前中央农业实验所在北碚曾举行一次秧田泡青的初步实验，泡青的肥效很好。1950 年，我们在武汉大学农事试验场也举行了一次田间实验。1951 年，与中南农业科学研究所合作，又举行了一次田间实验，并且配合了化验工作。这篇报告这两年的实验研究结果，并且提出关于秧田泡青的实际应用的讨论。

二、1950 年水稻秧田肥料试验

本试验的目的以不施肥为对照，以施用棉籽饼为标准，比较厩肥、猪粪灰和蚕豆茎叶绿肥（泡青）等三种施肥法的优劣。

1. 试验地点：武汉大学农学院前水稻田。

2. 试验设计：5×4 随机排列，每区 20 平方尺（4×5 方尺，1/300 亩），4 区团，每区团分 5 区。

3. 施肥方法：O 区，不施肥。I 区，施厩肥，农场牛粪褥草厩肥，每亩施用 2000 斤。于播种日施肥，施肥后即行播种。II 区，施猪粪灰，猪粪加草灰（约等量），每亩 1000 斤。施肥法同 I 区。III 区，施蚕豆茎叶绿肥，从蚕豆地割下来施用，时蚕豆正开花，生长旺盛，每亩 3000 斤。于播种前一星期（3 月 28 日）施用（茎叶稍切割），泡水到播种时放水播种。IV 区，施棉籽饼（土法木榨），每亩 300 斤。施肥法同 I 区。

水稻品各栽培方法：胜利秈，每亩播种量 15 斤（每区 0.05 斤），于 4 月 2 日泡种，4 月 4 日播种，5 月 16 日采取秧苗样本。秧田管理如常法。

4. 试验结果

于 5 月 16 日采取秧苗样本，比较秧苗生长情形。取样法分两种。一种每区采取 1 平方尺面积的秧田两方；另一种每区随机采取秧苗 100 株。采取后，洗去淤泥，吸干表面水分，称青重。称青重后，风干，称风干重。各区青重，风干重结果是表 1 和表 2。因表面水分不易吸干，青重不及风干重更能代表实际生长情形。O，I，II 三区间无显著差异。III，IV 两区间无显著差异。O，I，II 三区和 III，IV 两区间有显著差异。本试验结果说明秧田施用蚕豆茎叶绿肥（秧田泡青法）的肥效可以和施用棉籽饼相比。厩肥和猪粪灰肥效很慢，在秧苗生长期中（42 天）未产生肥效。

*原载于《土壤学报》，2（6）：46~54，1951.

表 1 1950 年水稻秧田肥料试验

		区团				平均数	平均数标准差
		甲	乙	丙	丁		
施肥处理区	O	222.5* 37.7+	162.0 45.8	295.2 56.4	159.0 22.5	209.7 40.6	±7.141
	I	247.7 38.4	227.0 30.7	266.2 42.7	165.3 27.7	226.5 35.0	±4.123
	II	180.5 29.8	188.0 40.3	374.7 35.6	223.0 34.3	241.6 37.5	±2.596
	III	395.9 76.6	379.5 80.5	317.4 53.2	307.0 60.4	350.0 67.7	±6.403
	IV	289.1 51.7	177.0 38.4	215.5 76.1	280.6 57.2	240.5 55.9	±7.874

每区取 2 方尺秧苗（42 日）的青重和风干重[*为青重（克），+为风干重（克）]

表 2 1950 年水稻秧田肥料试验

		区团				平均数	平均数标准差
		甲	乙	丙	丁		
施肥处理区	O	24.5* 4.9+	24.0 5.5	26.2 5.1	27.0 5.1	25.4 5.2	±0.125
	I	25.0 4.7	28.5 5.1	44.9 7.8	31.0 5.5	32.4 5.8	±0.622
	II	32.5 5.7	24.5 5.0	28.5 5.1	64.7（？） 5.3	37.6 5.3	±0.160
	III	42.5 8.9	48.0 5.5	52.0 9.1	61.3 9.3	51.0 8.2	±0.866
	IV	40.5 7.7	31.0 6.0	45.8 8.4	40.5 7.1	39.4 7.3	±0.115

每区取 100 株秧苗（42 日）的青重和风干重[*为青重（克），+为风干重（克）]

三、1951 年水稻秧田肥料试验

本试验目的是以不施肥为对照，以施用棉籽饼为标准，比较蚕豆茎绿肥（泡青）的肥效。

1. 试验地点：武汉大学农学院前水稻田。

2. 试验设计：3×7 区随机排列，每区 20 平方尺（4×5 方尺，1/300 亩），7 区团，每区团分 3 区。

3. 施肥方法：O 区，不施肥。III 区 施蚕豆茎叶绿肥，从蚕豆地割下来施用，时蚕豆正开花，但本年蚕豆生长不很好。每亩 3000 斤。于 3 月 20 日（播种前 22 日）施用秧田中，灌水泡青。IV 区施 棉籽饼，机榨。每亩 300 斤。于 3 月 20 日施于秧田中。

4. 水稻品种及栽培方法

胜利籼，每亩播种量 15 斤（每区 0.05 斤），于 4 月 3 日泡种，4 月 11 日播种，5 月 18 日移植，5 月 22 日取样。秧田管理如常法。

5. 试验结果

于 5 月 22 日取样。取样后，测定青重，风干重和含氮量。取样法仍分每区 2 平方尺和每区 100 株两种。测定结果见表 3，表 4 和表 5。由于表面水分不易吸干，青重误差仍大，风干重正确性高。由于丙区团 O 区 2 秧苗奇少，2 方尺样风干重奇小，以致不施肥（O）处理区平均数标准差特别大。但该区每株生长正常，100 株重不特殊。不论 2 方尺样本或 100 株样本，不施肥（O）处理区和施肥处理区（III 或 IV）间的差异均显著。就田间观察，施棉籽饼区（IV）比施绿肥区生长较好，但统计比较，无显著差异。这点和 1950 年实验相反（该年绿肥区比棉籽饼区好，亦无显著差异），可能由于本年棉籽饼施用得早，在田中腐熟得较好，可以更好的产生肥效。

各种施肥处理间相比，含氮量无显著差异。不施肥丙区的含氮量特别低，该区生长不正常，从 2 方尺重量亦可看出。

总结 1950 年及 1951 年结果，可以断论，采用蚕豆茎叶绿肥泡青（每亩 3000 斤）作为秧田肥料，可以与棉籽饼（每亩 300 斤）相比。也就是说，秧田泡青法可以基本上解决秧田肥料问题。

表 3 1951 年水稻秧田肥料试验

		区团							平均数	平均数标准差
		甲	乙	丙	丁	戊	己	庚		
施肥处理区	O	695* 142.0+	750 179.0	235 35.0	680 124.8	700 93.0	614 120.0	615 119.2	611.1 116.1	±16.823
	III	915 157.0	935 185.0	1040 143.5	865 140.4	720 128.0	752 167.0	1070 137.0	899.5 151.0	±7.681
	IV	785 137.3	1195 184.1	1000 187.0	990 165.0	1200 212.2	803 143.0	1125 168.5	1026.9 171.0	±9.3274

每区取 2 方尺秧苗（41 日）的青重和风干重[*为青重（克），+为风干重（克）]

表 4 1951 年水稻秧田肥料试验

		区团							平均数	平均数标准差
		甲	乙	丙	丁	戊	己	庚		
施肥处理区	O	49.7* 11.1+	47.0 7.0	52.0 9.4	29.0 5.0	36.1 6.0	37.3 8.1	37.6 7.0	41.2 7.7	±0.762
	III	62.5 14.5	52.2 10.2	65.3 11.0	102.0 21.8	50.5 9.0	60.8 12.6	71.5 12.8	66.3 13.1	±1.619
	IV	102.8 27.2	124.8 24.0	73.0 11.8	42.5 8.0	105.7 21.5	61.3 9.2	83.5 15.2	84.8 16.7	±2.086

每区取 100 株秧苗（41 日）的青重和风干重[*为青重（克），+为风干重（克）]

表 5 1951 年水稻秧田肥料试验

			区团							平均数	平均数标准差
			甲	乙	丙	丁	戊	己	庚		
施肥处理区	O		0.97 0.98	1.41 1.25 1.28	0.51 0.60	1.17 1.10	1.03 1.05	0.83 0.75 0.96	1.09 1.16		
		平均	0.975	1.313	0.555	1.135	1.040	0.847	1.125	0.9978	±0.093
	III		0.95 0.85	1.06 1.02	1.38 1.43	1.39 1.30 1.25	0.82 0.85	0.97 0.96	0.94 0.95		
		平均	0.900	1.040	1.405	1.313	0.835	0.965	0.945	1.058	±0.083
	IV		1.04 1.08	1.09 1.05	0.84 0.86	1.02 1.09	1.18 1.07	1.30 1.32	1.30 1.33		
		平均	1.060	1.075	0.850	1.055	1.125	1.310	1.315	1.1122	±0.062

秧苗（41 日）含氮量（烘干重%）

四、1951 年水稻本田试验

将施肥处理不同（不施肥，施蚕豆茎叶绿肥，施棉籽饼）的秧苗移植本田，观察比较秧苗移植后的生长情形及收获量。

1. 试验地点：武汉大学农学院农艺系水稻试验区。

2. 试验方法：将 1951 年水稻秧田肥料试验中甲、乙、丙、戊、己、庚六区团的秧苗，移植本田，得 3×6 随区排列，每区 5 行，每行 12 尺，行株距各 1 尺（1/100 亩），6 区团每团 3 区。

因选择地点不好，甲区团接近肥水浸灌，其生长情况不能和其他区团相比，而且一区团三区受肥水的影响又不相同。因此，只采取乙、丙、戊、己、庚五区团比较收获量，即 5 区团，每区团 3 区：

O 区　移植不施肥秧田秧苗

III 区　移植施用蚕豆茎叶绿肥青秧田秧苗

IV 区　移植施用棉籽饼秧田秧苗

5 月 18 日移植，8 月 20 日收获，本田未施肥。

3. 试验结果

籽实产量及叶杆产量见表 6 和表 7。经统计计算，籽实产量处理间有显著差异，叶杆产量处理间无显著差异。施绿肥秧苗（III 区）比施棉籽饼秧苗（IV 区）和不施肥秧苗（O 区）所生稻禾籽实产量大（O：III：IV：100：116：102）。本实验虽不能断论施蚕豆茎叶绿肥秧苗确优于施棉籽拿秧苗。但最低可以说采用蚕豆茎叶绿肥作为秧田肥料（秧田泡青法）无不良后果。

表 6　1951 年水稻本田试验

		区团					平均数 标准差
		乙	丙	戊	己	庚	
秧苗来源	O	428	347	347	419	390	386.4
	III	516	431	447	447	403	448.8
	IV	382	413	391	394	391	394.2

籽实产量（每亩斤数）

变量分析

变异	自由度	平方和	平均方	*F* 值
重复	4	5228	4124.5	6.04
处理	2	11879	5939.5	
机误	8	7870	983.7	
总数	14	24937		

n_1=2；n_2=8；*F*（5%）=4.46

表 7　1951 年水稻本田试验

		区团					平均数 标准差
		乙	丙	戊	己	庚	
秧苗来源	O	387	319	275	344	300	360.5
	III	462	387	369	400	356	412.3
	IV	344	350	350	362	350	391.7

叶秆产量（每亩斤数）

五、泡青秧田土壤氨态氮的增加情况

在举行 1951 年水稻秧田肥料试验时，在试验秧田紧邻，另开 12 区（每区 20 方尺），其中 6 区每区施用蚕豆茎叶绿肥（泡青与秧田同，每亩亦为 3000 斤），另 6 区不施肥。于 3 月 21 日泡青，4 月 5 日，4 月 12 日，4 月 23 日，5 月 10 日，分别采取表土样本（表土 4 寸）测定土壤含氨态氮量。在泡水情形之下，土壤不含硝酸，含氨态氮量即等于含无机氮量。土壤中含氨态氮量的增加情形亦即代表绿肥腐解进行情形。于 5 月 7 日并测定不施肥区的土壤含氨态氮量，比资比较。

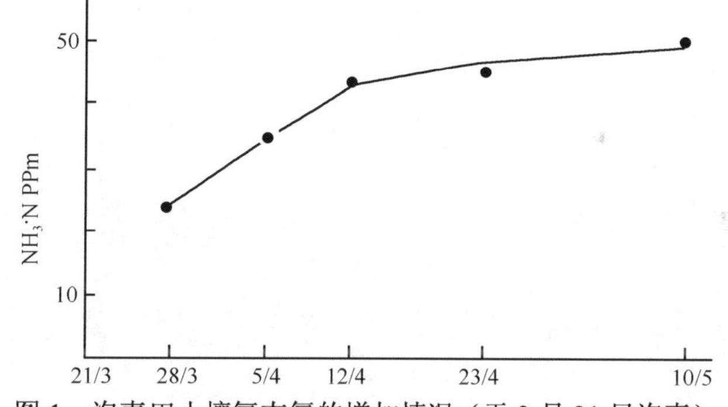

图 1　泡青田土壤氨态氮的增加情况（于 3 月 21 日泡青）

测定结果见表 8 及图 1。经统计分析，4 月 12 日以后，施用绿肥区中的氨态氮有显著增加，与秧田实验比照，说明秧苗需要氮肥时，秧田泡青法可以供应其要求。

从这试验体会，秧田泡青，在水稻播种前十天左右举行，可以发挥肥效。

表 8　泡青田土壤氨态氮的增加情况

分析次数	日期	一区	二区	三区	四区	五区	六区	平均数	5 月 7 日分析不施肥田 6 区，结果
1	3-28	25.7	20.4	22.6	21.1	23.4	20.6	23.8	16.0
2	4-5	26.2	57.0	39.9	15.2	45.5	28.2	34.8	22.3
3	4-12	41.0	63.3	47.4	41.8	39.1	26.7	43.1	20.7
4	4-23	35.1	65.9	36.5	40.3	55.2	35.1	44.7	23.2 33.2 24.8
5	5-10	39.2	67.8	40.9	55.0	58.8	37.0	49.8	平均 23.4 ppm

开始泡青时期：1951 年 3 月 21 日

测定数字单位：风干土重百万分数（p.p.m）

六、讨 论

水稻田肥料一半用在秧田里。农民为培植好秧苗，不惜花费很多本钱，很多时间去购买和搬运肥料。一担粪尿水，含量高的不过含有0.5%的氮，0.25%的五氧化二磷，0.20%的氧化二钾，含量低的，只有0.2%的氮，0.04%的五氧化二磷，0.03%的氧化二钾。这样稀释的粪尿，不能惜花费二三升米，走二、三十里路去挑。有些地方，在清明前后，粪尿水的价格会长到斗米一担。农民买籽饼做肥料，含量大约在5%的氮，1~5%的五氧化二磷，0.5%~2%的氧化钾左右。一般价格是五合到一升米一斤。一分秧田（1/10亩）用10斤籽饼和2担粪尿水就要一二斗米的肥料本钱，所费采购人力还不能计算。

从上面报告的实验结果估计，300斤蚕豆茎叶可以当得30斤籽饼，施用于一分秧田上，可以解决秧田的肥料问题。在长江流域和长江以南，一般土地上，冬季种蚕豆，到春分，每分地足有300斤以上的茎叶。如果就在秧田种蚕豆，只需及时犁翻，灌水泡青，所得绿肥价值还可包括它的根系部分。如果在另一块地上种蚕豆，一分秧田也只要一分地的蚕豆。种蚕豆不需好地，田边，路角都种得。（我们认为没有一家农民办不到。）普通种蚕豆，等它成熟收获，所得籽实，一分地不到一斗，连茎叶做猪食，总共价值，估计得高，不及两斗米价。根据调查，实验和计算，我们认为秧田泡青确是一种优良农业技术，在长江流域和长江以南可以普遍推广的。

我们是用蚕豆做实验的。蚕豆是一种豆类植物，作为泡青材料，冬季生长的别种豆类植物都有同样的效果。紫云英（红花草籽）、苕籽（蓝花草籽）、豌豆等等都可以用。经我们调查，不少地方用紫云英或苕籽泡青，结果都很好。

硫质对“噤水田”水稻生产量的影响*

陈华癸　庄正德　何殿元　雷自强

（武汉大学农学院）

一、水稻田施用石膏

在长江流域和长江以南，有许多地方种植水稻需要施用石膏。各地一致的基本情况是这样的：在正常田中，当秧苗移植本田以后，短期黄萎，3、5 日内回青复活，而在有些田中，黄萎时期可以延长到一、二十天。黄萎时期太长，则回青复活得太迟，影响水稻的正常发育，减少收获量，甚至无收。这种现象，有些地方称为“发秋”或称“犯噤”。“犯噤”的田地称为“噤水田”。施用石膏可以防止或纠正“犯噤”现象。在移植前后，施用少量石膏，秧苗就可以正常的回青复活。已经“犯噤”的秧苗，施用石膏后 3、5 日内也就回青复活。根据农民经验，施用石灰也可以，但需用量要很大。石膏每亩只要 5~10 斤，而石灰每亩需要 50~100 斤。

施用石膏对“噤水田”土壤的性质起什么变化？施用石膏对“犯噤”的水稻秧苗有什么作用？武汉大学农业化学系正在研究中。这篇是在一个农家的田地上举行的田间实验的报告。

二、实 验 经 过

这实验是在武昌湖泗区，何堰乡，刘均堡农民刘昌训的一丘田上举行的。这丘田是“噤水田”，每年要施用 3~4 斤石膏（应城石膏）。如果不施石膏，秧苗移植后就“犯噤”，一、二十天不回青复活。1951 年 5 月 12 日秧苗移植，于 5 月 15 日分 6 区（这丘田面积为 6 分，每区 1 分田，图 1 示实验田的形状和各区排列次序），随即施用下列药剂：

第 1 区　　施用硫黄 93 克
第 2 区　　施用硫酸镁 715 克
第 3 区　　施用氯化钙 323 克
第 4 区　　施用硫酸钠 936 克
第 5 区　　施用碳酸钙 291 克
第 6 区　　施用硫酸钙 500 克

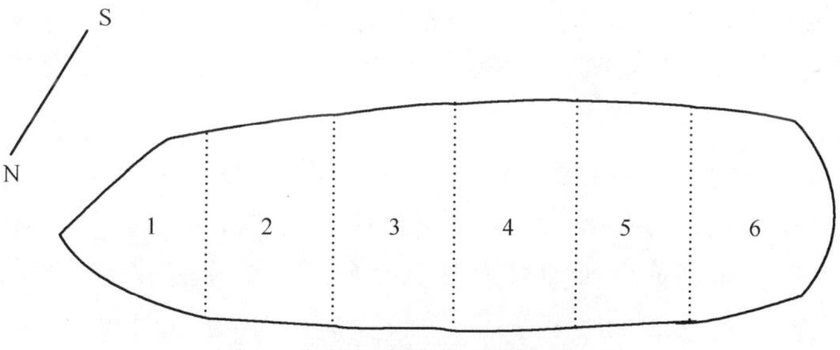

图 1　实验田的形状和各区排列

在施用药剂时秧苗黄萎，5 月 21 日观察秧苗回青复活情形。当时第 4，6，2，三区已全部回青。第 1 区只有极少数秧苗未回青。第 3，5 两区还是半青黄。按回青复活优差的程度排列，6 区的回青复活情况是：4—6—2—1—5—3。

7 月 30~31 日收获（当地早熟稻种）。收获时在每区中心收割一方。（以一窝为定点，一行为定线，收割 3.4~3.6 公尺见方一块。）得各区籽实产量如表 1。

*原载于《土壤学报》，2（1）：34~36，1952.

表 1 各区籽实产量比较表

区号	处理	收割面积	籽实产量	每亩籽实产量
1	硫磺	11.56	7.10	443.5
2	硫酸镁	12.25	8.10	473.2
3	氯化钙	12.25	7.00	379.4
4	硫酸钠	12.25	8.90	468.2
5	碳酸钙	12.25	6.40	340.7
6	硫酸钙	12.96	8.13	422.3

三、结果与讨论

将石膏分拆为钙质和硫质两部分，分别比较下列各区间的同异：

（1）施用钙比不施用钙（用钙量相同）

（2）施用钙或镁比不施用钙或镁（用钙或镁量相当）

（3）施用硫比不施用硫（用硫量相同）

第 1 项比较，施钙三区平均籽实产量每亩 373.7 斤；不施钙三区平均籽实产量每亩 461.0 斤。两类间无统计学显著差异。（施用钙质的比不施用钙质的产量反要小些。）这说明石膏中的钙质没有增加生产量的效果。

第 2 项比较，施用钙或镁四区平均籽实产量每亩 403.8 斤；不施钙或镁两区平均籽实产量每亩 455.5 斤。两类间也没有统计学显著差异。

第 3 项比较，施硫（硫黄或硫酸盐）四区平均籽实产量每亩 451.5 斤；不施硫两区平均籽实产量每亩 360.0 斤。经计算费雪尔氏“t”值（小样计算），得“t”=4.33，有统计学显著差别（n=4；P=0.05，t=2.776；P=0.02，t=3.747；P=0.01，t=4.604）。施用硫质增加籽实产量。据刘昌训和当地农民观察，施硫四区出穗整齐，不施硫两区出穗不整齐。

从这实验结果，可以初步结论如下：施用硫质（硫黄或硫酸盐类，硫黄在土壤中也硫化为硫酸）可使“犯噤”的秧苗早回青复活，及时生长发育，出穗整齐，丰收籽实。施用石膏的功用可能是由于其硫质成分，而不是由于其钙质成分。至于这实验是说明实验地一带的“噤水田”的情形，这是代表一般的需要施用石膏的水稻田的情形，则有待于更深更广的研究。

这实验，就写者所知，是水稻田缺硫现象的第一个实验报告。水稻缺硫的病象使移栽后的秧苗不回青，及时施用硫黄或硫酸盐类可以治疗。

致谢：本实验是依靠武昌县湖泗区何堰乡乡农会主席刘克炎同志和农民刘昌训同志的积极协助下举行和完成的。从开始到收获这两位同志都下田协助，谨此致谢。刘昌训同志自愿以自己的田地做实验（并在下药后将武汉大学农业化学系所写的产量保证书退回），这样爱科学、拥护科学工作的精神是特别值得表扬和感激的。武昌县、区、乡负责同志均协助推动这研究工作，统此致谢。

水稻田土壤中占优势的微生物物种*

陈华癸

（武汉华中农学院）

我们观察到，虽然土壤中微生物的种类很多，在一定条件下只有少数种类占优势。制备一定土样的营养冻粉平面培养，常见到在最高稀释盘中只有一种或少数几种占 80%以上。如果用选择培养基来分离培养一定生理群，也常见到一种或少数几种在最高稀释盘中占优势。这些事实指出，一定土壤在一定条件之下，只有少数几种微生物对于土壤中相应的生物化学活动起决定性作用。进一步认识到，正因为占优势的种类对于土壤中相适应的活动起着决定性的作用，对这些种类应当调查清楚，并详加研究。对它们了解得愈多，就愈能够掌握它们的活动，为农业生产服务。

这里报告的是对于长江流域水稻田土壤中占优势的微生物种类的初步研究。研究工作还在继续进行中。

在长江流域和长江以南，主要的水稻田土壤是水旱两作的，即夏天蓄水种水稻，冬天排水种冬季旱作物（小麦、油菜、蚕豆等）或绿肥作物（如苕籽、紫云英等）。

在种植水稻季节，田间蓄水，土壤和地表空气间隔着一层水。在排水种冬作时期，土壤的通气条件有显著的改善。这种情况的改变，鲜明地影响着土壤中微生物学的过程。我们研究工作的重点放在水稻田蓄水种水稻情况下的微生物学过程。

一、水稻田土壤在蓄水种水稻时期占优势的腐生细菌种类

虽然说，在蓄水条件下的生态因素是嫌气性的，但在蓄水初期，能在牛肉汁蛋白胨冻粉培养基上生长的，主要是兼性的微生物。并且其他条件相同，用好气性方法所得细菌数量，远比用嫌气性方法所得为多（表 1）。

我们用牛肉汁蛋白胨冻粉培养基分离土壤中的腐生细菌，在嫌气条件下培养（以期与嫌气性的土壤条件相符）。将在最高稀释盘中出现的种类称为在数量上占优势的种类。共得到 16 种，其中兼性嫌气性的孢子细菌 9 种，专性嫌气性的孢子细菌 3 种，兼性嫌气性革兰氏负反应的无孢子杆菌 4 种，如：*Bacillus megatherium*，*Bac. mycoides*，*Bac. subtilis*，*Bac. oligonitrophilus*，*Bac. flexus*，*Bac. danicus*（?），*Bac. fissus*（严格嫌气性），*Bac.* sp.，*Bac.* sp.，*Bac.* sp.，*Clostridium butylicum*（严格嫌气性），*Cl.* sp.（严格嫌气性），*Bacterium agile*，*Bact.* sp.，*Bact.* sp.，*Bact.* sp.

表 1　水稻田土壤中细菌的数量

日期（1956 年）	土壤情况	细菌数量（千/克土）	
		好气培养	嫌气培养
4 月 23 日	绿肥压青	62820	731
4 月 30 日	土壤蓄水	60896	884
5 月 9 日	插秧蓄水	45410	918
5 月 28 日	分蘖蓄水	12623	1410
6 月 11 日	分蘖蓄水	17384	431

我们观察到，在水稻生长初期，以兼性细菌种类占优势，而在水稻生长的后期，则专性嫌气性的优

*原载于《土壤学报》，5（1）：111~116，1957.

势提高了些。这现象说明，在水稻田蓄水条件下，虽然兼性嫌气性的和严格嫌气性的种类都能生长，由于生态因素是嫌气性的，对严格嫌气性的种类更为适宜的严格嫌气性的种类的优势逐渐提高。

在占优势的种类中，*Bacillus megatherium* 和 *Bac. mycoides* 特别值得重视，因为它们不仅在数量上占优势，而且在有机质（蛋白质和含磷物质）的矿物质化过程中，它们也是重要的种类。

二、分解纤维素的微生物

在水稻田蓄水时期，只有 *Bacillus omelianskii* 活跃的纤维素分解微生物。在绿肥压青、稻田蓄水以后，用培养法、埋片法（埋滤纸）或用显微镜直接检查在分解中的植物纤维，所得到的都只是这一种。

三、蛋白质分解过程中占优势的微生物种类

我们用豌豆粉作为唯一的氮素来源制备培养基。用这种选择培养基分离培养水稻田土壤（蓄水）中分解蛋白质的微生物种类。分离出在最高稀释皿中生长的菌落，加以纯化培养。纯化后再试验它们水解明胶和胨化牛奶的性能。我们将在豌豆粉培养基上数量最多的，而且具有水解明胶，胨化牛奶的特性的微生物种类称为蛋白质分解过程中占优势的微生物种类，得下列四种兼性嫌气性的孢子杆菌：*Bacillus megatherium*、*Bac. mycoides*、*Bac. rotans*、*Bac. glutinosus*。

四、氨化作用、硝化作用和反硝化作用

与所意料到的相符合，无论在排水或蓄水条件下，氨化作用都是不断进行着。而只有在排水情况下，才进行有硝化作用。表 2 示土壤在排水和蓄水时期中氨化作用和硝化作用的盛衰。

表 2　在排水和蓄水季节中氨态氮和硝态氮的存亡

日期	通气情况	作物	NH_3-N	NO_3-N
1945 年 11 月 30 日	排水	小麦	23.6p.p.m	5.3p.p.m
1946 年 2 月 11 日	排水	小麦	15.8	10.5
1946 年 6 月 14 日	蓄水	水稻	11.3	微量
1946 年 7 月 15 日	蓄水	水稻	7.1	微量

在种小麦时期，硝化作用是在进行着的；而在蓄水种水稻时期，硝化作用就停止了。如果土壤常年蓄水，则完全没有硝化作用（表 3）。

表 3　水稻田土壤全年蓄水条件下氨态氮和硝态氮的情况

日期	NH_3-N（p.p.m）	NO_3-N
1943 年 12 月 14 日	10.6	无
1944 年 1 月 10 日	10.7	无
1944 年 2 月 12 日	5.9	无
1944 年 3 月 1 日	3.6	无

表 4　旱地蓄水后土壤中的反硝化作用（实验室结果无淋失现象）

时间	含硝酸态氮量（土壤蓄水深 3 厘米）（p.p.m）	含硝酸态氮量（土壤湿润）（p.p.m）
开始	7.5	7.5
第 2 天	1.5	5.0
第 4 天	1.0	9.0
第 6 天	微量	11.0
第 8 天	微量	16.5
第 11 天	微量	13.0

表 5 绿肥压青，土壤蓄水后硝酸态氮消失

时间	NH_3-N_1（p.p.m）	NO_3-N_1（p.p.m）
开始	9.7	9.7
第 3 天	7.6	7.6
第 6 天	17.2	0.0

冬季作物（旱作）收获后，水稻田中多少含有一些硝态氮，这时开始蓄水，准备种水稻。在初蓄水的几天，土壤中反硝化作用十分旺盛，硝态氮很快消失（表 4，表 5）。

五、固氮细菌（azotobacter）的分布

虽然在蓄水条件下，土壤中固氮细菌（azotobacter）还是相当旺盛的。用小土块法和泥盘法研究土壤中的固氮细菌，在小土块矽酸胶培养盘上固氮菌出现的频率为 40%~70%（表 6）值得注意的，是固氮细菌生活在根际，而且是在根的表面。

表 6 水稻田蓄水种水稻，土壤中 azotobacter 的分情况

（在矽酸胶盘上证明小土块或小根段中含有 azotobacter 的频率百分数）

日期	土壤情况	根间土壤（小土块）%	根周围土壤（小土块）%	根麦面（小根段）%
1956 年 4 月 23 日	绿肥压青	46	—	—
1956 年 5 月 9 日	插秧	57	—	—
1956 年 5 月 28 日	分蘖期	57	40	0.5
1956 年 6 月 3 日	分蘖期	62.5	62	1.5
1956 年 6 月 11 日	分叶盛期	69.5	59	2.0

六、卵膦酯及其他含磷有机物质的磷素矿物质化过程中占优势的种类

表 7 列举各菌株水解卵膦酯能力的强弱

（水溶性磷占所加卵膦酯中全磷量的百分数，在 30℃培养两周）

种类	菌株	水溶性磷百分数/%
Bac. megatheriem	NO.3	8.89
	NO.3	16.11
	NO.13	10.25
	NO.20	4.71
	NO.23	8.33
	NO.27	5.27
	Ba 5	2.00*
Bac. mycoides	NO.10	5.83
未接种		微量

*这是东北农业科学研究所供给的磷细菌，其矿化磷负的能力经测定证明是比较强的。

我们用新鲜鸡蛋黄（用无菌手术操作，不加热，以避免可能的分解作用）或用纯化的卵膦酯作为唯一的磷素来源，制备选择培养基，分离培养在卵膦酯和其他含磷有机化物过程中占优势的微生物种类，得到下列两种：*Bac. megatherium*、*Bac. mycoides*。

七、总 结

1. 对于这个问题的研究工作，还在继续进行着。这篇报告是片断的。鉴于占优势的种类对于土壤中相应的活动起着决定性的作用，我们应该调查清楚这些种类，并加以详细的研究。对于它们了解得愈多，就愈加能够掌握它们的活动，为农业生产服务。

2. 在蓄水种植水稻的时期，水稻土壤是嫌气性的，共得 16 种在数量上占优势的细菌，其中 9 种是兼性嫌气性的孢子杆菌。

3. *Bacillus omelianskii* 是在蓄水条件下唯一活动着的纤维素分解微生物。

4. *Bac. megatherium* 和 *Bac. mycoides*（较次要）在我们研究的水稻田土壤中占显著重要的地位。它们不仅在数量上占优势，而且也在蛋白质分解过程和含磷有机物质分解过程中占优势。

5. 与估计相符，水稻田土壤在蓄水和排水季节中氨化作用都在进行着，而硝化作用只在排水季节进行。在排水季节中形成的硝酸，在蓄水种水稻的头几天就因反硝化作用而消失了。

6. 在我们所研究的蓄水的水稻田土壤中，固氮菌相当多。固氮菌不附生在根表面，而存在于根与根之间或根周围的土壤中。

土壤生物活性和土壤有机态磷的来源、积累和矿化——主要研究成果和问题*

陈华癸

（华中农学院，中国科学院武汉微生物研究所）

一、绪　　言

在十九世纪中叶，Mulder（1844）最先报道了土壤有机成分中含有磷。在十九世纪末和二十世纪初，Костылев（1890）和 Stoklasa（1911）先后探讨过土壤微生物的生命活动对于土壤的磷素循环和植物磷素营养的重要作用。在二十世纪初期，Aso（1904）和 Shorey（1913 及其他）等人在对于土壤中有机磷化合物的成分及其生物来源进行了初步的研究工作。Schollen Berger（1918，1920）最先对于土壤中有机磷含量的分析方法以及土壤有机磷含量进行了较为严格的研究工作。在本世纪的二十年代和三十年代，关于这个问题的研究成果逐渐地在积累着。然而，关于土壤有机磷的化学和微生物学性质的深入研究，则是在 1940 年前后才开始的[Wrenshall and McKibbin，1937；Chang （张信诚），1939，1940；Pearson and Simonson，1940；Yoshida，1940；Wrenshall and Dyer，1941；Dyer and Wrenshall，1941a，b；Pearson，Norman and Ho，1941]；而在 1950 年前后，这个问题的理论和实际才得到较广泛地重视，研究工作向着开阔和纵深发展（Pierre，1948；Сокозов，1948；Kaila，1948，1949；Bower，1949，Менкина，1950）；Pierre（1948），Black，Goring（1953），Ulrich，Benzler（1955），Bremner（1956）和 Barrow（1961）等人写了几篇总结性的文章。

每亩地面植物每年合成 500~1000 斤干有机物质，含磷 1~3 斤，其中有一半处于有机化合物状态。这些有机态磷的一部和全部就地转化为土壤的有机磷成分并进一步分解为无机状态。在表土中，有机磷的含量约占全磷量的 1/4~1/2，或每亩约含有机磷 20~80 斤。成土母质中是不含有机磷的。土壤中有机磷的形成和积累完全是土壤形成过程的产物，亦即在相当长的时间内，由于有机磷化物的生物合成（主要是地面植物的作用）和矿化（主要是土壤微生物的作用）的循环发展而形成和积累的。土壤有机磷含量随深度而递减，在根系达不到的地方就没有有机磷（图 1）。

土壤中有机磷化物的积累和矿化（磷素的生物循环）是土壤肥力的一个重要方面，它对于植物磷素营养具有十分重要的意义。然而，在 20 年前，农学家、土壤学家和农业化学家们对于土壤有机磷的重要性的认识确实是很不够的。可以这样说：在本世纪的四十年代以前，关于元素的生物循环及其对于土壤肥力和植物营养的重要意义的科学研究工作主要是针对着碳、氮两种营养元素进行的（在一定程度上也包括硫），而对于土壤中的磷素问题，一向只着重于无机磷化合物的固定和释放（主要是非生物学的变化）。Pierre （1948）的文章对于转变这个局面起了重要的作用，他总结了有关的研究成果并着重的探讨了土壤有机磷对于植物磷素营养的重要性。以后的研究工作进一步丰富和发展了这项认识。现在的认识是：土壤中植物营养的供给水平既受无机磷化合物的固定和释放作用的制约，又受有机磷的合成、固定和矿化的制约，两者几乎是同等重要的（图 2）。如果说，测定土壤可给性磷的速测方法主要是指出它的“强度价值”（McDonnell and Welsh，1957），这种“强度价值”有时和土壤无机磷含量的关系比较密切，有时却和土壤有机磷含量的关系比较密切（McDonnell and Welsh，1957）。

*原载于《微生物专题报告集》，219~234，1964，北京：科学出版社

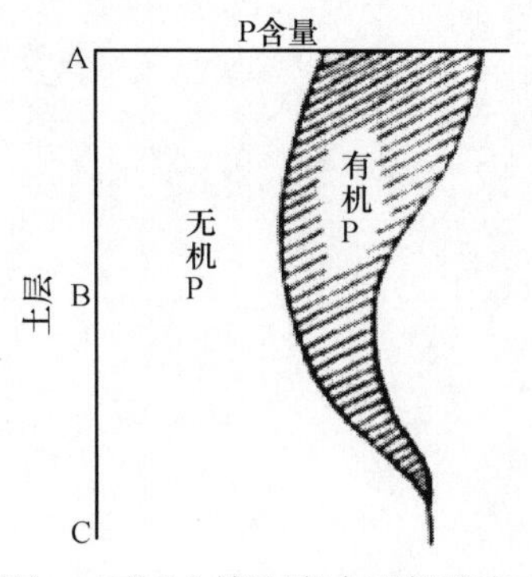

图 1 不同土壤深度中无机磷和有机磷含量(Thompson，1958，246)

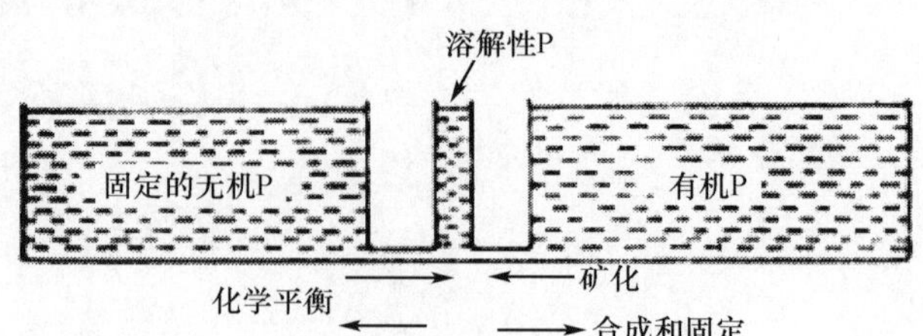

图 2 土壤中三类磷化物的关系(Thompson，1958，244)

二、一些植物化学和生物化学的基础知识

植物和微生物中的主要有机磷化物有：

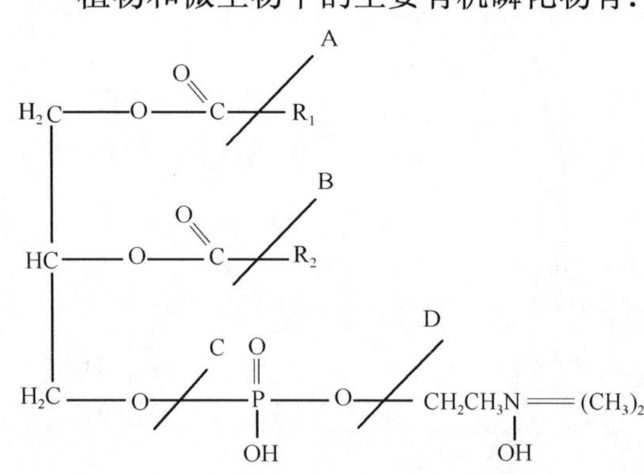

1. 磷酸化糖类：在细胞代谢过程中，糖类的合成和裂解以及相应的能量转化主要是通过磷酸化糖类的合成和裂解来完成的。糖类的合成和裂解包含着一系列的磷酸化作用，形成许多种磷酸酯类物质。这些磷酸化作用主要作用主要是在特定的酶的作用下进行的。

2. 磷脂：植物和微生物体中的主要磷脂是卵磷脂，它是由甘油、脂肪酸、胆碱和磷酸结合成的类脂化合物：

广义的卵磷脂酶有四种（A、B、C、D）。卵磷脂酶 A 分解 R_1，卵磷脂酶 B 分解 R_2，卵磷脂酶 C 分解磷酰胆碱，卵磷脂酶 D 脱磷酸产生无机磷酸盐。在本文中所提出卵磷脂酶，只是指对分解卵磷脂并产生无机磷酸盐的酶的作用，它是能分泌到体外的水解酶。

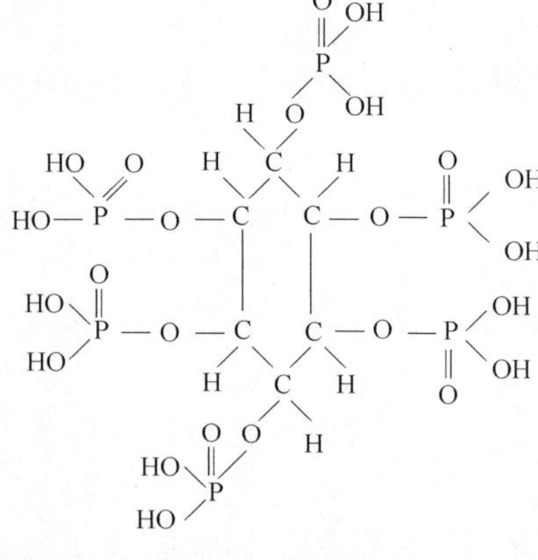

3. 肌醇膦酸脂：植酸是肌醇的六磷酸酯。植素是植酸的钙、镁盐。广义地说，植素包括植酸的各种盐类，如一价的铵盐和钠盐，两价的钙盐和镁盐和三价的铁盐和铝盐等。在讨论中也包括植酸的部分脱磷酸的衍生物，如肌醇的三磷酸酯和四磷酸酯等。

植素酶分解植素，产生无机磷酸盐，它也是能分泌到体外的水解酶。植素是植物（特别是种子）中的成分，迄今微生物产生植素的直接证据还不十分确切（参考 Caldwell and Black，1958)，但很多微生物却分泌植素酶。

4. 核朊、核酸和核苷酸：核朊是复合朊，包括朊和核酸两部分。核酸是分子量极大的核苷酸的多缩合物。核苷酸由核苷和磷酸两部分组成；核苷由氮基和核糖或去氧核糖两部分组成：

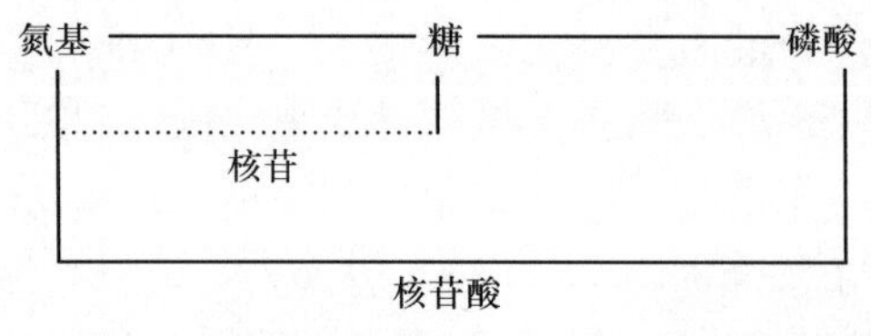

核苷或核苷酸的氮基可以是下列各种嘧啶或嘌呤。即：

含腺嘌呤的腺苷和腺核苷酸；

含鸟粪的鸟粪核苷和鸟粪核苷酸；

含胞核嘧啶的胞核苷和胞核苷酸；

含尿嘧啶的尿核甙和尿核苷酸；

含胸腺嘧啶的胸腺核甙和胸腺核苷酸；

含 5-甲基-胞核嘧啶的相应的核苷和核苷酸。

生物体中除构成核酸的核苷酸外：还有些自由核苷酸。这些核苷酸上的磷酸可以是一个、二个或三个，例如腺核苷酸：

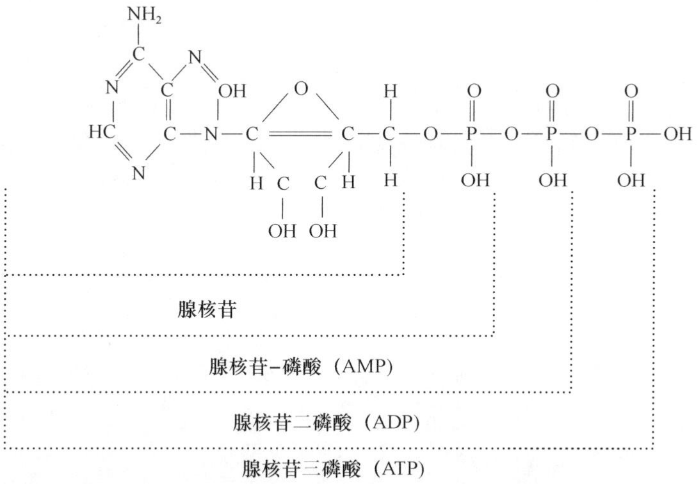

腺核苷

腺核苷–磷酸（AMP）

腺核苷二磷酸（ADP）

腺核苷三磷酸（ATP）

核酸、核苷酸和核苷按其所含糖的不同而分为含核糖的核糖核酸、核糖核苷酸和核糖核苷以及含去氧核糖的去氧核糖核酸、去氧核糖核苷酸和去氧核糖核苷两类。核酸酶将核酸水解为核苷酸。核苷酸在各种核苷酸酶的作用下脱磷酸，将核苷和磷酸分开。很多种微生物分泌核酸酶和核苷酸酶，在体外水解核酸和核苷酸，产生无机磷酸化合物。

在本文中，讨论含磷有机化合物的矿化作用，实际上是指脱磷酸作用，亦即产生无机磷酸化合物的作用。

三、土壤有机态磷含量和土壤有机质中 C∶N∶S∶P 的比例

迄今为止，分析土壤中有机态磷含量的方法仍旧是很不准确的（Bremner，1956；Barrow，1961）。现有的方法主要可以归结为两类：一类是灼烧法，将土壤样本中的有机磷灼烧成无机磷，然后采取同样的方法抽提灼烧过的土壤样本和未灼烧的对照土壤样本中的无机磷，两个抽提物中含磷量的差别被认为是土壤的有机磷含量。另一类是抽提法，先将土壤中含磷化合物抽提出来，再分别测定抽提液中氧化部分与不氧化部分，两部分含磷量的差别被认为是土壤的有机磷含量。两类方法都只是约测法。

土壤有机磷含量的变化幅度很大（表 1），最低可以只含 10~20ppm（Pearson，1940），最高可以超过 1000ppm（Kaila，1948）。一般而论，在腐殖质含量低，矿化作用强的土壤中，例如在多数热带和亚热带的红壤中和温带的灰钙土中，有机磷的含量较少，占全磷量的百分比比较低。Хейфец（1948，1950）报道苏联灰钙土表层的全磷量为干重的 0.062%，而有机磷只占全磷量的 11%。在腐殖质较丰富、矿化作用较慢的土壤中则相反。例如，Kaila（1948）根据 100 个芬兰土壤的分析结果，在矿质土中有机磷占土壤全磷量的 40%，在富含腐殖质的土壤中则占 60%。根据中国科学院土壤研究所的分析资料，我国含有机质 2%~3%的耕作土壤的有机磷含量为全磷量的 25%~50%（表 2）而有机质不足 1%的红壤的有机磷含量则不到全磷量的 10%。

表 1 全世界各地土壤中有机磷、无机磷含量和比例（自 Nye and Bertheur，1957 年及其他）

国家	土壤	地点数	pH	无机磷（ppm）	有机磷（ppm）	全磷（ppm）	有机 P/全 P	有机 C/全 P	资料来源
马来西亚	残积壤土和粘壤土	26	3.8-5.6	67	27	94	28	—	Owen，1953
	残积砂土和砂壤土	—	—	35	38	73	52	—	Owen，1953
锡兰	红壤（6）；非红壤（5）	11	4.3-8.0	630	312	942	33	83	Kandiah，1948
安哥拉	铁铝土	6	5.1-6.8	583	234	817	29	—	Almeid and de Mirmnd，1954

续表

国家	土壤	地点数	pH	无机磷（ppm）	有机磷（ppm）	全磷（ppm）	有机 P/全 P	有机 C/全 P	资料来源
马来西亚	滨海沉积土	—	3.8-5.6	146	105	251	42	—	Owen，1953
印度	中性和酸性水稻土	20	4.2-7.0	327	189	516	37	50	Ghani and Aleem，1943
英国	各种矿质土	18	4.9-8.0	827	359	1186	30	101	Dean，1938
美国	—	25	5.2-8.1	333	246	579	42	182	Thomposon，1954
南澳	红褐土	15	6.2-8.9	159	61	220	26	—	Williams，1950
加纳	—	67	4.6-7.5	107	27	134	20	247	Nye and Bertheur，1957
苏联	各种土壤	—	—	973	536	1616	33	45	Хейфец，1950

表 2 长江流域土壤中有机磷量的一些数据（中国科学院土壤研究所供给）

成土母质	分析样本数	有机质（%）	全磷（P_2O_2）（%）	有机磷（P_2O_2）（%）	有机磷、全磷（%）
长江中游老冲积物	8	1.82	0.059	0.020	34
长江下游沉积物、湖积物	19	2.87	0.108	0.006	33
第四纪红色黏土上红壤	23	2.04	0.069	0.061	45
石灰岩、黄色页岩石上土壤	16	2.03	0.156	0.044	28

在绪言中已经指出，土壤有机磷含量随着深度而渐减（Pearson and Simonson，1940）；Хейфец，1948，1950；Fuller and McGeorge，1951），Хейфец 的部分结果在表 3。

显然，土壤中的有机磷含量和有机碳、氮、硫等元素成分是有一定比例关系的（Schllen berger，1920；Pearson and Simonson，1940；Ghani and Aleem，1943；Kaila，1948；Днмтрепко，1948；Хейфец，1948，1950；Fuller and McGeorge，1951；Thompson and Black，1950；Thompson，Black and Zoellner，1954；Walkor and Adams，1958；Williams，Williams and Scott，1960）。Black and Goring（1953）总结前人的研究结果，指出土壤中 C∶N∶P 的比例平均为 110∶9∶1。

表 3 苏联土壤中有机磷和无机磷的分层含量（Хейфец，1948，1950）

土壤	层次（厘米）	全磷量（P_2O_5，%）	有机磷（干物量）	无机磷
深灰化土	4~12	0.160	0.055	0.105
	12~40	0.130	0.016	0.114
	40~62	0.110	0.004	0.106
深厚黑土	0~10	0.208	0.094	0.114
	20~30	0.138	0.052	0.085
	40~50	0.130	0.043	0.087
	60~70	0.130	0.042	0.088
	90~100	0.100	0.025	0.075
	100~110	0.000	0.013	0.077
典型灰钙土	0~5	0.156	0.017	0.130
	13~18	0.149	0.021	0.128
	20~25	0.135	0.024	0.111
	43~50	0.117	0.018	0.099
	70~75	0.138	0.010	0.128

Williams，Williams 和 Scott，（1960）研究各种苏格兰土壤的有机碳∶全氮∶全硫（在钙质土中除去 SO_4-S）：有机磷的比例，平均为 61.4∶4.3∶0.6∶1。他们根据自己和别人的研究结果，计算出全氮和有机磷之间的相关性较差（P±0.64）。有机碳∶全氮∶有机硫三者之间的相关性要比它们和有机磷之间的相关性大。似乎，土壤有机磷和含氮的土壤腐殖质成分之间有一定的独立性，这可能主要是由于植素及其衍生物的影响（见后）。

四、土壤中的磷脂类及其微生物分角作用

研究工作（Aso，1904；Stoklasa，1911；Shorey，1913；Wrenshall and McKibbin，1937；Сокозов，

1948）证明土壤中含有一些溶解于乙醚的磷脂类物质，但含量很少，不是土壤有机磷化物的重要成分。Wrenshall 和 Mckibbin （1937）报道，土壤有机磷化物中只有 0.3%是溶解于乙醚的。Сокозов（1948）报导，在灰壤和红壤中只有痕迹，在其他土壤中也只占全有机磷含量的 0.4%~2.6%。看来土壤生物残体中磷脂类物质分解得很迅速，积累不起来。

在土壤中不只有生物残体中的磷脂及其分解过程，并没有土壤磷脂的积累和分解过程。值得看重地提出这个问题。因为，有些土壤微生物学家并没有吸收土壤化学的研究成果来指导自己的研究工作。近十多年来，有些研究者们努力于用卵磷脂做选择加富媒介来分离培养能够加速分解土壤有机磷的“有机磷细菌”，并且企图用这种有机磷细菌来做提高植物土壤磷素营养的细菌肥料（Менкина，1950；张宪武和刘期松，1955；板野新夫和甘扬声，1956）。在原理上，用卵磷脂做选择加富媒介，所得到的微生物菌株是最能分解卵磷脂的，而和分解其他有机磷化物的功能并无本质上的联系。施用这种细菌肥料来加强土壤中并无积累、无需加强的卵磷脂分解作用，不能说不是有些无的放矢。有意识地选择分解卵磷脂能力强的菌株，作为细菌肥料来加强其他有机磷化物（不是卵磷脂）的分解作用，则不能说不是有些南辕北辙。

尽管如此，研究者们提出的科学资料对于土壤微生物学还是有一定的贡献的。Менкина（1950）以卵磷脂为唯一磷源进行选择加富培养，从黑土、灰壤和泥炭土中分离出几株分解卵磷脂能力较强的细菌，它们分别属于 *Bacillus* 和 *Serratia* 属。经鉴定，命名为 *B. megatherium* var. *phosphaticum* 和 *Serratia corallina* var. *phosphorica*，*B. megatherium* var. *phosphaticum* 不仅分解卵磷脂脱磷酸的能力强，分解核酸脱磷酸的能力也很强（表 4）。

表 4　有机磷化物磷酸矿化试验（砂培 21 天，加葡萄糖）（自 Менкина，1950）

	P_2O_5 量（卵磷脂）			P_2O_5 量（核酸）		
	开始（毫克）	结束时矿化量		开始（毫克）	结束时矿化量	
		毫克	%		毫克	%
对照	5	5	—	5	痕迹	—
芽孢杆菌 2 号	5	2.50	50	5	4.07	81
芽孢杆菌 6 号	5	1.59	31	5	4.01	80
芽孢杆菌 7 号	5	0.84	17	5	4.30	85
无芽孢杆菌 K	5	.2.50	50	5	1.14	22

板野新夫和甘扬声（1958）用类似方法在东北黑土中也分离出了分解卵磷脂能力强的 *B.megatherium* var. *phosphaticum*。张宪武和刘期松（1955）也用同样方法从东北淋溶黑钙土、森林灰化土和泥炭土中分离出 20 个分解卵磷脂能力较强的菌株，其中能力最强的四株均为 *Pseudomonas* spp.。

胡正嘉（1957）用卵黄从湖北省水稻田中分离出的分解卵磷脂脱磷酸的优势种类主要是 *Bacillus cereus*，有些 *B. mycoides* 也有一定的分解能力。Szember（1960）从土壤中得到 20 株分解卵磷脂脱磷酸能力强的微生物，其中有一株是放钱菌，其他多数是短杆菌，没有做详细交代。Bottcher（Bettxep，1961，1963）研究了土壤中芽孢杆菌的卵磷脂酶的活性，指出 *Bacillus cereus*，*B. mycoides* 等有较强的酶活性，而 *B. megatherium* （包括 Менкина 的 *B. megatherium* var. *phosphaticum*）的酶活性都很弱。胡正嘉（未发表数据）比较 *Bacillus cereus* 和板野新夫的 *B. megatherium* var. *phosphaticum*，前者有卵黄反应，后者没有卵黄反应。

Colmer（1947）采用有无卵磷脂酶作为芽胞杆菌的分类特征之一，他指出 *B. subtilis*，*B.megatherium*，*B. pumilus*，*B. alvei*，*B. circulans*，*B. bravis*，*B. macerans*，*B. polymyxa*，*B. sphaericus*，*B. mesentericus* 等种类都不含卵磷脂酶，而 *B. cereus*，*B. mycoides*，*B. albalactis*，*B.pranssnitzii*，*B. anthracis* 等则具有卵磷脂酶。Smith 和 Gordon（1952，自 Bergey’s manual，1957，pp.613-634）总结 Colmer（1949），McCaughey 和 Chu（1948）和 Knight 和 Proom（1950）的研究工作，认为 *Bacillus* 属中，*B. cereus* 和 *B. cereus* var. *mycoides* 是富于卵磷脂酶的种类，而 *B. megatherium* 和 *B. subtilis* 等则不含卵磷脂酶。

五、土壤中的核酸及其脱磷酸作用

早期的研究工作证明土壤中有嘌呤基和嘧啶基等核酸的组成成分以及 2，6-二羟基嘌呤，6-羟基嘌呤等核酸的分解产物（Shorey，1912，1913；Schreiner and Lathrop，1912），从而指出了土壤中含有核酸。以后，Wrensball 和 McKibbin（1937）报导土壤的有机磷中有 47.5%~65%是核酸；Сокозов（1948）的报导为 38%~57%，Bower（1949）的报道是 17%~33%。

然而，Yoshida（1940）和另一些研究者们却没有在土壤中找到核酸等成分（见 Adams，Bartholomew and Clark，1954）。Adams. Bartholomew 和 Clark（1954）采用现代方法分析土壤有机磷，发现含 575ppm 有机磷的 Corrington 土壤中只有不到 1ppm 核酸磷，而含 327ppm 有机磷的 Webster 土壤只有不到 6ppm 核糖核酸磷。他们的估计是土壤含核酸磷很少，不会超过总有机磷的 5%。Anderson （1958）采用离子交换树脂和紫外线刻印等方法测定三个耕作土壤的胡敏酸水解产物中的核酸的组成成分，计算出土壤中去氧核糖核酸中的磷含量只占土壤有机磷的 0.06%。

由于土壤有机磷的很大一部分还未能鉴定，土壤有机磷中究竟有多少核酸磷，是百分之几十还是百分之几，还不能定论。植物、动物和微生物的核酸成分在土壤中分解得很快。但是在土壤中究竟有没有比较稳定的“土壤核酸”，也不能定论。

研究工作证明（Dyer and Wrenshall，1941 b；Pearson，Norman and Ho，1941）将核酸或核苷酸加于土壤中，在不多几天内大部分核酸就被分解掉，脱去磷酸。Dyer 和 Wrenshall （1941 b）将酵母核酸和四种核苷酸（鸟核苷酸、腺核苷酸、胞核苷酸和尿核苷酸钙）加于土壤中，培养一周，就有 70%~80%有机磷变成为无机磷。Pearson，Norman 和 Ho （1941）的试验得到同样的结果（表 5）。在两家的研究工作中，核酸核苷酸的分解（脱磷酸）都是不完全的，在第一个星期的旺盛脱磷酸作用以后，进一步的脱磷酸作用进展很慢，或者处于停滞状态。Dyer 和 Wrenshall（1941b）认为这是由于在土壤培养过程中包含着一种特殊的“土壤核苷酸”的形成过程，这种“土壤核苷酸”抗微生物分解的能力强，比较稳定。他们用提取核苷酸的方法从土壤中抽提出“土壤核苷酸”，将这种“土壤核苷酸”加到土壤中去，进行土壤培养试验，结果这种“土壤核苷酸”分解得很缓慢，凡是加核酸或四种核苷酸的，在一周之内脱磷酸 68.8%~80.26%，而加“土壤核苷酸”的在第一周内只产生了 6.96%的无机磷，在以后的 12 周中进展也很慢。

表 5 土壤培养加核酸和核苷酸后有机磷变化（Pearson，Norman and Ho，1941）

处理	增加的有机磷（ppm）	有机磷（ppm）			
		0 天	5 天	45 天	60 天
对照	—	82	88	88	82
	—	82	86	86	76
核酸	56	138	108	100	106
	56	138	112	98	98
腺核苷酸	95	176	114	94	100
	71	153	116	92	96
鸟类核苷酸	54	136	136	106	90
	54	136	142	96	96
胞核苷酸	51	131	110	92	90
（胞核苷酸+尿核苷酸）钙	61	143	98	104	96
	61	143	124	106	—

Pearson，Norman 和 Ho（1941）不承认有什么稳定的“土壤核苷酸”，认为不完全分解只是由于不断进行着的微生物合成作用的结果。然而，Goring 和 Bartholomew（1950，1952）、Flaig， Kuron 和 Kaul （1955）都证明了黏土矿物和核酸互相吸附。Goring 和 Bartholomew （1951）的试验是这样进行的：他们是 *Aerobacter aerogenes*，*Bacillus sultilis* 和 *Rbizobium* sp. （豇豆族）的细菌体和高岭土或本脱土混合，用同样的分析方法分析没有和黏土混合的细菌体和已和黏土混合的细菌体，分析结果有显著区别；与黏土混合之后，水溶性的有机磷含量和“磷脂”含量都减少了，“核酸”含量却提高了，而磷朊

和残余有机磷没有显著变化。对比没有和黏土混合的细菌体和已和黏土混合的细菌体进行培养试验，培养七天以后，水溶性有机磷、“磷脂”和“核酸”的脱磷酸作用都由于黏土矿物的吸附作用而减弱，本脱土的影响大于高岭土。

似乎，土壤中存在着比较稳定的“土壤核苷酸”。

大多数微生物具有分解核酸或核苷酸脱磷酸的能力。Менкина （1950）和张宪武、刘期松（1955）用卵磷脂做媒介进行选择加富而分离出来的细菌种类分解核酸的能力也很强。Менкина 的三株 *B. megatherium* var. *phosphaticum* 在 21 天内矿化了 80%~86%核酸磷。张宪武、刘期松的 20 株细菌中有 19 株在 21 天内矿化了 62.23%~93.43%核酸磷。Joffries，Holtmen 和 Guse（1957）报导 *Bacillus*，*Serratia*，*Achromobacterium*，*Stroptococus*，*Streptomyces*，*Penicillium* 和 *Aspergillus* 等属的种类分解核酸的能力较强，而 *Flavobacterium*，*Alkaligenes* 和 *Micrococcus* 等属的种类没有这种能力。*Bacillus megatherium*，*Streptomyces rimosus*，*Str. albus*，*Serratia marcescens*，*S. rubidaea* 等种分解核酸的能力很强。Кварадхетия 和 Бетяева （1962）认为所有的 *Bacillus megatherium* 分解核酸的能力都很强，在他们的试验中，所有的菌株都不弱于 Менкина 和 *B. megatherium* var. *phosphaticum*。

六、植素及其衍生物在土壤中的转化和脱磷酸作用

Yoshida （1940）、Wrenshall 和 Dyer （1941）等比较确切地证实了土壤中含有植素及其衍生物。Yoshida （1940）从土壤中分离出了植酸的钠盐，加酸水解所产生的肌醇和磷酸相当于植素。Wrenshall 和 Dyer （1941）证明灰壤中含有植酸的铁盐。Bower（1945，1949）指导三个 Tows 土壤中的有机磷的 67%~83%在化学性质上像植素及其衍生物，后者最近似肌醇三磷酸。Anderson（1956）的研究结果指出，土壤中的肌醇磷酸主要是肌酸六磷酸（植酸），肌醇三磷酸和四磷酸不到 10%。Pederson（1953）报导一种生草灰化土含植酸磷 163ppm，合全有机磷量的 46%，他指出，在酸性土壤中的植酸最可能是铁盐，在碱性土壤中最可能是钙盐。这些植酸盐的溶解性比同类的磷酸矿盐还要低（Jackman and Black，1951，1952 a，b）。Caldwell 和 Black（1958a，b，c）的分析结果，49 个土壤样本的平均含植酸磷量为每克 39 微克，占全有机磷量的 17%（3%~52%）。森林土壤中的含量比草地土壤多，在两类土壤中，植酸磷的含量都随着 pH 的升高而降低。

综合各家研究结果，植酸盐作为土壤中有机磷化物的主要成分的见解是比较可靠的。就现在的主要证据而言，只有植物含植酸，微生物不含植酸。因此，土壤中的植酸只能来源于植物残体，并没有土壤微生物再合成的植酸物质。但这并不等于说，在土壤中只有植物残体中植酸成分的分解过程。正如上面指出的，植酸及其衍生物在土壤中转化为溶解性很低的成分，并且还和黏土矿物结合。因此，在土壤中实际在进行着的是：①植物残体中植酸的生物学分解过程；②植酸的非生物学固定和积累，形成“土壤植素”；③“土壤植素”的生物学分解作用。

然而，Smith 和 Clark （1951）加放射性 P^{32} 于土壤中，经培养后，分析土壤中的植酸磷，发现它也含有放射性。最近，Caldwell 和 Black （1958a，b，c）从土壤微生物的混合培养体中找出了植酸盐及它的异构体。根据他们的研究结果，土壤中的植酸盐有 40%左右是微生物来源的。如果这项研究结果是确实的，那么在土壤中实际进行着的过程就包括：①植物残体供给土壤植酸；②土壤微生物合成植酸；③植酸的生物学分解；④植酸的非生物学固定和积累，形成“土壤植素”；⑤“土壤植素”的生物学分解作用。

在本世纪初就有关于微生物分泌植素酶，能水解植酸脱磷酸的报导（Dox， 1911）。似乎很多种微生物都含有植酸酶，而以真菌的作用为强（Jackman and Black，1952a；Jackman and Black，1952a）。土壤中常遇见的真菌，如 *Penicillium chrysogenum*、*Arpergillus niger*、*Cunninghamella tlakcsleuna*、*Actinomocur* sp.，和细菌，如 *Arthrobactor helvolum*、*Bacillus cereus*、*B. subtilis*、*B.* sp.等都分泌或强或弱的植素酶。

分泌植素酶力较强的这些微生物分别属于不同的形态类型（四株酵母菌，一株放线菌，22 株细菌中包括球菌、生芽孢的杆菌和不生芽孢的杆菌）。另外，用卵磷脂选择加富得到的菌株之中也有五株能

水解植素。

土壤中加植素进行培养试验，在培养时期内，植素的脱磷酸作用比核酸或核苷酸的脱磷酸作用要缓慢得多。Dyer 和 Wrensball （1941b）的试验结果，在中性土壤中，8 星期脱磷酸 50%左右（核酸在一星期内脱磷酸 70%以上，）在酸性土壤中则进行得十分微弱。Pearson，Norman 和 Ho（1941）的结果相同，而在酸性土壤中加石灰能显著地提高植素的矿化速度。Jackman 和 Black（1952a，1952b）的研究工作表明，在土壤中，限制植酸盐矿化的因素不是生物学的，也就是说，不是酶的活性不足，由于很多土壤微生物都分泌植素酶，土壤中植素酶的活性总是比较充分的。土壤中植素分解缓慢主要是由于它处于不溶解状态的原因。土壤类外加植素酶并不能提高植素的脱磷酸作用，而加溶解性较强的植素却能显著地提高植素的脱磷酸作用。他们认为土壤中植酸盐处于高度的不溶解状态，因而限制了植素磷的快速水解。应用 Michalis 和 Menton 关于酶作用的公式来推算试验结果，植素溶解作用的限制性比植素酶活性强度的限制性大 71 倍。Jackman（1951）的试验证明在酸性溶液中植酸的铁、铝盐的溶解性很低。Anderson 和 Arlidge （1962）指出 pH 在 3~4 时微晶高岭土和 Insch 土壤黏粒吸收植酸的能力较高，但在同样条件下，氧化铁胶体却没有什么吸收能力；在更强的酸性条件下，氧化铁又能吸收溶液中的植酸。他们认为后者可能是：在强酸性条件下氧化铁胶体破坏并因此而产生了植酸铁盐沉淀。

七、植物残体分解过程中磷成分的生物合成和矿化

植物残体中既含有机磷成分，也含有无机磷成分，大致各占一半（张信诚，1939；Pearson，Norman and Ho，1941；Kaila，1949；Birch，1961）。因此，在植物残体的分解过程中，既包含有机磷的矿化作用，又包含无机磷的生物合成作用。这两个方向相反的作用的相对强度决定植物残体在土壤中磷素变化的总方向。张信诚（1939，1940）的研究工作指出，在植物物质分解的初期，磷素变化的总方向是有机磷的合成和积累，随着有机质分解过程的继续进行，磷素的变化终于达到一个转折点，在这个转折点以后，磷素变化的总方向是有机磷的矿化，亦即有机磷逐渐减少，无机磷逐渐增加。磷素转化的总方向与能源物质的供应情况有显著的关系，特别是较易分解的能源物质，如糖类、纤维素和类纤维素等。Kaila（1949）指出，微生物每分解 1000 份有机物质，大致要合成含有 3 份磷素的有机化合物，这比例随着有机物质和磷素的可给性不同而稍有进出。在她的研究工作中，牛粪、真菌菌丝和麦麸的含磷量分别为干物重的 0.455%、0.577%和 0.882%，在这些物质的分解过程中，有机磷的矿化过程一开始就是磷素转化的总方向。黑麦草和小麦根的含磷量分别为干物重的 0.052%和 0.097%，在整个试验时期内（90 天和 150 天）有机磷都是不断累积，无机磷不断减少，小麦根的变化幅度极小。在青饲料的 90 天分解过程中，全磷量占干物质的百分比从 0.411%提高到 1.020%，在这个时期中，先是有机磷增加，以后则是无机磷增加。在马粪的 90 天分解过程中，全磷量占干物质的百分比从 0.485%提高到 0.864%，磷素转化的总方向也先是有机磷增加，再是无机磷增加，但变化幅度很小。

Birch （1961）认为由于植物物质同时含有有机磷和无机磷，后者占全磷量的一半或更多些，因而，在植物物质分解的前期（至少三个月），只是它的无机磷成分在起着合成和分解的交迭作用（微生物的合成和分解，）原本植物物质中的有机磷成分则处于稳定的，不变化的状态。然而，报告老并没有提出植物有机磷处于不变化状态的直接证据，而把植物有机磷的分解和微生物再合成的有机磷的分解作用割裂开来，似乎也是证据不足的。

进行土壤培养试验，研究土壤有机磷的矿化作用，Thompson， Black 和 Zoellner（1954）将 25 对生荒和耕作土壤放在 40℃温箱内培养 25 天，有机磷的矿化量从 4~45ppm。有机磷的矿化强度与土壤有机碳和氮的矿化强度成正相关。在相等的有机磷条件下，土壤 pH 与有机磷矿化强度成正相关，即土壤 pH 愈大，矿化愈多。

Thompson 和 Black （1948）报导，土壤有机磷矿化随温度而递增。Eid，Black 和 Kempthorne （1951）报导，土壤有机磷矿化和土壤温度有明显的关系，在 35℃时比在 20℃时强很多倍。植物磷素营养的可给性，在 20℃时和土壤无机磷部分成正相关，在 35℃时却和土壤有机磷部分成正相关。

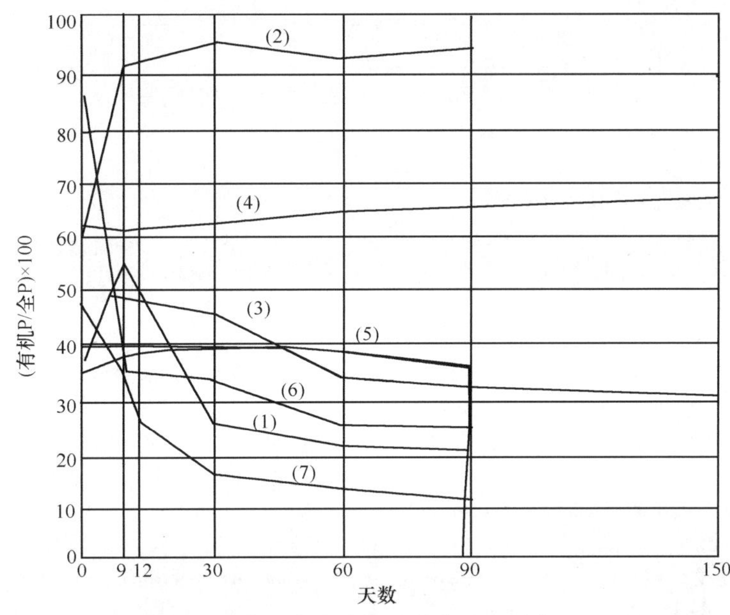

图 3　各种有机磷体在分解过程中有机磷的合成和矿化（从 Kaila，1949）

（1）青饲料；（2）黑麦草；（3）小麦根；（4）牛粪；（5）马粪；（6）真菌物质；（7）麦麸

土壤有机磷的积累和矿化除随土壤的类型，土壤地带分布的不同而变化外，不同的利用情况也产生显著的影响。Thompson，Black 和 Zoellner （1954）对比相邻接的生荒土壤和耕地土壤，生荒土壤中的有机磷含量要大些。生荒开垦为耕地后，有机磷含量的减少与土壤腐殖质含量的减少成正比例。de Turk（1938）报道耕作土壤中有机磷含量逐渐减少。Корабтева（1951）报道，经常施厩肥使得土壤有机磷和无机磷含量都提高了。由于施厩肥而提高的有机磷既包含在酸溶性部分中，也包含在比较稳定的碱溶性部分中，Kaila（1948）报道，施石灰能加强土壤中有机磷的矿化作用，经常施石灰降低土壤中的有机磷量。Thompson（1958，p.250）指出，施石灰提高酸性土壤中的磷素养料的可给性的生产经验在很大的程度上可能是由于加强了土壤有机磷的矿化作用。

八、小　　结

（1）随着研究工作的逐渐深入，人们对于土壤有机磷对土壤肥力，植物土壤磷素营养的重要性的认识日渐加强。

（2）由于研究方法的不完善，目前对于土壤中有机磷的含量和成分的知识还是很不确切的。

（3）卵磷脂分解快，在土壤中积累很少。关于土壤中有没有比较稳定的“土壤核酸”的积累和矿化过程，还不能定论。植素及其衍生物肯定是土壤中主要的有机磷成分之一，植素的脱磷酸作用进行得较慢，关键问题可能不是由于土壤中缺乏植素酶活性，而是由于植酸钙盐和铁盐的高度不溶解性（特别是铁盐）。还不知道土壤中有没有比较重要的其他有机磷化物成分。

（4）关于土壤中分解有机磷化物脱磷酸的主要微生物种类的科学资料还积累得很少，待充实。

（5）植物物质翻耕入土壤后（它既含有机磷，又含有无机磷，大致各占一半），磷素转化的总方向决定于无机磷的生物合成作用和有机磷的矿化作用的相对强度。这个相对强度受植物物质成分，特别是能源的影响。能源丰富，生物合成作用在一定时期内大于矿化作用（无机磷减少）。能源不足，矿化作用可以在开始时就大于生物合成作用（无机磷增加）。

（6）提高 pH（酸性土上施石灰）和提高温度一般能加强有机磷的矿化作用。

（7）进行土壤微生物学的研究工作应重视有关学科的研究成果，帮助我们较正确地提出研究任务，从而做出更有科学价值的生产意义的研究成果。

参 考 文 献

[1]板野新夫，甘扬声，1956，转化土壤中有溶性有机磷和无机磷化合物为可溶性磷酸盐的细菌，土壤学报，3：91-95.

[2]胡正嘉，1957，水稻田苕子翻耕后土壤中强烈分解有机磷的细菌（摘要），华中农学院学报，2：166-168.

[3]张宪武，刘期松，1955，转化有机磷细菌的研究，土壤微生物学集刊（中国科学院林业土壤研究所），11-23.

[4]Borrxep，K.，1961，Hecsap meeipsnsfijt6 73-678

[5]Berrep，k.，1963，3depho ienph-ahreihheon peahi cnpehi 32：231-238

[6]IMMrpehko，II，A.，1948，O coiepshih 1948， 105-501

[7]Kapaxha，M. T. u Bereba，B.I.，1962，Mhh6psieaa end hjhlehonek khchoru cuopoewxe mhrpocpra-hhmamk.Mhrpeohoiorha，31：216-220

[8]Hhmamh. Mhhrpocaoiorha，31：216-220.

[9]Roctayes，H. O.，1890，O hesoropbit ceoicrnax h cocrabe meperbox cacroe ruahctbo h ieconoleroo. ch6 ctp. 115.

[10]Mehknda，P. A.，1950，Batrephh，muepsinbyhmhe opragegeceob Meco

[11]Adama，A. P.，Bartholomew，W. Y. & Clark，F. E.，1954. Measurement of nucleic acid component in soil. Soil Sci. Soc. Am. Proc.，18：40-46.

[12]Anderson，G.，1956. The identification and estimation of soil inositol phosphate. J. Sci. Food and Agr.，7：437-444.

[13]Anderson，G.，1958. Identification of derivatives of deoxyribonucleic acid in humic acid. Soil Sci.，86：169-174.

[14]Anderson，G. C.，Arlidge，E. Z.，1962. The adsorption on inositol phosphates and glycerophosphate by soil clays，clay minerals，and hydrated sesquioxides in acid media. J. Soil Sci.，13：216-224.

[15]Aso，K.，1904. On organic compounds of phosphoric acid in soil. Tokyo Imp. Univ. Col. Agri. Bulletin，6：277-284.

[16]Barrow，N. J.，1961. Phosphorus in soil organic matter. Soils and Fertilizers，24：169-173.

[17]Birch，H. F.，1961. Phosphorus transformation during plant decomposition. Plant and Soil，15：347-366.

[18]Black，C. A. and Goring，C. A. L，1953. Organic phosphorus in soils. Soil & Fertilizers Phosphorus in Crop Nutrition，in Pierre and Norman（ed）：Academic Press，New York，U. S. A.，pp. 123-152.

[19]Bower，C.A.，1945. Separation and identification of phytin and its derivatives from soils. Soil Sci.，59：277-285.

[20]Bower，C. A.，1949. Studies on the forms and availability of soil organic phosphorus. Iowa Agr. Expt. Stn. Res. Bull.，362.

[21]Bremner，J. M.，1956. A review of recent work on soil organic matter. J. Soil Sci.，11：67-82.

[22]Caldwell，A. G. and Black C. A.，1958. Inositol hexrphosphate. I. Quantitative determination in extracts of soils and manures；II. Synthesis by soil microorganisms；III. Content in soils. Soil Sci. Soo. Amer. Proc.，22：291-293-296-298.

[23]Chang，S. C. （张信诚），1939. The transformation of phosphorus during the decomposition of materials. Soil Sci.，48：85-100.

[24]Chang，S. C.（张信诚），1940. Assimilation of phosphorus by a mixed soil population and by pure cultures of soil fungi. Soil Sci.，49：197-210.

[25]Colmer，A. R.，1947. The use of the enzyme leoithinase in grouping some member of genus *Bacillus*. J. Bact.，54：11-12.

[26]De Turk，E. E.，1938. Changes in the soil of the morrow plots which have accompanied long continued cropping. Soil Sci. Soc. Amer. Proc.，3：83-85.

[27]Dor，A. W.，1911. Phytase in lower fungi. J. Biol. Chem.，10：183-186.

[28]separation of organic P from soil. Soil Sci.，51：150-170.

[29]Dyer，W. J. and Wrenahall，C. L.，1941 b. Organic phosphorus in soil. IV. The decomposition of some organic phosphorus compounds in soil culture. Soil Sci.，51：323-330.

[30]Eid，M. T，Black，C. A. and Kempthorne，O.，1951. Importance of soil organic and inorganic phosphorus to plant growth at low and high soil temperatures. Soil Sci.，71：361-377.

[31]Flaig，von W.，Kuron，H. u. Kani，R.，1955. Uber die Sorption von Nncleinstoifen an Tonkolloiden，Zeit. F. phlanzenernehr. Dung. u. Bodunk.，71：141-154.

[32]Fuller，W. H. and McGeorge，W. T.，1951. Phosphates in calcarious Arizona soils. III. Distribution in some representative profiles. Soil Sci.，71：315-323.

[33]Ghani，M. O. and Aleem，S. A.，1943. Studied on the distribution of different forms of phosphorus in some India soils. I. Surface distribution. Indian J. Agr. Sci.，13：283-288.

[34]Goring，C. A. I. and Bartholomew，W. V.，1950. Microbial products and soil organic matter. II. The effect of clay on the decomposition and separation of the phosphorus compounds in microorganisms. Soil Sci. Soc. Amer. Proc，14：152-156.

[35]Goring，C. A. I. and Bartholomew，W. V.，1951. Microbial products and soil organic matter. III. Adsorption of carbohydrate phosphates by clays. Soil Sci. Soc. Amer. Proc. 15：189-194.

[36]Goring，C. A. I. and Bartholomew，W. V.，1952. Adsorption of mononaeleotides，nucleic soil and nucleoprotein by clays. Soil Sci.，74：149-160.

[37]Jackman，R. H. and Black，C. A.，1951. Solubility of iron aluminium，calcium and magnesium inositol phosphates at different pH values.

Soil Sci.，72：179-186.

[38]Jackman，R. H. and Black，C. A.，1952 a. Phytase activity in soils. Soil Sci.，73：117-125.

[39]Jackman，R. H. and Black，C. A.，1952 b. Hydrolysis of phytate phosphorus in soils. Soil Sci.，73： 167-171.

[40]Jeffreis，C. D. Holtman，D. F. and Guse，D. G.，1957. Rapid method for determining the ability of micro organisms on nucleic acids. J. Bact.，73：590-591.

[41]Kaila，A.，1948. Viljelysmaan orgaanisesta fosforista. Valtion maatalouskoetoiminnan Julkaijuja，129.

[42]Kaila，A.，1949. Biological absorption of phosphorus. Soil Sci.，68：279-289.

[43]Knight，B. C. I. G. and Proom，H.，1950. A comparative survey of the nutrition and physiology of mesophilic species in the genus *Bacillus*. J. Gen. Micribiol.，4：508-538.

[44]McCaughey，C. A. and Chu，H. P.，1948. The egg-yolk reaction of aerobic sporing bacilli. J. Gen. Microbiol.，2：334-340.

[45]McDonnell，P. M. and Walsh，T.，1957. The phosphate status of Irish soils with particular references to farming systems. J. Soil Sci.，8：97-112.

[46]Mulder，1844. Uber die Bastandheile der Ackererde. J. prakt. Chem.，32：321-344.

[47]Pearson，R. W.，1940. Determination of organic Pin soils，Ind. Eng. Chem. Anal ed.，12：198-200.

[48]Pearson，R. W.，Norman，A. G. and Ho. C.，1941. The mineralization of the organic phosphorus of various compounds in soils. Soil Sci. Soc. Amer. Proc.，6：168-175.

[49]Pearson，R. W. and Simonson，R. W.，1940. Organic phosphorus in seven Iowa soil profiles： Distribution and amounts as compound to organic carbon and nitrogen. Soil Sci. Soc. Amer. Proc.，4：162-167.

[50]Pearson，N. E.，1953. On phytin phosphorus in soil. Plant and Soil，4：252-266.

[51]Pierre，W. H.，1948. The phosphorus cycle and soil fertility. J. Amer. Soc. Agron.，40：1-14.

[52]Schollenberger，C. J.，1918. Organic phosphorus of soil：Experimental work on methods for extraction and determination. Soil Sci.，6：365-395.

[53]Schollenberger，C. J.，1920. Organic phosphorus content of Ohio soils. Soil Sci.，10：127-141.

[54]Schreiner，O. and Lathrop，E. C.，1912. The chemistry of steam-heated soils. U. S. Dept. Agr.，Bur. Soil Bull，Bull.，39.

[55]Shorey，E. C.，1912. Nucleic acid in soils. Science，39：390.

[56]Shorey，E. C.，1913. Some organic soil constituents. U. S. Dept. Agri. Bur. Soils.，88.

[57]Smith，D. H. and Clark，F. E.，1951. Anion exchange chromatography of inositol phosphates from soil. Soil Sci.，72：353-360.

[58]Smith，D. H. and Clark，F. E.，1952. Chromatographic separation of inositol phosphorus compounds. Soil Sci. Soc. Amer. Proc.，16：170-172.

[59]Smith，N. R. and Gordon，R. R.，1952. From Bergey's Manual of Determinative Bacteriology，7th ed. 1957，pp. 613-634.

[60]Stoklssa，J.，1911. Biochemischer Kreilauf des phosphate-ions in Boden Cbt. Bakt. II，29：358-519.

[61]Szember，A.，1950. Influence on plant growth of the breakdown of organic phosphorus compounds by micro-organisms. Plant and Soil，13：147-158.

[62]Thompson，L. M.，1957. Soils and Soil Fertility. McGraw-Hill Book Company，Inc.，New York，U. S. A.

[63]Thompson，L. M. and Black，C. A.，1948. The effect of temperature on the mineralization of soil organic phosphorus. Soil Sci. Soc. Amer. Proc.，12：323-326.

[64]Thompson，L. M. and Black，C. A.，1950. The mineralization of organic phosphorus， nitrogen and carbon in Clarion and Webster soils. Soil Sci. Soc. Amer. Proc.，14：147-151.

[65]Thompson，L. M. and Black，C. A. and Zoellner，J. A.，1954. Occurrence and mineralization of organic phosphorus in soils，with particular reference to associations with nitrogen. Soil Sci.，77：185-196.

[66]Ulrich，von B. u. Benzler，J. H.，1955. Der organisch gebundene Bhosphor in Poden，Zeits. f. Phlanzenernahr.，Dung. u. Bodonkund.，70：220-249.

[67]Walker，T. W. and Adams，A. F. R.，1958. Studies on soil organic matter. I. Influence of phosphorus content of parent materials on accumulations of C，N，C，and organic Pin grassland soils. Soil Sci.，85：307-318.

[68]Williams，C. H. Williams，E. G. and Scott，N. M.，1960. Carbon，nitrogen， sulphur and phosphorus in some Scottish soils. J. Soil Sci.，11：334-346.

[69]Wrenshall，C. L.and Dyer，W. J.，1941. Organic phosphorus in soils. II. Nature of the organic P. compounds A. nucleic acid B. phytin. Soil Sci.，51：235-248.

[70]Wrenshall，C. L.and McKibbin， R. R.，1937. Pasturo studies. XII. The nature of the organic phosphorus in soils. Canadian J. Res.，15 B，475-479.

[71]Yoshida，R. K.，1940. Studies on organic P compounds in soils； isolation of inositol. Soil Sci.，59：81-89.

水稻土中植物营养元素的生物循环*

陈华癸

（华中农学院）

首先应该指出，本文所讨论的水稻土对象是以长江流域的水稻土为主，这地带的水稻土，按其植被和水文来说，可以分为两个主要类型：

1. 夏季蓄水种水稻，冬季蓄水休闲的冬水田，按照地形的特点，冬水田又可以分为塝田和低地冬水田两亚类，塝田冬水田是指在坡地上的冬水田，做成冬水田的原因不是因为排水困难而是因为不敢排水，如果秋季排水种冬作，第二年春天不能保证插秧时所需要的水分，低地冬水田一般是排水困难，或无法排水的水田，有些低地冬水田是湖田，不断地接受湖积物。

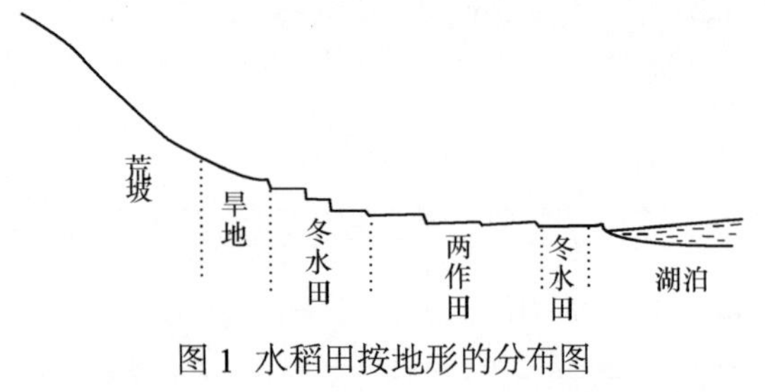

图 1 水稻田按地形的分布图

2. 夏季蓄水种水稻，冬季排水种旱地作物的水旱两作田，旱地作物可以是粮食作物（以麦为主）或技术作物（以油菜为主），也可以是绿肥作物（豆类绿肥如苕子和紫云英；非豆类绿肥如肥田萝卜）；还有许多地方种蚕豆或者是油菜和豆类绿肥间作。

按地形，水稻土的分布如图 1：本文以水旱两作田的生物循环为讨论的主要对象。

水稻田的氧化还原势及相应的生物学特性

水稻土的氧化还原势和一般旱作土壤有显著的差别，在蓄水条件之下，水层隔绝空气，水稻土的氧化还原势大都在 rH 10~25，这比旱地土壤要低得多（于天仁、李松华，1957；Неунылов 1948，图 2），旱地土壤的氧化还原势一般在 rH 30 以上。

影响水稻土氧化还原势的主要因素是：①氧气的供给情况；②植物根系的代谢作用；③土壤微生物的代谢作用，和④无机质氧化还原体系的作用。无机质氧化还原体系虽然有一定的影响，但是它们的摆幅较小，而且可以认为是被动的。因此，水稻土的氧化还原势表现如图 2（Неунылов，1948）的式样，根系附近氧化还原势高的原因是氧气通过水稻的输导组织从根表面排出的结果。于天仁、李松华（1957）测定在蓄水条件下，八种水稻田土壤的氧化还原势平均值为 rH23.8。黄东迈、李锡泾（1955）测定，在长期灌水条件下水稻土表土的氧化还原势 Eh 值为 45mv，排水晒田后则在三五天内可以提高到 520~575mv。

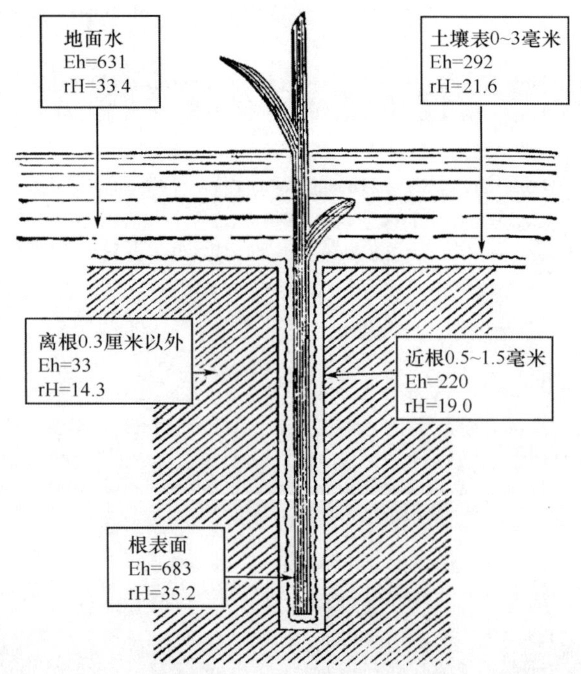

图 2 水稻土的氧化还原势（自 Неунылов，1948）

*原载于《稻作科学论文集》，167~182，1959，北京：农业出版社

上述的水稻土氧化还原势允许大多数种类的好气性微生物生活，但是不是最适宜的，受到一定的抑制作用，高度好气性的微生物不能活动。对于嫌气性微生物来说，比较旱地土壤更适宜些。对于兼性的微生物来说，当然是适宜的。正因为如此，水稻土中氧化还原势的变化对于土壤中微生物的动态有显著的影响，蓄水种水稻的土壤，短期排水晒田大大地增加了土壤中好气性微生物的数量和活动，如果说在排水以前每克土壤含好气性细菌不到1000万，那么排水后立即暴长10倍（曹燕珍，1956），再度灌水以后又回缩下降，即使在蓄水条件之下，好气性微生物的数量也远比嫌气性微生物的数量为多，用嫌气培养法分离培养水稻土中的微生物，所得种类包括兼性的和专性嫌气性的，春季，旱地灌水成为水田后，首先是好气性和兼性的种类为主，以后，嫌气性种类的优势逐渐上升，逐渐地占绝对优势（曹燕珍，见前）。

水稻土的碳素循环和腐植质含量

如果说，水稻土的形成过程可以认为是一种草甸沼泽过程，那么它和典型的草甸沼泽过程的主要差别在于它没有累积腐殖质的趋势，表1列举七个不同的无石灰性水稻土样本的有机碳含量，它们并不比在同一区域内的旱地土壤有机碳含量高。水稻土和同一气候地带的沼泽土壤，芦苇沼泽土壤相比，两者有很大的区别，芦苇沼泽土壤一般都有一层15~30厘米厚的腐殖质层，由腐殖物质、半腐殖化的物质和尚未分解的植物残体所组成。但是，一旦芦苇沼泽土壤开垦为水稻田以后，这一层腐殖质层便在三、二年内消失掉了，在同一地点，芦苇沼泽土和水稻土有这样重要的区别，使人很难承认水稻土的形成过程是一种草甸沼泽过程。

腐殖质累积量低的原因是两方面的，一方面由于收获物的移去，每年供给土壤的有机质原料比较少；另一方面由于高度的农业技术措施，使得分解作用也进行得比较快。

水稻土有机质原料的主要来源有下列三类：①水稻及冬季作物的根系；②有机质肥料，和③绿肥植物的茎叶和根系。在长江流域，冬季绿肥作物一般在4月上中旬翻耕，翻耕后土壤灌水，绿肥作物的茎叶和根系就在灌水的情况下腐解。在这种条件下，豆类绿肥植物腐解得很快，一般叶子及嫩根在三、四天内变黑，两星期内细胞组织破坏消失，较老的组织，细胞壁分解得较迟，一个月内还可以看得见完整的细胞壁物质。

在水旱两作田中，冬季旱（10月~3月），夏季蓄水（4月~9月），在冬季，有机物质是在通气条件较好而温度较低（在长江流域冬季土温一般在0℃以上）的条件下腐解，在夏季、有机物质是在通气条件不良而温度较高的条件下腐解。观察的结果，温度的影响比通气条件更重要，富于细胞壁物质的禾本科植物根在冬季分解得很慢，在4月初，绿肥翻耕的时候，一般的还残留很多尚未分解的水稻根残体，前一年的残体要在第二年春耕灌水后，水稻土蓄水种水稻的季节中才逐渐腐解完全。

表1　用 Ter Meulen 干烧法测定水稻土有机碳含量

土壤	重复		平均
	1	2	
武汉，水旱两作用	0.760	0.760	0.760
武汉，同上	0.922	0.928	0.925
武汉，同上	1.189	1.189	1.189
武汉，常年蓄水，种莲藕	0.779	0.779	0.779
南京，水旱两作田	0.491	0.504	0.498
南京，同上	0.877	0.883	0.880
长沙，同上	1.345	1.349	1.347

绿肥翻耕，蓄水种水稻后，占优势的腐生性微生物是细菌，以孢子杆菌（主要是 *Bacillus*

cereus-megatherium 群，*B. mycoides* 等），孢子梭菌（*Clostridium* spp.），革兰氏负反应的顶毛杆菌（*Pseudomonas* spp.）和周毛杆菌（*Bacterium* spp.）为主，放线菌和真菌不占重要地位（曹燕珍，1956），分解纤维素的细菌种类则是 *Bacillus omelianskii*（赵文洪，1956）。

如果按照 Robinson（1949，p.203）所说，将土壤的腐殖质化过程分为四个类型：①好气条件下的腐殖质化过程；②嫌气条件下的腐殖质化过程；③酸性条件下的腐殖质化过程；④潮湿热带条件下的腐殖质化过程；那么水稻土的腐殖质化过程是属于好气嫌气条件之间的。我们的研究，用好气性培养法所得的细菌总数远比用嫌气性方法所得为多，前者的数量在每克土千万与万万细菌的范畴，后者的数量只在每克土十万与百万的范畴，虽然如此，好气性微生物的生命活动是受到抑制的，高度好气性的微生物不能生活（如硝化作用就不能进行）；真菌的活动也受到抑制，数量很少，一般的好气性微生物活动也受到阻碍。如果短期的排水，改良通气条件，好气性微生物数量大增（见前）。从水稻土呼吸所产生的气体的性质来看，水稻土又表现了很显著的嫌气性活动特点，Acharya（1935）的研究工作指出，在轻度、中等和严格的嫌气性条件下，稻草分解，产生气体的情况是：嫌气性的程度愈高，所产生的沼气愈多。De 和 Mandel（1956）的研究工作指出，水稻土蓄水种水稻的前 3~5 星期内，土壤空气中没有沼气，以后有沼气，沼气的产生说明丁酸类型的发酵作用占一定地位，也说明嫌气性条件在逐渐加强。他们的研究结果和我们的结果是一致的，我们的研究结果指出，插秧一个月后，嫌气性孢子梭菌的数量占优势（曹燕珍，1956）。

水稻土的氮素循环

氨化作用：由于氨化作用是多种多样的微生物的作用，在水稻土中，无论在旱季或水季都畅行无阻，豆类绿肥翻耕后，氨化作用立即开始，土壤中的氨态氮不断增加（陈华癸，1952；王家玲，1956），由于土壤的阳离子吸收性，氨主要是吸收在土壤胶体表面，溶解在土壤中氨很少（陈华癸和萧泽宏，1948，表 2），这对于土壤的含氮量和植物的氮素营养都有好处，我们在长江流域各地的研究结果，水稻土的氨态氮量是百万分之几到几十之间，在水稻旺盛生长的季节，由于植物的旺盛吸收作用，土壤中氨态氮的含量可以很低。

豆类绿肥翻耕以后，在蓄水条件下，氨化作用进行得很旺盛，几乎没有停滞期或逆转期（见图 3，王家玲，1956），从图中还可以看出，绿肥翻耕后，土壤中氨态氮的增加比所加入的绿肥总含量还要高（压入绿肥 2500~3000 斤/亩，含氮 12.5~15 斤/亩，土壤氨态氮量增加 50~100 百万分，合 16.5~33 斤/亩×半尺），这结果指出，绿肥翻耕后，不但其本身的含氮物质氨化，并且带动了土壤中原有的含氮物质氨化，这结果和 Breadbent（1948）的结果相符合，他用 C^{14} 和 N^{15} 研究苏丹草绿肥压青后二氧化碳和无机质氮的形成过程，也指出了绿肥带动了土壤中腐殖质的分解作用和含氮物质的矿物质化作用。

在湖泊地带，水草广泛地被利用为绿肥，在插秧前一、二星期施入水稻田中，这时的水草大多处于幼嫩状态，腐烂分解很快，氨化作用立即开始，供给水稻吸收，丝草（*Potamogeton moachianus*）、芦苇（*Phragmites communis*）、标草（一种莎草科植物）、茭草（*Zizania caduciflora*）四种水草在水稻土中的氨化作用进度见图 4。

水稻土中推动氨化作用的主要微生物种类也就是占优势的腐生性微生物种类（*Bacillus mycoides*、*Bacillus cereus-megatherium* 群、*Bacillus* spp.，王家玲，1956）。

硝化作用：由于水稻土在蓄水种水稻期间的 rH 值在 10~25 之间，硝化微生物不能活动，基本上可以认为没有硝化作用。日本的研究者们 Shingo Mitsui（1956）认为水稻土可以分为两层，氧化层和还原层，在氧化层中有硝化作用，这观察是正确的，但同时也要指出，氧化层只有几毫米厚，苏联 Неуныпов（见前）也指出，在土壤表面和水稻根表面，rH 值可达到 35 以上，因而也有硝化作用，显然的，这样局限的硝化作用层不能对水稻土的氮素循环起显著作用，在蓄水条件下氨态氮是主要的，可以说是唯一重要的无机氮素成分。

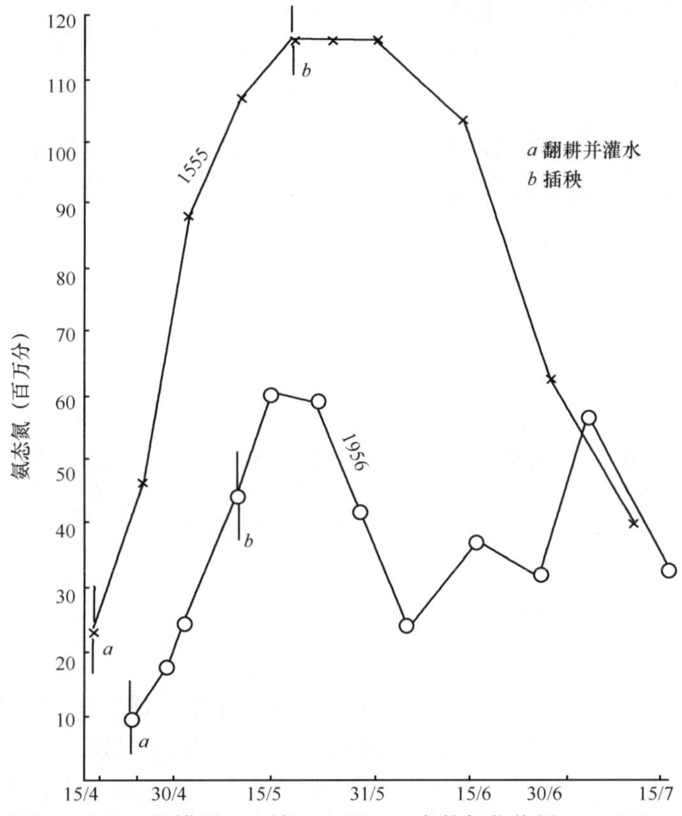

图 3 绿肥（苕子）翻耕后，土壤（水稻田）中的氨化作用（王家玲，1956）

表 2 冬水田中的无机态氮（陈华癸和萧泽宏，1948）

日期	土壤		表面水	
	HN_3-H	（$NO_3^-+NO_2^-$）-N	HN_3-H	（$NO_3^-+NO_2^-$）-N
1943-12-24	10.6ppM	微量	—	—
1944-1-10	12.7	”	0.3	微量
1944-2-16	5.9	”	—	—
1944-3-1	3.6	”	0.3	—

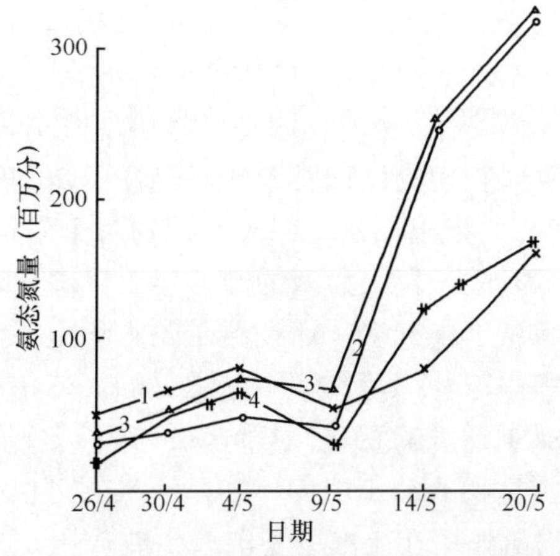

图 4 水草压青后，土壤中氨态氮的变化（程见尧材料）

1. 丝草 4 月 14 日压青；2. 芦苇 4 月 20 日压青；3. 标草 4 月 20 日压青；4. 茭草 4 月 19 日压青；4 月 24 日插秧

在水旱两作的水稻田中，冬季排水种旱地作物的时期，由于通气条件的改善，硝化作用有显著的增强，这时土壤中的硝态氮和氨态氮占同等重要的地位（表 3，陈华癸和萧泽宏，1948），在冬季蓄水的冬水田中只有氨态氮（见表 2）。

表 3 水旱两作的水稻田（四川北碚）**在冬季排水种小麦，夏季蓄水种水稻时土壤中的氨态氮和硝酸态氮含量**（自陈华癸和萧泽宏，1948）

	第 13 区		第 14 区		第 15 区	
	HN_3-H	NO_3-N	HN_3-H	NO_3-N	HN_3-H	NO_3-N
1946-2-11（1）	3.4	4.2	15.8	10.5	8.5	8.2
1946-7-15（2）	5.9	微量	7.1	微量	5.0	微量

注：（1）排水种小麦季节；（2）蓄水种水稻季节。

反硝化作用：水稻土中的反硝作用可以很强，前节中指出，两作水稻田中，冬季排水种旱地作物时，有相当旺盛的硝化作用，土壤中有硝酸盐类，到春季，一旦灌水，反硝化作用立即使硝酸消失（陈华癸和萧泽宏 1948；黄东迈和李锡泾，1955，见表 4；王家玲，1956），我们查不出，究竟硝酸还原后主要是变为氨或者变为氮气，Shigo Mitsui（1956，64 页）认为在水稻田中硝酸还原的结果是变为氮气，如果这样，旱地灌水对于土壤含氮量和植物氮素营养很不利，鉴于硝酸盐类肥料不适宜于水稻田的经验，这种看法可以认为是正确的。

固氮作用：共生性的固氮作用在实行水稻绿肥耕作制的土壤中起着十分重要的作用，在长江流域，两作田冬季种豆类绿肥（以苕子和紫云英为主），每亩可以收获地上部分 1000 斤到 8000 斤（鲜重），一般优良的农业技术可以保证 4000~5000 斤，约含氮 20~25 斤，地下部分不在内，种一亩豆类绿肥可以肥田三亩，这是一个专门的问题，不在这里详细讨论（参阅陈华癸，1955）。

表 4 水稻田灌水前后 NH_3-N 和 NO_3-N 的变化（自黄东迈和李锡泾，1955）

地点	深度（厘米）	HN_3-H（微克/克土）		HN_3-H（微克/克土）	
		灌水前	灌水后 4 天	灌水前	灌水后 4 天
练湖农场	0-5	10	67.5	6	痕际
3 号水田	5-15	30	90	6	痕际
练湖农场	0-5	15	52.5	3	痕际
1 号水田	5-15	20	67.5	—	—

根据我们的初步观察（陈华癸，1956）冲积物上的水稻土和水稻根际有相当旺盛的固氮菌（azotobacter），Tschapek 和 Rouco（1956）指出，固氮菌并不适宜于过高的氧化还原势，这正是水稻土所具备的，至于嫌气性的固氮菌，水稻土中当然比旱地土壤中更旺盛，关于好气性和嫌气性固氮细菌对于水稻土的氮素经济的意义，和发展它们为细菌肥料的可能性亟待详细研究。

印度的学者们（见 Singh，1942）指出固氮蓝绿藻类植物对于水稻土地的氮素经济起着重要作用，在有些水稻田中，固氮蓝绿藻类可以成一层很显著的漂浮层。日本学者 Wanatobe（1950，1951）指出，固氮蓝绿藻类在热带和亚热带的水稻土中很多，而在温带的水稻土中则很少，在长江流域的水稻田中，根据过去的研究资料，已知的一些固氮蓝绿藻类都不是常见的（陈华癸，1951）。

关于藻类植物对于水稻土的氮素经济的意义，对于藻类与固氮细菌的共生性固氮作用并没有予以应有的重视，而实际上，在各种藻类植物的初步培养体中，经常的可以分离出固氮菌来。Stokes（1940）的结论，认为与藻类植物共生的固氮作用在土壤中不能起主要作用，这并不是针对着水稻土而言的，应该做更深入的研究。

水稻土的硫素循环

自然界的硫素循环和氮素循环相伴着进行，水稻土中硫素循环的规律我们了解得很少。

蛋白质分解产生硫化氢，这是一般腐生性微生物共有的性能，硫化氢氧化为硫酸盐则是一些无机营养型的微生物的性能。我们对于硫磺细菌和硫化细菌的知识有一些，但是对于它们在土壤中的具体活动情况了解得很少。

我们在水稻土中培养出有色硫磺细菌、无色硫磺细菌和硫化细菌，但是对于它们在水稻土中的作用强度了解得很少。鉴于硫化氢在水稻土中有累积的趋势，甚至于有毒害作用（Shingo Mitsui，1956；Mitsui and Hashimoto，1949）的研究说明硫化氢累积能破坏水稻根的细胞组织，只有水稻根表面沉淀有铁锈层时才能防止硫化氢的毒害作用。

虽然说，水稻土中的硫素养料一般是能够满足水稻的需要的，无需施用硫素肥料，但是在有些水稻田中（发秋田、冷水田），不施石膏，插秧后秧苗不回青，或回青很迟，影响产量，施用少量石膏后（每亩 5~10 斤）可以治愈这现象，根据陈华癸等（陈华癸等，1952；程学达，1956），这是石膏中硫酸根的作用（表 5）。

无疑的，微生物生命活动对于上述几项问题都起着一定的作用的，但还缺少深入的研究。

表 5　用含硫物质处理，对冷水田秧苗回青及籽实产量的影响（陈华癸等，1952）

区号	处理	秧苗回青早迟秩序	收割面积（平方米）	籽实产量（斤）	每亩籽实产量（斤）
1	硫磺	4	11.56	7.10	443.5
2	硫酸镁	3	12.25	8.10	473.2
3	氯化钙	6	12.25	7.00	379.4
4	硫酸钠	1	12.25	8.90	468.2
5	碳酸钙	5	12.25	6.40	340.7
6	硫酸钙	2	12.96	8.13	422.3

水稻土的磷质循环

有机磷素的无机质化也是和蛋白质的分解作用密切地联系着的，含磷蛋白质的分解是有机磷素无机质化的主要方面，在长江流域，豆类绿肥翻入水田中后，有机磷素的无机质化也是比较快的，但比氨化作用稍慢些。前面指出，豆类绿肥翻入水田中后，氨化作用无停滞期或逆转期，但胡正嘉（1956）的研究工作指出，在一星期内，土壤可给性磷素略有下降，以后再上升（图 5），他同时指出，在长江流域，水稻田中分解有机磷化物的微生物活动很强盛，无机质化畅盛，从农业技术的要求来说，没有施用磷细菌肥料，加强有机磷素分解的需要。

在我们这里的水稻田中，分解有机磷素化合物的主要种类是 *Bacillus cereus-megatherium* 群，*Bacillus mycoides* 和 *Bacillus asterosporus* 等孢子杆菌，有些菌株分解卵磷脂的能力很强（表 6）。

另外，我们的初步探索，不论在水稻田或旱地土壤中，矽酸盐细菌都很多，它们能分解难溶性的磷矿石粉，产生可溶性磷素养料，现尚缺乏深入研究。

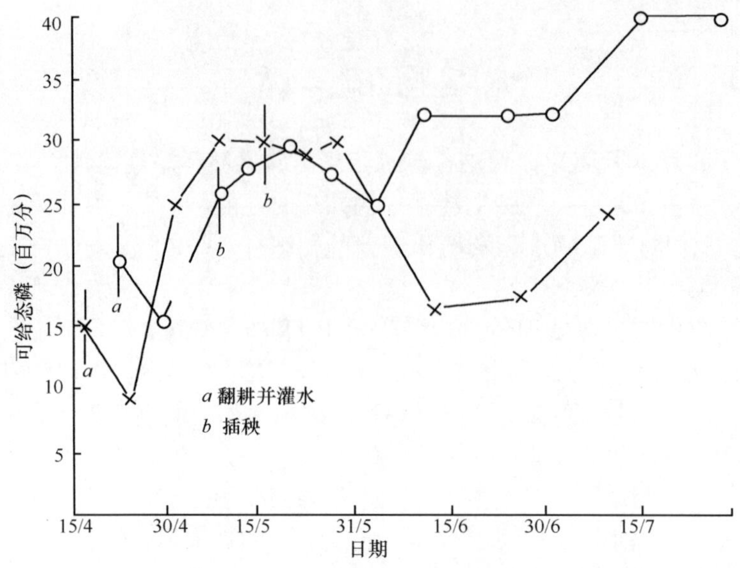

图 5 绿肥（苕子）翻耕后，土壤（水稻田）中可给性磷的变化（胡正嘉 1956）

水稻土的铁素循环

水稻土形成过程的特点之一是铁素的淋洗和淀积，它和灰壤化过程不同，在水稻土中铁移动而铝不移动(见表 7)，在灰壤中铁和铝伴随着移动，日本研究者 Keizoburo Kawaguchi 和 Yoshiro Matsuo（1956）的研究工作指出，同一土壤，种植水稻愈久，铁和锰的下移愈深，而铝和钛则很少移动，因此，铁素淋洗程度可以看作为水稻土形成程度的标志。

黄东迈和李锡泾（1955）的研究工作指出：水稻土中亚铁的变化与灌排有密切关系，土壤在干燥状态下，表土亚铁消失，在灌水后并保持水层时，亚铁在数日内即增高（图 6），亚铁含量的增加大大地增加了铁素的运动性，创造了淋溶的条件，在蓄水条件下亚铁含量的增加应该看作为氧化还原势降低的结果而不是原因，而氧化还原势降低的原因则主要是由于①隔绝了空气；②微生物的代谢作用。

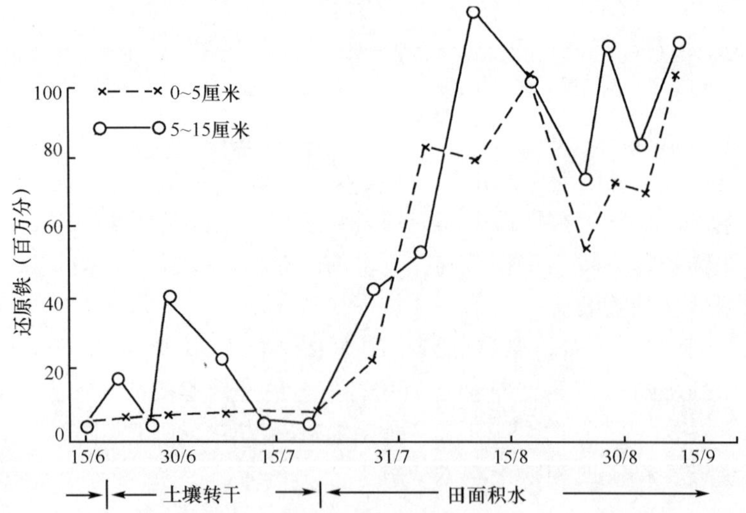

图 6 水稻土排水和积水情况下，土壤中还原铁的含量（黄东迈和李锡泾，1955）

表 6 用卵磷脂做磷液，几种从水稻田中所得的细菌的矿化程度（0.1N 酸溶解性；胡正嘉，1956）

菌 类	*bacillus cereus*				*bacillus mycoides*	*bacillus asterosporus*	对照
菌株号	3	7	13	23	10	65	不接种
矿化量%	44.5	47.5	45.4	38.0	36.9	40.2	痕际

表 7　水稻土各剖面层次粘土部分的矽、铝、铁含量（自梭颇，1936）

	土层	SiO_2	Al_2O_3	Fe_2O_3
喇堡粘壤土	0-15	0.10	37.61	1.34
	15-25	7.41	40.97	1.54
	25-50	9.21	35.15	8.58
	50-100	9.82	36.01	12.44
陆墓镇粘土	0-15	1.56	25.75	9.27
	15-40	1.88	27.83	7.93
	40-60	2.20	27.25	7.99
	60-80	4.65	26.49	7.02
	80-95	9.60	27.50	8.16
	95-100	1.17	25.86	8.72
	100-130	5.10	21.16	18.72
	131-	2.88	26.48	7.84

关于铁素在底土中淀积的原因，我们了解得更少，在水稻土中可以找到各种铁细菌，但我们并不知道它们对于在底土中铁素淀积起什么作用，英国研究者 Crawford（1956）的工作指出，腐生性微生物对铁的淀积作用（尤其是 *Corynebacterium simplex*）可能比铁细菌的作用更重要些。

总　　结

本篇总结了过去几年中关于水稻土中各种植物营养元素的生物循环的研究工作(主要是中国的研究工作)，讨论了水稻土的氧化还原势，碳素、氮素、硫素、磷素和铁素的微生物学变化，讨论对象以长江中下游为主。

现有的资料很少，还不能对水稻土的生物循环提供一个明确的概念，但已有的资料确指出了水稻土生物循环的特点，如果说，水稻土形成过程属于草甸沼泽过程，那么它是一类十分特殊的草田沼泽过程，其特点在于：①腐殖质的累积量很低，没有腐殖质累积层，这点和同一地带的典型草甸沼泽土壤（芦苇沼泽土）有显著的不同；②有机物质分解速度强，元素的生物循环十分旺盛；③水稻土的氧化还原势在好气性和嫌气性之间：高度好气性的微生物作用受到抑制，嫌气性的微生物活动比一般旱地为强，这种特性对水稻土的生物循环起着十分重要的作用，突出的表现在氮素循环特点和铁素的还原和淋洗特点上；④铁素的淋洗程度是水稻土形成程度的标志；⑤在铁素高度淋失的水稻土中，由于嫌气性分解作用所产生的硫化氢对于水稻有毒害作用，在有些水稻土中，施用少量的硫酸盐类对于秧苗回青有良好作用。

参 考 文 献

[1]于天仁、李松华，1957，土壤学报，5 卷 1 期，97-110 页。
[2]王家玲，1958，水稻田绿肥（苕子）翻耕后土壤中主要嫌气性蛋白性分解细菌以及土壤无机态氮质动态的研究，华中农学院论文。
[3]黄东迈、李锡泾，1955，土壤学报，3 卷 2 期，83-89 页。
[4]胡正嘉，1956，水稻田中绿肥（苕子）翻耕后水稻生长期间有机磷的转化及强烈分解有机磷的细菌，华中农学院论文。
[5]陈华癸、萧泽宏，1948，武汉大学理学季刊，9 卷 1 期，79-88 页。
[6]陈华癸，1951，中国植物学杂志，5 卷 2 期，46-49 页。
[7]陈华癸、庄正德、何殿元、雷自强，1952，土壤学报，2 卷 1 期，34-36 页。
[8]陈华癸，1955，土壤学报，3 卷 2 期，90-106 页。
[9]曹燕珍，1956，水稻田绿肥（苕子）翻耕后水稻生长期间土壤耕层中占优势的有机营养型微生物类群的研究，华中农学院论文。
[10]程学达，1956，土壤学报，4 卷 1 期，51-58 页。
[11]赵文洪，1956，水稻田绿肥（苕子）翻耕后，纤维素分解细菌种类的研究，华中农学院论文。
[12]梭颇（J.Thorp），1936，中国之土壤，188、195 页（中文）。
[13]Неуныпов Б. А.，1948.

[14]Сборник научных работ сельскохозяйственных опытноисследовательских учреждений Приморского края. Вып. 1，стр. 61-112.
[15]Acharya，C. N.，1935，Biochem. J. Vol. 29，p.528-，953-，116-，1459.
[16]Breadbent，F. E.，1947，Proc. Soil Sci. Soc. Amer. Vol. 12，pp. 246-249.
[17]Crawford D. V.，1956，Rapports 6e I. C. S. S. Vol. C，pp.197-202.
[18]De，K. A. And L. N. Mandel，1956，Soil Sci. Vol. 81，pp. 453-458.
[19]Kawaguchi K. And Y. Matsuo，1956，Rapports 6e I. C. S. S. Vol. C. pp.533-537.
[20]Mitsui，S. 1956，Inorganic Nutrition， Fertilization and Soil. Amelioration for Lowland Rice pp. 21-85.
[21]Mitsui，S.and Hashimoto 1949，p. 21.
[22]Robinson G. W.，1949， Soils，3 ed. pp. 203.
[23]Stokes J. L.，1940，Soil Sci. Vol. 49，pp.265-275.
[24]Tschapek M. And I. Rouco，1956，Rapports 6e I. C. S. S. Vol. C，pp.177-180.
[25]Shingh，R.N.，1942，Indian J. Agr. Sci. Vol.12，pp.743-561.
[26]Wanatobe，A.，1950，Miscellaneous Reports of Research Institute of Natural Resources Nos.17-18，pp.61-68，见 Nature Vol.168，748-49（1951）.

第二项 硝化作用和硝化细菌

Inorganic Compounds of Nitrogen in Rice Field Soils*

H. K. CHEN AND T. H. SHIAO[1]

INTRODUCTION

The state of inorganic nitrogen in the water-logged rice field soils is not comparable with soils which are naturally or artificially drained. In the drained, hence aerated, soils the chief inorganic compound of nitrogen is nitrate. Only very small amount of nitrite and ammoniacal nitrogen are found in neutral, well aerated soils. The decomposition of organic nitrogen in aerobic soils is accomplished in two steps, both steps being microbiological: The first step is the decomposition of organic nitrogen compounds by heterotrophic microorganisms with the production of ammonia. This step is known as ammonfication or deamination. It is either aerobic or anaerobic, depending on the state of oxygen tension in the soil. The second step is the oxidation of ammonia to nitrate with nitrite as intermediate product. This step is known as nitrification which is, mainly, if not all, conducted by the strictly aerobic, autotrophic nitrifying bacteria. The rice-field soils are either permanently flooded or are flooded during the whole, or the greater part of growth season of rice crop. Under the flooded condition, the soil air space is extremely restricted, the oxygen tension of the soil is very low, and the decomposition of organic matter is predominently anaerobic. No nitrification takes place, and the ammonia is, consequently, the end-product of decomposition of organic nitrogen compounds.

The data presented in this article are meant to show the order of concentrations of the three forms of inorganic nitrogen compounds (nitrate, nitrite and ammonia) and certain factors relating to the dynamic equilibrium of the three compounds in typical rice field soils, which may be interesting to the students of rice-field soils and rice culture. The data presented herein are not at all systematic. A number of determinations were done primarily for purposes not discussed in this article. A systematic study of the periodicity of inorganic nitrogen compounds in different types of rice field soils is being carried put by the authors at the time of writing this article.

FORMS OF INORGANIC NITROGEN IN ATER-LOGGED RICE FIELD SOILS

To some extent, the microbiological condition of the rice field soils resemble to the naturally occurring ponds and swamps. But knowledge concerning the nitrogen cycle in ponds and swamps is also not extensive. The better known reports on the nitrogen problem of rice field soils were based mostly on results obtained from artificially made waterlogged soil, placed in laboratory vessels. Subrahmanyan (1927) conducted experiments by adding water to air-dried soils in glass vessels, so were Itano and Arakawa (1932) and De and Bose (1938); and Janssen and Motzger (1928) carried out experiments by growing rice in flooded stone jars

*原载于《武汉大学理科季刊》，9（1）：79~88，1948

[1] Professor and assistant respectively, College of Agriculture, National Wuhan University, Wuchang, Hupeh. Dr. Chen was formerly of the National Agriculture Research Bureau, Ministry of Agriculture and Forestry, and the College of Agriculture, National University of Peking. Mr. Shiao was formerly of the National Agriculture Research Bureau, Ministry of Agriculture and Forestry. The experiments mentioned in the present article were carried out when the authors were members of N.A.R.B. and the National University of Peking.

containing soil from a "dry" arable field never been used for rice culture. These and other works have undoubtly contributed much to our understanding of soils under water-logged condition. It is obvious, nonetheless, that merely flooding a sample of soil does not make it a representative of rice field soil. Bamji (1938) determined the total, ammoniaeai, and nitrate nitrogen contents of soil samples taken from a rice field of several years standing, which will be discussed later.

Table I gives a few representative data on the concentrations of inorganic nitrogen compounds in permanently flooded rice field soil. The soil samples were taken from rice field of many years standing at the Tien-sen-chiao (天生桥) Experiment Station of the National Agricultural Research Bureau at Pei-peh, Szechuan (四川北碚). The soil was developed from purple coloured sandstones and shales. It was on the acid side of neutrality, pH 6.5, containing 1.2 per cent organic carbon and 0.09 per cent total nitrogen (C/N ratio 13: 1). Determinations were carried out on fresh soil samples taken to the depth of 4 to 5 inches by the method described by Leng (1944). Ammoniacal and nitrate-and-nitrite nitrogen were determined by Olsen's method.

TABLE Ⅰ Inorganic nitrogen in rice field flooded all the year round

	Soil			Flood water	
	NH_3-N	NO_2-&-NO_3-N	NO_2-N	NH_3-N	NO_2-&-NO_3-N
24, 12, '43	10.6	tr.	…	…	…
10, 1, '44	12.7	tr.	0.05	0.3	tr.
16, 2, '44	5.9	tr.	…	…	…
1, 3, '45	3.6	tr.	…	0.2	…

Remark: Figures are given in ppm of soil or water. The NO_2-&-NO_3-N was determined by Olsen's reduction method which was significant only to first decimal digit if the concentration was expressed in parts per million parts of soil. The differences between comparable figures are due probably to seasonal variation. For a brief description of the rice field soil see text.

Nitrate nitrogen and nitrite nitrogen have also been determined by phenol-disulphonic acid method and by Griess' method respectively. The ammoniacal nitrogen in the flood water was determined by A.O.A.C. official method of water analysis. It is clearly shown in Table I that the ammoniacal nitrogen is the chief form of inorganic nitrogen in rice field soil. The highest nitrate nitrogen found during the course of study was 0.7 ppm. The highest nitrite nitrogen found, when determined separately, was 0.05 ppm. Bamji (1938) gave 2-7 ppm of nitrate nitrogen for an unfertilised Indian rice field soil. The figures are many times higher than the authors' result. Not knowing much of the rice fields in India, the authors do not attempt to explain the difference, but to point out that while the authors' determinations were made on fresh soil samples, Bamji's determinations were made on soil samples dried at 55 ℃ for 26-28 hours. Leog (1944) found that rice field soil samples oven-dried at 110 ℃ always gave two or more times ammoniacal and nitrate nitrogen than fresh soil samples.

INORGANIC NITROGEN COMPOUNDS IN RICE FIELD FLOODED ONLY DURING THE GROWTH SEASON OF RICE

A number of determinations of inorganic nitrogen contents have been made on soil samples taken from a field of long term fertilising experiment, during the period between November, 1945 and September, 1946. The field experiment was laid down in 1942. A rotation of wheat and rice were grown each year. During the rice seasons the field was flooded. The field was drained every autumn for wheat. Before establishing the fertilising experiment, the field was a permanently flooded rice field. Among the experimental plots, three plots were selected for making periodical determinations of inorganic nitrogen compounds. The plots were given the following fertilising treatments：

Plot 13 1000 shih-king per mou（ca. 3 tons per acre）horse manure each year given before planting rice.

Plot 15 40 shih-king per mou（140 pounds per acre）ammonium sulphate twice a year given once before each crop. No ammonium sulphate was given in the spring of 1946.

Plot 14 Plot recieiving both fertilising treatments given to Plot 13 and Plot 15. No ammonium sulphate was given in the spring of 1946.

Eight sets of determinations were made covering ten months from Nov. 30，1945 to Sept. 15，1946，during which period a crop of wheat and a crop of rice were grown. Wheat was sown on Oct. 17， 1945 and reaped on May 6，1946. The field was flooded on May 11. Rice was planted on May 16 and reaped on Aug. 17 the same year. Table II and Figure I give the ammoniacal and nitrate （including nitrite）.

TABLE Ⅱ Ammoniacal and nitrate nitrogen in rice-wheat fields flooded during the rice season

	PLOT-13			PLOT-15			PLOT-14		
	NH_3-N	NO_3-N	Total	NH_3-N	NO_3-N	Total	NH_3-N	NO_3-N	Total
30，11，’45	4.2	2.7	6.9	23.6	5.3	28.9	7.3	7.9	15.2
11，2，’46	3.4	4.2	7.6	15.8	10.5	26.8	8.5	8.2	16.7
9，3	1.5	Tr.	1.5	16.0	5.1	21.1	3.0	Tr.	3.0
11，4	1.3	Tr.	1.3	3.5	2.3	5.8	2.2	1.7	3.9
11，5	2.2	5.1	7.3	9.1	7.7	16.8	5.8	9.5	15.3
14，6	2.6	Tr.	2.6	10.3	Tr，	10.3	6.1	Tr.	6.1
15，7	5.9	Tr.	5.9	7.1	Tr.	7.1	5.0	Tr.	5.0
15，9	1.5	Tr.	1.5	1.1	Tr.	1.1	1.3	Tr.	1.3

Remark： Figures are given in ppm. Wheat was sown on 17，10，’45， was given ammonium sulphate on 20，10，’45 and was harvested on 6，5，’46. Rice was planted on 16，5，’46 and harvested on 17，8，’46.T he field plots were flooded on 11，5，’46，and drained on 6，8，’46. For descriptions of fertilizer treatments see text.

Nitrogen contents found during the period. In December and January，the wheat crop was in a state of dormancy and the microbiological activity in soil was also negligible. A not very strong process of nitrification occurred during the period between last drainage and winter season. The ammoniacal nitrogen，and consequently the total inorganic nitrogen，was highest in Plot 15，which received a dressing of ammonium sulphate. There was probably a temporary microbiological synthesis of nitrogen in Plot 14，which received both horse manure and ammonium sulphate. Active growth of wheat took place in February，March and early April. The crop absorbed much nitrogen and markedly reduced the soil inorganic nitrogen content. Nitrate nitrogen decreased faster than the ammoniacal nitrogen， due probably to different rates of absorption of the two forms of nitrogen. In Plot 13 nitrate nitrogen disappeared during the later part of the growing season. Nitrification must have taken place during this period as judged by what happened in the later period. But the rate of nitrification was slower than the rate of absorption of nitrate by the plant. The growth period of wheat stopped around April 11. After that and before the field was flooded on May 11，there was active nitrification in all three fields. Nitrate disappeared after flooding during the whole rice season.

REDUCTION OF NITRATE IN FLOODED SOIL

The disappearance of nitrate in flooded soil is due to a process known as reduction of nitrate. The process is also microbiological. The end-product of nitrate reduction is either nitrite，ammonia or nitrogen gas，depending upon the microorganisms involved. On flooding a rice-field soil，nitrate reduction was found to proceed further than the stage of nitrite. In 1947，a block of rice field soil of many years standing was taken from the College Farm of the College of Agriculture，National University of Peking，Loe-Tau-Tsong， Peiping（北平罗道庄）. The soil was dried in shade. Air-dry soil contained 7.5 ppm nitrate nitrogen. The nitrate

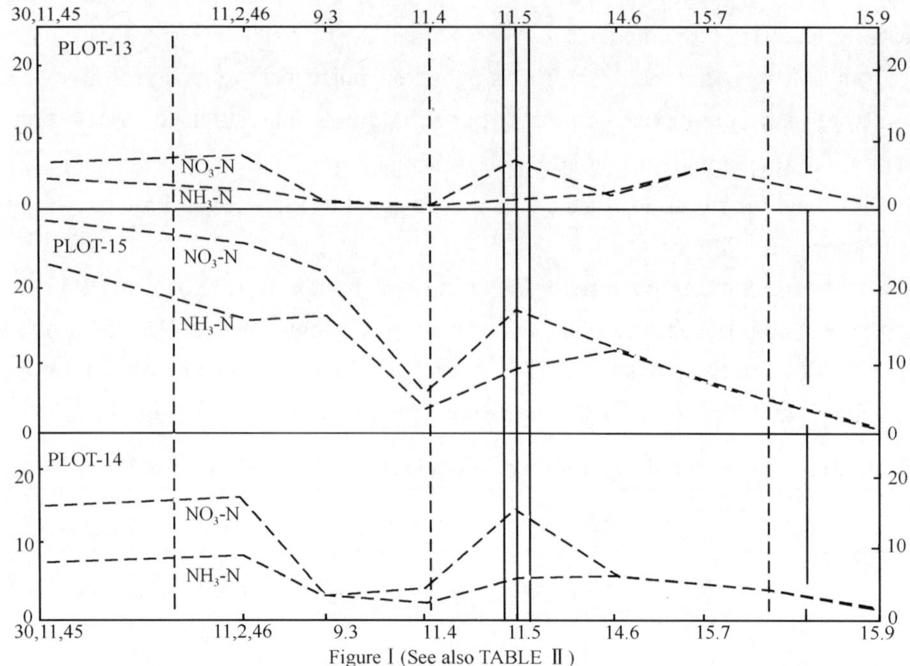

Figure Ⅰ (See also TABLE Ⅱ)

nitrogen was produced during the period of drying, which took more than two weeks. An experiment on nitrate reduction was conducted by treating subsamples of this soil in small flasks, 50 g soil being placed in each of 24 flasks. To 6 flasks of Set A, 245 cc water was added which covered up the soil to a depth of 3 cm. To 6 flasks of Set B, 47 cc potassium nitrate solution containing 3 mg nitrate-nitrogen and 198 cc water were added. To 6 flasks of Set C, 47 cc potassium nitrate solution, 10 cc glucose solution containing 125 mg glucose and 188 cc water were added. To 6 flasks of Set D, 15 cc water was added. Flasks of Sets A, B and C were rubber stoppered. Flasks of Set D were stoppered with cotton wool. The amount of nitrate-nitrogen in the flasks was determined by phenol disulphonic acid method. The flasks were kept at room temperature. Determinations were made on the second, fourth, sixth and nineth days following treatments. The result is presented in Figure II. Figure II shows clearly that reduction of nitrate follows immediately after flooding the soil. In flasks without extra nitrate (Set A), no nitrate was found 24 hours after flooding. In presence of glucose, 107.5 ppm nitrate was reduced to trace amount within 5 days. Without glucose, an equilibrium was reached also in 5 days when 80 percent of added nitrate was reduced. No increase of nitrite was found to accompany nitrate reduction. Apply the result found in this experiment to field conditions, the amount of nitrate nitrogen as ordinarily occurring in the drained period of rice field soils would be reduced within one day after flooding.

SUMMARY

The article presents a collection of data obtained in the past few years concerning the state of inorganic nitrogen in rice field soils. The data presented substantiate the expected state of inorganic nitrogen as deducted from the soils' microbiological conditions. Under flooded condition, the chief inorganic nitrogen in soil is ammoniacal, amounting to a few milligrams of ammoniacal nitrogen per 100 grams of soil. Nitrite and nitrate nitrogen present only in trace amounts. When the water-logged soil is drained for winter crop, a slow nitrification takes place. Between two flooded periods, which covers the winter crop season, the nitrate produced, though significant, is still second in concentration to ammonia. When the field is reflooded, the nitrate is reduced immediately, probably within a single day.

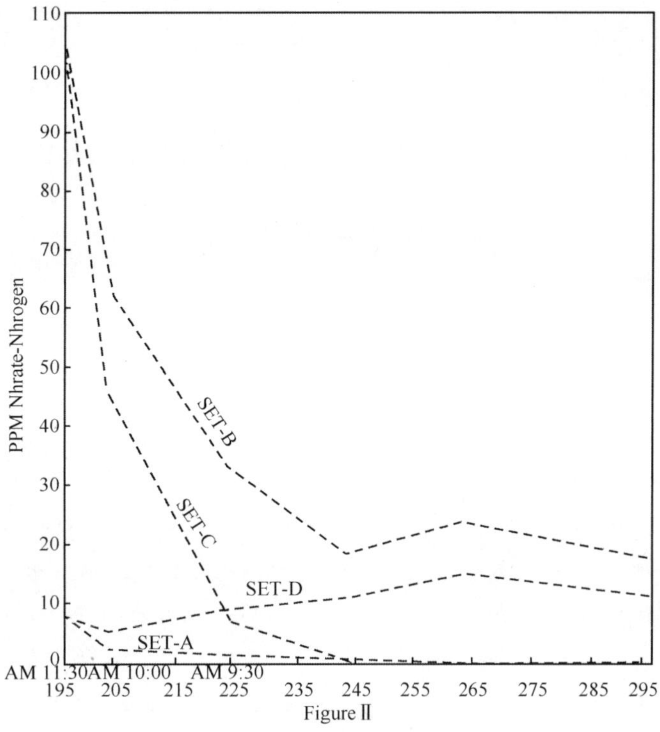

Figure Ⅱ

REFERENCES

[1]Bamji，N. S.（1938）Indian J. Agric Sc. vol. 8：pp. 839-847.

[2]De，P. K. and N. M. Bose（1938）Indian J. Agric. Sc. vol. 8：pp. 487-498.

[3]Itano，A. and S. Arakawa（1932）Berichte des Ohara Institute fuer landwirtschaftliche Forschungen vol.5：pp. 427-446.

[4]Janssen，G and W. H. Metzger（1938）J. Amer. Soc. Agron. vol. 20：pp. 259-476.

[5]Leng，L. T.（冷福田，1944）Nung Pao（农报）vol. 9：pp. 291-294.

[6]Subrahmanyan，V. （1927）J. Agric. Sc. vol. 17：pp. 449-467.

硝化细菌生理学特性的初步研究*

陈华癸　周　啟

（华中农学院）

（一）前言

自从 C.H.维诺格拉斯基（1890）发现硝化细菌以后，由于获得纯培养的困难以及研究方法的不完善，硝化细菌的生理学及其生物化学过程的研究，一向进展不大。

根据现有的硝化细菌生理学特性的研究资料，认为硝化作用只有在充分通气的条件下才能进行（维诺格拉斯基，1890），让空气通过培养基（里斯，1952）或者将培养基在滴漏器中巡回（里斯，1952）都会加强硝化作用。如果以氮气冲淡进入土壤中的空气，则土壤条件中的硝化作用就被抑制（阿墨和巴梭罗墨，1951）。由此认为硝化细菌对于环境通气程度的反应是非常敏感的，证明它是一种高度好气性的微生物。

此外，在水稻土方面，由于蓄水种水稻期间的 rH 值是在 10~25 之间，同时从水稻田土壤的农化分析中又发现冬季排水种旱地作物的时期，土壤中的硝态氮和氨态氮占有同等重要的地位。但是在蓄水种水稻的季节或冬季蓄水的冬水田中又没有硝态氮的存在，或仅仅含有极其微量的硝态氮（陈华癸和肖泽宏，1948），由此就得出结论：水稻田只有在排水的情况下，才进行有硝化作用；而在蓄水种水稻时期，硝化微生物不能活动，基本上可以认为没有硝化作用。（陈华癸，1957，1959）

当我们在分析水稻田与旱地土壤生物区系的过程中，却获得了与上述结论完全相反的结果，发现在冬季蓄水的冬水田土壤中，不仅有硝化细菌的存在，同时它的数量远远大于旱地土壤。因此，在我们的面前就产生了一个问题：硝化细菌究竟能不能在缺氧的条件下繁殖？在缺氧的条件下是不是能够进行硝化作用？这亦就促使了我们关于这个问题的进一步探索，下面就是我们一年多来初步研究的结果：

（二）试验部分

试验分两部分，即①水稻田土壤与旱地土壤中硝化细菌数量的测定；②水稻田硝化培养体在不同通气条件下的移植试验。

一、水稻田土壤与旱地土壤中硝化细菌数量的测定：

实验方法：在 1958 年 1 月到 4 月我们从武昌狮子山华中农学院农场的冬水田及小麦试验地上采取上样，分别进行硝化细菌数量的测定，继之又在 1959 年 5 月到 7 月在农化教研组的水稻试验地和棉花试验地上分别取样，进行硝化细菌的测数。

硝化细菌测数的培养基成分：$(NH_4)_2SO_4$ 0.5 克，$CaCO_3$ 1 克，1∶20 标准盐溶液[1] 1000 毫升。

硝化细菌测数的方法是采用稀释测定法，每试管（1.8×15）装培养基 5 毫升，每管接种土壤稀释液 0.5 毫升，每稀释度四个重复，在 30℃左右的温度中培养，接种后 15 天用 Griss 溶液检验各级稀释管的亚硝酸反应，以推算硝化细菌数量，现将一部分实验结果列表（表 1）如下：

表 1　水稻田土壤与旱地土壤中硝化细菌的数量　　（单位：千/每克干土）

时间	水田	旱地
1958-1-1	11.32	0.03
1958-3-24	13.46	0.09

*原载于《华中农学院学报》，1（3）：42~46，1959.

[1] 标准溶液的成分　K_2HPO_4　0.5 克，$MgSO_4$ 0.25 克，NaCl 0.25 克，　$FeSO_4$ 0.005 克，$MnSO_3$ 0.005 克，蒸馏水　100 毫升。用 1/10N KOH 调剂 pH 至 7.2，在 121℃　25 分钟灭菌后待用。

续表

时间	水田	旱地
1958-4-1	22.22	0.96
1959-5-26	40.35	1.85
1959-6-26	23.08	0.69
1959-7-11	33.98	0.94

注：1958-1-1，3-24，4-1，分析的水田为冬季蓄水的冬水田，旱地为小麦试验地；1959-5-26，6-26，7-11，分析的水田为蓄水种水稻的时期，旱地为棉花试验地。

从上表结果中可以看出，不仅在蓄水种水稻的土壤中含有硝化细菌，同样在冬季蓄水的冬水田中也含有硝化细菌；而且，不论在冬季蓄水的冬水田或蓄水种水稻的土壤中，其中所含有硝化细菌的数量远比旱地土壤要高，根据我们的试验相差可达 23 倍到 377 倍之多。

此外，在上述的实验材料中可以说明硝化细菌在缺氧的条件下可以繁殖；为了更进一步的证实这一点，并查明硝化细菌能不能在缺氧的条件下进行硝化作用，我们就布置了下面的一个实验。

二、水稻田土壤的硝化培养体在不同通气条件下的移植试验：

实验方法：将 59-5-26 水稻土接种的 10^{-3} 培养体（用 Griss 溶液检验证明有 NO_2 反应）移植至装有 5 毫升及 15 毫升硝化细菌培养基（成分与测数同）的试管中，然后将实验分成两组：

第一组：在接种后的培养基上，立即加入灭菌的液状石蜡，石蜡高度为 2.6-3.0 公分。下分四个处理：即①5 毫升培养基，②对照（5 毫升培养基，不接种），③15 毫升培养基，④对照（15 毫升培养基，不接种）。

第二组：在接种后的培养基上，不加灭菌液状石蜡，下分四个处理：即①5 毫升培养基，②对照，③15 毫升培养基，④对照。

每组每处理用二个重复。

在 32℃~35℃的温度中培养，五天后用 Griss 溶液检验结果，并将第一组及第二组的培养体各一支作为接种材料，分别移植至 5 毫升及 15 毫升硝化细菌的培养基中，实验分成二大组：

第一大组：用加灭菌液状石蜡的培养体接种：①接种后的培养基，立即加入灭菌液状石蜡：下分四个处理：即（I）5 毫升培养基，（II）对照；（III）15 毫升培养基；（IV）对照。②接种后的培养基，不加液状石蜡，下分四个处理：内容与上同。

第二大组：用不加灭菌液状石蜡的培养体接种。

（1）接种后的培养基，立即加入灭菌的液状石蜡；下分与上述相同的四个处理。

（2）接种后的培养基，不加灭菌液状石蜡，分四处理，与上述相同。

每处理都用二个重复。

在 32℃~35℃的温度中培养，五天后检查结果；并将第一大组接种后加灭菌液状石蜡的和第二大组接种后不加液状石的培养体作为接种材料，再移植至新的培养基中培养，实验仍旧分成与上述相同的二大组。

以后每隔五天除检查结果外，都按以上处理进行移植一次。

现将连续移植八次的实验结果列表（表 2）如下：

表 2 水稻田土壤的硝化培养体在不同通气条件下的移植结果

移植次数	移植日期	移植材料	培养结果								检查日期
			加灭菌液状石蜡				不加灭菌液状石蜡				
			5C.C.	对照	15C.C.	对照	5C.C.	对照	15C.C.	对照	
第一次	59-8-18	不加液状石蜡培养体	++	--	++	--	++	--	++	--	59-8-23
第二次	59-8-23	第一次移植加液状石蜡培养体	++	--	++	--	++	--	++	--	59-8-28
		第一次移植不加液状石蜡培养体	++	--	++	--	++	--	++	--	

续表

移植次数	移植日期	移植材料	培养结果								检查日期
			加灭菌液状石蜡				不加灭菌液状石蜡				
			5C.C.	对照	15C.C.	对照	5C.C.	对照	15C.C.	对照	
第三次	59-8-28	第二次移植加液状石蜡培养体	++	--	++	--	++	--	++	--	59-9-2
		第二次移植不加液状石蜡培养体	++	--	++	--	++	--	++	--	
第四次	59-9-2	第三次移植加液状石蜡培养体	++	--	++	--	++	--	++	--	59-9-7
		第三次移植不加液状石蜡培养体	++	--	++	--	++	--	++	--	
第五次	59-9-7	第四次移植加液状石蜡培养体	++	--	++	--	++	--	++	--	59-9-12
		第四次移植不加液状石蜡培养体	++	--	++	--	++	--	++	--	
第六次	59-9-12	第五次移植加液状石蜡培养体	++	--	++	--	++	--	++	--	59-9-17
		第五次移植不加液状石蜡培养体	++	--	++	--	++	--	++	--	
第七次	59-9-17	第六次移植加液状石蜡培养体	++	--	++	--	++	--	++	--	59-9-22
		第六次移植不加液状石蜡培养体	++	--	++	--	++	--	++	--	
第八次	59-9-22	第七次移植加液状石蜡培养体	++	--	++	--	++	--	++	--	59-9-27
		第七次移植不加液状石蜡培养体	++	--	++	--	++	--	++	--	

注：检验培养结果 Griss 溶液

从上表的实验结果中，进一步的证实了硝化细菌可以在完全缺氧的条件下繁殖，同时亦可以从这个实验中说明，硝化细菌不仅可以在有氧的条件下进行硝化作用，在缺氧的条件下，同样的也可以进行着硝化作用。

从以上的两个实验中说明了水稻土壤中，不论在冬季蓄水的冬水田或蓄水种水稻的时期，都有硝化细菌的繁殖，同时也有硝化作用的进行。

（三）讨论和小结

（1）根据实验结果，证明水稻田土壤，不论是在冬季蓄水的冬水田或者蓄水种水稻的时期都有硝化细菌的繁殖，而且其中硝化细菌的数量远比旱地土壤要高，根据我们的试验可相差 23~377 倍之多。这些结果与过去认为蓄水的水稻田中硝化微生物不能活动的结论是相抵触的。

（2）水稻田土壤中的硝化细菌，不仅能在有氧的条件下繁殖和进行硝化作用，而且在缺氧的条件下同样的能够繁殖和进行硝化作用。这一点与 C.H.维诺格拉斯基发现硝化细菌以来，科学家们一直到最近（里斯，1952）还认为硝化细菌是高度好气性的微生物的见解是相抵触的。

（3）从这次初步研究工作中证明了蓄水的水稻田土壤中有硝化细菌的繁殖，并有硝化作用的进行。但是，根据很多农化分析的资料，在蓄水状态下，水稻田土壤中却又没有硝态氮的累积或很少累积，这样，对水稻田土壤中氮素养料的转化、损失和施肥法，就提出了一项重要的值得研究的问题。

过去，由于认为在蓄水条件下，基本上没有硝化作用，而只有强盛的反硝化作用。因此，也就认为水稻田不宜施用硝酸盐肥料，而只宜施用硫酸胺。由于硫酸铵在蓄水条件下不能硝化，而使氮素不至于造成严重的损失。或者，正如三井进午所提供的，水稻田在灌水状态下，土层中有氧化层和还原层的分化。而其中只是在表面少于 1 厘米的氧化层中有硝化作用的进行，同时由此所产生的 NO_2^- - N 当进入还原层后，即由于反硝化作用的活动而使之还原成为 N_2 或 NO 而损失。因此，他认为，为了避免水稻田土壤中氮素的损失，氮肥必须要进行深施。

但是，根据我们现在研究工作所指出的，在水稻田表层以下的土壤中，仍旧有强烈的硝化作用。因此，与反硝化作用相结合，尽管采用了全层施肥法，依然会导致氮素的大量损失。这与文献中所记载的，水田中施用氮肥如硫酸铵、豆饼及鱼粉作为水稻的基肥时，其利用率往往较旱地为低，一般只在40%~60%左右。同时从蓄水的水稻土表层所逸出的气泡的分析中证明，含有的气体主要是甲烷和氮气，而其中氮气约占总量的10%~95%（Hamien，W.H. and Auiyec，P.A.S，1915）。这些情况，正是说明了水稻田在蓄水状态下，耕层中不仅有强盛的反硝化作用，同时亦有硝化作用的进行，结果就会使氮素遭到了大量损失。过去我们研究工作中所指出的，水稻田在冬季排水时有硝化作用，原先的认识是以为在排水条件下硝化作用加强了，现在看来，主要是反硝化作用减弱了。

由此，进一步深入研究水稻田中的硝化作用以及控制其中氮素养料的损失和加强氮素的有效利用率是水稻生产技术中的一项主要问题。

（4）在这项研究工作中，对于硝化细菌生命活动的基本理论上也提出了一项必须钻研的问题：即硝化细菌能够在缺氧的条件下进行硝化作用，那么硝化作用氧化 NH^{+}_{4} 所需要的氧究竟从何而来的呢？它的有机合成的能源又从哪儿来的呢？这些都是值得深入研究的。

参 考 文 献

[1]Jane Meiklejohn.Some Aspects of the Physiology of the Nitrifying bacteria，AUTOROPHIC MICRO.ORGANISMS 1954，pp.6

[2]Г.ДчС，ВЙОХИмИЯ Автотрофнщх БАнтЕрий Моса-1958

[3]Е.Д.鲁邦，硝化微生物的生理学，土壤学译报，1957 年 2 期

[4]陈华癸，1957 年，水稻田土壤中占优势的微生物种类，土壤学报，5 卷 1 期

[5]陈华癸，1959 年，水稻土中植物营养元素的生物学循环，稻作科学论文选集，167 页

[6]三井进午，（日）水稻无机营养，施肥和土壤改良，上海：上海科学技术出版社，1959

[7]Нсунщлов В.А.，1948，Окислитеlьно восаповительнмо процесщ в почвах Рдсовмх долей и метоцщ управления ими с целыо повышения Уржайности .СБ .научн работ сепбс кохозяйственнщх Опщтно-нседовательских Учреждений приморското краявщпуск вып. Ⅰ владивосток Hamien，W.H. and Auiyec，P.A.S.，Theganeo of Sulamp rice Soils. Ⅱ. Ihrir utryotion for the aeration of the root of the crop. Mem.Qept.Agric.qnd.4：1-18，1915

水稻田土壤中硝化作用和硝化微生物的研究

Ⅱ.水稻田土壤中的亚硝酸细菌加富培养条件下的繁殖和亚硝酸化强度*

陈华癸　周　啟

（华中农学院）

在前一报告中报道了在水稻田土壤中，不能在氧化势较高的氧化层，或还原势较高的还原层都有较旺盛的硝化细菌和硝化作用。这种硝化细菌能在缺氧条件下繁殖，并进行硝化作用。这种硝化细菌还和反硝化细菌紧密伴生，因而导致土壤氮素的损失：

$$NH_3 \rightarrow HO_2NO_3 \rightarrow N_2\uparrow$$

在这一报告中报导以下的研究成果：

一、用维诺格拉斯基硝化培养的加富培养体。在这个培养体中已经没有硝酸细菌存在了，硝化作用只进行到 NO_2^- 为止。但在这个培养体中仍有伴生的反硝化细菌存在，未能完全清除，但是类型减少。

二、接种这种亚硝酸细菌的加富培养体至麦克江（Moiklejohn，1950）硝化培养基上，在 28~30℃下培养，不论在有氧或缺氧的条件下，都能将培养基中的 NH_4^+ 盐几乎全部的氧化为 NO_2^- 盐。这项氧化作用，在有氧条件下于 13 天内完成，在缺氧条件下于 16~22 天内完成。NO_2^- 盐，也不进一步还原为所态氮。

在培养的过程中，亚硝酸细菌的数量一般在两星期内达到最高量，以后逐渐下降。通常在前两星期内，亚硝酸细菌的增长比伴生的异养细菌快，到转折点以后，亚硝酸菌数量开始下降，伴生异养细菌的数量则不断增值。

三、如果在亚硝酸盐产生到最高量以后，加入灭菌的葡萄糖溶液（按 1.5%左右葡萄糖用量计算），则有反硝化作用进行，NO_2^- 盐在 5~7 天内完全消失。

在培养过程中，亚硝酸细菌数量的动态规律，与亚硝酸细菌加富培养体接种的很相类似。这种现象在有氧和缺氧条件下都能进行，强度的差别不大。

四、在麦克江硝化培养基中接种少量水稻田土壤，进行同样的培养试验，则既有 NH_4^+ 盐氧化为 NO_2^- 盐的作用，又有 NO_2^- 盐氧化为 NO_3^- 盐的作用在进行着。在有氧条件下，培养后 10 天，NO_2^--N 达到最高量，随后逐渐下降，在 16~22 天内全部消失。在缺氧条件下，NO_2^--N 量在培养 13~16 天时最高，至 22~30 天全部消失。

五、水稻田土壤接种的硝化培养体，在生产 NO_2^- 盐以后，加入灭菌的葡萄溶液（按 1.5%左右葡萄用量计算），则不能在有氧或缺氧的条件下，都产生较旺盛的反硝化作用，NO_2^--N 在 36 小时内完全消失，并在培养物的表面形成一层较厚的气泡层。

如果水稻田土壤接种的硝化培养体，经一个月的培养，其中 NO_2^- 盐已全部氧化成为 NO_3^- 盐时，加入灭菌的葡萄糖溶液，同样，在 36 小时内可以使 NO_3^- 盐全部还原，并在培养物表面形成气泡层。（图表略）

---

*原载于《中国土壤学会 1963 年学术年会论文摘要集》，46~47，1963.

水稻田土壤中的硝化作用和硝化微生物的研究
Ⅳ.硝化细菌和反硝化细菌在水稻田土壤中的活性*

周　啟　陈华癸

（华中农学院）

引　言

在土壤植物营养元素生物学循环中，硝化作用和反硝化作用是其中两个重要的环节。首先通过硝化作用将氨态氮氧化成亚硝酸态氮和硝酸态氮；然后经过反硝化作用又将亚硝酸态氮和硝酸态氮还原成分子态氮。这样一个氧化和还原的过程就使土壤中大量无机态氮遭到流失。

自维诺格拉斯基首先发现硝化细菌后，证明它是一种化能自养型细菌，它在氧化氨态氮时必须要有氧气作为受氢体并产生能量，同化二氧化碳成为有机碳化合物以组成它的细胞物质。所以一直来认为硝化作用和硝化微生物是绝对需要氧气的；同时，有的学者在检测蓄水水稻田土壤时，没有发现有无机态氮，或者只有痕迹。从而认为蓄水稻田中没有硝化微生物的结论（陈华癸和肖泽宏，1948），随后，日本的Mitsai 1956年提出蓄水稻田的土壤可分为氧化层（0~1厘米）和还原层，其中氧化层有硝化细菌存在和活动，而还原层土壤则没有硝化细菌存在。此外，有些研究者利用示踪元素N^{15}技术检测有机氮和无机态氮施入稻田后，证明可利用的氮素有35%~52%遭到损失（Broadbent，1971；朱兆良等，1977）。所以这里就提出一个问题。如果，蓄水稻田只有局部地方有硝化细菌在活动，那么为什么有大量的无机态氮遭到损失呢！这些损失的无机态氮是从何而来的呢？

为此，我们在上世纪50年代中期开展了水稻田土壤中的硝化作用和硝化微生物的研究。通过三年多时间的研究所得到的结果是：①蓄水水稻田土壤中的硝化细菌和反硝化细菌的数量远比旱地土壤要高（周啟，1959）；②不论栽水稻或不栽水稻的蓄水田，在水稻不同生长阶段中，不论是氧化层或不同深度的还原层均有硝化微生物存在和活动；③首先发现水稻田中的硝化微生物是兼嫌气性的（陈华癸，周啟，1961；Cnen，H.K. and Zhou Qi，1964）；④分离到亚硝酸细菌的纯培养（陈华癸，周啟，1964）；⑤证实亚硝酸细菌纯培养与异养型的一株反硝化细菌紧密伴生在一起，是造成水田土壤中无机态氮大量损失的原因（Chen，H.K.and Zhou Qi，1964）。

下面就将上述的这些结果进行详细的报道。

实 验 结 果

一、水稻田土壤中亚硝酸细菌和反硝化细菌的数量和分布

水稻田亚硝酸细菌和反硝化细菌的数量是在硝化培养基和反硝化培养基中分别采用稀释法和平面计数法测定（周啟，1959），结果表明蓄水水稻田土壤中的亚硝酸细菌不仅存在于0~1厘米的氧化层中，而且也存在于1~25厘米的耕作层中。反硝化细菌的数量则在氧化层中还明显高于1~25厘米的耕作层（表1和表2）。

表2的结果说明水稻不全生长阶段的整个耕作层中都存在着一定数量的亚硝酸细菌和反硝化细菌，除了水稻成熟期，亚硝酸细菌的数量略有降低，使它与反硝化细菌之间的比例增大外，在其他生育阶段，亚硝酸细菌和反硝化细菌都保持着比较稳定的比例。

---

*中文版未正式发表；英文版发表美国在*Soil Sci.*135（1）：31-34，1983

表 1 在水稻不全生长阶段中亚硝酸细菌和反硝化细菌的数量（×1000/克干土）

深度cm	取样期								
	插秧前蓄水			水稻反青期			水稻分蘖盛期		
	NB*	DNB	NB/DNB	NB	DNB	NB/DNB	NB	DNB	NB/DNB
0-1	112	10700	1：96	76	7980	1：105	99	7810	1：79
1-15	70	1640	1：23	87	1470	1：17	109	3880	1：36
15-25	20	1470	1：74	62	1890	1：31	104	4350	1：42

* NB，亚硝酸细菌；DNB，反硝化细菌；NB/DNB，亚硝酸细菌：反硝化细菌

表 2 亚硝酸细菌和反硝化细菌在中稻生长中的数量

取样时间（月-日）	水稻生长期	亚硝酸细菌（$\times10^3$ / 克干土）	反硝化细菌（$\times10^3$ / 克干土）	亚硝酸细菌：反硝化细菌
6-14	反青期	184.8	4368.0	1：24
6-23	分蘖盛期	218.4	5460.0	1：25
7-10	抽穗期	186.0	5586.0	1：30
8-15	成熟期	102.0	5145.0	1：50

二、水稻田土壤中的硝化细菌和反硝化细菌在好气和嫌气条件下的消长

1.实验使用Meiklejohn's硝化培养基，将不接种的对照试管和接种一小块水稻土的试管放在28℃中分别进行好气性培养和嫌气性培养。嫌气培养或者是用一层灭菌液状石蜡复盖在液体培养基表面上（1厘米厚），或者是将培养试管放置在一个装有碱性没食子酸的大试管中，用塞子塞紧管口，以吸收氧气来达到嫌气条件。结果说明水稻田硝化细菌不论在好气条件下，或者在嫌气条件下都能进行硝化作用。而且两者的作用趋势也很相似，所不全的只是作用强度的差异而已。结果见表3和图1。

表 3 接种水稻土的培养物在好气和嫌气条件中产生的亚硝酸态氮 NO_2-N，mg/L

培养时间（天数）	好气		嫌气			
			液状石蜡覆盖		没食子酸处理	
	接种	不接种	接种	不接种	接种	不接种
0	0.0	0.0	0.0	0.0	0.0	0.0
5	3.3	0.0	3.1	0.0	2.1	0.0
10	20.2	0.0	19.6	0.0	10.7	0.0

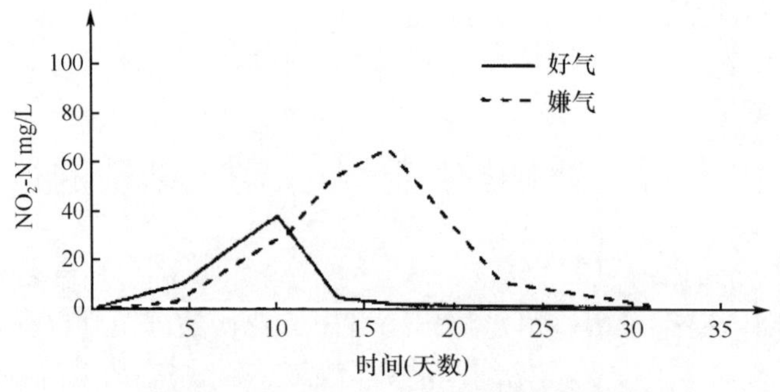

图 1 水稻田土壤接种在 Meikilejonn's 培养基中硝酸态氮的产生过程

2. 另一个实验也是将一小块水稻土接种到Meikilejonn's培养基中，在28℃中分别在好气和嫌气条件下培养37天，此时，培养物中的氨态氮已全部氧化成硝酸态氮（123mg/L），相当于原有培养基中含有140mg/L的87.5%。然后将灭菌葡萄糖液加入到这些培养物中并继续培养。结果在一天半的时间内，硝酸态氮全部还原成气态氮并在培养物表面产生一层气泡层（表4和图2）。

表 4 接种水稻土的培养物加葡萄糖继续培养的实验结果

培养时间（天数）	NO_2-N，mg/L				NO_3-N，mg/L			
	好气		嫌气		好气		嫌气	
	CK	葡萄糖	CK	葡萄糖	CK	葡萄糖	CK	葡萄糖
37+0	123	123	123	123	0	0.0	0	0.0
37+1/2	127	n.d.[a]	126	n.d.[a]	0	1.1	0	1.1
37+3/4	127	n.d.	123	n.d.	0	25.0	0	22.3
37+1	127	n.d.	126	n.d.	0	97.4	0	83.5
37+1 1/4	127	n.d.	124	n.d.	0	0.4	0	0.4
37+1 1/2	127	0	124	0	0	0.0	0	0.0

[a] n.d.——没有测定

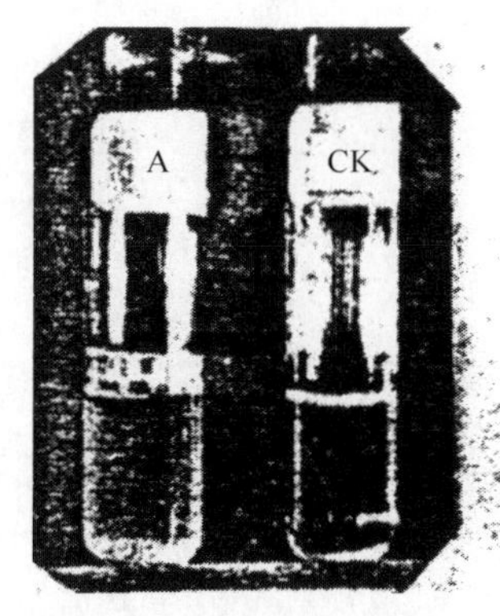

Aerobic

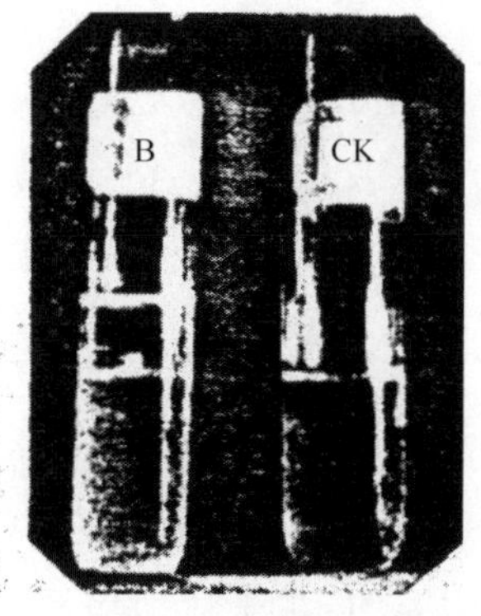

Anaerobic

图 2 在接种水稻土的培养物中加入葡萄糖液再继续培养后产生的反硝化作用（示氮泡层），试管上面的 A 和 B 表示加有葡萄糖

三、水稻田亚硝酸细菌纯培养的分离

在水稻田亚硝酸细菌分离过程中，要淘汰互相依存的硝酸细菌并不困难，最不容易淘汰的倒还是与亚硝酸细菌紧密伴生的异养型细菌。在我们的实验中，通过不断加富和连续移植的手段（陈华癸、周啟，1963），进行了两年多时间，才使培养物中亚硝酸细菌数量占有绝对优势，然后将此培养物进行大量稀释，并对所有的稀释管进行严格的筛选，最后获得了两株亚硝酸细菌纯培养。

四、水稻田亚硝酸细菌的伴生菌是一类反硝化细菌

从近乎纯培养的亚硝化培养物积聚培养中分离获得了伴生菌纯培养，它是一种异氧型的极毛杆菌，它不仅在好气和嫌气条件下生长和繁殖（表5）而且也能在硝化培养基和反硝化培养基中生长和繁殖。但是这种伴生菌纯培养在硝化培养基中不能进行硝化作用，而在反硝化培养基中才能引起反硝化作用（表6）。

表 5 伴生菌接种在 Meiklejohn 培养基中的亚硝酸细菌和反硝化细菌

接种时间（天数）	NO_2-N，mg/L		亚硝酸细菌*（A）		反硝化细菌*（B）		（A）/（B）	
	好气	嫌气	好气	嫌气	好气	嫌气	好气	嫌气
0	0.0	0.0	0.2	0.1	41.6	33.5	1∶208	1∶335
4	11.1	9.3	23	50	506	2540	1∶22	1∶51
10	117	56	120	120	346	3260	1∶29	1∶27

续表

接种时间（天数）	NO_2-N，mg/L		亚硝酸细菌*（A）		反硝化细菌*（B）		（A）/（B）	
	好气	嫌气	好气	嫌气	好气	嫌气	好气	嫌气
15	130	105	465	230	506	11700	1∶1.6	1∶51
22	131	131	700	260	396	17100	1∶0.6	1∶66
30	131	131	260	230	435	8490	1∶1.7	1∶37

*×10^4/ml

表 6 伴生菌在硝化培养基和反硝化培养基中的消长和生化反应

培养时间（天数）	硝化培养基		反硝化培养基	
	菌数（×10^4/mL）	NO_2-N 反应	菌数（×10^4/mL）	NO_2-N 反应
0	0.8	--	0.8	--
2	1.4	--	7100.0	++
4	192.7	--	138600.0	++
6	54700.0	--	108700.0	++
8	8500.0	--	1895.0	++
10	22.0	--	68.2	++

五、伴生菌的硝化作用和反硝化作用

将伴生菌接入到Meiklejohn的亚硝酸培养基中，在28℃进行好气和嫌气条件培养，到第10天和第22天，亚硝酸态氮的含量已达到131mg/L（相当于加入原有硝化培养基的140mg/L的94%）（见图3）；然后加入灭菌葡萄糖液并继续培养16天表明，反硝作用从第四天开始至第七天已全部结束（图4）。

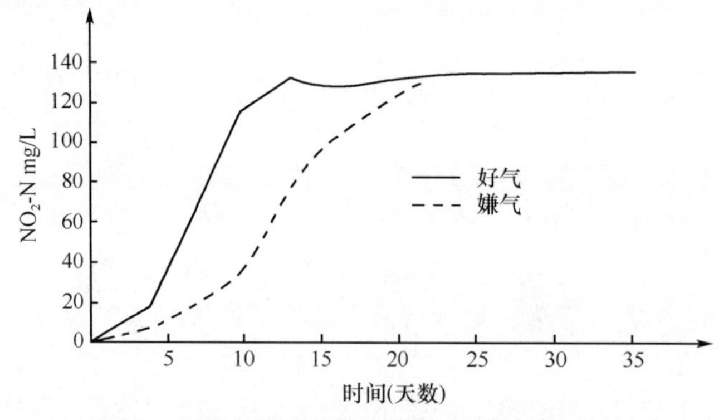

图 3 用伴生菌接种在培养基中产生的 NO_2-N

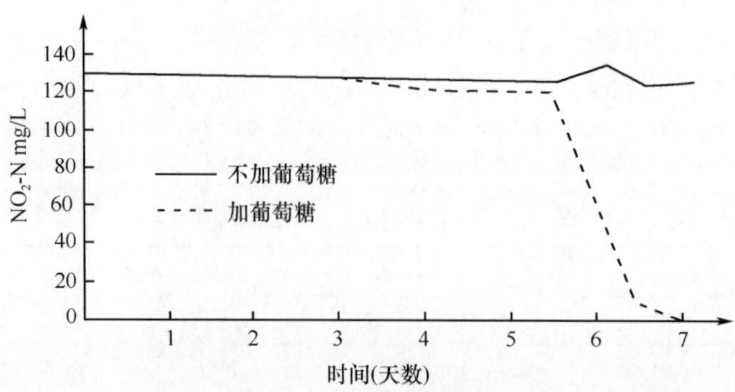

图 4 伴生菌接种培养 16 天后加入葡萄糖产生的反硝化作用

讨 论

实验结果对硝化微生物和硝化作用的绝对好气性的普遍性提出了异议.首先证明水稻田土壤中有兼嫌气性的硝化微生物，它们在淹水条件下，在整个耕作层的土壤中依然大量存在并进行硝化作用。我们已从这种土壤中获得亚硝酸细菌的纯培养（陈华癸、周啟，1964），这种纯培养在形态上和旱地土壤中的绝对好气性的Nitrosomonas相似。但是对于氧气的要求上，两者却不相同，这可能是由于随着生态条件的不同出现的一种特定的生态类型。因为，水稻田亚硝酸细菌长期生活在淹水条件下的适应结果，使它在极低的Eh值条件下仍然可以存在并进行生命活动。正如有些科学家从深海底部的淤泥中发现有硝化细菌并证明它们是处在生活状态的一样（ Carey & Waksman，1934；Zobell，1935）。

实验结果还指出，水稻田土壤中不仅存在有大量的反硝化细菌，而且有的反硝化细菌还能和亚硝酸细菌紧密伴生在一起。这种伴生的异养型反硝化细菌不仅在嫌气条件进行反硝化作用，而且还能在好气性条件下进行。同时它即使在完全没有有机碳化合物的硝化培养基中也能单独正常生长和繁殖，但不能进行硝化作用。这种伴生菌纯培养在反硝化培养基上在好气或嫌气条件下都能引起强烈的反硝化作用。所以，当接种有水稻土或近乎纯培养的亚硝酸细菌的硝化培养物，在累积有一定量的NO_2-N或NO_3-N时，加入灭菌的有机碳化物，如葡萄糖液就能使伴生的反硝化细菌将NO_2，-N或NO_3-N迅速还原成为气态氮而流失。由此可以认为，水稻田土壤中硝化过程所产生的无机氮素的损失，实质上是由硝化作用和反硝化作用紧密活动的结果。在水稻田的具体条件下，在耕作层土壤中首先由硝化作用将氨态氮氧化成NO_2-N，或进一步氧化成NO_3-N，接着，就由伴生菌引起的反硝化作用，而使得亚硝酸态氮或硝酸态氮还原成气态氮（N_2）而损失。在蓄水条件下，反硝化作用是如此强烈，致使稻田土壤的化学检测不能发现NO_2-N和NO_3-N，或者只有痕迹。

因此，从水稻田土壤肥力的角度来说，这种硝化作用本来是一个有利的过程，但当它与反硝化作用紧密结合在一起就造成为一种有害的微生物学过程。所以在1960年前后出现一种硝化抑制剂，企图抑制土壤中的硝化作用，以防止无机氮素的大量流失。实际上这是一个行不通的措施。只能有待于后人的继续研究来解决了。

参 考 文 献

[1]Chen，H.K.，Shiao，T.H.，Inorganic compounds of nitrogen in the field soils. 武汉大学理科季刊（英文版），9（1）：79-88，1948

[2] Mitusi，S.（三井进午），水稻无机营养；施肥和土壤改良（中译本），上海：上海科学技术出版社，1959

[3] Broadbent，H. Nitrogen-15 in soil studies，AAEA Vienna，pp.47-54，1971

[4]朱兆良，蔡贵倍，俞金洲.稻田中 N^{15} 标记硫酸氨的氮素平衡的研究初根.科学通报，第 22 期，503 页，506 页，1977

[5]陈华癸，周啟，水稻田土壤中的硝化作用和硝化微生物的研究.I.水稻田土壤中的硝化作用，土壤学报，9（1-2）：56-64，1961

[6]Chen H.K，Zhou Qi，Facultatively Anaerobic Nitrification and Nitrite. Forming organisms 第八届国际土壤学会议论文摘要集，土壤生物学，pp.761-768，罗马尼亚，布加勒斯特，1964

[7]陈华癸，周啟，水稻田土壤中的硝化作用和硝化微生物的研究.II.水稻土壤的亚硝酸细菌加富培养物在培养条件下的繁殖和亚硝酸化的强度。中国土壤学会 1 963 年学术年会论文摘要集（第一部分）46-47 页，1963

[8]周啟，土壤微生物区系的分析方法，微生物通讯，1（1）：31-39，1959

[9]Meiklejohn J.，The Isolation of *Nitrosomonas Quropaea* pure culture. J. Gen. Microbiol. 4：185，1950

[10]Carey，C.L. and S.A. Waksman. The presence of nitrifying bacteria in deep seas. Science，79：349，1934

[11]Zobell，C.E. Oxidation-reduction potentials and activity of matrine nitriftiers. J. Bact. 29. 1935

[12]Wartz，B. and Jean-Marie Schmastz. Culture de betemtrophes an mitieu mineral C. R. Acad. Sci. 255：1036-1038，1962

水稻田土壤中硝化作用和硝化微生物的研究
V.亚硝酸细菌伴生菌的生物学特性及其对亚硝酸细菌的影响*

陈华癸　周　啟

（华中农学院）

硝化作用是土壤中植物营养元素生物循环的一个重要环节。它是由两种化能自养型细菌分两个阶段进行的。第一阶段是由亚硝酸细菌 （*Nitrosomonas*） 将氨态氮氧化成亚硝酸态氮；第二阶段则由亚硝酸细菌 （*Nitrobacter*） 进一步讲亚硝酸态氮氧化成硝酸态氮。这个氧化过程一直认为是绝对需氧的。植物营养元素生物循环的另一个重要环节是反硝化作用，它是由一种异样型细菌将亚硝酸态氮和硝酸态氮还原成分子态氮（N_2）而流失。这一过程通常认为是在嫌气性条件下进行的。

日本研究者 Shingo Mitsui 在 1956 年提出，将蓄水田土壤分成氧化层（0~1 厘米）和还原层两层。氧化层中有好气性微生物活动，如硝化细菌在这里将铵态氮氧化成亚硝酸态氮和硝酸态氮，当它们渗漏到还原层后，由反硝化细菌将它们还原成分子态氮而流失[7]. 这个典型图示，当时受到很多研究者的重视。

此外，苏联学者表明除了水田表面的氧化层中有硝化作用外，由于水稻根部能吐出氧气而使根表周围 5 毫米以内的土壤中也有硝化作用，在此以外就没有硝化作用进行了[1]。

20 世纪五十年代中期，我们在研究水稻田土壤中的硝化作用和硝化微生物时得到的结果表明。①水稻田土壤中的硝化微生物数量远高于旱地土壤中的硝化微生物，两者相差可达 23 倍到 377 倍[1]。②不论栽有水稻或不栽水稻的蓄水田土壤中，以及在水稻不同生长阶段中，在氧化层和不用深度的还原层土壤中均有硝化微生物存在并进行硝化作用[3]。这一结果表明：Shingo Mitsui 和苏联学者认为只有氧化层和水稻根表周围一薄层土壤中才有硝化作用的说法是不全面的。③首次发现水稻田土壤中的硝化微生物是兼厌气性的。它在好气性条件下可以进行硝化作用，在嫌气条件下也能进行硝化作用[2]。与旱地土壤中的硝化微生物只能在好气条件下进行的情况不同。实际上早在 1934 年 Carey 和 Waksman 在深达四千多米的海底中分离到硝化细菌，并认为这些硝化细菌不是由大陆土壤污染的，而是海底所固有的[3]。此外，Zobell 也证明在 Eh 值在−100 到−300 毫伏的海洋底部沉积物中有硝化细菌[5]。表明生活在不同条件下的硝化细菌和其他微生物一样具有多样性特征。所以，从 1890 年以来一直认为硝化作用和硝化微生物是绝对需氧的定论是不确切的。④分离获得亚硝酸细菌纯培养。实验首先通过加富培养和连续移接的方法[4]。逐步消除或淘汰培养物中的其他微生物，创造亚硝酸细菌的数量在培养物中占有绝对优势的条件。但是在连续移接分离过程中，发现与亚硝酸细菌始终生活在一起并共同进行硝化作用的硝酸细菌却比较容易消除，而且它被消除之后，对亚硝酸细菌的活动也没有发生什么影响；而是在亚硝酸细菌纯培养的分离过程中，最不容易消除的倒是与亚硝酸细菌紧密生活在一起的一种异养型伴生菌（它也是一种引起反硝化作用的细菌）。由此，必须要选择一种有利于亚硝酸细菌迅速生长而不利于伴生菌生长的培养基。试验结果证明 Lewis ＆ Pramer 培养基最为合适。它比常用 Meiklejohn 培养基要好。在两周时间的培养中，不仅亚硝酸细菌的菌数增加很快，而且对异养型伴生菌的生长还起到一定的抑制作用。使亚硝酸细菌的数量和伴生菌数量的比例达到 1∶0.02，也就是说，每有 50 个左右的亚硝酸细菌只有一个伴生菌，从而使亚硝酸细菌在培养物种占有绝对优势。这样再继续连续移接，就较容易的获得亚硝酸细菌纯培养。⑤通过加富和连续移接很多次而获得的积聚培养物。其中只含有一定数量的亚硝酸细菌和异养型伴生菌以及亚硝酸态氮（NO_2-N），如果在这种积聚培养物种加入少量无菌的有机碳化合物（如葡萄糖）后，分别在好气和嫌气条件下培养，很快就会发现培养物上产生一个很厚的气泡层，同时在培养

*本文未正式发表

物中完全检测不到NO_2-N，说明NO_2-N 已全部被伴生菌还原成分子态氮（N_2）而逸出[3]。

上述水稻田土壤中硝化作用和硝化微生物的研究结果清楚表明，水稻田土壤中硝化微生物数量远高于旱地土壤、它们不论在氧化层和不同深度的还原层土壤中都有分布，并进行硝化作用，而且这些微生物均具有兼厌气性的特征以及与异养型的反硝化细菌（伴生菌）紧密生活在一起。这一切因素都是造成水稻田土壤中无机氮营养元素大量流失的主要原因。其中特别是亚硝酸细菌和引起反硝化作用的伴生菌生活在一起。所以，进一步的研究就必须要搞清伴生菌的生物学特性及其对亚硝酸细菌的影响。

实验方法和结果

一、 亚硝酸细菌伴生菌的生物学特性

（一） 伴生菌的分离和纯化

从只含有亚硝酸细菌和伴生菌的积聚培养物接种到牛肉膏蛋白胨斜面上。28℃~30℃中培养，然后将斜面菌苔接至无菌水中做成菌悬液，在牛肉膏蛋白胨平板上进行单细胞分离。如此反复多次，并结合显微镜进行形态观察，直至获得伴生菌纯培养。结果证明伴生菌是假单胞菌属（*Psudomonas* sp.）的一个成员。

（二）伴生菌的生物学特性

（1）实验证明该伴生菌是一种能引起反硝化作用的异养型细菌，将它接种在反硝化培养基中，并分别在好气和嫌气条件下培养。结果见表 1，从表中清楚表明，伴生菌不仅在嫌气条件下将培养基中的硝酸盐还原为NO_2-N，而且在好气条件下也能进行同样的还原作用。

表 1 伴生菌接种在反硝化培养基中在好气和嫌气条件下进行的反硝化作用（NO_2-N mg/liter）

接种时间（天）	接种			不接种		
	好气	嫌气*		好气	嫌气*	
		A	B		A	B
0	0.0	0.0	0.0	0.0	0.0	0.0
5	3.3	3.1	2.0	0.0	0.0	0.0
10	20.2	19.6	10.7	0.0	0.0	0.0

*A. 加灭菌液状石蜡；B. 用焦性没食子酸钠吸收氧气；

（2）在只含有亚硝酸细菌、伴生菌和NO_2-N 的积聚培养物中加入葡萄糖后，在好气和嫌气条件下均能将NO_2-N 还原成分子氮逸出，并在培养液表层形成气泡层。结果见表 2。

表 2 积聚培养物中加入葡萄糖后产生的反硝化作用

接种时间（天数）	积聚培养物加葡萄糖 NO_2-N mg/liter		积聚培养物不加葡萄糖 NO_2-N mg/liter	
	好气	嫌气	好气	嫌气
12+0	55.7	55.7	55.7	55.7
12+1/2	41.7	41.7	55.7	55.7
12+3/4	34.8	33.4	55.7	55.7
12+1	11.7	13.0	55.7	55.7
12+1 1/4	0.4	0.6	55.7	55.7
12+1 1/2	0.0	0.0	55.7	55.7

（3）伴生菌纯培养分别接种在亚硝酸培养基和反硝化培养基中，菌数的消长，实验结果表明，随着培养时间增加，两种培养基中的菌数逐渐上升，随后菌数又缓慢下降。见表 3。

表 3 亚硝酸细菌的伴生菌在亚硝酸培养基和反硝化培养基中的消长

接种时间（天）	亚硝酸培养基		反硝化培养基	
	伴生菌数（个）×10000ml	NO_2-N	伴生菌数（个）×10000ml	NO_2-N
0	0.8	--	0.8	-
二	1.4	--	7100	++
四	193	--	139000	++
六	54700	--	109000	++
八	8500	--	1900	++
十	22	--	68	++

上表结果表明两种培养基中的菌数，都有一定的生长曲线规律。由此表明，伴生菌能够在完全没有有机碳化物的培养基中良好生长和繁殖。

二、 伴生菌对亚硝酸细菌的影响

实验结果表明，与亚硝酸细菌生活在一起的异养型伴生菌。一旦与亚硝酸细菌分离后，对亚硝酸细菌有明显的影响。单独生活在无机培养基中的亚硝酸细菌与积聚培养物相比有很大区别。首先，亚硝酸细菌纯培养形成的 NO_2-N 比积聚培养中形成的数量要少 12~100 倍[5]；另外，单独生活在无机培养中的亚硝酸细菌通过多次连续移植，硝化作用的能力会逐渐衰退，同时菌体也很快会导致死亡。但是，亚硝酸细菌的积聚培养物，经过保存二年半以上的时间，再多次移植也不会失去它的活性。我们还在好气和嫌气条件下培养获得的积聚培养物，连续移植 20 次，硝化作用始终不变，也毫不衰退。同时这种积聚培养物在嫌气条件下保存三年以上也没有失去它的活性。实验结果还表明，当亚硝酸细菌单独生长在无机培养基中而开始衰退前接入异养型伴生菌，硝化作用又会开始加强，其作用强度几乎和积聚培养物的强度完全相同。

小结与讨论

一、 验结果表明，异养型伴生菌不论在嫌气条件下还是好气条件下都能将硝酸盐还原成 NO_2-N，以及将 NO_2-N 还原成分子态氮而流失。但是通常的反硝化细菌虽然在好气和嫌气条件下都能生长，但它只有在嫌气条件下才能进行硝酸盐的还原反应，在好气下却不能进行。现在这种异养型伴生菌（*Pseudomonas* sp.）不仅在嫌气条件进行硝酸盐还原作用，同时在好气条件下也能进行硝酸还原作用。这是非常特殊的。

二、 验还表明，异养型伴生菌纯培养单独接种在亚硝酸培养基中也能正常生长和繁殖，培养六天时间，每毫升培养基的菌数可由 0.8×10000/ml 个上升到 54700×10000/ml 个，显然这些异养型伴生菌能够利用培养基唯一碳源-碳酸盐作为碳素养料[2, 3]。此外，Wurtz 和 Schmaltz （1962）的实验也证明，这些和亚硝酸盐细菌紧密伴生的异养型细菌，在完全没有有机碳化合物中也能生长和繁殖[11]。从而说明，异养型伴生菌的营养特性也是兼性的。也就是说，它既可以在有机培养基中生长繁殖，也可以在无机培养基中生长和繁殖。这也是非常特殊。

三、 从异养型伴生菌和亚硝酸细菌纯培养之间的关系来看，伴生菌对亚硝酸细菌是有利的，特别是对亚硝酸细菌所引起的硝化作用有非常有利的影响。如果伴生菌脱离了亚硝酸细菌，那么亚硝酸细菌的消化作用就会受到明显的抑制。两者之间的这些关系，许多研究硝化细菌的学者，如 Gibbs；Nelson；Kingma Boltjes；Meiklejohn；Gundersen 等在实验中也都得到了类似的结果[5]。但是，异养型伴生菌对亚硝酸细菌的这种紧密关系的作用机制至今还不了解，目前仅仅只是一些推测，有的认为伴生菌可供给亚硝酸细菌所需要的维生素和生长因素以促进亚硝酸细菌的良好发育[5]；有的伴生菌可能改善培养基的氧化还原点位，或者去除亚硝酸细菌的代谢产物，从而使它能够良好生长[5]。总之没有得到确切的、具有说服力的实验数据。所以，只能待于后人的研究和解决了。

综合异养型伴生菌的生物学特性及其对亚硝酸细菌纯化培养的影响所获得实验结果来看。很明显，由于能在好气和嫌气条件下都能产生强烈反消化作用以及在有机和无机培养基中都能生长繁殖的异养型伴生菌兼厌气性的亚硝酸细菌紧密结合在一起，这就是造成蓄水稻田土壤中无机氮素大量流失的原因。文献资料表明，水田中施用氮肥，如硫酸铵、豆饼、鱼粉作为水稻基肥时，其利用率明显比旱地土壤要低，一般只在40%~60%左右[5]；De 和 Digar 用不同的有机氮和无机氮施入水田的实验中指出，呈气体状态散发出来的氮素，占施入氮总量的 30%~40%；我们在实验室中的实验表明，有硝化作用产生的 122.5 mg/liter 的 NO_2-N 在 36 小时内全部还原为分子态氮而流失[6]。因而，在旱地土壤中对植物氮素营养有利的硝化作用。而在蓄水的水稻田土壤中却成为一种有害的过程。因此，进一步找出防止或减弱水稻田土壤中的硝化作用和反消化作用进行的途径，以防止土壤中氮素养料的损失和加强氮素营养的有效利用率是水稻生产技术中一项重要的农业科学技术任务。

此外，在这项研究工作中，对亚硝酸细菌和伴生菌生命活动的基本理论上也提出了一项必须要钻研的问题。即亚硝酸细菌能够在缺氧的条件下进行硝化作用，那么它氧化 NH_4^+ 所需要的受氢体究竟从何而来呢？它的有机合成所需要的能源又从哪里而来呢？同样，伴生菌在有氧条件下进行反硝化作用，那么它的作用机制又是怎样呢？也不清楚。同时，伴生菌是一种异养型细菌，它却能在完全没有有机碳化合物的无机培养基中正常生长和繁殖，这又是如何解释呢！这些问题都是值得深入研究的。

另外，亚硝酸细菌和伴生菌的关系是如此的密切的不能分开。当它们共同生活在一起时，却能顺利地完成上述这些特殊条件下的内容。这是不是两者之间具有一种互补作用？当然，这只是一种推测而已。

参 考 文 献

[1]陈华癸，周啟. 水稻田土壤中的硝化作用和硝化微生物的研究Ⅰ. 水稻田土壤中的硝化作用. 土壤学报，9（1-2）：56-64. 1961

[2]Chen H. K & Zhou Qi. Facultatixely Anaerobic Nitrfication and Nitrite-forming Oraganisms 第八届国际土壤学会议论文文集，土壤生物学，761-768，罗马尼亚布加勒斯特

[3]Zhou Q. & Chen H. K. The Activity of Nitrifying and Denitrifying Bacteria in Paddy Soil. Soil Science. 135（1）：31-34，1983

[4]陈华癸，周啟. 水稻田土壤中的消化作用和消化微生物的研究III. 亚硝酸细菌纯培养的分离. 中国土壤学会、中国微生物学会. 1964 年土壤微生物专业会议专题报告及研究报告摘要集，武汉，研究报告 45-46 页，1964

[5]周啟. 亚硝酸细菌的生物学（综述）中国土壤学会、中国微生物学会 1964 年土壤微生物专业会议专题报告及研究报告摘要集，武汉，研究报告 28-41 页，1964

[6]陈华癸，周啟. 水稻田土壤中的消化作用和消化微生物的研究Ⅱ. 水稻田土壤的亚硝酸细菌加富培养物在培养条件下的繁殖和亚硝化强度. 中国土壤学会 1963 年学术年会论文摘要集. 南京：中国土壤学会. 1963

[7]三井进午. 水稻的营养、施肥和土壤改良. 上海：上海科学技术出版社. 1959

[8]Meiklejohn J. Myxooacteria mistaken for nitrifying bacteria. Nature. 168，561，1951b

[9]Meiklejohn J.The pure culture isolation of *Nitrosomonas europaea*. Transact 4th Int. Congr. Soil Sci. Vol.1. p195，1950

[10]De & Digar. Loss of nitrogen gas from water logged soils. J. Agri. Sci. 44（part 2）：129-132. 1954

[11]Wurtz, Schmaltz. 在无机培养基上异养细菌的培养. 生物学文摘，第三分册，微生物学.1964 年第二期

第三部分　其　　他

提　要　这是不属于第一部分和第二部分内容的课题。研究内容有两项，共发表论文 31 篇[注]。

第一项

陈华癸教授独立撰写和由他指导研究生、青年教师，以及与同事合作撰写的学位论文。

第二项　教学改革研究

①国家教委立项的课题——高等农林院校本科生物系列课程教学内容和课程体系改革的研究和实践。1996 年在牵头单位华中农业大学召开了一次研讨会，由陈华癸教授主持，其中有华中农业大学和山东农业大学校长参加，其他均为参与这项研究课题的各院校代表。②经过四年多时间的研究，各院校关于这项课题，发表在不同杂志上的论文总共有 47 篇。2000 年因该项教改成果突出获得华中农业大学教学成果奖特等奖；2001 年又获得湖北省高等学校省级教学成果奖一等奖，同时在全省第三届教育科学研究优秀成果评选中又评上一等奖。

注：包括已在《微生物与农业》上刊登的论文，见附件 1。

第一项　陈华癸教授独立撰写和由他指导研究生、青年教师，以及与同事合作撰写的论文

Notes on A Collection of Freshwater Algae from CHANGLI，HOPEI.*

HUA-KUEI CHEN

（Department of Biology）

The materials for this report were a collection made in august，1932，by Mr. Y. C. Wang，formerly of the Tsing Hua University. Samples were taken from different localities of Changli，a district in the northeastern part of Hopei Province. It may be remarked here that this paper is to be regarded as a simple systematic report on the Algae existing in that district in a single summer season rather than as a distributional study.

The following field notes are Mr. Wang's:

Vial 651. Free floating in a pool of Nan-men. Aug. 22th，1932.

Vial 653. Free floating in a ditch by the road of Pei-men. Aug. 22th，1932.

Vial 657. In stream of Ngoh-shu Hill. Aug. 22th，1932.

Vial 658. On the stone of a stream near Ngoh-shu Hill，yellowish brown in color. Aug. 22th，1932.

Vial 668. From same locality as vial 658.

Vial 675. In a marshy ditch along the road；young globose，solid，showing olive green color. Aug. 23th，1932.

Vial 674. From same locality as vial 675，but showing blue green color.

Vial 676. In stream of Fun-huang Hill. Aug. 24th，1932.

Systematic Enumeration of the species observed.

CLASS ISOKONTAE

Group Volvocales

Series Tetrasporales

Family Palmellaceae

Genus. Palmella，Lyngbye.

Palmella mocosa kutz. （?） Vial 676.

Group Chlorococcales

Series Zoosporinae

Family Hydrodictyaceae

Genus. Hydrodictyon，Roth.

Hydrodictyon reticulatum L. Vial 653.

Family Coelastraceae.

Genus. Scenedesmus，Meyen.

S. incrassatulus Bohl. Vial 653.

---

*原载于北京大学《自然科学季刊》，251~256，1934.

S. bijuga Lag.	Vial 653.
S. arcuatus var.capitatus Smith.	Vial 653.
S. curvatus Bohl.	Vial 653.
Group Oedogoniales	
Family Oedogoniaceae	
Genus. Bulbochaete，Agardh.	
B. monile Wittr，and Lund.	Vial 651.
B. minor var，germanica Hirn.（?）	Vial 651.
B. affinis Hirn.	Vial 651.
Genus. Oedogonium，Link.	
O. paucocostatum Trans.	Vial 651.
O. curtum Wittr，and Lund.	Vial 651.
O. tapeinosporum Wittr.	Vial 651.
O. pringsheimii Cramer.	Vial 651.
O. varian Wittr. & Lund.	Vial 651.
O. vaucherii （Le Clerc.）Al. Br；Wittr.	Vial 674.
O. fragile Wittr.	Vial 674.
Group Conjugatae	
Series Euconjugatae	
（a）Mesotaeniaceae	
Family Mesotaeniaceae	
Genus. Cylindrocystis，Menegini.	
C. diplosora Lund.	Vial 668.
C. crassa De Bary.	Vial 668.
（b）Zynemales	
Family zygnemaceae	
Genus. Spirogyra，Link.	
S. porticalis （Muller.）Cleve.	Vial 676.
S. varian（Hass）Kutz.	Vial 676.
S. subsalina Cederc.	Vial 676.
S. jurgensii Kuetz.	Vial 651.
S. farlowii Trans.	Vial 651.
Genus. Zygnema，Agardh.	
Z. cruciatum（Vauch.）Agardh.	Vial 651.
Series Desmidiaceae	
Family Desmidiaceae	
Subfamily Closterieae	
Genus. Closterium，Nitz.	
C. lanceolatum Kutz.（?）	Vial 668.
C. sigmoidium Laguh.（?）	Vial 668.
C. moniliferum（Bory）Ehrenb.	Vial 668.
C. Ehrenbergii Menegh.	Vial 674.
Subfamily Cosmarieae	
Genus. Euastrum，Ehrenb.	

E. insulare（Witter.）Boy. Vial 651.

Genus. Cosmarium，Corda.

C. subarctoum Racib.（?） Vial 676.

C. garrolense Roy. and Biss. Vial 668.

C. obtusatum Schm. Vial 676.

C. subtumidum Nordst. Vial 676.

C. aptogonum Breb. Vial 674.

Genus. Sphaerozosma，Corda.

S. excavatum Ralfs. Vial 674.

Genus. Hyalotheca，Ehrenb.

H，dissiliens（Sm.）Breb. Vial 651.

CLASS HETEROKONTAE

Group Heterotrichales

Family Tribonemaceae

Genus. Tribonema，Derb. and Sol.

T. bombycinum Agardh. Vial 676.

T. minor Klebs. Vial 676.

CLASS CYANOPHYCEAE

Group Chroococcales

Family Chroococcaceae

Genus. Aphanocapsa，Naeg.

A. rivularis（Carm.）Rab. Vial 651.

Genus. Merismopedia，Meyen.

M. punctata Meyen. Vial 653.

Group Hormogoneales

Family Oscillatoriaceae

Genus. Oscillatoria，Vaucher.

O. amphibia Agarch. Vial 653.

O. anguina Bory. Vial 658.

O. formosa Bory. Vial 658.

O. Agardhii Goment. Vial 658.

O. limosa Agardh. Vial 674.

Family Nostocaceae

Genus. Nostoc，Vaucher.

N. punctiforme（Kuetz.）Har. Vial 674.

N. pruniforma（Lin.）Agardh. Vial 674.

Genus. Cylindrospernum，Kutz.

B. minutissimum Collins. Vial 675.

The following table summarizes the number of genera and species found from this region

		Genera	Species and varieties
chlorophyceae	Volvocales	1	1
	Chiorococcales	2	5
	Oedogoniales	2	10
	Conjugatae	8	20
Heterokontae		1	2
Cyanophyceas		5	10
Total		19	48

The writer is indebted to Dr. L. C. Li，Fan memorial institute of Biology for his kind asistance during this investigation. Acknowledgment is also due to Mr. Y. C. Wang for the use of his valuable collections.

***NOTE* 1.** The classification followed is that of West and Fritsch，“British Freshwater Algae.” Cambridge，1927.

***NOTE* 2.** The diatoms are not here included.

LITERATURE CITED

[1]Collins，F. S.：Green algae of North America，1928.

[2]Pascher，A.：Susswasserflora. Heft 4，5，6，7：Chlorophyceae；Heft 9：Zygnemales.

[3]Smith，G. M.：Phytoplankton of the inland lakes of Wisconsin. Part 1 and 2，Bulletin of Wisconsin Survey No. 57 Part 1，1919；Part 2，1922.

[4]Smith，G. M.：A monograph of the algaegenus Scenedesmus，based upon pure culture studies. Transactions of the Wisconsin Academy of Sciences，Arts and Letters. Vol.18，Part 2，1916.

[5]Smith，G. M.：The fresh water algae of the United States 1933.

[6]Tiffeny，L. H.：Oedogoniaceae.1930.

[7]Tilden，J.：Minnesota algae. Vol.1 Report of the Survey Botanical Series 8，1910.

[8]Transeau，E. N.：Key to Spirogyra（unpublished）

[9]West，G. C. & Fritsch，F. E.：A Treatise on the British Fresh-water Algae. 1927.

[10]West，W.：A Monograph of the British Desmidiaceae. Vol. 1-5. 1904-1923.

陈华癸在八十岁生日庆祝会上的致谢词*

陈华癸

（华中农业大学）

感谢各位领导、各位来宾、各位同学、同事和亲友：

做生日，尤其是做整生日是中华民族的传统，也是全世界各族人民的传统。

在生日聚会上互相祝福。

在今天的聚会上，各位的发言和寄来的贺辞，在祝福之中有不少过誉之词，其实不符，我很惭愧。

回忆我 1947 年到武汉以来，始终在这个圈子里生活和工作，缺点和错误很多，对事业造成不少损失，对和我一起生活和工作的人们造成不少伤害，我感激组织上对我的教导和指正，感激同志们对我的帮助和宽容。

历史的进程是曲折的，但终归是不断进步的。华中农学院，华中农业大学，土壤农化系也不例外，这些进步，这些成绩的取得，首先归功于中国共产党，归功于国家和人民的支持，是全校全系教职员工和同学们的集体贡献。成绩是集体的，功过是个人的。在生日祝福的欢乐中，我谨向支持、关怀、爱护我的战友们表示衷心感谢。

我趁此，还要向农业出版社，向文集的主编，著者和赞助者们致谢，这本文集，封面没有我的名字，里面也没有我的文章，但明白地是为我而出版的一本纪念文集，文集的撰稿人都是武大农化系和华中农学院的毕业生，文集涉及的题目范围很广，追索这些文章的内容，与著者们当时在学校学习的关系，显示出两个明显的特点：

一个特点是，著者们在学校学到的知识对他们仅仅起到了个专业启蒙的功能，文章中反映的广度和深度是著者们毕业后长时期在不同岗位上工作和学习的成就。

另一个特点是，无论在广度上多么分散，无论在深度上多么精深，却都没有脱离他们所接受的本科教育的基础，大学本科教育为以后的事业打下基础，没有这个基础是很难取得这些成就的。

得出一个结论，一定要把大学本科办好。本科教育的任务就是为以后的发展打下基础，而每一个毕业生的发展则是万紫千红，多姿多态的。

当前，我们正面临着改革开放，建设有中国特色的社会主义市场经济的新时期，学校教育正在进行着同步的改革，华中农大和全国高等学校一样，已经从较单一的本科教育为主的学校发展为专科，本科、研究生多层次教育的学校，学校日益兴旺发达。一个倾向可能掩盖另一个倾向在发展多层次教育的同时，切不能放松本科教育这一大学教育的支柱层次，一定要坚持办好，办出质量。

在改革大潮中，在国家兴旺发达的势头上，学校正面临着时代的机遇和冲击。

机遇要抓住，能否抓住机遇在于我们自己的实力和努力；冲击要顶住，顶住冲击在于我们对教育事业的坚贞和对祖国的挚爱。

借此机会，我愿和全体师生员工们互相勉励，并寄主要的希望于中年和青年。

最后，再一次表达我的感谢和祝福。

---

* 本文未正式发表。

如何使土壤肥料更好地为农业持续跃进服务*

陈华癸

（华中农学院土化系）

一、土 壤 肥 力

土壤肥力就是土壤生育植物的性能。

①土壤是由疏松的大小颗粒搭成的，因此它具备植物生根立株的性能。

②土壤有保水透气的性能。植物根又需要水，又需要空气才能生活，土壤满足了这两项条件，既能保水，又能同时排水透气。

③土壤含有植物所需要的各种养料，能够基本上满足植物的养料需要，土壤不仅只是植物养料的死仓库，如果仅只是养料的死仓库，迟早养料会用光，不能再生育植物了。土壤是生活的自然体系，在这体系中进行着活跃的物质运动，由于活跃的物质运动，植物养料也不断地在循环周转着，使得有限的植物养料能够无限的满足植物生长的要求。

土壤是一种很复杂的，活跃的物质体系。

（1）土壤矿物质：土壤固体成分中 95%~99%是矿物质，这些矿物质由大小不等的颗粒组成，大些的是砂粒，小些的是粉砂粒，最小的是粘粒。砂粒的成分主要是石英石（氧化硅），对于植物营养无直接意义，但是对于土壤的排水透气性很重要。粉砂粒和粘粒的矿物质成分复杂，含有各种元素成分，如磷、钾、钙、镁、铁等，这些元素成分是以不溶解于水的化合物状态存在的，它们不能直接被植物吸收，只有当风化以后，变成为水溶性的化合物状态才能被植物吸收。在土壤中进行着的风化作用，逐渐地将它们转化为植物可以吸收的状态，例如，麻石中含有正长石，正长石中含有很多植物需要的钾素养料。但是，未经风化的正长石里的钾不溶解在水里，植物不能够吸收利用。只有经过风化后，正长石遭到了物理的，化学的和微生物的分解作用，其中的钾素转变为水溶解性的化合物，植物才能够吸收利用。风化作用是缓慢的，对于植物营养料的供应只能起比较次要的，虽然是不可忽视的作用。

（2）土壤有机质和生物循环：除矿物质外，土壤还含有机质成分，本省农业土壤中有 1%~2%的有机质成分（以腐殖质为主），其中含有植物营养所需的各种营养元素成分。土壤有机质成分不断地在形成着和分解着，它是土壤中生物循环的一个环节。土壤中的生物循环可以下列图式说明：

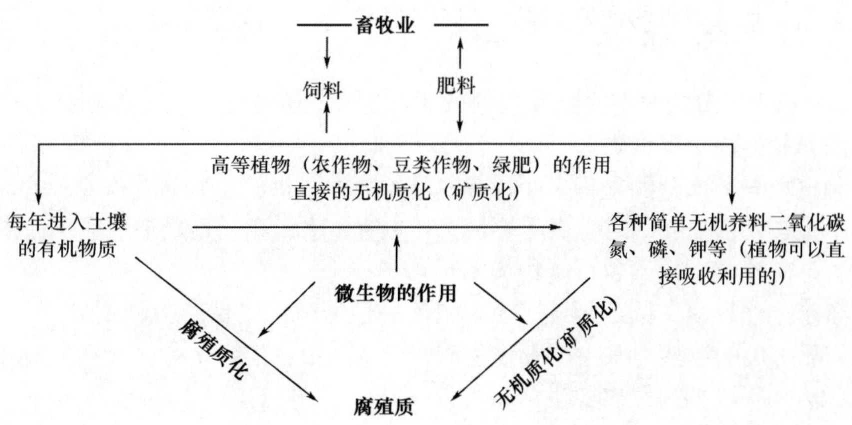

*原载于《中共湖北省地县委农业书记会议参考资料》，1959.

在这生物循环中，植物（农作物）和微生物是对立的合成者。植物将无机化合物合成为有机化合物（植物的体质和它的分泌物）。微生物将有机化合物分解成无机化合物。由于这样的循环周转，土壤中有限的植物养料成分可以无限的，一年又一年的满足植物的需要。

土壤的生物循环也包括腐殖质的形成和分解作用。土壤腐殖质对于土壤的保肥，保水，排水透气和耕作性质都起主要作用。

生物循环是土壤中最活跃的因素。

（3）土壤胶体：土壤黏胶是胶体物质，土壤腐殖质也是胶体物质，它们能合在一起，形成土壤胶体系，在土壤的胶体表面吸附着很多的植物营养物质（主要是阳离子，如氢、钙、镁、钾、铵、钠等），这些吸附物质是植物可以直接吸收利用的养料。

由于土壤黏粒的多少，黏粒的性质和腐殖质的多少，土壤胶体的吸附性能不等，对于土壤保肥性有重要作用，砂土黏粒少，吸附性小，保肥性差。黏土吸附性大，保肥性比较强。腐殖质的吸附性最大，因此，培育提高土壤的腐殖质含量，对于改善土壤的保肥性质有很重要的作用。

为什么胶体吸附性和保肥性有关？这是因为吸附在胶体表面的养料既能够供植物直接吸收，却又能对抗水的淋洗作用，不会因过多的雨水淋洗掉。如果胶体吸附性很弱，养料不吸附在胶体表面上，而是溶解在土壤水中。虽然也能供植物吸收，但是很容易被雨水淋洗，损失掉，例如白善土黏粒比较少，吸附性弱，保肥性差，施肥见效快，可是肥性不持久，经多年的施用大量有机肥料，培育成为乌白善土（油白善土），腐植胶体增加了，吸附性提高了，土壤的保肥性也改善了。

（4）土壤酸碱度：我省土壤可以分为石灰性和非石灰性两大类。石灰性土壤含碳酸钙（石灰），微碱性，大多数农作物都适合，少数植物不能在石灰性土壤上生长，例如，马尾松、茶等，在种这些植物时要特别注意，非石灰性土壤中不含碳酸钙，又可分为两类，一类是中性和微酸性的，也适合于大多数农作物生长。另一类是强酸性的，对植物生长有一定的妨碍。在酸性土壤上施用石灰可以中和酸性，增加生产。例如江汉平原的冲积土主要是石灰性。丘陵地主要是中性或微酸性，鄂西黄土和江南红黄土主要是酸性或强酸性的。

（5）土壤水和空气：土壤水和空气都存在于土壤空隙中，因此它们是矛盾的，水多气少，水少气多。然而，要植物能够旺盛生活，要土壤中的微生物能够旺盛生活，水和空气都是必要的，因此，如何解决水和空气的矛盾是培肥土壤营养植物的一项主要问题。

土壤水中溶解着一些植物的养料。这些养料植物可以吸收，但也容易淋洗损失。

（6）土壤中的植物养料：从上面的叙述，可知，土壤中的植物养料分别的存在于：矿物质颗粒中，有机质成分中，胶体表面上，和溶解在水中。矿物质颗粒中的植物养料必须经过风化作用才能变为植物可以吸收的形态，风化作用是很缓慢的，对植物营养来说只起次要的作用。有机成分中的植物养料经微生物分解后变为植物可以吸收的形态。这是植物吸收养料的最主要的来源，胶体表面吸附的和土壤水中溶解的植物养料是植物能够直接吸收利用的，但是存量不多，它依靠有机物质的矿物质化作用不断的补给。

形象化地解说：有机成分的分解是植物养料的生产工厂，而胶体表面吸收则是养料的临时库存和供应站，水里溶解的养料是街头零售点。

土壤中含各种植物养料成分的量有多少。有些种类能充分供给植物（农作物）的需要，因此在农业生产中不是问题，有些元素成分则时常供应不足，其中氮、磷、钾三元素最为突出，称为植物营养三要素，在某些情况下，其他营养元素也可能供应不足。

由于土壤具备了上述性质，因此具备了生育植物的性能——土壤肥力。

但是，仅仅依靠土壤的自然力来生产农作物，生产量是很低的，只有在人的控制和改造之下，才能不断地提高土壤肥力，不断的提高农业生产量。

二、施肥是提高土壤肥力、提高农作物生产的手段

前面讲过，生物循环是土壤中最活跃的因素，培肥土壤和营养植物的最主要的问题之一就是加强生物循环的进度，控制它的转化方向和相对强度，要求能够做到腐殖质含量逐渐增长（改善土壤的保水、保肥、排水透气和耕作性质），而不是逐渐下降，或停滞在目前含量过低的水平上，同时还要求旺盛而适当的矿物质化过程，保证植物得到充分的养料。

土壤培肥和施肥的主要任务就在于如何把这个生物循环管理好，并且进一步提高它。以氮素的生物循环为例，空气中含 80%氮气，可是大多数农作物都不能利用它作为氮素养料，只有土壤中的一些微生物（固氮菌等）和与根瘤菌共生的豆类植物才能够转化氮气为植物能够利用的氮素化合物，它们首先把氮气合成有机氮化物（固氮作用）。这些有机氮化物还不能被其他农作物直接吸收利用，必须经过氨化微生物和硝化微生物的作用，把有机氮化物转化为铵盐（氨化作用）和硝酸盐（硝化作用）才能够被多种农作物吸收利用。土壤中的硝酸盐一方面是农作物的优良氮素养料，另一方面它既容易被淋洗损失，也能被反硝化微生物转化为植物所不能利用的氮气，因而损失掉。

由此可见，如何通过农业技术措施来管理好这个循环，例如种豆类作物加强固氮作用，改善耕作和灌排技术加强氨化作用和硝化作用，防止淋失和反硝化作用，是培肥土壤，提高农作物生产的一个重要方面。

施肥是培肥土壤，提高农作物生产的另一个重要方面，因为，施肥能在土壤原有的肥力基础之上增加肥力条件，把土壤肥力提高到更高的高度，满足高额丰产的需要。因此不断地施肥，土壤肥力不断地提高，农作物产量也就可以不断的上升。

肥料可以分为有机肥料和矿质肥料两大类。

施用大量的有机肥料增加了土壤中的有机质材料，因而可以同时加强矿物质化，改善农作物的养料条件；也加强了腐殖质化，改善土壤的保肥、保水、排水透气和耕性。当然，必须同时满足和改善微生物的生活条件，因为只有在旺盛的微生物生命活动中，有机物质才能腐殖质化和矿质化，才能产生肥效。

施用化学肥料直接改善农作物的营养条件。

有机肥料和化学肥料的配合使用是培肥土壤和营养农作物、提高生产的重要技术措施。有机肥料和化学肥料的使用不能偏差，偏差了那一面，都对高速度的发展农业生产不利。

提高有机肥料质量和数量的主要措施是养猪积肥，实行农牧兼营的农业生产制度，猪多，肥多，粮多，收入多，循环增长；种绿肥和牧草。增施化学肥料主要是实现农业生产化学化，大小并举，土洋并举的生产化学肥料。

以下对粪肥，绿肥和化学肥料，它们在土壤中转化的规律和增产意义作简单介绍。

粪肥　粪肥不宜单独堆积，单独堆积时，所含的养料成分损失很大，不合理的堆积可能损失植物养料成分一半以上。粪肥一般是和土或干草烂在一起堆积，制成混粪，为了分别说明，这里将和土的称为土粪，和草的称为烂粪。

土粪是一种优良的积肥方法，一般将粪吸收在三倍左右的土壤中，这样做，由于土的吸收作用，养分不至于损失。

粪吸收于土壤中后，一面腐殖质化，一面矿物质化，经过几个月，粪土相融，形成为高度腐熟的肥料。将腐熟的土粪集中的施用于种子附近，为农作物创造了优良的生活条件，多施腐熟土粪不会伤苗伤根，肥效很长，很稳，大量使用土粪，结合深耕，分层施肥，对改良土壤起很大的作用。每亩土壤，一尺厚约重 60 万斤，一般含腐殖质 1.5%，合 9000 斤。土粪一般含有机成分（腐殖质及其他）5%~7.5%，一次一亩施用 20000 斤，合 1000~1500 斤，提高了土壤腐殖质含量 1/4~1/6。虽然其中一部分要矿质化，变为植物的无机质养料，连续几年施用大量土粪，必然的会提高土壤腐殖质含量，因而提高了土壤的肥力水平。

烂粪在堆积过程中，由于旺盛的微生物生命活动，所含的有机物质逐渐的腐殖质化，形成优良的肥

料，按照施用时的腐熟程度而定，烂粪可以分为腐熟的和半腐熟的两种状态，腐熟的烂粪，粪草都失去原来的形状，成为均匀一致的，疏松的黑色腐殖质肥料，半腐熟的烂粪，粪草都有一定的变化，但还没有完全失去原来的形状（尤其是草），农业生产中实际使用的，一般是半腐熟的烂粪。

施用烂粪必须做到粪土相融，才能充分发挥肥效，烂粪和土粪不同，由于没有先和土融合，不可直接与种子或作物根接触，尤其是半腐熟的烂粪。如果直接接触，则产生的热，毒质或过稠的养料会伤害幼苗或植物根。正确的施用烂粪，务必充分散开，翻耕入土，反复耕耙，做到肥料和土壤充分混合，然后由于土壤微生物的生命活动，将肥料进一步腐殖质化和矿物质化，粪土相融，成为肥沃的土壤。不正确的施用烂粪，将半腐熟的、甚至于新堆积的烂粪在播种或插秧前不久施用，施用时不散开，不耕耙，或耕耙不够，形成不均匀的肥料堆和空白地相间的田地状态，这样，作物生长必然很不一致，直接在肥料堆上的由于受毒害而死亡或严重的热烧，在肥料堆附近的生长十分旺盛，一部分过旺倒伏，在没有施到肥料的地方严重脱肥，植株矮小，这种现象，各地仍相当普遍，急待改进。

烂粪的肥效也很慢，很稳。大量施用烂粪，结合深耕，和大量施用土粪一样，能提高土壤腐殖质含量，因而提高了土壤的肥力水平。

绿肥 种一季绿肥。生产几千斤到一万斤以上有机物质，然后在适宜的条件下，翻耕入土，这些有机质（肥料）逐渐矿物质化，供下一季农作物养料的需要。同时，由于腐殖化也提高了土壤腐殖质含量，另外，绿肥根在土壤中的生命活动，对于土壤构造和耕性也有改良作用。

如果将豆类绿肥和非豆类绿肥相比，豆类绿肥（如红花草籽，苕子等）有一项显著的优越属性，即它和根瘤菌共生，产生根瘤，引起共生性的固氮作用，从而将空气中的氮气转化为土壤中的氮素养料。种一季豆类绿肥，亩产绿肥6000斤，从空气中获得的氮素养料可达20斤，相当于硫铵100斤，例如石首远安公社计划建立绿肥基地20万亩，仅只它的固氮能力就相当于年产一万吨硫铵的合成氮工业。

绿肥一般采取当有机质生产量最大而且青嫩的时期翻耕入土；尤其是豆类绿肥，在土壤中腐烂分解很快，产生肥效很快。

绿肥和烂粪或土粪相比，绿肥的矿质化作用强而且快，而腐殖质化相对的比较弱。

最合理的施肥法是粪肥和绿肥配合使用。粪肥的腐殖质化作用比较强，肥效慢而稳，后期肥效好；前期肥效差。绿肥的肥效快而猛，前期肥效好，后期肥效差。配合使用则同时提高了矿质化作用和腐殖质化作用，前后期肥效均匀。

化学肥料 仅只是粪肥和绿肥的配合使用还不能满足作物生长的最大要求。①有机肥料在土壤中矿物质化，供给植物的规律并不能和植物生长发育期间的需肥规律完全相同，在农作物生长时期，有时需肥多，有时需肥少，不但需肥少，这时如果速效肥过多反而有害，因此，施用有机肥料只宜于保证植物在一般需肥时的需要，而在需肥多时，就要施用速效性的追肥补充，化学肥料是最好的速效性追肥（人粪尿虽好，但对生产需要来说，是远远不够的）。②植物在不同时期对于肥料的数量和质量的要求不完全相同，不同种农作物的需肥情况也不相同。粪肥和绿肥都是完全肥料，我们能够调节它的数量，不能改变它的养分比例，因此，就需要成分比较单一的化学肥料来调节养分比例。③有机肥料主要是地里生长出来的。绿肥是直接从地里长出来的，粪肥是间接从地里生出来的。因此，施用大量有机肥料，其中所含的磷、钾等养料成分，还是地里来，地里去，增加了周转率，提高了利用效率，但并没有提高绝对数量。化学肥料是工业产品和矿产品，施入土壤，还增加了土壤养料的绝对数量。

因此，化学肥料使用，农业生产的化学化是不断提高农业生产量的迫切需要。合成氮工业的发展，磷矿的开采和加工，钾矿的开采和加工，社会主义工矿业生产的巨大力量对于农业建设的重要性会一天比一天显著。

随着国家的工业化和农业化学化，今年提供农业生产的化学肥料从品种到数量愈来愈多，如何正确地、合理地、经济地、有效地使用，充分发挥化学肥料的优越性，将是一个很重要的问题。

这里不能深入到施肥方法问题中去，只能对于施用氮肥和磷肥的个别问题加以解释。①凡是含硝酸成分的氮肥不宜用于水田，因为在水田中，反硝化作用强，施用硝酸肥料会因为反硝化作用而损失，同时土壤胶体吸附硝酸的能力很弱，硝酸肥料几乎全是溶解在土壤水中，容易淋失。因此，施用含硝酸的肥料必须按“少吃多餐”的方法使用，施用后，如果农作物不能立即吸收，一场雨水都会淋洗掉。②凡是含铵成分的氮肥。水田旱地都适宜，土壤胶体吸附铵的能力比较强，因而淋洗损失的现象不像含硝酸肥料那样严重，然而含铵的肥料也有缺点，例如硫酸铵，它是生理酸性的，不正确的使用，会使土壤酸性加强，因此最好配合施用石灰。③过磷酸钙本身是水溶解性的，可以被植物直接吸收，但施入土壤后，可能起不利的变化，在石灰性土壤中，它会变成磷酸三钙，从水溶解性变为不溶解性，因而降低了磷的有效性。在强酸性的红黄土中，它会变成磷酸铁和磷酸钙，它们都是不溶解性的，这种作用称为固结作用。为了防止固结作用，应该将过磷酸钙与腐熟烂粪混合，或做成颗粒肥料使用，避免和土壤直接接触。根外喷施磷肥也是很好的办法。

三、土壤的保水性和排水透气性

作为土壤的肥力条件来说，有必要将旱地土壤和水稻土分开来讨论，因为旱地作物和水稻土对于水和空气的要求的差别比较大。

1. 旱地土壤的保水性和排水透气性

土壤的保水性满足植物（农作物）对水分的要求，排水性也就是透气性，如果土壤不能排水，大雨以后，土壤中就充满了水分，隔绝了空气，旱地作物就会渍死，植物要呼吸需空气，土壤中微生物的生命活动也要空气，前面已经提到过，微生物的生命活动是土壤中水分重要的肥力因素。

土壤为什么同时有保水性和排水透气性？这是和它的颗粒性分不开的。

土壤是由大小不同的颗粒搭成来的，颗粒之间有空隙，空隙状态和土壤的保水性和排水透气性有密切关系。

土壤空隙大小，分为两级：非毛细管空隙和毛细管空隙。

非毛细管空隙是大空隙，在大孔隙中，水分受地心吸力的作用而下流，如果地下水位很低，下面没有不透水的层次。大孔隙中的水分很快地就漏掉。这种水对农作物没有什么用，而漏掉以后，改善土壤透气条件，则是农作物十分需要的。因此，土壤大空隙或非毛细管空隙是土壤肥力的重要因素。

毛细管空隙是小空隙，它和土壤的保水性密切相关，水在毛细管空隙中流动得比较慢，而且，决定水在毛细管空隙中的流动的规律的主要因素不是地心吸力，而是比较复杂的毛细管运动的规律；从湿往干处流，从比较大的空隙往更小的空隙中流；雨水或灌溉水落在干旱土壤中，先满足最小的毛细管空隙，再满足中等的毛细管空隙，再满足较大的毛细管空隙。下场小雨，土壤上层湿，下层干，毛细管水从上向下行，大太阳晒，土壤表面干，下部湿，毛细管水向上行，土壤水分不匀，左右干湿不一致，毛细管水总是从湿流到干。

毛细管空隙中的水——毛细管水——是作物吸收的主要来源。不仅如此，植物根吸收根周围的水，使得根周围比较远的地方更干些。这样，在毛细管运动规律的支配下，较远处的毛细管水就向根周围流动。因此，根可以获得土壤中全部的毛细管水。

从上述可知，同时提高土壤的毛细管空隙和非毛细管空隙，是提高土壤肥力的重要方面，什么条件决定土壤的毛细管空隙和井毛细管空隙，如何改善这种条件？

（1）土壤质地，或称泥沙比例，是一项重要因素。前面提到，土壤是大小不等的颗粒搭起来的，按大小区分，土壤颗粒可以分为砂、粉砂、黏粒三等，砂粒大，粉砂粒中等，有些像灰面样，黏粒最小，是胶体状态的小颗粒，土壤中所含大小颗粒的比例称为质地，或泥砂比例。砂土地砂粒多，非毛细管空隙多，排水透气性好，但是毛细管空隙少，保水性差，不耐旱，三天没有雨或不灌溉就遭旱，黏重土地，或称胶泥土地。黏粒多，毛细管空隙比较多，而非毛细管空隙少，下雨后，排水不畅，土壤不透气，农作物的根部受水渍害（例如胶板马肝土）。同时黏重土壤的保水性虽比砂土地好些，也不很强，雨后天晴，地面不断蒸发，毛细管不断送水到地面，帮助蒸发，土壤干得也很快。

泥砂比例适宜的土壤——油砂土，两合土——的毛细管空隙和非毛细管空隙性质都比较好，毛细管空隙比砂土多，比胶泥土也不少，非毛细管空隙比胶泥土多，比砂土少。同时，由于毛细管空隙和非毛细管空隙的互相间隔，地面蒸发的强度也比较小，因此，从保水性来说，油砂土或两合土比砂土好，比胶泥土不坏或还要好些；从排水透气性来说，比胶泥土好，比砂土差。

因此，改造砂土或胶泥土为两合土，是土壤改良的重要措施，用客土法，“砂加泥好得稀奇，泥加砂象糖黏粑”，改善泥砂比例。这几年，有不少先进的改土成绩值得学习，参考。

在泛滥冲积土地带，客土法不一定要从别处运土来，在泛滥冲积土中，往往是一层砂一层胶泥的交叠沉积的，摸清土壤的层次底细，往往可以通过深耕翻土，就地改良土壤的泥砂比例。今年土壤普查工作中就发现了几处这样的土层情况，替土壤改良找到了办法。

（2）土壤构造：土壤里的大小颗粒并不完全是单独存在的，而是许多大小颗粒在一起，形成土壤构造单位，许多构造单位再搭成土壤构造。土壤形成构造后，土壤空隙状态就起了变化，单个颗粒与单个颗粒之间有空隙，而构造和构造之间也有空隙，构造和构造之间的空隙一般都比较大，属于非毛细管空隙，因此土壤构造形成增加了耕作中的非毛细管空隙，对于排水透气有好处，然而，构造形状不同，起的作用不一样，有时也能起坏作用，要具体研究，加以培育或改造。

土壤构造形成团粒构造？目前我们还没有掌握培育土壤团粒构造的全部规律，有几条是知道的：①施用大量的有机肥料，提高土壤腐殖质含量，②在无石灰性的土壤上施用石灰；③植物（农作物）的须根长得旺，④土壤微生物活动旺盛，这四条是培育土壤团粒构造的重要条件，肥沃的菜园土壤具有优良的团粒构造，保水透气性比一般大田土壤要好得多，土壤耕作园田化的一项指标就是把大田土壤培育成为团粒构造优良的，肥沃的菜园土壤。

（3）土壤的层次对于保水透气性质也有重要的意义，土壤从上到下，并不是完全一致的，形成各种层次，泛滥沉积的土壤表现为由各种大小颗粒形成的土壤层次，粗砂层，细砂层，胶泥层相间的排列着，砂层排水透气性强，胶泥层保水性强，并且可以减慢上面砂层的排水速度，层次排列得适宜，例如蒙金土，对于土壤的保水透气性质有很大的好处，层次排列得不适宜，就需要把它搅乱，和匀，改造成为两合土。

深耕和优良的耕作方法是改良土壤保水透气性质的重要技术措施。

不合宜的土壤层次可以用深耕的方法改良。例如，山东寿张县的深翻改土工作是做得很有成绩的，那里是黄河泛滥沉积的土壤，表面砂土层下面盖着胶泥层，通过深耕，改造为两合土。

在一般土壤上深耕可以加深耕层，耕层土壤比下层土壤泡松，保水透气性都有改善。

深耕结合施用大量的有机肥料（在无石灰土壤上加石灰），不仅加深了耕层，培育了土壤团粒构造。

“死土改活土，活土改油土”的意义是多方面的，其中主要的一方面是改善土壤的保水和排水透气状况，翻耕死土，经过日晒冬凛结，成大块的死土，就疏散成为活土，初步具备了植物扎根，保水和透气性的性质，结合施用大量有机质肥料，充分做到肥土相融。

土壤构造团粒化，保水透气性进一步改善，腐殖质胶体增加，微生物生命活动旺盛，供肥和保肥性质都提高了，活土就变为油土。

当然，深耕必须采取优良的耕作方法，耕作方法不好，火色不当，不但不能改良土壤的保水透气性质。还会带来很大的损失。

（4）地下水位和土壤下面不透水的层次，对于土壤的保水性和透水性有关，对旱地作物来说，地下水位高，或是较长期的渍水，总是不利的，必须挖深排水沟，降低水位，排除渍水。

以上简单地介绍了旱地土壤的保水性和透气性，以及改良的途径。

2. 水稻田土壤的保水性和透气性

水稻根也需要氧气，进行呼吸，水稻田土壤汇总的微生物也只有在提高了空气条件下才能够更旺盛的生活，由于水稻植物体内有导气管，能从叶面将空气传导到根部区，因此，水稻根浸水条件下还能够呼吸。但是如果要求更旺盛的呼吸，要求土壤中有益微生物的旺盛活动，仅只从叶面经导气管传下去的空气就很不充分了，需要改善土壤的透气条件。湿润栽培，干干湿湿，灌跑马水的增产意义就在于此。

必须在充分满足水稻对水的要求下同时提高土壤的透气条件，才是水稻高额丰产的基础。

水稻田表面上的水层主要是起蓄水的作用。如果只靠本田表面蓄水，就隔绝了空气在隔绝了空气，水稻根呼吸不旺。土壤中的有益微生物的活动也不旺，有害微生物的活动反而旺盛了。例如，有机肥料分解，产生植物能够吸收的氮素养料的微生物活动在透气情况下比淹水情况下高，相反的，产生硫化氮的作用，在淹水情况下比透气情况下高。硫化氮对水稻根有毒害作用，尤其是在有机肥料下得多的情况下，这现象更严重，在深根，多肥，密植的条件下，排水晒田，在保证水分供应条件下，改善透气条件是增产关键。

烂泥田，青泥田，冷侵填不能排水，是限制生产量的关键因素。降低水位是改良烂泥田，青泥田和冷侵田的基本措施。

当然，不论是水田或旱地，彻底解决农作物的给水问题，不是单纯的改善一块土地本身的保水透气的性质所能完全解决的。大面积的水利化才能够彻底解决农作物的供水排水透气问题，然而，这又不是说土壤的保水透气条件是不重要的，水利化只有在优良的土壤保水透气条件下最能发展它最高的作用。优良的土壤保水透气条件是高效率的水利化的基础。

水利化要求省水、省工，要求尽量的利用雨水，改良土壤的保水透气性质是达到这些要求的土壤基础，可以举一个简单计算说明，原来一块沙土地，一亩地一尺厚的毛细管水只能保存 45 000 斤，改良后，提高到 70 000 斤，那么原来三天要灌一次的，现在五天灌一次就可以了。雨水多存些，灌溉次数可以减少些，就显著地提高了水利化的增产效率。

尤其是，只有在深耕多肥，肥、水、气条件优良之下，才提供农作物高额丰产的条件，水利化才能发挥它最大的作用。

以上简单地介绍了土肥水的一些自然规律和改良的途径，提供参考，至于具体的技术措施只有针对着一定的土壤，有的放矢的设计和执行。

钾细菌的形态、生理及其对磷钙矿物的分解能力*

陈廷伟　陈华癸

（华中农学院）

钾细菌（原名硅酸盐细菌）在苏联和我国已开始制成为细菌肥料应用到农业生产中。但是，关于钾细菌的形态和生理特性的研究材料还很缺乏，特别是它对磷钾矿物的分解能力还缺乏系统的实验材料证实。本项研究的目的即为在研究钾细菌形态和生理特性的基础上，重点研究对磷钾矿物的分解能力。

一、钾细菌的分离

采用以土壤矿物为钾源的无氮培养基**作为分离用的选择性培养基。钾细菌在这种培养基上生长得很好，形成圆形，有光泽，凸起的无色半透明菌落；菌落浓稠而有弹性。挑起这种菌落经过初步镜检后即可纯化作研究用的菌株。

二、形 态 特 征

钾细菌是能产生肥大荚膜的孢子杆菌。菌体为长杆状，大小（1~1.2）×（4~7）微米；荚膜为椭圆形，大小（5~7）×（7~10）微米。孢子高亦为椭圆形，比菌体粗大，大小为（1.5~1.8）×（3~3.5）微米。菌体末端圆形，常成单个不成链状；菌体中往往有1~2个颗粒体。革兰氏染色负反应。

在含氮的查贝克和无氮的阿息比培养基上，皆能形成圆形、凸起的无色半透明菌落；如同半颗玻璃珠一样。但是，在无氮培养基上的菌落凸起度更大（与培养基的切线角度在45°~90°之间），并且更为丰满和浓稠而有弹性。在无氮培养基上产生厚大的荚膜，但是在含氮的查贝克培养基上没有荚膜或者只有黏液层；在牛肉汁琼脂上只能生长少量的无荚膜菌体。在淀粉铵琼脂或马铃薯培养基上，皆不形成荚膜而易于形成孢子，同时，在这种培养基上的菌落形态变为混浊而失去弹性。

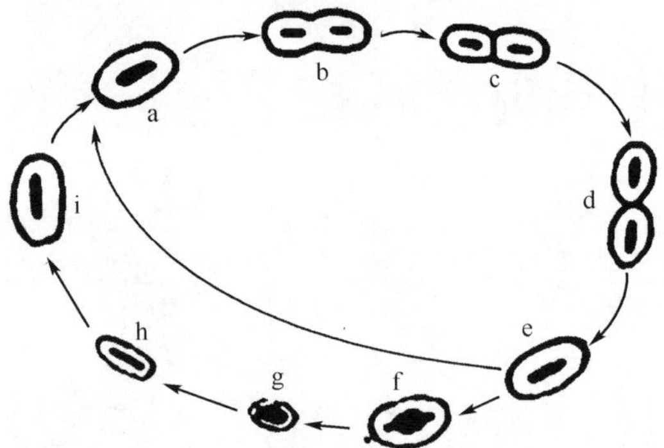

图1　钾细菌的生活史，a-e 营养体发育世代；f-i 孢子发育世代

钾细菌菌体和荚膜的发育很为特殊。在无氮培养基上生长一昼夜的幼年培养体每一荚膜中可以有2~4个杆菌；然后荚膜随同一起分裂，培养三天后的每一荚膜中只有一个杆菌。培养7~10天后的老年培养体中，可以见到荚膜内产生椭圆形的厚膜孢子，培养时间更久时，孢子脱离开荚膜，只剩下一个个空的荚膜。培养时间很久的培养体中还会出现畸形的菌体，荚膜中的杆菌体伸长，弯曲成弧形，C形甚至S形的不正常形态（图1）。

---

*原载于《微生物》，2（3）：104~112，1959.

**无氮培养基：

蔗糖	5克	$FeCl_3$	0.05克
Na_2HPO_4	2克	$CaCO_3$	0.1克
土壤矿物		蒸馏水	1000毫升
$MgSO_4$	0.5克	琼脂	20克

三、生 理 特 性

钾细菌的生理特性主要表现在对环境条件的适应性强。对碳源利用范围广，可以很好地利用各种简单糖类和淀粉。能利用各种无机氮源，但是对有机氮源利用力不强。不能在牛肉汁琼脂上生长；在石蕊牛奶上微弱冻化牛奶；明胶穿刺只成黏液生长在表面，不液化明胶，对灰分元素有特殊的利用能力，能在以磷灰石为唯一磷源的培养液中生长，也能在以长石，云母和土壤矿物为钾源的培养液中生长，它甚至还能在只含有葡萄糖和磷灰石的蒸馏水中生长得很好。在灰分元素缺乏的培养中，钾细菌产生大量荚膜黏液，紧紧包被在矿石上并从其中夺取生长发育所必需的养分（图 2）。在这种条件下的菌体往往产生一层以上的荚膜。在养分愈缺乏的情况下，钾细菌产生的荚膜愈大，而荚膜黏液量也俞多，同时粘结矿物也愈紧密。

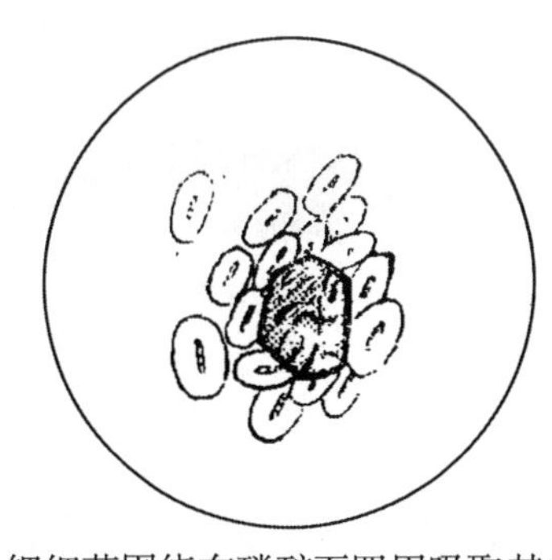

图 2 钾细菌围绕在磷矿石四周吸取其中养分

钾细菌为兼性需氧性细菌，生长的最适宜温度是 25º~30ºC 最适 pH 是 7.0~7.2。这种细菌利用各种碳源时不产生明显的酸。

四、钾细菌对磷灰石的利用和分解能力

在无磷培养液*中，以海州磷灰石（通过 100 孔筛）为唯一磷源，接种钾细菌培养 3 天后用平面培养法测定钾细菌数量；并用奥斯特瓦德黏度计测定培养液的黏度。结果是在不加磷灰石的无磷培养液中，钾细菌不能生长发育，培养液黏度与蒸馏水相同（以蒸馏水为 1 比较）；在以磷灰石为磷源的培养液中，钾细菌每毫升数量达到 630 万个，黏度增加 3 倍左右（表 1）。在固休培养基上还可以直接观察比较。在无磷琼脂中，钾细菌落很小（平均直径不到 1 毫米），而加有磷灰石的培养中生长起丰满的大菌落（直径平均 7.9 毫米）。

表 1 钾细菌对磷灰石的利用能力

培养基	比黏度	钾细菌数（个/毫升）	比-P 培养液增加倍数	备注
-P	1.00	50	——	
-P+磷灰石	1.92	25×10^4	5，000	包括粘结在磷灰石上的
-P+磷灰石	2.21	630×10^4	12，600	黏液

在以磷灰石为唯一磷源的培养液中，应用化学分析方法测定了钾细菌对磷灰石的分解能力。培养 10 天后的培养体先用 6%的 H_2O_2 处理，破坏荚膜黏液和菌体后使其与磷灰石分开，然后按一般化学分析方法测定溶液中的水溶性磷。分析结果是接种钾细菌的比不接种对照水溶性磷增加 223%~250%，释出磷素为磷灰石原有不溶性磷的 0.12%~0.18%。试验的每一处理有两个重复，先后进行了二次试验，所得结果很相近（表 2）。

*无磷培养基

蔗糖	5 克	$CaCO_3$	0.1 克
（NH_4）$_2SO_4$	1 克	$FeCl_3$	0.005 克
$MgSO_4$	0.5 克	蒸馏水	1.000 毫升
$FKCl_3$	0.3 克	pH 调至 7.2	

表 2　钾细菌对磷灰石的分解能力（光电比色法）

培养基	处理	水溶性磷（P_2O_5）ppm			比对照	不溶性磷分解
		I	II	平均	增加%	率%
-P+磷灰石	对照	0.15	0.2	0.17		
同上	接种钾细菌	0.70	0.40	0.55	223%	0.122%
-P+磷灰石	对照	0.2	0.2	0.2		
同上	接种钾细菌	0.8	0.6	0.7	250%	0.177%

五、钾细菌对含钾矿物的利用和分解力

在无钾培养液中，分别以长石（通过 100 孔筛），云母（通过 60 孔筛）及土壤矿物为钾源，接种钾细菌培养 3 天后测定其中的菌数和黏度。结果是在缺钾培养液中，钾细菌也有少量生长，因为一般试剂常含有微量钾素。但是在添加长石，云母和土壤矿物的处理中钾细菌生长得更好。这说明钾细菌是能从上述矿物中取得钾素的。

表 3

培养基	比黏度	钾细菌数（万/毫升）
-K	1.05	4.1
-K+长石	1.10	27.0
-K+云母	1.10	15.0
-K+土壤矿物	1.11	68.0

同样，以化学分析方法测定了含钾矿物中钾素释出的数量。样品处理方法同前。试验进行二次，分别用重量法和容量法测定培养液及培养物中的钾含量。

结果是在以长石为钾源的培养液中，接种钾细菌的比不接种对照水溶性钾，最高增加 87%；云母粗的增加 65%；土壤矿物粗的增加 45%（表 4 及表 5）。

表 4　钾细菌对含钾矿物的分解能力（重量法*）

培养基	处理	水溶性钾（K_2O）P.P.m.			比对照增加%
		I	II	平均	
+长石	对照	3.96	3.52	3.75	
同上	接种钾细菌	5.82	8.14	6.98	87
+云母	对照	3.41	3.19	3.30	
同上	接种钾细菌	7.04	3.85	5.44	65

表 5　钾细菌对含钾矿物的分解能力（容量法**）

培养基	处理	水溶性钾（K_2O）ppm			比对照增加%
		I	II	平均	
-K+长石	对照	8	8	8	
同上	接种钾细菌	10	10	10	25
-K+云母	对照	7	7	7	
同上	接种钾细菌	8	11	9.5	35.7

续表

培养基	处理	水溶性钾（K_2O）ppm			比对照增加%
		I	II	平均	
-K+土壤矿物	对照	10	10	10	
同上	接种钾细菌	15	14	14.5	45

*重量法为卡塔纳耶夫亚砂酸钴银法。
**容量法为光电比色法。

六、 关于微生物对岩石矿物破坏作用的机制问题

细菌在含有磷灰石或正长石等矿物的培养液中生长时，大量的菌胶团缠裹矿石颗粒，并同时分解这些矿物，取得磷素或钾素养料。这类细菌不产酸，它分解磷灰石或正长石等矿物，并不是由于产酸而导致的酸性分解。用火棉胶将钾细菌和磷灰石分开，使得培养液中的水分和溶质可以通过胶膜与细菌拉触，但磷灰石粉微粒和细菌都不能透过胶膜接触，在这样的条件下，细菌不能生长，磷灰石不被分解，因此，钾细菌分解利用岩石矿物中养分的途径可能是：①细菌与矿石接触并产生特殊的酶的作用，破坏矿石结晶结构，并释放出其中养分；②表面的物理化学的接触交换作用。

从以上研究结果证明，钾细菌不但有特殊的形态和生理特性，而且具有利用大气氮素和分解磷灰石和长石、云母中磷钾元素的能力。这是在土壤中广泛分布的一种有益细菌，应研究和利用这种细菌为农业生产服务。

施　　肥*

陈华癸

（华中农学院 中国科学院武汉微生物研究所）

第一节　历史经验

我国农业生产历史悠久，祖辈相传，积累了十分丰富的经验。我们对于农业生产技术的发展史还没有比较详细的了解，关于积肥、施肥的历史经验也知道得很少。但尽管如此，在周代的农业生产中就已使用肥料，这一点则是在古书中可以查考的。

《礼记月令》中载有："土渊溽暑，大雨时行，烧剃行水，利以杀草，可以粪田畴，可以美土疆。"提到刈除杂草，烧成灰，可以肥田。《孟子》："百亩之粪，上农夫食九人也。"明代王象晋《群芳谱》内对此作了解释："积地莫若积粪，地多无粪枉费人工，故孟子不曰百亩之田，而曰百亩之粪。"礼记的成书时代尚难确定，而孟子的年代是可以查考的，也就是说，至少二千四百年前，我国农民已经使用肥料，并且认识到它的重要意义。

西汉《氾胜之书》著于公元前一世纪，开篇第一句："凡耕之本，在于趣时、和土，务粪泽，早锄、早获"，已经指出了土、肥、水、耕作管理等是农业生产的重要环节；对于种植各种作物都强调施用肥料，例如："区种粟二十粒，美粪一升，合土和之"；："区种大豆法，……；其坎成，取美粪一升，合坎中土，搅和，以纳坎中，临种沃之，坎三升水，坎内豆三粒"。《氾胜之书》中的肥料包括蚕矢和洞中熟粪。

后魏贾思勰的《齐民要术》著于公元六世纪，书中述及肥料的种类已很广泛，包括蚕矢、熟粪、坏墙垣、绿肥、豆粪等。并且对于施肥法和肥效有较具体的叙述。如说："凡美田之法，绿豆为上，小豆、胡麻次之，悉皆五六月中穗种，七月八月犁掩杀之，为春谷田，则亩收十石，其美与蚕矢熟粪同"，"凡谷田，绿豆、小豆底为上"，"故墟、新粪、坏墙垣乃佳"。

元《王祯农书》著于公元十四世纪，其中粪壤篇开篇写道："田有良薄，土有肥硗，耕农之事，粪壤为急。粪壤者所以变薄田为良田，化硗土为肥土也。"明确地指出了施肥与培肥土壤的关系。书中还指出，从古代的土地轮休制到后来的土地常年利用，施肥是保持土壤肥力的必要措施。书中写道："古者分田之制，上地家百亩，岁一耕之；中地家二百亩，间岁耕其半；下地家三百亩，岁耕百亩，三岁一周。盖以中下之地，瘠薄硗确，苟不息其地力，则禾稼不蕃，……所有之田岁岁种之，土敝气衰，生物不遂。为农者必储粪朽以粪之，则地力常新壮而收获不减。"这部书中所记述的肥料种类很多，说明从后魏到元，积肥施肥的技术有很大的发展。书中的肥料种类包括踏粪（厩肥）、苗肥（绿肥）、草粪（青草堆肥和稻田压青草）、火粪（即熏土）、各种动植物残体，石灰和泥肥（沟泥）等；书中并指出施用石灰的作用是"下田水冷，亦有石灰为粪，则土暖而苗易发"，证明当时已开始使用矿质肥料。

但在上述古农书中，都没有提到追肥法。直到明季，即公元十七世纪，王象晋著的《群芳谱》才开始有底肥与追肥相结合的施肥法的叙述。如写种稻："犁田须犁耙三四遍，青草或粪穰灰土厚铺于内，䆉烂打平"（施底肥）。"耘稻，扬稻后，将灰粪或麻豆饼屑撒田内，用水耘去草，……浇粪时，待雨弗致缺水，则稻苗不竭"（中耕、灌溉和追肥）。又如写种白菜："五月上旬撒籽，用灰粪盖，粪水频浇，密则芟之。"书中对各种作物都明白地写出施底肥，但只是对个别作物，才写施追肥，并没有对每一种作物都指出了底肥和追肥的配合使用，可见当时候施追肥还不是普通的措施。

*原载于《中国农业土壤论文集》，上海科学技术出版社. 275~296，1962.

与《群芳谱》同时的《沈氏农书》对水稻施肥法有比较详细的记载，写到水稻施底肥和追肥技术，称底肥为“垫肥”、追肥为“接力”。对于追肥认为要按水稻生长发育季节施用，而且要看苗追肥。记载如下：

“下接力须在处暑后苗做胎时，在苗色正黄之时，而苗色不黄，断不可下接力，到底不黄，到底不可下也。暑苗茂密，度其力短，俟抽穗之后，每亩下饼三斗”。

十八世纪中叶编写的《授时通考》中引用了《劝农书》一则，有如下记载：“用粪时候亦有不同，用之于未种之先，谓之垫底，用之于既种之后，谓之接力。垫底之粪在土下，根得之而愈深；接力之粪在土上，根见之而反上。故善稼者皆于耕时下粪，种后不复下也。大都用粪者要使化土，不徒滋苗。化土则用粪于先，而使瘠者以肥。滋苗则用粪于后，徒使苗枝畅茂而实不繁。故粪田最宜斟酌得宜为善。”这里提出两点：①底肥的重要性；②肥料不仅肥苗，而且肥土。对于追肥似乎还没有肯定它在施肥措施中的重要意义。《劝农书》的写作年代不详，明末的几本农书中都没有提到，可能是明末清初的著作。

由上所述，可见我国施追肥可能开始于十五、十六世纪。从那时候起，我国肥料学说已经形成下述几项特点：①广搜肥源，肥料种类繁多，主要是有机肥料，但也已经使用石灰等矿质肥料。②肥料经过一定的腐熟过程后才施用。③以底肥为主，追肥为辅。④追肥按生长发育阶段，看苗施肥。⑤施肥不仅仅是直接营养作物，而且是培肥土壤。

自十五、十六世纪以后，四五百年内通过反复实践，肥料学学说不断地丰富和提高，这是我们的祖先遗留给我们的宝贵遗产，是我国社会主义农业建设的科学技术基础的一个重要方面。

从清末到解放以前，农业科学特别是肥料科学方面，主要是把资本主义国家的一套理论搬了过来，脱离生产实际，不能反映我国具体情况，以致我国农民的施肥经验，未能总结提高。

解放以后，党号召农业科学研究工作者深入农业生产实践，把总结丰产经验作为农业科学工作者的重要任务。对农业科学技术采取两条腿走路的方针，专业研究与群众性科学研究相结合，开展了群众性的农业科学研究活动，推动了农业科学技术的迅速发展。在党的正确领导下，全国各地分级地总结了先进生产经验和专业研究成果，通过提炼，使之上升为理论，进一步指导生产。在施肥法方面，经过十年来的总结和提炼，形成了一套以有机肥料为主、无机肥料为辅，基肥为主，追肥为辅的农作物施肥方法，并且不断地发展着，提高着。

第二节 土壤中植物营养料的转化和施肥对于提高土壤肥力的作用

（一）植物的营养料

植物在土壤上生长，从种子发芽起，到成熟结实止，在整个生长和发育的过程中，必须吸收足够的光、热、水、空气和各种营养物质。通过光合作用，植物将吸收来的二氧化碳和水合成糖类，这是植物物质合成过程的第一步。植物需要的二氧化碳，主要靠叶面从空气中吸收；而空气中的二氧化碳，则主要是由于生物（动物、植物和微生物）的呼吸和各种有机物质的分解所产生的，其中微生物分解各种有机物质（动植物的尸体、各种肥料和土壤中的有机物质）起着十分重要的作用。在施用大量有机质肥料的土地上，由于微生物腐烂分解作用而产生的二氧化碳，增加了空气中的二氧化碳浓度，有利于地面农作物的光合作用。此外，最近的研究工作证明，植物根还能够从土壤中吸收碳酸根（碳酸根和重碳酸根是二氧化碳在土壤溶液中的存在形式），从而部分地供给了光合作用所需要的二氧化碳。

植物合成糖类物质以后，再进行吸收作用，吸收氧气（从地表面空气和土壤孔隙中的气体吸收），将糖类分解、转化，并且与氮、磷、硫、钾、钙、镁、硅、氯、钠、铁、硼、锰、铜、锌、钼 ，以及某些现在还在探索中的放射性元素等，经过十分复杂的变化，合成为植物体内细胞组织和器官的各种成分。这些成分在不断的新陈代谢中形成，长大、转变和进行着各种生命活动

氮、磷、硫、钾、钙、镁、铁等元素需要量大，称为植物营养所需的大量元素（还包括二氧化碳、水、氧气中所含的碳、氮、氧三种大量元素）。硼、锰、铜、锌、钼等元素的需要量很少（植物体只含

百万分之几或不到百万分之一），以及一些现在尚不明确的、需要量更少的放射性元素，称为微量元素。这些元素主要是由植物根从土壤中吸收。

土壤中所含上述元素的量有多有少。有些能充分供给植物的需要，因此在农业生产中不成问题。有些元素则常成不足，必须适当补给；其中尤以氮、磷、钾三元素最为突出，称为植物营养三要素。在某些情况下，其他大量元素或微量元素也可能供应不足，需要适当补给。

植物利用这些营养成分是以溶解性的或气态的无机化合物状态吸收的，因此，植物营养称为无机营养，和动物不一样，动物主要是利用有机化合物养料。

但最近的研究工作又提出了，植物营养只能说基本上是无机营养，土壤中还供给植物一些维生素类和胡敏酸等复杂有机物，这些有机物对于作物丰产也起着一定的作用。

（二）土壤中各种植物营养物质及其转化

土壤含有各种植物营养成分，它们以十分复杂的情况存在于土壤各个组成成分中。就一般的农业土壤而言，百分之九十五至百分之九十九的土壤固体是矿物质，包括砂粒，粘粒和碳酸盐等。

砂粒的成分主要是石英粒（氧化硅），对于植物营养无直接意义，而对于土壤的排水透气性质却十分重要，土壤的砂粒愈多，砂性愈强，排水透气性也愈强。粉砂粒和粘粒的矿物质成分复杂，含有多种植物营养成分，这些成分是以不溶解于水的化合物状态存在，不能直接被植物吸收；只有在土壤中经过物理的、化学的和生物的风化作用，才逐渐转化为植物可以吸收的溶解性化合物状态。这种风化作用是缓慢的，对于植物营养料的供应，是比较次要的，但也是一种不可忽视的作用。在土壤中还进行着与风化相反的作用，称为固结作用。例如：南方红土中，溶解性的磷酸盐转化为不溶解的、不能被植物吸收的磷化合物；在北方石灰性土壤中，部分钾素被矿质粘粒所吸收固结，成为不能被植物所吸收利用的状态。这种相反的作用，在土壤中不是主导的物质变化方向，然而在类如上述的一定情况下，对植物营养起着不利的作用。

土壤黏粒除供给植物营养成分外，对于土壤的排水透气性也有重大的影响。土壤中粘粒愈多，土壤愈黏重板结，排水透气性也愈差。土壤中砂粒和粘粒的比例决定土壤的质地，对于土壤的耕性具有十分重要的作用。

一般农业土壤中含百分之一至百分之五的有机质（以腐殖质为主），其中也含有多种植物营养成分。有机质含量虽少，但对植物营养所起的作用很大，因而土壤有机质或腐殖质含量在很大程度上决定着土壤的肥力。土壤有机质在土壤微生物和高等植物（农作物）的生命活动下进行着旺盛的生物循环，而在这循环周转中，不断地在很大程度上满足了植物营养的要求。关于生物循环对于植物营养和土壤肥力的意义，下面还要作较详尽的讨论。

土壤有机质除供给植物营养物质外，对于土壤的物理性状和耕性也具有重要的作用；特别是土壤有机质中的主要成分腐殖质，对于土壤构造（土壤团粒的形成）、耕性和排水透气性起着重要的调节作用。由于以不同比例的砂粒和粘粒（质地）所形成的土壤耕性可受腐殖质含量的影响而得到调节，因此提高土壤腐殖质含量，可以使砂土变黏，使黏土疏松，改善土壤耕性和保水保肥的性质。

土壤黏粒是胶体物质，土壤腐殖质也是胶体物质，它们结合在一起形成土壤的胶体体系。在土壤的胶体表面吸附着很多的植物营养物质（主要是阳离子，如氮、钙、镁、钾、铵、钠等等）。这些吸附物质是植物可以直接吸收利用的。实际上，土壤中速效性的钾素养料主要就是胶体上吸附的钾素离子。

土壤黏粒的多少、黏粒的性质和腐殖质的多少，以及土壤胶体吸附性能的不等，对于土壤保肥性质有重要作用。砂土粘粒少，吸附性弱，保肥性差。黏土吸附性强，保肥性佳。北方土壤的黏粒主要是伊利石（又称云泰石或水化云母），吸附性、保肥性比较强。南方红黄土的黏粒主要是高岭石、三水铝矿和赤铁矿，吸附性保肥性比较差。腐殖质的吸附性最强，以等重量比较，每一百克高岭石的吸附当量只有几毫克，每一百克伊利石的吸附当量为几十毫克，每一百克的腐殖质的吸附当量达二百至三百毫克。因此，土壤腐殖质含量，即使仅增加百分之一，对土壤的保肥性能就有很大的改善。

另外，土壤水中也溶解着一些植物营养物质吸附在胶体表面。一般的农业土壤水中含百分之零点一左右的溶解性物质，溶解在土壤水中的营养物质是植物可能直接吸收的。

在土壤中、土壤水中的溶解性植物养料与胶体吸附的植物养料由于物理化学的作用而相互转化着。它们又与有机质中和矿物质颗粒中的植物养料由于生物作用和风化作用而相互转化着。因此，土壤是物质转化（运动）十分活跃的自然体。

前面提到、土壤中物质的生物循环对于植物营养起着最为重要的作用。这种循环可用下图式说明：

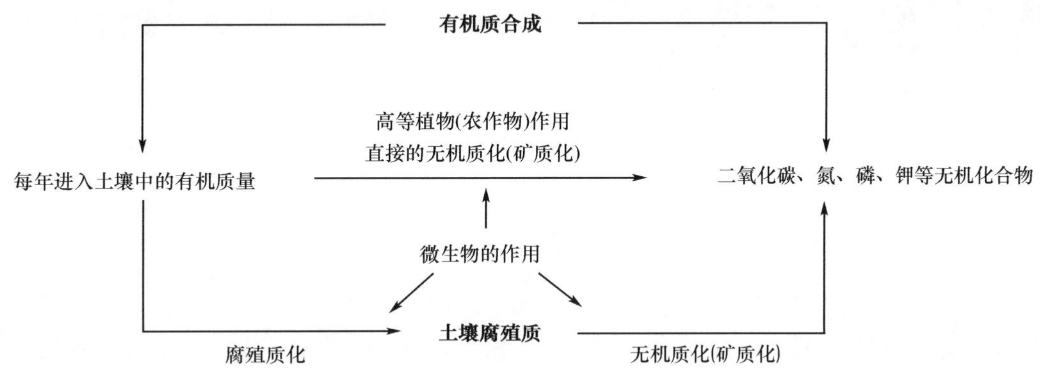

要提高农作物产量，必须加强这个生物循环的过程，控制它的转化方向和相对强度，使土壤腐殖质的含量逐渐增长，从而改善土壤的结构、耕性、保水、保肥性质；同时要求矿质化过程不太快，也不太慢。矿质化太慢，会使植物得不到必要的无机养料；矿质化太快，在短时期内产生多量的无机养料，植物一时吸收不了，会招致淋溶损失，或使植物营养失调，引起徒长、倒伏。并且矿质化太强也意味着土壤腐殖质含量的相对损失，造成土壤肥力水平的降低。

解放以来，农作物产量不断增长，这与土壤肥力不断提高有着密切的关系。土壤肥力提高的原因是多方面的，加强生物循环，并使它向着有利于生产的方向进展是主要的原因之一。轮作制度的改善，土壤耕作管理技术和作物栽培技术的提高，带动了整个生物循环的加强；大力积肥、造肥和合理用肥，不仅将大田中生长的有机物质的绝大部分直接或间接地归还了土壤，而且把山青、湖草、塘泥和商品肥料（包括饼肥、矿质肥料和化学肥料等）施入土壤，显著地增加了土壤中的有机质和矿物质成分。这些成分进入土壤生物循环中，更加强了土壤的生物循环。土壤肥力和农业生产，就在不断加强的生物循环过程中逐年提高。

以下对于生物循环中的某些各别方面作简单的阐述，以便更具体地理解土壤肥力与施肥的关系。

（三）各地腐殖质的形成和分解

腐殖质是土壤微生物分解各种动植物有机物质（腐烂分解）后，再合成的特殊物质。它是胶体状态的高分子有机化合物。微生物分解各种有机物质，一方面把它们直接矿质化，产生各种植物营养料；一方面进行腐殖质化而合成腐殖质。同时腐殖质本身也不断地被土壤微生物所分解和矿质化，因此土壤中的腐殖质是在下断地形成和分解之中而存在的。土壤中腐殖质含量的增多或减少，决定于每年土壤有机质转化成为腐殖质的量和每年腐殖质被矿质化的量的差额；它受每年加入土壤的有机质数量、土壤的水热条件和土壤微生物活动的影响。

每年加入于土壤中的有机质数量决定于自然条件和生产条件。一般说来，南方加入的有机质虽比北方多，但土壤中腐殖质的含量并不比北方多，甚至于还少些。在气候较冷、湿度较大、土壤透气性较差的地带，腐殖质化和矿质化的相对强度比在气候较暖、湿度较低、土壤透气性较好的地带要大些。因此在东北地带，土壤腐殖质含量比较高。在东北地带的生荒草地，由于每年产生的有机物质全部归还于土壤，荒草的覆盖度大，土壤未经耕作；比较紧密，排水透气性差，每年腐殖质化的总量，腐殖质化和矿质化的相对强度都比开垦的土地为大，因此腐殖质的含量也比耕作的熟地高。显然，合理的耕作制度和耕作方法，尽一切可能积肥、保肥和施用大量有机肥料，就能使耕作土壤熟化中的腐殖质含量保持在较高的水平上。保持东北耕地土壤的腐殖质含量在百分三至百分之五是完全可能的。

华北地区气温较高，湿度较低，在久经耕作的土地上，土壤腐殖质含量比较低，一般在百分之一点五以下。华北土壤中腐殖质含量的高低，在很大程度上影响着土壤肥力和生产量的高低。因此，在华北采取各种措施，大量施用有机肥料，是提高土壤肥力和争取农业高产的根本问题之一，一九五八年丰产田的生产实践充分地证实了这一点。

西北地区气候更为需干旱。在最干旱的季节，表层土壤水分很少，微生物活动微弱，植物有机质（如植物根、绿肥等）在土壤中或保持原状，或分解得十分缓慢，这对土壤肥力的提高作用很小。在湿季，土壤透气情况仍较好，有机质的矿化较强而腐殖质化则较弱，尤其是在灌溉地区，矿质化的强度更大，土壤腐殖质含量比华北更低。因此，在西北地区大力积肥、保肥和施肥更是提高土壤肥力和农业生产量的重要措施。栽培多年生牧草也有十分重要的意义。根据绥德水土保持科学试验站的研究资料，种草木樨地土壤腐殖质提高百分之零点二五以上，种苜蓿地提高百分之零点五五，后作物的产量显著提高。

长江以南，从北到南，基本规律相似，只是在数量上所区别。在雨水足、气温高、自然植被茂盛的情况下，每年土地上有机质的生成量大，腐殖质化和矿质化的强度也大，土壤中含有一定的腐殖质量，但不很高（百分之二至百分之四）；若自然植被破坏，每年有机质生成量降低，微生物活动继续旺盛，土壤腐殖质含量消耗很快，在红土荒地上，腐殖质含量可以降低到百分之零点五以下。在久经耕作的土地上，如何大量积肥、保肥、施肥，尽可能地将有机物质加到土壤中去，这是我国南方保持和提高土壤肥力，争取农业丰产的基本措施。实际上，南方耕地的腐殖质含量一般比荒地高，但进一步的增产还需更大的努力。在提高腐殖质含量的同时，还要加强生物循环，一方面达到较高的腐殖质含量水平，一方面进行旺盛的有机质化和矿质化，在循环周转之中，达到高额丰产。

水稻田在泡水情况下，由于透气性较差，一般比旱地土壤的腐殖化程度高些。但是水稻田大都只是季节性的灌水，处于干湿相间的状态。经年泡水的烂泥田，腐殖质含量一般较高。在水稻的生长时期，短期的落水晒田能加强矿质化作用。

南方水稻田种植冬季绿肥，对于培育和提高土壤腐殖质的作用比较弱，然而对于加强生物循环、加强有机质的合成和矿质化、更多地供应植物营养料，却具有十分重大的意义。大量施用猪牛粪、堆肥和塘泥，对于提高土壤腐殖质含量有较大的效果。村落住屋周围的肥料来源多，因而土壤的培肥和改良效果是比较显著的。

（四）植物营养元素的生物循环

空气中约有百分之八十的氮气，但它不能被绝大多数动植物吸收利用。一般岩石矿物中不含氮素成分。地面上生物体中的氮素成分主要是直接（植物）或间接（动物）从土壤中吸收的，而土壤中的氮素成分则主要是靠土壤微生物的生命活动而逐渐累积。有些微生物如根瘤菌、固氮菌、固氮蓝藻等，在适宜的条件下能够吸收利用空气中的氮气，先使之转化为它们身体中的蛋白质成分，然后进入土壤氮素的生物循环中去，使土壤成为含有氮素养料的自然体。一般农业土壤约含氮百分之零点零五至百分之零点二。土壤含氮量与土壤的腐殖质含量有较稳定的比例关系，一般为二十比一。

土壤中氮素的生物循环大致如下：植物和土壤微生物在其生长和发育过程中将土壤氮素合成蛋白质——蛋白质是含氮的有机化合物，约含氮百分之十六，是生物体的基本成分。在这些生物体死亡以后，有的残留在土壤中，有的通过施肥归还到土壤中，受到其他微生物的腐烂分解作用，一部分转化为腐殖质中的氮素成分，一部分矿质化而成为铵盐。蛋白质分解产生铵盐的作用称为氨化作用，是氨化微生物生命活动的结果。铵盐形成后，绝大部分以离子形态吸附在土壤胶体上，不致淋洗损失。不论是存在土壤水中或吸附在土壤胶体上的铵离子，都是农作物可以直接吸收的良好氮素养料。除蛋白质直接氨化外，腐殖质中的氮素成分，或早或晚也要氨化，产生铵盐。

铵盐形成后，一部分直接被农作物吸收，一部分在土壤中被另一类微生物——硝化细菌——的硝化作用而氧化为硝酸盐。硝化作用分两个步骤：先形成亚硝酸盐，然后形成硝酸盐。在透气良好的旱作土壤中，硝化作用的速度比氨化作用快，因此土壤中主要的矿质氮化物是硝酸盐。硝酸盐也是农作物的良好氮素养料，它的缺点在于它绝大部分溶解在土壤水中而不吸附在土壤胶体上，淋洗损失的比例比铵盐大得多。

由于土壤透气良好是各种旱性作物旺盛生长所必需的土壤条件，所以硝化作用的旺盛进行不认为是无益的微生物转化过程，一方面它同样的能为农作物吸收，一方面它也是土壤透气条件良好的标志。

水稻土在泡水条件下只含微量的硝酸。过去一向认为，硝化作用是氧化作用，在泡水缺氧的水稻田中，只有在表面一薄层（不到一厘米厚的氧化层）和水稻根的表面进行，然而最近的研究工作（华中农学院，一九五九）指出，在整个水稻土根层土壤中有很多的硝化细菌进行硝化作用，但是由于反硝化作用的旺盛进行，形成的硝酸很快地就损失了。因此，水稻土中的硝化作用实际上对土壤肥力是不利的。

反硝化作用是反硝化细菌将硝酸盐还原为亚硝酸盐、氮气和铵盐的作用。氮气是呈不能为农作物吸收利用的氮素形态，所以这种作用对于土壤肥力是不利的。反硝化作用在泡水的水稻田中不透气、缺氧的条件下进行十分旺盛，但反硝化细菌只能将硝酸盐还原为氮气，不能将铵盐转化为氮气，故水稻土中的铵盐比硝酸盐稳定，肥效高。

前面提到，根瘤菌、固氮菌和固氮蓝藻等有固氮作用，能将空气中的氮气（一般农作物不利用的）转化为土壤中氮素成分（农作物可以利用的），这不仅是土壤中氮素成分最根本的来源，而且它们的活动和在农业生产中有意识地加以利用，对于提高农业生产有十分重要的现实意义。豆类植物与根瘤菌共生固定氮素，在农作物中占有特殊地位，被认为是肥田的农作物。不仅豆类绿肥作物有重要的肥田作用，各种豆类作物都有或多或少的肥田作用，对后作物有显著的好处。固氮菌生活在土壤中，固氮蓝藻生活在水田中，它们的生命活动提高了土壤的氮素含量。根瘤菌肥料和固氮菌肥料的研究、生产和使用已取得不少的成就。固氮蓝藻也已成为科学研究的一个重要对象。

土壤中的磷素原本存在含磷的成土母质中，一般酸性母质含磷量低，碱性母质含磷量高。因此，由酸性岩石母质形成的土壤比从碱性岩石母质形成的土壤含磷量低些。这些母质中所含的磷素成分，植物不能直接吸收利用，只有风化以后才转化为植物能吸收的状态（称为可给性磷）；形成可溶性磷以后，进入生物循环，植物吸收可给性磷，成为植物体中的有机质状态的磷。有机质状态的磷归还到土壤中后，由于微生物的分解作用，再转化为植物可以吸收的可给性磷供下一季植物吸收利用。农业土壤中约有五分之一至二分之一的磷素是以有机质状态存在，这种磷素的生物循环是对农作物生长最重要的磷素养料条件。

土壤中的钾素原本也存在于成土母质中（正长石、云母等），经风化作用转化为能溶解在土壤水中和吸附在土壤胶体上的钾盐，成为植物可以吸收的形态（可给性），植物吸收可给性的钾素以后，以钾盐的状态存在于植物体中，当植物体归还土壤中，细胞组织破坏以后，钾盐就再转化为溶解在水中或吸附在土壤胶体上的状态，供后一季植物吸收利用。

以上是植物营养三要素在土壤中循环周转的情况，其他植物营养元素也以各种不同的形式在土壤中循环周转者，农业技术措施的任务就在于调节和加强植物营养元素的循环周转中的各个重要环节，满足农作物的需要。

现代的工业技术力量对于农业生产，对于增加植物营养元素的供应，起着十分重大的作用。氮、磷、钾等化学肥料的制造和施用，已经成为提高和调节植物营养条件的主要力量之一。解放以来，党和政府十分重视化学肥料的制造和推广施用，社会主义工业化的高速度发展，更为化学肥料的生产提供了有利的条件。

（五）施肥对于改善土壤养料供应条件的意义

根据生产经验，采用各种农业措施来调节和补充土壤供给植物养料的条件，充分而适当地满足农作物生长和发育各个时期的营养需要，是争取作物丰产的必要前提，其中施肥是一项最重要的技术措施。施足底肥（主要是肥效较缓慢的有机质肥料，或矿质肥料），是在原来的土壤肥力基础上提高土壤的供肥水平；施种肥和分期追肥，是在农作物需肥最多而土壤供应不上的时期予以及时的补充。底肥、种肥和追肥合理配合，可以收到更大的效果。施肥不仅可以在总量上和生长季节上更好地满足农作物的需要，而且可以调节各种肥料之间的比例。

（六）施肥对于改善土壤理化生物性质和耕性的意义

施肥的任务不仅在于营养植物，对于改善土壤的理化生物性质和耕性也都起着重要的作用。施用大量有机肥料，一方面由于矿质化而供给植物以养料，另一方面由于腐殖化而增加了土壤腐殖质成分。对于绝大多数耕作土壤来说，土壤的腐殖质含量对于土壤肥力有决定性的意义。提高腐殖质含量可以改善土壤的胶体特性和土壤的保水保肥性能，尤其能改善土壤的结构性，从而改善土壤的排水透气性能和耕作性能和耕作性质。施用热性的有机肥料还能够提高土温。施用石灰可以调节土壤酸性，并能使冷土变热。石灰和腐殖质胶体相结合也能够改善土壤结构。施用石膏能改良苏打盐土，并防止水稻热烧（其作用实质尚待阐明）。

施用含泥质或砂的肥料有改善土壤质地的作用。在砂土上施用塘泥不仅供给了植物养料。还改善了土壤质地。连续施用大量的土粪或塘泥还有厚土层的作用。根据农民经验，每亩施四万斤塘泥或土粪约可增厚土层一寸。

施用细菌肥料的意义在于人为地改变土壤微生物的种类和数量，加强有益微生物的活动，抑止有害微生物的活动，从而更有效的调节和加强土壤的生物循环。

第三节　以有机肥料为主、无机肥料为辅，基肥为主，追肥为辅的施肥体系

我国农民在长期的农业生产实践中，积累了十分丰富的积肥、制肥、保肥和用肥的经验。解放后，在党的正确领导下，专业的科学研究与群众的科学研究相结合，紧密地联系生产实践，在全国范围内逐年总结经验，加以提高，已经有了一个比较完整的施肥方法。它的特点表现在：①有机肥料为主、无机肥料为辅；②基肥为主、追肥为辅；③肥料种类多和多种肥料的配合施用；④各种施肥技术措施的高度科学性。

（一）以有机肥料为主

以有机肥料为主体、配合施用无机肥料，对我国农业生产具有十分重要和多方面的意义。首先，它保证了土地肥力稳定而不断的提高。世界各国的农业生产历史都充分的证明了施用有机肥料，能保证土地肥力的不断提高。每年从土壤上收获的农作物，以各种形式几乎全部的归还土壤；另外，施入塘泥、草炭、山青、湖草等非农田生长的有机物质，使每年加入土壤的植物营养物质不少于或者大于收获物的总量。这些营养物质通过土壤中微生物的生命活动和农作物的生长，进行着旺盛的有机质化和矿物质化过程。

以有机肥料为主不仅能供给农作物生长所需的营养成分，而且还能供给土壤微生物所需的有机养料，保证土壤微生物的高度活跃进性。在施用大量有机肥料的土壤中，要比没有施用肥料或只施化学肥料的土壤中的微生物数量多而生命活动也强得多。微生物生命活动的加强，与土壤中植物养料的供应水平有着密切的关系。而丰富的有机质供应，是土壤微生物进行旺盛的矿质化和产生农作物所需的植物养料的重要条件。同时，土壤微生物在进行生命活动的过程中。还制造了与土壤肥力有密切关系的土壤腐殖质。世界各地的研究工作都是证明了，大量而连续地施用厩肥可以提高土壤腐殖质含量，并相应地提高土壤肥力水平。

在这问题上我国的生产实践经验更为丰富。把黄（红）土变为黑土，实质上就是把腐殖质含量少、肥力低的土壤培育成为腐殖质含量多、肥力高的土壤。例如长江下游的白鳝土，按其腐殖质含量的多少而可以分为乌白鳝土、普通白鳝土和死白鳝土三等。普通白鳝土的耕性比较好，干湿都可以耕作；但由于粉砂性强，灌水把田后土粒下沉，土质不柔，插秧比较困难（不能插清水秧），且土壤保水抗旱性、保肥性较差，植物养料含量也比较低，虽然施肥见效快，但需肥较多。死白鳝土是普通白鳝土因长期不培育、施肥量过少、土壤腐殖质消耗而形成的。乌白鳝土则是种水稻的一等土壤，它是普通白鳝土经过不断的较大量的施用有机肥料培养而成，腐殖质含量较高（大于百分之二），土壤的保水、保肥性状已

经得到了改善，因此没有土粒下沉、难以插秧等缺点。

施肥以有机质肥料为主，与以化学肥料为主相比，前者的优点是肥料中含有完全的而且是平衡的植物营养成分，因为各种有机质肥料基本上是完全肥料，而化学肥料只含一种或少数几种植物营养元素，由于肥料供给的不平衡，容易导致农作物的营养失调。

（二）基肥为主、追肥为辅

以有机肥料为主的施肥原则，势力必导致以底肥为主、追肥为辅的施肥方法。因为大多数种类的有机肥料的肥效是比较迟缓的，施入土壤中后，在土壤微生物的作用之下，逐渐而不断地产生农作物所能利用的营养物质，这样就成为农作物营养的经常供应基础。在这经常供应基础之上，再按照农作物各个生长发育时期的不同需要，在需要最多的时期，及时地追施速效性肥料，以满足农作物对养料的需求。

根据一九五八年全国水稻丰产科学技术交流会议的资料，基肥占总施肥量的百分比：长江流域双季早稻以百分之八十左右为宜，双季晚稻以百分之七十至百分之八十为宜，一季晚稻以百分之六十左右为宜；北方及长江流域的中熟稻种，以百分之七十至百分之八十为宜；华南地区高温多雨，肥效快而不能持久，双季早稻以百分之七十左右为宜；双季晚稻以百分之六十左右为宜。

华北地区种植冬小麦，某肥施用土粪，一九五七年每亩为二千至四千斤，一九五八年略有增加。山东农业科学研究所总结一九五九年大面积丰产经验，施基肥、追肥与产量关系如下：

每亩小麦产量范围	基肥（圈肥）	追肥
100~200 斤	0~1000~3000 斤	人粪尿 1000 斤
200～300 斤	1000～5000 斤	硫酸铵 10～20 斤
300～400 斤	5000~10 000 斤	硫酸铵 20～30 斤
400～600 斤	10 000~25 000 斤	人粪尿 1000～3000 斤
		硫酸铵 30～40 斤
600～800 斤	15 000~35 000 斤	硫酸铵 20～60 斤

棉花地一般基肥用量点总施肥量百分之六十至百分之八十。例如四川简阳试验站一九五八年的施肥试验，每亩收籽棉三百五十七至四百五十三斤，施用猪粪尿二千斤，硫酸铵零至六十斤，其中以猪粪尿作为基肥。

必须肯定，以有机肥料为主、无机肥料为辅，基础为主、追肥为辅施肥方法，完全不降低追肥或矿质肥料和化学肥料在提高农业生产上的重要意义。

没有适当的追肥，要完全靠基肥逐渐产生肥效来满足农作物生长的最大需要是不可能的。基肥逐渐产生的肥效不会和农作物在整个生长发育过程中的需要完全符合。根据华中、华东四十八个水稻试验田的平均数据：不施肥的每亩产量四百三十六斤；只施有机质基肥的亩产五百四十一斤；有机质基肥加耖口肥的，亩产五百八十二斤；有机质基肥加分蘖肥的，亩产五百八十九斤；有机质基肥加分化肥的，亩产六百一十六斤。据华中农业科学研究所一九五七年一季晚稻的试验，以一千六百斤厩肥作为底肥，三十五斤硫酸铵作为耖口肥，不另施追肥，亩产七百零八斤，而同量的厩肥、某肥和硫酸铵用量，但将硫酸铵分别作为耖口肥、分蘖肥、分化肥和孕穗肥，亩产量达八百四十七斤。实际上，解放以来随着基肥施量的不断提高，各种农作物的追肥措施不是更简单化，而是更复杂更细致了。

国家肥料工业的迅速发展，化学肥料施用量的不断提高，充分地说明了施肥以有机肥料为主，丝毫不减弱无机肥料和化学肥料在农业生产上的意义。相反地，为了争取更大的丰收，土壤中现有植物养料不是太多，而是很不够，凡是能够改善土壤养料供应的手段，不论是施用有机肥料、无机肥料或化学肥料都是相辅相成，而不是相互排斥的。这里，国家雄厚的矿产资源和巨大的工业技术力量，都是提高农业生产的最好保证。

（三）肥料种类多，多种肥料混合施用

要保证充足的有机质肥料，必须千方百计挖掘肥源。我国广大农民在积肥、制肥、保肥和用肥方面积累了丰富的经验。凡是可能获得的有机物质材料几乎都可作为肥料，或者直接施用、或者沤制后施用。以牲畜粪尿堆积沤制成的土粪和厩肥（栏粪），人粪尿、山青和湖草，以植物物质为主的堆肥和氹肥，草炭和塘泥，海肥和各种杂肥，一年生和多年生的绿肥和牧草，在我国分布很广，在农业生产上起着极其重要的作用。

我国施肥技术的特点，不仅在于施用的肥料种类多，而且采取同时混合施用几种有机质肥料的办法。多种肥料混合施用是广搜肥源的必然结果，它对于土壤肥力和农作物生产有哪些特别的意义，还有待进一步研究。根据生产实践，多种肥料混合施用比单纯施用一种有机质肥料效果更好。目前我们体会到的，至少有下列两点好处：①多种肥料混合施用，比单施一种肥料，营养物质更为完全，更为平衡。②虽然大多数有机肥料都是迟效性肥料，但其产生肥效的早迟快慢还是各不相同的。例如，在长江流域，晚春施用绿肥和青嫩湖草，分解很快，肥效猛而不能持久；用量少，后期肥效不佳，用量多，前期肥效过猛。相反的，使用塘泥或冬季沤制的堆肥，分解很慢，肥效迟而持久，对后期肥效有保障，而前期肥效很差。因此，为了争取高额丰产，保证肥效足、稳、持久，不仅在基肥、种肥和追肥之间要适量选择适合的肥料，配合施用，而且基肥也要多种肥料混合施用。

（四）施肥措施的高度科学性

我国农民祖代相传，逐渐积累和丰富的施肥技术措施具有高度的科学性，这不仅表现在针对各种农作物形成了一套比较完整的施肥制度，而且在具体实施施肥制度时能与一定的地点和时间条件相结合，例如看天、看地、看苗的施肥原则就是一定的施肥制度和特殊条件有机结合的体现。

看天施肥：是针对一定的气候条件（温度和湿度）施肥。例如，湖北省江汉平原一季晚稻和插秧较迟的中稻，用青嫩湖草做基肥，在插秧前十天到半个月将湖草压入土中，这时天气已暖，分解很快，到插秧时，湖草已经半腐熟，既能产生肥效，又避免了初施湖草时过旺的发酵阶段。早稻田或插秧期早的中稻田，一般不用湖草做基肥。因为开始时气温低，分解慢，肥效很小，接着气候急剧变暖，如插秧后气温猛烈上升，湖草不是在插秧前而是在插秧后进入猛烈发酵阶段，产生大量的气体和一些对于植物有毒害的物质（如硫化氢等），会伤害秧根，使秧苗坐蔸。早稻和插秧期早的中稻一般以厩肥为主。厩肥经过冬天堆制，初期过旺的腐解阶段在堆制时期度过，施入土壤中后，肥效比较稳定。

又如山西运城棉花劳模郭震西的丰产施肥技术措施是：为了既要保证充足的养料，又要避免在高温多雨的季节（六月下旬到七月下旬）因施肥不当而造成徒长，一九五八年在以土粪和过磷酸钙为基肥的基础上，苗期少施氮肥，只在五月末和六月初施少量硫酸铵，这种肥料肥效快，不持久，用量少，只满足了苗期的需要，而没有多余的肥效带进雨季去。到七月上旬（仍在雨季中）施用了少量肥效不很猛的腐熟马粪、棉饼和过磷酸钙，这样既适当地补充了肥料，又不致使氮肥肥效过猛。到七月末，雨季将结束，雨水渐少，正是棉株开始结桃的时候，猛施了四十斤硫酸铵，充分地满足了棉株的需要，又躲开了雨季，避免了徒长，结果结桃很多。

看地施肥：是针对一定的土壤性质和水热情况施肥。在砂性土壤中，有机质养料的分解较快，肥效较速，但由于保肥性差，肥分容易流失，因此基肥可施半腐熟的肥料，追肥次数宜较多，每次追肥量宜较少。相反的，在黏重的土壤中，基肥宜施腐熟程度较高的肥料，追肥次数可以少些。在砂性土壤中施用大量的塘泥，不但供给农作物以所需的养料，而且有改善土壤质地的作用（砂加泥）。在黏重的土壤中，将种肥混合在砂中做盖籽肥，可以保证幼苗出土，并且长久使用后，有一定的改善土壤质地的作用（泥加砂）。深翻土壤与分层施肥相结合，做到层层有肥，不仅保证了全部根系范围的肥料供应，而且有加速熟化新翻底土的作用。一九五八年全国展开深耕改土运动，各地有经验的农民都强调深耕必须与多施有机肥、分层施肥相结合，加速生土熟化，才能获得增产。

各种厩肥按其成分、含水多少、腐解时产生热量的多少，可以分为凉性和热性两类。例如马粪、驴

粪、羊粪是热性，猪、牛粪是凉性。低湿背阴地带，早春地温低，宜施用热性肥料。而朝阳温热地带，可以施用凉性肥料。此外，农民施用凉性、热性两类肥料时，还根据某些目前还不大理解的性能，例如，湖北种植水稻，有热烧现象时施用凉性的石膏，有冷烧现象时，施用热性的石灰，都有显著的肥效。就目前的初步研究，冷烧是脱肥的现象，大都因气温低，有机质肥料矿化慢，因而养料矿化慢，因而养料供应不足，加施石灰有加速微生物生命活动、加速矿化、改善植物营养条件的作用。一般冷烧也可以施用少量速效性肥料如尿水、硫酸铵等解决。热烧大都是由于有机质肥料施用不当（过多或腐熟不够）而造成的毒害作用，施石膏防治热烧的实际效果、原因和机制，尚待研究。

肥料的品质和数量，肥料的施用方法和土壤耕作情况对于发挥肥料作用和相应的施肥技术有着密切的关系，这里首先谈肥料的数量与浓度的关系。在用肥量不大的情况下，为了保证一定的肥料浓度和靠拢根系的吸收范围，多采取集中施肥的原则。华北农谚“施肥一大片，不如一条线”，生动地说明了这个道理。这一农谚主要是指在条播地上施用土粪。土粪一般是高度腐熟的粪土混合物，直接接触根系，不会伤根，而且能更好地为植物所吸收利用。如果把用量不大的土粪与全部耕层土壤混合，则由于肥料过于稀释，反而不能使作物根系充分利用。而在长江流域，一九五七至一九五八年大量施用半腐熟的厩肥和堆肥，有经验的农民强调使肥料充分散开，并且反复拖耙，做到水土相融。厩肥或堆肥是不能与根系直接接触的，在气温高、腐解旺盛的情况下，直接接触一方面产生的营养成分过浓，会损伤根系，一方面还产生一些对根系有毒的物质（如硫化氢）。散开而反复拖耙后，肥料与土壤充分混合，能做到供肥充分而养分又不过于浓稠。而且，在水稻密植的情况下，作物根系实际上满布耕层，它们从耕层土壤全部而不是从耕层土层局部吸收养料。

深耕如果不相应地增施肥料，则可能由于肥料受深耕的稀释作用不能充分发生肥效。显然的，从耕深四寸加深到耕深八寸，如果肥料用量不变，在分层施肥、土肥相融的原则下，肥料浓度实际上降低了一半，增施一倍肥料，才能保证相同的浓度。中国科学院植物生理研究所（一九五八年）在江苏金山调查六十一块一季晚稻田，在每亩施肥量相当于猪粪肥二十至六十担的范围内，加深耕层并不增产，甚至可能减产。十六块田浅耕小于二十厘米，平均亩产八百一十四斤，五块田深耕二十至三十三厘米，平均亩产六百九十三斤。在每亩施肥量大于一百担的范围内，十一块田浅耕小于二十厘米，平均亩产九百八十八斤。可见深耕必须相应地增施肥料，才能获得增产效果。

同一田块的不同部位，土壤肥力情况也不是完全一致的，尤其是新经改良或整理的田块，差异更大，因此，不同部位的施肥量也应当有所区别。在多年经营同一块田地，充分了解田地的肥力布情况，有区别地施用基肥是完全可以做到的。再进一步，植株在一块田地上生长，由于遗传禀性的差异，和田地肥力的不匀，各植株的生长情况和需肥情况也往往不完全相同。在同一田块上，适时追施肥料，提小苗、赶大苗；在同一地带的不同田块上，对三类苗田重施追肥，消灭三类苗，这对于几年来的农业生产丰收，单位产量不断提高，起着十分重要的作用。

看苗追肥：看苗追肥在于正确诊断农作物在各个生长发育期的健旺形态，而不是简单地追求植株深绿肥大，不适当的深绿肥大不是植株健旺的征象，反而时常是徒长的表现。例如水稻劳模陈永康指出，一季晚稻的生长发育期中有“三黄三黑”的表现，在“三黄”时期健康的色泽是金黄色（即不是深绿色），这时期如果追肥过多，变成为深绿色，势必导致徒长或倒伏。在“三黑”时期，健康的色泽是深绿色，如果表现为金黄色，则是脱肥的象征，势必影响产量。除颜色之外，植株的形态，叶片的大小和竖直状态，都与农作物的营养条件有表里关系，可以作为追肥时期和用量的指示。

在同一田块中，或相邻的几个田块之间，植株生长情况不会完全一致。详细比较，正确掌握健旺植株的形态、色泽、是看苗施肥的工作基础。

少吃多餐：我国农民施肥的高度科学性还表现在“少吃多餐”的追肥原则上。与世界各国的追肥制度对比，我国一般追肥次数多，每次追肥用量少。例如长江流域种棉花采用五施两补的施肥原则，即分基肥、种肥、提苗肥、现蕾肥和结铃肥等五次施肥，另外在苗期后期和开花期再根据植株生长情况适当补施两次。

“少吃多餐”不仅符合农作物不同生长时期的需要，而且更能够发挥看天、看地、看苗施肥的灵活性。同时对于减少速效性肥料的淋失或固结作用也是有利的。在排水畅顺，雨水较多或灌溉农田中，速效氮素肥料，不论是尿水、硫酸铵或硝酸铵，一方面被植物吸收利用，一方面由于微生物的氨化作用和硝化作用，形成硝酸盐，硝酸盐溶解性大，容易随水淋失。“少吃多餐”能够相应地减少淋失。溶解性的磷酸盐肥料（如过磷酸钙）在土壤中的动态和硝酸盐相反，它不会淋失，但容易固结而成为不溶解性的磷酸化合物，因而失去了可给性，植物不能够吸收利用，“少吃多餐”地施用磷酸盐肥料，可以相应地提高吸收率，降低固结率。

速效性化学肥料最宜采用“少吃多餐”的施肥方法。近年来发展的喷射施肥法，使“少吃多餐”可以做得更为细致，受到广大农民的欢迎，推行得很快。

第四节　肥肥土、土肥苗的原理和实践

这里提出肥肥土、土肥苗的原理，并不排斥施肥直接供给农作物吸收的办法，例如施用溶解性的速效追肥和喷施肥料等，而是说在以有机肥料为主的施肥体系中，施用大量的有机质基肥需要按照肥肥土、土肥苗的原理进行，这是生产经验反复证明了的。

在前面已经简要地阐明了，有机肥料不是直接对农作物产生肥效，而是要经过土壤微生物的生命活动，腐烂分解，一方面由于矿质化产生农作物能够直接吸收的养料，另一方面由于腐殖质化产生土壤腐殖质，从而改善土壤的理化生物性质和耕性，换言之，有机肥料只有在它进入土壤的生物循环之后，才产生肥效。在这意义上发展了肥土相融的施肥原则。也就是说，肥土相融不只是指肥料和土壤的充分混合，而且意味着肥料进入土壤的生物循环，与土壤融合成为一个生活的整体。有机肥料并不是直接肥苗，而是先肥土，然后土再肥苗。

为了有利于讨论各种有机肥料在土壤中的转化规律和使用中存在的问题，首先对有机物质的成分及环境条件对于有机物质的矿质化和腐殖质化的影响做简要的阐述。

有机肥料的成分

不同的有机质肥料在土壤中转化的速度是不一样的。一般而论，细胞壁厚，纤维质化、木质化程度高的植物质有机肥料，例如稻草、麦稭、秋后的杂草等，腐烂分解的速度慢，而细胞壁薄，纤维质化、木质化程度低的植物质有机肥料，例如苜蓿、豆类绿肥、春夏青嫩的山青、湖草等，腐烂分解的速度快。同一种植物，老熟的藁杆比青嫩的茎叶难于腐烂分解。从化学元素成分来说，含碳素成分高、氮素成分低的有机肥料腐烂分解和产生肥效比较慢；含碳素成分较低、氮素成分较高的有机肥料，腐烂分解和产生肥效比较快。例如大麦绿肥的肥效比苕子绿肥的肥效迟，后者的含氮量较高。有机肥料的碳氮比例对于肥效快慢有十分重要的意义，如豆类绿肥、腐熟粪肥、豆饼、血粉等肥料的碳氮比例小，这类肥料施入土壤后，可以立即产生肥效。至于所产生的肥效是否猛烈，则要看它们的物质成分、粉碎情况及土壤气候情况而定。相反的，稻草、麦稭、未腐熟的堆肥等的碳氮比例大，这类肥料施入土壤后，不能立即产生肥效。而要经过一段时期，在这时期中，由于微生物的生命活动，碳素成分逐渐减少（不断地放出二氧化碳气），终于在碳氮比例下降（转折点约为碳氮比例二十五比一）的情况下，才开始产生肥效。至于从施肥到开始产生肥效之间究竟要经过多少天数，则要看它们的物质成分、分散情况及土壤气候条件来定。为了缩短这个时间，有些肥料如堆肥、凼肥等，应先行堆积沤制，在堆积沤制期中，由于微生物的生命活动，逐渐腐热，成为可以直接产生肥效的肥料。腐熟程度愈高，施入土壤后产生肥效愈早。

碳氮比例高的有机肥料，在初步腐烂阶段（即碳氮比例大于二十五比一的阶段）对农作物生长不仅没有好处，还时常发生有害的作用：①在短期内使土壤的可给性植物养料元素成分（氮、磷等）为微生物所吸收而转化为有机质状态，因而暂时失去肥效；②在透气不良的条件下，产生一些有毒物质；③腐烂发热，靠着根时烧根，因此，春季在播种前施用有机肥料，应施用腐熟程度高的、已经经过了初步腐烂阶段的肥料，避免有害作用，提早产生肥效。在秋耕、冬耕时期施用有机肥料，施肥后土壤有几个月的休闲时期，可以施用腐熟程度较低的有机肥料。

气候、土壤条件是影响有机肥料腐烂分解的重要因素。在失水干燥、或温度在零度以下时，微生物

生命活动停止、有机物质不腐烂。水热条件逐步提高，微生物的生命活动逐步加强。当土温在摄氏三十度左右，饱和含水量在百分之八十左右时，土壤微生物的生命活动最强，有机物质的矿质化作用也最强。因此，在南方湿热地带，有机肥料的腐烂分解和矿质化作用比北方寒冷干燥地带要强得多；同一地带，多雨的热天比寒冷天气或干旱天气作用强得多；干旱地带的灌区比非灌区作用强得多。

土壤的透气条件对于有机肥料的转化也有重要影响。上面讲到，当土壤水分条件在饱和含水量百分之八十左右时有机物质的矿质化最快；当水分过多时，空气被隔绝，微生物生命活动就受到一定的阻碍作用。水分充足的旱地与泡水土壤相比，前者有机肥料腐烂分解得快；在泡水不透气情况下，腐殖质化作用相对地强些，矿质化作用比较弱些。黏重透水透气性差的土壤与透水透气性好的砂性土相比，前者有机质肥料腐烂分解得慢些。将有机质肥料深埋在土壤底层比在土壤表层腐烂分解得慢些。因此，黏性土一般肥效比较慢而持久，砂性土肥效快而短暂。在东北寒冷多湿的地带，过深地施用有机质肥料会导致草炭化而不是矿质化。

以下分别讨论我国主要肥料的使用经验及其对于培肥土壤、营养植物的意义；并比较具体地阐明肥肥土、土肥苗的原理和实施。

土粪：粪肥不宜单独堆积。单独堆积时，所含植物养料成分损失很大。不合理的堆积可能损失植物养料成分一半以上。粪肥也不可直接施于作物根际，这样做，由于微生物的生命活动、腐烂分解所产生的热、有毒物质和过于浓稠的溶解性植物养料物质，都会伤害作物。土粪是我国北方积制和使用的优良方法。一般将粪肥吸收在三倍左右的细土中，这样做，由于土壤的吸收作用，肥分不致损失。如果用土量过少，则吸收不完全，会损失一部分肥分；如果用土量过大，则体积过大，运肥困难。

粪肥吸收于土中后，一面腐殖质化，一面矿物质化，经过几个月的腐熟，粪土相融，形成为高度肥沃的“腐殖质土壤”。华北农业科学研究所分析华北二百五十六个土粪样品的结果如下：水分（23.2±0.55）%，石灰（3.56±0.06）%，有机物（7.46±0.16）%，全氮（0.321±0.006）%，氨态氮（0.0101±0.0009）%，硝酸态氮（0.026±0.0011）%，磷酸（0.306±0.067）%，氧化钾（0.733±0.012）%，碳氮比例（14.11±0.17）%。含有这样成分的腐熟土粪，就能够生长出好庄稼。将腐熟土粪集中施用于种子附近，为作物创造了优良的生活条件。多施土粪不会伤根，而施用后能够不断地供给作物以植物养料。大量积制和施用土粪，每亩施用一万斤以上，结合深耕，分层而充分地与土壤混合，可起改良土壤、培肥土壤的作用。

施用厩肥、堆肥、沤肥和凼肥（草塘泥等）的经验，厩肥是指用褥草垫厩，粪肥和杂草混合堆积而成的肥料，或指用土垫厩，粪肥和土混合堆积而成的肥料。后者与土粪同物异名，讨论见前，这里只指前者而言。堆肥是用少量粪肥与大量的植物物质混合堆积而成。不论是厩肥或堆肥，在堆积的过程中，由于旺盛的微生物生命活动，所含有机物质逐渐腐殖质化和矿物质化而形成优良的肥料。按照施用时的腐熟程度，厩肥和堆肥都可以分为腐熟的和半腐熟的两种。腐熟的厩肥和堆肥、粪草都失去原来的形状，成为均匀一致的，疏松的暗褐色腐殖质肥料；半腐熟的厩肥和堆肥，粪与草虽都有一定的变化，但还没有完全失去原来的形状（尤其是草）。农业生产中实际使用的一般是半腐熟的厩肥和堆肥。

施用厩肥和堆肥必须做到土肥相融，才能充分发挥肥效。厩肥和堆肥与土粪不同，由于没有和大量的土壤混合，不可直接与作物根系接触，尤其是半腐熟的厩肥或堆肥。如果与作物根系接触，则产生的热，毒质或过稠的养料成分会伤害植物。正确的使用厩肥和堆肥，务须使之充分散开，耕翻入土，反复耕耙，做到肥料和土壤充分混合，然后经过土壤微生物的生命活动，将肥料进一步腐殖质化和矿物质化，土肥相融成为肥沃的土壤，供给植物养料成分。长江流域，以厩肥或堆肥为基肥，最好与冬耕相结合，一方面利用翻耕和以后的多次耕耙，使充分与土壤混合，一方面有充分的时间使在土壤中进一步腐殖质化和矿物质化，真正达到土肥相融的目的。

不正确地施用厩肥或堆肥，将半腐熟的甚至于新堆的厩肥或堆肥在播种或插秧前不久施用，施用时不散开、不耕耙或耕耙不够，会使田地形成不均匀的肥料堆和空白地相间的状态。在这样的田地上，作物生长必然很不一致，处在肥料堆上的植株由于受毒害而死亡，或遭受严重的热烧；在肥料堆附近的植株生长十分旺盛，一部分发生倒状；在没有施到肥料的地方，因严重脱肥，植株矮小。这种现象各地仍相当普遍，急待改进。

为了达到一定的技术要求，如果需要厩肥或堆肥作为种肥，甚至于作为追肥使用，应该事先制造充分腐熟的厩肥或堆肥，在施用时拌和适量的土壤。

凼肥或草塘泥是在泡水情况下腐烂分解的有机肥料，凼肥或者在田头挖凼积制，或者就在田中挖凼积制。凼子挖好后，加入各种有机物质，以植物物质为主，适当加入一些粪尿以促使腐熟，加水浸泡，经过一冬或相当时间，沤制成肥料。田头凼肥一般的使用状态是半腐熟的，凼中肥液可以做追肥使用。田中的凼肥则在耕作整地时散开使用，这样做不仅得到了半腐熟或腐熟的有机肥料，也利用了在凼底形成的肥泥。四川的田块肥料比一般凼肥更进一步，在积制过程中加以耕耙和水分管理，因而做到了一面积制沤肥，一面土肥相融。

绿肥：绿肥在农业中占十分重要的地位。在农业生产中，人为地控制植物营养物质在土壤中转化的方向和速度是培育土壤肥力、争取丰产的农业技术的任务，栽种绿肥就是一项专门为这种任务而进行的农业技术。种植一季绿肥作物、再把它翻耕入土，实质上是完成了一次植物营养物质的生物循环。种植一季绿肥，将这一季之中可能利用的植物营养物质转化为绿肥作物有机质，然后在适宜的条件下，翻耕入土，这些有机质（肥料）逐渐转化成为植物能够吸收利用的状态，可以满足下一季农作物的需要。此外，绿肥在土壤中的腐殖质化，绿肥根系在土壤中的穿透松土作用，绿肥植物生长时期对土壤水热条件的影响等等，也改善了土壤的耕作、保水、保肥等性质。

我国主要的绿肥作物是生长期较短的冬季豆类绿肥（籽籽、紫云英、黄花苜蓿等），夏秋季豆类绿肥（绿豆、泥豆等），草木樨，以及大麦、萝卜菜、荞麦等。多年生的苜蓿一方面是饲料，一方面也具有绿肥的作用。在南方丘陵地带也有利用荒地种植绿肥作物，再把它搬到田地中去做绿肥的。

豆类绿肥比非豆类绿肥，有显著的优越属性，即它和根瘤菌共生，产生根瘤，引起共生性的固氮作用，从而将空气中的氮气转化为土壤中的氮素植物养料。种一季豆类绿肥，亩产绿肥六千斤（包括地上和地下部分），从空气中获得的氮素成分可达二十斤，相当于硫酸铵一百斤，或豆饼三百斤。仅以豆类绿肥所固定的氮素养料的货币价值计算，就可以当得一季，冬季作物或短期的夏秋作物的收获物的货币价值。

绿肥作物一般是在有机质生产量最大而且是青嫩的时期翻耕入土，尤其是豆类绿肥，茎叶嫩，碳氮比例低（一般在二十比一上下），在土壤中腐烂分解得很快，在适当的水分和温度条件之下，它的腐解速度比厩肥或堆肥快得多。在长江流域的水稻田中，春四五月翻耕豆类绿肥，三五天内叶子就开始腐烂，很快地产生肥效。

最近的研究工作和农民的生产经验都证明，豆类绿肥、青嫩湖草与厩肥、堆肥、土粪等肥料相比，有一重要的差别。连年地较大量地使用厩肥、堆肥或土粪，不仅不断地供给农作物以植物养料成分，而且能逐渐提高土壤的腐殖质含量，这是由于它们矿物质化比较慢，相对的腐殖质化程度比较高。也正因为如此，这些肥料必须配合速效性的追肥使用。相反的，在水温条件较高的地带（长江流域和长江以南，以及北方的灌溉农田），豆类绿肥或青嫩湖草翻耕入土后，矿质化很快，腐殖质化程度比较低，因此，对于供给植物养料的作用很大，而对于培育土壤腐殖质含量的作用则比较小。就长江流域和长江以南来说，豆类绿肥或青嫩湖草不是迟效性肥料而是中等的或相当速效性的肥料。

农业化学的研究结果与农民生产实践经验是一致的。长江下游一带的稻田，使用豆类绿肥或湖草，在目前的技术水平下，用量不宜过多，一般每亩在一千五百至三千斤新鲜物质之间，如果简单地计算其中所含的元素成分，含氮不过五至十五斤，含磷酸不过一至三斤。由于施用绿肥或湖草过量而造成的徒长、倒伏或其他不良现象，不能从施肥量本身去理解，而必须研究施肥后引起的土壤性质（包括溶解性植物养料）的变化。最近的研究工作还指出，绿肥或青嫩湖草翻耕入土后，不仅它本身所含的植物营养物质很快就矿质化，而且由于它供给了土壤微生物以大量的优良有机养料，大大加强了土壤微生物的生命活动，因而不仅使绿肥或湖草本身很快的矿物化，而且推动了土壤中原有的有机成分的加速矿物质化（中国水稻科学会议论文，一九五七年）。

这里再次指出多种有机质肥料配合使用的必要性，不要一块田完全上厩肥或堆肥，另一块田完全用绿肥，而是要把腐殖质化作用比较强的厩肥（堆肥）与矿物质化作用比较强的绿肥配合使用。

稻草回田（种双季稻，早稻稻草田田）稻草回田在两广是比较普遍的农业技术措施，不仅可将稻草中所含的植物营养物质全部归还土壤，供下季农作物的利用，而且提供了南方土壤中微生物十分需要的有机质养料。在水热条件优越的地带，加施有机质养料能加强土壤中微生物的生命活动，从而改善土壤中植物养料的供应状况。

稻草回田在长江南北（北亚热带）也会试用，效果不好。这是由于稻草的成分和这一带的水热条件所决定的。稻草是纤维质化、木质化比较高的物质，碳氮比例很大（约三百比一），这样的物质翻耕入土，在一定的时期内碳氮比例维持在二十五比一以上，此时没有肥效。在南亚热带，水热条件比较好，在两季稻之间，空闲的时间比较长（约一个月），稻草及时腐烂，对于晚稻可以产生优良肥效。在北亚热带，虽然夏季的水热条件也很好，但两季水稻之间空闲的时间很短（半个月以内），稻草不能及时腐烂，在双季晚稻插秧后，不能及时产生肥效。南方用稻草肥田，肥效很猛，这和上面讨论的绿肥相同，不能仅比稻草本身所含植物养料的多少来理解，亩产稻谷五百至一千斤的早稻，连稻草带地下部分不过八百至一千五百斤左右，含氮不过二至四斤，而必须从由于稻草回田所引起的土壤微生物的高度活跃性来理解。

这是一项众所周知的农业生产经验。在温带一般不能用成熟的禾本科植物的藁杆茎叶直接翻耕入土地，作为肥料，必须事先堆积腐熟，制成堆肥，才能使用；而在亚热带和热带则很多地方都直接将成熟的禾本科植物的藁杆茎叶翻入土中，作为肥料。

塘泥（沟泥、湖泥、河泥）塘泥对于培肥土壤、营养作物的意义是多方面的。首先，塘泥含有较多的有机质成分，它是接受了水中的动植物残体，在水底下经过微生物的腐烂分解作用而形成的。在水底下进行腐烂分解作用，腐殖质化的强度大，矿物质化强度小，因而腐殖质可以逐渐累积，形成富含腐殖质的塘泥。质量较好的塘泥、沟泥、湖泥、河泥含有机质一般在百分之二至百分之十，含氮百分之零点二至百分之零点四，磷百分之零点一六至百分之零点五六。显然，塘泥质量的好坏对于肥效有很大的影响。塘泥所含基本上是腐殖质而不是半腐熟的动植物有机质，因此用量不嫌其多，每亩使用一万至二万斤很普通，五万至十万斤也不会造成用量过多的损害。

正因为塘泥使用量很大，它对于土壤的物理性质和耕性具有很大的改造作用。一亩地一尺厚的土壤约重六十万斤，一次施用二万斤上下可能不会产生显著的影响，但是用到五万斤，不仅显著地改变了土壤的理化性质和耕性，而且使土层厚度增加一寸以上，在砂性大的土壤上，使用较黏重的塘泥有改善土壤质地的显著作用。在质地优良的油砂土上，施用过多的黏重塘泥，显然会对土壤质地起不好的作用。

在湖北省土壤普查工作中，农民区划出“杂土”一类，并指出“杂土”是由长年地、较大量地施用塘泥、绿肥、堆肥和湖草而形成的土类，与原本的土壤在形态上和肥力上都有质的区别。杂土被认为是第一等的土壤。

如何将大量的塘泥妥当地施入土中，使产生土肥相融的效果，这是一个问题。塘泥挑起晒干后，一般成为坚实的大块，铺在田面，对耕性和田面会起恶劣的作用，因此，必须在春节以前挑好塘泥，铺在冬闲田上，让它经过冬冻，风化、松散成碎块，在开春时才便于打碎犁翻入土，反复耕耙后，与土壤充分混合。开春以后再挑塘泥，一般不能收到土肥相融的效果。

草炭：草炭可分为高位草炭和低位草炭两类。高位草炭含植物养料成分少，酸度大，不宜做肥料，因其吸水性强，一分高位草炭可吸水三至六分，可以作为粪尿的吸收剂。低位草炭在我国分布很广，含植物养料成分较多，中性或接近中性，适宜于作肥料，肥效很好。草炭是腐殖质化了的有机物质，施入土壤后，转化较慢，尤其是生产和使用草炭最多的东北地区，天气寒冷，草炭的腐烂分解更慢。因此，草炭最好作为粪尿的吸收剂。草炭与或多或少的粪肥相混合，一起堆积腐熟，制成草炭厩肥或草炭堆肥，肥效更好。草炭吸收粪尿，对于粪肥肥分的保存作用很大。据东北农业科学研究所的试验，纯马粪堆积两个半月，损失氮素百分之三十二，而用草炭与马粪混合，堆积两个半月后，氮素反而增加了百分之二点九。

低位草炭是土壤微生物的优良养料，施入土壤中后，能加强土壤中微生物的生命活动，提高土壤的肥力水平。用草炭制成混合细菌肥料，施用于含有机质成分较多的土壤中，肥效很显著。

肥肥土、土肥苗的施肥原则不仅充分体现在施用大量有机质肥料及相应的施肥方法上，而且还体现在各种用理化方法来提高土壤肥力，改善作物营养条件的技术措施上，例如熏土、晒土和晒田等。

熏土（火粪、火土、烧土）熏土是用柴草等燃料在温度不很高、少氧的情况下，将土加以熏热，用作肥料（或称化土为肥）。这是我国独特的制肥和用肥经验。熏烧土壤，温度不宜过高（摄氏二百五十至三百度），要暗火熏烧而不可明火大烧。根据东北学业科学研究所一九五八年的试验，加热到摄氏一百度对土壤有机质含量无显著影响，加热到摄氏二百度，有机质有少量分解，加热到摄氏三百度，有机质含量损失四分之三，加热到摄氏五百度，全部有机质消失。一般熏土温度在摄氏二百五十到三百度左右，在这样温度熏烧处理之下，土壤有机质成分有些损失，而可给性养料成分有显著增加。例如，福建农业科学研究所一九五八年的研究结果，田土熏过以后，比不熏土的增产百分之七点六至百分之三十九点六，熏土后，有机质含量损失百分之十点二三，氮素很少损失，硝酸态氮略降低，速效性磷略提高，而速效性钾有显著提高，比不熏土的增加十二至十八倍。

仅从养料成分来理解熏土的增产意义，显然是不够全面的，但我们还没有从各个可能方面（如病虫害、杂草、土壤物理性质及耕性和土壤局部灭菌等方面）来研究熏土的增产意义。

晒垡：熏烧土壤是在燃烧加热的条件之下“化土为肥”，而晒垡则是在天然曝晒下“化土为肥”。长江流域很多水稻田，冬季翻耕后晒垡，有显著的增产作用；晒垡以后，土壤中的速效氮、磷、钾都有显著增加。同样的，对于晒垡的增产意义也须从多方面来理解。

水稻栽培期间的落水晒田，除对植物的根系发育有重要的影响外，对于土壤肥力条件的改善也有显著的作用。短期的落水晒田，可以改变土壤的水和空气条件，对于土壤微生物的生命活动有很大的影响，微生物的总量和生命活动强度大大增加了；对植物营养有重要意义的固氮作用、氨化作用（产生速效性氮的作用）、有机磷化物的分解作用、硫化作用（产生硫酸盐的作用）都显著地增加了，而对作物有害的反硫化作用（产生硫化氢的作用）则显著地降低了（华中农学院，一九五九年）。

细菌肥料：细菌肥料的施用也只能够从肥肥土、土肥苗的原理来理解和正确实施。施用细菌肥料并不会直接增加土壤什么成分，也不能直接改造土壤的理化耕作性质，细菌肥料只是在它对于土壤中的微生物生命活动起了重要影响以后，才能发生肥效。例如：根瘤菌肥料只有在适当的豆类植物根上形成根瘤，起共生性的固氮作用时才有肥效；固氮菌只有在土壤中大量繁殖，起固氮作用或产生植物生长刺激素时才有增产作用；磷细菌肥料只有在土壤中引起有机磷化合物的水解作用时才产生肥效。没有肥肥土、土肥苗的概念，而要求细菌肥料对作物产生直接的效果是不正确的。

石灰和石膏：施石灰是改善土壤肥力条件的技术措施。石灰能中和土壤酸性，在农业生产实践中已广为应用，它的作用本质也早经阐明。在酸性土上施石灰的意义是多方面的：①调节土壤酸性，使更适合一定的农作物的要求；②减轻活性铝和锰的毒害作用；③中和由于有机肥料分解所产生的酸性物质；④供应酸性土壤中比较缺乏的钙质养料。施石灰中和土壤酸性，可以一次较大量地施用，三、五年乃至于十年、八年用一次，也可以小量地每年每季使用。我国农业生产实践主要是采取后一种方法，每次每亩施用生石灰五十至一百斤。

正因为石灰对于改良酸性土壤有重要的作用，因而也产生了一种偏见，即只从化学中和来考查和评价施用石灰的意义，这是不够全面的。我国农民施用石灰的目的是多方面的。农民的产生经验和现代微生物学的研究工作证明，施用石灰除改变土壤酸度外，还显著地改善了土壤的微生物活动。在堆肥堆中，加施石灰可加速它的腐熟作用，这种现象只能够部分地用中和酸性来解释，即在有机物质分解的过程中，产生一些有机酸，这些有机酸的累积对于微生物的进一步活动有毒害作用，施用石灰可以将这些有机酸转化为钙盐，中和它的酸性，并且降低它的溶解性，因而消除了有机酸的毒害作用。在接近中性的土壤中，施用五十至一百斤石灰，对于土壤的酸碱反应并没有重大的影响，然而在施用绿肥或稻草回田的同时，施用石灰可以显著地加速绿肥或稻草的腐烂分解作用。

施石灰治水稻冷烧和脱肥很可能是由于促进了土壤微生物的矿质化作用。有些地方还有用石灰作为追肥的经验，如湖北省松滋县在施用绿肥的基础上，在水稻生长后期，绿肥中分解得比较快的成分已经耗尽，难于分解的部分和已经腐殖质化的部分的进一步矿质化比较迟缓，因而形成了脱肥现象，这时施少量石灰，有显著的追肥作用，即可以加强土壤中有机成分的矿质化作用，改善作物的营养条件。

施石灰还有改善土壤结构性的作用，石灰中的钙质和新形成的腐殖质能够与土壤中的粘粒和砂粒胶凝起来形成团粒构造，从而改善土壤的保水、排水、透气性和耕性。

施石膏可以改善苏打盐土，这是在实践和理论上都已经比较明确地证明了的。除此以外，在我国长江流域和长江以南的有些水稻田，也有施用少量（每亩三至五斤）石膏的技术措施。在有些水稻田（发噤）中，水稻插秧以后，秧苗返青迟缓，施石膏可以提早返青，效果很快，三五天即可见效。初步的研究表明，它与作物的硫素营养有关。除石膏外，硫酸镁、硫酸钠都有同样作用（武汉大学，一九五二年）。浙江农业科学研究所（一九五六年）的研究指出，含硫石灰有作用，不含硫的石灰就没有作用。另外，水稻热烧，也可以施用少量石膏防治。石膏的作用也可能在于它能使泡土变实，有利于根土密结。确实地阐明水稻田施用石膏的生产经验，尚须进行更多的研究工作。

对开展菌根研究的几点意见*

陈华癸

（华中农学院）

我们一向把真菌分为两个大的生理群：一群为腐生真菌；一群为寄生真菌。寄生真菌中，当然，最引人注意，而且研究最深的是那些致病的寄生真菌，这也是植物病理学主要的研究对象，但在自然界中，确实存在着更多种类的寄生真菌，它们在寄主体上营寄生生活，对寄主植物并无显著危害作用。对这些种类繁多、并无显著危害作用的寄生关系，人们重视的程度就不很够了，研究得就不太多了。形成菌根的菌根菌同寄主的关系就是其中的一部分，它们是许多种真菌同许多种寄主植物发生的相当普遍的达式关系。

菌根这个名称是1885年才提出来的，到现在只有100年的历史。当然，在某些植物根上长有菌丝套，对植物生长有某些作用，这些知识在100年以前也并不是完全不知道的，而是说，对菌根真菌进行科学研究只有100年的历史。在这100年中，由于研究工作并不多，而且寄生物和寄主又都是多种多样的，积累的知识有限。外生菌根对寄主的好处认识的比较早一些，而内生菌根，特别是VA菌根(注1)对寄主起着有益的作用，认识得更晚一些，只有在本世纪30-50年代才逐渐认识，到现在不到50年的历史。这是一大群真菌和一大群植物之间存在的错综复杂的关系。

为什么最近一些年来，对菌根菌的研究逐渐重视起来了呢？这主要是因为这些年进行的虽然为数不多的研究已产生了某些对国民经济有意义的科学技术，引起了人们的注意。主要在以下几方面：某些外生菌根菌同木本植物之间形成外生菌根的关系，例如让松树苗在苗圃中形成菌根，不仅在苗圃期，而且在定植后，都对生长起着显著作用。又如VA菌根的形成对柑橘的生长起到显著促进作用。再如某些豆类和根瘤菌以及VA菌根菌形成三者联合体，对于豆类的生长，对于固氮作用，对于提高土壤中磷素的利用率都起到显著作用。这些现象的被发现及其在生产中的利用使我们对菌根菌的研究工作越来越重视了。

谈到研究菌根，由于菌根菌种类多，多半是广泛寄主性，能形成菌根的寄主种类也多，两者间形成的共生关系也是多种多样的，表现这些共生关系的特定条件也是多种多样的。因此，我们的研究工作只能是某个特定的菌根菌同某个特定的寄主之间的相互关系。它们的共生体是如何形成的，形成后有何特定功能，这些功能有多大强度，在什么条件下这些功能才能表现出来，所有这些都不能把菌根菌作一方面，把寄主植物作一方面，加起来得出一个简单的、笼统的认识。尽管多种菌根菌和构成菌根关系的多种植物之间的知识是可以互相参考，互相启发的，但不能简单地以一概全。至于某些特定的关系究竟是什么，在国民经济上有多大意义，值不值得重点发展，我们不能在研究之先得出答案，而只能在研究之后得出答案。只能先做研究，后下判断，而不是先判断有无价值再去研究，这些年研究菌根的历史已表明：有些菌根的形成是有重要国民经济意义的，是应该重点研究、重点发展的。

目前，菌根研究存在一个重要的技术关尚未突破，特别是对VA菌根菌来说，直到现在还不能进行*in virto*培养(注2)，在基础研究上就很难进行严格的试验，在生产应用上就不能快速地、廉价地、大量生产菌根菌的菌种，供接种应用。对于某些多年生的树木和经济植物，在它们的生产过程中，有一个环节是在菌圃上集中育苗，然后定植，目前已掌握的取得菌根菌的方法是可以应用的，尽管目前使用的方法在繁殖上比较迟缓，效率不高、成本较高，但可集中在苗圃上使用，使苗木长好菌根后再移栽定植，

---

*原载于《土壤肥料》，（5）：3~4，1982.

（注1） 内生菌根分成两类：有隔菌丝的内生菌根和无隔菌丝的内生菌根，后者又称为藻菌菌根，但是人们根据它在根内形成大量的泡囊和丛枝状的构造，更习惯称它为泡囊-丛枝状菌根简称VA菌根（Vesjcular-arbuscular mycorrhiza）。

（注2） *in Vitro*培养，即人工条件下离体培养。

一亩地可育几千株苗，定植后可以长几年，几十年甚至几百年。因此，目前生产菌种的方法虽然原始、效率不高，但在生产上还是现实的。大田生产的情况则不一样了，一亩地要播种几万粒种子，长出几万株幼苗，每株都保持一个独立的菌根体系，而且只长一年，因此，必须有一种生产技术，保证能廉价地大量生产菌种，每年使用，而我们目前还没有这种技术。例如，尽管某些豆科植物可以同根瘤菌和菌根菌形成三者联合体有显著增产作用，显著提高固氮效率，显著提高利用土壤中磷素的作用，但这一技术如应用于大面积生产，就遇到如何大量生产廉价的菌种的问题。因此，要么攻下菌根菌在 in vitro 条件下纯培养这一关，大量地、廉价地制备菌根菌种；要么采取另外的也可以大量繁殖菌根菌的办法，达到同样可以大量应用菌根菌的效果。目前，我们想到的还是通过努力突破 in vitro 培养，既然它能在活的寄主上生长、繁殖，我们就不难设想，我们也可以创造同活的寄主所提供同样的供菌根菌生长与发育的条件。

菌根研究的另一个问题是尽管菌根菌是多寄主的，但并不是说某一特定菌株同所有能形成菌根的寄主都起一样的作用，试验已表明：同一菌株在某一寄主上起良好作用，在另一寄主上可能不起有益作用。因此，菌根菌研究的另一问题是要进行一方面是菌根菌，一方面是寄主，一对一对地研究，一对一对地筛选比较，找出一对一对能产生显著良好作用的菌根，作为我们的研究对象。这一工作的工作量显然是庞大的，但必须这样做。如果我们没有找到一对具有高效果的共生关系，我们就谈不上利用这种关系。我们的研究任务必须落实到具体的菌株和具体的寄主植物上，一个研究者或一个研究单位只能针对一种或少数几种植物来遴选高效的菌株，这是研究菌根的第二项比较困难的任务。

以上这些工作都必须有人做，我们这个全国菌根学习班正好提供了这一条件，可以组织一批同志开展这项工作，可以通过这个班掌握基础的研究方法。我认为这个班开得很有意义，对我国开展菌根研究会起重要作用，参加的同志在这个班学习的基础上，结合各自的研究对象开展工作，从具体工作中发现并找出哪些菌根菌在国民经济中可以起一定作用，针对特定的菌根关系，进行深入的研究，去取得有一定国民经济价值的科学成果，对我们来讲是十分必要的。

祝同志们在这个班的学习中，取得满意收获。

Overcoming Restriction of *Streptomyces hygroscopicus* 10-22 by the Modification of *S. fradiae* — An Attempt to Develop a Transformation System for *S. hygroscopicus* 10-22*

QIN ZhONGJUN　DENG ZIXIN　ZHOU QI　CHEN HUAKUI

(Agricultural Antibtotics Laboratory，Huazhong Agricultural University，Wuhan 430070)

ABSTRACT　No transformant was obtained when pIJ702（*tsr*，*mel*$^+$） from *S. lividans* TK24 was used to transform *S. hygroscopicus* 10-22. pIJ702 isolated from *S. fradiae* ATCC 10745，however，was transformed into 10-22 at a frequency of 10^3-10^4 transformants/μg DNA. Among the transformant colonies，only 1/1000 of them were black in colour （*mel*$^+$） while a great majority of them remained white （*mel*$^-$）. Plasmid pIJ702 band was only visualized on agarose gels from the black colonies but not from the white colonies. However，when pIJ702 isolated from both black and white transformants were used to transform *S. lividans* TK24，the mel gene was expressed normally in the recipients. The preparation was also successful in transforming *S. hygroscopicus* 10-22，and again gave rise to 1/1000 of black colonies only. When the 10-22 （pIJ702） black colonies were plated on non-selective medium，among the majority of black colonies grown，there were a few white colonies，which were proved to be host mutants of 10-22. These mutants were transformable by pIJ702 and homogenous black colonies were obtained.

KEY WORDS　Restriction-modification system；pIJ702；*Streptomyces fradiae*；*Streptomyces hygroscopicus*；Transformation

Streptomyces hygroscopicus var. *yingcheinensis* 10-22 produces three important agricultural antibiotics useful in combating rice sheath blight disease，leaf spot of corn and cotton wilt，respectively [1]. In the attempt to clone antibiotic biosynthetic genes from this strain，we have tried to use many Streptomyces plasmid vectors，including pIJ702，to transform the protoplast of this strain，but met no success.

Streptomyces are producers for many restriction enzymes. Restriction is often encountered when foreign DNA is transferred from different *Streptomyces* species. In this paper，we report a new method to introduce pIJ702 into *S. hygroscopicus* 10-22 and some unusual properties of pIJ702 in this strain. The use of some of the properties is also described.

MATERIALS AND METHODS

1. Strains and Plasmids

S. hygroscopicus 10-22，90-11 （*str*^{r}，*leu*$^-$） [2] and *S. fradiae* ATCC 10745 are the stock of our lab. The blocked mutant N-103 was obtained by NTG mutagenesis （Liang Rongfang et al.，unpublished）. *S. lividans* 1326[6]，JT46 （pro-2，str-6） [9]，TK24 （str-6） [6] and plasmid pIJ702 （mel$^+$，tsr） are all provided by the Streptomyces group of the John Innes Institute.

* This work received support from the Natural Science Foundation of China, the Ministry of Agriculture and International Foundation for Science. We thank Dr. J. Lucania form ER Squibb and Sons, New Brumswick, NJ, USA for the gift of thiostrepton. We also thank Prof. D. A. Hopwood and Dr. T. Kieser from John Innes Institute, Norwich, England for giving us various help in several aspects.

原载于《遗传学报》，20（2）：180~184，1993.

2. Media

2 CM [3] is the medium for the sporulation of 10-22 and its derivatives. R2YE [6] is the medium for protoplast transformation and regeneration. The liquid medium for 10-22 and 90-11 is YEME （sucrose concentration is 10% instead of 34%）.

3. Methods

Methods for NTG mutagenesis，protoplast preparation，fusion and transformation as well as for plasmid isolation are as described in [6]. The curing of pIJ702 from 10-22 and its derivatives was simply done by plating spore suspension at appropriate dilutions on 2CM plate，and then selecting for white colonies from black populations. If the loss of the black colour is accompanied by the loss of the thiostrepton resistance，pIJ702 is considered to be cured. The loss of pIJ702 is further confirmed by plasmid isolation.

RESULTS

1. pIJ702 isolated from TK24 is difficult to be introduced into *S. hygroscopicus* 10-22

Three routine methods were used to overcome the restriction barrier of 10-22 to foreign plasmid DNA of pIJ702. Firstly，more than 1 μg of pIJ702 from *S. lividans* TK 24 was used to transform 10-22，no transformants were obtained. Secondly，*S. hygroscopicus* 90-11（str^r，leu^-） was protoplasted and fused with the protoplast of *S. livdans* 1326 （carrying pIJ702）.Thiostrepton resistant （$thio^r$），streptomycin resistant （str^r） fusants were picked. pIJ702 isolated from these fusants were used to transform 10-22，again it was not successful. Thirdly，the spore suspension of 10-22 was mutagenized by NTG，and was used for protoplast preparation. pIJ702 （more than 1 μg isolated from TK24） was used to transform such protoplast，but still no thiostrepton resistance transformants visible.

2. pIJ702 isolated from ATCC 10745 can transform *S. hygroscopicus* 10-22

Some strains of *S. fradiae* are reported to possess several restriction-modification systems [8]，but pIJ702 are reported capable of being introduced by transformation into some of the *S. fradiae* strains. Thus an attempt was made to introduce pIJ702 into *S. fradiae* ATCC 10745. Indeed some thiostrepton resistance （$thio^r$） and black （mel^+） transformants were obtained. pIJ702 isolated from such transformants was used to transform the protoplasts of *S. hygroscopicus* 10-22，thiostrepton resistant transformants were obtained at a frequency of 10^3-10^4/μg DNA used （Table 1）.

Table 1 Transformation of *S. hygroscopicus* 10-22 and *S. lividans* by pIJ702

Host of pIJ702	Receptor			
	10-22		TK24	
	Transformants/μg DNA	B / B+W（%）	Transformants/μg DNA	B / B+W（%）
TK24	0		10^6	100
S. fradiae ATCC 10745	10^3-10^4	0.1	10^6	100
B	10^5	0.1	10^6	100
W	10^5	0.1	10^6	100

B and W are black and white transformants of 10-22 by pIJ702，respectively.

3. The unusual expression of mel gene of pIJ702 in *S. hygroscopicus* 10-22

Among 10-22 transformants obtained after transformation of pIJ702 purified from S. fradiae only 1/1000 of the transformant colonies produced melanin（mel^+）on R2YE plate and majority of transformants were seen white. Plasmid DNA from black and white colonies were isolated respectively，only pIJ702 from black colonies but not from white colonies is detectable on agarose gel and remains at high copy number. pIJ702 from both black and white colonies can all transform 10-22 at high frequency （ca. 10^5 transformants/μg DNA），but the ratio of black to white is still constant （1：1000）. Both plasmid preparations gave rise to uniform black colonies at high frequency （ca. 10^6 transformants/μg DNA） in *S. lividans* TK24 （Table 1）. This experiment indicates that pIJ702 is still present in white colonies without detectable changes，but somehow *mel* gene can not be expressed normally.

4. The normal expression of mel gene of pIJ702 in host mutant

The black 10-22 transformants of pIJ702 （mel^+，$thio^r$） was plated unselectively for rare white colonies. Such colonies were found to have lost thiostrepton resistance. Agarose gel electrophoresis further confirmed that pIJ702 is eliminated in white colonies. 6 of such colonies were protoplasted and were transformed by pIJ702 either from *S. fradiae* 10745 or from *S. hygroscopicus* 10-22. Almost all thiostrepton resistant transformants produced melanin. This experiment suggested that tsr and mel genes are expressed normally in these rare white colonies which prove to be the host mutants of 10-22.

5. The transformation of pIJ702 into 10-22 mutant blocked for antibiotic biosynthesis

S. hygroscopicus 10-22 produces three important agricultural antibiotics. Chemical mutagenesis had been used to isolate mutants blocked for antibiotic biosynthesis. N-103 is one of them （Liang Rongfang et al.，unpublished）. When pIJ702 was used to transform the protoplast of N-103，the same situation as the transformation of 10-22 （described in results section 2） was observed. The transformation of N-103 is not only influenced by the modification status，the expression of the mel gene also seems to be abnormal （the ratio of black to white colonies remains 1：1000）. Similarly，white colonies cured of pIJ702 among black populations were also obtained after a step of non-selective growth. One such strain，Q3030（$thio^s$，mel^-） was protoplasted and transformed by pIJ702 again，almost all of the thiostrepton transformants produced melanin.

6. pIJ702 becomes potentially useful vector for *S. hygroecopicus* 10-22

In order to investigate the possibilities of pIJ702 as cloning vector in the previously described host mutants of 10-22 （e.g.Q105 or Q303），we first examined the plasmid status of pIJ702 in these two strains，and no deletion in any part on pIJ702 was found. The presence of the intact circular form of pIJ702 tends to suggest that it can replicate autonomously. Secondly，we tested the genetic stability of pIJ702 in such strains and found variations in different strains. Judging from the retention of the thiostrepton resistance in the population of spores after one round of growth under non-selective condition，the stability of pIJ702 in Q303 approaches 100%，whereas less than 5% is found in Q105.

DISCUSSION

The common producer of three agricultural antibiotics-*S. hygroscopicus* 10-22 expresses strong restriction to foreign DNA. Restriction to externally isolated cloning vectors derived from the indigenous replicons is at present circumvented by the modification of such DNA in native host [4]. It had been attempted to use widely useful vectors such as pIJ702 （derived from pIJ101） to transform this strain but had not been successful [4].

We noticed that SfrI and ShyI are two restriction endonucleases produced respectively by *S. fradiae* ATCC 3355 and *S. hygroscopicus* which recognize the same nucleotide sequences. We also noticed that pIJ702 can transform some *S. fradiae* strain at low frequency [5]，which are known to possess several restriction-modification systems [8]. This prompted us to think if there were some similarity in modification specificity between *S. fradiae* 107145 and *S. hygroscopicus* 10-22，then the restriction of *S. hygroscopicus* 10-22 to pIJ702 could be circumvented by in vivo modification of pIJ702 DNA by *S. fradiae* 10745. Therefore，we first introduced pIJ702 into *S. fradiae* 10745 by transformation and established a transformation system of *S. hygroscopicus* 10-22 using pIJ702 isolated from *S. fradiae* 10745. Thus，this result not only provided complementary vector to the present gene cloning systems in *S. hygroscopicus*，but also provided useful information that *S. hygroscopicus* 10-22 and *S. fradiae* 10745 must have similar restriction-modification system（s）.

pIJ702 was chosen for this experiment because it carries a mel gene which is able to confer host colonies black in colour，making transformants visually distinguishable. Abnormal expression of the mel gene of pIJ702 is *S. hygroscopicus* 10-22 reflects genetic interference of the host to the mel gene expression or to the replication of the whole pIJ702. Whether it is due to specific repression or due to plasmid incompatibility which might be present between indigenous plasmids and pIJ702 is unknown. It is known that *S. hygroscopicus* 10-22 carries at least four plasmids （Qin Zhongjun et al.，unpublished），plasmid incompatibility seems to be a good explanation but not yet proved at present study.

Apparently，to what extent in different *Streptomyces* species such an attempt could be useful is unknown，but there is a good reason to believe that the principle of this technique should be widely applicable.

Remarkable progress had been made using pIJ702 as cloning vector in *Streptomyces* [7]. Establishment of this plasmid as useful cloning vector for *S. hygroscopicus* 10-22 could be very useful for the study of the molecular biology of the primary and secondary metabolism in this important producer for three agricultural antibiotics.

ACKNOWLEDGEMENTS

This work received support from the Natural Science Foundation of China，the Ministry of Agriculture and International Foundation for Science. We thank Dr. J. Lucania form ER Squibb and Sons，New Brumswick，NJ，USA for the gift of thiostrepton. We also thank Prof. D. A. Hopwood and Dr. T. Kieser from John Innes Institute，Norwich，England for giving us various help in several aspects.

REFERENCES

[1] Ling R. and Zhou Q.：1987. *Chinese J. Biotechnology*，3（2）：130-136.

[2] Liang R. and Zhou Q.：1989. *Chinese J. Biotechnology*，5（3）：226-231.

[3] Yuan R. and Zhou Q.：1989. *Chinese J. Antibiotics*，8：380-387.

[4] Deng Z. and Zhou X.：1989. Agric. Biotechnology in China，The Ministry of Agriculture，pp.142-151.

[5] Cox，K.L. et al.：1984. *J. Bacteriol.*，159：499-504.

[6] Hopwood，D.A. et al.：1985. in：Genetics Manipulation of Streptomyces—A Laboratory Manual，the John Innes Foundation，Norwich，England.

[7] Hopwood，D.A. et al.：1986. in：The Bacteria，S.W. Queener and L.E. Day，eds.，Academic Press，New York，Vol.9，pp.159-229.

[8] Matsushima，P. et al.：1987. *MGG*，206：393-400.

利用果类资源生产果醋的研究*

陈华癸　阎淳泰　梁运祥

（华中农业大学）

本文报告了利用果品加工上的下脚料及果品生产上的残次、落果等原料，并采用了固态发酵法生产食醋的工艺，并对工艺加以适当改造，用于果醋生产。简化了生产工艺、降低了生产投资和生产成本，对于合理利用资源，增加农民收入，繁荣市场，将起到一定的促进作用。报告中介绍了工艺流程，选用的菌种、果醋成品的成分含量。

一、前　　言

我国果树栽种面积约 7600 万亩，总产量近 1700 万吨，平均亩产 400 余斤，人均年占有量仅 15 公斤。尽管如此，而目前我国果品加工能力却只有 350 万吨，占总产量的 21%。其中果核、果皮、果渣及附带的果肉、汁作为工业下脚料而被废弃的约 100 万吨；除此，加上水果自然腐烂（烂果率为 20%）约 300 万吨；水果生产中落地果、残次果 5%~10%约 150 万吨；和由于采收粗放、分级、包装运输水平低，设备落后，造成果品碰压、刺伤严重，从产地到销地损失率约在 20%，严重的达 30%~40%，约 100~200 万吨。

野生果类资源在 100 万吨以上，除少部分利用外，其中很大部分自生自灭了。综合上述损失累计约 600~700 万吨，相当于近 2000 万亩果园的产量，资源的浪费却是惊人的。

这些不能鲜食的水果及水果加工厂下脚料，除霉变腐烂者外，仍具有一定的营养价值，而且有其独特的果香、天然色素等。如果能将其就地加工为果醋，不但可以大大节约粮食，为国家创造财富，也是发展农村商品经济的有效措施之一。

我国年产食醋 100 多万吨，耗粮 20 多万吨，产值 3 亿元以上。假若能将我国果类资源每年累计损失量的一半用来生产果醋，出品率为原料的 50%，其果醋产量也达 150 万吨，相当于我国食醋行业全年产量，则可节粮 20 多万吨。

果汁醋作为一种健身饮料，在欧美及日本各国发展很快。我国果汁醋的生产自 80 年代初开始起步，目前已具备一定批量生产能力。果汁醋一般含有丰富的维生素及无机盐类，因此不但能调味，还具有一定的保健功能。并且风味独特，口感新鲜，比起粮食醋来，更为消费者接受。据已报道的果醋生产，有的是采用水果榨成果汁采用静止表面发酵法。这种方法发酵周期长，设备占地面积大，卫生条件不好控制。近来报道的多为果汁液体深层发酵法。这种工艺比较先进，目前还只局限于山东、四川、上海、河北等几个大城市进行批量生产，而且主要是以鲜果汁为原料，对菌种、设备要求都很高，建厂投资大，其产量销量都没有形成局部气候，其影响及优越性尚未充分发挥。由于生产成本较高等原因，目前果汁醋液体深层发酵及静止表面发酵法仍不可能在全国普遍推广。

基于以上种种原因，我们主要利用了果品加工上的下脚料及生产上的残次、落果等为原料，并采用了固态发酵法生产食醋的传统工艺，并对该工艺加以适当改造，用于水果醋的生产。采用新技术生产果醋，一是简化了生产工艺；二是降低了生产投资和生产成本。果类一经破碎直接进入酒精发酵、省去榨汁工艺及设备，直接利用当地调味品厂食醋生产车间原有设备和技术力量，不需增加许多投资即可生产。即便果品加工厂新建果醋车间，投资也不大，这对于合理利用资源，增加农民收入，繁荣市场，将起到一定的促进作用。现将研究结果予以报告。

*原载于《中国调味品》，11：16~20，1990.

二、试验材料与方法

（一）菌种

酒精酵母，黑曲霉 uv-11，黑曲霉 F_{27}，黑曲霉 N_{343}。

（二）培养基

查氏培养基、麦芽汁培养基、麸皮培养基。

（三）主料

1.水果

（1）广柑来源于湖北省秭归县。根据运输破损程度分为完好、机械损伤两大类，有明显霉烂变质的水果弃之不用。

（2）橘皮：来源于湖南省麻阳县。

（3）苹果：市销果光苹果。

（4）苹果皮、果核：市销国光苹果取其皮核。

（5）苹果园落地果、残次果：来源于华中农业大学果园。

（6）野生果金樱子：来源于湖北省英山县。

（7）野生猕猴桃：湖北五峰县，由于高温，麻袋包装长途运输至武汉，有部分变软，破损，挤破，属等外级野果。

（8）桃子。

2. 碎米 华中农业大学业畜牧站。

3. 麸皮、谷壳 华中农业大学畜牧站，校附属农场。

4. 食用酒精[95%（v）]：武汉酒精厂生产。

5. 偏重亚硫酸钾（$K_2S_2O_5$）分析纯或液态 SO_2。

（四）化验分析

1. 一般化学分析由华中农业大学酿造教研室及武汉调料公司一综合厂化验室承担。

2. 芳香成分，无机离子含量，各种氨基酸含量由华中农业大学基础部无机分析化学教研室协助检测。

3. 猕猴桃醋大生产样品由武汉市质检所协助检测。

（五）试验规模

1. 小型试验：每批 10kg/塑料桶。

2. 中型试验：每批 50kg/小缸。

3. 大型生产试验：3500 公斤猕猴桃。

（六）果醋试制的工艺设计及操作要点

1. 原料要求

1）野生果、残次果、果皮、果核：新鲜、无腐烂变质、无药物杂质。

2）生产用水：应符合 $TJ_{20\text{-}76}$ 生活饮用水卫生标准。

3）添加剂：应符合国家标准 GB_n50-77 规定。

4）淀粉质原料及辅料：无霉变。

2. 工艺流程

1）工艺流程根据原料不同采用 I 和 II 两种类型。

工艺流程 I：

新鲜水果或落地果、残次果可采用本工艺流程。如广柑、苹果等。

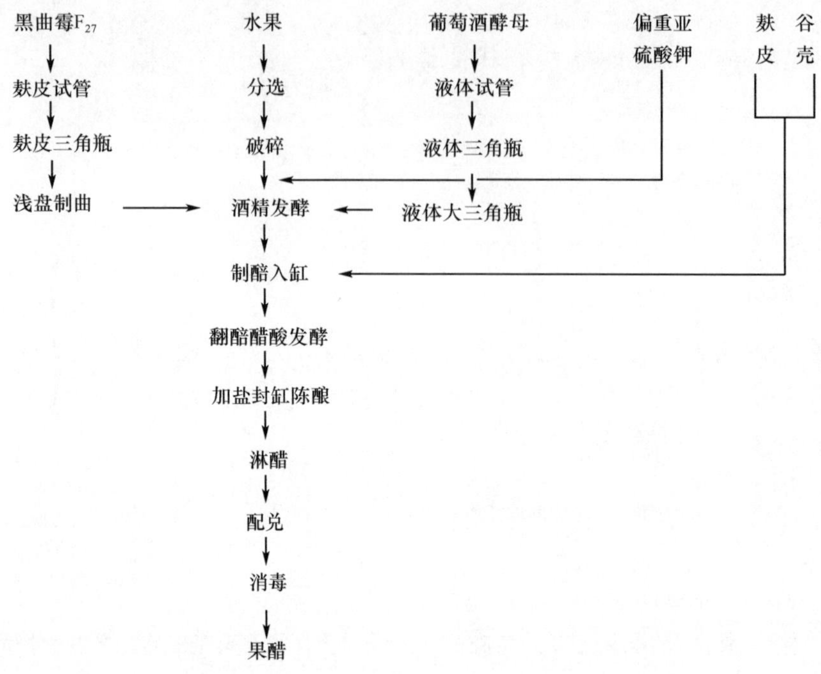

工艺流程 II：

野果、果皮、果核、果渣、残次果、落地果等果汁含量较少的原料可采用此工艺。如橘皮、金樱子、苹果皮、核等。

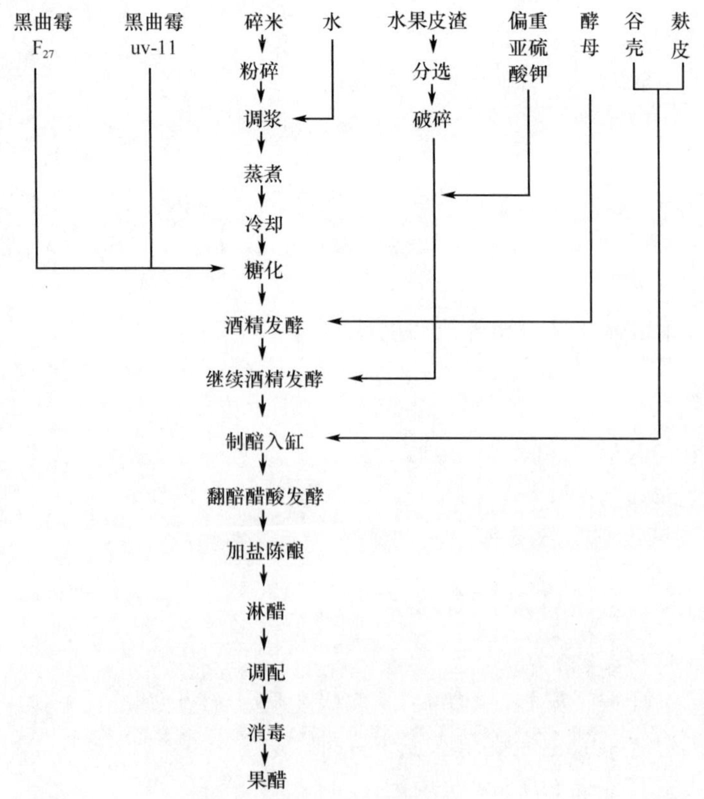

2）偏重亚硫酸钾用量表

	每百公斤水果偏重亚硫酸钾用量
1. 果子清洁、良好、品温低于 20℃，酸度在 0.8%以上（以酒石酸计）	10g
2. 果子清洁，完全成熟，品温低于 20℃，酸度在 0.6%~0.8%（以酒石酸计）	13g
3. 果子破裂、果皮、果核、果渣、无霉烂	20g

三、试 验 结 果

（一）使用纤维素酶提高糖化率及水果出汁率的研究

1. 采用纤维素酶产生菌黑曲霉 N_{343} 和糖化酶产生菌 UV_{-11} 分别制曲，加酶液配比（N_{343}：UV_{-11}）选择为 3∶7，其糖化率提高 12.22%。

2. 采用纤维素酶产生菌黑曲霉 F_{27} 和糖化酶其混合糖化，可使山芋粉糖化率提高 13.86%。

3. 采用纤维素酶产生菌黑曲霉 F_{27} 固体曲提高水果出汁率 12.16%~22.51%。

（二）利用野生资源落地果、残次果、果品加工厂下脚料制果醋的研究其工艺是可行的。感官指标，理化指标，卫生指标达到部颁标准，其产品具有本品种特有果香及一定营养成分。

（三）本工艺在基本不增加设备的情况下，只要把原食醋生产工艺稍加改变便可利用野生资源、落地果、残次果、果品加工厂下脚料，不但可以部分代替主粮，而且变废为宝，成本也比较低，适合大、中、小型工厂推广应用。

（四）果醋感官指标

1. 色泽　浅黄至琥珀色。

2. 香气　具有本品种特有果香。

3. 滋味　酸味柔和、绵长、适口、无异味。

4. 体态　澄清、无悬浮物、无沉淀物。

（五）果醋理化指标及卫生指标

表 1　武汉产品质量监督检查所猕猴桃醋检验报告

检验项目	标准要求	检验条件	样本检验结果	单项评定	备注
总酸 g/100ml（以乙酸计）	≥3.50		5.47	合格	
不挥发酸 g/100ml（以乳酸计）	≥0.70		4.26	合格	
还原糖 g/100ml（以葡萄糖计）	≥1.00		1.32	合格	
游离矿酸	不得检出		未检出	合格	
砷 mg/L（以 As 计）	≤0.5		0.1	合格	
铅 mg/L（以 Pb 计）	≤1		0	合格	
细菌总数（个/ml）	≤5000		<10	合格	
大肠菌群（个/ml）	<3		<3	合格	

表 2　醋的一般性质及维生素 C 含量

样品（产地） 项目	果醋 （华农）	香醋 （镇江）	熏醋 （山西）	蒜醋 （山西）	香醋（K 型） （镇江）	陈醋 （山西）	白醋 （上海）	老陈醋 （山西）
pH	3.48	3.28	3.21	3.64	3.40	3.52	2.40	3.62
总酸 g/100ml	5.68	5.72	3.96	5.87	6.43	6.35	5.69	9.92
还原糖 g/100ml	0.96	4.80	7.75	5.40	5.40	9.01	—	19.45
VC mg/100ml	0.72	—	0.48	—	—	—	0.12	—

（六）果醋芳香成分、无机离子含量、各种氨基酸含量

1. 芳香成分　用日立 163 气相色谱仪测定，以高分子微球为担体，直接进样可检出多种芳香成分，见表 3。

2. 无机离子　用原子吸收光谱仪测定 Ca、Fe、Cu、Zn、Pb 离子，用火焰光度计测定 K、Na 离子，结果见表 4。

3. 各种氨基酸　用日立 835-50 型氨基酸自动分析仪测定 17 种氨基酸，结果见表 5。

表 3　芳香成分含量（ppm）

成分	果醋	香醋	熏醋	蒜醋	香醋(K 型)	陈醋	白醋	老陈醋
甲醛	8.1	28.0	23.0	21.4	24.7	22.8	26.9	21.6
甲酸	692.3	305.4	251.9	588.1	—	224.9	—	317.9
乙醛	15.3	108.9	161.7	50.2	258.5	149.1	15.4	188.1
乙醇	18831.0	4033.4	1279.6	1815.3	10711.4	1043.7	95.8	810.0
异丙醇	17.9	137.3	244.8	116.9	103.6	187.3	14.0	340.0
异丁醇	434.1	188.4	202.9	13.0	184.6	263.5	—	291.7
双乙酰	—	20.3	186.8	53.5	56.5	56.5	—	34.9
乙酸乙酯	1571.8	2206.9	2626.1	134.8	1495.9	1908.1	7.7	2433.0
正丁醇	34.6	193.0	172.8	86.2	108.3	473.9	—	—
丙酸乙酯	794.3	1010.5	2757.8	408.7	1103.1	1572.4	—	2731.8
第二戊醇	987.3	716.9	4701.5	1398.0	759.6	2419.7	—	1446.6
异戊醇	310.8	193.6	47.5	7.1	121.6	315.4	—	222.1
糠醛	32.9	575.5	647.9	249.6	936.8	683.2	—	552.3
乳酸	700.1	4418.8	4939.6	1902.8	7142.2	6619.9	+	4210.7
乳酸乙酯	394.0	8019.6	7046.1	1745.0	11677.2	9591.2	—	6692.5
乙酸异戊酯	37.4	—	15.6	54.2	31.7	42.5	—	63.3

“+”表示痕量，“—”表示未检出。

表 4　无机离子含量（ppm）

成分	果醋	香醋	熏醋	蒜醋	香醋（K 型）	陈醋	白醋	老陈醋
Ca	2250	350	838	1213	58	1048	38	350
Fe	345.7	153.8	332.5	96.0	100.5	430.0	3.5	76.9
Cu	0.21	0.55	0.30	0.66	0.96	0.61	0.19	0.85
Zn	1.22	0.88	0.93	1.43	1.24	1.51	0.22	0.36
Pb	1.20	1.03	0.98	0.96	1.21	1.09	0.22	1.71
K	1158.3	3110	2400	3675	3775	6100	10.7	10400
Na	102	6990	1760	950	4530	6323	7000	12550

综上所述，华农果醋与其他名醋相比有以下特点：

（1）果醋含有丰富的 Vc，其含量显著地高于其他食醋。果醋总酸度为 5.68mg/100ml，pH=3.48。与其他食醋酸度相近。故果醋酸度适中，开胃生津，既是营养丰富的调味品，又可作为新型酸性保健饮料。

（2）芳香成分单纯，乙醇含量最高为 18831ppm 比其他食醋高出许多倍。异味物甲烯、乙醛、双乙酰、糠醛等含量低，所以色泽清澈、果香浓郁、滋味醇厚、口感纯正。

（3）矿物质含量丰富。营养元素 Ca、Fe 含量比大多数其他醋要高；Cu、Zn、K 与其他醋相当；而钠的含量低仅为 102ppm，比其他醋低 1~2 个数量级；且 K-Na 比值大，故作为防衰老、防治高血压和冠心病的低钠调味品，或具有防病健身功效的保健饮料是很适宜的。

（4）华农果醋是以水果加工的下脚料为主要原料，故氨基酸含量较少，但仍高于白醋，而与山西蒜醋相近。

表 5 各种氨基酸含量（mg/ml）

成分	果醋	香醋	熏醋	蒜醋	香醋（K 型）	陈醋	白醋	老陈醋
天门冬氨酸	0.64	1.35	0.97	1.02	1.77	2.27	0.15	3.78
苏氨酸	0.38	1.12	0.73	0.43	1.27	1.16	0.07	1.81
丝氨酸	0.50	1.59	0.89	0.48	1.67	1.43	0.09	2.19
谷氨酸	2.21	8.17	3.49	2.34	8.19	6.63	0.28	11.3
甘氨酸	0.55	1.64	0.89	0.58	1.89	1.57	0.06	2.43
丙氨酸	0.86	2.78	2.52	1.01	3.38	2.70	0.11	4.48
胱氨酸	0.19	0.44	0.18	0.15	0.35	0.25	0.14	0.35
缬氨酸	0.60	1.58	1.09	0.66	1.87	1.71	0.16	2.74
蛋氨酸	0.12	0.26	0.39	0.17	0.34	0.32	0.17	0.41
异亮氨酸	0.49	1.23	0.88	0.51	1.41	1.29	0.14	2.03
亮氨酸	0.71	2.08	1.58	0.63	2.27	2.08	0.09	3.44
酪氨酸	0.26	0.51	0.37	0.27	0.37	0.54	0.13	0.74
苯丙氨酸	0.46	0.93	0.47	0.33	1.43	1.52	0.15	1.29
赖氨酸	0.48	1.34	0.65	0.44	1.52	1.40	0.05	2.30
组氨酸	0.23	0.65	0.31	0.13	0.44	0.47	—	0.68
精氨酸	0.14	1.80	—	0.07	0.67	0.34	—	0.83
脯氨酸	1.05	2.89	1.71	0.90	2.93	2.98	—	4.95
总和 Σ	9.87	30.4	17.12	10.12	31.77	28.66	1.79	45.75

四、问题与讨论

1. 关于原料贮藏试验，经一年时间未变质，制醋工艺不变。
2. 果醋消毒过程中，芳香成分损失较大，应考虑密闭回收。

利用原生质体融合技术提高黑曲霉产酶活力的研究*

郝　勃　阎淳泰　陈华癸

（华中农业大学微生物系）

摘　要　以黑曲霉N343和UV-11为出发株，分别经亚硝酸诱变处理，再以制霉菌素浓缩处理，从中筛得3株维生素缺陷型和1 株氨基酸缺陷型的突变株。选取来自N343的突变株NB_3（B_1^-，PP^-，FA^-）和来自UV-11的突变株UBI（Pro^-）为融合亲本，在研究其原生质体释放和再生条件的基础上，以PEG诱导进行了原生质体融合。用直接选择法检出98个融合子，经过筛选得到两株稳定的融合子：FNU32和FNU38。它们除了在形态、培养特征等方面具有明显杂种性能以外，都能将两个亲本各自出发株的酶活特性（高活力的纤维素酶或淀粉糖化酶）囊括于一体，因而大大提高了菌株在诸如制糖与粗饲料精制等方面的应用价值。

关键词　黑曲霉；营养缺陷型；原生质体融合；纤维素酶；淀粉糖化酶

自1974年Ferenczy报道[1]以离心法诱导白地霉（*Geotrichum candidum*）营养缺陷型突变株的原生质体融合以来，国内外陆续报道[2]了一些在其他丝状真菌应用原生质体融合技术的研究。在我国，以米曲霉为材料开展的原生质体融合的研究较为多见[3-5]，在改良酱油酿造的米曲霉菌种方面做了大量有关的研究。

作者从野生黑曲霉筛到一株纤维素酶活力很高的菌株F27[6]，并成功地将它应用于酱油酿造，与米曲霉3042混合制曲，对提高酱油原料蛋白质利用率有显著的效果。随后在研究薯干淀粉糖化时，发现将黑曲霉UV-11与黑曲霉F27的衍生菌株N343制成混合曲，其糖化效率明显优于由UV-11制成的单一曲[7]。这表明欲提高薯干等农副产品的糖化效果，必须要有一株酶系丰富、活力高和比例适当的新型菌株。为此，作者采用原生质体融合技术并且获得了这一类新型菌株。

1　材料和方法

1.1　菌种

黑曲霉（*Aspergillus niger*）N343由华中农业大学微生物系发酵工程室提供，黑曲霉UV-11购自中国科学院微生物研究所。

1.2　培养基

基本培养基（MM）为察氏培养基，如加0.2%酵母膏则为补充培养基（SM）。完全培养基（CM）为米曲汁培养基。AM-1为菌丝体制备用培养基[8]。PD、PM和YM分别为土豆浸出汁培养基、蛋白陈培养基（蛋白胨0.6%，pH7.0）和酵母浸膏培养基（酵母膏0.6%，pH7.0）。

1.3　试剂

效价为 10000u/ml 制霉菌素，由 70%～75%乙醇溶解口服片配制成。PB 试剂，为 1/15mol/L 的 KH_2PO_4/K_2HPO_4 加 0.6mol/L 的 NaCl 配成，pH5.0。PBA 试剂，由 0.2mol/L Na_2HPO_4/NaH_2PO_4 加 0.4mol/L NH_4Cl 配成。20%PEG-6000，以 pH7.5 的 PBA 试剂溶解，另补加 0.01mol/L $CaCl_2$ 和 0.05mol/L 甘氨酸。工具酶为蜗牛酶（购自中国科学院生物物理研究所）、溶菌酶（SERVA，N.Y. ，USA ）和纤维素酶（日本，ONOZUKA，12-10 型）按不同比例混合后，以 PB 或 PBA 试剂溶解，最终 pH 值为 6.0。

1.4　营养缺陷型突变株的获得

1.4.1　原养型株诱变处理：按照常规方法[9]选择致死率为95% 以上的处理时间。

1.4.2　浓缩处理：诱变处理后的分生孢子悬液倒人CM平板上28℃培养至下一代孢子长出。然后按文献[9]的方法以制霉菌素进行浓缩处理。

1.4.3　营养缺陷型突变株的检出与鉴定，均按常规方法[9] 进行。

*原载于《微生物学通报》，23（6）：332~335，1996.

1.4.4 原养型株等的固体曲酶活力测定：淀粉糖化酶按照DNS比色法[10]测定活力；纤维素酶分别以DNS比色法[11]测定C_L酶活和C_X酶活。

1.5 营养缺陷型突变株的原生质体的制备

参照文献[8]的方法。纯化原生质体以无菌的双层高级镜头纸（广州耀华仪器用品厂）过滤酶解液，然后以500r/min离心沉淀原生质体。

1.6 原生质体融合处理

等量混合两亲本纯化原生质体各2.0×10^6个以上，加入PEG助融，振荡均匀后置28℃处理20min，以适量冷的PBA试剂（pH7）中止融合。趁冷离心（1000r/min，5min） 洗涤一次。经稀释后，取1.0×10^4个/ml以上稀释液涂布于高渗MM上，取$10^2\sim10^3$个/ml稀释液涂布于高渗SM上。

融合频率（Ff）：

$$\frac{\text{在MM上出现菌落数}\times\text{相应稀释倍数}}{\text{在SM上出现菌落数}\times\text{相应稀释倍数}}$$

1.7 融合子的检出与鉴定

综合文献[12，13]的方法进行。

2 结果和讨论

2.1 营养缺陷型突变株的获得

从菌株N343筛到3株维生素缺陷型突变株，编号NB1（B_1^-，PP^-），NB2（B_1^-， FA^-）和NB3（B_1^-，PP^-， FA^-）；从菌株UV-11筛到1株氨基酸缺陷型突变株，编号UB1（Pro^-）。

以上突变株的显微形态和培养特征均与各自的原养型菌株基本相同，但它们的产酶活力均比相应的原养型株明显下降。例如，突变株UB1的淀粉糖化酶活力仅为UV-11的50%，突变株NB3的纤维素酶活力仅为N343的45%（C_L酶）和35%（C_X酶）。

回复突变率的测定结果还表明，以上突变株均低于10^{-6}水平，故将它们作为融合亲本不会干扰以直接法检出融合子的可靠性。

2.2 原生质体制备和再生的适宜条件

以突变株UB1和NB3为材料，比较了不同的培养方法、菌龄、工具酶配比和渗透压等因素对原生质体制备和再生的影响，以找出最适宜的条件。

2.2.1 培养方法对原生质体制备的影响：选用6种菌丝体生长培养基进行摇瓶培养，结果表明，无论是菌株UB1或NB3，生长在CM与AM-1两种培养基中的菌丝体，经溶菌后，原生质体的释放量都最高；如果将CM改成固体CM并在其平板上放一块无菌玻璃纸进行培养，则效果比CM摇瓶培养更佳（表1）。并且玻璃纸培养法在随后的操作中具有简便和快速的优点。

表1 菌丝体培养方法对原生质体释放的影响*

菌株		NB3	UB1
摇瓶培养	PD	$<10^4$	5.0×10^4
	SM	2.0×10^4	1.0×10^5
	PM	$<10^4$	$<10^4$
	YM	$<10^4$	$<10^4$
	AM-1	2.0×10^4	3.0×10^5
	CM	3.2×10^4	4.5×10^5
CM 玻璃纸培养		1.6×10^5	6.6×10^5

* 表中为镜检的每 ml 原生质体数

2.2.2 菌丝体的菌龄对原生质体释放和再生的影响：以突变株UB1为材料，用玻璃纸培养法，在12～24h之间每2h取一块生长的菌丝体进行酶解。结果表明，菌龄14～16h的菌丝体，酶解3h释放的原生质体量最高，可达1.1×10^6个/ml，这些原生质体在CM上再生率为54.4%～59.2%。生长16h以后的菌丝体，随着菌龄延长，在同样的酶解时间内，原生质体释放量有逐渐下降的趋势，但其再生率则变化不大。

2.2.3 工具酶对原生质体释放的影响：将纤维素酶、蜗牛酶和溶菌酶以不同比例混合进行比较，结果表明，它们分别为0.8%、0.4%和0.2%时，两个突变株（UB1和NB3）的原生质体释放量均为最高（表2）。

表 2 工具酶的配比对原生质体释放最（个/ml）的影响

突变株	A*	B	C	D	E
NB3	$<10^4$	$<10^4$	4.0×10^4	3.5×10^4	5.0×10^4
UB1	5.0×10^4	1.0×10^4	8.0×10^5	1.2×10^6	7.4×10^5

* A=1.0%纤维素酶+0.5%溶菌酶；B=1.0%蜗牛酶+0.5%溶菌酶；C=1.0%纤维素酶+1.0%蜗牛酶；D=1.0%纤维素酶+0.5%蜗牛酶+0.25%溶菌酶；E=0.8%纤维素酶+0.4%蜗牛酶+0.2%溶菌酶

结果还表明，酶解温度以31～34℃最合适。当温度升至37～40℃则明显不利于原生质体释放。酶解时间在适温下以2～3h为宜。

2.2.4 稳定剂对原生质体释放和保存的影响：结果表明，突变株NB3和UB1的原生质体对半高渗溶液（如0.3mol/L NaCl）和高渗有机质溶液（如0.6mol/L蔗糖）都比较敏感，即它们在上述溶液中，分别在20～60min内就会破裂。但在高渗无机盐溶液中这些原生质体较为稳定，其中尤以在pH6.0的PBA试剂中最好。在此试剂中，NB3和UB1的原生质体释放量分别可达2.5×10^5个/ml和9.0×10^5个/ml，而且将它们保存于4℃中，7～8h内不会破裂。

2.3 原生质体融合和融合子的性状

2.3.1 融合处理时间与融合频率（F_f）的关系：融合处理时间对融合频率有明显的影响，表3显示，15～20min处理时间最为合适。如果继续延长时间，融合频率会迅速下降。这可能与PEG对原生质体的毒性和原生质体对有机质稳定剂的敏感性有关。

表 3 融合处理时间（t）与融合频率（F_f） 的关系

t/min	5	10	15	20	25	30
$F_f\times10^{-3}$	1.10	3.73	4.98	4.46	2.32	1.85

2.3.2 融合子的稳定性：以直接选择法从MM平板上检出融合子98株。其中有50 株分生孢子颜色与菌株UV-11相似，呈深褐色；另外48株分生孢子颜色与菌株N343的相似，呈黑色。但前50株融合子经传代2～3代后，其中29株的菌落出现镶嵌黑色与褐色的分生孢子，这表明是异核体融合子出现的分离现象。其余融合子在MM或CM平板上继续传代达10次，性状都稳定。

表 4 融合子 FNU32 和 FNU38 的酶活特征

酶活力*	出发菌株		融合子	
	N343	UV-11	FNU32	FNU38
C_L酶	108.4	28.5	102.1	93.7
C_X酶	1051.0	365.2	872.5	1122.0
淀粉糖化酶	835.4	7938.0	6664.0	5468.0

* 酶活力单位：C_L酶与C_X酶为 mg 葡萄糖/g（干曲）·h；淀粉糖化酶为 mg 葡萄糖/g（干曲）·min

2.3.3 融合子的酶系特征：从以上性状稳定的融合子中挑出融合子FNU32和FNU38，发现它们所含的高活力的酶系均为淀粉糖化酶和纤维素酶（表4），这分别是菌株UV-11和N343的酶系特征。说明通过融合后所得到的这两个融合子都同时包含双亲的基因组，因此将两个出发株具有的优良性能都分别囊括于一体。

参考文献

[1]Ferenczy L，Kevei F，Zsolt J. Nature，1974，248：793.
[2]辛明秀，马玉娥. 微生物学通报，1995，22（6）：365-370.
[3]邢来君，张军，孙光，等.真菌学报，1987，6（4）：242-247.
[4]邢来君，温廷益，罗会文，等. 真菌学报，1989，8（3）：227-232.
[5]辛明秀，蒋亚平. 微生物学通报，1994，21（3）：143-147.
[6]阎淳泰. 中国调味品，1985，11：26-31.
[7]郝勃，宋士良，邓岑华. 中国调味品，1987，3：11-13.
[8]曹军卫. 武汉大学学报（自然），1984，4：95-102.
[9]《微生物诱变育种》编写组. 微生物诱变育种，北京：科学出版社，1973.
[10]阎淳泰，朱火堂. 酿造学实验指导，华中农业大学，1987.
[11]齐义鹏. 纤维素酶及其应用，成都：四川人民出版社，1986.
[12]朱岛重臣. 日本酿造协会会志，1984，79（10）：721-723.
[13]梁平彦，刘宏迪，陈开英，等. 遗传学报，1981.8（4）：287-293.

2株黑曲霉原生质体的形成和再生过程观察*

郝 勃 阎淳泰 陈华癸

（华中农业大学微生物学系）

摘 要 在相差显微镜下详细地观察了2株黑曲霉在最适条件下形成原生质体及其再生的过程。结果表明，黑曲霉的原生质体均能从菌丝体的顶端及其他部位释放出来。释放的原生质体呈球形，大小不一。大多数原生质体内均能清晰地看见含有1个巨大的液泡。原生质体再生同时存在2 种形式：一是先从原生质体产生1个突出物，随后突出物逐渐增大并延伸成1条旋绕的无隔菌丝，继续发展成有隔的正常菌丝；第二种方式是由酵母状细胞短链发育成正常菌丝。

关键词 黑曲霉；原生质体形成过程；原生质体再生过程

自1974 年Ferenczy首次报道将细胞融合方法成功地引入丝状真菌领域以来[1]，原生质体融合技术已陆续在丝状真菌各种群中展开，并被证实为一种有效的育种手段。笔者曾通过原生质体融合技术，从中筛选到了兼备2个有关融合亲本优良性状的融合子[2]，在进一步的研究确认该融合子的稳定性之后，也就验证了原生质体融合在丝状真菌中工业用途菌株改良方面是重要的手段之一。但是欲使本技术取得效果，获得大量生活力强的纯原生质体是一个先决条件。因此，我们在上述的利用原生质体融合技术来提高黑曲霉产酶能力的研究中，详细地观察了2株黑曲霉在最适条件下的原生质体释放与再生过程。

1 材料与方法

1.1 材料

1）黑曲霉（ *Aspergillus niger* ）。营养缺陷型突变株N343-HB3 和UV-11-HB1，分别为笔者以亚硝酸常规诱变处理黑曲霉N343和UV-11得到[2]。

2）米曲汁培养基（CM）。由米曲霉（ *Aspergillus orzye*） 沪酿3.042的米饭曲水解液配制而成[2]。

3）工具酶。为蜗牛酶（ 购自中科院生物物理所）、溶菌酶（ N.Y.，USA，SERVA出品，效价不低于24 000 U/mg）和纤维素酶（日本Yakult Honsha Co.出品） 以2∶4∶1重量混合以后，用pH 6.0的PBA试剂（即1/15 mol/ml的KH_2PO_4-K_2HPO_4缓冲溶液加上0.6mol/ml的 NaCl）溶解得到。

1.2 方法

1）原生质体制备。在CM平板上铺1张无菌圆形（ 与平皿直径相当） 的玻璃纸，将供试菌株的分生孢子悬液涂上，于28℃倒置培养16～18h后，将生长在玻璃纸上的菌丝体与玻璃纸一同取下，放入15ml工具酶液中，在31℃～34℃水浴中保温3h，同时以50～100r/min速度振荡。用冰浴中止酶解，然后趁冷经双层无菌镜头纸过滤，滤液以500r/min 离心10min，沉淀以0.6 mol/ml 的 NaCl液悬浮。

2）原生质体再生。将上述悬浮液中原生质体培养（28 ℃）在等渗麦芽汁（4~ 5Balling）中摇瓶（50～100r/min），每2～4 h 取样于相差显微镜下观察。

2 结果

2.1 原生质体形成过程

供试菌株的原生质体和其他许多常见丝状真菌一样，可以从菌丝顶端和其他部位释放。如图1-A～B显示，在菌丝顶端和其他部位均同时出现菌丝膨大变形，说明这都是由于细胞壁已被水解掉，细胞膜

*原载于《华中农业大学学报》，16（1）：48~51，1997.

凸出将要成为原生质体。原生质体释放时，通常是在菌丝体的释放部位先膨大形成1个小球体（ 如图1-A菌丝顶端所示） ，然后该球体逐渐增大，最后脱离菌丝；有的还可以在原有的释放部位紧接着又形成一个或几个原生质体，此种现象，可能是由于菌丝体有关泡囊形成与细胞膜融合的机制[3]所导致的。有报道[4, 5]认为，在菌丝上缓慢膨大脱落的原生质体较为稳定，很快形成的原生质体则很脆弱。但我们未曾发现很快破裂的原生质体，这表明我们选择的条件，不仅释放的原生质体数量多（镜检每ml含原生质体个数：N343-HB3为1.7×10^5，UV-11-HB1为6.6×10^5），而且质量也很好，从而使再生率很高（在CM平板上再生率为54.4%～59.2%）。

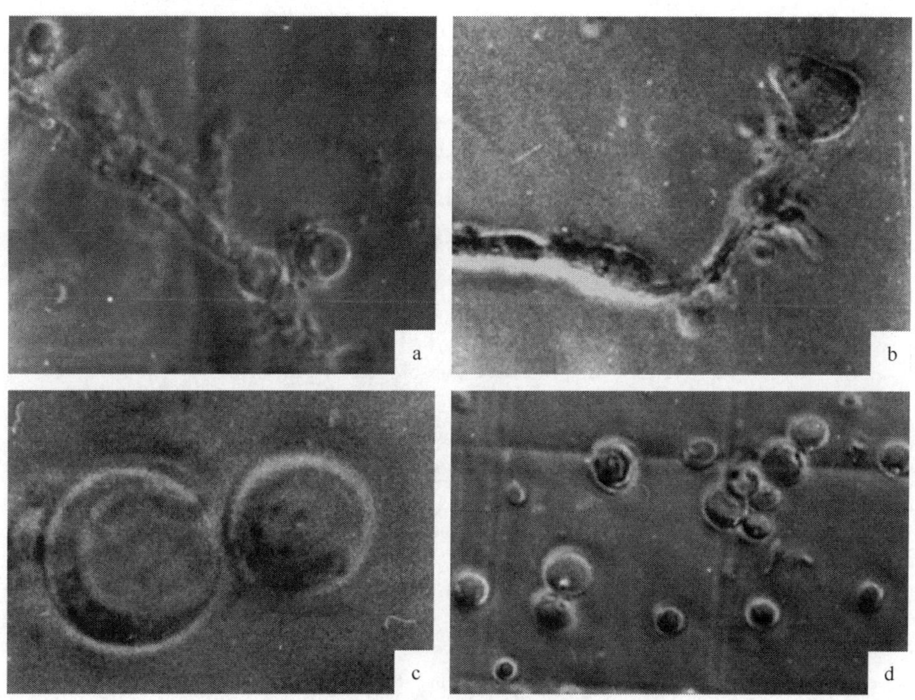

图 1 原生质体形成过程

Fig. 1 The procedure of protoplast formation

a. N343-HB3 菌丝顶端释放出原生质体（ × 700）；b. UV-11-HB1 菌丝顶端和中部释放出原生体（ × 700）； c. N343-HB3 释放出的原生质体呈球形（ × 1100）； d.UV-11-HB1 释放出的原生质体呈球形（ × 700）

释放的原生质体多呈球形，大小不一，直径在1～20μm之间，但以5～10μm直径的中等大小的原生质体最多，其原生质中一般含1个巨大的液泡，其余的内含物呈月牙形，透明性较液泡差，如图1-C～D所示。有报道[4]认为，在丝状真菌早期形成的原生质体才含有液泡，但我们在观察供试菌株的原生质体释放过程中，却未见到含颗粒状细胞质的原生质体。

2.2 原生质体再生过程

供试菌株的原生质体在液体培养基中有2种再生形式：一种是首先在原生质体距液泡的近端伸出1个突起物而使其失去球形（ 图2-A），随后该突起物或称“芽”逐渐增大（ 图2-B）并延伸发展成1条旋绕的无隔菌丝（ 图2-C）；随着培养时间延长，该菌丝发育成正常菌丝；另一种形式是原生质体先发育成酵母状细胞短链（ 图2-D），而后形成正常菌丝。根据我们观察，在本方法下从原生质体到正常菌丝体的整个再生过程需要15～17h。此外，在丝状真菌中除了上述两种再生形式外，尚有第3种方式[4]，即从原生质体上多个部位发育出芽管，而后长出再生的菌丝体。但在本观察中没有此种再生形式。

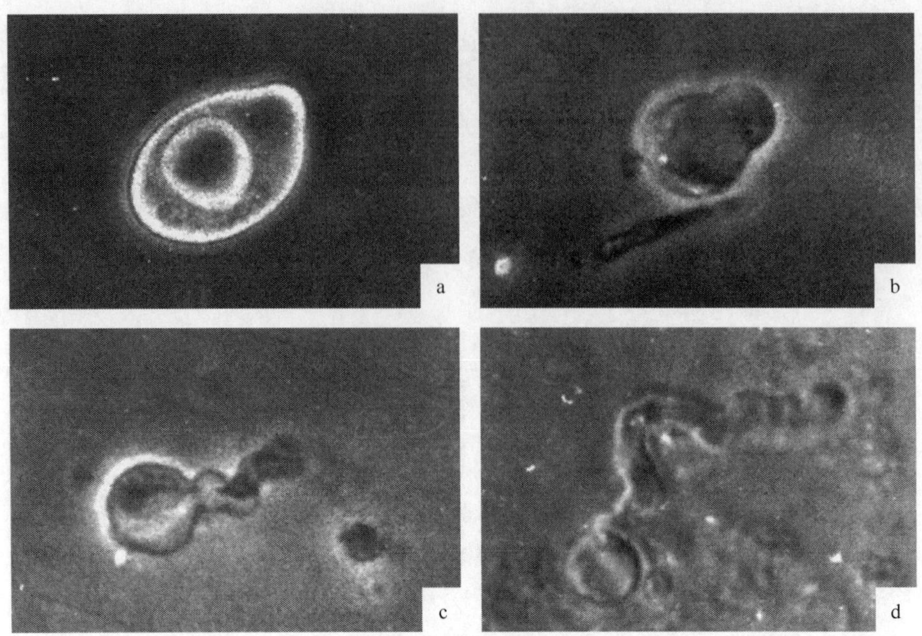

图 2 原生质体再生过程

Fig. 2 The procedure of protoplast regeneration

a. 原生质体液泡近端伸出的突起物（× 1100）；b～c. 原生质体发育成旋绕的无隔菌丝（ × 1100）；
d. 原生质体发育成酵母状细胞短链（ × 1100）

参 考 文 献

[1] Ferenczy L，Kevei F，Zsolt J. Fusion of fungal prot oplasts. Nature，1974，248：793～794.

[2]郝勃. 黑曲霉 N343 和黑曲霉 UV-11 营养缺陷型突变株原生质体融合的研究：[硕士学位论文] . 武汉：华中农业大学图书馆，1989

[3]周与良，邢来君. 真菌学. 北京：高等教育出版社，1986. 40～44.

[4]陈漱，曹军卫. 几种丝状真菌原生质体的形成与再生. 真菌学报，1986，5（ 2） ：117～123.

[5]曹军卫. 黑曲霉原生质体的制备及再生. 武汉大学学报（ 自然科学版） ，1984. 4：95～102.

绿衣观音土曲的特征和制作工艺*

廖美德[1]　陈华癸[2]　许耀才[2]

（1. 湖北农学院农业工程系）

（2. 华中农业大学土化系）

绿衣观音土曲又名“绿衣小曲”、“绿曲”、“南曲”。是近年来在湖南常德地区、湖北江汉平原和鄂东南地区普遍用来酿制小曲白酒的曲种。该种小曲有不同于其他小曲曲种的表现特征，具有独特的功能微生物组成和制作工艺。现就其特征和制作工艺报道如下。

1　绿衣观音土曲的特征

绿衣观音土曲包括引曲（曲母、种曲）和土曲（成曲）两部分。此曲以大米为原料，成品引曲圆饼形，直径5.0～7.0cm，厚1.5～3.0cm，表面淡黄、多皱纹、有酒香。土曲以观音土、大米、米糠等为原料，圆球形，直径5.5～6.5cm，重9.0～130g，有深裂缝，曲表密被菌丝和绿色分生孢子，故表面为绿色。

2　绿衣观音土曲的制作工艺

2.1　引曲的制作

2.1.1　工艺流程

大米→浸泡→洗净白浊→和水磨浆→滤干→草木灰吸干→拌引曲粉→制坯→入箱→开箱→倒盘→一烧→二烧→三烧→烘干（晒干）→成品引曲。

2.1.2　原料配比和加工

大米（m）：引曲粉（m）=30：25

取新鲜早稻30kg，以清水浸泡10～16h ，洗去白浊后和水磨浆，滤水后用草木灰吸干，拌入2.5kg引曲粉，制坯。

2.1.3　培菌

保温培养分两步进行，即曲箱培养和曲架培养。

曲箱培养　先在箱底铺一层蒸煮过的谷壳，然后将曲坯单层置于谷壳上。曲坯先为圆球形，放置后自然下塌成圆饼状。于28℃~30℃室温中培养，当箱内品温达34℃~36℃时开箱（一般曲箱内培养12~18h）。

曲架培养　开箱后将曲饼倒盘，置于竹匾上，第一天先单匾置于曲架上培养（称之为一烧），经22～24h后，将两匾叠放继续于曲架上培养（称之为二烧，以下类推），再经22～24h进入三烧（22-24h），三烧结束，将曲饼晒干或烘干即得成品引曲。

2.1.4　引曲制作过程中表观变化

曲坯入箱后，经8h品温升至室温，坯内微生物生长旺盛，可听见有冒气泡声，曲表有少量短绒毛状菌丝突起，曲表白色。曲箱培养结束时，表面有细密皱纹。经三烧后，曲表逐渐由白变为淡黄，皱纹加深。成品引曲泡松有酒香。

2.2　土曲制作工艺

2.2.1　工艺流程

（观音土粉、黏米糠、米粉、引曲粉）拌合→加水和匀→制坯→入箱→开箱→倒盘→一烧→二烧→三烧→四烧……→六烧→烘干→成品土曲（成曲）。

---

*原载于《湖北农学院报》，14（2）：61~63，1994.

2.2.2 原料配比和加工

观音土（m）：黏米（m）：黏米糠（m）：引曲粉（m）=65：10：25：15：1.5～2.5

观音土、大米用前先磨碎过筛，将原料充分混合后，加水拌匀，加水量以捏成形不散及放置不变形为度。

2.2.3 培菌

土曲的培养与引曲相似，亦分为二步培养。曲箱培养时，土曲坯可堆积培养，不必单层放置培养，曲架培养经一烧、二烧、三烧、四烧、六烧后即可出房烘干。

2.2.4 土曲制作过程的外观变化

土曲入箱后经8h微生物生长旺盛，品温上升到室温，经15~20h后，品温上升到34℃~36℃时开箱。在曲箱中曲表无明显变化，可听到冒气泡声和闻到酒香味。一烧结束时，曲表有一层白色菌丝膜、并且有白色点状菌丝密集区，二烧结束时，曲表有少量绿色分生孢子形成，至三烧，分生孢子逐渐增多，并开始出现浅裂缝，四烧结束时曲表分生孢子层加厚，裂缝加大，六烧则无明显变化，曲表水分潮湿感降低。

绿衣观音土曲制作过程中的温度控制用焚烧谷壳以暗火保持室温在28℃~30℃，湿度控制在80%~90%之间，若温度或湿度过大，则以开门或开天窗来调节。

3 讨论

绿衣观音土曲是众多小曲中的一种，同其他小曲相比，绿衣观音土曲的独特之处表现在以下几点：

第一，表观特征独特。绿衣观音土曲的引曲为淡黄色，曲表多皱纹；土曲表面被绿色分生孢子，有深裂缝，而其他小曲曲表白色，光滑。

第二，制作原料不同。该曲引曲以大米为原料，土曲以观音土、大米、米糠为原料，且观音土占原料的70%。其他小曲多以大米为原料，少数混有少量米糠。

第三，留种方式不一样。绿衣观音土曲包括引曲和土曲两部分，土曲专门用于酿酒，引曲则专门用于繁殖土曲和引曲。其他小曲则以上一次成品小曲为下一次种曲。

第四，培菌方式不同。绿衣观音土曲培菌分曲箱培菌和曲架培菌两步，出箱后利用竹匾叠放保持曲的品温，蒸发水分。而其他小曲多先用缸或其他容器培养，待收汗后，堆积于箩筐中保持温度，促进吐湿，且不断翻曲以防升温过高。

第五，接种方式不同。绿衣观音土曲将引曲粉与原料充分混匀后制坯。而其他小曲多先踩曲、切曲、制坯，然后用裹粉的方法接种。

第六，酒曲中微生物组成特别。绿衣观音土曲中主要糖化菌为棒曲霉（*Aspergillus clavatus*），而其他小曲中的糖化菌为根霉（*Rhizopus* sp.）和毛霉（*Mucor* sp.）。

参考文献

[1]赵学慧. 华中农学院学报，1983，2（2）：26-28.
[2]景泉. 酒曲生产实用技术.北京：食品工业出版社.1983.

棒曲霉 *HD*-11 的形态特征的研究*

廖美德[1]　许耀才[2]　陈华癸[2]

（1. 湖北农学院农业工程系，荆沙市，434103）

（2. 华中农业大学土化系，武汉，430070）

摘　要　从酒曲中分离得到菌株*HD*-11，经鉴定属棒曲霉（*Aspergillus clauatus*）。本文报道了其形态特征，并将*HD*-11 与棒曲霉群的几个菌株进行了比较。

关键词　棒曲霉*HD*-11；形态特征；菌株

棒曲霉*HD*-11 是从绿衣观音土曲中分离得到的一个具有较高糖化酶活性的菌株。本文研究了其形态特征并将它与棒曲霉群其他几个菌株进行了比较。结果报道如下。

1　材料与方法

1.1　材料

HD-11（*Aspergillus clavatus*），从酒曲中分离得到；*As*3.724（*Aspergillus clacatus*）、*As*3.7909（*A.clavatus*）、*As*3. 4298（*A.clavatu nanica*）、*As*3.98（*A. gigantus*），中科院北京微生物所提供。

1.2　培养基

察氏培养基　其配方为：硝酸钠3g，磷酸氢二钾1g，硫酸镁0.5g，氯化钾0.5g，硫酸亚铁0.01g，蔗糖30g，水1000ml。

马铃薯培养基（PDA）其配方为：20% 马铃薯浸汁1000ml，葡萄糖20g，琼脂20g。

1.3　扫描电镜样品制备

在察氏平板上培养供试菌种→挑取分生孢子头粘台→锇酸熏蒸→真空干燥→离子溅射法喷镀→观察→照相。

2　结果

2.1　棒曲霉*HD*-11 形态特征（图版Ⅰ～图版Ⅴ）

HD-11 在 PDA 培养基上保温30℃培养7d，菌落直径为50～60mm；在察氏培养基上保温30℃培养7d，菌落直径为30～40mm。菌丝体白色，有隔。气生菌丝毛毡状，紧贴基质生长，孢子头丰富，从基内菌丝生出，近环状排列（在PDA 上尤为明显）。顶囊棒状，表面呈灰蓝绿色（幼时为白色，在察氏培养基上有分化）。菌落边缘有宽1～5mm 的薄层气生菌丝紧贴基质。在察氏培养基上，菌落表面微皱，菌落明显向上隆起。在PDA 上，菌落平坦，无皱纹，顶囊和气生菌丝的顶端均有无色透明黏液状分泌物。菌落背面暗黄色，有近同心状环和放射状深皱纹。

分生孢子产孢结构长0.7～3.0mm，棒状，成熟时顶端开裂呈3 瓣、4 瓣或5瓣，分生孢子梗光滑，（18～29）μm×（750～2470）μm，顶囊棒状，（29～90）μm×（120～250）μm。分生孢子小梗单层、密集全面着生，生于基部小梗：（2.5～3.0）μm×（2.0～3.0）μm，生于顶端小梗：（2.5～3.0μm）×（7.0～9.0）μm。分生孢子卵圆形或近球形，表面光滑或有浅凹纹，呈灰蓝绿色或白色，直径大小为（3.0～4.5）μm。分生孢子串生于小梗顶端，以孢间连桥联结，紧密呈柱状排列在顶囊上，成熟时裂开呈现3 瓣、4 瓣或5 瓣，如短柱状。

*原载于《湖北农学院报》，15（2）：102~104，1995.

2.2 *HD*-11与棒曲霉群几个菌株的比较

将*HD*-11 同中科院北京微生物所保存的棒曲群的*As*3.3909、*As*3.724、 *As*3.4298、*As*3.98从形态和生理上进行比较，结果见附表，并参见图版Ⅵ。

附表 *HD*-11 同棒曲霉群几菌株的比较（PDA 培养基）

项目	*HD*-11	*As*3.3909	*As*3.724	*As*3.98	*As*3.4298
中文名	棒曲霉	棒曲霉	棒曲霉	巨大曲霉	矮棒曲霉
拉丁名	*Asprgllus clavataus*	*A. clavatus*	*A. clavatus*	*A. glgantcus*	*A.clavatu nanlca*
菌落	菌落先期白色，毛毡状，后期密被绿色分生孢子头，背面有放射状纹	菌落先期白色后期密被蓝绿色分子孢子头，久置后孢子头呈灰色，背面一般无皱纹	菌落先期白色后期密被蓝绿色分生孢子头，久置后呈铁灰色，背面多平整	菌落白色，久置后不变色，气生菌丝发达，背面平坦	菌落先期白色扩展慢，后期产生较稀少深色分生孢子头，气生菌发达，背面平坦有分泌物
分生孢子头	分生孢子头近环状排列，先为白色后变为蓝绿色	分生孢子头近环状排列，先白色经蓝绿色变铁灰色	分生孢子头散生，由白色、蓝绿色变为铁灰色	分生孢子头散生，白色	分生孢子头散生，先白色后为暗褐色
分生孢子梗	分生孢子梗比菌丝体宽，由基内菌丝生成	分生孢子梗比菌丝体宽。由基内菌丝生成	分生孢子梗比菌丝体宽。由基内菌丝生成	分生孢子梗与菌丝同宽，菌丝或气生菌丝生成	分生孢子梗比菌丝体宽，由基内菌丝或气生菌丝生成
顶囊	棒状 小梗全面着生基部光滑	棒状 小梗全面着生基部光滑	棒状 小梗全面着生基部光滑	近球形 小梗全面着生基部有疣状物	棒状 小梗着生 2/3 区域，基部光滑
分生孢子	卵圆形，光滑或浅纹，分生孢子在顶囊上紧密排列，老熟时孢子团开裂成短柱状	卵圆形，光滑或浅纹，分生孢子在顶囊上紧密排列，老熟时孢子团开裂成短柱状	卵圆形，光滑或浅纹，分生孢子在顶囊上紧密排列，老熟时孢子团开裂成短柱状	圆球形，有 状物，分生孢子顶囊上呈链状，排列松散	长圆柱形，一端较大，一端较小，较大的一端有一环状纹，分生孢子在顶囊上呈链状，排列松散。
生活周期（30℃）	36～40 h	24 h	24 h	24 h	120 h

3 小结

HD-11是从酒曲中分离得到的菌株，同*As*3.3909、*As*3.724相比，它们在分生孢子、分生孢子头形态上没有明显差异，三者仅在菌落形态和某些生理特征上有差异。*HD*-11明显不同于*As*3.4298和*As*3.98。根据Raper[3]的分类标准，HD-11属棒曲霉（*Aspergillus clavatus*）。

参 考 文 献

[1]《常见与常用真菌》编写组. 常见与常用真菌. 北京：科学出版社，1978.
[2]柏内特 J H. 真菌学基础. 北京：高等教育出版社，1989.
[3]Raper. The Gene Aspergillus. Willams Wilkins Balt imore Manryland，1965.

湖北省十二个磷矿的磷矿粉肥试验研究*

陈华癸　李学垣　尹名济　徐凤琳　麦月珠等

（华中农学院磷矿粉肥有效化研究组）

湖北省磷矿资源丰富，矿点多，储量大，但矿床成因和矿石性质比较复杂。为了查明不同矿出产的磷矿粉肥的性质和肥效以及影响肥效的因素，有利因矿制宜，合理利用，多、快、好、省地满足农业生产大上、快上对磷肥的需要，我们收集了十二个磷矿的 40 多个矿石样品，进行了：①磷矿粉肥的当季肥效；②磷矿粉肥的持续肥效，③磷矿性质和当季肥效的关系等三方面的试验研究。现将已有结果整理如下。

一、当季肥效的盆栽试验

1. 五种磷矿粉肥对红花草子当季肥效的比较试验

1972 年秋，我们用远安、荆襄、黄陂、广济、黄梅五个磷矿出产的磷矿粉肥样品进行了盆栽试验。磷矿粉肥细度为 80%过一百目筛（下同）。供试土壤是本院实习农场死黄土，土壤 pH5.5（KCl），有机质含量 0.6%（丘林法），速效磷（P_2O_5）含量 0.9 毫克/100 克土（0.5 M $NaHCO_3$ 法）。

土壤锤细后过 1 厘米筛。用 20×20 厘米白瓷钵，装土 12 斤。妹钵施尿素 1.8 克，硝酸钾 1.2 克。对照钵不施磷肥。施磷矿粉肥处理钵，每钵施磷矿粉肥 10 克。各处理重复 4 次。1972 年 10 月 10 日播种，1973 年 3 月 28 日收获，产量以全植株地上部分鲜重计算。

试验结果（表 1 和图 1）表明，在严重缺磷的死黄土上，各种磷矿粉肥对红花草子都有显著的增产效果，但差别较大。远安磷矿粉肥比对照增产 446%，荆襄、黄梅、广济、黄陂磷矿粉肥肥效依次下降，黄陂磷矿粉肥比对照增产 65%。

表 1　五种磷矿粉肥对红花草子的当季肥效

处理	红花草产量（克/钵）	产量比率（%）
不施磷肥	41.3	100
黄陂磷矿粉肥	68.3	165
广济磷矿粉肥	97.3	236
黄梅磷矿粉肥	111.3	269
荆襄磷矿粉肥	131.0	317
远安磷矿粉肥	225.6	546

图 1　四种磷矿粉肥对红花草子的当季肥效

*原载于《湖北农学院报》，15（2）：102~104，1995.

2. 十种磷矿粉肥对黄豆当季肥效的比较试验

1973年春，用黄陂、郧西、广济、荆襄、孝感、黄梅、大悟、通山、鹤峰、远安10个磷矿出产的磷矿粉肥样品进行了盆栽试验。黄豆品种为六月爆。供试土壤和施用的氮、和磷矿粉肥量同前。设两种对照：①不施磷肥，②施过磷酸钙（每钵3克）。各处理重复4次。1973年4月4日播种，7月6日收获，称子粒重和总干物质重。结果见表2和图2。

试验结果同样表明，在严重缺磷的死黄土上，10 种磷矿粉肥直接施用，对当季黄豆都有显著的增产效果，各种磷矿粉肥肥效的差别也较大，大致可分为两类：

第一类，远安、鹤峰、通山和大悟四个磷矿的磷矿粉肥。它们的肥效比不施磷肥的增产 167.2–267.2%，与过磷酸钙的肥效相近，甚至超过。

第二类，黄梅、孝感、荆襄、广济、郧西出产的磷矿粉肥。它们的肥效，比不施磷肥的增产 57.8 %~101%，相当于过磷酸钙肥效的57.7%~73.7%。

黄陂出产的磷矿粉肥比不施磷肥的只增产20.3%，相当于过磷酸钙肥效的44%。其肥低的原因可能是由于试验用的矿粉代表性的偏差（试验用黄陂磷矿粉肥样品含全磷量22%，有效磷量0.86%。但我们从黄陂磷矿收集了各种类型矿样8种进行化学分析，样品的平均含全磷量为26%，有效磷量为1.86%。见本文第四部分）。

表2 十种磷矿粉肥对黄豆的当季肥效

实验处理	总干物质重（克/钵）	子粒重（克/钵）	肥效率（%）	
			（一）	（二）
不施磷肥	19.7	6.4	100	-
黄陂磷矿粉肥	23.3	7.7	120.3	44.0
郧西磷矿粉肥	30.4	10.1	157.8	57.7
广济磷矿粉肥	31.5	10.9	170.3	62.3
荆襄磷矿粉肥	34.8	11.9	185.4	68.0
孝感磷矿粉肥	35.1	12.2	190.1	69.7
黄梅磷矿粉肥	37.9	12.9	201.6	73.7
大悟磷矿粉肥	49.9	17.1	267.2	97.7
通山磷矿粉肥	47.7	17.2	268.8	98.3
鹤峰磷矿粉肥	62.9	21.7	339.1	124.0
远安磷矿洞肥	65.6	23.5	367.2	134.3
过磷酸钙	52	17.5	273.4	100

注：0.05差异显著标准1.3克/钵，0.01差异显著标准5.4克/钵。

图 2 十种磷矿粉肥对黄豆的当季肥效

二、当季肥效的田间试验

在 1972 年黄陂县五岭公社，武昌县流芳公社磷矿粉肥试验示范基础上，1973 年秋季在本院实习农场的青泥田（18 丘）、胶泥田（18 丘）和黄泥田（小垅）上，进行了磷矿粉肥对红花草子当季肥效的田间试验。

供试土壤每百克土含速效磷量（方法同前）：青泥田为 1.6 毫克，胶泥田 1.8 毫克，黄泥田 2.1 毫克。供试磷矿粉肥；荆襄磷矿粉（全磷 29.5%，2%柠檬酸浸提有效磷 2.2%）和宜昌磷矿粉（全磷 28.3%，2%柠檬酸浸提有效磷 4.5%）。

小区面积 0.1 亩，重复 3 次，单行顺序排列。各处理每亩施磷矿粉肥 150 斤，设不施磷肥和施过磷酸钙（每亩 50 斤）两个对照。磷肥于 1973 年 7 月 26 日双季晚稻第一次中耕时施下。红花草子于当年 9 月 21 日播种，1974 年 4 月 11 日收割，称地上部分鲜重。结果见表 3。

表 3 两种磷矿粉肥在三种土壤上对红花草子的当季肥效

土壤	处理	鲜重（斤/区）				肥效率（%）
		1	2	3	平均	
清泥田	不施磷	267.2	283.9	270.2	273.8	100
	荆襄磷矿粉肥	308.9	312.3	325.7	315.6	115.3
	宜昌磷矿粉肥	355.7	367.4	362.4	361.8	132.6
	过磷酸钙	367.4	384.1	375.8	375.8	137.3
胶泥田	不施磷	250.5	267.2	342.2	253.3	100
	荆襄磷矿粉肥	255.6	292.3	308.9	288.6	112.8
	宜昌磷矿粉肥	308.5	325.1	317.3	316.9	125.7
	过磷酸钙	317.3	334.0	359.1	336.8	132.9
黄泥田	不施磷	267.8	269.6	265.9	267.8	100
	荆襄磷矿粉肥	305.0	300.2	292.1	299.1	111.7
	宜昌磷矿粉肥	323.6	318.3	320.7	320.8	119.8
	过磷酸钙	338.7	332.7	335.0	335.4	125.3

注：试验资料用综合法进行统计处理，两处理平均数差试验误差 2 倍以上，表明处理差异显著。实验误差：青泥田为土 66 斤，胶泥田为土 15.2 斤，黄泥田为土 3.2 斤。

试验结果表明，两种磷矿粉肥在三种土壤上的肥效差别规律是一致的：①两种磷矿粉肥均有增产效果；②荆襄磷矿粉肥的肥效相当于过磷酸钙的 50%左右，和盆栽试验结果相符；③宜昌磷矿粉肥的肥效与过磷酸钙接近。参照化学分析和田间试验的结果，可以认为宜昌磷矿粉肥的当季肥效与远安，鹤峰磷矿粉属于同一类型。

三、十种磷矿粉肥持续肥效的盆栽试验

在十种磷矿粉肥对黄豆当季肥效比较试验（见表 2）的基础上，又连续进行了三茬作物（第二茬荞麦；第三茬红花草子；第四茬水稻）的持续肥效比较试验。

表 4 十种磷矿粉肥对第二茬作物荞麦的肥效

实验处理	干物质重（克/钵）					肥效率（%）
	1	2	3	4	平均	
不施磷	10.3	9.0	10.2	9.5	9.8	100
黄梅磷矿粉肥	32.9	35.8	34	39.1	35.5	362.2
广济磷矿粉肥	35.8	36.1	39.9	38.0	37.7	384.7
大悟磷矿粉肥	40.4	36	38.3	39.6	38.6	393.9
通山磷矿粉肥	39.1	42.6	36.9	40.9	39.6	404.1
孝感磷矿粉肥	40.9	37.0	39.0	41.1	39.5	403.1
黄陂磷矿粉肥	37.4	35.7	35.9	39.8	37.2	379.6
荆褒磷矿粉肥	40.2	43.0	39.8	33.4	39.1	399.0
郧西磷矿粉肥	37	40	37.8	35.9	37.7	384.7
远安磷矿粉肥	40.3	36.9	41.8	39.6	39.7	405.1
鹤峰磷矿粉肥	38	40.5	42.8	41.9	40.8	412.1
过磷酸钙	33.2	37.1	36.4	37.5	36.5	372.4

注：0.05 差异显著标准 0.3 克 / 盆，0.01 差异显著标准 7.2 克 / 盆。

1．第二茬作物——荞麦试验

1973 年 6 月 20 日播种，8 月 4 日每盆追硝酸钾 1 克，8 月 20 日每盆再追硝酸钾 1 克，9 月 24 日收获。因花而不实，只得每盆干物质重量。试验结果见表 4。荞麦生长期间有如下表现：

（1）出苗后 20 天，各种施磷肥处理的长势都比不施磷肥旺盛。远安，鹤峰、通山，大悟四种磷矿粉肥表现最好，过磷酸钙次之，其它磷矿粉肥又次之。出苗一个月以后，十种磷矿粉肥肥效的差别日益缩小，但都超过了过磷酸钙。

（2）出苗后现第 1 片真叶时，不施磷肥处理即表现缺磷象徵，叶色暗绿，幼茎紫红，生长缓慢。但出苗一个月以后，长势转旺，两个月后，施磷处理的叶色都退淡，下叶枯死，长停滞，但不施磷的仍然青枝绿叶。

2. 第三茬作物——红花草子试验

1973 年 9 月 25 日播种，10 月 12 日和 12 月 14 日每钵分别追施硝酸钾 1 克和少量微量元素混合液。1974 年 8 月 26 日收获，测鲜重。试验结果见表 5。

表 5 十种磷矿粉肥对第三茬作物红花草子的肥效

实验处理	背草鲜重（克/钵）					肥效率（%）
	1	2	3	4	平均	
不施磷	27	37	30	31	28.7	100

续表

实验处理	背草鲜重（克/钵）					肥效率（%）
	1	2	3	4	平均	
黄梅磷矿粉肥	130	125	130	130	129	449.5
广济磷矿粉肥	125	135	135	125	130	453.6
大悟磷矿粉肥	215	210	225	225	219	763.1
通山磷矿粉肥	185	150	190	150	169	588.8
孝感磷矿粉肥	170	170	180	185	176	613.2
黄陂磷矿粉肥	135	130	130	135	132	459.9
荆褒磷矿粉肥	210	220	225	215	217.5	756.1
郧西磷矿粉肥	210	210	215	195	207.5	723.0
远安磷矿粉肥	420	410	400	395	406.5	1416.4
鹤峰磷矿粉肥	230	220	230	235	228.7	796.9
过磷酸钙	85	90	95	85	88.7	309.1

注：0.05 差异显著标准 12.1 克/钵，0.01 差异显著标准 16.3 克/钵。

红花草子的生育状况有如下表现：

（1）在二叶期，各处理均表现缺磷微象，处理间差异不明显，五叶期后（播种后一个月），所有施磷矿粉肥处理的红花草子长势都超过了不施磷和施过磷酸钙的处理。两个月后，各种磷矿粉处理间的差异持续显著。

（2）在各种磷矿粉肥对红花草于当季肥效的盆栽试验中，年前肥效表现都不够明显，但在第三茬肥效试验中，各种磷矿粉肥在年前都表现出明显肥效。足见在第一茬作物施的各种磷矿粉肥中的磷，到第三茬作物时，已明显的向有效形态方面转化。

（3）与不施磷和施过磷酸钙的相比，施磷矿粉肥的在第二，第三茬作物上的增产幅度有显著提高。例如远安磷矿粉肥的肥效率（以不施磷肥为 100%），在第一茬作物黄豆上为 367.2%（表 2），第二茬作物荞麦为 405.1%（表 4），第三茬作物红花草子则为 1416.4%（表 5）。又如施黄陂磷矿粉肥的，第一茬作物增产不显著，而第二，第三茬作物却增产四倍左右。

3．第四茬作物——水稻试验

将不施磷，施过磷酸钙，施荆襄磷矿粉肥和远安磷矿粉肥等处理，已经种过三茬作物的盆栽土壤移至不漏水的盆钵中，进行水稻试验。1974 年 5 月 9 日插秧（品种为华矮 15 号），每钵 3 穴，每穴 5 株。6 月 21 日每钵追施硫酸铵 3 克。7 月 22 日收获。试验结果见表 6。

表 6　两种磷矿粉肥对第四茬作物水稻的肥效

处理实验	子粒干重（克/钵）					肥效率（%）
	1	2	3	4	平均	
不施磷	8.0	7.9	7.0	7.4	7.6	100
过磷酸钙	12.4	14.6	12.3	12.8	13.2	173.9
荆襄磷矿粉肥	19.4	19.5	17.8	18.9	18.9	248.7
远安磷矿粉肥	20.6	18.9	20.2	23.3	20.7	272.5

施远安，荆襄磷矿粉肥的处理，经过三茬作物以后，在第四茬作物水稻上产生了显著肥效，比不施磷肥的分别增产 172.5%和 148.7%，比施过磷酸钙的分别增产 56.7%和 43.1%。

1973 年我们用荆襄磷矿粉肥在死黄土上对水稻做的当季肥效试验。并未表现出增产效果。可见，磷矿粉肥在第四茬作物水稻上的肥效，是磷矿粉肥在前三茬作物中，产生了向有效形态方面转化的结果。

4．磷矿粉肥肥效的持续性和有效化

将四茬作物的试验结果（表 2、表 4、表 5 和表 6）结合起来考查，可以看出，各种磷矿粉肥的后效都超过过磷酸钙的后效，而且过磷酸钙的后效主要是速效养分残效，磷矿粉肥的后效则应主要归结为其中迟效成分向有效形态转化。例如，在表 1 中，黄陂、广济，黄梅、荆襄和远安五种磷矿粉对红花草子的肥效，分别比不施磷肥的处理增产 65%、136%、169%、217%和 446%；在表 5 中，同样五种磷矿粉肥对第三茬作物红花草子的肥效，分别比不施磷肥的增产 349.9%、353.0%、349.5%、656.1%和 1316.4%（应指出，表 1 和表 5 的结果是在同一种土壤而不是同一钵土壤上得到的。表 2、表 4、表 5 和表 6 的结果则是来自同一盆钵土壤的持续性肥效试验）。

（待续）

湖北省十二个磷矿的磷矿粉肥试验研究（续）*

陈华癸 李学垣 尹名济 徐凤琳 麦月珠 等

（华中农学院磷矿粉肥有效化研究组）

四、矿石性质和肥效关系的探讨

1. 矿石类型、全磷量、有效磷量与肥效的相关性

我们对进行肥效试验的十种磷矿粉肥的矿石类型（薄片矿物鉴定）、全磷量（酸溶法）和有效磷量（2%柠檬酸浸提）进行了鉴定和分析（见表 7），结果可以看出：

（1）对红花草子和黄豆的当季肥效，基本上与矿石有效磷含量成正相关，与全磷量不呈数量相关。如供试样品中，远安、荆襄磷矿粉的全磷量基本一样（分别为 29.8%和 28.5%），但有效磷含量差别较大（分别为 3.60%和 1.86%），对红花草子和黄豆的增产效果几乎相差一倍。而供试样品中郧西、广济磷矿粉的全磷量虽差别大（分别为 26%和 10.1%），但有效磷含量相近（分别为 1.1%和 1.04%），故对黄豆的当季肥效接近（参看表 1、表 2）。由此推想，亩施 100~150 斤远安、鹤峰或大悟磷矿粉肥（有效磷含量为 3.6%~4.28%），与亩施 30~50 斤过磷酸钙，由于两种磷肥含有效磷的总量相差不远，所以其肥效也应相近，盆栽和田间试验的结果证实了这一点。

表 7 十种磷矿粉肥的矿石类型和含磷量

磷矿产地	矿床成因类型	矿石类型*	全磷量（%）	有效磷（%）	枸溶率**（%）	第一茬作物黄豆当季增产顺序	第三茬作物红花草子增产顺序
黄陂	前震旦纪沉积变质	晶质磷灰岩	22.0	0.86	3.90	10	8
郧西	震旦纪沉积	胶磷矿	26.0	1.10	4.23	9	5
广济	前震旦纪沉积变质	晶质磷灰岩	10.1	1.04	10.30	8	9
黄梅	前震旦纪沉积变质	晶质磷灰岩	26.0	1.60	6.15	5	10
孝感	前震旦纪沉积变质	晶质磷灰岩	27.0	1.90	7.03	6	6
荆襄	震旦纪沉积	胶磷矿磷灰岩	28.5	1.86	6.50	7	4
大悟	前震旦纪沉积变质	晶质磷灰岩	26.2	4.06	15.49	4	3
通山	志留纪沉积	胶磷矿磷块岩	10.9	2.95	27.01	3	7
鹤峰	震旦纪沉积	胶磷矿	20.4	4.28	21.09	2	2
远安	震旦纪沉积	胶磷矿磷块岩	29.8	3.60	12.08	1	1

注：• 磷矿结晶形态系根据矿石薄片鉴定结果
• • 枸溶率即有效磷量占全磷量的百分数

（2）各种矿源的磷矿粉肥后效的差异，与矿粉的全磷量有一定的关系。例如，各种矿粉在第三茬作物的增产效益顺序，通山磷矿由第一茬作物的第三降到第七，荆襄磷矿由第七升到第四，郧西磷矿由第九上升到第五。这可能主要是由于通山磷矿全磷量较低，仅 10.9%，荆襄矿为 28.5%，郧西矿为 26%。又如，鹤峰、远安磷矿粉肥对第一茬作物的肥效相近（见表 2），但在第三茬作物上的肥效，鹤峰矿只

*原载于《湖北农业科学》，12：16~19，1974.

有远安矿的一半（见表5），可能主要是因为鹤峰矿全磷量比远安矿低（9.4%）。

（3）不同矿产地的磷矿粉肥的有效磷含量和其矿石成因类型、结晶形态密切相关。供试样品中除郧西、大悟两矿是例外情况外，沉积矿床的胶质磷矿（微晶质或非晶质）的有效磷含量高；沉积变质矿床的晶质磷矿的有效磷含量低。

2. 同一矿产地、不同矿层、不同矿样含磷量的差异

我们在孝感地区三个沉积变质磷矿（孝感四方山磷矿、大悟阳坪、黄麦岭磷矿，黄陂团山沟磷矿）进行了调查，并采集了23个矿石样品进行了全磷量和有效磷量分析（见表8、表9），结果可以看出：

表8 孝感、黄陂两矿区磷矿石样本的分析结果

矿区	矿样	全磷量（%）	有效率（%）	枸溶率（%）
孝感四方山	富矿样	31.00	2.60	8.4
	贫矿样	3.75	0.28	7.5
黄陂团山沟	1号	36.00	2.7	7.5
	2号	31.00	2.45	7.9
	3号	28.80	2.20	7.6
	4号	25.80	2.45	9.5
	5号	24.80	0.30	1.2
	6号	26.30	2.25	8.6
	7号	25.50	2.25	8.8
	8号	9.75	0.33	3.4

·含电气石

表9 大悟县三个矿区磷矿石样本的分析结果

矿区	矿样	全磷量（%）	有效率（%）	枸溶率（%）
阳坪河东	第一层	33.5	3.62	10.8
	第二层·	2.8	0.67	23.9
	第三层	24.5	3.15	12.9
	第四层	28.5	2.57	9.0
	第五层·	5.5	0.90	16.8
	第六层··	36.0	2.38	6.6
阳坪河西	1号	30.5	2.54	8.3
	2号	28.5	2.84	10.0
	3号	14.0	2.25	16.1
	4号	18.5	2.20	11.9
	5号	15.0	2.60	17.3
黄麦岭	厚层矿样	29.8	2.75	9.2
	薄层矿样	25.0	2.27	9.1

注：·不属采矿岩层
··含石英脉

（1）同一矿区出产的磷矿石的全磷量和有效磷量的变化是相当大的。例如，大悟磷矿11个矿样的全磷量从14%~36%，有效磷量从2.2%~3.62%；黄陂磷矿8个矿样的全磷量从9%~36%，有效磷量从0.3%~2.7%。

（2）同为沉积变质矿床，矿石的有效磷含量和枸溶率也有明显的差异。如同为沉积变质矿床，大悟

磷矿的有效磷含量就比黄陂磷矿高得多。

（3）矿石的全磷量和枸溶率之间呈明显的负相关。一般趋势是全磷量愈高，枸溶率愈低，见图 3。

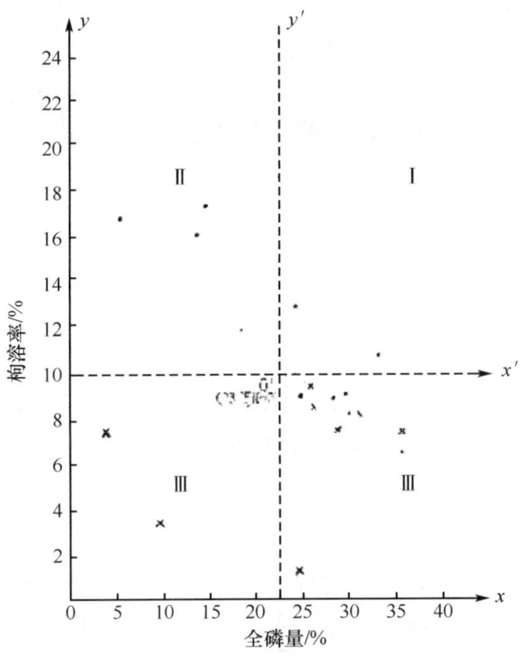

图 3　孝感、黄陂、大悟三县 23 个矿样全磷量与枸溶率相关图

相关系数 r_{xy}=0.5117；0.05 相关显著标准 0.3809；0.01 相关显著标准 0.4869

同一矿区不同矿层，不同矿样全磷量和有效磷量变化较大的原因，我们认为有以下几点：①矿体形成时的沉积过程本身是不一致的，这就造成了不同矿层、不同部位含全磷量的不一致。②矿层露头由于风化和淋洗，产生次生富集作用而提高了全磷量、有效磷量和枸溶率。③在区域变质过程中，变质结晶状况不同而造成有效磷量的变化。变质程度深的有效磷量降低。④孝感地区三个磷矿成矿后都受岩浆活动的影响，接触变质作用使局部地段磷矿富集，全磷量增高，有效磷量降低。例如大悟县阳坪磷矿相连接的两个采矿台阶，上台阶第一层全磷量 33.5%，有效磷量 3.62%；下台阶一大层磷矿由于岩浆活动（矿层中有石英脉），使全磷量增高至 36%，而有效磷量却降低到 2.38%。同样，黄陂团山沟磷矿相邻近的 3 号、4 号矿样，虽同属一个采矿台阶，3 号矿样受岩浆活动（矿石中夹有大量柱状电气石晶体）影响较大，4 号矿样受岩浆活动（不含电气石）影响较小，致使 3 号矿样的全磷量比 4 号矿样增高，有效磷量降低。

五、孝感县槐河店磷矿粉肥的肥效

1973 年孝感地区燃化局和孝感磷矿，在孝感县小河区槐河店发现一种磷矿，为黄褐色土块状，火焰试验无萤光反应，经地质队化验，有效磷含量高达 9.14%~14.8%。

1974 年春，我们取样进行了含磷量的化学分析、油菜幼苗法试验和水稻田间试验，结果如下：

1. 化学分析

样品一（油菜幼苗法试验用的样品）含全磷量 17.5%，有效磷量 10.7%，枸溶率 61%。

样品二（五岭公社重光大队水稻、田菁田间试验用的样品）含全磷量 18%，有效磷量 12%，枸溶率 66%。

化验结果指出，槐河店磷矿粉肥的枸溶率比现知枸溶率最高的摩洛哥磷矿粉肥（枸溶率通常在 30%左右）还高出一倍以上。

2. 油菜幼苗试验（一）

供试土壤为本院实习农场死黄土。

试验处理；在满足氮、钾的基础上，①不施磷肥；②施过磷酸钙每钵1克，③施槐河店磷矿粉肥每钵2.5克；④施荆襄磷矿肥粉每钵2.5克。重复4次。试验于1月2日播种，2月2日收获。

化验结果（见表10）可以看出，油菜在五叶期前，对荆襄磷矿粉肥的吸收利用能力很低，植株因严重缺磷而生长受到抑制，收获时只有二叶一心，与不施磷肥的相同。可是，施槐河店磷矿粉肥的与施过磷酸钙的，油菜植株都已长出四叶一心，且前者的株高和鲜重都超过了后者，肥效率分别为不施磷肥的10倍和9倍。

表10 槐河店磷矿粉肥肥效的油菜幼苗试验（一）结果

处理	株高（厘米）	叶片数	鲜重（克/钵）	肥效率（%）
不施磷肥	3.7	二叶一心	0.9	100
施过磷酸钙	8.7	四叶一心	8.3	922.2
施槐河店磷矿粉肥	9.5	四叶一心	9.6	1066.8
施荆襄磷矿粉肥	2.9	二叶一心	1.1	122.2

3. 油菜幼苗试验（二）：槐河店磷矿粉肥和过磷酸钙等有效磷量的肥效比较试验

在前一试验中，槐河店磷矿粉肥施用量和过磷酸钙施用量的比例为2.5∶1（干重）。本试验在等有效磷量基础上进行比较。

供试过磷酸钙的有效磷含量为13.8%，槐河店磷矿粉肥的枸溶性磷为10.7%。因此，施用的槐河店磷矿粉肥与过磷酸钙的比例为1.3∶1（干重）。

供试土壤为本院实习农场的死黄土（含有效磷0.9毫克/100克土）和面黄土（含有效磷2.1毫克/100克土）。重复4次。试验于3月20日播种，4月10日收获。试验结果（见表11）表明，在两种土壤上，等有效磷量的槐河店磷矿粉肥和过磷酸钙的肥效相近。

4. 槐河店磷矿粉肥对水稻、田菁当季肥效的田间试验

1974年春，在黄陂县五岭公社重光七队和重光大队农科所进行了槐河店磷矿粉肥当季肥效的田间试验。

表11 槐河店磷矿粉肥肥效的油菜幼苗试验（二）结果

土壤	处理	株高（厘米）	鲜重（克/钵）	肥效率（%）
死黄土	槐河店磷矿粉肥	14.7	15.1	93.2
	过磷酸钙	13.9	16.2	100
面黄土	槐河店磷矿粉肥	17.8	18.9	96.4
	过磷酸钙	18.0	19.6	100

重光七队试验田的土壤为白散土。施肥处理：在每亩施碳酸氢铵50斤的基础上，分①不施磷；②施过磷酸钙（有效磷14%）44斤，③施槐河店磷矿粉肥（有效磷12%）58斤。水稻品种为二九青，小区面积0.6亩，3次重复。试验结果如表12。

表12 槐河店磷矿粉肥对水稻的当季肥效

处理	小区产量（斤）				亩产（斤）	肥效率（%）
	1	2	3	平均		
不施磷	31.4	31.8	32.8	32.0	711	100
施过磷酸钙	36.8	33.6	35.8	35.4	787	110.6
施槐河店磷矿粉肥	33.6	34.8	34.6	34.0	764	107.4

在重光大队农科所小麦后茬早稻间种田菁进行的槐河店磷矿粉肥当季肥效试验。试验在每亩施菜饼180斤的基础上，设①不施磷；②施过磷酸钙50斤；③施槐河店磷矿粉肥100斤。试验结果如表13。

表13　槐河店磷矿肥对早稻及间种田菁的当季肥效

处理	早稻产量（斤/亩）	肥效率（%）	田菁产量（斤/亩）	肥效率（%）
不施磷	552.4	100	192.6	100
施过磷酸钙	600.0	108.6	429.6	223.1
施槐河店磷矿粉肥	609.5	110.3	492.6	255.8

这些试验结果表明，槐河店磷矿粉肥是一种当季肥效很高的磷矿粉肥，可作速效磷肥使用。

1974年春季，我们在黄陂县五岭公社重光七队100多亩新平整的土地上，施用6000多斤槐河店磷矿粉肥作为早稻秒口肥和追肥，也普遍取得较好肥效。

六、小结和讨论

1. 本试验报告是三年来对湖北省磷矿粉肥有效化研究的一部分，内容包括对湖北省12个矿区、34个矿样当季肥效的生物研究、化学分析和矿石薄片的矿物鉴定。

2. 盆栽试验表明，黄改、郧西、广济、荆襄、孝感、黄梅、大悟，通山、鹤峰、远安10处出产的磷矿粉肥，对红花草子、黄豆都有当季肥效，但肥效高低不等，按上述顺序，黄陂的最低，远安的最高。黄豆当季肥效试验结果是：远安、鹤峰、大悟、通山四种磷矿粉肥肥效最高，接近或超过过磷酸钙的肥效，为第一类（施肥量，磷矿粉肥10克，过磷酸钙3克），黄梅、荆襄、孝感、广济、郧西五种磷矿粉肥次之，相当于过磷酸钙肥效的57.71%~73.71%，属第二类；黄陂磷矿粉肥的肥效较低，只相当于过磷酸钙肥效的44%。但是，应该指出：供试验的黄陂磷矿粉肥样品有效磷含量，不到该磷矿平均含量（1.87%）的一半，而大悟磷矿粉肥样品的有效磷含量，比该磷矿平均含量（2.65%）几乎高一倍。

在三种土壤上进行田间试验，宜昌磷矿粉肥（代表第一类）和荆襄磷矿粉肥（代表第二类）的肥效试验与盆栽试验结果相符合。盆栽试验虽没有包括宜昌磷矿粉肥，但从矿床成因、矿石类型、含磷量及肥效试验来看，宜昌磷矿粉肥和远安磷矿粉肥属于同一类型。

3. 10种磷矿粉肥的当季肥效与其有效磷含量成正相关，有效磷含量愈高，当季肥效愈显著，但和全磷量高低没有明显关系。

4. 各种磷矿粉肥对第二茬作物荞麦、第三茬作物红花草子和第四茬作物水稻，均表现出愈来愈高的后效。磷矿粉肥后效的高低与其全磷量有一定的关系。在同一成因类型、有效磷含量相近的磷矿中，全磷量高的，后效一般也高。

5. 矿石鉴定和化学分析表明，沉积变质矿床的品质磷矿的有效磷含量比沉积矿床的胶质磷矿低，当季肥效也较低。供试样品中，郧西沉积矿床胶磷矿和大悟沉积变质矿床品质磷灰岩例外。

6. 从大悟、孝感、黄陂三县的21个矿样的化学分析表明，同一矿源的不同矿层、不同矿样的全磷量和有效磷量的差异相当大。全磷量枸溶率呈明显的负相关。

（1）优先选择有效磷含量高的矿石制作磷矿粉直接施用，在就地开采、就地加工、就地利用时，更应注意这一点；而有效磷含量低、全磷量高的磷灰岩应加工，作成化学磷肥，或配合其他有效化措施后再作磷矿粉肥施用。从有效磷含量看，远安、宜昌、鹤峰、通山等磷矿适宜制作磷矿粉肥。对同一矿床的不同矿层，也应根据情况分别利用，以达到经济合理地利用磷矿资源。

（2）从磷矿粉肥当季肥效与持续肥效看，目前每亩一次施用量一般不宜低于100斤。至于在一年多熟的情况下，磷矿粉肥是每季都施，或一年施一次，抑或几年施一次，每年施多少，哪一种方法好？我省还缺乏较长期定点的试验资料，对这些问题尚难确切回答。建议有关部门加强领导，开展这方面的试验研究。

（3）孝感县槐河店磷矿的出现绝非偶然、孤立的地质现象。希望有关部门组织力量进行勘探，研究其成因类型、分布规律，更好地满足农业大上、快上对磷肥的需要。

第二项　教育改革研究

农业生产和农业科学*

陈华癸

（华中农学院）

一

现代农业科学的产生和发展，是由生产所决定的。毛泽东同志在《实践论》中曾经指出："马克思主义者认为人类的生产活动是最基本的实践活动，是决定其他一切活动的东西。人的认识，主要地依赖于物质的生产活动，逐渐地了解自然的现象、自然的性质、自然的规律性、人和自然的关系"[1]。无疑地，人们获得农业生产知识的最基本的活动是农业生产实践。人们在长期的生产实践中，积累了解决生产问题的农业生产技术知识。

在原始的农业生产中，人们只会种植少数几种作物。这些作物都是些半野生的品种，用火耕杖种。那时，人们还不知道如何管理作物，保护土壤，而是边开垦、边丢荒，经营着生产水平很低的农业生产，过着不断迁移的半定居的农业社会生活。在这样艰苦的生产和生活条件下，人们逐渐地积累了较多的生产经验和生产知识，也改进了一些生产工具。农作物的种类丰富了，半野生的植物品种逐渐地为人们自己在生产中培育出来的植物品种所代替。金属农具代替了木石器。无计划的丢荒农作变成为固定的、有计划的休闲轮作。定居下来的农业社会，要求保持地力长期不衰，也要求土地每年提供较多的农产品。在这种情况下，对农业生产技术的要求提高了，人们逐渐地取得了精耕细作的知识和技术。随着农业生产的发展，农业生产的技术水平也不断提高。从原始农业发展到近代的精耕细作的农业，全部农业生产技术知识的获得，都是人们依赖于生产活动，逐渐地认识自然的现象、自然的规律性的成果。

与此同时，人们从工、农业生产实践以及其他实践活动的各个方面，逐渐地了解了自然界的种种联系和种种过程的体系，从而产生了关于各种自然现象的统一的自然科学知识。从欧洲的文艺复兴运动起，自然科学从神学的束缚下解放出来，发展为近代和现代的自然科学。近代和现代的自然科学，一方面同生产和生活实践密切结合，相伴地发展着；一方面又走着相对独立发展的道路，自然科学的各个分科遂逐渐形成。人们吸取和利用自然科学各个分科的成就，去研究农业生产中的各种技术问题，促使了现代农业科学的产生。因此，现代农业科学，是农业生产和自然科学相结合的产物。

现代农业科学的形成和发展，有力地提高着农业生产的技术水平，促进农业朝着现代化方向前进。现代农业科学对发展农业生产的重大作用，可以用农业化学的发展及其对于农业生产作出重大贡献的事实来说明。

在两千年以前，人们已经认识到肥料对于提高农作物产量的重要意义。公元十四世纪我国王祯著的《农书》中，已经记述了多种多样的有机肥料和矿质肥料，包括踏粪、苗粪、草粪、火粪、石灰和泥粪等。[1]在这以后，西欧在十七世纪发现了硝土的肥田作用，十八世纪由于智利硝土的开发，而使硝土成为一种肥料。这个时期，也正是化学从炼金术中解放出来，成为真正的科学而飞速发展的时期。到十九世纪，化学已经伸入生物学的领域，促使了植物生理学和农业化学的形成和发展。十九世纪农业化学的主要成就之一，是阐明了植物从土壤和肥料中吸取的营养料主要是一些简单的无机化合物，而在这些简单的无机化合物中，含有植物所需要的各种营养元素。这些营养元素之中，有些是土壤中含量比较丰富

*原载于《红旗》，（7-8）：10~18，1963.

1《毛泽东选集》第1卷，人民出版社1952年版，第271页。

的，植物用之不竭，有些是在土壤中比较贫乏的，不能满足农作物高额丰产的需要。在农业生产中施用肥料，主要就是补给植物所需要的在土壤中比较缺乏的营养元素。农业化学的这项重要发现，同当时正在兴起的化学工业相结合，产生了化学肥料工业。早期的化学肥料工业，除直接开发和利用富含硝酸钠的智利硝矿外，还从骨粉和海鸟粪中提制过磷酸钙。接着，开发了各种磷矿，用来制造更多的过磷酸钙，并且开发了钾矿，制造硫酸钾和氯化钾。在二十世纪初，化学合成氮肥的技术产生了，巨大的合成氮工业也就诞生了。化学肥料工业提供了在农业生产中大量施用化学肥料的物质条件，大大提高了农业生产水平。这正如马克思曾经指出的："在自然丰度相等的各种土地上，这相等的自然丰度能被利用到何种程度，部分地要看农业化学的发展如何，部分地要看农业力学的发展如何。"[2]二十世纪以来，农业化学和为农业服务的化学工业继续发展，农业生产的化学化，对于提高农业生产水平的作用日益增强，已经成为农业生产现代化的一项重要内容。

同样，现代农业科学的其他分科的发展，对于提高农业生产的技术水平也都起着重大的作用。

生物学是农业科学的基础理论，它是研究生物的遗传变异，生长发育和物质代谢的科学。自从达尔文提出生物进化的学说以后，生物遗传和变异的知识逐步形成为一门独立的科学。遗传学的形成和发展，主要是二十世纪的事情。可以这样说，一直到十九世纪，人们主要是依靠自然的恩赐和朴素的配种和培育知识得到一些较好的农、牧业品种。而随着遗传学和有关科学的发展，创造和选育新品种，才进入了实验科学的阶段。通过系统选种，杂交育种，以及化学引变和辐射引变等专门的选种育种的科学实践，不断地为农业生产提供了一些优良的新品种。

关于植物的生长发育和物质代谢的知识——植物形态学和生理学——是栽培农作物的科学基础。农作物的生长和发育，一般地可以分为营养生长和生殖生长两个阶段，两阶段又各分为若干分段。农作物在每一阶段和分段各有特点，要求不同的环境条件。因此，在阐明各种农作物的生长发育阶段和各分段以及它们对于环境条件的特殊要求之后，人们就能够比较正确地总结先进的栽培技术经验，揭发其实质，同时还能够相应地提出新的栽培技术措施来。现代饲养家畜的先进技术，也是生产实践经验和动物生理学知识相结合的成就。

植物病理学、昆虫学和微生物学的研究，阐明了植物病虫害的本质，摸清了各种病虫害发生发展的规律。这些研究成果和有机化学在合成农药方面的成就相结合，创造了许多防治病虫害的有效技术，将农业生产从对病虫害无能为力的状态中解放出来。在家畜、家禽的病害及其防治的研究方面，也取得了显著的成就。

人们从研究陆地表层地质运动的规律和地面生物的生命运动的规律，以及二者之间的相互关系中，揭发了土壤和土壤肥力的形成和发展的规律，并且发现了成土母质、气候、植被、地形和水文等是影响土壤肥力的形成和发展的主要自然因素。合理的轮作换茬，可以创造优良的植被条件；平整土地，可以改善地形条件，深耕和施肥，可以改善成土母质条件；灌溉、排水和保墒耕作，可以改善水文条件。研究土壤肥力的形成和发展以及有关的农业生产技术措施的理论和应用，是土壤学和有关科学的任务。土壤学和有关科学的科学成就是用地和养地相结合，合理利用土地的科学基础。

化学、物理学和数学是生物科学的基础知识，从而也是农业科学技术的基础知识。农业生产的化学化，除化学肥料的应用外，还包括杀虫剂、灭菌剂和除莠剂等的应用。最新的物理学和数学的成就，对农业科学技术的提高，也起着更直接的作用。

地理学和气象学的知识，对于农业生产的区域化和农作物的抗逆防灾，均起着巨大的指导作用。

建立在现代技术科学和工业技术力量基础上的农业生产的机械化、电气化、水利化和化学化，不仅大大地提高了农业劳动生产率，完成人们用双手和陈旧落后的农业生产工具所无法完成的生产技术活动，而且也大大地提高了农作物的单位面积产量和总产量。

由此可见，农业生产和农业科学的发展，是密切结合，互相推动的。农业科学来自农业生产实践，它反过来又对农业生产的发展起重要的促进作用。同时，农业科学还从起源于人类生产和生活的整个的自然科学领域中吸取丰富的滋养。

二

现代农业科学所以能够迅速地提高农业生产的技术水平，促进农业生产的发展，促进农业现代化，是因为在农业科学研究中，可以充分利用现代自然科学的各项成就，通过特定的科学研究的方法，更准确、更迅速地总结农业生产经验，研究特定的生产技术措施的增产效益，探讨同农业有关的客观世界的各种规律。

农业科学技术产生和发展的历史证明，我们的祖先在生产实践中积累生产知识的过程是十分艰难的，农业生产技术水平提高的速度很慢。在农业生产实践中，探讨特定的技术措施的增产效益，只能从观察这项技术措施对于农业生产所产生的可见形态和最终产量来判断。但是，这样的判断是很难作为定论的。因为，同样的技术，应用在不同的土地上或者在不同的年景里，对于农作物所产生的影响往往是不一致的。只有经过人们大量的、多次的、甚至几代的生产实践，才能从错综复杂的现象中比较准确地判断这个特定的技术措施的增产效益及其适用的条件。

现代农业科学研究则不同。它除了从农业生产实践经验中吸取丰富的滋养外，还从整个的自然科学领域中吸取滋养。农业生产过程是十分复杂的，它包含自然界的种种运动形态，而且这些运动形态互相关联和互相转变着。这些运动形态包含生物学的变化，也包含含蓄在生物学的变化之中的化学、物理的变化，以及造成生物生活环境的土壤、水和气候的变化。自然科学的每一种学科，都是分析单个的运动形态或一系列互相关联和互相转变的运动形态的。农业科学利用自然科学的各项专门成就，进行专门的农业科学研究，不仅能够观察农作物整体的变化，以及周围环境对农作物所起的综合影响，还能够更深入地分科研究包含在农作物整体变化和环境的综合影响之中的各个因素的变化，它们之间的相互影响，以及它们对于农业丰产所起的实际作用。

在农业科学研究中，可以通过特定的研究方法，把研究对象——特定的运动形态、特定的增产因素或环节——相对地孤立起来加以研究。我们也可能有效地控制环境条件，一方面使得研究对象在已知的环境条件中生活和运动，尽量减少未知因素的干扰，另一方面又可以研究特定的研究对象和特定的环境条件之间的制约关系。因此，在科学研究中，对于原因和结果之间的本质联系，比较在复杂多变的生产实践中更容易理解，从而能够较快地、较准确地作出科学的判断。

例如，种植豆科植物，一般需要肥料较少，而且能提高土壤肥力，对后茬作物有增产效益。早在公元六世纪，我国贾思勰著的《齐民要术》中就写道：“凡美田之法，绿豆为上”[1]。农民从生产经验中认识到，豆科植物只有在它的根上生长根瘤时才产生肥效。各种豆科植物种植在不同的土壤里，有的生根瘤，有的不生根瘤。在原本不生根瘤的土地上种植豆科植物，如果要它生长根瘤，可以采用土壤接种法，即将原本能生根瘤的土壤和种子拌在一起播种。这是长时期农民在生产实践中积累的宝贵经验。而农业科学研究对于总结这一生产经验，并加以提高，起了显著的作用。科学研究发现，在上述生产实践中，最本质的因素是一种在土壤中生活的细菌（根瘤菌）侵入豆科植物，使后者形成根瘤，根瘤菌在根瘤中生活，发生共生固氮作用。共生固氮作用将空气中的氮气（这是一般高等植物不能利用的氮素物质）变为植物的氮素营养料，直接地营养了豆科植物，还间接地改善了后茬植物的营养条件。对共生固氮作用的本质的阐明，导致根瘤菌肥料的制造和应用，这是现代农业生产技术的一个组成部分。这项科学研究专门地研究了包含在复杂的生产过程中的特定的微生物的运动规律。虽然通过科学研究探索这一微生物的运动规律也不简单，但比起通过种植豆科植物长期的生产实践过程来总结这一经验，是简单得多了。而且，由于根瘤菌是肉眼看不见的细菌，培养根瘤菌需要农业生产环境所不能提供的工作条件。如果完全依靠农业生产实践的直接经验，没有专门的科学研究工作（微生物学研究），是无法创造出制造和应用根瘤菌肥料的现代技术来的。

又例如研究合理施肥，为了判断特定的农作物在特定的条件下，究竟从土壤中取得多少磷素养料，从施用的肥料中取得多少磷素养料，从而判断特定的肥料或施肥方法的增产意义，研究工作者们就利用含有较高的磷的放射性同位素的肥料进行科学试验，从植物所吸收的放射性物质的比例来判断这种肥料或施肥方法的实际效果。经过研究，发现许多植物在幼嫩时期吸收土壤中的磷素养料的能力很弱，主要依靠吸收优质磷肥（如过磷酸钙）中的养料来满足它的生活需要；植物长大以后，吸收土壤中磷素养料的能力加强，可以完全或大部分靠土壤中的磷素成分满足生活需要。这样，就阐明和发展了一项丰产技术措施，即在不少情况下，施少量过磷酸钙作为种肥，是十分有效的施肥方法。

这些都说明，利用农业科学去总结农业生产经验，探讨和掌握与农业有关的客观规律，可以大大简化问题的复杂性，加强人们的洞察力，易于比较迅速而准确地作出判断，有效地促进农业生产的发展。

三

农业科学对发展农业生产的作用，是很明显的。要迅速发展农业生产，必须十分重视和加强农业科学研究工作，使农业科学更好地为农业生产服务。

农业科学研究要更好地为农业生产服务，必须同时在两方面开展工作。一方面是用现代农业科学知识总结和提高群众的生产实践经验，解决当前迫切需要解决的农业生产技术问题，充分发挥农业科学对农业生产的指导作用。另一方面是进行长远的、基础农学的研究工作，不断地揭示新的科学规律，掌握新的农业科学知识，从而使农业科学对原本不能够起指导作用的问题变为能够起指导作用。

充分发挥农业科学对当前农业生产的指导作用，是农业科学工作者的重要任务。我国的农业生产，已经积累了丰富的经验，同时还存在着不少需要解决的问题。在农业生产技术的各个方面，现代的农业科学已经积累了许多确切的知识。这些知识的应用，是高产量、高劳动生产率的现代化农业生产的组成因素。农业科学研究工作者应当面对生产实际，运用这些知识研究和总结丰富的农业生产经验，解决生产中迫切需要解决的问题。目前，这些知识的大部分还没有在我国的农业生产中广泛应用，发挥其应有的作用。例如，在种子方面，现代的良种繁育科学已经掌握了利用杂种优势的科学技术。这项科学技术的实际运用，能够提高玉米、高粱等农作物的产量，在有些国家已经被广泛应用。而在我国，目前还处于技术准备阶段。又例如，在植物保护方面，现代的植物病理学和昆虫学已经揭发了许多病虫害的发生规律，并且创造了不少有效的防治方法。但是还需要经过大量的、针对着各地不同情况的试验研究工作，才能使这些防治方法在广泛的农业生产实践中充分发地发挥作用。因此，在农业科学研究的这一方面，需要投入大量的科学技术力量。这些，经过相当长时期的努力，就能够使我国的农业生产达到现代化的水平。

然而，这还不是农业科学研究的全部任务，还必须进行另一方面的研究工作。包含在复杂的农业生产过程中的各种规律，有很多是我们现在还没有确切掌握而又急待掌握的，这些也是我们农业科学研究的重要课题。这方面的研究工作，就其对提高当前农业生产水平的作用来说，不一定就立竿见影。但是，就其对农业科学技术的发展来说，却是十分重要的。这方面的研究工作，蕴含着农业科学研究推动农业生产发展的巨大潜力。这也可以用关于豆科植物和根瘤菌的共生固氮作用的研究的历史经验来说明。当人们通过科学研究，发现了根瘤菌所起的特殊作用，以及人工培养根瘤菌的技术以后，接着也就创造了制造和利用根瘤菌肥料的生产技术，保证豆科植物结根瘤，从而更有效地发挥共生固氮作用的增产效果。然而，另外有一些非豆科的植物（如杨梅、木麻黄、桤木等）也结根瘤，根瘤中也生活着特定的微生物（但不是根瘤菌），也发生共生固氮作用。对于这种特定的微生物，目前我们还没有掌握人工培养的方法，因此也就不能够制出相应的微生物肥料应用于生产实践。也就是说，在这个问题上，科学知识的不足，在一定程度上限制了农业生产技术水平的提高。因此，我们除了动员足够的科学技术力量来充分研究、运用农业科学知识于当前农业生产实践外，还必须保证以一定比例的力量来进行许多新的农业科学研究工作。

农业科学研究要更好地为农业生产服务，充分发挥它的作用，在农业科学研究工作中，必须保持实验科学的严谨精神。

在农业科学研究工作中，总结先进的农业生产经验，使之上升到理论，再用来指导农业生产实践，这是集中广大群众的智慧，不断地丰富和发展农业科学，不断地提高农业生产水平的一项最基本的科学研究活动。

先进的生产经验包括某些自然规律的正确运用。总结和提炼先进生产经验的科学任务，就在于说明在这个或那个先进生产经验中究竟是如何地运用着自然规律的。进行这项科学活动，当然首先要调查研究应用这些先进生产技术所引起的农作物在生长发育过程中的各种变化和有关的环境条件特点，用现有的自然科学知识来说明这些现象，并且去粗取精，加以提炼，揭发出这些先进生产经验的科学实质。这样做是必需的。但从实验科学的要求来说，这样做，只达到了科学活动的假说阶段。进一步还需要通过一定的试验研究来考察假说的正确性，并作必要的修正。这样得出来的科学结论，才能够有力地指导生产实践，发挥先进生产经验促进农业生产不断发展的巨大作用。

在科学研究活动中，坚持田间试验的方法，是贯彻实验科学精神的中心环节。田间试验，是农业科学研究中最主要的研究方法。一切农业科学理论和技术措施都需要通过田间试验的考察。只有在农业科学理论或特定的技术措施为田间试验结果所证明时，才能认为它是反映了自然的和生产的客观规律。

进行田间试验，主要是用对比的方法。用田间试验的术语来说，采用某种特定技术措施的田块称为处理区，不采用这种特定技术措施的田块称为对照区。田间试验的目的，就是对比处理区和对照区农作物的生长情况和产量的差别，对这种特定技术措施的增产效果作出判断。

当然，在田间试验的实施中，也还需要排除一系列的实际困难，才能够从试验中得出清楚的结果，作出比较正确的判断。因为两试验区的土地条件（土质、肥力、地形、水文等）和环境条件（如气候、光照、病虫害等），总难免有些差别。即使在两区中所采用的农业技术措施完全相同，两区中农作物的生长情况和产量也不会完全一样。田间试验的经验指出，在一般的试验条件下，由于两区田地本身的不一致性所导致的产量差别，通常在百分之五至十之间，甚或更大些。如果特定的技术措施的增产效果达到百分之三十、百分之五十或一倍以上，不论是通过一般的生产实践或田间试验，都不难作出肯定的判断。然而，更多的技术措施的增产效果并没有这样大的幅度，而是增产百分之五、百分之十或百分之二十左右。这是十分宝贵的不可忽视的增产幅度。农作物单位面积产量的逐步提高，主要是依靠这类幅度不很大的、这样和那样的增产技术措施的综合增产效果的逐步积累。如何能够将增产幅度不很大的特定技术措施的效果，和由于两区田地本身的不一致性所导致的幅度差不多的产量差别区分开来，从而能够作出比较正确的判断，就成为做好田间试验的一个重要问题。在田间试验的科学实践中，已形成了一门专门的学问，即田间试验设计和统计分析。采用这些方法，就能够有效地解决这项困难的，但是重要的科学技术问题。

田间试验不仅能够考察单一的增产因素的增产效益，还可以在一定范围内考察多种增产因素的综合效益。特定的技术措施的增产作用不是孤立地发生的，它要求一定的条件。在许多情况下，特定的技术措施要和另一些特定的技术措施结合起来，才能产生显著的增产效果。如果我们在生产实践中或科学研究中，把这些增产技术措施一个一个孤立地加以比较，有时往往得不出什么有益的结果，甚至会作出不正确的判断。针对这种情况，在基本上阐明了有关技术措施的增产作用的原理的基础上，进行多因素的田间试验，能够帮助我们以较小的力量，在较短的时间内，作出比较正确的科学判断。例如，在很多种土壤上，单施氮肥能够增产，单施磷肥却不能增产，只有在施氮肥的基础上加施磷肥，才能获得施用磷肥的增产效果（亦即在氮肥增产的基础上施磷肥进一步提高产量）。这样，如果只进行单施氮肥和单施磷肥的田间试验，就只能得出施氮肥能增产、施磷肥不能增产的错误结论来。而只有正确地进行研究氮磷联合效用的多因素田间试验，才会得出符合上述客观实际情况的正确结论。

任何田间试验都是在特定的条件下进行的，所得的结果，严格而论，只是在这特定的试验条件下发生作用的规律。由于农业生产的实际条件变化很大，如果把特定的农业生产条件下得到的田间试验的成果，直接地在很大范围中推广应用，是有一定的盲目性的。为了尽可能地避免由于这种盲目性而造成损失，还必须进行区域试验（即区域性的田间试验）。根据土壤、地形和气候的区域性特点，在各个区域选择确有代表性的地点进行区域试验，是将科学研究成果多、快、好、省地推广到农业生产实践中去的重要环节。区域试验既是鉴定性的，也是试探性的。即使是区域试验的成果，也还需要通过千百万人在生产实践中最后地加以考验和发展，才能够成为提高农业生产的技术因素。

还需要指出，在农业科学研究中，我们必须同等地重视包含在农业生产过程中各种物质运动的实质和这些物质运动在特定的时间、地点条件下的具体作用。只有这样，才能发挥农业科学研究对农业生产实践的指导作用。例如，一般而论，同样的规律决定着土壤中磷素的变化和植物的磷素营养。但是，在不同的土壤中种植不同种类的农作物，对于磷肥的需要量和施用方法却是有很大出入的。在长江流域的红壤上，施磷肥的增产效果很显著；在它附近的冲积土上，施磷肥的增产效果就比较小些。在同样的冲积土上，种稻、麦，施磷肥的增产效果就比较小，或者不增产；种油菜、紫云英，施磷肥的增产效果就比较大。显然，关于植物磷素营养的科学知识，只有这样具体化之后，才能起指导合理施肥的实际作用。研究农业生产过程中特定的物质运动在特定的时间、地点条件下的具体作用，也主要是采取田间试验和区域试验的方法进行的。

要做好农业科学研究工作，使其更好地为农业生产服务，农业科学研究工作者，在解决农业生产中的特定问题时，需要掌握有关的自然科学知识，吸取和利用自然科学中一切有用的成果。只有这样，才能使农业科学研究和现代自然科学的成就紧密联系起来，更好地为农业生产服务。现代自然科学的发展，特别是物理、化学、生物及各种边缘学科的发展，对于农业科学的发展产生了很大作用，并将产生更大的作用。如从植物营养的研究来看，在十九世纪中期，微生物学还处于大发展的初期，还没有伸入农业科学的领域。当时，在初步揭发植物的无机营养规律的基础上，人们对于土壤供给植物养物的能力（或称为土壤的化学肥力），理解为土壤中含有植物营养元素总量的多少，以及它们溶解性的大小问题，并且主要是以化学变化的规律性来解说土壤中植物营养元素变化的规律性。实际上，并不完全是这样。土壤中植物营养元素的变化，还在很大程度上决定于土壤中微生物的生命活动。因此，只有在十九世纪末叶，微生物学伸入农业科学领域以后，农业化学和土壤微生物学的知识相结合，才有可能比较正确地理解土壤中植物营养元素的运动规律。显然，对当时的农业科学研究工作者来说，微生物学知识就成为研究植物营养元素的运动规律所必须具备的自然科学知识。

然而，这还不是解决土壤中植物营养元素的运动规律所必需的全部自然科学知识。到二十世纪初，生物化学又诞生了，而且很快地伸入到农业化学的领域中来。生物化学有力地阐明了各种营养元素在植物体中的变化规律，和它们对构成植物体质和经济性状所起的实际作用。接着，生物物理学又诞生了。作为生物物理学的组成部分之一的放射性同位素示踪方法，是揭发植物、土壤和肥料三者之间的物质运动规律的有力手段。因此，要研究关于土壤中植物营养元素的运动规律，现代的农业科学研究工作者，就必须同时具备现代化学、微生物学、生物化学和生物物理学等方面的自然科学知识。不掌握这些方面的知识，就不可能将农业科学研究工作做好。

同时，也可以看到，随着科学的发展，农业科学研究日趋专门化。在这种情况下，要使我国农业科学的发展得到更好的保证，以便更好地为农业生产服务，就必需在全国范围内逐步地建立起农业科学研究的各处专门学科，并且要求农业科学研究工作者深透地掌握一门科学的理论和实际，以及相应的试验研究方法。这样，才能更好地实行农业科学研究工作的分工与协作，在积年累月、锲而不舍的科学实践中，积累丰富的科学研究成果，解决我国农业技术改革中所提出的综合技术任务。

教学、科研、推广是农业院校的三项基本任务*

陈华癸

（华中农学院）

党的十一届三中全会以来，由于采取了一系列正确的农村政策，极大地调动了农民的积极性、创造性，我国的农业出现了欣欣向荣的景象。近四年中，全国农业生产总值以每年递增5.6%的速度持续增长，连续三年创造了建国以来的最高产量。随着生产的发展，出现了竞相学科学、用科学的科学种田热。可以预测，这种重视科学、学习科学、应用科学的可喜局面，必将带来生产更大发展，是达到翻两番战略目标的一个重要保证。

农业的大变革向高等农业院校提出了更多更高的要求。在这一形势下，我们必须遵循党中央所提出的“搞四化要进行一系列改革”的指示，冲破不符合新的历史任务和革命实践要求的老框框，旧套套，钻研新情况，解决新问题，创立新章法。这是摆在农业高等院校，特别是重点院校面前一项紧迫的任务。

尽管高等学校的改革头绪多端，但首先要统一对高等农业院校任务的认识，在正确认识的基础上制定相应的政策，提出适当的改革措施。我认为，重点高等农业院校办成教学、科研两个中心的提法有进一步发展的必要，重点高等院校应以教学、科研和推广为三项基本任务。

（一）

在五十年代，高等农业院校的任务只有一项，即培养人才。第一次全国高等教育会议明确规定：“以理论与实际一致的方法，培养具有高度文化水平，掌握现代科学技术的成就，全心全意为人民服务的高级建设人才。”在制订第一次全国科学技术规划时，分析全国的科学技术力量，提出我国科学技术力量有五个方面军，其中之一是大专院校，而且在人数上，科学素养上，大专院校当时是占第一位的。尽管如此，当时并没有对农业高等院校提出既要搞好教学，又要搞好科研的两重任务。虽然许多院校在完成教学任务的同时，也有过不同规模的科研活动，但多为教师的“自发经营”，或为受有关单位之托而应邀完成的“额外任务”，并没有列为学校必须承担的本职工作。一位教师，一个教学组织可以开展科学研究工作，也可以不开展。我院在1956~1965年的十年间，科研项目20项，参加科研的教师只占全体教师的13%，有一些研究项目在生产中发挥了较大的作用，为了适应科研工作的需要，我院还成立了相应的科研管理部门。

粉碎“四人帮”后，邓小平同志多次指示，要把重点大学办成既是教学中心，又是科研中心。方毅同志在科学大会的报告中阐述了这一方针。至此，教学和科研才明确地规定为重点大学的双重任务。任务明，行动的步伐就加快。以我院为例，1978年以来，参加科研的教师每年为30%左右，在教学任务继续加重，基本上没有专职科研编制的情况下，承担了中央和地方的70个研究项目，并且取得了一批有一定水平的研究成果，其中有9项受到中央的奖励，有22项受到省、市的奖励，开创了建国以来的新局面。

与此同时，随着研究生教育发展成为高等院校的一项基本任务，研究生的培养既要求学校有相应的科学研究基础，也为科学研究提供了新生力量，重点高等农业院校既是教学中心又是科研中心的形势已初步形成。

*原载于《高等农业教育》1：42~44，1984.

（二）

党的十二大和五届人大第五次会议，提出了到本世纪末工农业总产值翻两番的战略目标。赵紫阳总理和国家其他领导人多次向科学技术人员提出要对经济生产做出贡献的号召，也就是说不仅要在科学技术发展的前沿上做贡献，而且也要在科学技术知识转化为生产力上做贡献。用“科研中心”这个术语概括上述科研、推广两项任务不够明确，在实际工作中也发生了一定的障碍。障碍的一个方面是只将已经能够在生产实际中应用的科学技术成就作为科学研究的成果，对科学技术的应用要求过急。障碍的另一方面是只把实验室试验和田间试验的成果做为科学研究的任务，而忽略了将这些成果转化为生产力，在生产上实际应用，也即技术推广工作的必需性。实际上，农业高等院校目前还没有主管技术推广的部门。在评定教师的工作成绩中，时而强调发表论文的数量和质量条件，时而强调对学生的教学工作量和质量条件，而对教师在将科学技术转化为生产力方面的努力，往往不予重视，甚至不加理睬。

因此，有必要将重点高等农业院校的根本任务明确为教学、科研、推广三项，即既是教学中心，又是科研中心，又是推广中心。制定与之相应的各项政策，成立相应的推广工作部门，在努力将科学技术转化为生产力方面做出应有的贡献。

建国三十多年来，也曾多次地讨论农业院校应担负教学、科研、推广三项任务，近年来也常提教学、科研、生产三结合，但生产这个概念不够明确，主要是指为学校本身从事经济生产呢，还是以推广科学技术为农业生产服务呢？作为学校，无异议地应以后者为重，因此还是教学、科研、推广三项任务更为明确。

但是由于过去提到教学、科研、推广三项任务的同时，大多以美国的制度为模式，想照美国那样，农学院担负一个州或一个地区的农业技术推广的领导工作。这实际上是不符合我国的实际情况的。因此，要么是只要教学和科研两项任务，要么就是把一个省或地区的推广领导任务全担下来。这两者都是不合适的。

我的建议是，学校要有推广任务，但不担任领导一个省、一个地区的推广任务，而是：①将本校的研究成果尽快地转化为生产力，有什么成果推广什么成果。例如，我院农化组的教师们在硼肥对于棉花的生长发育和增产的作用方面取得了卓有成效的研究成果。那么，顺理成章地努力推广这项研究成果并应用到农业生产中去，让它产生应有的经济效果。②充分发挥教师的专长，促进现实的农业生产，有什么专长，推广什么专长。例如，我院王就光教授发挥专长指导郊区菜农防治病虫害，增产效果十分显著。

（三）

承认高等农业院校有教学、科研、推广三项任务，首先要反映在有关的政策上，鼓励教师正确对待和积极完成这三项任务。在政策上承认这三项任务都是教师的本职工作，在计算工作和评价工作成绩中都应有相应的对待。

首先是工作问题。毫无疑问，计算工作量是实行责任制的必要因素。目前，教育部只对教学工作规定工作量的计算标准，而没有对科研和推广工作提出相应的计算标准，特别是1982年职称提升工作中规定教师必须有面对学生的最低教学工作量。这样，如果教师蹲一年点，在点上开展科研和推广工作，就自然而然属于因为没有完成教学工作量而不能评职提升了。这里，可以采取美国大学教师工作量的计算方法，即职责当量（FTE）的计算方法。首先承认教学、科研、推广都是本职工作，再根据教师工作的实际情况，规定承担其中的一项、两项或三项，然后规定每项工作的工作量份额，对工作量的考核就是考核各项工作量份额的实际完成情况。

其次是工作成绩问题，也就是一定工作量的实效问题。对于教育工作，这是教书、教人、教学内容和学生实际收获的综合度量，对于科学研究，以论文的数量和水平，或者成果的鉴定为尺度。对于推广工作，如何衡量成绩？在学校工作中，实际上没有经验。承认推广是农业院校的基本任务之一，对教师工作的评价上就需要有相应的衡量尺度。

再次是组织机构问题。有学校管理教学的行政部门，有管理科研的行政部门，至少，我们学校还没有管理推广的行政部门，科研处代管一些推广工作，有不少困难，迫切需要建立相应的推广管理部门。

既然教学、科研、推广都是教师的本职工作，那么，系和教研室也就同时是教学、科研、推广的实际执行机构。

最后是物质条件问题。前面提到，粉碎“四人帮”以后六、七年中，重点农业院校已形成了既是教学中心，又是科研中心的局面。但是，科学研究还缺乏稳定的经费保证，学校的科研队伍仍然处于是“雇佣军”的状态。

关于推广工作的经费条件和科学研究不尽相同，可能需要支出，也可能有一定收入，需要什么，如何解决，还需要进一步实践和积累经验。

改革农业教育的两点意见*

陈华癸

（华中农学院）

我国的农业教育要“面向现代化，面向世界，面向未来”，迎接世界新的技术革命的挑战，就必须大力进行改革。这里提出两点改革的意见：一点是宏观的农业教育体系的改革；一点是微观方面的，即对如何办好一个学校的改革意见。

办好我国的农业教育，首先要建立适合我国国情的农业教育体系，本科、专科、中专、农业中学或普通中学的农业教育等，需要有合理的层次、人数、布局和规格。

农业大学教育（这里指本科和研究生教育），要有一个全国的合理布局。这需要中央和各省协调规划，中央搞一套，各省搞一套，很难形成一个整体性的合理布局。以目前的师资、财力来说，也是很不够的。师资严重欠缺，校舍、设备和消耗性开支又严重不足。又要马儿跑，又要马儿不吃草，是不符合实事求是精神的。

鉴于我国的农业本科教育主要是四年制的，三年制的专科学校似可不办，将必需办的个别专业合入本科中办。

为了层次合理、分明，三年制的专科学校应改为二年制，与四年制的本科形成一个清楚的层次区别。

从现实情况看，二年制的专科和三、四年制的中等专业学校是要并举的。这是由不同的学生来源决定的。城市普及高中，农村普及初中，在这个教育发展的过程中，普高毕业生读二年制农专，初中毕业生读三、四年制中专是两朵并蒂花。二年制专科和三、四年制中专可以在全省统一规划下分地区建立，以本地区的学生为主要来源，以服务于本地区农业发展为毕业后的主要去路。

在农村的普通中学中设农业教育课和农业中学的格局应该如何处理？由于我国幅员广大，各省、区的发展程度相差也很大，不能一概而论。对于中等发达的省区，趋势是中等城市设普通中学，在乡、小镇设农业中学，以初中为主。有条件读普通中学的乡、小镇的中学生可以入中等城市普通中学住读。

当然，干部培训、函授、业余教育、技术推广等也都是农业教育的组成部分，篇幅所限，就不多写了。

关于如何把一个学校办好，就任何一个学校而言，可以罗列出许多应兴应革的具体事项。但是，归根结蒂，任何事都要人来办，关键是在有改革精神的领导班子的有力领导下，如何发挥人的主动性，培养人的进取精神。铁饭碗、大锅饭是人的进取精神的消蚀剂，不打破，什么事都做不好。教职工和学生的铁饭碗、大锅饭都要打破。

现在，各条战线都在试行干部的聘任制、工人的合同制，且已经有了一定的经验，其实聘任制也就是合同制。学校的教师、职员、工人也都可以搞合同制。我国的老经验，学校的老师一年一聘，这不好，一方面教职工没有安全感，另一方面不利于专业教学和研究计划的实行，不利于熟悉岗位工作和积累经验。可试行在一年试用期以后，可以订三年、五年、十年的合同。订了合同，就要按合同的要求考察工作表现，而且在合同期满以前，学校和教职工本人都要为对方创造可以继续签订下一期合同的条件，教职工表现不很出色，可能得不到继续聘任。学校当局不为教职工创造比较满意的工作和生活条件，也很可能留不住优秀的教职工，合同期满后，很可能另行高就。一个大学教师，试用一年，订两个三年合同，两个五年合同，两个十年合同，一辈子也就差不多了。

---

*原载于《高等农业教育》，2：35~39，1984.

现在大学生入学后可以说铁饭碗已经基本到手，这对在大学学习期间形成一个你追我赶、生动活泼、富于进取精神的学风是十分有害的。学生的政治修养、资质不同，努力不同，在四年的学习期间，自然有学习效果的差异，“全五分”的思想，本身是不符合实际的。应该实行淘汰制，一个大学，若每年入学一千，毕业八百，淘汰二百。毕业生中有二百考取研究生，进入更高层次的学习，剩下六百有的得到学士学位，有的得不到学士学位，差别是智愚贤不肖的客观反应，必将在德、智、体三方面都形成一个努力进取、不甘落后的生动局面。

高等农林院校本科生物系列课程教改设想*

李合生　陈华癸

（华中农业大学）

一、面向21世纪的高等农林院校生物系列课程教学改革的重要意义

科学家预言21世纪是生命科学的世纪。这就意味着现代生命科学在理论上将会出现巨大的飞跃，在农业、医学、工业和国防上，也将引起巨大的变革。生命科学将成为21世纪自然科学发展与社会进步的关键学科。“科技兴国，教育为本”，高等学校生物学教育是发展生命科学与培养人才的基地。因此，生物科学的教学、科研日益受到人们的重视。而我国高等农林教育中生物系列课程教学内容和课程体系虽然经多次改革、多次反复，也取得了一定的成绩，但基本上仍沿袭50年代初的模式，在教学计划、课程设置、人才培养等方面，还保留着计划经济的烙印，与国家社会主义市场经济的建立、科学技术的迅速发展很不适应，与经济发达国家相比，有较大差距，亟待改革，其核心就是进行教学内容和课程体系的改革。

在高等农林院校，生物学按照研究生物类型、对象的不同，建立了微生物学、植物学、动物学，由于所研究的生命特征及其规律不同，又建立了细胞学、形态解剖学、植物生理学、遗传学、生态学等分支。此外，由于物理学、化学、数学等不同学科的相互渗透和深入，还建立了生物化学，生物物理学、分子生物学、电生理学等边缘分支科学。目前，生物科学正从生物个体研究分别朝着微观（分子生物学）和宏观（生态生理学）两个方向发展。

在国外高等农林院校里，大多是把生物学各分支科学，划分为若干类别，分别列为不同层次（研究生、本科生、专科生）、不同专业（农学、植物学、园艺学、林学、畜产学、植物保护学等）学生的必修课和指定选修课、任意选修课。日本、美国、加拿大等国比较典型，例如日本名古屋大学农学院，把生物化学分为Ⅰ、Ⅱ、Ⅲ、Ⅳ，其中Ⅰ、Ⅱ为普通生物化学，分两学期开出，规定为农业化学和食品工业化学专业的必修课，为农学、畜产、林产专业的相关选修课；而Ⅲ、Ⅳ为高级生物化学，规定为农业化学和食品工业化学专业的相关选修课。另外，为农业化学专业硕士生开出了生物化学专题Ⅰ和生化大实验Ⅰ，为食品工业化学、生物化学专业，开设了生物化学专题Ⅱ和生化大实验Ⅱ。从动物生理学来看，名古屋大学农学院的畜产专业规定家畜生理学、繁殖生理学及其实验课（单列）为必修课，而以比较生理学、泌乳生理学、产卵生理学、植物生理学、森林植物学为选修课。在植物生理学方面，美国明尼苏达大学农学院为全校各专业开设了11门有关植物生理课程（如：植物生理学概论、植物细胞生理学、实验原生质学、植物代谢、植物生长与发育、光合微生物生理学、植物分析方法等），这些课程以植物生理学概论为必修课，其他均为选修课。威斯康辛大学为植物系学生开设了7门植物生理学，既注意到分子水平和机理研究（膜、营养、分化），又注意到生态方面的问题（生理生态学），课程设置反映了该校和教师的专长，选修课多，学生可以跨系、跨专业、甚至跨学校选课，而且选修课比例也高。

在国内，植物学方面已进行了初步改革，中国农业大学、北京林业大学、北京农学院三院校，根据不同专业设置不同的必修课和选修课，如农化专业必修课为植物学Ⅰ（形态解剖），植物学实验Ⅰ，选修课为杂草学；植病专业的必修课为植物学Ⅱ（分类学）和植物学实验Ⅱ。微生物学方面，各农林院校共开出了7门微生物学，一个是植物生产类（农业、环境微生物等），一个是动物生产类（畜牧、水产微生物），一个是微生物生产类（普通微生物），它们都含有共同的基础微生物学部分。国内各农林院

*原载于《教育与教材研究》，6：16~18，1996.

校，在植物生理学、动物生理学、遗传学、基础生物化学方面都基本上是按统一的教学计划作为必修课的，多数学校实验课未单列。

与国外相比，国内农林院校生物系列课程在教学内容、方法、课程结构等方面存在的主要问题如下：

（1）课程设置上，在80年代初制定的统一的教学计划指导下，植物生理学、植物学、遗传学、微生物学、动物生理学等都列为相关专业的必修课，实验课未单列，而且大部分专业是以同一模式上课，都讲求学科的完整性、系统性，按统一教学大纲实行模板式教学，内容庞杂，未紧密结合专业，而有些课程如鱼类生理学、禽类生理学、油菜生理学、棉花生理学等又分得过细，缺乏综合性和比较性。此外，选修课开设很少，选修课所占总学分的比例也很低。美国依阿华州大学农学院，农学专业总学分159，其中必修课和指定选修课学分为128，任意选修课学分为31，即任意选修课所占总学分比例为20%。日本名古屋大学农学院畜产专业必修课学分99，选修课学分33，即选修课占25%。而华中农业大学四年制各专业学生总学分为160，五年制兽医专业为197，其中必修课学分一般占总学分85%左右，选修课只占15%左右，而且在选修课中，大部分为指定选修课，任意选修课只占总学分的2%~4.5%。与国外相比，差距甚大，这对学生的学习主动性是一个束缚。学生要预测四年后的市场人才需求非常困难，带有相当大的盲目性和不可预测性，在发展学生个人理想、愿望方面又是一个不折不扣的“紧箍咒”。

（2）在教学内容方面，有相当一部分陈旧、落后，教材更新速度慢。目前国内生物系列课程教材几乎是每8~10年才修订再版一次。而美国的微生物学、植物生理学、植物学等都是每两年左右更新一次。教师讲课内容贴近时代，新的内容不断补充。分子生物学的内容在植物生理学中占有显著的地位。

（3）在实验课方面，基本是处于理论课的从属地位，重视不够，学生做实验是“照单配药”，动手能力差，被动地学习做某些实验。实验课尚未形成一个独立的课程体系，实验经费不足，实验方法落后，验证性的多，实验设备仪器陈旧。国外实验室，像欧美、日本等国的大学实验室经费充足、设备不断更新，实验内容先进，实验室是全天开放，教师稍加指导，主要由学生独立完成实验，实验室是多用型、通用型，实验课是分段集中上，实验室利用率非常高。

（4）在教学方法上，多年来一直沿用“以教师为中心、以课堂为中心、以教材为中心”的传统教学模式。只重视知识的传授，注重授课计划的完成，而忽视学生能力和素质的培养。教学手段方面，生物学各类课程如微生物学、植物学、植物生理学、动物生理学等都没有配套的电视教学片，更没有CAI、多媒体、试题库等先进手段。在国外，教师挂牌教学，学生自由选择课堂，教师结合科研、专长讲课，学生听起来有兴趣；同时，教学手段先进，CAI、多媒体、试题库、教学录像片比较普及，对学生有吸引力，可调动学生学习的积极性和主动性。

二、面向21世纪的高等农林院校本科生物系列课程教学改革的设想

1. 高等农林院校本科生物系列课程改革的思路

（1）在教育思路上来一个破旧立新的大转变：改变过去40年来把生物学课程划分过细的局面，加强对学生的综合性的、整体性的、开拓性的素质教育，一定要打好基础，拓宽知识面，提高自身素质，迎接新世纪、新科技及市场经济的挑战。面向21世纪的生命科学，必然是各学科相互渗透与相互交融的“大学物学”时代。因此，无论是植物生产类，还是动物、微生物生产类专业的学生都应当学习普通生物学（或叫生物学基础）、学习遗传学（含动、植物遗传学内容）等。

（2）改革课程教学的主要任务就是传授知识的传统教育观念，加强对学生的分析、解决问题的能力和创新精神及适应市场竞争能力的培养。对生物学各类实验课和实验室要进行改革，让学生多学一些“实用知识”。

（3）改变过去以教师、教材、教室为中心的灌输式的教学方法，注重发挥学生的积极性和主动性，使学生具有自我开拓和获取知识的能力。

（4）改变过去一进大学某专业学习，就定了“终身”的状况，注重学生个性发展，让学生在毕业前1~2年有预测市场，设计未来，自主选修课程的机会。

2.高等农林院校本科生物系列课程教学改革的目标

以邓小平同志的“教育要面向现代化、面向世界、面向未来”为指导思想，转变教育思想，更新教育观念。经过5年时间的研究与实践，优化本科生物系列课程结构，推出一批新的教学改革方案和新教材、新软件，使生物系列的理论课、实验课的教学内容、课程体系、教学方法及教学手段，能够适应我国21世纪社会主义市场经济、农林科技发展和人的个性发展的需要，使生物系列课程的教育水平、学术水平、人才素质逐步接近国际先进水平。

3. 生物系列课程的教学内容和课程体系改革的主要框架

（1）为了达到上述目标，根据不同专业类群的要求，本着打好基础，拓宽知识，整体综合，先进实用，结合专业，加强实践的原则，将生物系列课程划分为两个系列，两种类型。

1）理论教学系列

基础型课程　应该设置一些知识面广、内容新，可为农林类所有专业奠定宽厚基础的课程；对分得过细的课，合而为一，还其本来面目，且要内容精练，减少学时，对曾经为青年人奠定生物学基础起过重要作用的普通生物学应该恢复其应有的位置。为此，我们对生物系列基础型课程提出“3+3”的设想：基础课为生物学基础（或普通生物学）、分子生物学基础、植物学或动物学；专业基础课为遗传学（含动物、植物、微生物遗传）、基础生物化学（含动物、植物、微生物生物化学）、基础生物生理学（植物生理学导论或动物生理学导论或微生物生理学导论）。

选修型课程　在打好基础之后，从微观或宏观两个方面，设置一些有一定深度的交叉学科和新学科作为学生深造之用，如：作物生理学、林木生物化学、生态生理学、植物细胞分子生物学、环境微生物学、分子遗传学、发育生物学等。

2）实验教学系列

基础型：如植物形态解剖实验，普通生物学实验，遗传学实验等。

选修型：如林木生理生化研究技术，微生物生理生化研究技术，分子生物学实验技术，基因工程技术等。

（2）增加选修课的门数，提高选修课占总学分的比例，为学生在毕业前充分提供短期（1~1.5年）预测市场、选修课程的机会。

（3）编写基础课和选修课的教学大纲和新教材，缩短原有教材的修订、更新周期。

（4）改进教学方法，改灌输式为启发式、讨论式，制作CAI软件和电视教学录像片，改革考核办法。

（5）开展实验室建设管理改革研究，建设开放式、通用型的现代化实验室，使其成为培养智能型人才的重要基地。

转变教育思想和观念，深化本科生物系列课程教学内容和课程体系的改革*

陈华癸　李合生

（华中农业大学）

一、转变教育思想和观念是生物系列课程教学内容和课程体系改革的先导

当今世界正处在大变革的历史时期，科学技术和经济建设迅速发展，各种科技领域相互渗透，新学科和交叉学科不断涌现，科学知识的更新速度和科学技术转化为生产力的速度加快，科学技术发展综合化趋势加强。而我国高等农林教育中教学内容和课程体系虽经多次改革、多次反复，也取得了较大的成绩，但基本上仍沿袭着50年代初的模式，在教育计划、课程设置、人才培养模式等方面，还保留着计划经济的烙印，设置专业过多、过细、过窄、部分教学内容陈旧，教学方法老一套，课程体系结构不够合理，模式单一，必修课过多，选修课偏少，课程之间由于追求各自的完整性、系统性，也出现了重复脱节现象。从而导致了所培养的学生基础不扎实，专业面过窄，综合素质不高，适应性不强。为适应形势发展的需要，改革高等农林教育的教学内容和课程体系势在必行。然而要进行一项成功的改革，没有正确的指导思想是不可能实现的。当今站在教学第一线，担负着教学改革任务的教师和管理干部多数是在旧模式下培养出来的，在改革实践中自觉或不自觉地会受到一些传统的教育思想和教育观念的束缚，因此在这场教学改革实践中必须加强自身主观世界的改造。结合我们农林高等院校生物系列课程的现状，我们认为要抓好思想和观念的三个大转变。

（一）从单纯向学生传授知识、技能为中心的传统教育，转变为着重培养学生分析、解决问题的能力和创新精神的现代教育。学校教育的重要任务就是培养学生的创新精神和独立生活、工作及获取知识的能力。可是在传统的大学教育中，教师反复强调学生要学习知识，但未重视教学生如何用所学到的知识创造性的解决问题，学生的思维和动手能力较差。在课程设置上我们认为不能把实验课当做理论课的附属品，无足轻重，要把生物实验课单列出来，提高它应有的地位，引导学生重视动手能力、正确分析实验结果和解决疑难问题的能力的培养。在教学方式上要从根本上改变目前普遍存在的“满堂灌”、“照本宣科”的现象，不能讲得过细过碎过全，要通过启发式和讨论式教学，杜绝死记硬背，培养学生积极思维和创新精神。在教学内容上要尽可能地多传授有效的知识，动态的知识，不仅要介绍新的科技成果，而且要多向学生讲述在取得新成果的过程中是如何思维和实践，如何从失败中找到原因并取得成功的。教学过程要与课程论文、毕业论文及第二课堂、科研工作相结合，建立新型的平等和谐的师生关系，这些都是培养学生创造能力和创新精神的有效措施。此外，还要重视学生个性发展，逐步实行学分制，增加选修课，提高选修课的比例，加强因材施教要为不同层次和不同水平，不同就业去向的学生开设不同类型的课程。生物系列课程可分为基础必修型课和选修型课，前者，面向动植物生产类各专业百分之百本科生；后者，面向动植物生产类各专业10%~50%的本科生，而生物学基础课（普通生物学）则面向农林院校各专业所有的本科生。这样可以避免一刀切、一个模式，既可为优秀人才拓宽学习的空间“让腿长的跑得更快些”，又可为一般的学生节省一些宝贵时间，选修自己喜爱的任何一门课程，发展个人的特长、爱好，增强自己在市场经济中的竞争能力。

（二）改革过去生物科学知识和课程设置越分越细的分割知识教育，转变到强调综合性、整体性的素质教育。分析与综合是认识客观事物的两种途径，要使学生既能从微观方面深入认识生命现象的本质，

*原载于《国家教委高等教育面向二十一世纪教育改革经验交流论文汇编》，325~330，1997，北京：高等教育出版社.

更应学会从客观方面进行综合，了解生命现象、生物活动在整个地球生态环境中的地位与作用。对生物界来说，无论是动物、植物、微生物，从起源到进化，从基本结构到遗传代谢变化都是彼此紧密联系，不可截然分割开来的，因而在培养本科学生时，不能让动物生产类的学生只学习动物遗传、动物生化，而把植物遗传、植物生化拒之门外，严格分割开来，那是不利于学生综合性整体性素质培养的。应该看到现代的科学技术，是既高度分化又高度综合，而以高度综合为主的整体化趋势，这就要求课程设置的综合化，要重视和加强基础科学的综合教学。同时，现代科学与现代社会相互渗透，使得社会问题和技术问题的解决都要依靠多门学科，多种技术的综合应用。因此教学上要由以学科为中心的统一的教学计划、统一的教材、统一的学识，即统一的知识教育阶段进入到交叉学科的综合知识教育阶段。还有一个观点也是值得我们注意的，就是我们加强基础科学教学，并不是单纯为专业课服务的，而是为提高本科生整体素质服务的。因此我们对农林类动、植物生产系列的本科各专业，都设置了相同的普通遗传学和基础生物化学（含动物、植物、微生物）生物学基础、生态学概论（大生态）等。此外，作为生物系列课程来讲，过去由于过分强调学科特点，忽视了相关性，片面追求各门课程的大而全，因此在相关课程之间存在着教学内容相互重叠，教学过程相互隔离，甚至出现相互矛盾的现象，这样既浪费了宝贵的学时，又影响教学效率的提高，而且也不利于系列课程的整体性和综合性。像细胞减数分裂和DNA结构等内容一直在生物学、植物学、普通生物学、细胞生物学、遗传学、生物化学、分子生物学等课程中进行了重复地讲授。这就要求我们在新的生物系列课程教学大纲编写中要多进行相关课程的传阅、协调，达到有机的结合、衔接，避免在面向21世纪的生物系列课程中出现新的重复。我们认为课程之间的衔接是必要的，而重复脱节是应当避免的。

（三）转变以追求增加学时数进行教学改革的教育思想，建立集约观点，确立减少学时，充实内涵、引进先进教学手段，提高教学质量的思想。在过去的多次教学改革中，都出现过各门课程固守山头，有些人认为只有增加教学时数，才可以提高教学质量。实际上这样改革的结果，学时数虽有增加，但教学质量并未相应提高，而且增加了课程间的重复现象，课堂上从古到今满堂灌，在未增加学时数，甚至减少学时的情况下，教学内容有增无减， 便出现了“压缩饼干”的现象，加上教学手段又是老一套，结果教学改革效果并不好。这次面向21世纪教学改革中，我们项目组提出了每门课程要减少学时，精炼内容，提高水平，并注意引进新的科技成果，不断更新课程教学内容。编写新的教材，适当控制教材篇幅但不要浓缩，要便于学生自学，还要通过课程之间的相互贯通融合以及相互渗透的办法，尽量避免课程之间重叠现象。节约下来的学时，一是用于增加个别新兴的课程，如分子生物学基础、生态学概论、生物学基础等。二是用于单列生物系列实验课，提高实验性环节占总学时的比例达到30%以上，三是用于发展学生个性，因材施教，增加选修课程的“活动空间”，让选修课占总学分比例，逐步达到25%左右。同时在学时减少，新内容增加，知识面拓宽，水平提高的情况下，引进先进教学手段是不可缺少的，因此在生物系列课程改革中，CAI软件、电视录像教学都要相伴而行。这方面，我们项目组的微生物课在教材修改、试题库建设、CAI软件等方面都取得了阶段性的成果。

二、加强领导、加强协作是教学改革取得成功的重要保证

在教学改革中，要加强领导，项目组之间思想要相互沟通，行动上要相互配合，搞好整体安排非常重要。国家教委[96]12号文件指出，各有关学校要将本校承担的项目作为学校整体教学改革规划中的重中之重，要与学科专业建设，课程建设，师资队伍建设和教材建设等结合起来。[96]62号文件又指出，各项目组要从项目研究工作的实际出发，主动邀请相关项目组及教学指导委员会学科组参加本项目的研讨会，相互配合，团结协作，各司其职，共同完成研究任务，这是很正确的。加强领导、加强协作是完成面向21世纪教学改革的重要保证。本项目第一主持单位华中农业大学的党委书记、校长多次主持召开面向21世纪教学改革项目组与院、系、处行政领导的联席会议；主持单位山东农业大学和北京林业大学的校领导及教务处也是多次召开会议，经常指导、部署、督促、检查项目组的教学改革工作。华中农业大学还在6月上旬召开全校本科教学工作会议，其指导思想是动员和组织全校师生员工转变教育思

想和教育观念，落实教改核心地位，深入推进教学内容和课程体系改革全面提高教学质量，促进学校教学工作再上新台阶。由于领导重视，因此在研究力量，改革力度，具体实施，研究经费等方面就比较落实，工作有起色。否则，单靠几个项目组的研究人员（教师）的活动是无法把教改工作推动起来，落到实处的。此外，我们还体会到面向21世纪的各个项目组之间必须及时经常沟通信息，相互协调，如农、林类本科专业目录修订的研究就制约着各系列课程体系的建立。如果各项目组背靠背，各干各的，各吹各的号，各说各重要，最终还会使一场改革落空的。我们很赞成工科力学课程项目组总负责人范钦珊先生的意见，参加面向21世纪项目的院校和老师，要按照选优联合的原则，切实实现院校的联合，项目的联合，课程的联合，为国家创建出最优化的系列课程设置和教学大纲及教材。在项目组之外，还要与国家教委批准的理科生物学基地建设及农业部、林业部教学指导委员会生物学科组紧密结合，共建未来，一定可以收到事半功倍的效果。

三、面向21世纪教学改革必须是理论与实践的紧密联合

教学改革不能从理论到理论，为改革而改革，为发表几篇论文向国家教委交差了事，而是要脚踏实地从我国现状出发，面向现代化、面向世界、面向未来，研究出一个符合我国国情的高水平的整体优化的教学内容和课程体系改革方案，编写出高水平的新大纲和新教材，同时要在各个学校的本科教学班进行教学改革实践，至少重复一次，以通过实践检验这场改革的先进性和科学性。华中农业大学已在本校园艺系专业和畜牧兽医学院畜牧专业选择了97级本科生教学班进行从基础到专业，从人文素质到专业素质的全方位的教学改革试验，并已制定了教学改革试验方案，进行了多次讨论，下一步是落实到人，编写教学大纲，边教边改，目前正在实施之中，可望在今年年内取得一些实质性的进展。中国农业大学学习讨论了国家教委副主任周远清同志5月13日在该校所作的关于高等教育情况及发展思路报告的基础上，以华中农业大学 提出的关于动物生产类和植物生产类系列课程教学计划试验方案为蓝本，结合中国农业大学的实际情况，提出了初步修改意见，并分别召开了动物生产类（畜牧、兽医、食品、气象）和植物生产类（农学、园艺、植保）有关教师座谈会，提出了理科生物学基地教学计划和非生物学专业的生物系列课程的门类，课程的教学内容及教学大纲（初稿），并计划1997年秋季在部分专业本科生中进行教学改革方案试验。

21 世纪农林本科生物系列课程改革的研究与实践（上）*

李合生　陈华癸

（华中农业大学）

一、背 景 分 析

生物系列课程包括现有的植物生理学、动物生理学、生物化学、植物学、动物学、普通生物学、微生物学、遗传学、细胞生物学、分子生物学。这些课程都是农林类各专业的重要专业基础课或基础课。科学家预言，21 世纪是生命科学的世纪，生物科学肩负着特殊的使命。由于分子生物学和生物技术的迅速发展，使生命科学成为当代最活跃的领域之一。农业经济、农村社会和农业科学技术的发展以及大量新型农林科技人才的培养，对生物科学提出了更高的要求。而我国高等农林教育中生物系列课程教学内容和课程体系虽经多次改革取得了一定的成绩，但一部分教学内容仍然陈旧落后，重复脱节，课程设置模式单一，选修课程很少，知识面窄，对学生能力培养重视不够。这与我国社会主义市场经济的建立、科学技术的迅速发展很不相适应；与经济发达国家相比，有较大的差距。这种状况亟待改革，其核心就是进行教学内容和课程体系的改革，目的是使高等农林本科生的生物学素质、知识、能力结构能适应 21 世纪的需要。

二、目前农林类本科生物系列课程和学生生物学素质的现状

我国农林院校生物系列课程的教学近十多年来取得了长足的进步，教学水平有所提高。但是，由于近几年普通高校招生入学考试取消了生物课，学生不重视生物课学习，导致高中生物学教学质量下降。在大学本科阶段，教学计划和课程设置模式比较单一，结构不太合理，选修课少，学生选课的自由度小，从而限制了学生生物学素质提高的机会、空间及深度。在农林院校本科学习阶段，生物课程安排较少，也是造成学生生物学素质较低的重要原因。如植物生产类中农学、园艺专业开设的生物系列 4 门课程占总学时的 13%左右。据 4 所大学本科林业专业统计，4 年平均总学分为 165 分，生物系列课程为 17.7 分，占总学分的 11%。

其次，从生物系列课程的教学内容来看，有相当一部分陈旧落后，缺乏综合性，知识面窄。如动物生产类专业不学植物，只学动物遗传、动物生化；植物生产类专业也类似。这就使得学生的生物学素质的培养大打折扣。另外，在实验课方面，生物实验课大多未单列，实验经费不足，实验设备更新慢，学生独立进行实验操作机会少，出了问题也不善于分析和解决，学生动手能力较差，存在着高分低能现象。

三、中外高等农林院校生物系列课程设置的比较研究

近 20 年来，生命科学的飞速发展特别是分子生物学的突破性成就，使人们毫不怀疑地相信 21 世纪将是生命科学的世纪，其中生物技术将会对农林业的发展作出巨大的甚至是决定性的贡献。为顺应生命科学飞速发展形势，许多西方国家不仅对生命科学研究加大了资助强度，而且还对农、林、医学的高等教育大力扶持。以美国为例，在近年 48 万博士学位获得者中，学习生命科学的占 51%，表明优秀青年科学家的主流已属意专注生命科学前沿。各大学也纷纷将生物类专业的设备及课程安排作了相应调整。

---

*原载于《中国农业教育》，（4）19~22，2000.

而我国农林院校的专业设置及课程安排还基本保持着50年代初的苏联模式，这显然与国内农业发展需要、国际农业发展趋势、特别是21世纪对高层次农林业科技人才的需求极不相称。改革高等农林院校生物学课程设置已刻不容缓。

我国大多数农林院校开设的生物系列课程主要有植物学、植物生理学、生物化学和遗传学4门课。

欧、美、日、俄等经济发达的国家在生物系列课程设置上有如下几个共同特点：①注意打好基础，如基础生物化学导论综合了植物生物化学与动物生物化学的基本内容；②分子生物在课程中占有明显的地位。如设置植物生理学与分子生物学；③环境生理、生态学受到普遍重视；④结合各专业生产实际开设的新课程和选修课多，有作物生理学、果树生理学、树木生理学、应用生物学、树木生物化学等；⑤重视学生动手能力培养，实验技术课单列，占有重要位置，设置有植物生理实验、生物化学实验技术、基础工程技术。下面具体介绍几所大学的生物系列课程开设情况。

1.新西兰 Massey 大学

Massey 大学是新西兰的一所著名大学，所开设的生物系列课程如下：①植物生物学；②植物生物技术导论；③植物王国；④植物分类方法；⑤基因及其产物的生物化学；⑥生物化学；⑦微生物与遗传学；⑧基因操作；⑨自然植物遗传；⑩植物生理学。

2.美国普度大学农学院、肯塔基大学农学院、戴维斯加州大学农学院

美国大学的农学院一般包括农学系、蔬菜系、果树系、环境园艺系、植病系、昆虫系、农业工程系、农经系、食品系、土壤系等，动物、兽医等独立为兽医学院。在植物科技学科中，其主要基础与应用基础课如下：①生物学；②植物生理学；③遗传学；④植物分类原理；⑤生物化学；⑥农业生物技术；⑦植物生长和发育；⑧植物生物物理学；⑨植物解剖；⑩作物生理生态。

四、高等农林院校本科生物系列课程改革的思路

1.在教育思想上来一个破旧立新的大转变，改变过去四十多年来把生物学课程按生物四大系统（动物、植物、微生物和人）划分过细的局面；加强对学生综合性、整体性、开拓性的素质教育。面向21世纪的生命科学，必然是各学科相互渗透与相互交融的“大生物学”时代。因此，无论是植物生产类还是动物生产类专业，学生都应当学习普通生物学（或生物基础）、普通遗传学（含动、植物遗传学内容）等。

2.改变过去不合理的课程体系和陈旧的内容，适应科学技术发展和市场经济的需求，建立新的课程体系，充实新内容，协调好课程之间的关系，避免新的重复、脱节现象。如分子生物学与遗传学，普通生物学与动、植物学等。

3.改变过去认为课程的主要任务就是传授知识的传统教育观念，加强对学生的分析、解决问题的能力和创新精神及适应市场竞争力的培养。对生物学骨干课实验应单列，实验室要改为通用型和开放型。

4.改变过去以教师、教材、教室为中心的灌输式教学方法，引进新的教学手段，开展讨论式教学，注重发挥学生的积极性、主动性和创造性，使学生具有自我开拓和获取知识的能力。

5.改变过去学生一进大学就定专业“终身”的状况，注重学生个性发展和市场需求的变化，增加选修课的门数和比例，让学生在毕业前有预测市场、设计未来以及自主选修课程的机会。

五、面向21世纪高等农林类本科人才应具备的生物学素质

概言之，21世纪的高等农林类本科学生应当具有丰富的生物学理论知识、过硬的生物学实验设计、测试分析技能及创新能力。

（一）生物学基本理论知识的要求

作为面向21世纪农林类本科人才的生物学理论知识应具有综合性、整体性、系统性和先进性。这

就需要对原有生物系列课程的教学内容进行更新，设立新课程，优化课程体系。无论是动物生产类还是植物生产类的本科，都应从宏观到微观地掌握生物类（包括动物、植物、微生物、病毒）的形态分类、生化、生理、遗传以及分子生物学基础、生态学基础等。要求学生掌握和了解从动物到植物、微生物、病毒，从生物大分子到细胞、组织、器官、个体，从田间、盆栽到地球、宇宙，从作物生态学到农业生态学、大生物圈，从生命起源到生物进化，从现象到机理等等一系列综合性的科学知识，改变过去对生物学知识的分割状况；还要改变过去那种农林类本科生只学习生物大分子、细胞、组织、器官、个体而忽视宏观知识的状况，以提高农林类本科生的整体生物学素质。

（二）生物学基本能力的要求

为了适应现代科学技术迅速发展和高新技术的广泛应用及社会主义市场经济的需要，面向 21 世纪高等农林类本科人才生物学能力的基本要求是：

1.室内的仪器分析测试能力，即以植物、动物、微生物及病毒为材料，进行生理生化、遗传、生长发育、分类等研究时必不可少的技术能力。如：可见光、紫外光比色技术，红外线 CO_2 气体分析技术，电泳技术，高速冷冻离心技术，染色体检测技术等。

2.在田间进行生理生化、生长发育、遗传变异、生态试验的实验设计、取样、前处理、物质含量及活性检测、统计、分析的能力。在野外识别动、植物种类、采取和制作标本的能力。

3.具有对室内的实验结果或田间试验结果进行整理、分析并写出规范的实验报告或田间试验小结、科技论文的写作能力。

4.识别植物细胞亚微结构、三维结构电镜照片的技能等。

六、各专业类群中生物系列课程类型的划分、要求、主要内容和差异

根据农林院校专业类群的需要制定不同类型生物系列主干课程体系：

1.对于植物生产类专业（农业、园艺、农业化学、植物保护等）应加强综合性课程，以加强基础，拓宽知识面，加强动手能力、创新能力培养。开设课程如下：植物学、植物学实验、基础生物化学、植物生理学、植物生理生化实验、普通遗传学、动物生物学（必修或指定选修）、普通微生物学（必修或指定选修）。

2.对于动物生产类专业（动物科学和动物医学）应加强综合性课程和特需专业基础课，还应开设植物生物学实验课系列。开设课程如下：动物学、动物生理学、动物生理学实验、基础生物化学、基础生物化学实验、普通遗传学及实验、微生物学和植物生物学（必修或指定选修）。

3.对于生物科学类专业（生物科学、生物技术等）应加强现代生物学各学科的设置，体现基础厚、知识面广、内容新的特色。开设课程如下：植物生物学及其实验（单列）、动物生物学及其实验（单列）、微生物生物学及实验（单列）、生物化学、生态学、普通遗传学、细胞生物学、分子生物学、发育生物学、免疫学、生物信息、生物工程概论、生化技术等 16 门。

4.经管、人文、工程类专业的学生应当学点自然科学，开设普通生物学或现代生物学基础，作为必修的公开基础课或指定选修课，主要介绍生物学基础知识，以进化论为主线，微观方面介绍分子、遗传学和生物工程的基础知识，宏观方面介绍生物与环境、生物与人类共存持续发展等。

5.食品科技专业应当了解动物、植物和加强生化基础知识及生化分析技术，开设普通生物学、生物化学、生化实验 3 门课。

七、不同模式生物系列课程体系改革方案及实施情况

华中农业大学、山东农业大学、中国农业大学等校分别制定了不同模式的生物系列课程体系改革方案，并开始教改试点的筹划和实施工作。

1.制定教改方案的基本原则

（1）紧扣人才培养目标，把“厚基础、宽口径、高素质、强能力、广适应”作为制订方案的重要原则。

（2）重组课程体系，避免重复脱节，加强课程综合性，以适应科技的综合化、整体化发展趋势和市场经济发展的需要。

（3）压缩必修课学时，多开小课；增加选修课门数及其占总学分的比例，适应学生个性发展的需要。

（4）重视学生创新能力和实践能力的培养。

2.教改方案内容

根据以上基本原则，将生物系列课程分为两个系列、两种类型。

（1）理论教学系列课程设置

①基础必修型课程

设置一些知识面广、内容新而精、可为农林类所有专业奠定宽厚基础的课程；对分得过细的课程则合而为一，精炼内容，减少学时，增设生物基础课程。为此提出“5+4”的设想：

5 门基础课——植物学、动物学、生物学基础或普通生物学、动物生物学、植物生物学。

4 门专业基础课——普通遗传学（含动物、植物、微生物）、基础生物化学（含动物、植物、微生物）、生物生理学（植物生理学或动物生理学）、普通微生物学。

②选修型课程

在学习必修课的基础上，从微观和宏观两个方面设置一些有深度的交叉学科新课程，作为有兴趣的学生（含考研学生）深造选修之用。如分子生物学基础、作物生理学、果树生理学、蔬菜生理学、林木生物化学、细胞生物学、生态生理学、生物化学与分子生物学、病原微生物学、分子遗传学等。

（2）实验教学系列课程设置

①基础必修型：如植物生理生化实验（或生化实验、生理实验）、植物学实验等。

②选修型：如植物组织培养技术、微生物检测技术、特种作物生理生化分析、分子生物学实验技术。

3.几种教改推荐方案（见下页附表）

21世纪农林本科生物系列课程改革的研究与实践（下）*

李合生　陈华癸

（华中农业大学）

八、有关生物系列课程教学内容改革的说明

（一）植物学

教学内容及基本要求　是植物生产类专业重要的必修基础课，其主要内容包括植物个体发育过程中各器官的形态建成过程及其解剖结构，植物系统演化过程中植物界几大类群的特征，被子植物分类概要及一些重点科、属、种的特征，系统演化及经济价值，同时对现代植物学的单科前沿问题有一个基本的认识。

新教学大纲的特点　在内容上力求先进，贯彻少而精的原则，删去一些与其他课程重复的内容；在编排上，从静止地观察植物发展到动态观察和研究植物的一生，进而从观察植物个体发展到植物群落直至整个植物生态系统。因此将植物学中形态解剖部分内容简化，充实植物发育生物学和生态学内容，植物分类部分则从单纯识别植物转变为植物系统演化过程中形态进化和对环境的适应。在学时上，原植物学理论教学为44学时，改革后减少为36学时。

（1）删去与其他课程重复的内容及繁琐的内容，如细胞膜系统、藻类植物生活史等。

（2）调整的内容，如将营养器官变态、茎、叶、花、果形态学术语、种子种苗等调整到植物学实验中。

（3）增加的内容：①植物研究新成就如克隆概念、叶片衰老、人工种子等，植物个体发育形态发生研究方法，植物分类新概念，新的分类方法，分类学发展动态等；②植物学名形成，以属名为基础的种名、科名等；③濒危植物、抗污染植物等介绍；④与生产有关的内容衔接，如芽的异质性。

（二）植物学实验

教学内容及基本要求　是一门综合性、技能性、理论联系实际的实践课，主要内容包括：植物实验基本理论、基本知识、基本方法，如临时装片法、生物绘图法，不同方式徒手切片制成永存片、压片，离析制片，种子、花粉离体萌发方法，组织化学鉴定法，胚整体压挤法，植物分类基本方法，标本采集、制作，植物实验室一般操作等。同时适当拓宽学生知识面，如了解石蜡制片过程，识别细胞亚微结构，掌握显微镜种类、用途及测微尺的使用等。

新教学大纲的特点　更新了60%~70%教学内容，重新安排的内容有：①徒手切片和用支持物进行徒手切片制成永存片；②根尖压片方法；③种子萌发；④花粉离体萌发计算花粉生活力；⑤检查花粉在柱头上萌发和花粉管在花柱中生长制片方法；⑥离析方法；⑦植物学研究中各种常用药品配制；⑧珍稀濒危及国家重点保护植物、抗污染植物、监测植物、水土保持植物等的识别；⑨孢子植物标本采集与制作。

新大纲减少了绘制植物结构、细胞图的次数。用电视显微镜投影方法，减少学生对一些难以观察的结构耗费过多的时间，如椴树茎次结构、蚕豆和大豆根次生结构观察等。

*原载于《中国农业教育》，5：16~18，2000.

（三）普通动物学

教学内容及基本要求　是水产养殖、动物科学、植物保护、生物技术等专业的基础课，主要研究动物各类群的形态结构和有关的生命活动规律。具体内容包括：动物各主要类群的形态、生态、生理功能、行为及其统一性；动物的分类、起源和进化。要求学生掌握动物学的基本理论体系，研究动物学的方法及经济动物的综合利用途径，以指导今后的研究工作。

新教学大纲的特点　在内容上删去了与生物学课程及组织胚胎学等课程的重复，进行了恰当分工与衔接。由于压缩了学时（过去总学时 80 学时，理论课 50 学时，实验课 30 学时；现在总学时 50 学时，理论课 30 学时，实验课 20 学时），理论教学改革采取“糅合”方式，例如过去扁形动物的理论教学，先讲代表动物的形态结构，再讲门的特征，现在不单列代表动物这一节，而直接从代表动物的形态结构中抽提出门的特征。在实验安排上高度“综合”，例如过去软体动物、节肢动物分别开设，现在综合在一个实验中。新大纲中还增加了一些有关动物的结构、机能、分类、起源进化方面的研究新进展以及有关现代新科技的基础知识。

（四）生物学基础

教学内容及基本要求　是农林类和文科类专业学生的生物学入门课程，内容包括：生命起源、细胞形态结构及功能、物质和能量代谢、生殖和发育、遗传基本规律及其分子基础、基因工程、生物进化、生物分类、生物多样性（植物、动物、微生物等）、生物与环境等，系统地介绍现代生物学各主要分支学科，使学生获得较全面的生物学基础知识，并从宏观上了解现代生物学的概貌。

新教学大纲的特点

（1）作为一门基础课起桥梁作用，既是高中生物学课程的补充、提高和深化，又为以后的专业课、专业基础课奠定基础。

（2）以生物体的基本结构和生命活动基本规律为重点，以生物进化为主线，贯穿于教学的始终，让学生在了解整个生物界的同时树立进化的、辩证的、发展的和相互联系的观念，有利于学生的整体素质的提高。

（3）以植物、动物、微生物的具体生物例证来讲生物体的基本结构和共有的生命活动基本规律，体现个性与共性的统一。

（五）普通遗传学

教学内容及基本要求　是生物类各专业的基础课，主要内容包括：高等真核生物遗传的细胞学基础、孟德尔遗传规律、真核生物的遗传作图、染色体结构变异和数目变异、性别决定、细菌及病毒的遗传作图、细胞质遗传、数量性状的遗传特点及分析方法、基因突变的一般规律及突变体的分离与鉴定、近亲繁殖与杂种优势利用、群体遗传平衡以及物种形成与进化。要求结合实验，使学生掌握遗传学的基本原理，掌握对动物、植物和微生物遗传分析的一般方法和基本实验技术，为进一步学习有关专业课程和分子遗传学奠定较好的遗传学知识基础。

新教学大纲的特点　本课程多年来被人为地分割为动物遗传学、植物遗传学，分别为动物生产类和植物生产类学生开设；改革后合成为一门“普通遗传学”，为全校动物、植物生产类各专业学生的共同必修专业基础课，实行全校一个教学大纲，授课时动物、植物生产类专业稍有侧重。总体设想是把教学改革的重点放在加强学生进行遗传分析的思维方法训练特别是常规分析方法训练方面，放在加强学生分析问题和解决问题的基本能力的培养方面。在教学内容调整方面：①精简本课程教学范围内讲不透、讲不清的章节；②精简与其他课程相重复的内容；③避免各章节之间的脱节；④充实遗传分析方法的学习；⑤更新过时的旧概念。

本大纲精简的内容占原大纲的 20%，增加的内容约占原大纲的 25%，对过时的概念予以更新、更正的占原大纲的 3%左右。在实验课方面，植物和微生物实验更新的内容占原大纲 14%，动物类实验更新内容占原大纲 7%。

（六）基础生物化学

教学内容及基本要求 是农林类和生物类专业的专业基础课，目的是阐明生物体内物质结构与功能关系、物质与能量的转化原理、物质转化与信号传导的一般原理。内容包括：糖类化学、脂类化学、蛋白质化学、核酸化学、酶学、维生素、糖类代谢、脂类代谢、蛋白质代谢、代谢调控。结合必要的课堂实验，使学生掌握生物化学研究的基本方法及其基本理论与实验技能。

新教学大纲的特点 本课程是生物类各专业的必修专业基础课，系各种专业生物化学和组织生物化学内容合并而成。

（七）动物生理学

教学内容及基本要求 是动物类专业（含动物医学）必修的专业基础课，内容包括细胞的功能、血液、血液循环、消化、呼吸、泌尿、内分泌、生殖、神经及感觉各器官系统的生理特征、生理功能、功能调节及其控制规律。通过本课程的学习，要求学生建立器官系统机能整体性、机体机能整体性观点，得到生理学逻辑思维方法的训练，为专业课学习打下基础。

新教学大纲的特点 一方面与相关学科衔接好，避免内容重复脱节，保证内容精炼，同时在原有大纲的基础上，扩展教学内容以适应传统养殖模式向市场经济养殖模式转变。教学内容调整落实在哺乳类、鸟类动物生理两大块上：增设禽类生理特点的章节，适应特种经济禽类养殖业发展的需要；增设毛皮动物生理特点的章节，适应毛皮动物养殖发展的需要。在实验课中，加强动手能力培养，实验内容以涉及重点章节内容的经典实验为主，同时也编写了一些可供选择的实验内容。

1995 年学时数为 129 学时，1996~1997 年为 100 学时，本大纲削减到 90 学时（理论 55 学时，实验 35 学时），动物生理学实验单列一门课。

（八）植物生理学

教学内容及基本要求 主要内容包括：植物细胞生理、水分生理、矿质营养、光合作用、呼吸作用、有机物质运输、植物生长物质、植物生长生理、生殖生理及衰老、抗性生理。要求学生掌握植物生命活动基本规律，为学习专业课奠定良好的基础。

新教学大纲的特点 过去的植物生理学教学中存在的主要问题是课程设置模式单一，无层次，无选修余地；教学内容陈旧，教材更新周期太长，缺乏综合性和针对性，联系农业生产实际不够；缺乏特色，与其他课程有重复现象。近年来，植物生理学在微观方面正在与分子生物学紧密结合，在宏观方面则与生态学相结合，与农村生态关系更密切，因此在植物生理学教学内容改革中反映了这两方面的新进展，精减内容占原大纲的 15%左右，增加的新内容占原大纲 15%左右，教学总学时控制在 50 学时以内。

（九）植物生理生化实验

教学内容基本要求 是一门单列的植物生产类各专业的专业基础课，其主要内容包括两大部分：一是植物生理生化实验原理，如样品的分离、提纯、离心技术、层析技术、电泳技术、光学分析技术、气体测压技术、免疫化学技术等原理；其次是植物生理生化实验技术，如细胞生理、水分生理、矿质营养、光合作用、呼吸作用、植物激素、生长生理、抗性生理、核酸、蛋白质含量、酶活性等的测定技术。要求学生掌握主要植物生理生化指标测定原理和方法。

新教学大纲的特点 是将原来附属于植物生理学和生物化学两门课的实验内容相合并，单列一门必修课，单独记学分，以加强对学生实验操作能力的培养。内容划分为两大部分，第一部分为实验基本原理，第二部分为实验操作。在实验项目选择上，删去了一些验证性的实验内容，增加了一些定量测定的新技术方法，目的在于提高学生动手能力和开展科学实验的测试能力。原学时数为 52 学时，现为 50 学时。

（十）普通微生物学（或微生物学概论）

教学内容及基本要求　微生物学作为生物学的重要组成部分，是农业院校植物生产类专业以及生物科学、生物技术等专业的重要专业基础课。内容包括：微生物细胞的结构与功能，多样性的微生物类群，微生物的营养、代谢和生长，微生物的遗传与变异，微生物形态及微生物在自然界物质循环中的作用。结合实验，要求学生掌握真细菌、古细菌、蓝细菌、放线菌、真菌等主要微生物类群的形态特征，掌握微生物生理、遗传和生态学方面的基础知识，为后续课程的学习打下基础。

新教学大纲的特点　本课程应有很宽的涵盖面，包括微生物的主要类群及形态、分类的基本原则、微生物的营养与代谢、微生物生态区系；与应用相联系，应涵盖食品与工业微生物、食用微生物、土壤微生物、农用微生物（固氮及其他共生微生物、微生物肥料、微生物与饲料、植物病虫害微生物防治等）。要注意介绍微生物在分子生物学和生物工程发展中的作用，微生物代谢部分要注意勿与生物化学内容相重复。

（十一）普通生物学

教学内容及基本要求　本课程以生命活动的共同规律来介绍生物科学的基本知识和基础理论，力求反映现代生物学的新成就及应用。主要内容包括：生物的物质基础，生命的基本单位——细胞及组织的结构与功能，生物多样性及其保护，生物体的结构与功能（被子植物与哺乳动物为主），生殖与发育，遗传与变异，生物的进化，生物与环境等。

新教学大纲的特点

（1）知识范围较广：不仅着重介绍植物、动物的形态结构与生理功能，而且从生命的化学组成、个体生物学、生物的多样性、生物的遗传、进化及生态等角度，从微观到宏观地介绍生物科学的基础理论和基本技能知识。

（2）注重生物学知识的综合性、系统性：借鉴了国内外现代生物学教材新体系的特点，注重生物界的全貌和关于普遍规律的知识及一些新的研究进展。

（3）涉及的专业较广：目前畜牧、食品、资环、农业工程、生物物理等院系开设此课，由于各专业的培养目标及后续课程设计不同，各专业可根据自己的要求选择相应的内容重点讲授。课程总学时为81学时，其中理论课50~54学时，实验课27~30学时。

（十二）动物生物学

动物生物学的研究特别是动物生理生化的研究一向在整个生物学研究中处于领先地位，植物生理生化中许多新研究领域的开拓，多数是受到动物生物学研究成果的启发，有些甚至是直接照搬动物生物学的研究成果，将其应用到植物中来，并且得到了验证（当然，某些植物特有的生理功能，如光合作用、氮素及其他无机元素的同化等应属例外）。考虑到这些特点，我们认为，为植物生产类专业开设的动物学课程应突破传统动物学内容的束缚，除对动物的类群、形态结构、生理功能作系统介绍外，在宏观方面应突出讲授动物生态学内容，特别是动物生存环境的变迁与生态平衡、濒危物种的保护等内容；在微观方面应突出介绍动物细胞学、胚胎学、发育生物学及动物生物技术的研究进展，以使学生对新世纪生物学的迅速发展建立起完整的概念。

（十三）植物生物学

目前的动物生产类专业不开设植物类课程，不利于培养“宽口径、广适型”人才。绿色植物的生产属于地球上的初级生产力，动物生产则属于次级生产，后者必须依赖于前者。本课程为动物生产类专业开设，其内容涵盖面广泛，除包括植物的形态结构、系统分类、生理功能等传统内容外，应更多地注重宏观方面的内容，如生态学、植物学等，强调绿色植物在地球物质循环、生态平衡、环境保护和人类社会可持续发展中的作用。同时，也要注意介绍植物分子生物学基础和生物工程研究的现状和展望。

（十四）生物化学实验

把生物化学实验内容从原生物化学课中单列出来建立一门新课，单独记学分。本实验课可作为植物生产类、动物生产类及生物专业的必修专业基础课，改革重点在于使学生由单纯验证性实验、分散的小实验转向系列实验操作，掌握基本生化技术和原理，能灵活应用这些技术进行科学研究和产品开发。重点掌握生物大分子（核酸、蛋白质）分离、提纯的一般原理（讲授6学时）及分离、提取、层析、电泳、冷冻、离心等基本生化分析技术。

（十五）生物学概论（Ⅰ、Ⅱ）

本课程是从微观和宏观两个方面介绍生命科学的基础理论和一些研究新进展，使学生了解生物界概貌和普遍规律，了解生命科学对人类的重要贡献以及对未来社会发展的重要作用。

新教学大纲的特点

（1）知识范围较广，从分子、细胞、个体、群体及生物与环境等方面介绍生物科学的基础理论，涉及较广泛的生物学基础知识。

（2）注重生物学知识的综合性、系统性，借鉴了国内外现代生物学教材新体系的特点，注重生物界全貌和普遍规律的知识及一些新进展。

（3）注重介绍一些新的研究进展和应用意义，设置专题讲座介绍生命科学研究领域（特别是农业领域）一些新的研究进展和应用前景。

概论Ⅰ面向非生命科学专业类，学时为30~60学时（不同专业可根据需要选择不同学时）；概论Ⅱ面向植物生产类、动物医学类等专业，学时为30学时。

农林院校本科生物系列课程教改取得丰硕成果*

李合生　陈华癸

（华中农业大学）

本刊讯　由原国家教委下达和资助的“高等农林院校本科生物系列课程教学内容和课程体系改革的研究与实践”项目，牵头主持单位为华中农业大学，主持单位为北京林业大学、山东农业大学，参加单位有中国农业大学、西北农业大学、南京农业大学、东北林业大学、南京林业大学等8所农林院校，牵头主持人是陈华癸院士和李合生教授，参研人员85人（其中教授、副教授70人，中级人员15人）。研究的生物系列课程主要有植物学、动物学、微生物学、植物生理学、动物生理学、生物化学、普通生物学、遗传学等。该项目于1999年通过鉴定、验收，并荣获华中农业大学2000年教学成果特等奖和湖北省教学成果一等奖。

该项目研究和实践的主要成果有：

1.根据科学技术发展和人才素质培养及市场的需求，增设了起点高、内容新的课程，如现代生物学基础、生物学概论、普通生物学、动物生物学、植物学实验、植物生物学、植物生理生化实验，加强了对面向21世纪高素质人才的培养。

2.改变了过去的课程体系和陈旧的教学内容，建立了新的课程体系，并进行了优化重组，将生物系列课程划分为两个系列，两种类型，即：理论教学系列和实验教学系列（实验课单列）；基础必修型和选修型。根据农林院校专业类群的需要，制定了不同类型生物系列主干课程体系的改革方案：

（1）文法、经管及化学、农机类等专业，开设普通生物学或现代生物学基础（生物学概论）课程，作为必修课或指导性选修课。

（2）植物生产类专业，开设植物学、植物学实验、基础生物化学、植物生理学、植物生理生化实验、普通微生物学（微生物学概论）、普通遗传学、动物生物学等，计8门生物学类课程。

（3）动物生产类和动物医学类专业，开设普通动物学、动物生理学、动物生理学实验、植物生物学、基础生物化学、基础生化实验、普通遗传学、微生物学等，计8门生物学类课程。

（4）食品科学与工程专业，开设普通生物学、生物化学、生化实验，计3门生物学类课程。

（5）生物科学类专业，开设植物生物学及实验、动物生物学及实验、微生物生物学及实验、生态学、生物化学、细胞生物学、普通遗传学、分子生物学、生物信息、免疫学、生物工程概论、生化技术等16门课程。

上述生物系列课程教改方案，已在华中农业大学、中国农业大学、山东农业大学、南京农业大学、西北农业大学97、98级本科学生班中进行了1~2年的教改试点，试验效果良好，并已列入新制定的教学计划中。

3.编写了有特色的生物系列课程的新教学大纲20种以上。在农林类本科各专业新的教学计划的总学时普遍压缩200~300学时的前提下，改革后的生物系列课程学时数都普遍减少了10~20学时，同时内容进行了更新，反映了近代新成果、新进展，删去了重复和繁杂内容。如：植物学和植物学实验由100学时减少至90学时，生物化学、植物生理学、遗传学都减少了10学时，动物学由原来的70学时减少到50学时。在内容上，新的遗传学教学大纲，在原有教学大纲的基础上改动的内容占30%，新大纲增加的分子水平的新内容占原大纲的25%，删掉的陈旧内容占原大纲计划的20%。植物学新的教学大纲和原有的教学大纲相比，增删内容占1/5左右。这些变动，体现了生物课程的教学时数少了，但起点高了，内容新了，更加精炼了。

---

*原载于《中国大学教育》，5：46~47，2001.

4.编写出版了一批生物系列课程新教材，为培养新世纪人才提供了优质“硬件”。在生物系列课程教学内容改革中，改变了过去人为地把遗传学、生物化学等课程划分为动物、植物、人类、微生物四大分支，越分越细的局面，而新编写了综合性强，内容新的普通遗传学、基础生物化学的教学大纲，还其学科本面目，并有创新。新编了面向 21 世纪本科人才培养的起点高、知识面广、内容新、精的面向 21 世纪生物系列课程教材 12 本。

5.在教学方法和手段及考核方法方面进行大力度的改革，改灌输式教学为启发式、讨论式即参入式教学，已在本科生物系列课程教学中普遍采用，有力地调动了学生的学习积极性和主动性，在学生学习知识的同时，还培养了他们获取知识的能力。本项目组参研学校在教学改革试验中，都普遍采用了投影片、幻灯片、电化教学录像片、VCD 及 CAI 课件，还研制和试用了试题库等，从而提高了教学质量。

6.为了加强本科学生动手能力、分析能力及创新能力的培养，该项目将部分生物学实验课单列，并且探索从过去验证性实验转为综合性、研究性、应用性实验，加强实物识别，让学生参加实验课准备和预试以及开展第二课堂活动，实施实验操作考试等一系列措施，有利于学生创新能力的培养。如植物学实验从植物学课中独立出来。在教学内容和方法上，以前是把早已制好的永存切片给学生用作观察器官的显微结构，验证课堂讲授内容，而现在是采用学生自己动手，将需要了解的代表性植物材料由学生自己动手制切片，然后再去观察器官的结构，使学生既了解了研究植物所用的方法，又在技能上得到了训练，也巩固了课堂讲授内容。在实验课改革的同时，对实验室管理体制进行了深入改革，加大投资改善实验条件，建设开放型、通用型的实验室，直接服务于本科生创新能力的培养。实验课单列后，学生对实验课重视起来，实验操作考试效果好。

7.为适应学生个性发展和市场需求的变化，各专业在必修课程学习的基础上，从微观和宏观两个方面设置了一些有深度或交叉学科的选修课程，如分子生物学基础、作物生理学、果树生理学、林木生物化学、生态生理学、病原微生物学、分子遗传学、植物组织培养技术、特种作物生理生化分析等，有利于学生个性的发展和知识面的拓宽以及实验操作技能的加强。

8.项目组成员积极参与了生物系列课程优质课程建设与改革，取得了显著成绩，获奖多项。据不完全统计，中国农业大学、北京林业大学、东北林业大学、华中农业大学的植物学、动物生理学、植物生理学、生物化学课程教学改革体系建设都荣获过教学成果奖或优质课程证书。

下 篇 著 作 集

第四部分　教材

第五部分　专著

第六部分　译著

第四部分　教　　材

《土壤微生物学》

《土壤微生物学》
陈华癸编著
民国二十六年（1947 年）初版
民国二十七年（1948 年）第二版
国立编译馆出版，正中书局印行

该书为部定大学用书。全书共十章，并附有参考文献。

全书内容

《微生物学》（初版）

《微生物学》（初版）

主　　编　陈华癸

副 主 编　娄隆后　张天伏

参编人员　叶维青　刘萝筠　陈华癸　周　啟　娄隆后　袁永生

　　　　　张天伏　张元龙　许光辉　黎尚豪

本书是农业部委托各农业院校根据1955年高等教育部颁发的教学大纲基础上，结合8年来的教学经验和1958年教学改革的精神集体编写的，整个编写是在华中农学院进行的。

本书可作为高等农业院校土化专业用“土壤微生物学”和农学、果蔬、植保等专业用“微生物学”的教学参考书。全书共有二十四章。

该书由高等教育出版社1959年出版（初版），本书为三十二分之一开本；字数421 000；印数2000（精装本）+（平装本）2500册。

全书内容

第一章　绪　　论

第二章　细菌和放线菌的形态（附超显微镜微生物）

第三章　酵母菌、霉菌和其他真菌类的形态学

第四章 土壤藻类和微小动物

第五章 微生物的营养

第六章 酶和酶的作用

第十一章 微生物分解己糖的基础理论

第十二章 微生物在不含氮有机物分解过程中的作用

第十三章 微生物与发酵及酿造

第十四章　沼气及其利用

第十五章　微生物在自然界氮素循环中的作用（一）（氨化、硝化、反硝化作用）

第十六章　微生物在自然界氮素循环中的作用（二）（非共生固氮作用）

第二十三章 菌肥的制造和使用

第二十四章 农业技术措施对土壤中微生物生命活动的影响

《微生物学》（修订第二版）

《微生物学》（修订第二版）

主　　编　陈华癸
副 主 编　娄隆后　张天伏
参编人员　华中农学院　陈华癸　李阜棣　周　啟　曹燕珍
　　　　　北京农业大学　娄隆后
　　　　　东北林学院　张天伏
　　　　　中国科学院水生生物研究所　黎尚豪

本书第二版分为三篇共二十二章（绪论除外）。第一篇，微生物学基础；第二篇，自然界物质转化的微生物；第三篇，土壤微生物学和土壤肥力。每篇各有七至八章。有些章只是在第一版的基础上进行了一些小修小改，增添或删去了一些内容。有些则改变较大，不仅内容的增删修改较大，在章节的形式上也采取了必要的变动。第一篇各章除了删去“酶和酶的作用”一章和增添“细菌的变异”一章外，其他章节的改变主要是在编写的形式上。第二篇主要是在内容细节的增删、修改方面。第三篇各章从内容到形式改变都较大。第一版中原有的“堆肥和厩肥的微生物学”和“沼气及其利用”两章在第二版中合并为一章。

本书为全国高等农业院校试用教材，供农学、果蔬、植保和土壤农化专业使用。

1962年农业出版社出版。书为16分之一开本；字数457 000，上海第五次印刷，印数24 001~27 500册。

全书内容

第三篇 土壤微生物与土壤肥力

《微生物学》(修订第三版)

《微生物学》(修订第三版)

主　编　陈华癸　樊庆笙

编写者　万金精　方宇澄　刘梦筠　叶维青　陈华癸　李阜棣
　　　　胡正嘉　钱泽澍　樊庆笙

审稿人　万淑婉　王毓庆　余　苹　吴文礼　张　扬　张梦昌
　　　　李志超　李季伦　李政祥　李淑高　杨洁彬　郝余祥
　　　　郝锡宓　胡美玲　娄无忌　郭恒聪　秦　京　黄怀琼
　　　　温琼英

本版教材与修订第二版对照有下列几点做了较大的修订：①绪论中简述微生物学发展史部分增添了关于微生物分子生学发展时期的介绍；②细菌、真菌和病毒三大类微生物分章阐述；③关于微生物的遗传和变异着重于在分子水平上阐述基础理论知识；④由于生物固氮研究在近二十年中有重大发展，本版中做了重点介绍；⑤编写了《土壤生物活性》一章，从整体上阐述土壤微生物的生命活动及其对土壤肥力和农业生产的影响，以及环境的微生物净化问题；⑥关于《微生物与肥料》、《微生物与饲料》、《抗生素和植病防治》、《微生物与害虫防治》和《沼气发酵》等问题都分章阐述，以利于不同专业的分别取舍；⑦编写术语索引和微生物中名、学名对照表。

本书为全国高等农业院校试用教材，供农学、土壤专业用。

1979 年农业出版社出版。该书为 16 分之一开本；字数 430 000 字，印数 25 000 册。

全书内容

《微生物学》（修订第四版）

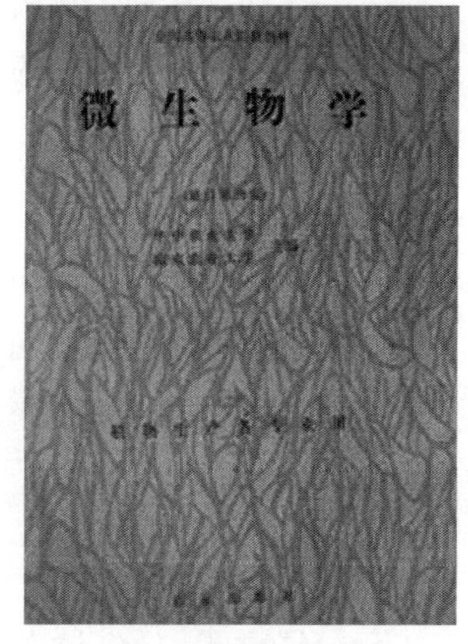

《微生物学》（修订第四版）
主编　陈华癸　樊庆笙

李阜棣、周敞、胡正嘉、喻子牛、周俊初提供了素材或对草稿提出了修改意见，全书由何绍江通篇清稿。此外，尚有周俊初、张淑芳、黎若娅、刘孔鑫、魏四洪等参与抄写和制图工作。

本书为全国高等农业院校教材。第四版修订的指导思想与前三版基本上是一致的，即作为农业高等院校植物生产各专业（以农学和土壤农化专业为主要对象）的专业基础课的教材。全书仍为十五章约三十万字。将第三版的绪论一章删去，只保留一篇短的绪言；将《微生物的营养、环境条件和纯培养》和《微生物的代谢和生长》各写成一章；将第三版中《微生物生态系》一章的内容分散到有关的章节中去；将《血清学，抗原抗体的体外反应》、《细菌的分类和鉴定》及《菌根和菌根菌》独立出来，写成三个小章。有关甲烷细菌和沼气发酵的基础知识写进了有关章节，删去了独立的一章，每一章节都做了一定程度的修订。

农业出版社 1989 年出版，书为 16 分之一开本，字数 302 000 字；印数 1–11 700 册。

全书内容

《微生物学实验》

《微生物学实验》

主编　陈华癸

参编　李阜棣　周　啟　赵学慧　曹燕珍　胡正嘉　陈华癸

本书既是一本实验课的教材，又是一本土壤微生物学工作手册。这本书的服务对象是：①农学院农业系统各专业的微生物学实验课；②土壤农化专业土壤微生物学实验课、教学实习和学生的科学研究活动；③微生物学和土壤微生物学教师和研究生的备课活动和科学研究活动。

本书共分八个部分和附录，共有109个实验。该书由农业出版社在1962年出版（初版）。书为16分之一开本；字数202 000；印数1−4 000册。

全书内容

写给实验课的指导教师

作为微生物学实验课的教材，建议：①由指导教师根据特定的实验课的教学内容，预先选择一定节目作为指定的学习文件；②在每次实验课开始时，由指导教师将具体要求、工作程序、必要的修改和注释写在黑板上，并简明扼要地做一些说明（或在每学期初编出详细的教学计划）。这样做，既可以较充分地利用这本书，不用另发成套的讲义或实验指导书，并可以结合当时当地的实验课的具体条件。

例（1）特定实验课的实验X，题目为“细菌荚膜的衬托染色法”。

阅读本书实验六“细菌荚膜衬托染色法：复习实验四”“细菌的简单染色法”。

实验工作程序：

1.制涂片；简单染色。

2.做黑素处理。

3.镜检。

4.书面作业：绘图并说明；提问解答。

说明：

1.本次实验用固氮菌进行。

2.本实验用1/5石碳酸复红染色液染色。

例（2）特定实验课的实验Y，题目为：“用嫌气培养法培养巴斯德梭菌”。

阅读本书实验三十八“用焦性没食子酸吸收氧气培养嫌气性细菌”；实验十一“酵母菌水浸片的制备”；实验十二“酵母菌细胞中肝糖粒染色法”。

实验工作程序：

1.接种。

2.实习用焦性没食子酸吸收氧气法。

3.保温培养3-5天。

4.制水浸片，加碘液处理；镜检。

5.书面作业：绘图并报实验结果。

说明：本班实验集体采用标本瓶代替大试管。1000 毫升标本瓶，用 10 克焦性没食子酸和 50 毫升 20% NaOH。瓶口用真空封闭油膏封闭。

华中农学院微生物学教研组

1962 年 4 月

《微生物学进展》

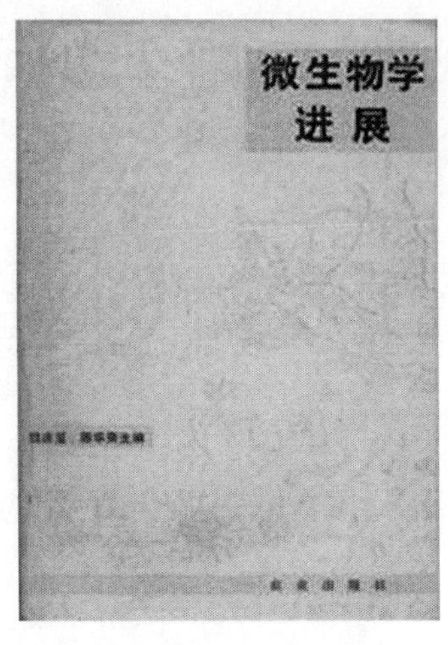

《微生物学进展》

主编　樊庆笙　陈华癸

1980年农业部委托南京农学院举办全国高等农业院校微生物学讲习班。邀请高等院校和微生物学科研单位的教授和专家二十余人讲授微生物学基础理论和研究技术的近代成就。使参加讲习班的各院校教师得以深入学习，系统地掌握现代微生物学的基础知识和基本技术，在各院校微生物学课程的教学实践中，能充实统编教材《微生物学》(农业出版社1979年出版)的内容，提高教学质量，发挥教学效果。为了使这些专题讲授内容能供微生物学工作者同行，兹将其中十六个专题内容，汇编成《微生物学进展》一书，由农业出版社作为统编《微生物学》的主要参考书出版。可供各大专院校微生物学教师、微生物学科研工作者和微生物学专业学生参考。

本书由农业出版社1984年出版；书为32分之一开本；字数为372 000字；印数1−6800册。

全书内容

第五部分　专　　著

《土壤学》（下册）

《土壤学》（下册）

主编　陈华癸

参编人员：侯光炯（西南农学院）、黄瑞采（南京农学院）、陆发熹（华南农学院）、唐耀先（沈阳农学院），其他有华中农学院的王开凤、杨补勤、庄正德、李学垣、张光远和陈华癸。

该书为高等农业院校试用教材，供农学类各专业及土壤农化系用，全书共有十三章。

农业出版社 1962 年出版（初版）。本书为 16 分之一开本；字数 281 000 字；印数 1–2400 册。

全书内容

《土壤微生物学》

《土壤微生物学》
陈华癸　著

这是一本专著。全书共有十九章，并附有参考书，适合于土壤农化专业作为教学参考书。也可以作为自学土壤微生物学者阅读。

高等教育出版社 1957 年出版（初版）；书为 32 分之一开本；字数为 198 000；印数 1–2300 册。

全书内容

《土壤微生物学》

《土壤微生物学》
陈华癸　李阜棣　陈文新　曹燕珍　编著

这是一本专著，可供农学类各专业和土壤农化专业参考使用。全书共有十章；上海科学技术出版社 1981 年出版（初版）；书为 32 分之一开本；字数 232 000；印数 1−8000 册。

全书内容

《中国共生固氮研究五十年》

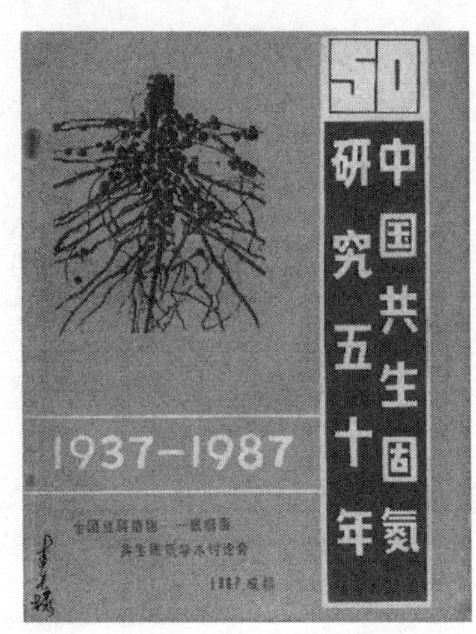

《中国共生固氮研究五十年》

主编　陈华癸　樊庆笙

顾问　张宪武

编委　丁　鉴　陈廷伟　李阜棣　周湘泉　杨苏声　姚惠琴

南京农业大学　1988 年编辑出版（未公开发行）

这本书将五十年来（1937 年第一篇：大豆根瘤菌菌株的研究，张宪武）我国各研究单位的沿革和主要工作整理出来，汇集了全国共生固氮研究者们的工作成绩、经验和见解。为我国共生固氮研究继往开来、增进学术、加强合作起到了重要的促进作用。

全书内容

江苏省农科院土肥研究所
江苏省淮阴农科所微生物室
山西省农科院土肥研究所
广西省

第二部分 应用推广研究

第三部分 附录

《农业哲学基础》

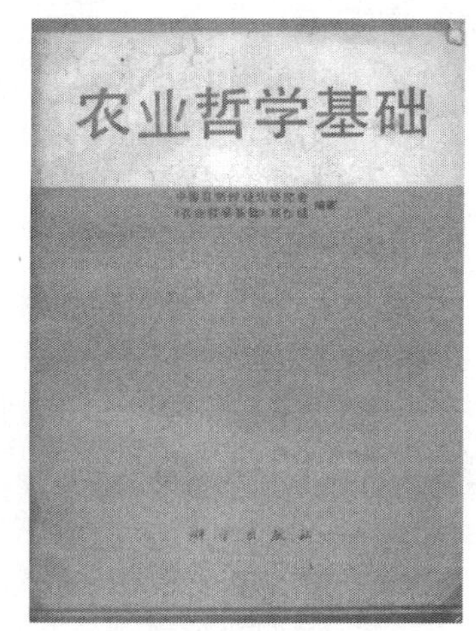

《农业哲学基础》
主编　金善宝　沈其益　陈华癸
科学出版社　1991 年出版

这本书是我国部分高等农业院校自然辩证法工作者在中国自然辩证法研究会以及农业专业组关怀和支持下集体编撰写。全书共分六篇。论述了农业的本质、特征及其发展规律、农业发展中的若干关系、部门农业中的辩证法、农业的发展前景、农业科学的体系和发展规律、农业科学方法论、农业发展战略以及农业决策和管理等。这样从哲学的角度比较系统地论述农业生产、农业科学、农业教育与农业管理问题的书，在我国还是第一本。

我国农村正处在一个伟大的历史转变时期，正在由传统农业向现代农业转变由自给半自给经济向社会主义商品经济转变，农村的生产体制和多种经济的内涵，也在突破着历史的近 30 多年的种种与四化建设不相适应的束缚。《农业哲学基础》一书，在这个伟大的历史转变时期问世，它必将经受实践的检验，接受群众和专家的鉴评，这无疑地又将促进农业哲学研究工作的深化与发展，而这正是本书编写者们衷心期望的。

全书内容

第四篇 农业决策和管理

第五篇 农业发展前景和发展战略

第六篇 农业子系统中的辩证法概述

《生物资源再利用原理与技术》

《生物资源再利用原理与技术》
陈华癸　蔡泽民　编著

本书是国家“九五”重点图书出版规划项目。生物资源包括可利用的生物种类、生物量及其再生资源，它是生物工程学的一个重要分支。全书共分 8 篇 29 章 138 节，其中同济大学环境生物工程系张恭勤教授和华中农业大学真菌研究室罗信昌教授分别撰写蚕蛹的综合利用和菌蕈与农林副产品的利用等两篇章。本书可供从事食品、发酵、抗生素生产、农林副产品加工、乡镇企业的广大科技工作者和管理人员以及农林、轻工院校的有关专业和综合性大学生物系广大师生参考，也可以作为有关院校学生和企业职工的培训教材。

本书为 16 开本，由湖北科学技术出版社 1999 年 8 月出版发行。

全书内容

第一篇　绪论

第二篇　生物资源再利用的基础知识

第八篇 附录

第六部分　译　　著

《豆类–根瘤菌的共生关系及其农业利用》

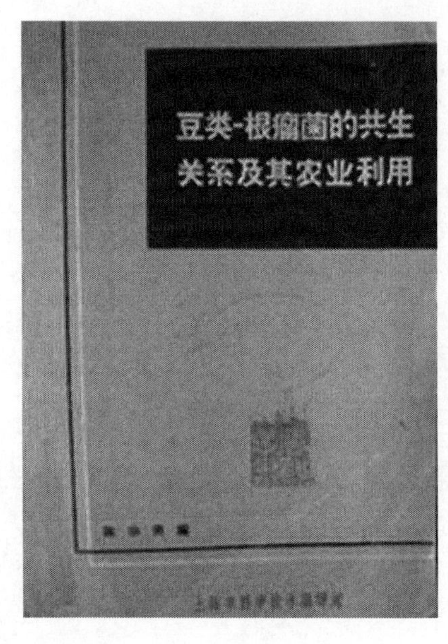

《豆类-根瘤菌的共生关系及其农业利用》
陈华癸　编译

豆类和根瘤菌的共生固氮作用是自然界营养元素生物循环的一个重要的环节，而且对培肥土壤，提高农业生产也很重要。但是，豆类和根瘤菌的共生关系的研究，不论是在科学理论或生产技术方面，都还有许多根本性的和关键性的问题没有得到解决。因此，关于豆类和根瘤菌的共生关系的研究仍是农业微生物学的一项重要课题。为此编译了这本译著，全书共有九篇文章。本选译的文献都发表于 1947–1962 年，有两篇是属于全面评论性质的（Allen 1958；Raggiotn. Raggio M. 1962）。其他几篇则侧重于不同的问题和方面。

本书由上海市科学技术编译馆于 1965 年出版第一版）; 32 分之一开本；字数 225 000；印数 1–2000 册。

全 书 内 容

第一篇　共生固氮作用的生物学　原著 Allen，E.K；Allen，O.N.《Eneyclopedia of plant Physiology》8：48-118，1958.（谢寿长、赵学慧译，陈华癸校）

第二篇　豆类植物结瘤情况调查　原著 Allen，O.N；Allen，E.R.《Soil sci.soc.Amer. Proc.》12：203-208，1947.(周啟译，陈华癸校)

第三篇　豆科植物和根瘤菌的共生关系　原著 Norris，D.O.《Empire，J.Exptl.Agric.》24：247-270，1956（曹燕珍译，陈华癸校）

第四篇　根瘤菌和豆科植物的共生关系以及细菌菌株对共生关系的影响　原著 Thomton，H.G.《Proc.Roy.Soc.》B，189：171-176，1951-1952（王子芳译，陈华癸校）

第五篇　豆科植物在根瘤共生作用的影响，寄主的决定因素和功能的比较研究。原著 Nutman，P.S.《Biol.Rov.Cambridge Phil.Soc.》32（2）：109-151，1956（胡正嘉译，陈华癸校）

第六篇　根瘤菌固氮作用的生物学和化学　原著 Virtanen，A.I.《Biol.Rov.Cambridge Phil.Soc.》22：239-269，1947（胡正嘉译，陈华癸校）

第七篇　影响豆科植物摄氮的若干因素　原著 Van Schreven，D.A.《Nutrition of the Legumes》137-163，1958(吴仁慧译，陈华癸校)

第八篇　根瘤菌的存活情况　原著 Vincent，J.M.《Nutrition of the Legumes》108-123，1958.（吴仁慧译，陈华癸校）

第九篇　根瘤　原著 Raggio，M. Raggio， N.《Ann， Rev. Plant Pnysio.》13：109-128，1962（王家玲译，陈华癸校）

《微生物世界》

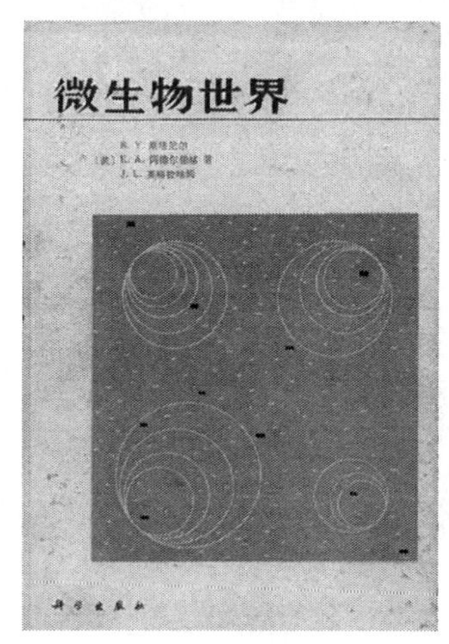

《微生物世界》

美国 R.Y.斯塔尼尔、E.A.阿德尔伯格：J.L.英格拉哈姆著《微生物世界》翻译组译　陈华癸　　校

本书是 R.Y.Stanier 等所著的《微生物世界》第四版。它是一本介绍普通微生物学基础知识的专著，内容全面，资料较新，与第三版相比有较大的增改，特别是细菌和微生物遗传部分几乎全部改写。对微生物的代谢作用及生理、生化方面也有较详细的介绍。为了进一步深入探讨，每章后面还附有重要的阅读文献，本书是从事微生物科学研究人员及有关大专院校师生较好的参考书。本书共有三十一章；并附有学名和人名索引，以及内容索引。

科学出版社 1983 年出版（第一版）；该书为 16 分之一开本；字数为 1 066 000；印数 1–4900 册

全书内容

附　　件

附件 I　在《微生物与农业》上已经登载，不再在本《论著集》上重复登载的论文目录

第一部分

1.胡传炯、周平贞、周啟、陈华癸、A.D.L. Akkermans. 科学通报 43（9）：969-973，1998

2.胡传炯、周平贞、周啟、陈华癸、马桑菌株内生菌的分离、回接及生物学特征. 中国农业科学 32（2）：72-77，1999

第二部分

3.陈华癸、周啟 水稻田土壤中的硝化作用和硝化微生物的研究 I. 水稻田土壤中的硝化作用. 土壤学报 9（1-2）：56-64，1961

4.陈华癸、周啟 农业技术措施对水稻田土壤中微生物生命活动的影响及其生产意义 I. 烤田对土壤微生物区系变化的影响.土壤学报 9（3-4）：133-139,1961

5.陈华癸、周啟 水稻田土壤中的硝化作用和硝化微生物的研究III. 亚硝酸细菌纯培养的分离. 中国土壤学会、中国微生物学会，1964 年土壤微生物专业会议专题报告及研究报告摘要集. 武汉，研究报告 45-46 页， 1964

6.Chen H. K & Zhou Qi. Facultatively Anaerobic Nitrification and Nitrite-forming Organisms 第八次国际土壤学会议论文集. 土壤生物学 761-768 页，罗马尼亚布加勒斯特，1964

7.Zhou Q. & Chen H. K. The Activity of Nitrifying and Denitrifying Bacteria in Paddy Soil. 美国 Soil Science 135 (1)：31-34,1983

第三部分

8.覃重军、邓子新、周啟、陈华癸. 利用弗氏链霉菌的修饰系统克服吸水链霉菌的限制性障碍——发展吸水链霉菌转化系统的尝试. 遗传学报 20（2）：180-184，1993

9.覃重军、邓子新、周啟、陈华癸. 吸水链霉菌应城变种的四个内源质粒及其逐个消除的研究. 微生物学报 35（1）：14-20，1995

10.覃重军、邓子新、赵国方、梁蓉芳、周啟、陈华癸. 建立质粒的“突变克隆”体系和获得农抗 5102 超量表达以及激活新抗表达.《微生物与农业》，林开春、史贤明、袁德军、万中义、覃重军主编，北京科学出版社 403-408 页，2004

附件Ⅱ 历届招收的研究生名单

序号	姓名	学位	学习时间	学位论文题目	合作导师
1	胡正嘉	研究生	1953~1956	水稻田中绿肥（苕子）翻耕后水稻生长期间有机磷的转化及强烈分解有机磷的细菌	
2	曹燕珍	研究生	1953~1956	水稻田绿肥（苕子）翻耕后水稻生长期间土壤耕层中占优势的有机营养型微生物类群的研究	
3	王家玲	研究生	1953~1956	水稻田绿肥（苕子）翻耕后土壤中主要嫌气性蛋白性分解细菌以及土壤无机态氮质动态的研究	
4	赵文洪	研究生	1953~1956	水稻田绿肥（苕子）翻耕后，纤维素分解细菌种类的研究	
5	王子芳	研究生	1953~1954	1954 赴前苏联留学	
6	周 啟	研究生	1957~1961	水稻田土壤中的硝化作用和硝化细菌以及农业技术措施对水稻土中微生物区系的影响	
7	陈廷伟	研究生	1957~1958	（离校返京实习）	
8	麦茂英	研究生	1958~1959	（生病退学）	
9	陶天申	研究生	1962~1965	胶冻样芽孢杆菌解钾机理的研究	
10	谭金华	研究生	1962~1966	（文化大革命期间未正常毕业）	
11	周俊初	研究生	1963	（文化大革命期间未正常毕业）	
12	喻子牛	研究生	1964	（文化大革命期间未正常毕业）	
13	罗运国	研究生	1965	（文化大革命期间未正常毕业）	
14	范业宽	硕士	1979~1982	稻棉轮作对改良次生潜育化水稻土的效果	李学恒
15	李仲贤	硕士	1979~1982	关于紫云英根瘤菌结瘤性能自发的和吖啶橙与热处理诱导的变异以及 Nod^-与 Nod^+分离系同寄主共生的比较形态研究	
16	郑玲	硕士	1982~1984	紫云英根瘤菌有效根瘤形成的遗传分析	李阜棣
17	王福生	硕士	1982~1985	大豆根瘤菌（*Bradrhizobium japonicum*）竞争结瘤的生态分析	李阜棣
18	阮小安	硕士	1982~1985	宜昌百里荒地区引种白三叶草改良草地过程中根瘤菌的应用和生态学研究	曹燕珍
19	吴捷	博士	1982~1988	*Coraria nepalensis* Wall.共生固氮根瘤的形态发育及其内生菌的研究	
20	胡学俊	硕士	1983~1986	紫云英根瘤菌结瘤基因的消除和 pRP4 质粒在根瘤菌和其他格兰氏阴性细菌之间的转移	李阜棣
22	王长霖	博士	1984~1987	豌豆根瘤菌共生质粒行为的研究	
23	韩素贞	硕士	1984~1987	转座子 Tn5-Mob 对紫云英根瘤菌 SR72 的诱变以及 Tn5-Mob 和 RP4 对 SR72 质粒的诱动	周俊初
24	张忠明	硕士	1986~1988	紫云英根瘤菌（*Rhizobium astagali*）基因文库的构建及结瘤基因的分离	
25	郝勃	硕士	1986~1989	黑曲霉 N343 与黑曲霉 UV-11 的营养缺陷型突变株的原生质体融合的研究	闫醇泰

续表

序号	姓名	学位	学习时间	学位论文题目	合作导师
26	彭文涛	硕士	1986~1989	慢生型大豆根瘤菌（*Bradrhizobium japonicum*）基因文库的构建及结瘤基因的分离	周俊初
27	覃重军	博士	1989~1992	吸水链霉菌应城变种载体宿主系统的发展及其抗生素生物合成的遗传操作	周啟
28	葛向阳	硕士	1989~1992	添加黑曲霉（*Aspergillus niger*）F27 固体曲提高啤酒生产原料糖化利用率及其对啤酒质量的影响	闫醇泰
29	廖美德	硕士	1990~1993	绿衣观音土曲及其主要糖化菌棒曲霉（*Aspergillus clavatus*）的研究	许耀才
30	农广	博士	1991~1994	紫云英根瘤菌（*Rhizobium huakuii* Chen）的共生质粒、共生基因克隆和早期共生关系的研究	周俊初
31	郭先武	博士	1994~1997	华癸根瘤菌（*Mesorhizobium huakuii* Chen）染色体基因和质粒基因群体遗传学比较研究	
32	马立新	博士	1996~1999	绿荧光（gfp）基因在根瘤菌中的表达及适用于革兰氏阴性细菌载体系列的构建	周俊初
33	阮堂	留学生（越南）	1959		
34	Stephen Dowdle	留学生（美国）	1982~1984		

附件Ⅲ　陈华癸简历

1914年1月11日	出生在北京市，原籍江苏省昆山县
1931年	毕业于北京大学预科
1935年	毕业于北京大学生物系
1935~1936年	任北京大学生物系助教
1936年	赴英国伦敦大学卫生和热带病学院学习一年
1937年	在英国著名的洛桑试站做博士论文
1939年	获英国伦敦大学哲学博士学位
1940年	获美庚款基金资助，在西南联大清华农业研究所从事研究工作
1941~1945年	中央农业研究所，任技正
1946~1947年	筹建北京大学农学院土壤系，任教授、系主任
1947~1952年	筹建武汉大学农学院农业化学系，任教授、系主任
1952年	任华中农学院（华中农业大学）教授、土壤农化系主任、生物固氮研究室主任、华中农学院院长
1953年	参加中国民主同盟
1956年	参与筹建和成立中国科学院武汉微生物研究所（后改为武汉病毒研究所）兼任研究员、副所长，参加中国共产党
1964~1983年	先后当选为第三届、第五届和第六届全国人民代表大会代表
1979~1983	华中农学院院长；先后当选为中国科学技术协会第二届全国委员会委员、中国农学会常务理事、副会长、顾问；中国农学会土壤肥料研究会第一届、第二届理事长；中国微生物学会常务理事、副理事长、名誉理事长
1980年	当选为中国科学院生物学部委员
1981年	被授予博士研究生导师资格
1981~1985年	国家科委六五计划固氮生物学重大问题研究组组长
1981~1990年	任国务院学位委员会第一届委员
1982年	任农业部组织的农业微生物考察团赴英国有关大学和科研单位进行一个月的考察。此外，先前他也任副团长和团长率领考察团赴美国和澳大利亚作短期考察
1984年	历任华中农学院（华中农业大学）学术委员会主任、农业部科学技术委员会委员以及中国农业科学院学术委员会委员
1991年	中国农业百科全书总编辑委员会委员
1991年	农业部华中农业大学农业微生物重点开放实验室学术委员会主任

编 后 记

《陈华癸论文著作集》的编辑工作已顺利完成，包括有论文集、著作集和附件三部分。编者参考了中国科学技术协会编、中国科学技术出版社 1993 年出版的《中国科学技术专家传略》农学编土壤卷 I 中陈华癸传略的内容，这篇传略是教授口述，本校农化教研室程见尧老师撰写的。在编写本书过程中，程老师还提供了一些当时未编入这篇传略的有关资料，他还多次热情而细致地提出修改和补充意见，这种认真负责的精神，颇使我们感动，特此向他致以衷心的感谢。

陈教授在大学还没有毕业期间（1934~1935）就开始发表论文，一直到 2001 年为止总共发表论文 115 篇，其中有 10 篇已在《微生物与农业》中刊登（见附件 I），另外尚有早期发表的 4 篇论文，经多方努力依然无法收集到。因此，在本书中实际刊登论文为 101 篇，包括第一部分 65 篇，第二部分 14 篇，第三部分 22 篇。

陈教授在一生中共编写了十五本著作，包括教材 8 本、专著 5 本和译著 2 本。

陈教授还担任较多具体的教学和学术工作。同时他还非常重视高等农业教育的改革，提出诸多宝贵意见。在本“论著集”中，采用影集的方式加以说明，其中由朱辉老师翻拍了一部分照片，随后由李合生老师继续将需要的照片拍摄完毕并编辑好电子版，李老师这种主动热情和认真负责的精神值得学习，这也是他尊敬陈教授的一种具体行动。编辑组在此向他致以亲切的谢意。

总的来说，这本“论著集”基本上汇集了教授的全部功绩，反映了他的高尚品德、特有的教学和学术思想、渊博的知识、严谨求实的治学和学风、重大的贡献。出版这本“论著集”的目的，一方面是为了纪念教授的一百岁诞辰，另一方面也是为了便于读者和后人学习与继承教授的崇高品格和严谨精神。

当然，本“论著集”中也有一些缺点，主要是“论著集”中的一些文章，特别是根瘤菌分子遗传方面的论文，很多电泳图片很不清晰，虽然有的已做了一些修正，但有的原始论文的图片就很不清晰，因此无法进一步改善，实在遗憾。

本“论著集”是全体编辑人员共同努力的结果。但由于经验不足，时间仓促，不能尽善尽美，由此可能还有一些错误和不完善的地方，恳切希望作者和读者批评、指正。